Computational Nanomedicine and Nanotechnology

Renat R. Letfullin • Thomas F. George

Computational Nanomedicine and Nanotechnology

Lectures with Computer Practicums

Springer

Renat R. Letfullin
Rose-Hulman Institute of Technology
Terre Haute, IN, USA

Thomas F. George
University of Missouri-St. Louis
St. Louis, MO, USA

ISBN 978-3-319-43575-6 ISBN 978-3-319-43577-0 (eBook)
DOI 10.1007/978-3-319-43577-0

Library of Congress Control Number: 2016955820

Printed on acid-free paper

This Springer imprint is published by Springer Nature
The registered company is Springer International Publishing AG
The registered company address is: Gewerbestrasse 11, 6330 Cham, Switzerland

Preface

The field of nanomedicine as combined with nanotechnology is new and exciting, with broad applications that are most relevant to our everyday lives and well-being. This field, which combines multiple disciplines including medicine, biotechnology, physics, optics/radiology, engineering, and pharmacy, has matured to the extent that the time is ripe for this current textbook aimed at advanced undergraduate and graduate students. In addition to its pedagogical nature, it is also of value to research scientists who want to enter the field or even those who are already in the field.

The book is organized according to chapters, and within each chapter is a set of lectures in the form of both text for the student and PowerPoint for use by the instructor, followed by homework exercises. Selected chapters also contain computer practicums, including Maple codes and worked-out examples.

A special emphasis in the book is on how the combined use of chemistry and light with nanoparticles can serve as treatments and therapies for cancer. This includes nanodevices, nanophototherapies, nanodrug design, and laser heating of nanoparticles and cell organelles.

Nanomedicine is a new science. Because of this, the material of this book is mostly based on our own original journal papers, in some cases reproducing them in the text of the chapters. Also, we provide very few references in the textbook. The detailed list of references and reviews can be found in those original papers listed in the book.

The first three chapters of the book are general introductory chapters introducing the basic knowledge required for nanomedicine and nanotechnology study. Chapter 1 presents the definition of nanomedicine and nanotechnology, introduces the functions and properties of medical nanodevices, the design and fabrication of nanorobots and nanoparticles, and discusses current and potential applications of nanomedicine.

Chapter 2 gives a brief introduction to the structure of cancer cells and their properties and how cancer cells differ from normal cells. The causes of cancer, types of cancer, techniques for selective targeting of cancer, and gene therapy are explained in detail in this chapter.

Chapter 3 introduces cancer therapy and detection techniques based on utilizing the plasmonic nanoparticles. We discuss here the design and methods of activation of the nanodrugs selectively delivered to the tumor site and the plasmon resonance detection techniques using fiber optics.

The remaining chapters of the book are devoted to the modeling and computer simulations of the complex phenomena of electromagnetic radiation interaction with nanostructures, resulting in various accompanying effects around the nanoparticles and nanoclusters. The results of these simulations are then discussed in relation to the effective implementation of nanotherapy treatments. For example, in Chap. 4 we model the light absorption properties of nanoparticles in the X-ray, optical, and RF ranges of the spectrum.

In Chap. 5 we introduce a computer software and perform Lorenz-Mie diffraction simulations of absorption, scattering, and extinction efficiencies of different types of nanoparticles as a function of particle size and the wavelength of radiation in different surrounding biological media. Also, we calculate and then use the differences in optical properties of normal and cancer cell organelles in order to design new cancer detection and treatment methods.

Chapters 6 and 7 first introduce the thermodynamic model for nanoheat transfer and then focus on the heating and cooling kinetics of nanoparticles by short and ultrashort laser pulses in the femtosecond, picosecond, and nanosecond regimes. Time dynamic simulations are performed for metal nanoparticles in a surrounding biological medium as well as for healthy and cancerous cell organelles heated by optical, RF, and X-ray pulses. Chapter 8 is devoted to spatial thermal fields modeling and 3D simulations within and around a plasmonic nanoparticle and cell organelles surrounded by a biological medium heated by radiation.

In Chap. 9 we perform optical, thermal, kinetic, and dynamic modeling to develop and test new ideas and new dynamic modes in selective nanophotothermolysis, involving nanocluster aggregation in living cells, nanobubble overlapping around heated intracellular nanoparticles/clusters, and the laser thermal explosion mode of single nanoparticles—"nano-bombs"—delivered to the cells.

A goal of plasmonic laser nanosurgery/nanoablation is to deliver extremely precise heat blasts in programmed patterns to cancer cells, abnormal cell organelles, or mutated DNA molecules in order to destroy them. In Chap. 10, we perform modeling of soft and hard biological tissue ablation by plasmonic nanoparticles heated by short and ultrashort laser pulses. Short and ultrashort laser pulses hold key interest in precise nanoablation as they can allow for deposition of thermal energy to the target area by many orders of magnitude faster than the heat diffusion process that leaks heat to adjacent healthy cells.

This book, written for teachers and students and for research scientists, is on the cutting edge of a most exciting realm of science and medicine. It should not only help the reader to become more knowledgeable and versant in nanomedicine and nanotechnology but also stimulate the reader to be creative and go beyond the topics and ideas presented here.

Terre Haute, IN Renat R. Letfullin
St. Louis, MO Thomas F. George

Contents

Author Bios

Renat R. Letfullin is a faculty member in the Department of Physics and Optical Engineering at Rose-Hulman Institute of Technology and holds an appointment as full Professor of Physics and Nanotechnology at Radiological Technologies University. He is also Founder and Director of the Nanomedicine program at Radiological Technologies University.

Dr. Letfullin has made seminal contributions to the fields of wave and quantum optics, lasers, aerosol physics, and nanoscience, including nanooptics and nanobiotechnology. He is an outstanding theoretician and experimentalist in the fields of optics and kinetics of chemical pulsed lasers and, during the past decade, has branched out into biophotonics and nanomedicine. He has discovered new phenomena and scientific effects in these fields, where examples include effect of diffractive multifocal focusing of plane and spherical waves, effect of giant laser energy gain, the autowave of a new type—self-supporting photon-branched chain reaction, and effect of focusing of de Broglie matter waves. He has carried out an extensive array of studies in nanoscience, including laser heating and evaporation of nanostructures and optical properties of nanoparticles. He has developed and designed new self-contained laser systems with multi-mega-joule output energy per pulse, optical reactor for efficient laser processing of dispersed materials, phase technique for optical diagnostics of small particles, and new diffractive atom lens. His latest achievements are in the field of nanophotonics and nanomedicine, including discoveries of a plasmon explosion of nanoparticles, new dynamics modes in selective nanophototherapy, and RF and X-ray optics of nanoparticles.

Dr. Letfullin is a world-class physicist and an expert in the areas of optics, laser physics, nanophotonics, and nanomedicine. He has been recognized nationally and worldwide, such as being elected as a SPIE (Society of Photo-Optical

Instrumentation Engineers) Fellow and named an OSA (Optical Society of America) Senior Member for achievements in optics, photonics, laser physics, nanophotonics, biophotonics, and nanomedicine. A well-respected international researcher, Dr. Letfullin has authored over 150 papers and conference proceedings, edited 4 books, and coauthored 12 chapters in 7 different books. His work has garnered several grants and awards including 4 patents in laser technology and optical engineering.

Dr. Letfullin has made numerous and important contributions to designing and teaching physics, optics and biomedical optics courses, and developing a national first Master of Science degree program in nanomedicine. He has led the development of several new research laboratories on biophotonics and nanomedicine, created innovative teaching tools including online video courses and webinars, and supervised numerous masters and doctoral students.

Dr. Letfullin actively serves the research community as an editor for the research journal *International Journal of Theoretical Physics, Group Theory and Nonlinear Optics* published by Nova Science. He is an editorial board member and reviewer for a number of high-ranking scientific journals, and he has chaired numerous conferences.

Born in Russia, Dr. Letfullin received a Bachelor of Science degree in Physics and earned a Master of Science degree in Optics and Spectroscopy from Samara State University. He received a Doctor of Philosophy degree in Laser Physics from Saratov State University, followed by junior/senior research appointments at P. N. Lebedev Physics Institute of Russian Academy of Sciences (Samara Branch). He then served as a visiting faculty at Mississippi State University, and since 2004 joined the faculty at Rose-Hulman Institute of Technology. He also serves as a faculty member and program director at Radiological Technologies University.

Dr. Letfullin has been married since 1992 to Tatiana Letfullin who is an accomplished classical pianist. They are proud to have two daughters, Alla and Mary. Alla is Nanoscience Ph.D. student at Joint School of Nanoscience and Nanoengineering—a collaborative project of North Carolina A&T State University and the University of North Carolina. Youngest daughter Mary is a cheerful girl who enjoys art.

Thomas F. George is chancellor and professor of chemistry and physics at the University of Missouri—St. Louis. As chancellor, he oversees academic and administrative operations of a university with 17,000 students, 2500 faculty and staff, 40 academic buildings, and a $200 million annual operating budget. In addition, $150 million of new capital construction, renovation, and landscaping on campus is taking place during 2014–2016. The annual amount of philanthropic contributions during his tenure as chancellor has quadrupled to over $26 million, reaching as high as $31 million in 2014.

Chancellor George is active in the St. Louis community. He has received the Distinguished Higher Education Award from the Dr. Martin Luther King, Jr., Missouri State Celebration Commission, the Outstanding Community Service Award from the St. Louis County Branch of the National Association for the Advancement of Colored People, the Lifelong Vision Award from the Lifelong Vision Foundation of St. Louis, and the Silver Beaver Award and Good Scout Award from the Greater St. Louis Area Council of the Boy Scouts of America; and he has been named St. Louis North County Chamber of Commerce Citizen of the Year. He serves on numerous civic boards and currently is chair of the Higher Education Consortium of Metropolitan St. Louis, and he served for 4 years as chair of the Great Lakes Valley Conference (athletics). In addition to his role as campus and community leader and fund-raiser, Chancellor George contributes at the state and national levels, such as currently serving as vice chair of the Council on Public Higher Education in Missouri and president of the Coalition of Metropolitan and Urban Universities (comprised of 95 institutions across the USA and Canada).

Chancellor George is an active researcher in chemistry and physics, specializing in chemical/materials/laser physics and nanoscience, including nanomedicine. His work, supported by a wide variety of federal and international agencies, has led to 775 papers, 7 authored and 18 edited books, and 225 conference abstracts. He has been recognized nationally and worldwide, such as being awarded the Marlow Medal by the Royal Society of Chemistry in Great Britain and named a fellow in various professional organizations (New York Academy of Sciences, American Physical Society, Society of Photo-Optical Instrumentation Engineers, American Association for the Advancement of Science) and foundations (Dreyfus, Sloan, Guggenheim). He has been elected as a foreign member of the Korean Academy of Science and Technology, and he has held the title of visiting professor of physics at Korea University in Seoul. His research has been acknowledged by an honorary doctorate in physics (*honoris causa*) from the University of Szeged in Hungary. He also has received an honorary doctorate in education for locality development from Phranakhon Rajabhat University in Thailand. He has been awarded the Medal of Honor from Gulf University for Science and

Technology in Kuwait and the Diploma of Honour from Seinäjoki University of Applied Sciences in Finland.

An accomplished jazz pianist, Dr. George has studied with faculty at the Berklee College of Music in Boston and the Eastman School of Music in Rochester, New York. He has performed extensively in public, including recent performances at various venues in Missouri and Illinois, the University of Arkansas, Gettysburg College in Pennsylvania, Nanjing University and Northwestern Polytechnical University in China, city of Szeged in Hungary, West University of Timisoara in Romania, Gulf University for Science and Technology in Kuwait, University of Sarajevo in Bosnia, and University of Dubrovnik in Croatia, and the Cultural Club in Muscat, Oman. He has also performed in Russian Siberia.

Born in Philadelphia, Chancellor George became an Eagle Scout (with bronze, gold, and silver palms) and received his high school diploma from Friends' Central School, where he earned varsity letters in soccer and wrestling. He received a Bachelor of Arts degree (*Phi Beta Kappa*) with a double major in chemistry and mathematics from Gettysburg College in Pennsylvania. He earned a Master of Science degree and Doctor of Philosophy degree in theoretical chemistry from Yale University at age 23, followed by postdoctoral appointments at MIT and UC Berkeley. Joining the faculty at the University of Rochester, he was promoted by age 29 to full professor of chemistry. He then served as dean of natural sciences and mathematics at SUNY—Buffalo, provost at Washington State University, chancellor at the University of Wisconsin—Stevens Point, and since 2003 as chancellor at the University of Missouri—St. Louis.

Dr. George has been married since 1970 to Dr. Barbara Harbach, Curators' Professor of Music, director of the School of Fine and Performing Arts, and director of Women in the Arts at the University of Missouri–St. Louis. They are the proud owners (actually, servants) of four cats—two tortoise females (Luna and Stella) and two younger, mischievous males (Jeremiah and Jarrett).

Chapter 1
Introduction to Nanomedicine

Contents

Abstract This chapter presents the definition of nanomedicine and nanotechnology, describes the relation of nanomedicine to the basic sciences, introduces the functions and properties of medical nanodevices, the design and fabrication of nanorobots and nanoparticles, and discusses the current and potential applications of nanomedicine.

Electronic supplementary material: The online version of this chapter (doi:10.1007/978-3-319-43577-0_1) contains supplementary material, which is available to authorized users.

R.R. Letfullin, T.F. George, *Computational Nanomedicine and Nanotechnology*,
DOI 10.1007/978-3-319-43577-0_1

1.1 Lecture 1: Definition of Nanomedicine and Nanotechnology (Slide 1.1)

- New Technological Frontiers in Twenty-First Century
- Relation of Nanomedicine to the Basic Sciences
- Definition of Nanomedicine
- Definition of Nanotechnology
- Imaging and Treatment Modalities: Advantages and Disadvantages
- Conclusions

The last century was a renaissance time for many sciences as shown on Slide 1.2. In physics, the twentieth century began with the discovery of quantum mechanics and theory of relativity, breakthroughs in atomic and nuclear physics, and the invention of superconductivity and lasers. Microelectronics brought fast computers and high-speed communication, information technologies, and the invention of the Internet. Transportation also changed a lot during the twentieth century. But where is science and technology going today in the current century?

In the natural sciences, there are three main science and technological frontiers in the twenty-first century: nanotechnology, biology and information technology as shown on Slide 1.2. Nanomedicine is a new science which combines two new technological frontiers—nano and bio, i.e., applying nanotechnology to biology with medical applications. Nanotechnology is revolutionizing medicine, giving unprecedented opportunities to treat the human body at the atomic and molecular levels like we are recrafting conventional machines today.

What is nanomedicine? An efficient way to define this is to find its place among the established sciences. Slide 1.3 shows the interdisciplinary relation of nanomedicine to basic sciences and engineering. Biophysics itself is a

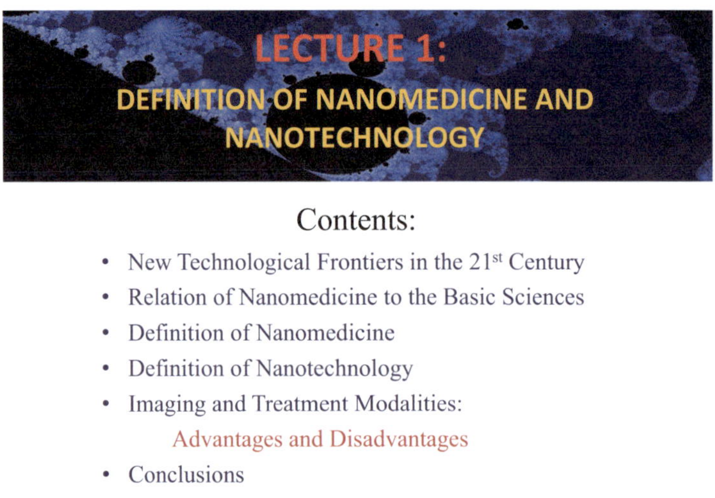

Slide 1.1 Contents of Lecture 1

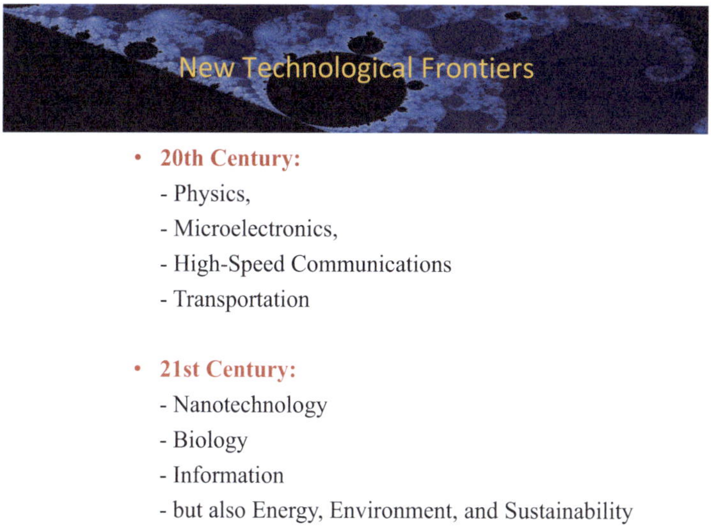

Slide 1.2 New technological frontiers

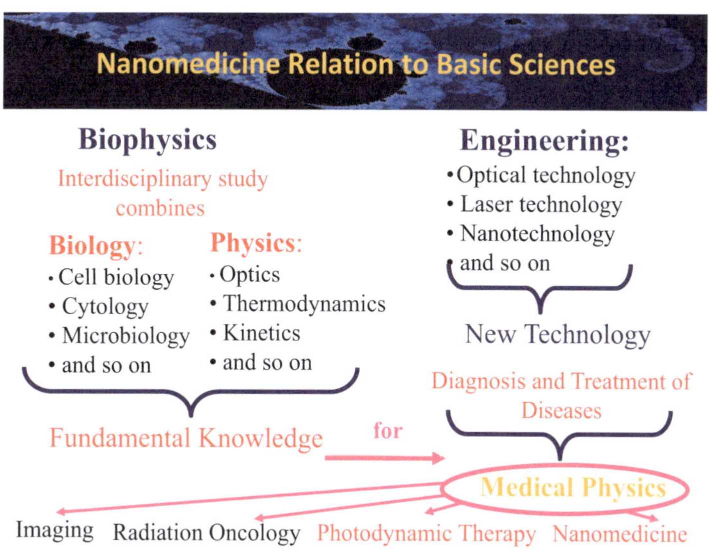

Slide 1.3 Relation of nanomedicine to the basic sciences

multidisciplinary science, combining biology and physics. From biology, nanomedicine involves disciplines like cell biology, microbiology, cytology etc. Physics provides nanomedicine with basic knowledge in optics, thermodynamics, kinetics, so on. The chemistry is needed for nanomedicine, for example, to target the cancer cells by nanodrugs, and to understand the chemical bonding between protein molecules of antibodies and the cell to the surface atoms of nanoparticles.

- Nanomedicine is an interdisciplinary field of science which involves physics, engineering biology, chemistry, medicine and other sciences for diagnosis and treatment of diseases by means of nanotechnology.
- The ultimate goal of nanomedicine is to treat the human body at the atomic and molecular levels, and to repair a body like we repair conventional machines today.

Slide 1.4 Definition of nanomedicine

Nanomedicine has strong application components as well, and it uses engineering tools like nanotechnology, laser technology, nanophotonics, etc. Thus, biophysics and chemistry provide the fundamental knowledge for nanomedicine, and engineering supplies new technologies and tools for nanomedicine. All this is needed and designed for *diagnosis and treatment of different diseases.*

The diagnosis and treatment of diseases by means of physics and engineering tools falls under the large area called *medical physics*, which has many branches like radiography, brachytherapy, nuclear imaging, fluoroscopy, etc. Photodynamic therapy and nanomedicine are parts of medical physics, where nanomedicine utilizes the established disciplines of biophysics, chemistry and engineering for the diagnosis and treatment of diseases.

Now we are ready to give a definition of nanomedicine (Slide 1.4):

Nanomedicine is an interdisciplinary science which involves physics, biology, chemistry, optics, medicine, and engineering for diagnosis and treatment of diseases by means of nanotechnology.

The ultimate goal of nanomedicine is to treat the human body at the atomic and molecular levels and repair the body like we are repairing the conventional machines today.

A key point in the definition of nanomedicine is nanotechnology (Slide 1.5). What is nanotechnology? The word "nano" comes from the Greek language and means "dwarf", i.e., something very small. When we say "nano," specifically we mean "one billionth" of something. For example, *one nanometer* is *one billionth of a meter*, i.e., 10^{-9} m. "*Any research and development at the atomic, molecular, or macromolecular levels, on the length scale of approximately 1–100 nm range, is called nanoscience* [1]".

The behavior and the properties of matter at this size range is quite different from the behavior and properties of bulk material. *Nanotechnology creates and designs structures, devices and systems that have novel properties and functions because of their small and/or intermediate size* [1]. Nanotechnology involves purposeful imaging, measuring, modeling, and manipulating matter at this length scale [1].

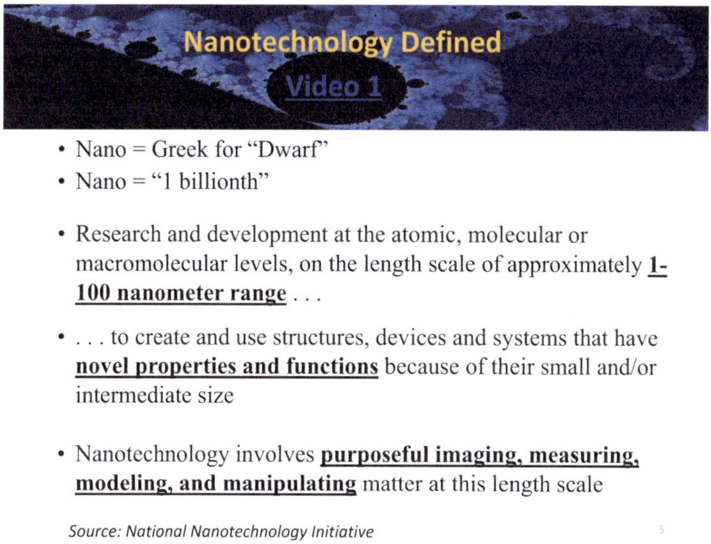

Slide 1.5 Definition of nanotechnology

Dr. Richard Feynman started a nanotechnology discussion in his lecture "There's Plenty of Room at the Bottom" in 1959 at a meeting of the American Physical Society. He stated that (1) it should be possible to build machines small enough to fabricate objects with atomic precision; (2) if information could be written on an atomic scale, the entire 24 volumes of the *Encyclopedia Britannica* could be placed on a pinhead; (3) an electron microscope could be made to see individual atoms; and (4) microcomputers could be built. There are many short videos available on YouTube presenting Feynman's vision of the small world.

How can the nanoworld actually help conventional medicine today? Current clinics already have many imaging and treatment modalities as listed on the Slide 1.6, like *radiography, fluoroscopy, mammography, computed tomography, nuclear medicine, positron emission tomography, magnetic resonance imaging, chemotherapy, magnetic field treatment, microwave and RF treatments.* Definitely they have some benefits as shown on the Slide 1.7. For example, conventional clinical modalities allow deep penetration into the body, with slice scans at specific regions, and with the fast scanning speeds, the entire body scan takes less than an hour. They provide accurate information on the position, density, structure and boundaries of internal organs. And most importantly, these modalities extend patients' lives.

However, besides the benefits, there are disadvantages of conventional imaging and treatment modalities listed on Slide 1.8. Today's clinical modalities are unable to detect and treat diseases at the single cell or DNA levels. But cancer, for example, is a DNA mutation disease. Missing the mutations of DNA results in late diagnosis and treatments leading to metastasis with the disease around the body. The spatial resolution of most modalities is about 1 mm, and a 1 mm tumor already contains thousands of cancer cells since a single cell has a size range of

Traditional Imaging and Treatment Modalities

- Radiography
- Fluoroscopy
- Mammography
- Computed tomography
- Nuclear medicine
- Positron emission tomography
- Magnetic resonance imaging
- Chemotherapy
- Magnetic field treatment
- Microwave and RF treatment

Slide 1.6 Traditional imaging and treatment modalities

Benefits of Conventional Clinical Modalities

- Deep penetration into the body
- Whole body scan
- Slice scan
- Fast scan
- Position accuracy
- Density information
- Organs structure and boundaries
- Extend patient life

Slide 1.7 Benefits of the conventional clinical modalities

10–20 μm. Those cells metastasize throughout the body via blood circulation. Because of the resolution limit, the today's imaging modalities can detect only tumors larger than 1 mm.

Finally, when we see a tumor, we have three main treatment methods available today: chemotherapy, radiation therapy and surgery. All of them are unselective, damaging healthy tissue and organs with many side effects. For example, X-ray radiation treatments at therapeutic doses can cause secondary cancer. Today's

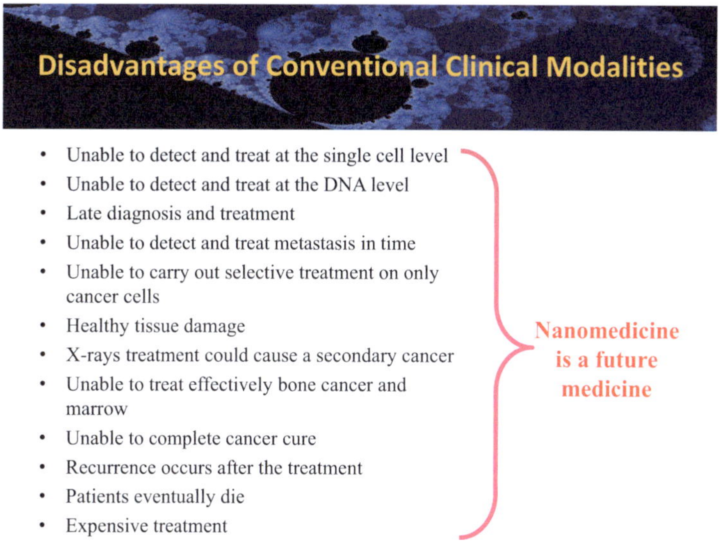

Slide 1.8 Disadvantages of the conventional clinical modalities

modalities are unable to treat effectively bone cancer and marrow. But marrow is responsible for the body's immune system to produce blood cells fighting any disease. For instance, when a cancer reaches the marrow the only treatment option we have today is marrow transplantation by shutting down an immune system of the patient first, i.e., sterilizing a patient's marrow by radiation. This procedure puts the patient at very high risk to be killed by any germs like a flu. Today's clinical modalities are unable to completely cure cancer, where recurrences can occur after the treatment, and many patients eventually die from the cancer. One more disadvantage is that chemotherapy, radiation therapy and surgery are very expensive with often unpredictable results.

The nanomedicine is able to resolve the abovementioned deficiencies of imaging and treatment modalities we have in clinics today. Nanomedicine can operate at the single cell and gene levels of detection, which provides early diagnosis and treatment of diseases, solving the metastasis problem. The nanomedicine is a selective treatment, with no side effects, and has capability for effective treatment of bone cancer and marrow.

In conclusion, nanomedicine is a smart medicine, and it is an exciting future medicine (Slide 1.9). It will provide fast and painless selective treatment without a side effects. We believe that via 3–5 year immunization cycle with nanodrugs, nanomedicine will defeat cancer for good and give life for cancer patients.

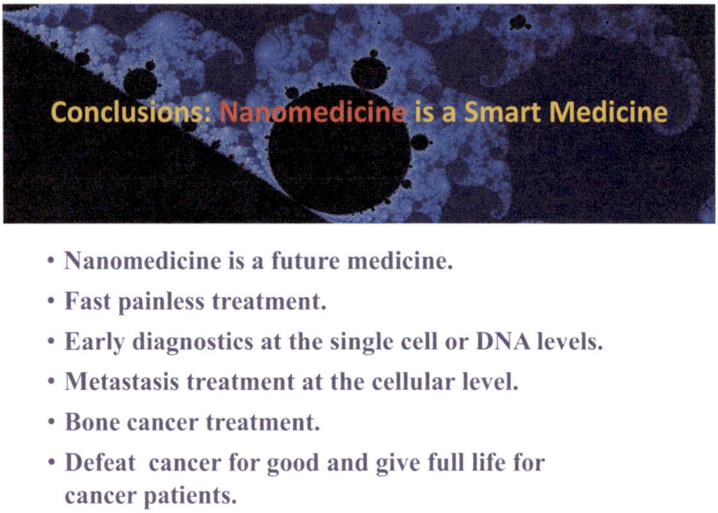

Slide 1.9 Conclusions

1.2 Lecture 2: Medical Nanodevices, Properties, and Functions (Slide 2.1)

- Medical Nanomachines and Nanorobots
- Nanoparticles in Medicine
- Functions and Properties of Medical Nanodevices
- Removing Used Nanorobots
- Conclusion

In recent years, there has been a tremendous increase in research at the nanoscale for materials. Nanomedicine utilizes several varieties of nanostructured materials and nanodevices currently being developed: *nanorobots, nanoparticles, nanochips, quantum dots, nanovalves,* etc. *Nanorobots* are defined as nanomachines capable of operating with nanosized objects and with nanoscale precision. *Nanorobots* are also known as *nanobots, nanoids, nanites, nanomachines* or *nanomites. Nanoparticles* are tiny particles of matter made of any material (metal, semiconductor, glass, polymer, organic, liquid or combinations of different materials) of any shape (nanosphere, nanoshell, nanorod, nanotube, fullerene) and structure (solid, porous, bubble, bilayer) in size range 1–100 nm designed to perform varies functions and manipulation of matter at this length scale. A *quantum dot* is a tiny piece of semiconductor material where the charge carriers are confined in all three dimensions. This lecture briefly describes medical nanodevices, nanoparticles, and their properties and functions.

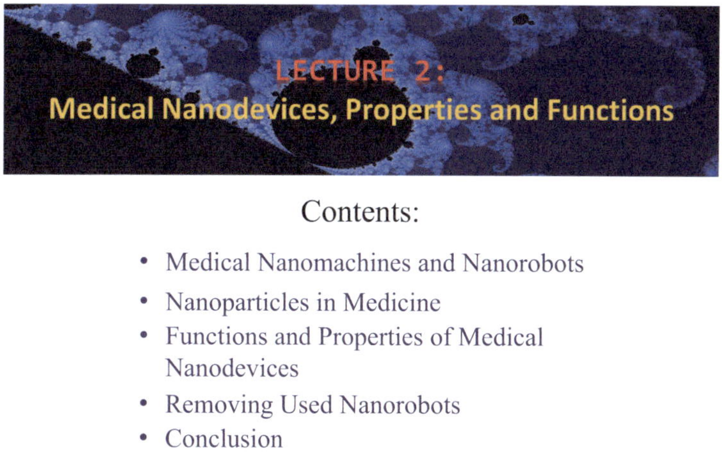

Slide 2.1 Contents of Lecture 2

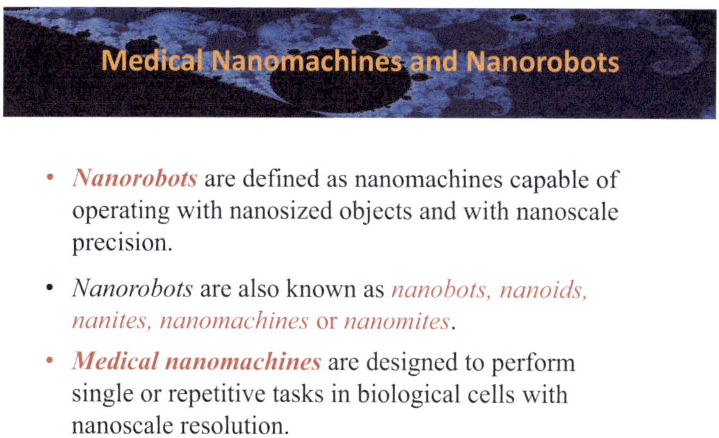

Slide 2.2 Medical Nanomachines and nanorobots

1.2.1 Medical Nanomachines and Nanorobots

Nanorobotics represent a new molecular technology field that has grown over the years, creating robots or machines made up of parts that are at or close to the scale of nanometers (Slide 2.2). Having evolved through several developmental stages, this field is still under research and development. Nanodevices have opened a new world of discoveries and feasibilities in many fields, such as space exploration, military defense, medicine, industry and energy. Applications of nanorobots in medicine are briefly discussed in Lecture 3.

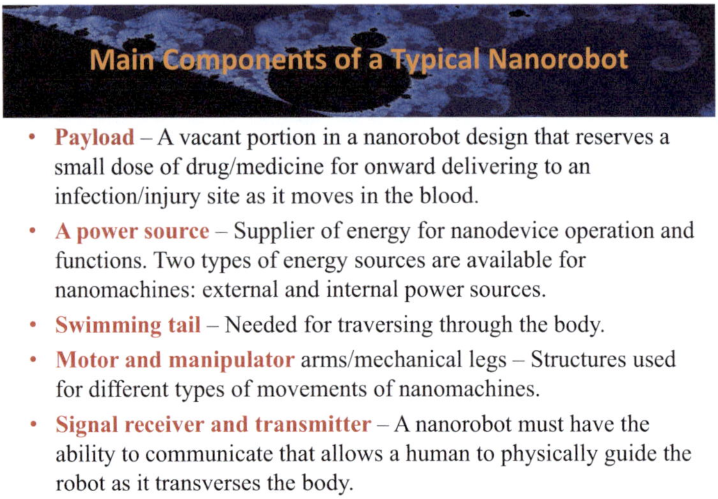

Slide 2.3 Main components of atypical nanorobot

Medical nanomachines are designed to perform single or repetitive tasks in biological cells with nanoscale resolution. Nanorobots could be designed as self-replicating machines. The main components of the typical nanorobot could include (Slide 2.3):

- Payload—This is a vacant portion in nanorobot design that reserves a small dose of drug/medicine for onward delivering to an infection/injury site as it moves in the blood.
- Power source—This supplies the energy for the nanodevice operation and functions. Two types of energy sources are available for nanomachines: external and internal power sources described below.
- Swimming tail—This is needed for traversing through the body.
- Motor and manipulator arms/mechanical legs—These are structures used for different types of movements of nanomachines.
- Signal receiver and transmitter—Nanorobots must have the ability to communicate, allowing humans to physically guide the robot as it transverses the body.

Currently built medical nanodevices include (Slide 2.4) *three-dimensional DNA crystals, nanovalves for drug delivery, molecular propellers, nanochips, DNA nanorobots, nanorobots treating hepatitis C*, etc.

A three dimensional DNA crystal is an assembly of synthetic DNA molecules forming 3D crystal structures capable of hosting guest structures (Slide 2.4). These porous crystals can be used as molecular containers to build biochips, nanorobots, biodegradable solid-state catalysts and biosensors, or for delivery of enzymes in enzyme replacement therapies.

Examples of Current Medical Nanodevices

- *Three-dimensional DNA crystal, nanovalves for drug delivery, molecular propellers, nanochips, DNA nanorobots, nanorobotstreating hepatitis C*, etc.
- ***A three-dimensional DNA crystal*** is an assembly of synthetic DNA molecules forming 3D crystal structures capable of hosting guest structures. These porous crystals can be used as molecular containers to build biochips, nanorobots, biodegradable solid-state catalysts and biosensors, or for delivery of enzymes in enzyme replacement therapies.

Slide 2.4 Examples of current medical nanodevices

Examples of Current Medical Nanodevices (continued)

- ***Nanovalves*** *for drug delivery* are tiny valves placed in the pores of nanostructures, for example mesoporous silica nanoparticles, filled by drug for controlled release of drugs. The nanovalves open and close in response to change of pH level in the cellular environment.
- ***Molecular propellers*** are special shaped molecules having several molecular-scale blades tilted at a different pitch angle around the rotational axis. The molecular propellers are able to push and pump fluids when rotate by using the molecular motors. ***Nanochips*** are used for DNA sequences analysis – DNA hybridization, which is a process of pairing of separated DNA strands with complementary strands of known sequence. These strands act as probes. Nanochips use electric current to separate DNA probes.

Slide 2.5 Examples of current medical nanodevices (continued)

Nanovalves for drug delivery are tiny valves placed in the pores of nanostructures, such as mesoporous silica nanoparticles, filled by drugs for controlled drug release (Slide 2.5). The nanovalves open and close in response to a change of pH level in the cellular environment. It is found that a pH factor of normal cells differs from tumor cells. This fact can be used for selective delivery and release of the drug to the abnormal cells only.

Molecular propellers are special shaped molecules having several molecular-scale blades tilted at a different pitch angle around the rotational axis (Slide 2.5). The molecular propellers are able to push and pump fluids when rotated by using the molecular motors. In nature, there are many examples of sophisticated molecular motors.

Nanochips are used for DNA sequences analysis, specifically DNA hybridization, which is the process of pairing separated DNA strands with complementary strands of known sequence (Slide 2.5). These complementary strands act as probes. Nanochips use an electric current to separate DNA probes. The current technique of DNA hybridization is tedious, time consuming and prone to errors, whereas the nanochip is quick, analyzing the DNA in minutes instead of hours.

1.2.2 Nanoparticles in Medicine

Nanoparticles are particles made of different materials and come in variety of shapes and structures in a size range 1–100 nm (Slide 2.6). Nanoparticles can be made from almost any material, including semiconductors, metals, and even structures that are organic in nature. Through various processing techniques, nanoparticles can be multi-layered, or even hollow, allowing for a variety of uses. These are divided into two broad types (Slide 2.6): inorganic material-based nanoparticles and organic material-based nanoparticles. The former have the primary advantage of being designed for the individual purpose, whereas the latter are less designable and are chosen for their inherent attributes. The body is quick to reject foreign substances, so the use of organic material as a protective coating on inorganic particles is a way to make the body accept nanoparticles. Nanoparticles

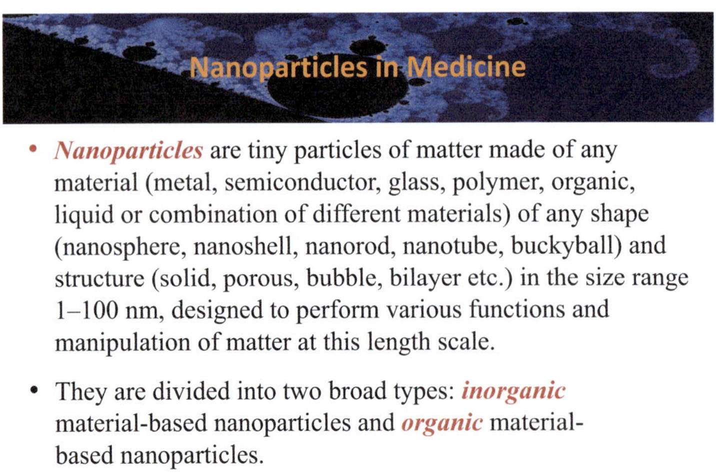

Slide 2.6 Nanoparticles in medicine

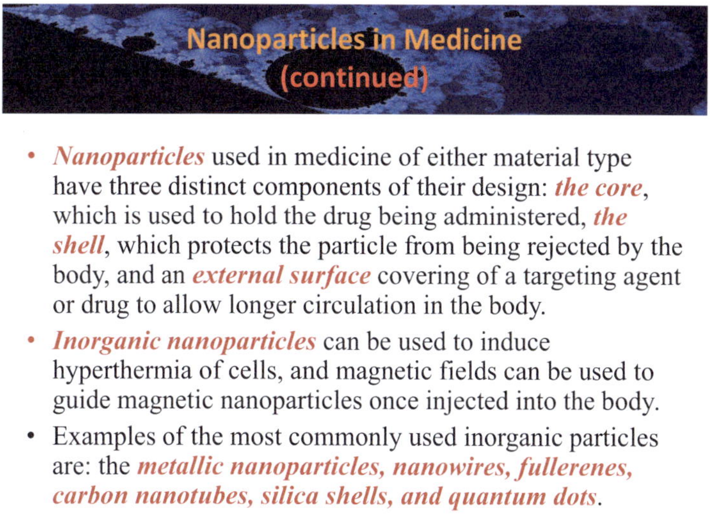

Slide 2.7 Nanoparticles in medicine (continued)

used in medicine of either material type have three distinct components of their design (Slide 2.7): (1) the core which is used to hold the drug being administered, (2) the shell which protects the particle from being rejected by the body, (3) and a surface covering of a targeting agent or drug to allow longer circulation in the body. Below we provide a few examples of nanoparticles of different types, which can be practical for nanomedical procedures.

Inorganic nanoparticles can be used to induce hyperthermia of cells, and magnetic fields can be used to guide magnetic nanoparticles once injected into the body (Slide 2.7). A hyperthermia treatment is the placement of nanoparticles on tumor cells followed by an energizing of the particles via electromagnetic waves. The nanoparticles effectively burn the tissue, and this can be used as either a method to kill tumor cells or as a means to increase the uptake of therapeutic drugs used in the treatment. The examples of most commonly used inorganic particles are: *metallic nanoparticles, nanowires, fullerenes, carbon nanotubes, silica shells, and quantum dots.*

Metallic nanoparticles are mostly made of noble metals, like gold, silver and platinum, or metal oxides, like iron oxide, nickel oxide and cobalt oxide, and have a wide variety of applications in diagnostics and therapy (Slide 2.8). For example, the imaging modalities used in clinics today—magnetic resonance imaging (MRI), computed tomography (CT), positron emission tomography (PET)—and optical imaging require contrast agents with specific morphological properties. By using metal nanoparticles of a specific design, we can create contrast enhancers that locate themselves strictly on cells of interest (through fine—tuned targeting agents) and also last longer, allowing us to watch the growth, death or movement of tumor cells over a longer period of time. Metal nanoparticles are used in the

Metallic Nanoparticles

- ***Metallic nanoparticles*** are mostly made of noble metals, like gold, silver and platinum, or metal oxides, like iron oxide, nickel oxide and cobalt oxide, and have a wide variety of applications in diagnostics and therapy. .
- ***Imaging modalities*** used in clinics today (MRI, CT, PET and optical imaging) require contrast agents with specific morphological properties. By using metal nanoparticles of a specific design, we can create contrast enhancers that locate themselves strictly on cells of interest (through fine-tuned targeting agents) and also last longer, allowing us to watch the growth, death or movement of tumor cells over a longer period of time.
- The metal nanoparticles are used in the therapy of cancer. Examples include ***nanophotothermolysis*** and ***nanophotohyperthermia***, where the metal nanoparticles delivered to the cancer cells are heated by radiation to ablate a tumor in the targeted area.

Slide 2.8 Metallic nanoparticles

Carbon Nanoparticles

- ***Fullerene*** is the fourth allotrope of carbon, consisting of 60 carbon atoms forming specific spherical shape, *buckyball*. Its unique properties, such as heat resistance and superconductivity, enable applications in high-temperature superconducting materials, high-power optical materials, high-performance catalyst, and high-strength composite materials.
- ***Carbon nanotubes*** are arrangements of planar carbon lattices into cylinders of various chiralities with sizes reaching several millimeters in length and only a few nanometers wide. In biology, the ability of nanotubes to get inside cell nuclei makes them invaluable in the diagnosis, imaging and treatment phases of cancer therapy. Carbon nanotube can be used to recognize target DNA sequences, as a drug delivery system, used in artificial heart valves, tooth roots, bones, etc.

Slide 2.9 Carbon nanoparticles

therapy of cancer (Slide 2.8). Examples include nanophotothermolysis and nanophotohyperthermia, where the metal nanoparticles delivered to the cancer cells are heated by radiation to ablate a tumor in the targeted area.

Fullerene is the fourth allotrope of carbon, consisting of 60 carbon atoms forming a specific spherical shape, *buckyball* (Slide 2.9). Its unique properties, such as heat resistance and superconductivity, enable applications in high-temperature superconducting materials, high-power optical materials, high-performance catalyst, and high-strength composite materials.

Carbon nanotubes are arrangements of planar carbon lattices into cylinders of various chiralities, with sizes reaching several millimeters in length and only a few nanometers wide (Slide 2.9). In biology, the ability of nanotubes to get inside cell nuclei makes them invaluable in the diagnosis, imaging and treatment phases of cancer therapy. Carbon nanotube can be used to recognize target DNA sequences, as a drug delivery system, used in artificial heart valves, tooth roots, bones, blood vessels, artificial ligaments and tendons.

Silica-based nanoparticles (SiNPs) made of silicon dioxide have uniform morphology, adjustable pore volume, controllable diameter, modifiable surface potential, easy functionalization and synthesis (Slide 2.10). There are two types of silicon nanoparticles—mesoporous (MSNs) and core/shell nanoparticles (C/S-SiNPs). MSNs have been successfully used for drug delivery because of unique mesopores and nanochannels which allow for a high payload of the drug and easy stimulated release. C/S-SiNPs, like silica-coated gold nanoparticles, are ideal for imaging agent delivery because the agents used, such as fluorophores, quantum dots and gold nanoparticles, can be easily doped into the C/S-SiNPs. The shell structures can protect imaging agents inside, while the agents give precise locations of the tumor tissue. Size is crucial to delivery and drug administration, pathway of cell uptake, biodistribution, and targeting efficiency. It has been found that C/S-SiNPs operate best in the range 10–100 nm, while MSNs are optimal with a 50 nm diameter for cell uptake. It has been found that the smaller the particles, the more toxic to the cells.

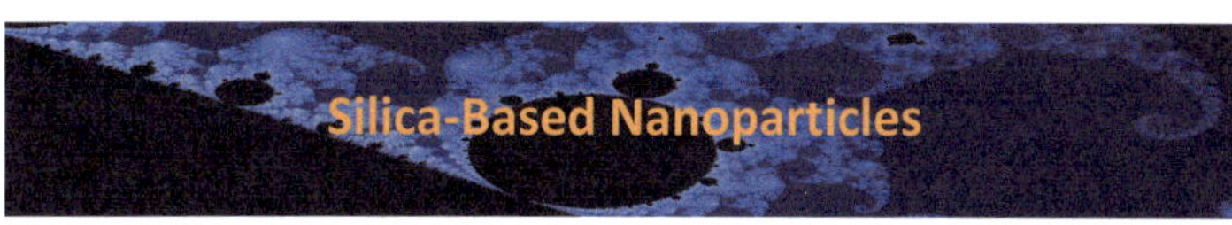

- ***Silica-based nanoparticles*** (SiNPs) made of silicon dioxide have uniform morphology, adjustable pore volume, controllable diameter, modifiable surface potential, easy functionalization and synthesis.
- There are two types of silicon nanoparticles, ***mesoporous (MSNs)*** and ***core/shell nanoparticles (C/S-SiNPs).***
- ***MSNs*** have been successfully used for drug delivery because of unique mesopores and nanochannels which allow for a high payload of the drug and easy stimulated release.
- ***C/S-SiNPs***, like silica-coated gold nanoparticles, are ideal for imaging agent delivery because the agents used, such as fluorophores, quantum dots and gold nanoparticles, can be easily doped into the C/S-SiNPs.

Slide 2.10 Silica-based nanoparticles

Quantum Dots

- ***Quantum dots*** are another example of inorganic nanoparticles that can be used for the purpose of increased imaging ability.
- These dots are ***small semiconductor crystals*** that we can be fine-tuned to have very specific spectral bands, allowing differentiation between the location of the quantum dot (on the tumor cell) and the healthy tissue beside it with a level of precision not possible before.
- The labeling of multiple disease markers with ***quantum dot-based barcodes*** could pave the way towards more sensitive and accurate disease detection systems. Quantum dot-based probes may be used to study the heterogeneity of disease markers and link them to prognosis for improved diagnostic strategies.

Slide 2.11 Quantum dots

Quantum dots are another example of inorganic nanoparticles that can be used for the purpose of increased imaging ability (Slide 2.11). The dots are small semiconductor crystals that can be fine-tuned to have very specific spectral bands, allowing differentiation between the location of the quantum dot (on the tumor cell) and the healthy tissue beside it, with a level of precision not possible before. The labeling of multiple disease markers with quantum dot-based barcodes could pave the way towards more sensitive and accurate disease detection systems. Quantum dot-based probes may be used to study the heterogeneity of disease markers and link them to prognosis for improved diagnostic strategies.

Organic particles are used primarily for drug delivery or as a coating on inorganic particles to assure they are not rejected by the body's immune system (Slide 2.12). The debate between using organic or inorganic particles isn't simply a matter of what works best, but also a matter of what hurts less—the toxicity level of nanoparticles vary widely and must be taken into consideration when determining what treatment/diagnosis route to take. Examples of organic particles include *polymer-based nanostructures, dendritic nanoparticles, liposome nanoparticles, polymersome nanoparticles, lipid formulations, micelle nanoparticles, nanoemulsions, alginates, biological nanoparticles,* etc.

Polymer-based nanostructures are created from conjugating several functional units and soluble macromolecules or by co-polymer self-assembly (Slide 2.12). Many of these structures are loaded with therapeutic or imaging agents, which through diffusion or stimulation of local environment control the release. Polymeric nanoparticles are a class of nanocarriers established for numerous drug delivery applications. Polymeric cores are shielded by materials, such as polyethylene glycol (PEG), to guarantee the structure's stability. Targeting molecules can easily be added to the surface of polymer-based nanoparticles.

Organic Particles

- ***Organic particles*** are used primarily for drug delivery or as a coating on inorganic particles to assure they are not rejected by the body's immune system.
- Examples include ***polymer-based nanostructures, dendritic nanoparticles, liposome nanoparticles, polymersome nanoparticles, lipid formulations, micelle nanoparticles, nanoemulsions, alginates, biological nanoparticles,*** *etc.*
- ***Polymer-based nanostructures*** are created from conjugating several functional units and soluble macromolecules or by co-polymer self-assembly. Many of these structures are loaded with therapeutic or imaging agents which, through diffusion or stimulation of the local environment, control the release. Polymeric nanoparticles are ***a class of nanocarriers*** established for numerous drug delivery applications.

Slide 2.12 Organic particles

Organic Particles (continued)

- ***A dendrimer nanoparticle*** is a type of polymeric nanostructure and can be synthesized as well-defined spherical structure over 1–10 nm in diameter. A single dendrimer is able to sense, image and perform therapeutic functions.
- ***Polymersome nanoparticles*** are drug delivery carriers having structural similarity to liposomes and made of synthetic polymer amphiphiles. These drug delivery systems have higher stability and higher fluidity on the sides/surface than liposomes.
- ***Lipid formulations.*** There are many types of lipid-based drug delivery systems. Such vehicles typically are comprised of a digestible lipid with a blend of surfactants, co-surfactants and potentially co-solvents.

Slide 2.13 Organic particles (continued)

Dendrimer nanoparticles are a type of polymeric nanostructure (Slide 2.13). They can be synthesized as well-defined spherical structures over 1–10 nm in diameter. Multiple types of these can be created based on the core structure from the polymerization process. A single dendrimer is able to sense, image and perform therapeutic functions. To achieve this, multiple function groups have to be coupled to the surface of a dendritic polymer.

Polymersome nanoparticles are drug delivery carriers having structural similarity to liposomes and made of synthetic polymer amphiphiles (Slide 2.13). When hydrated and extruded, they form polymer shell vesicles of about 100 nm by self-assembly. The ratio of hydrophilicity/hydrophobicity can be used to regulate the morphology of the nanoparticle. These drug delivery systems have higher stability and higher fluidity on the sides/surface than liposomes.

Lipid formulations. There are many types of lipid-based drug delivery systems (Slide 2.13). Such vehicles typically are comprised of a digestible lipid with (in the case of more complex self-emulsifying formulations) a blend of surfactants, co-surfactants and potentially co-solvents. The ability of a self-emulsifying drug delivery system to be diluted is essential for use as a drug delivery vehicle since, after administration, it will be diluted by intestinal media.

Micelle nanoparticles are drug delivery carriers made up of lipids (Slide 2.14). They are self-assembled into small nanoparticles having a hydrophobic core (drug) within size range of 10–200 nm.

Biological nanoparticles are unicellular microorganisms with different shapes and sizes (Slide 2.14). For example, a biological nanoparticle called "nanocell" consists of globular bacteria having no nucleus, which is impeccable, as it wouldn't lead to mutations. A nanocell can be filled by drugs and used as a drug delivery system.

Hybrid nanoparticles are nanocarriers made up of the metallic and polymeric materials for the central core, which is coated with single or multiple layers of lipid to create a protective (surrounding) membrane (Slide 2.14). The hybrid nanoparticles are used as delivery systems with considerable high drug-carrying capacity, and the lipid layer on the surface of a nanoparticle is able to direct drug-releasing kinetics. They are also able to serve as imaging agents.

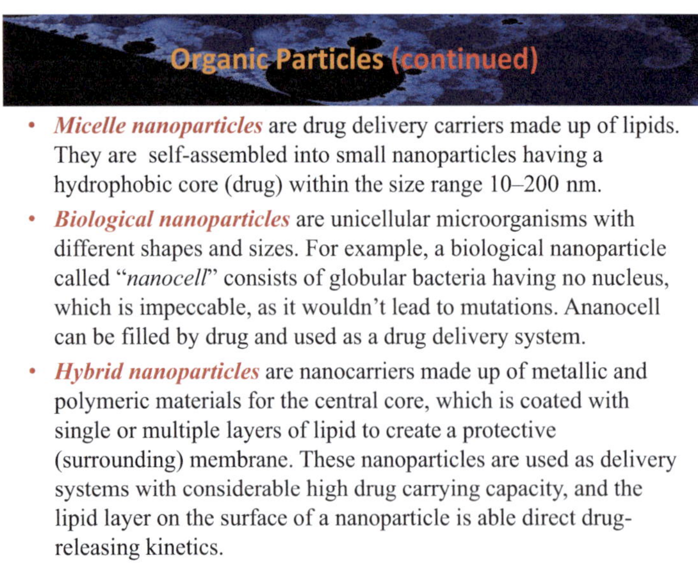

Slide 2.14 Organic particles (continued)

1.2.3 *Functions and Properties of Medical Nanodevices*

Medical nanomachines and nanoparticles are designed to perform several functions in the biological environment with nanoscale precision (Slide 2.15). They should be capable of actuating, sensing, signaling, information handling, and swarm behavior at nanoscale. These functions can either be demonstrated individually or in combinations. Some functions necessary for nanorobots include (Slide 2.15):

- Swarm intelligence—related to reorganization and distributive intelligence.
- Self-assembly and replication—refers to automated assembling and reproduction ability of a nanorobot.
- Information processing—refers to programming ability of a nanorobot.
- Nano to macro world interaction—relates to a framework which enables immediate access to nanorobots.

In order to perform these functions and activities in the body environment, the medical nanodevices should possess the following properties (Slide 2.16):

- Special morphological properties: structure, shape and size
- Surface chemistry
- Moving and swimming ability
- Biocompatibility
- Powering
- Communication
- Navigation

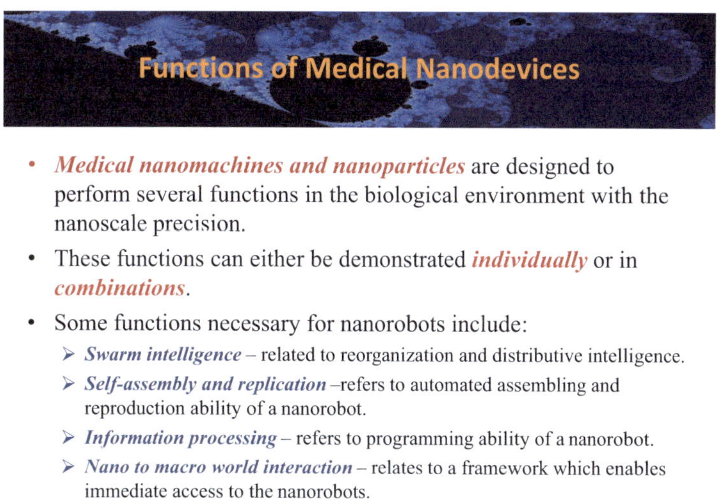

Slide 2.15 Functions of medical nanodevices

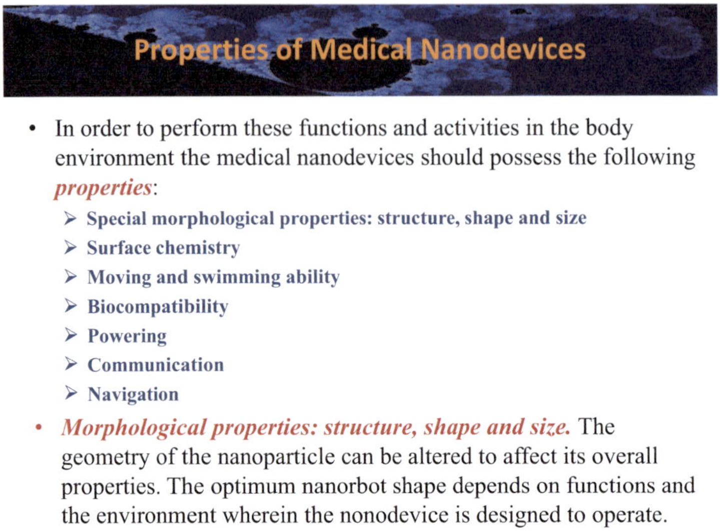

Slide 2.16 Properties of medical nanodevices

Morphological properties: structure, shape and size. The geometry of the nanoparticle can be altered to affect its overall properties (Slide 2.16). The optimum nanorobot shape depends on functions and the environment where the nonodevice is designed to operate. Below are few examples of nanomachine shapes and structures.

For example, *free-floating nanodevices* are nanorobots intended for material delivery, or storage devices operating independently with no preferred orientation (Slide 2.17). Thus, spherical shape offers the best volume efficiency for this type of nanorobot.

Swimming nanodevices are designed to move effectively though the blood stream (Slide 2.17). Less resistance to the surrounding fluids is needed for actively-swimming nanodevices. It can be achieved for smaller surface area in contact with fluid. The ellipsoid shape will be ideal for this type of nanomachine.

Intracellular nanodevices are designed to operate inside human tissues and cells to carry payloads like drugs, nanocomputers, power supplies, tools, communicational and navigational equipment (Slide 2.17). The spherical shape provides a maximum volume for volume storage efficiency. The carbon nanotubes with cylindrical shape are also considered to be effective nanocarriers.

Tessellating nanomachines involve cooperation by other nanodevices in close physical contact (Slide 2.18). They are designed to fit together in a mosaic pattern. The triangles, hexagons and polygonal prismatic shapes can provide a tight connection in designed order.

The shape of nanoparticles can be altered to affect its overall properties (Slide 2.18). For example, the aspect ratio (i.e., ratio of length to width) of a gold

Morphological Properties of Medical Nanodevices

- For example, **free-floating nanodevices** are nanorobots intended as material delivery or storage devices operating independently with no preferred orientation. Thus, the ***spherical shape*** offers the best volume efficiency.
- ***Swimming nanodevices*** are designed to move effectively though the blood stream. Less resistance to the surrounding fluids is needed for actively-swimming nanodevices. It can be achieved for smaller surface area in contact with fluid. The ***ellipsoid shape*** will be ideal for this type of nanomachine.
- ***Intracellular nanodevices*** are designed to operate inside human tissues and cells to carry payloads like drugs, nanocomputers, power supplies, tools, communicational and navigational equipment. The ***spherical shape*** provides a maximum volume for a volume storage efficiency. The carbon nanotubes with ***cylindrical shape*** are also considered to be effective nanocarriers.

Slide 2.17 Morphological properties of medical nanodevices

Morphological Properties (continued)

- ***Tessellating nanomachines*** involve cooperation by other nanodevices in close physical contact. They are designed to fit together in a mosaic pattern. The ***triangles, hexagons and polygonal prismatic shapes*** can provide a tight connection in designed order.
- ***The shape of nanoparticles*** can be altered to affect its overall properties. For example, the aspect ratio, i.e., ratio of length to the width) of a gold nanorod particle affects its absorbance spectra, which is much higher than the spherical-shaped gold nanoparticle of approximately similar size.
- The geometry of the nanoparticle has also been shown to affect its ***intake into the cells***.
- The nanoparticle geometry can also affect ***its half-life*** in the blood. The half-life is the amount of time it takes for the concentration of nanoparticles in the blood to be reduced by 50%.

Slide 2.18 Morphological properties (continued)

nanorod particle affects its absorbance spectra, which is much higher than the spherical-shaped gold nanoparticle of approximately similar size.

The geometry of the nanoparticle has also been shown to affect its intake into the cells (Slide 2.18). One method of cellular uptake of nanoparticles is receptor-mediated endocytosis, whereby the nanoparticle first attaches to a specific receptor on the cell surface that triggers its uptake. 50×14 nm gold nanorods have much less cellular uptake by receptor-mediated endocytosis than spherical colloids with diameters of 50 or 14 nm. This may suggest that a spherical object has a better chance of getting inside a cell than one with a high aspect ratio. However, carbon nanotubes have been taken up by cells through non-receptor-mediated endocytosis, regardless of the nanotube's surface functionalization or cell type. It has been suggested that this is due to the very high aspect ratio of the nanotubes.

Nanoparticle geometry can also affect its half-life in the blood (Slide 2.18). The half-life is the amount of time it takes for the concentration of nanoparticles in the blood to be reduced by 50 %. For example, research shows that in rats the filamentous micelles stay in circulation for over a week, while PEG-coated spherical vesicles are removed from the circulation within days.

The structure of the nanoparticle can affect its overall properties as well (Slide 2.19). Structurally, nanoparticles can be made as uniform core particles, porous nanoparticles, nanoshells, and single and multilayer nanoparticles with different surface chemistry. The nanoparticle core may be synthesized from a wide variety of materials in different geometries and sizes. Materials such as carbon in carbon nanotubes or gold in gold nanorods or silica and gold in gold nanoshells have been used to deliver thermal therapy to cancer cells, either through cytotoxic application of heat or in synergy with chemotherapeutics. For example, nanoshells have big advantages over uniform core nanoparticles: (1) nanoshells can contain strong drugs inside, and (2) their absorption spectrum can be tuned to different wavelengths of radiation by controlling the ratio of the thickness of the walls to the overall diameter of the shell (Slide 2.19).

- ***The structure of nanoparticles*** can affect their overall properties as well.
- Structurally the nanoparticles can be made as ***uniform core particles, porous nanoparticles, nanoshells, and single and multilayer nanoparticles with different surface chemistry***.
- ***The nanoparticle core*** may be synthesized from a wide variety of materials in different geometries and sizes. Materials such as carbon in carbon nanotubes or gold in gold nanorods or silica and gold in gold nanoshells have been used to deliver thermal therapy to cancer cells.
- ***Nanoshells*** have big advantages over uniform core nanoparticles: (1) nanoshells can contain strong drugs inside, and (2) their absorption spectrum can be tuned to different wavelengths of radiation by controlling the ratio of the thickness of the walls to the overall diameter of the shell.

Slide 2.19 Structure of nanoparticles

Surface Chemistry of Nanoparticles

- ***Surface chemistry*** on the surface of nanoparticles and nanomachines plays an important role in targeting abnormal cells, cell uptake, navigation of nanodevices, and selective drug delivery.
- The surface chemistry can ***increase the half-life time*** of nanoparticles inside the body and thereby increase the availability to the target cells.
- In addition, as the interface between the nanomaterial and the biological environment, the surface chemistry largely determines whether a nanoparticle ***suppresses or stimulates immune responses.*** Without surface modification, nanoparticles have been reported to be removed from the bloodstream within seconds by macrophages cells which remove foreign objects and stimulate other immune cells.

Slide 2.20 Surface chemistry of nanoparticles

Chemotherapeutic agents may be carried in the cores of hollow or porous nanoparticles, also called nanocarriers or nanovectors. Liposomes, dendrimers, fullerenes, polymer nanoparticles and mesoporous silica are examples of nanoparticles that have been used to carry drugs. These materials may also be programmed with mechanisms by which the drug can be released. For example, drug release may be triggered by heat, light, ion concentration, ultrasound, radiation or internal triggers, like reduced pH of a tumor interstitial or endocytosed vesicle.

Surface chemistry on the surface of nanoparticles and nanomachines plays an important role in targeting abnormal cells, cell uptake, navigation of nanodevices, and selective drug delivery (Slide 2.20). The surface chemistry can increase the half-life of nanoparticles inside the body and thereby increase the availability to the target cells. It can also affect the amount of cell uptake by target and non-target cells. In addition, as the interface between the nanomaterial and the biological environment, the surface chemistry largely determines whether a nanoparticle suppresses or stimulates immune responses. Without surface modification, nanoparticles have been reported to be removed from the bloodstream within seconds by macrophages cells which remove foreign objects and stimulate other immune cells.

The size of nanoparticles plays a crucial role in the overall properties of nanoparticles, their functions and applications (Slide 2.21). Particles under 100 nm in size can freely circulate around the body, penetrate into any organs, and effectively collected at the tumor sites. For example, the overall size of the nanoparticle drug delivery system may be tuned for delivery to the tumor site via tumor vasculature and then to an intracellular target. Nanoparticle size can mediate the efficiency of its uptake into cells. It has been shown that maximum cellular uptake is achieved with 50 nm particles. Finally, smaller sized nanoparticles undergo better renal clearance from the bloodstream after the targeted nanoparticle therapy (Slide 2.21).

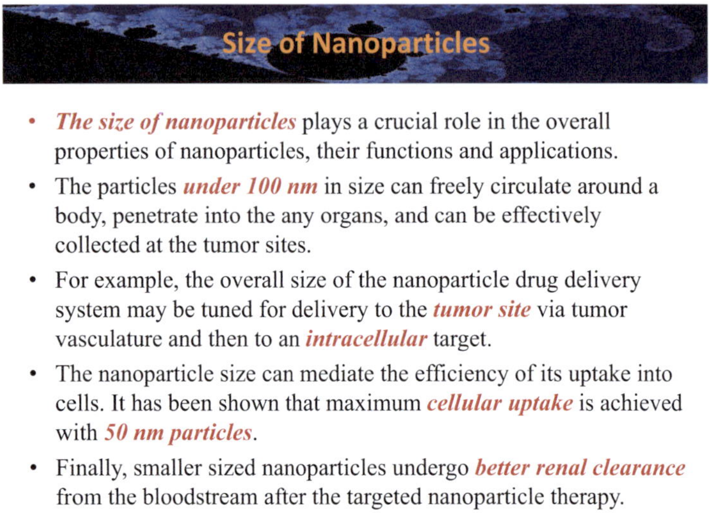

Slide 2.21 Size of nanoparticles

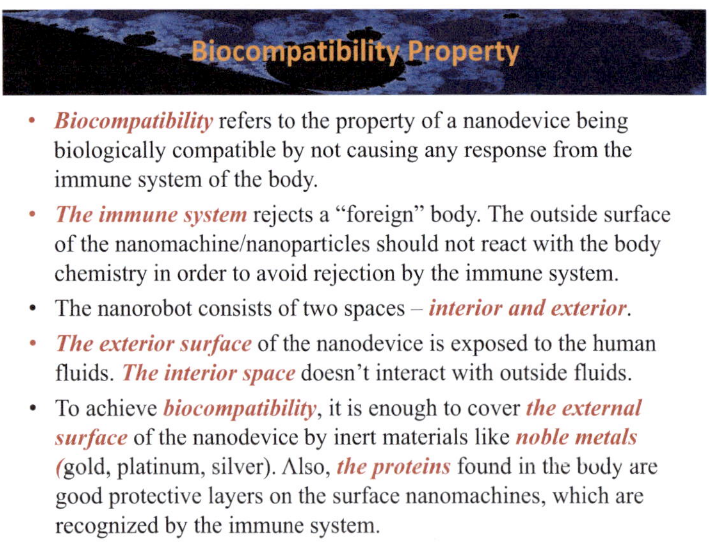

Slide 2.22 Biocompatibility property

Biocompatibility refers to the property of a nanodevice as being biologically compatible by not causing any response from the immune system of the body (Slide 2.22). The immune system rejects a "foreign" body. The outside surface of the nanomachine/nanoparticles should not react with the body chemistry in order to avoid rejection by the immune system. The nanorobot consists of two spaces—interior and exterior. The exterior surface of the nanodevice is exposed to the

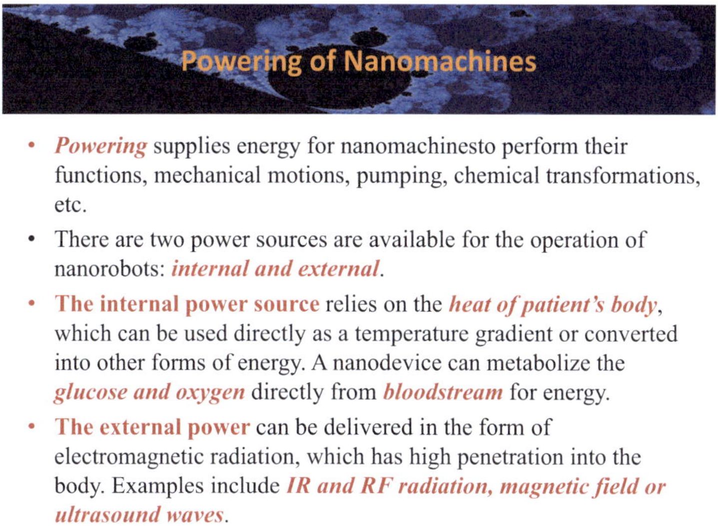

Slide 2.23 Powering of nanomachines

human fluids. The interior space doesn't interact with outside fluids. To achieve biocompatibility, it is enough to cover the external surface of the nanodevice by a material which doesn't cause any allergy reaction of immune system (Slide 2.22). Noble metals like gold, platinum and silver are traditionally used for nanomedicine applications. Also, the proteins found in the body are good protective layers on the surface of nanomachines, which are recognized by the immune system.

Powering supplies energy for nanomachines to perform their functions, mechanical motions, pumping, chemical transformations, etc (Slide 2.23). Two power sources are available for the operation of nanorobots: internal and external. *The internal power source* relies on the heat of a patient's body, which can be used directly as a temperature gradient or converted into other forms of energy (Slide 2.23). A nanodevice can metabolize the glucose and oxygen directly from the bloodstream for energy. *External power* can be delivered in the form of electromagnetic radiation, which has high penetration into the body (Slide 2.23). Examples include IR and RF radiation, magnetic field or ultrasound waves.

Communication with the outside world and between the nanorobots is required to perform various functions inside the body (Slide 2.24). One way to send messages into the body is RF messaging in the 1–10 MHz frequency range. Most of the cell communication uses chemical signals. One nanorobots will not suffice to do a job for several million of cells at once. The goal is to teach and make the nanorobots collaborate with each other. Making a bionano swarm on a tumor is a great example. Designing a communication network for the nanorobots is an extremely challenging task. We are still very far away, but again, nature has made systems or structures which can do exactly that—viruses. The creation of nanotubes, nanotransitors, nanophotonics, nanowires and nanowaveguides is accelerating this giant step toward fully operating nanorobots (Slide 2.24).

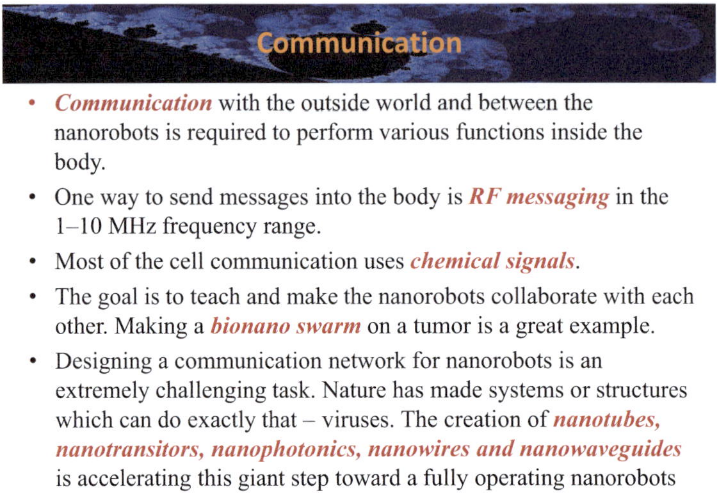

Slide 2.24 Communication

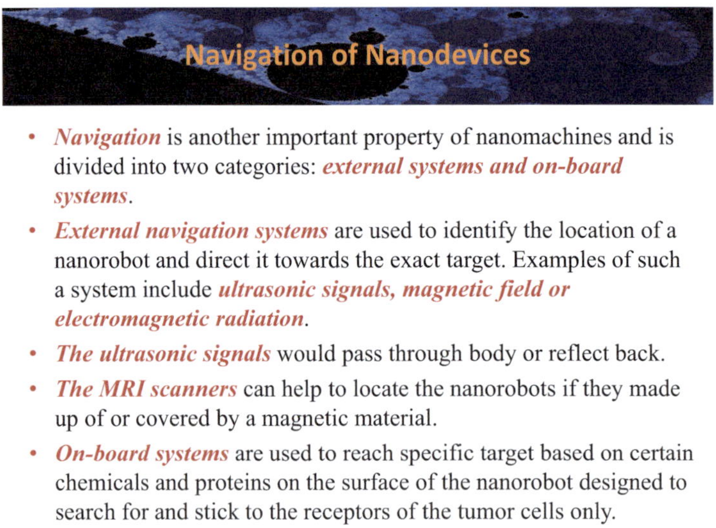

Slide 2.25 Navigation of nanodevices

Navigation is another important property of nanomachines and is divided into two categories: external systems and on-board systems (Slide 2.25). *External navigation systems* are used to identify the location of a nanorobot and direct it towards the exact target. Examples of such a system may include ultrasonic signals, magnetic field or electromagnetic radiation. The ultrasonic signals would pass through the body or reflect back. The MRI scanners can help to locate the nanorobots if they made up of or covered by a magnetic material. *On-board systems* are used to reach specific targets based on certain chemicals and proteins on the

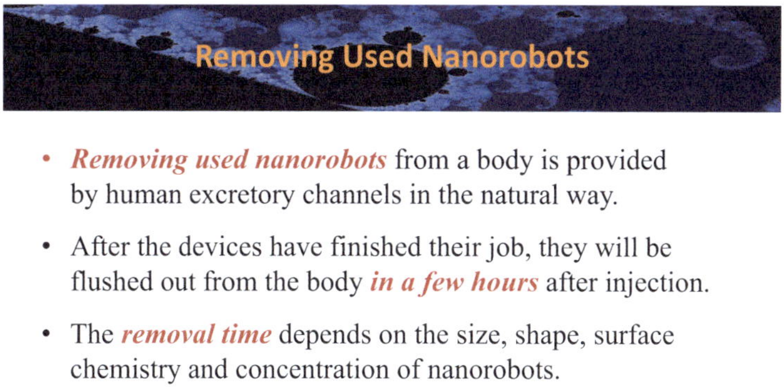

Slide 2.26 Removing used nanorobots

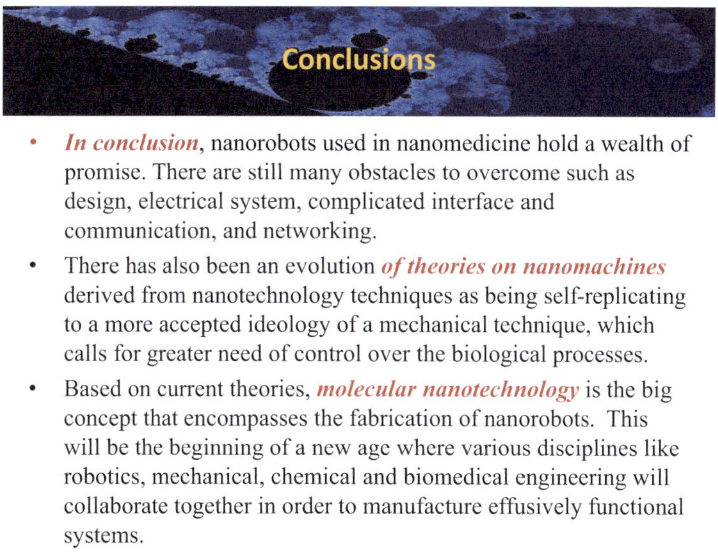

Slide 2.27 Conclusions

surface of the nanorobot designed to search for and stick to the receptors of the tumor cells only.

Removing used nanorobots from a body is provided by human excretory channels in the natural way (Slide 2.26). After the devices have finished their job, they will be flushed out from the body in a few hours after injection. The removal time depends on the size, shape, surface chemistry and concentration of nanorobots.

In conclusion, nanorobots used in nanomedicine hold a wealth of promise (Slide 2.27). There are still many obstacles to overcome, such as design, electrical system, complicated interface and communication, and networking. There has also been an evolution of theories on nanomachines derived from nanotechnology techniques as being self-replicating to a more accepted ideology of a mechanical technique, which calls for greater need of control over the biological processes. Based on current theories, molecular nanotechnology is the big concept that encompasses the fabrication of nanorobots. This will be the beginning of a new age where various disciplines like robotics, mechanical, chemical and biomedical engineering will collaborate together in order to manufacture effusively functional systems.

1.3 Lecture 3: Design and Fabrication of Medical Nanodevices (Slide 3.1)

- Design of Medical Nanodevices
- Methods of Nanomaterials Fabrication
  - Top-Down Approach
  - Bottom-Up Approach
- Synthesis of Metallic Nanoparticles and Applications
- Synthesis of Semiconductor Nanoparticles and Applications
- Synthesis of Oxide Nanoparticles and Applications
- Vapor Phase Reactions
- Solid-State Phase Segregation
- Heterogeneous Nucleation
- Kinetically Confined Synthesis
- Synthesis of Carbon Nanotubes and Applications
- Conclusion

Here we look at the design and fabrication of medical nanodevices and methods of nanoparticle synthesis (Slide 3.1).

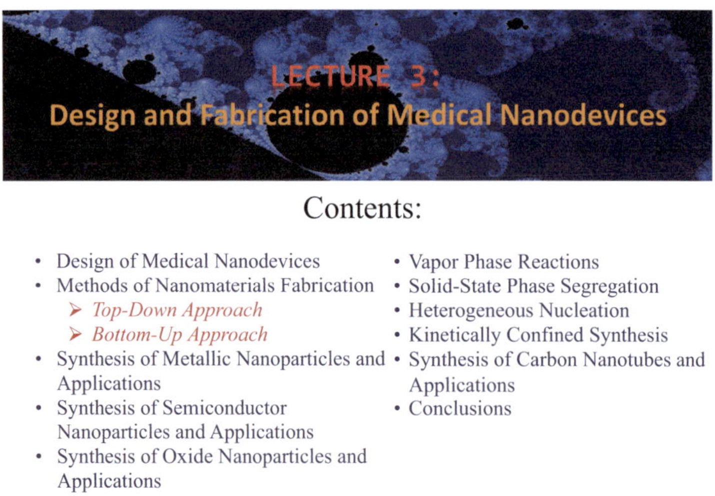

Slide 3.1 Contents of Lecture 3

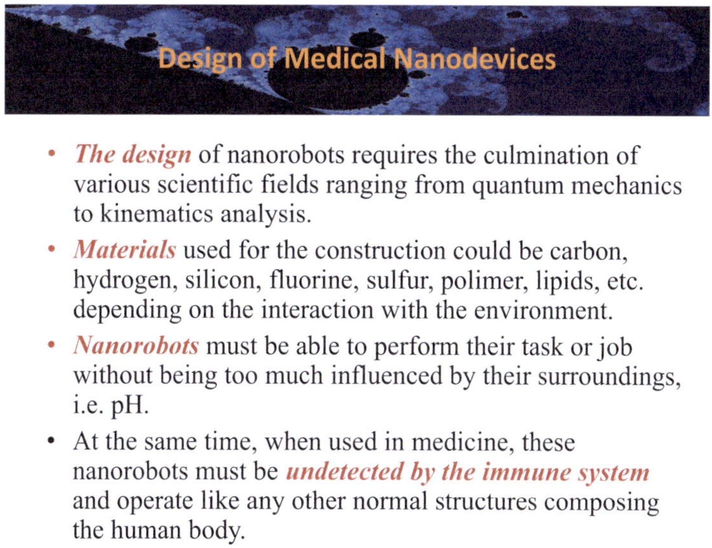

Slide 3.2 Design of medical nanodevices

1.3.1 Design of Medical Nanodevices

The design of nanorobots requires the culmination of various scientific fields ranging from quantum mechanics to kinematics analysis (Slide 3.2). Materials used for the construction could be carbon, hydrogen, silicon, fluorine, sulfur, polimer, lipids, etc., depending on the interaction with the environment. Nanorobots must be able to perform their task or job without being too much influenced by their surroundings, i.e., pH. At the same time, when used in medicine, these nanorobots must be undetected by the immune system and operate like any other normal structures composing the human body.

Like building a macroscale robot, nanorobots development must also follow a roadmap (Slides 3.3 and 3.4). Researchers have identified the following main steps in the development of medical nanodevices:

Step 1. Developing nanocomponents which include power supplies, fuel buffer tanks, sensors, motors, manipulators, on-board computers, pumps, pressure tanks and structural support.

Step 2. Assembling of nanorobots which includes making complex assemblies from the nanocomponents with nanoscale precision. Ideally the assembling would be performed by bionanorobots which have ability to build the nanomachines from available reserves or supplies, to repair damaged parts, to sense environment and respond appropriately, and so forth.

Step 3. Artificial intelligence would require development in basic computational capabilities, and rules would have to be established so that bionanomachines can make quick and sharp decisions at the nanoscale level (Slide 3.4). Nanocomputers, communicational and navigational equipment will have to be developed in order to establish communication between the nanorobots and outside world. This will allow us to program swarm behavior. All these capabilities will form the basic principles of a nanorobot with artificial intelligence.

Step 4. Self-assembling and self-evolution. Bioswarms will have the ability to self-assemble and self-evolve, and would manage energy efficiently.

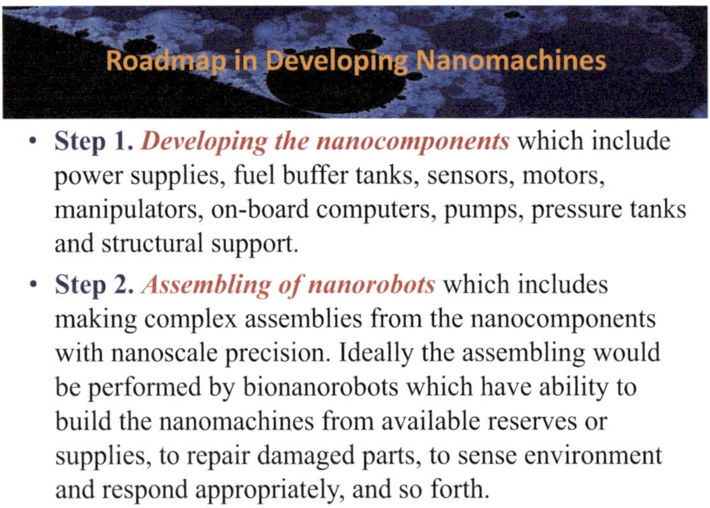

Slide 3.3 Roadmap in developing nanomachines

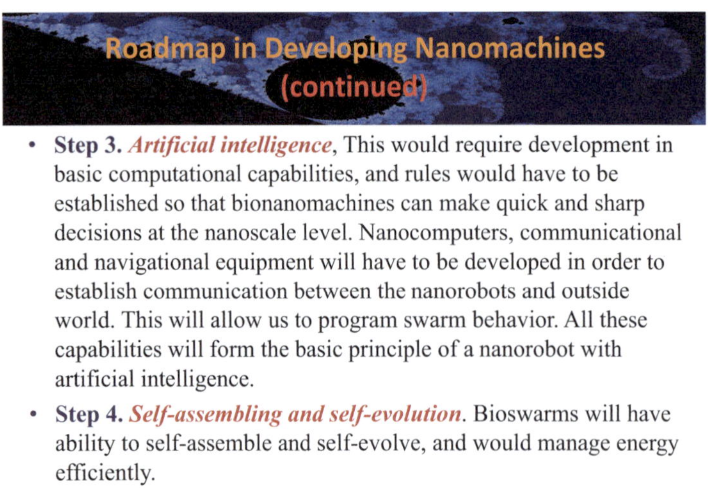

Slide 3.4 Roadmap in developing nanomachines (continued)

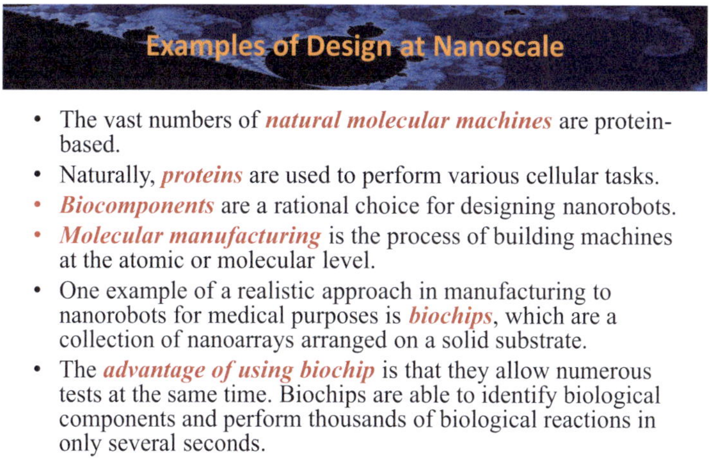

Slide 3.5 Examples of design at nanoscale

The vast numbers of natural molecular machines are protein-based (Slide 3.5). Naturally, proteins are used to perform various cellular tasks. Now, with the help of powerful crystallographic techniques, it is easier to understand the structure of the proteins. With the help of computers, protein-folding mechanisms can be demonstrated and perfected, and it is also possible to confirm and determine the structure of rarely known proteins. All these advancements can help us solve mysteries related to molecular machinery, and hence pave the way to manufacture and apply these machines in fields like medicine, electronics, military and space exploration. Biocomponents are a rational choice for designing nanorobots.

Molecular manufacturing is the process of building machines at the atomic or molecular level, and nanorobots would have the primary role in this process. One example of a realistic approach to manufacturing nanorobots for medical purposes is *biochips*, which are a collection of nanoarrays arranged on a solid substrate. The advantage of using biochips is that they allow numerous tests at the same time. They are able to identify biological components and perform thousands of biological reactions in only several seconds.

1.3.2 Methods of Nanomaterials Fabrication

Two techniques are used for the synthesis of nanomaterials (Slide 3.6). The first is the *top down approach* which uses nanoscale imprinting and milling techniques to create nanostructures out of bulk materials. The second is the *bottom-up approach* which uses basic chemistry techniques to build atom-by-atom and assemble nanostructures. For synthesizing nanoparticles, in particular, the top-down approach employs *milling* and *repeated thermal cycling* as primary

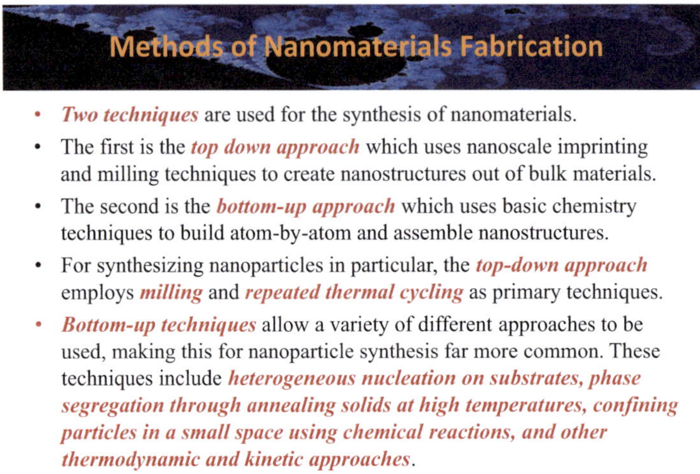

Slide 3.6 Methods of nanomaterials fabrication

techniques. Bottom-up techniques allow a variety of different approaches, making this for nanoparticle synthesis far more common. These techniques include *heterogeneous nucleation on substrates, phase segregation through annealing solids at high temperatures, and confining particles in a small space using chemical reactions, and other thermodynamic and kinetic approaches.*

1.3.2.1 Top-Down Approach

Milling produces nanoparticles from bulk materials at sizes larger than 20 nm (Slide 3.7). The particles created in this technique have a wide distribution of sizes, variation in shape, and impurities from the milling environment.

Applications of nanoparticles produced from this process are the fabrication of nanocomposites and nanograined bulk materials (Slide 3.7). One example uses a high-energy ball milling process for synthesis of nanocrystalline hydroxyapatite particles that can be used as bone filler for coating metallic prosthesis, in dental implants, and for making matrices for controlled drug release applications.

Repeated thermal cycling produces very fine particles by repeated quenching of the bulk material (Slide 3.7). As a result of heating, the material with low thermal conductivity can be broken into small pieces. This process has limitations: it can be difficult to design and control, and it is not applicable to all materials.

Applications. Metal nanopowders, consisting of 25 nm nanosilver or 80 nm nanoaluminum particles, are produced by heating the metals of interest by using repeated pulsed electrical arc discharge and plasma in a reactor (Slide 3.7).

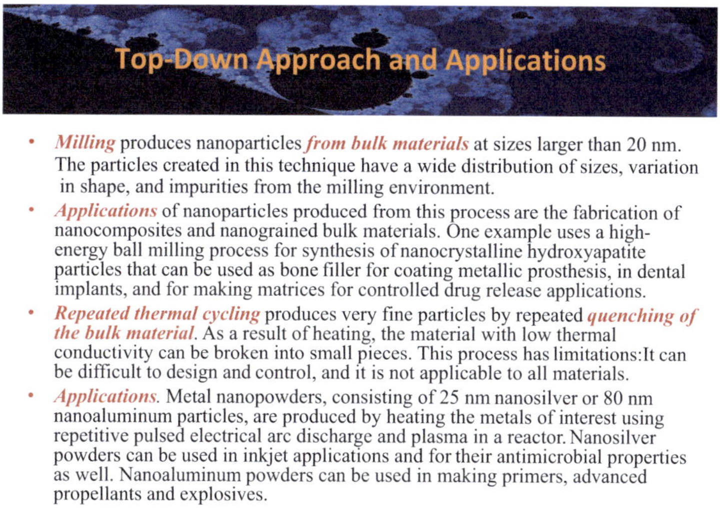

Slide 3.7 Top-down approach and applications

Nanosilver powders can be used in inkjet applications and for their antimicrobial properties as well. Nanoaluminum powders can be used in making primers, advanced propellants and explosives.

1.3.2.2 Bottom-Up Approach

Homogeneous nucleation is one method of the bottom-up approach (Slides 3.8 and 3.9). It can be used to synthesize nanoparticles in all the three states: liquid, solid and gaseous. This section will be divided further into the synthesis of metallic nanoparticles, semiconductor nanoparticles, oxide nanoparticles, etc., but the fundamentals behind synthesizing these particles are the same. Homogeneous nucleation is initiated in a saturated medium by disturbing the equilibrium through decreasing the temperature of the solution/medium. This will lead to supersaturation, nucleation and growth of the centers of nucleation in the medium. For example, when the temperature of a glass matrix is slowly reduced, quantum dots will be formed due to supersaturation in the matrix.

Of important note is that the above processes have a lower limit for size. This limit can be reduced further by reducing the surface energy of the newly created nanoparticle mixture using temperature, by increasing the supersaturation of the growth species, by using a different solvent, by introducing additives in the solution, and by incorporating impurities into the solid phase when applicable.

The process of supersaturation of the growth species is governed by two processes (Slide 3.9). First, the growth species have been generated, diffused in the mixture and adsorbed on the growth surface. Secondly, there will be a surface

Bottom-Up Approach

- ***Homogeneous nucleation*** is a one method of the ***bottom-up approach.***
- It can be used to ***synthesize nanoparticles*** in all the three states: ***liquid, solid and gaseous***.
- Homogeneous nucleation is initiated ***in a saturated medium*** by disturbing the equilibrium of saturation throught decreasing the temperature of the solution/medium.
- This will lead to ***supersaturation, nucleation and growth*** of the ***centers of nucleation*** in the medium. For example, when the temperature of a glass matrix is slowly reduced, quantum dots will be formed due to supersaturation in the matrix.

Slide 3.8 Bottom-up approach

Bottom-Up Approach (continued)

- The process of ***supersaturation*** of the growth species is governed by ***two processes***.
- ***First,*** the growth species have been ***generated, diffused in the mixture and adsorbed on the surface*** of the growth surface.
- ***Secondly***, there will be ***surface growth*** due to irreversible incorporation of growth species onto the solid surface.
- There are several ways of controlling the nanoparticle growth process, which can help control the size distribution. The growth control can be achieved by surface processes such as ***mononuclear growth, polynuclear growth*** or by ***diffusion-limited growth***.
- ***Mononuclear growth*** is layer-by-layer growth where one layer of growth is completed before the formation of another layer starts on the surface.
- In ***polynuclear growth***, the surface process is very fast because of high surface concentration, which causes the growth of the second layer to start before the layer is completed. Polynuclear growth is favored over mononuclear growth for formation of the same-sized nanoparticles, because as particles grow bigger, the radius difference gets smaller.
- **Diffusion-limited growth** is the most preferable process for formation of monosized nanoparticle (i.e., particles of the same size); it can be simply controlled through "drop-by-drop" addition of the reactant or a variation of the technique.

Slide 3.9 Bottom-up approach (continued)

growth due to irreversible incorporation of the growth species onto the solid surface [2].

There are several ways of controlling the nanoparticle growth process, which can help control the size distribution (Slide 3.9). The growth control can be achieved by surface processes such as *mononuclear growth, polynuclear growth* or by *diffusion-limited growth*. Mononuclear growth is layer-by-layer growth where one layer is completed before the formation of another layer starts on the surface. In polynuclear growth, the surface process is very fast because of high surface concentration, which causes the growth of the second layer to start before the layer is completed. Polynuclear growth is favored over mononuclear growth for formation of the same-sized nanoparticles, because as particles grow bigger, the radius difference gets smaller. The diffusion-limited growth process is the most preferable for formation of monosized nanoparticles (i.e., particles of the same size), since it can be simply controlled through "drop-by-drop" addition of the reactant or a variation of the technique.

1.3.3 Synthesis of Metallic Nanoparticles and Applications

For synthesis of metallic nanoparticles using homogeneous nucleation methods described above, metal complexes are reduced using reducing agents in dilute solutions (Slide 3.10). Precursors may include metals, inorganic salts and complexes. Some of the common reducing agents used are NaOH, H_2O_2, CO, P and H_2. And polymer stabilizers like polyvinyl alcohol and sodium polyacrylate are used as well. A combination of a diluted low concentration of the reactants in solution and polymer stabilizers forming a layer around the nanoparticles results in diffusion-limited growth and steric stabilization to produce monosized nanoparticles. The kind of reducing agent used can be a factor in deciding the size of the nanoparticle produced, as well as the size distribution. A strong reducing agent will promote the growth of smaller nanoparticles because the reaction will be fast. With a weaker reducing agent, the reaction will be slow, which might lead to formation of relatively larger

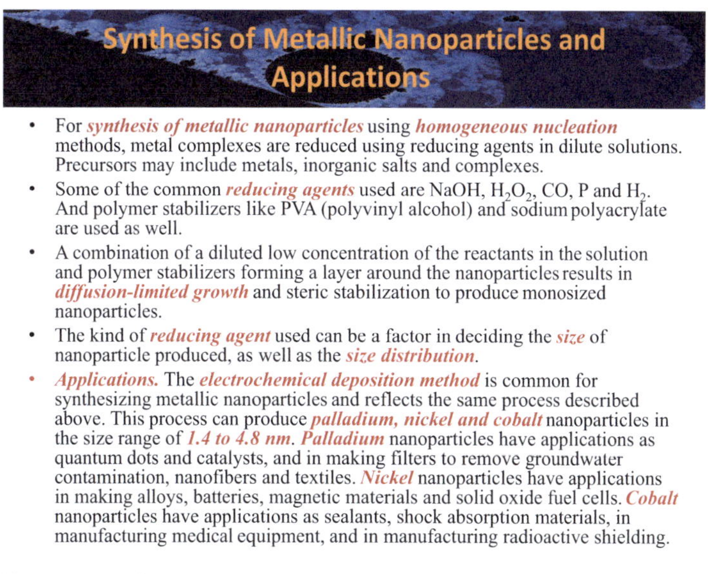

Slide 3.10 Synthesis of metallic nanoparticles and applications

nanoparticles. Also, if the slow reaction causes continuous formation of new nuclei, it will result in a wide size distribution, or if no further nucleation occurs, the reaction will favor diffusion-growth process, and the size distribution will be narrow.

Applications. The electrochemical deposition method is common for synthesizing metallic nanoparticles and reflects the same process described above (Slide 3.10). An electrochemical cell with a metal anode and a metal or glassy carbon cathode is set up with organic solutions of tetraalkylammonium halogenides as the electrolyte [2]. When an electric field is applied, the metal ions produced by the anode generate nucleation and growth of metallic nanoparticles in the solution. Tetraalkylammonium halogenides also acts as a stabilizer to prevent agglomeration of nanoparticles. This process can produce palladium, nickel and cobalt nanoparticles in the size range of 1.4–4.8 nm. Increasing the current density in the process can produce smaller nanoparticles. Palladium nanoparticles have applications as quantum dots and catalysts, and in making filters to remove groundwater contamination, nanofibers and textiles. Nickel nanoparticles have applications in making alloys, batteries, magnetic materials and solid oxide fuel cells. Cobalt nanoparticles have applications as sealants, shock absorption materials, in manufacturing medical equipment, and in manufacturing radioactive shielding.

1.3.4 Synthesis of Semiconductor Nanoparticles and Applications

This process uses pyrolysis of organometallic precursors to synthesize "non-oxide" semiconductor nanoparticles (Slide 3.11). Pyrolysis is a process of overcoming the activation energy by increasing the energy of the material using heat. Organometallic materials are organic compounds that have at least one metal-to-carbon bond and are used as catalysts in many applications. The organometallics are dissolved in an anhydrate ("waterless") solvent at high temperatures (200 °C and above) in a sealed airless environment with stabilizers present in the reaction space.

Applications. Cadmium selenide (CdSe) quantum dots are synthesized using trioctylphosphine selenide (TOPS) as a precursor (Slide 3.11). Trioctylphosphine/TOPS solutions are the polymer stabilizer and oleic acid is the solvent. CdSe quantum dots can be used in manufacturing polymer-based photovoltaic cells, chemical sensors, optical temperature probes and emitters for color displays.

1.3.5 Synthesis of Oxide Nanoparticles and Applications

This process uses sol-gel processing to synthesize metal oxides and organic–inorganic hybrid materials (Slide 3.12). Sol-gel preparation uses metal salts as precursors. Catalysts are then used to promote hydrolysis of these salts. This results

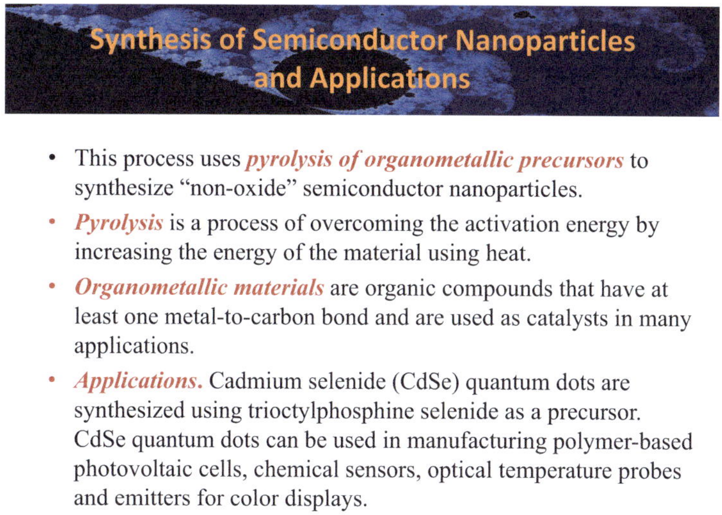

Slide 3.11 Synthesis of semiconductor nanoparticles and applications

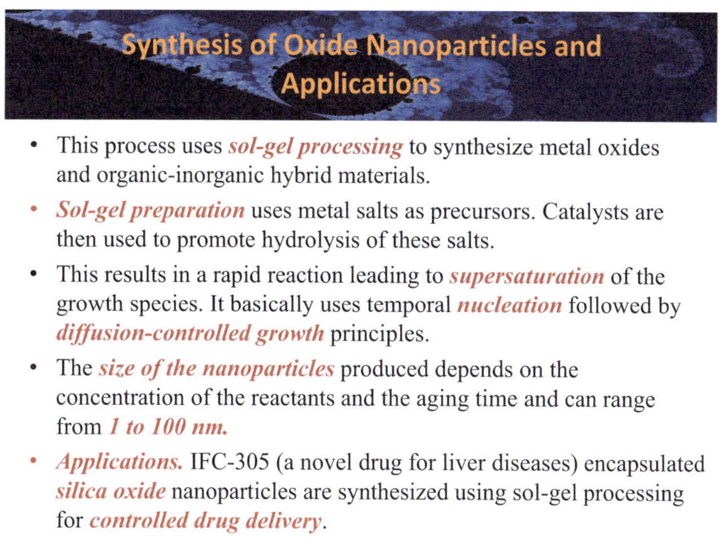

Slide 3.12 Synthesis of oxide nanoparticles and applications

in a rapid reaction leading to supersaturation of the growth species. It basically uses temporal nucleation followed by diffusion-controlled growth principles. The size of the nanoparticles produced depends on the concentration of the reactants, and the aging time and can range from 1 to 100 nm.

Applications. IFC-305 (a novel drug for liver diseases) encapsulated silica nanoparticles are synthesized using sol-gel processing for controlled drug delivery

(Slide 3.12). A SiO_2 sol-gel matrix is synthesized with various concentrations of the drugs through hydrolysis and polycondensation reactions using tetraethoxysilane (TEOS) and different concentrations of water as precursors. This matrix in particular was chosen for the application because of its biocompatible and bioresorbable properties. Also, this process can be conducted under mild processing conditions, which allows for the drug to be embedded in the matrix without changing its base properties. The release of the drug is dependent on the molar ratio of silica precursor to water, and on the concentration of the drug itself, which allows the development of the process to be controlled appropriately by changing these parameters.

1.3.6 Vapor Phase Reactions

Synthesis in the vapor phase usually happens in vacuum conditions at high temperatures (Slide 3.13). As a result, low concentrations of growth species are available as reactants in the diffusion-controlled growth process. The nanoparticles generated are then collected on a low-temperature non-stick substrate. The challenge with this process is that not all nanoparticles will settle on the substrate. Another challenge is that even the nanoparticles that are collected do not reflect a true representation of the size distribution. Stabilization to prevent these nanoparticles from agglomeration is also difficult (unlike the liquid phase where one can just add the polymer stabilizers to the solution).

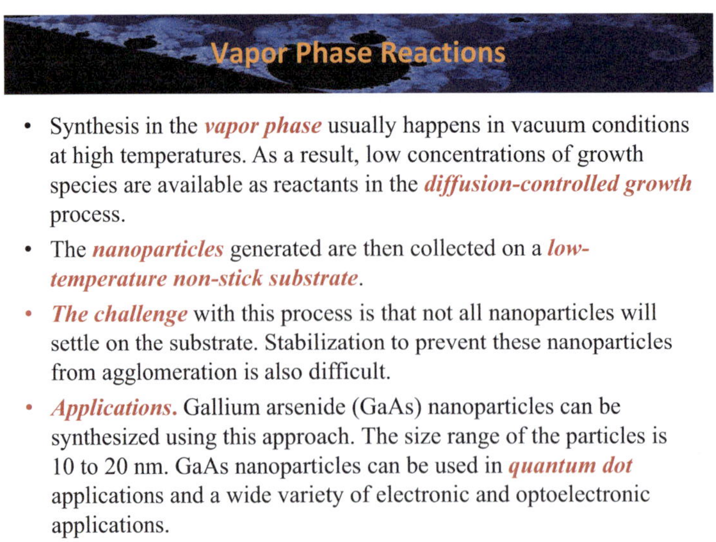

Slide 3.13 Vapor phase reactions

Applications. Gallium arsenide (GaAs) nanoparticles can be synthesized using this approach by nucleation of trimethyl gallium and AsH_3 precursors, with hydrogen gas acting as the reducing agent (Slide 3.13) [2]. Reaction occurs at 700 °C followed by nanoparticles being collected on a substrate at 350 °C. The size range of the particles is 10–20 nm. GaAs nanoparticles can be used in quantum dot applications and a wide variety of electronic and optoelectronic applications.

1.3.7 Solid-State Phase Segregation

Metals and semiconductor nanoparticles are synthesized in a glass matrix using this method (Slide 3.14). Precursors are mixed in a liquid glass melt at high temperatures during glass making [2]. Then the glass is cooled down to the phase transition temperature for a planned period of time. This causes supersaturated precursors to form nanoparticles by nucleation and growth through solid-state diffusion.

Applications. This particular example does not use a glass matrix, but nonetheless it is interesting, using a solid-state-diffusion growth process (Slide 3.14). $Al(OH)_3$ and $CoCl_2{\cdot}6H_2O$ solid reactants are mixed together at 450–650 °C in an agate mortar, followed by drying the mixture at 110 °C overnight. Then the mixture is heated to a planned temperature in a tube furnace with air flow. Finally, the mixture is cooled down to room temperature using liquid nitrogen. This results in formation of nanocrystalline cobalt aluminate ($CoAl_2O_4$) which can be used as an inorganic

Solid-State Phase Segregation

- ***Metals and semiconductor nanoparticles*** are synthesized in a glass matrix using this method.
- Precursors are mixed in ***a liquid glass*** melt at high temperatures during glass making.
- Then the glass is cooled down to the phase transition temperature for a planned period of time. This causes ***supersaturated*** precursors to form nanoparticles by ***nucleation and growth through solid-state diffusion***.
- ***Applications.*** This process is used in formation of ***nanocrystalline cobalt aluminate*** ($CoAl_2O_4$) nanoparticles which can be used as an inorganic ceramic blue pigment for applications in paint, glass, porcelain enamels and fiber. Its optical properties can also be used in manufacturing color filters for automotive lamps or luminescent materials in optical devices.

Slide 3.14 Solid-state phase segregation

ceramic blue pigment for applications in paint, glass, porcelain enamels and fiber. Its optical properties can also be used in manufacturing color filters for automotive lamps or luminescent materials in optical devices.

1.3.8 Heterogeneous Nucleation

Heterogeneous nucleation reactions take place on a substrate and for synthesis of quantum dots or nanoparticles (Slide 3.15); a contact angle of larger than 0° between atoms and the surface is required. The growth of the nanoparticles on the substrate can take place in the following manner:

- If the growth species are more strongly bonded to each other than to the substrate, they will form 'islands' on the substrate, with a group of atoms sticking together on the substrate. This process is referred to as *island growth*.
- If the growth species equally favor strongly binding to the substrate just as they do to each other, the distribution of the atoms will be layer-by-layer—one monosized particle layer at a time—before more atoms start to make another layer above it. This is referred to as the *layer growth*.
- *Island-layer growth* is a combination of the above two processes to make a continuous film. Some atoms will tend stick together and form islands, whereas others will individually directly stick to the substrate. This will give a larger size distribution.

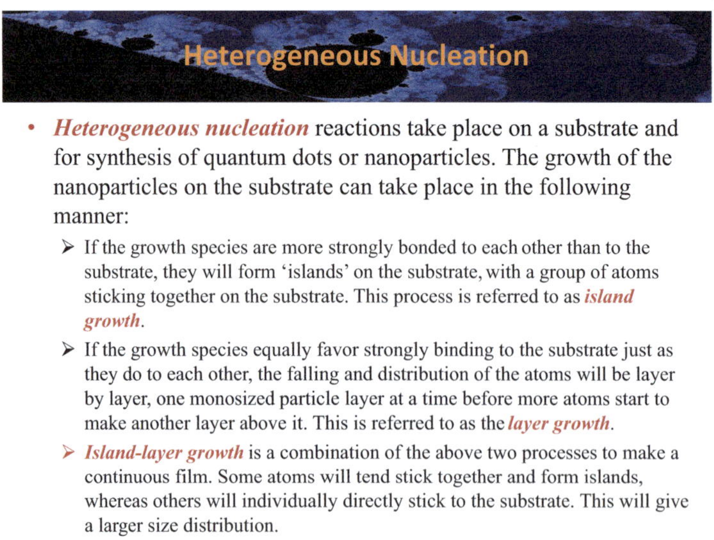

Slide 3.15 Heterogeneous nucleation

- ***Homogeneous surface defects*** that act as nucleation centers need to be created.
- This can be done using a process like ***thermal oxidation***.
- ***Oxygen*** is used in thermal oxidation to form oxide layers on the surface of the substrate.
- This can be done by ***heating the substrate*** such that the oxygen atoms can diffuse through the substrate surface and form defects.
- ***Applications. Nickel nanoparticles*** between in the size range 20-600 nm with a narrow size distribution can be formed using homogeneous nucleation process. As mentioned above, nickel nanoparticles have applications in making alloys, batteries, magnetic materials and solid oxide fuel cells.

Slide 3.16 Heterogeneous nucleation (continued)

Homogeneous surface defects that act as nucleation centers need to be created (Slide 3.16). This can be done using a process like thermal oxidation. Oxygen is used in thermal oxidation to form oxide layers on the surface of the substrate, which can be achieved by heating the substrate such that the oxygen atoms can diffuse through the substrate surface and form defects.

It should be noted that this process is different than vapor phase reactions. This is because in such reactions, the nanoparticles are generated in the vapor before depositing onto the substrate whereas in the heterogeneous nucleation process, the growth species strike the surface first, followed by formation of nuclei on the surface.

Applications. Nickel nanoparticles in the size range 20–600 nm with a narrow size distribution can be formed using the homogeneous nucleation process (Slide 3.16). As mentioned above, nickel nanoparticles have applications in making alloys, batteries, magnetic materials and solid oxide fuel cells.

1.3.9 Kinetically Confined Synthesis

This process deals with stopping the synthesis when we have achieved our desired products/results (Slide 3.17). This can be accomplished by supplying only a limited amount of precursor for the reaction, by physically filling up the reaction space; or by terminating the growth species' progress by having organic compounds or alien ions occupy the reaction site. One of these techniques is used in the *growth termination* method. Here, the stop growth method is used for synthesis of

Kinetically Confined Synthesis

- This process deals with ***stopping the synthesis*** when we have achieved our desired products/results.
- This can be accomplished by supplying only a ***limited amount of precursor*** for the reaction, by physically ***filling up*** the reaction space; or by ***terminating the growth species'*** progress by having organic compounds or alien ions occupy the reaction site.
- One of these techniques is used in ***growth termination*** method. Here, the stop growth method is used for synthesis of monosized nanoparticles, organic compounds or specific ions.
- ***Applications.*** This techniques is used to synthesize cadmium sulfide (CdS) colloidal nanoparticles. The size of the nanoparticles can be controlled by the concentration of organic ligands introduced to the system. CdS nanoparticles have applications is optoelectronics.

Slide 3.17 Kinetically confined synthesis

monosized nanoparticles, organic compounds or specific ions. These organic compounds can be introduced to the reaction at the right time to occupy the growth sites through organic bonding. The same concept applies to ions, because of their particular charge and attraction forces they can step in and take the place of the limited amount of precursor that is available in the reaction.

Applications. Cadmium acetate and sodium sulfide precursors are mixed together in a water/methanol solution, followed by filtering and suction drying using nitrogen to synthesize cadmium sulfide (CdS) colloidal nanoparticles (Slide 3.17). The size of the nanoparticles can be controlled by the concentration of organic ligands introduced to the system [2]. CdS nanoparticles have applications in optoelectronics.

1.3.10 Synthesis of Carbon Nanotubes and Applications

Carbon nanotubes were discovered in the soot of arc discharge in 1991 (Slides 3.18 and 3.19). Since then, researchers have successfully developed some manufacturing techniques for carbon nanotubes, such as *arc discharge, laser ablation, plasma torch,* and *chemical vapor deposition*. Below we briefly describe arc discharge and laser ablation.

Arc discharge is the most traditional manufacturing method, which consists of placing a certain amount of inert gas into a vacuum chamber between the two graphite electrodes (Slide 3.18). The discharging of electrodes leads to the formation of soot on the cathode containing the carbon nanotubes. This method produces

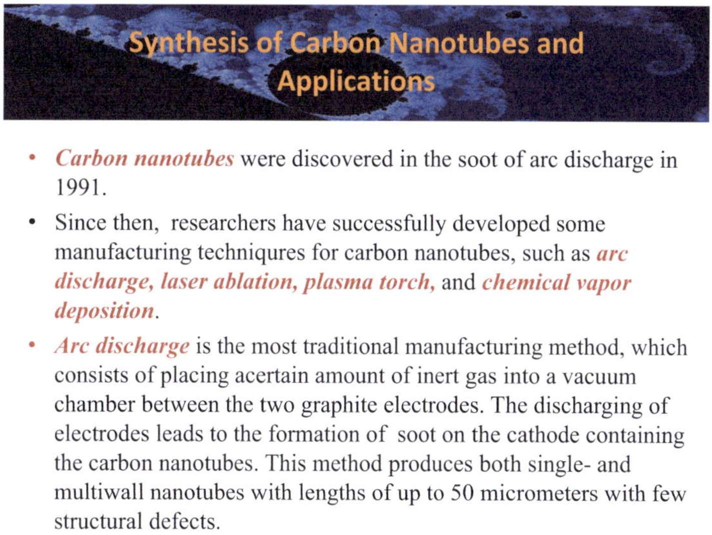

Slide 3.18 Synthesis of carbon nanotubes and applications

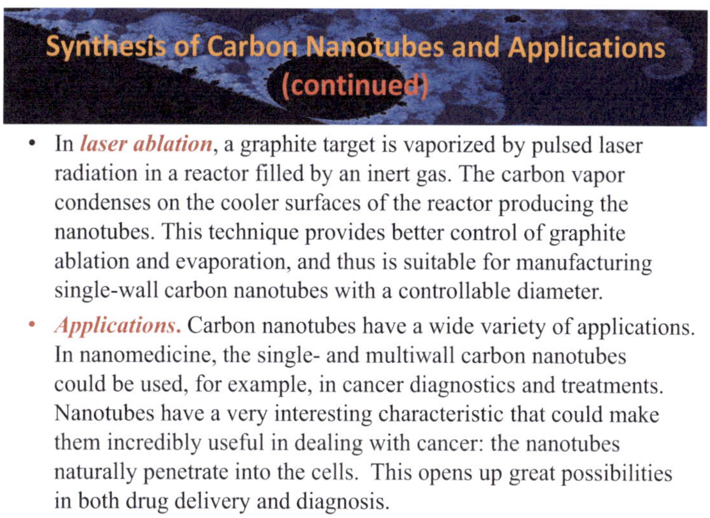

Slide 3.19 Synthesis of carbon nanotubes and applications (continued)

both single- and multiwall nanotubes with lengths of up to 50 μm with few structural defects.

In *laser ablation*, a graphite target is vaporized by pulsed laser radiation in a reactor filled by an inert gas (Slide 3.19). The carbon vapor condenses on the cooler surfaces of the reactor to procude the nanotubes. This technique provides better control of graphite ablation and evaporation, and thus is suitable for manufacturing single-wall carbon nanotubes with a controllable diameter. In addition, the method can be optimized to achieve high-yield and continuous production.

Applications. Carbon nanotubes have a wide variety of applications (Slide 3.19). They hold promise to be the new electrode materials in lithium-ion batteries and to increase the charge and discharge current density and capacity of the batteries. In nanomedicine, the single- and multiwall carbon nanotubes could be used, for example, in cancer diagnostics and treatments. Nanotubes have a very interesting characteristic that could make them incredibly useful in dealing with cancer: the nanotubes naturally penetrate into the cells. This opens up great possibilities in both drug delivery and diagnosis.

In conclusion, we have discussed design approaches and methods of fabrication of medical nanodevices with various types of nanoparticles (Slide 3.20). There are a wide variety of techniques to choose from for the synthesis of nanoparticles. We have also discussed different applications that these nanoparticles are being used in or are being considered for. Important points that need to be considered when designing a synthesis process are:

- Synthesis of nanoparticles is possible in all three phases: liquid, solid and gaseous.
- Stabilization processes to prevent the nanoparticles from agglomeration are very important.
- Making a same-sized distribution of nanoparticles is another important factor, which can be controlled by diffusion-limited growth processes or by confining the growth space.

Conclusions

- We have discussed ***design approaches*** and ***methods of fabrication*** of ***medical nanodevices*** with various types of ***nanoparticles***.
- There are a wide variety of techniques to choose from for ***synthesis of nanoparticles***.
- We have also discussed different ***applications*** that these nanoparticles are being used in or are being considered for.
- ***Important points*** that need to be considered when designing a synthesis process are:
  - Synthesis of nanoparticles is possible ***in all three phases: liquid, solid and gaseous***.
  - Stabilization processes to prevent the nanoparticles from ***agglomeration*** are very important.
  - Making a ***same size distribution*** of nanoparticles is another important factor, which can be controlled by ***diffusion-limited growth*** processes or by ***confining the growth space***

Slide 3.20 Conclusions

1.4 Lecture 4: Applications of Nanomedicine (Slide 4.1)

- Applications of Nanomedicine
  - Drug Delivery
  - DNA Analysis
  - Nanoparticles in Nanomedicine
  - Nanovalves for Drug Delivery
  - Improved Imaging
  - Gene Therapy and Delicate Surgeries
  - Aging, Diagnosis and Treatment of Diabetes and Kidneys
  - Cleaning Robots
  - Improving the Brain
  - Superior Implant Materials
  - Artificial Tissues and Organs
  - Body Surveillance
  - Conclusions

The application of nanotechnology to medicine promises to bring both incremental and breakthrough improvements. Most incremental improvements are the result of miniaturizing current processes which nonetheless can have a huge impact. Lab-on-a-chip technology, which has grown dramatically in the last two decades, now allows researchers to miniaturize multiple assays into a single small package. A single drop of blood can now be used to run multiple tests without the need for a fully stocked analytical laboratory, making the screening and diagnosis of

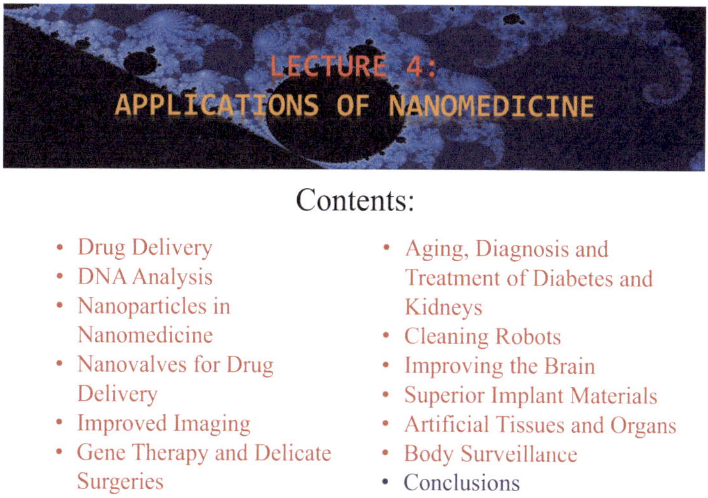

Slide 4.1 Contents of Lecture 4

Improving Imaging	Implant Materials
Drug Delivery	Artificial Tissue
DNA Analysis	Improving Brain
Selective Nanophotothermolysis of Cancer	Cleaning Teeth, Lungs and Arteries

Slide 4.2 Some applications of nanomedicine

diseases cheaper and easier, thus allowing doctors to make better decisions for the care of their patients. Breakthrough improvements create new avenues for prevention, diagnosis or treatment by capitalizing on properties or mechanisms unique to the nanoscale.

Nanomedicine is capable of treating any sites and organs of human body at the molecular level against (eventually) all diseases. Let us here briefly discuss a few current and potential applications of the nanomedicine listed on Slide 4.2. Some of them are still in the research-and-development stage.

Drug Delivery. Nanoparticles (nanoshells, nanotubes) can delivery very strong drugs directly to a tumor site without causing any side effects to the body (Slide 4.3). Using the nanoparticles as drug delivery systems decreases the toxicity and side effects of chemotherapy, since drug dosage is limited. The drugs used in chemotherapy, or other cancer-killing drugs, are dissolved and either put into capsules, entrapped or attached to a matrix of nanoparticles.

Nanoparticle drug delivery systems come in many shapes and sizes (Slide 4.4). For example, researchers have found that mesoporous silica nanoparticles are very effective for controlled drug delivery. The nanopore openings of these nanoparticles can be easily controlled. Three-dimensional DNA crystals can be used as molecular containers to build biochips, nanorobots, biosensors or drug delivery systems.

Polymersome nanoparticles are drug delivery carriers having structural similarity to liposomes and made of synthetic polymer amphiphiles (Slide 4.5). These drug delivery systems have higher stability and higher fluidity on the sides/surface than liposomes.

Drug Delivery

- ***Nanoparticles*** (nanoshells, nanotubes) can delivery very strong drugs directly to the tumor without causing any side effects to the body.
- Using the nanoparticles as drug delivery systems ***decreases the toxicity*** and side effects of chemotherapy, since drug dosage is limited.
- The drugs used in chemotherapy, or other cancer-killing drugs, are dissolved and either put into ***capsules, entrapped*** or attached to ***a matrix of nanoparticles.***

Slide 4.3 Drug delivery

Drug Delivery (continued)

- ***Nanoparticle drug delivery*** systems come in many shapes and sizes.
- For example, researchers have found that ***mesoporous silica nanoparticles*** are very effective for controlled drug delivery.
- The ***nanopore openings*** of these nanoparticles can be easily controlled
- ***Three-dimensional DNA crystals*** can be used as molecular containers to build biochips, nanorobots, biosensors or drug delivery systems.

Slide 4.4 Drug delivery (continued)

Lipid-based drug delivery systems include many different types (Slide 4.5). Such vehicles typically comprise a digestible lipid with (in the case of more complex self-emulsifying formulations) a blend of surfactants, co-surfactants and potentially co-solvents. The ability of a self-emulsifying drug delivery systems to be diluted is essential for its use as a drug delivery vehicle since, after administration, it will be diluted by intestinal media.

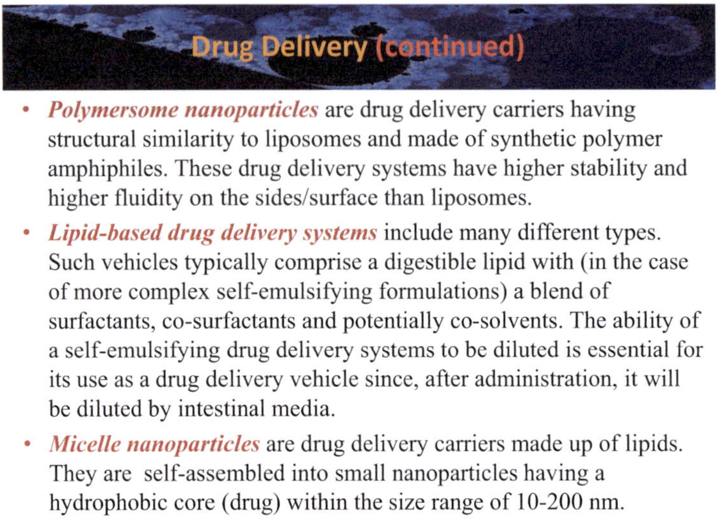

Slide 4.5 Drug delivery (continued)

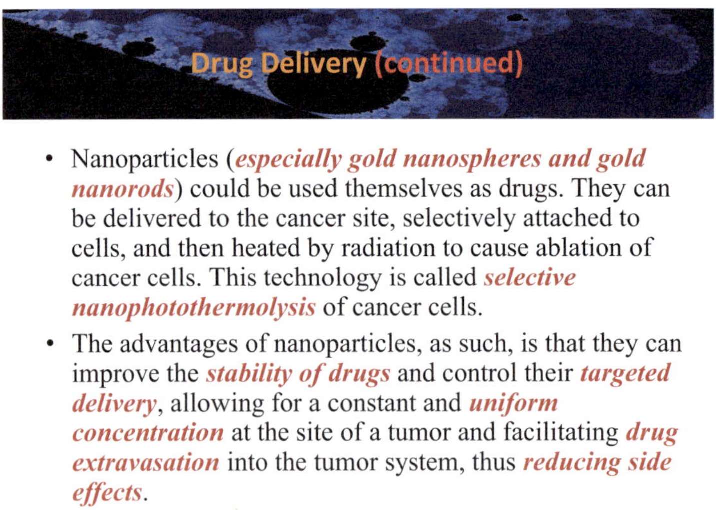

Slide 4.6 Drug delivery (continued)

Micelle nanoparticles are drug delivery carriers made up of lipids (Slide 4.5). They are self-assembled into small nanoparticles having a hydrophobic core (drug) within the size range 10–200 nm.

Nanoparticles (especially gold nanospheres and gold nanorods) could be used themselves as drugs (Slide 4.6). They can be delivered to the tumor, selectively attach to the cancer cells, and then heated by radiation to cause ablation of cancer

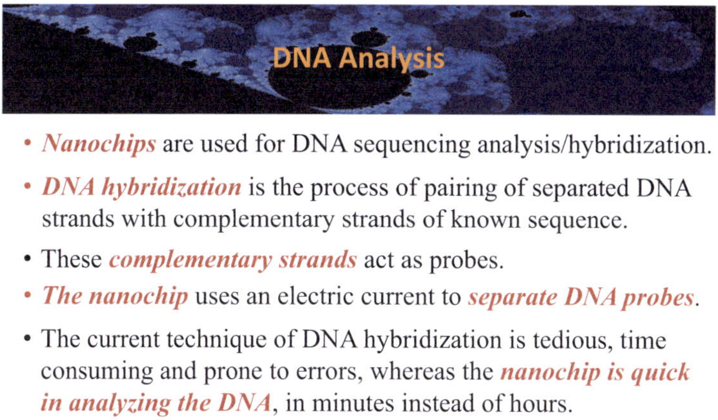

Slide 4.7 DNA analysis

cells. This technology is called selective *nanophotothermolysis* of cancer cells, which we will discuss in more detail later. There are many short videos available on YouTube demonstrating the nanodrug delivery technique.

The advantages of nanoparticles, as such, is that they "can improve the *stability of drugs* and control their *targeted delivery*, allowing for a constant and *uniform concentration* at the site of a tumor and facilitating *drug extravasation* into the tumor system, thus *reducing side effects*" (Slide 4.6) [3].

Metallic nanoparticles have a wide variety of applications in diagnostics and therapy (Slide 4.8). As we studied in Lecture 2, they are used as contrast agents for imaging modalities like MRI, CT, PET and optical imaging. They are also used in the therapy of cancer. Two therapies, nanophotothermolysis and nanophotohyperthermia, both utilize molecular-targeting to deliver nanoparticles to the localized area, and then use the unique light-absorption properties of the metallic nanoparticles to generate heat in the targeted region to ablate the tumor [4].

Silica-based nanoparticles have been successfully used for drug delivery because of unique mesopores and nanochannels which allow for a high payload of the drug and easy stimulated release, as we discussed in Lecture 2 (Slide 4.8). The silica-coated gold nanoparticles are ideal for imaging.

Carbon nanotubes have the ability to penetrate inside cell nuclei, making them invaluable in the diagnosis, imaging and treatment phases of cancer therapy (Slide 4.9). These nanotubes can be used to recognize target DNA sequences, as a drug delivery system, in artificial heart valves, tooth roots, bones, blood vessels, artificial ligaments, tendons, and so on.

Organic particles are used primarily for drug delivery or as a coating on inorganic particles to assure they are not rejected by the body's immune system, as shown in Lecture 2 (Slide 4.9). Examples of organic particles include *polymer-based nanostructures, dendritic nanoparticles, liposome nanoparticles, polymersome nanoparticles, lipid formulations, micelle nanoparticles, nanoemulsions, alginates, and biological nanoparticles.*

Nanoparticles in Nanomedicine

- ***Metallic nanoparticles*** have a wide variety of applications in diagnostics and therapy. For example, they are used as contrast agents ***for imaging*** modalities like MRI, CT, PET and optical imaging.
- They are used in the therapy of cancer. Two therapies, ***nanophotothermolysis*** and ***nanophotohyperthermia***, both utilize molecular-targeting to deliver nanoparticles to the localized area, and then use the unique light-absorption properties of the metal nanoparticles to generate heat in the targeted region to ablate the tumor.
- ***Silica-based nanoparticles*** have been successfully used for drug delivery because of unique mesopores and nanochannels which allows for a high payload of the drug and easy stimulated release. The silica-coated gold nanoparticles are ideal for imaging.

Slide 4.8 Nanoparticles in nanomedicine

Nanoparticles in Nanomedicine (continued)

- ***Carbon nanotubes*** have ability to penetrate inside cell nuclei, making them invaluable in diagnosis, imagin, and treatment phases of cancer therapy. Carbon nanotubes can be used to recognize target DNA sequences, as a drug delivery system, in artificial heart valves, tooth roots, bones, blood vessels, artificial ligaments, tendons, and so on.
- ***Organic particles*** are used primarily for drug delivery or as a coating on inorganic particles to assure they are not rejected by the body's immune system. Examples of organic particles include *polymer-based nanostructures, dendritic nanoparticles, liposome nanoparticles, polymersome nanoparticles, lipid formulations, micelle nanoparticles, nanoemulsions, alginates and biological nanoparticles.*

Slide 4.9 Nanoparticles in nanomedicine (continued)

Polymer-based nanostructures are a class of nanocarriers established for numerous drug delivery applications (Slide 4.10). Polymeric cores are shielded by materials such as polyethylene glycol to guarantee the structure's stability. Targeting molecules can easily be added to the surfaces of polymer-based nanoparticles.

Biological nanoparticles are unicellular microorganisms with different shapes and sizes (Slide 4.10) [5]. For example, a biological nanoparticle called "nanocell"

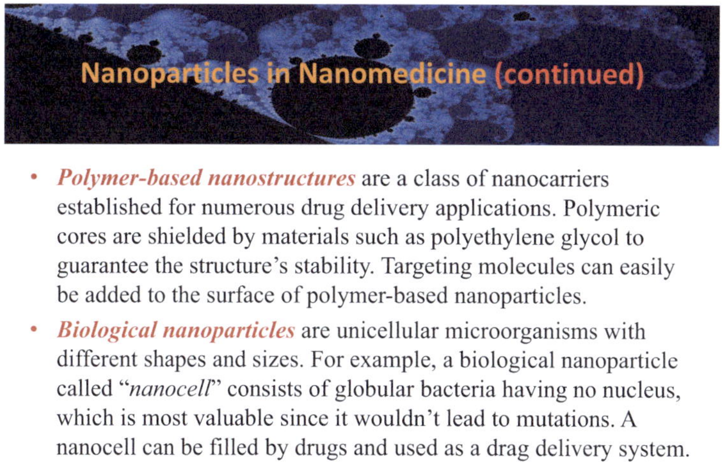

Slide 4.10 Nanoparticles in nanomedicine (continued)

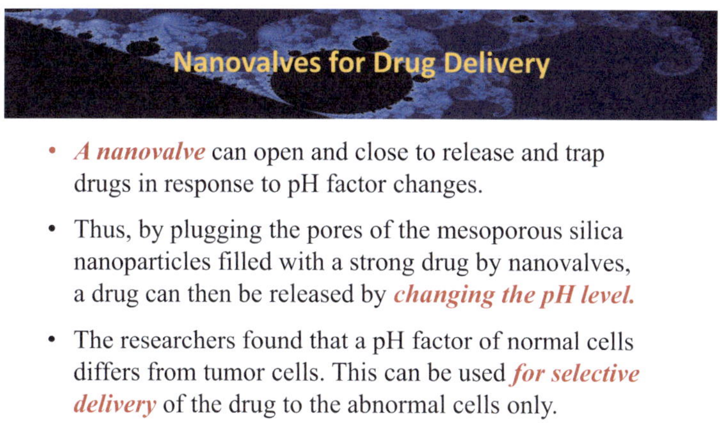

Slide 4.11 Nanovalves for drug delivery

consists of globular bacteria having no nucleus, which is most valuable since wouldn't lead to mutations. A nanocell can be filled by drugs and used as a drug delivery system.

Nanovalves for drug delivery are briefly described in Lecture 2 and shown on Slide 4.11. A nanovalve can open and close to release and trap drugs in response to pH factor changes. Thus, by plugging the pores of the mesoporous silica nanoparticles filled with a strong drug by nanovalves, a drug can then be released by changing the pH level. Researchers have found that a pH factor of normal cells differs from tumor cells. This can be used for selective delivery of drugs to abnormal cells only.

Improved imaging is discussed on Slide 4.12. The metal nanoparticles have the property of plasmon-resonance scattering at the visible and near-IR spectra of

Improved Imaging

- ***Hybrid nanoparticles*** are made up of metallic and polymeric materials for the central core, which is coated with single or multiple layers of lipid to create a protective (surrounding) membrane. The hybrid nanoparticles are able to serve as ***imaging agents***.
- ***Quantum dots*** are small semiconductor crystals that can be fine-tuned to have very specific spectral bands. The labeling of ***multiple disease markers*** with quantum dot-based barcodes could pave the way towards more sensitive and accurate disease detection systems.
- ***Quantum dot-based probes*** may be used to study the heterogeneity of disease markers and link them to prognosis for improved diagnostic strategies.

Slide 4.12 Improved imaging

radiation, which we will study later in this book. By attaching these particles to the cancer cells or DNA, they become visible, providing much earlier diagnostics at the single cellular and molecular levels. The metallic nanoparticles (gold nanoshels, nanospheres and nanorods; iron, cobalt and nickel oxides) are used as the contrast agents in CT and MRI scans and optical imaging techniques.

Hybrid nanoparticles are made up of metallic and polymeric materials for the central core, which is coated with single or multiple layers of lipid to create a protective (surrounding) membrane [5]. These nanoparticles are able to serve as imaging agents (Slide 4.12).

Quantum dots are small semiconductor crystals that can be fine-tuned to have very specific spectral bands (Slide 4.12). The labeling of multiple disease markers with quantum dot-based barcodes could pave the way towards more sensitive and accurate disease detection systems as shown in Lecture 2. Quantum dot-based probes may be used to study the heterogeneity of disease markers and link them to prognosis for improved diagnostic strategies.

Gene therapy. Nanorobots can be used to repair different mutations of DNA (Slide 4.13).

Delicate surgeries. A major application of nanomedicine would be in surgery. Cell repair machines can be used to perform genetic surgery (Slide 4.13).

Aging. The aging process involving cell damage could be repaired by nanorobots from the inside out (Slide 4.14).

Diagnosis and treatment of diabetes. Nanomachines are capable of controlling and monitoring glucose levels in diabetic patients (Slide 4.14).

Treatment of kidneys. Medical nanomachines could be designed for destroying kidney stones (Slide 4.14).

Gene Therapy and Delicate Surgeries

- Nanorobots can be used to introduce different ***modifications and corrections to the DNA*** or the proteins attached to the DNA in the right place.
- A major application of nanomedicine would be in ***surgery. Cell repair machines*** can be used to perform genetic surgery.

Slide 4.13 Gene therapy and delicate surgeries

Aging, Diagnosis and Treatment of Diabetes and Kidneys

- *Aging*. The aging process involving cell damage could be repaired by nanorobotsfrom the inside out.
- ***Diagnosis and treatment of diabetes.*** Nanorobots can help in controlling and monitoring glucose levels in diabetic patients.
- ***Treatment of kidneys.*** Medical nanomachines may be utilized for the targeting and destruction of kidney stones.

Slide 4.14 Aging, diagnosis and treatment of diabetes and kidneys

Cleaning nanorobots. *Artery-cleaning nanorobots* could remove extra fat and bad proteins from the walls of blood vessels and arteries (Slide 4.15). The size of these nanomachines is very small. They can easily swim in blood vessels and penetrate into any organs, and even be placed inside the cells through the porous membrane. Artery-cleaning nanorobots can be used for prevention of a heart attack and can be applied in the treatment of atherosclerotic plaques. Medical nanomachines can help find atherosclerotic lesions in blood vessels and help in their removal (Slide 4.15).

Lung-cleaning nanorobots are inhaled, where nanorobots can collect foreign particles like fibers of asbestos and toxic particles from the lungs (Slide 4.16).

Teeth cleaning nanorobots may collect harmful bacteria in the mouth and can help to detect different oral diseases (Slide 4.16).

Improving brain capability. A nanostructured data storage system may store a nanocomputer and an amount of information equivalent to an entire library

- ***Artery-cleaning nanorobots*** could remove extra fat and bad proteins from blood vessels and arteries
- Such nanorobots can be used for prevention of a ***heart attack*** and can be applied in the ***treatment of atherosclerosis***.
- Medical nanomachines can help find ***atherosclerotic lesions*** in blood vessels and help in their removal.

Slide 4.15 Cleaning robots

- ***Lung-cleaning nanorobots*** are inhaled, where nanorobotscan collect foreign particles like fibers of asbestos and toxic particles from the lungs.
- ***Teeth-cleaning robots*** may collect harmful bacteria in the mouth.
- ***Oral introduction of nanorobots*** can help also in the diagnosis and testing of different dental diseases.

Slide 4.16 Cleaning robots (continued)

(Slide 4.17). These nanochips/nanomemory then can be attached to brain cells to create an extra network of memory in the brain. Thus, in the future we may be able to browse our brain in searching information instead of browsing the Internet.

Superior implant materials. Creating new allows and discovering new nanomaterials can provide much stronger and lighter implant materials for dentistry and artificial body parts (Slide 4.18).

Artificial tissues and organs. Researchers hope to regenerate skin, bone and more sophisticated organs by means of nanotubes and nanomaterials (Slide 4.18). Nanorobots can be used as support devices for regeneration of skin, bone and injured organs.

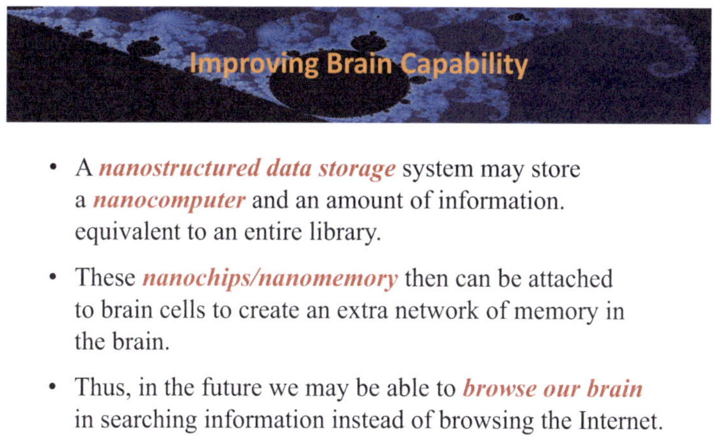

Slide 4.17 Improving brain capability

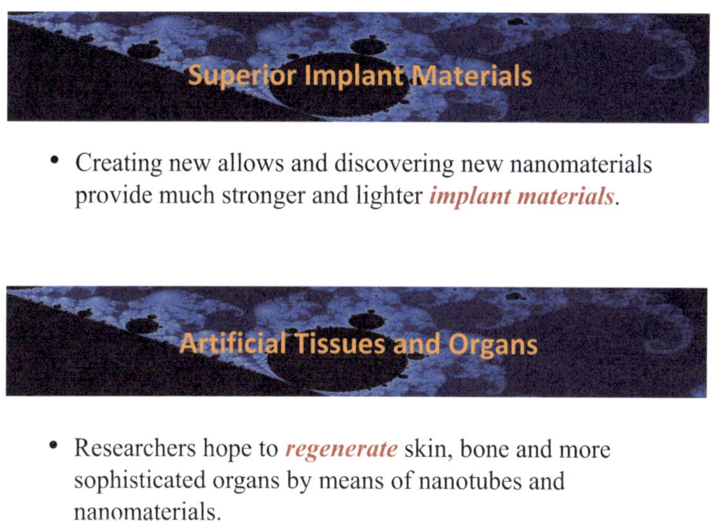

Slide 4.18 Superior implant materials, artificial tissues and organs

Body surveillance. Nanorobots can be used to continuously monitoring vitals inside the human body (Slide 4.19). This can be an effective tool for cancer prevention. If any bacteria or cancer cells invade the body, they will be destroyed by the nanorobots patrolling the body, since the bacteria or mutated DNA do not match the blueprint recorded in the nanomachine (Slide 4.19).

In conclusion, nanomedicine is revolutionizing the way we treat the patients (Slide 4.20). We have discussed here only a few current and potential applications

- Nanorobots can be used to ***continuously monitoring*** vitals inside the human body.
- This can be an effective tool for ***cancer prevention***.
- If any bacteria or cancer cells invade the body, they will be destroyed by the nanorobots patrolling the body, since the bacteria or mutated DNA do not match the ***blueprint recorded*** in the nanomachine.

Slide 4.19 Body surveillance

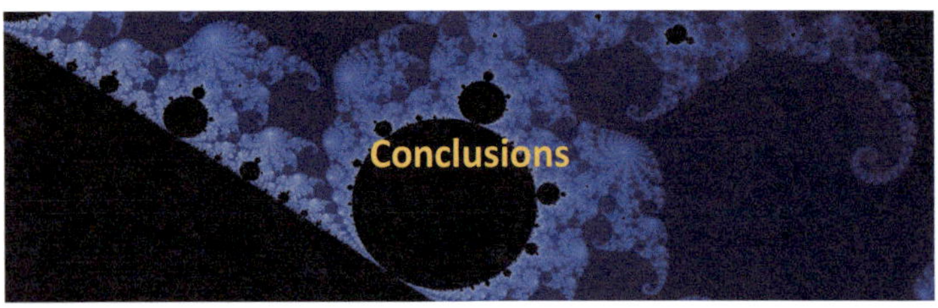

- ***Nanomedicine*** is revolutionizing the way we treat the patients.
- It enables rapid detection and treatment at the cellular level.
- ***Nanoparticles*** improve currently used methods.
- ***Nanorobots and nanomachines*** have opened a new world of discoveries and feasibilities in nanomedicine.

Slide 4.20 Conclusions

of nanomedicine just to give a sense of its capability. Rapid detection and treatment at the cellular level are very important to many life threatening diseases. For cancer, early detection would cure the cancer and save lives instead of just extending the patient's life. As seen in this lecture, nanodevices/nanoparticles will play a very large role in the future of medicine, taking part in many new advances in the fields of diagnostics and treatment of diseases. Nanoparticles can provide very good solutions for some medical problems we face and improve currently used methods. Nanorobots and nanomachines as discussed here are the new molecular technologies that have gradually developed over the years, creating robots or machines made up of parts that are at or close to the scale of nanometers. These new technology fields have evolved through several developmental stages and are still under research and development. These small devices have opened a new world of discoveries and feasibilities in nanomedicine.

Homework Exercises

Section 1.1: Definition of Nanomedicine and Nanotechnology

1.1. What are three main technological frontiers in the twenty-first century?
1.2. What is the definition of nanomedicine?
1.3. What is the definition of nanotechnology?
1.4. How does nanomedicine relate to the basic sciences?
1.5. What imaging and treatment modalities do you know?
1.6. List the advantages of today's imaging and treatment modalities.
1.7. List the disadvantages of today's imaging and treatment modalities.
1.8. How can nanomedicine help to improve conventional clinical modalities?
1.9. True or false? Nanomedicine is a part of medical physics.
1.10. Select one. Nanotechnology studies a matter at the length scale of:

(a) 1–1000 nm
(b) 1–100 pm
(c) 1–100 μm
(d) 1–100 nm
(e) All of the above
(f) None of the above

1.11. Select one. The size of a human cell is:

(a) 10 mm
(b) 1 μm
(c) 10 μm
(d) 1 mm

Section 1.2: Medical Nanodevices, Properties and Functions

1.12. What is the definition of nanorobots?
1.13. What are the main components of medical nanomachines?
1.14. Give examples of medical nanomachines.
1.15. What is a nanovalve?
1.16. Describe the structure and operation of a molecular propeller.
1.17. What is a nanochip?
1.18. What is the definition of a nanoparticle?

1.19. What are the two main types of nanoparticles?
1.20. What are the three distinct components of nanoparticle design?
1.21. Give examples of inorganic nanoparticles.
1.22. What are applications of metallic nanoparticles?
1.23. What is a fullerene?
1.24. What is a carbon nanotube?
1.25. What are two types of silicon nanoparticles?
1.26. What are applications of silica-based nanoparticles?
1.27. What is a quantum dot?
1.28. What are applications of quantum dots?
1.29. Give examples of organic nanoparticles.
1.30. What are applications of polymer-based nanostructures in nanomedicine?
1.31. What is a dendrimer nanoparticle?
1.32. What is a polymersome nanoparticle?
1.33. What is a key application of lipid formulation?
1.34. What is a micelle nanoparticle?
1.35. How does a biological nanoparticle work?
1.36. What is a hybrid nanoparticle?
1.37. What are the main functions of medical nanorobots?
1.38. Select one or more. Metallic nanoparticles have a property of:

(a) Free radicals
(b) Chemically active
(c) Plasmon resonance scattering
(d) High volume
(e) High surface area
(f) All of the above
(g) None of the above

1.39. Select one. Properties of medical nanodevices include:

(a) Shape and size
(b) Biocompatibility
(c) Powering
(d) Communication
(e) Navigation
(f) All of the above
(g) None of the above

1.40. What shape of nanorobot is used for free-floating nanomachines? Actively-swimming nanodevices? Intracellular nanodevice? Tessellating nanomachines?
1.41. How does the geometry of a nanoparticle effect its overall properties?
1.42. Select one or more. Structurally the nanoparticles can be made as:

(a) Uniform core particles
(b) Porous nanoparticles
(c) Nanoshells

(d) Single layer particle
(e) Multilayer nanoparticle with
(f) All of the above
(g) None of the above

1.43. What is a nanoshell?
1.44. What are the roles of surface chemistry on the surface of nanoparticles?
1.45. How does the size of a nanoparticle effect its overall properties?
1.46. True or false. Biocompatibility means that the immune system fights with foreign nanoparticles.
1.47. How is biocompatibility achieved?
1.48. What are two main sources for powering medical nanomachines?
1.49. Select one or more. Powering a medical nanorobots is provided by:

(a) Laser beam
(b) Chemical reactions
(c) RF wave
(d) Body heat
(e) All of the above
(f) None of the above

1.50. What is an internal power source?
1.51. What is an external power source?
1.52. Select one or more. Communication with nanomachines is reached by:

(a) Acoustic waves
(b) RF waves
(c) Magnetic field
(d) All of the above
(e) None of the above

1.53. What are two categories of the navigation system of nanorobots?
1.54. Select one or more. Navigation of nanomachines is reached by:

(f) Ultrasonic signals
(g) RF waves
(h) Magnetic field
(i) All of the above
(j) None of the above

Section 1.3: Design and Fabrication of Medical Nanodevices

1.55. What materials are used for construction of medical nanodevices?
1.56. List and describe the steps of a roadmap for developing of nanomachines.
1.57. What are two techniques for the synthesis of nanomaterials?
1.58. Select one or more. A top-down approach in nanoparticle fabrication includes:

(a) Heterogeneous nucleation on substrates
(b) Phase segregation through annealing solids at high temperatures
(c) Confining particles in a small space using chemical reactions
(d) All of the above
(e) None of the above

1.59. True or false. In a bottom-up approach, the nanostructures are created out of bulk material.
1.60. What is homogeneous nucleation?
1.61. What are the ways of controlling nanoparticle growth in a bottom-up approach?
1.62. What is an organometallic compound?
1.63. How is a cadmium selenide quantum dot synthesized?
1.64. Describe the synthesis of oxide nanoparticles?
1.65. How are nanoparticles are generated in a vapor phase reaction?
1.66. What is solid state segregation?
1.67. What is heterogeneous nucleation?
1.68. What is kinetically-confined synthesis?
1.69. What are carbon nanotubes manufacturing techniques?
1.70. Describe the arc discharge method for fabrication of carbon nanotubes.
1.71. Describe the laser ablation technique for production of carbon nanotubes.
1.72. What important points have to be considered when designing a synthesis of nanoparticles?

Section 1.4: Applications of Nanomedicine

1.73. Select one or more. Applications of nanomedicine include:

(a) Improving imaging
(b) Drug delivery
(c) DNA analysis
(d) Selective nanophotothermolysis
(e) New implant materials
(f) Artificial tissues
(g) Improving brain capability
(h) Cleaning teath, lungs and arteries
(i) All of the above
(j) None of the above

1.74. What are advantages of nanoparticle-based drug delivery systems?
1.75. What are applications of carbon nanotubes in nanomedicine?
1.76. What are applications of quantum dots in nanomedicine?
1.77. What is gene therapy?
1.78. Describe the operation of cleaning nanorobots.
1.79. What is body surveillance?

Topics for Presentations and Writing Assignments

Presentation #1. Nanoparticles in cancer therapy and diagnosis.

Presentation #2. Liposomes, lipid formulations, nanoemulsions and alginate nanoparticles in cancer therapy and diagnosis.

Presentation #3. Silica-based nanoparticles, polymeric nanostructures and dendrite nanoparticles in cancer therapy and diagnosis.

Presentation #4. Nanotechnology for biophotonics, i.e., bionanophotonics

Presentation #5. Methods and tools for cancer treatment and diagnosis in conventional medicine.

References

1. http://www.slideshares.net
2. G. Cao, Y. Wang, *Nanostructures and Nanomaterials: Synthesis, Properties, and Applications* (World Scientific, Singapore, 2011)
3. https://www.ncbi.nlm.nih.gov
4. R.R. Letfullin, A.R. Letfullin, T.F. George, Absorption efficiency and heating kinetics of nanoparticles in the RF range for selective nanotherapy of cancer. Nanomed. Nanotechnol. Biol. Med. 11, 413–420 (2015)
5. F.I. Alexis, E.M. Pridgen, R. Langer, O.C. Farokhzad, Nanoparticle technologies for cancer therapy. Handb. Exp. Pharmacol. **197**, 55–86 (2010)

Chapter 2
Introduction to Cancer Cells and Targeting

Contents

Electronic supplementary material: The online version of this chapter (doi:10.1007/978-3-319-43577-0_2) contains supplementary material, which is available to authorized users.

R.R. Letfullin, T.F. George, *Computational Nanomedicine and Nanotechnology*,
DOI 10.1007/978-3-319-43577-0_2

Abstract This chapter gives a brief introduction to the structure of cancer cells and their properties, and how cancer cells differ from normal cells. In order to provide a thorough understanding of how cancer cells are formed, a brief introduction about the different biomolecules present in cells—namely, carbohydrates, proteins, lipids, and nucleic acids—is described. In cancer cells, it is observed that there are distinct changes in the nucleus, mitochondria, ribosome, and cytoskeleton, which are the organelles in the cells, and hence the structure and function of these organelles in normal and cancer cells are discussed. The causes of cancer, types of cancer, techniques for selective targeting of cancer, and gene therapy are explained in detail in this chapter.

2.1 Lecture 5: Introduction to Cell Biology (Slide 5.1)

- Introduction to the Cell
- Biological Molecules
  - *Carbohydrates*
  - *Proteins*
  - *Lipids*
  - *Nucleic Acids*
- Cell structure
  - *Nucleus*
  - *Mitochondrion*
  - *Ribosome*
  - *Cytoskeleton*
  - *Other organelles*
  - *Plasma membrane*
- Cell Division
- DNA Mutations
- Causes of Genetic Changes
- Conclusions

This lecture briefly introduces the *biology of the cell* providing the basics of the *cell structure* and *functions* needed to understand the nature of *cancer cells*. Slide 5.2 gives a general introduction to the biological cell. *Cells* are building blocks of any organism. Our body is composed of trillions of cells. They are specialized to perform different functions and tasks. Cells are made of *organelles* sited in the *cytoplasm* protected by the *cell membrane*. Thus, the organelles are structural components of the cell that carry out cellular functions. Cells also carry *genetic information* that is unique to every individual. Cells are capable of *self-replicating* via a process called *mitosis*. The deviation from normal cell functions can occur in the *cell cycle*, or glitches can be found in the genetic information, and these can cause cancer.

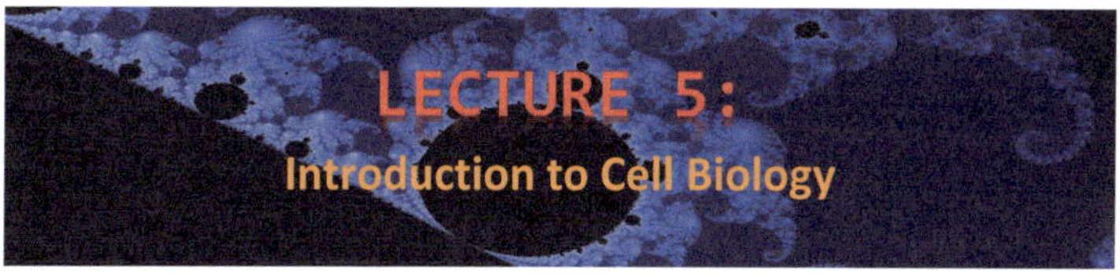

Contents:

- Introduction to the Cell
- Biological Molecules
  - *Carbohydrates*
  - *Proteins*
  - *Lipids*
  - *Nucleic Acids*
- Cell Structure
  - *Nucleus*
  - *Mitochondrion*
  - *Ribosome*
  - *Cytoskeleton*
  - *Other Organelles*
  - *Plasma Membrane*
- Cell Division
- DNA Mutations
- Causes of Genetic Changes
- Conclusions

Slide 5.1 Contents of Lecture 5

- Compose ***entire body.***
- ***Differentiated***, or specialized, to perform certain tasks.
- ***Organelles*** are the structures inside of cells that carry out cellular functions.
- ***Genetic information*** that is unique to every individual is enclosed.
- ***Self-replication*** via a process called mitosis.
- Problems can occur in the ***cell cycle*** or glitches can be found in the genetic information, and these can cause ***cancer.***

Slide 5.2 Introduction to the cell

In order to effectively treat cancer, the biology of a normal cell and cancer cell needs to be well understood. The cell itself and intracellular organelles are made of four main types of *molecules* (Slide 5.3):

- *Carbohydrates*
- *Proteins*
- *Lipids*
- *Nucleic acids*

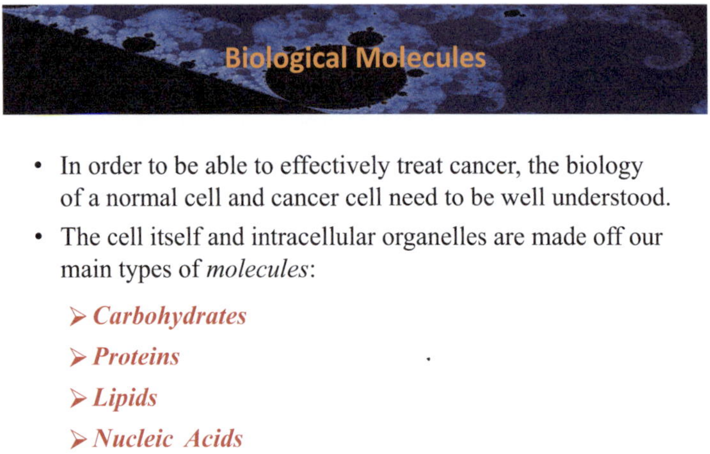

Slide 5.3 Biological molecules

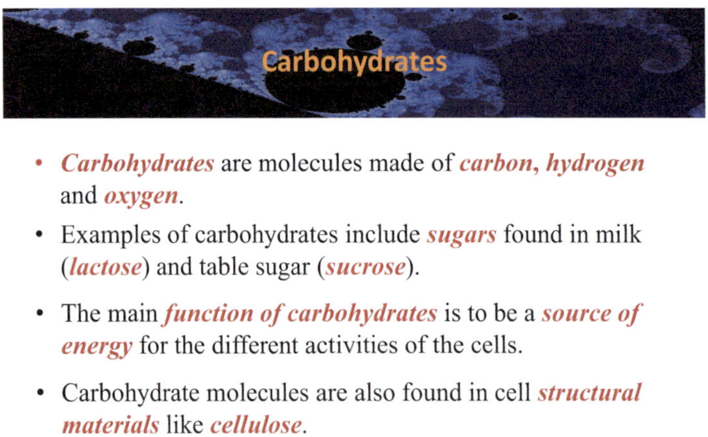

Slide 5.4 Carbohydrates

Carbohydrates are made of *carbon*, *hydrogen*, and *oxygen* (Slide 5.4), where examples include the *sugars* found in milk (*lactose*) and table sugar (*sucrose*). The main *function of carbohydrates* is to be a *source of energy* for the different activities of the cells. Carbohydrate molecules are also found in cell *structural materials* like *cellulose*.

Proteins are monomer molecules made up of *amino acids* (Slide 5.5). There are about 20 different amino acids used in the cells. Because of the wide variety of amino acids, proteins have numerous *functions* in the cells. They form *structural* features of the body, *transport* materials across cellular membranes, act as biological *catalysts*, maintain *contact between adjacent cells*, *control the activity of cells*, including decisions regarding cell division, etc.

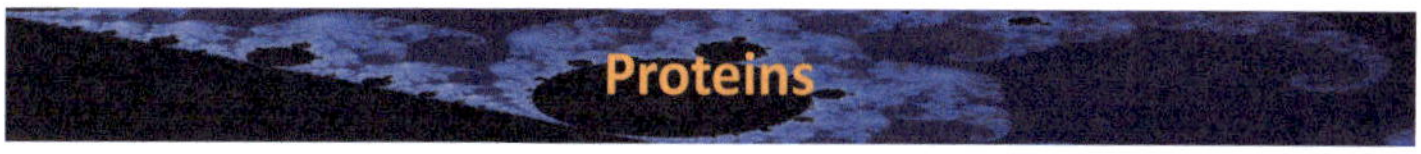

- ***Proteins*** are monomer molecules made up of ***amino acids***.
- There are about ***twenty*** different amino acids used in the cells.
- Because of the wide variety of amino acids, proteins have numerous ***functions*** in the cells.
- They form ***structural features*** of the body, ***transport*** materials across cellular membranes, act as biological ***catalysts, maintain contact between adjacent cells,* and *control the activity of cells***, including decisions regarding cell division.

Slide 5.5 Proteins

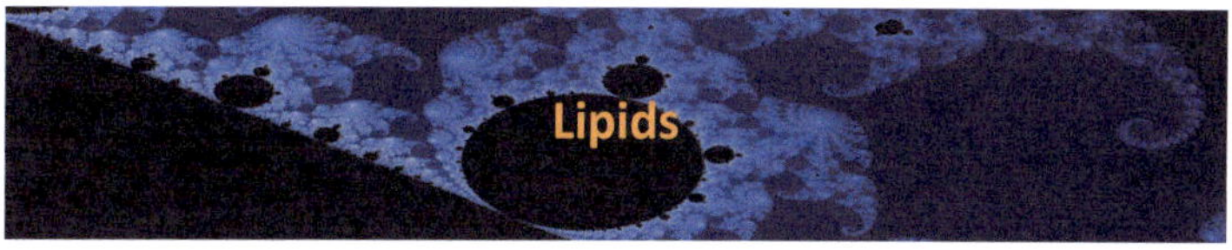

- ***Lipids*** include a variety of biomolecules such as ***fats, oils, waxes***.
- They are hydrophobic and are composed mainly of ***carbon, hydrogen and oxygen***.
- The main ***function of lipids*** is the formation of ***biological membranes***.

Slide 5.6 Lipids

Lipids include a variety of biomolecules such as *fats, oils, and waxes* (Slide 5.6). They are hydrophobic and are composed mainly of *carbon, hydrogen, and oxygen*. The main *function of lipids* is the formation of *biological membranes*.

Nucleic acids are the molecules used for storing all information about the cell (Slide 5.7). DNA (deoxyribonucleic acid) and RNA (ribonucleic acid) are two main types of nucleic acids. The monomers used to build nucleic acids are called *nucleotides*, which are labeled by the single letter abbreviations A, C, G, T, and U. RNA is a chain of nucleotides that carries information present in genes. The important functional differences between DNA and RNA are as follows:

- RNA is copied from DNA.
- DNA is blueprint for specific production of RNA.
- RNA is single stranded and DNA is double stranded.

- ***Nucleic acids*** are the molecules used for storing all information about the cell.
- ***Doxyribonucleic acid*** (DNA) and ***ribonucleic acid*** (RNA) are two main types of nucleic acids.
- The monomers used to build nucleic acids are called ***nucleotides***, which are labeled by the single letter abbreviations A, C, G, T and U.
- ***RNA*** is a chain of nucleotides that carries information present in genes. The important functional ***differences between DNA and RNA*** are:
  - ➢ ***RNA is copied from DNA.***
  - ➢ ***DNA is a blueprint for specific production of RNA.***
  - ➢ ***RNA is single stranded and DNA is double stranded.***

Slide 5.7 Nucleic acids

- ***The functions*** of individual cells are controlled by smaller structures within the cells called ***organelles***.
- There are ***four main*** important ***organelles*** which have a direct relevance to cancer:
  - ➢ ***Nucleus***
  - ➢ ***Mitochondrion***
  - ➢ ***Ribosome***
  - ➢ ***Cytoskeleton***

Slide 5.8 Cell structure

2.1.1 Cell Structure

The functions of individual cells are controlled by smaller structures within the cells called organelles. There are four main important organelles which have a direct relevance to cancer (Slide 5.8):

- *Nucleus*
- *Mitochondrion*

- *Ribosome*
- *Cytoskeleton*

The *nucleus* is the brain of the cell (Slide 5.9). It has a spherical shape and contains two membranes. It is home to 46 *chromosomes*. Chromosomes are made up of a complex between DNA and proteins called *chromatin*. *Genes* are certain sections of DNA which control cell activities. Changes in the nucleotide sequence of a gene alter the behavior of the cell.

Mitochondrion is the energy producer for the cell (Slide 5.10). Like the nucleus, it is surrounded by a double membrane. Mitochondria convert food into energy, which leads to the production of chemicals that may be toxic for the DNA and can cause genetic changes.

- The ***nucleus*** is the brain of the cell.
- It has a spherical shape and contains two ***membranes***.
- It is home to forty-six ***chromosomes***.
- Chromosomes are made up of a complex between DNA and proteins called ***chromatin***.
- The ***genes*** are certain sections of DNA which control cell activities.
- ***Changes*** in the nucleotide sequence of a gene alter the ***behavior*** of a cell.

Slide 5.9 Nucleus

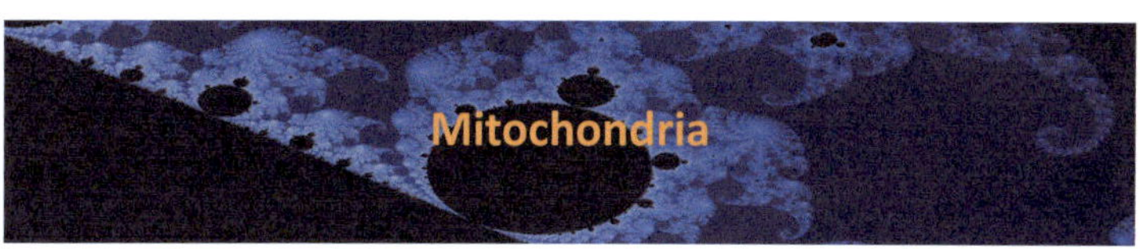

- ***Mitochondrion*** is the energy producer for the cell.
- Like the nucleus, it is surrounded by a ***double membrane***.
- ***Mitochondria*** convert food into energy.
- ***Energy production*** in mitochondria leads to the production of chemicals that may be a toxic for the DNA and can cause genetic changes.

Slide 5.10 Mitochondria

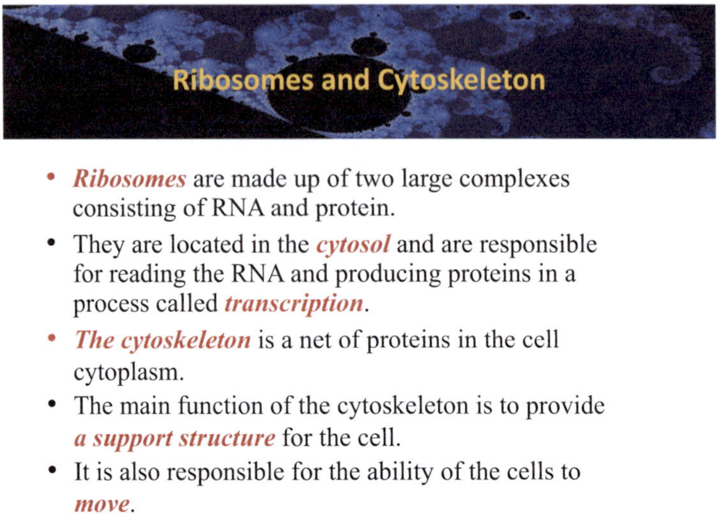

Slide 5.11 Ribosomes and cytoskeleton

Ribosomes are made up of two large complexes consisting of RNA and protein (Slide 5.11). They are located in the *cytosol* and are responsible for reading the RNA and producing proteins in the process called *transcription.*

The cytoskeleton is a net of proteins in the cell cytoplasm (Slide 5.11). Its main function is to provide a support structure for the cell. It is also responsible for the ability of the cells to move.

Other organelles are as follows (Slide 5.12):

- *Centrosome*—made of two *centrioles* which organize microtubules.
- *Lysosomes*—contain *hydrolytic enzymes* to digest macromolecules.
- *Peroxisomes*—produce *hydrogen peroxide* for oxidation processes.
- *Endoplasmic reticulum.*
- *Golgi apparatus.*

A *plasma membrane* encloses cytoplasm containing the organelles and forms the cellular boundary (Slide 5.13). The membrane is made of a *phospholipid bilayer*, which is an extracellular matrix containing *glycoproteins*, *integrins,* and *carbohydrate side chains.* These lipid molecules play roles in the process of cellular adhesion and cell-to-cell contacts. The plasma membrane is porous to allow for nutrients, waste, and other materials to pass through the cell.

Other organelles are:

- ***Centrosome*** – made of two ***centrioles*** which organize ***microtubules*.**
- ***Lysosomes*** – contain ***hydrolytic enzymes*** to digest macromolecules.
- ***Peroxisomes*** – produce ***hydrogen peroxide*** for oxidation processes.
- ***Endoplasmic reticulum*.**
- ***Golgi apparatus*.**

Slide 5.12 Other organelles

- ***The plasma membrane*** encloses cytoplasm containing the organelles and forms the cellular boundary.
- The membrane is made of a ***phospholipid bilayer***, which is an extracellular matrix that contains ***glycoproteins, integrins*** and ***carbohydrate side chains***.
- These lipid molecules play roles in the process of ***cellular adhesion and cell-to-cell contacts.***
- The ***plasma membrane*** is ***porous*** to allow for nutrients, waste and other materials to pass through the cell.

Slide 5.13 Plasma membrane

2.1.2 Cell Division

While many of the cells age and die, there is a process known as *mitosis* whereby a cell reproduces to create two identical copies of itself (Slide 5.14).

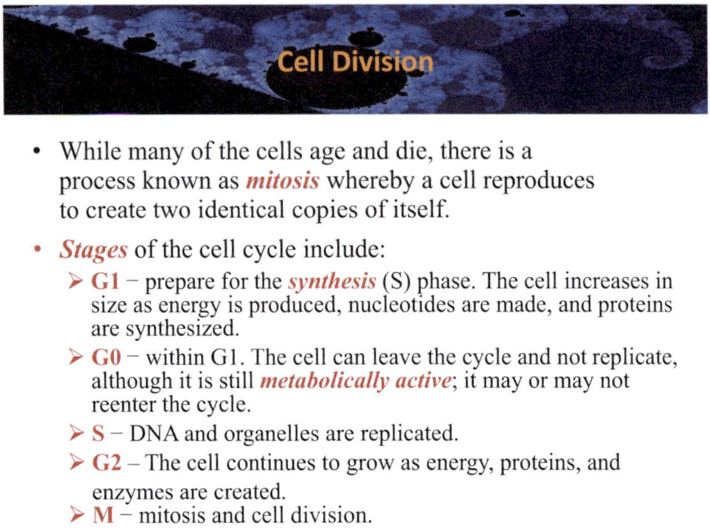

Slide 5.14 Cell division

Stages of the cell cycle include

- *G1*—prepare for the *synthesis* (S) phase. The cell increases in size as energy is produced, nucleotides are made, and proteins are synthesized.
- *G0*—within G1. The cell can leave the cycle and not replicate although it is still *metabolically active*; it may or may not reenter the cycle.
- *S*—DNA and organelles are replicated.
- *G2*—The cell continues to grow as energy, proteins and enzymes are created.
- *M*—mitosis and cell division.

There are a few *checkpoints* built into the cell division process to ensure that cells do not divide unless they have completed the replication process correctly (Slide 5.15):

- *G1/S*—The cell size is checked, and the DNA is inspected for damages before it replicates in the S phase.
- *G2/S*—Verifies that all of the DNA was replicated and confirms that it was replicated without any damages or mistakes.
- *M*—Checks that spindle fibers were formed and connected to the centrosomes to pull the chromatids apart.

Errors may occur during the replication process that could lead to changes in the nucleotide sequence of the DNA (Slide 5.16). If these changes occur within genes, they can alter the function of the cell. For example, the cancer cells behave differently as a result of a series of mutations (permanent changes) in the genes. Types of genetic changes are as follows:

- *DNA mutations*
- *Gene amplification*

Checkpoints in the Cycle

There are a few ***checkpoints*** built into the cell division process to assure that cells do not divide unless they have completed the replication process correctly:

- **G1/S** –The cell size is checked, and the DNA is inspected for damages before it replicates in the S phase.
- **G2/S** – Verifies that all of the DNA was replicated and confirms that it was replicated without any damages or mistakes.
- **M** – Checks that spindle fibers were formed and connected to the centrosomes to pull the chromatids apart.

Slide 5.15 Checkpoints in the cycle

Genetic Changes

- Errors may occur during the replication process that could lead to ***changes in the nucleotide sequence*** of the DNA.
- If these changes occur within ***genes***, they can alter the function of the cell.
- For example, the cancer cells behave differently as a result of a ***series of mutations*** (permanent changes) in the genes.
- Types of genetic changes are:
  - ➢ ***DNA mutations***
  - ➢ ***Gene amplification***

Slide 5.16 Genetic changes

DNA mutation is a process where incorrect '*reading*' of RNA happens during the *transcription* event (Slide 5.17). The altered RNA may lead to the production of a protein that no longer functions properly. There are three types of DNA mutations:

- *Nonsense mutations*
- *Missense mutations*
- *Frameshift mutations*

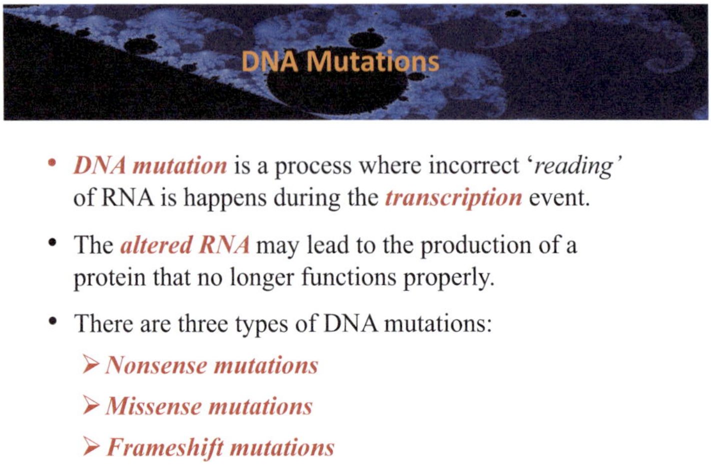

Slide 5.17 DNA mutations

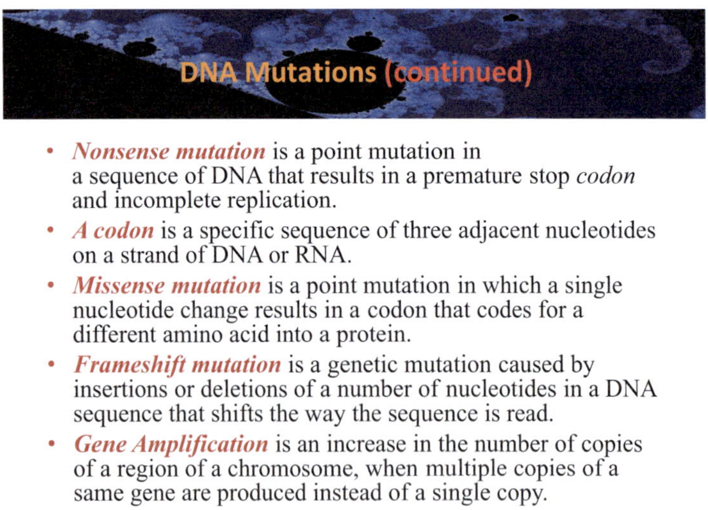

Slide 5.18 DNA mutations (continued)

Nonsense mutation is a point mutation in a sequence of DNA that results in a premature stop *codon* and incomplete replication [1]. A *codon* is a specific sequence of three adjacent nucleotides on a strand of DNA or RNA (Slide 5.18).

Missense mutation is a point mutation in which a single nucleotide change results in a codon that codes for a different amino acid *into a protein* (Slide 5.18) [2].

Frameshift mutation is a genetic mutation caused by insertions or deletions of a number of nucleotides in a DNA sequence that shifts the way the sequence is read (Slide 5.18) [2].

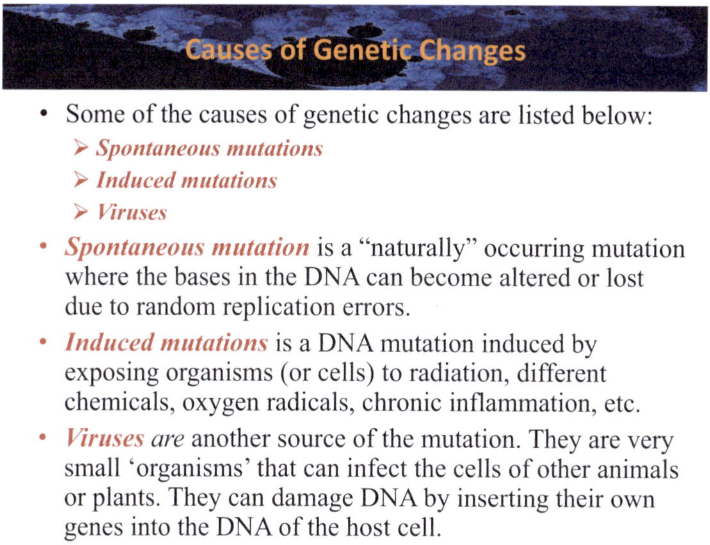

Slide 5.19 Causes of genetic changes

Gene amplification is an increase in the number of copies of a region of a chromosome, when multiple copies of the same gene are produced instead of a single copy (Slide 5.18).

Some of the *causes of genetic changes* are listed as follows (Slide 5.19):

- *Spontaneous mutations*
- *Induced mutations*
- *Viruses*

Spontaneous mutation is a "naturally" occurring mutation where the bases in the DNA can become altered or lost due to random replication errors.

Induced mutation is a DNA mutation induced by exposing organisms (or cells) to radiation, different chemicals, oxygen radicals, chronic inflammation, etc.

Viruses are another source of the mutation. They are very small 'organisms' that can infect the cells of other animals or plants. They can damage DNA by inserting their own genes into the DNA of the host cell.

In conclusion, the cell is categorized as both the smallest structural and functional unit of any living organism (Slide 5.20). Depending on the organism itself, cells can potentially have different compositions in which varying organelles or components are included, but almost all cells contain a nucleus which is suspended in the cytoplasm and further enclosed by a plasma membrane. The genetic material, stored in the form of chromatin and chromosomes in the cell nucleus, programs the cells to carry out necessary functions to sustain life and ensures that both hormones and enzymes are represented in an appropriate amount and balanced fashion. Normal cells have the ability to replicate in an orderly and consistent manner. Cancer cells also contain DNA, but with mutation to represent an abnormality

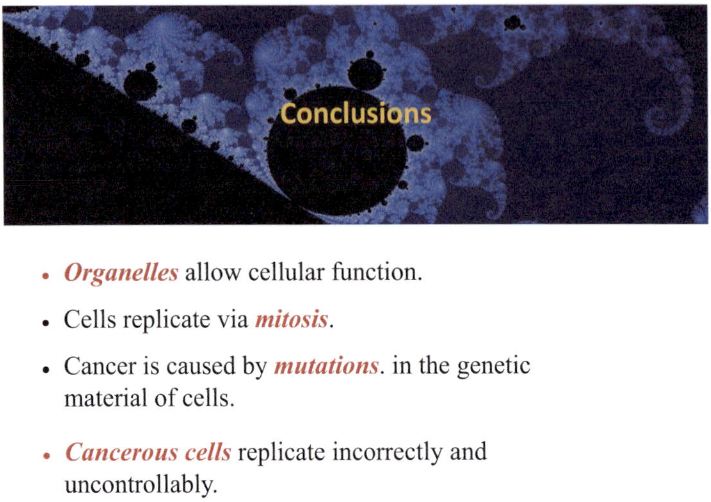

Slide 5.20 Conclusions

which creates impaired and defective functioning. Cancer cells differ from normal cells as they hold the ability to metastasize and infect other tissues in the body as they grow uncontrollably, whereas normal cells do not.

2.2 Lecture 6: Introduction to Cancer Biology (Slide 6.1)

- Some Causes of Cancer
- Stages of Cancer
- Categories of Cancer
- Types of Tumors
- Cancer Formation
- Normal Cells versus Cancer Cells
  - Normal Cells
  - Cancer Cells
- Cancer Cell Structure
- Properties of Cancer Cells
- Conclusions

Cancer is defined as a disease in which cells in a part of the body begin to grow out of control (Slide 6.2). It is the second most common cause of death in the United States. Tumors can be benign in which case they are not cancerous; also, tumors can be malignant in which case they are cancerous. There are different types of cancer which include lung cancer, breast cancer, leukemia, and many more. In order to be

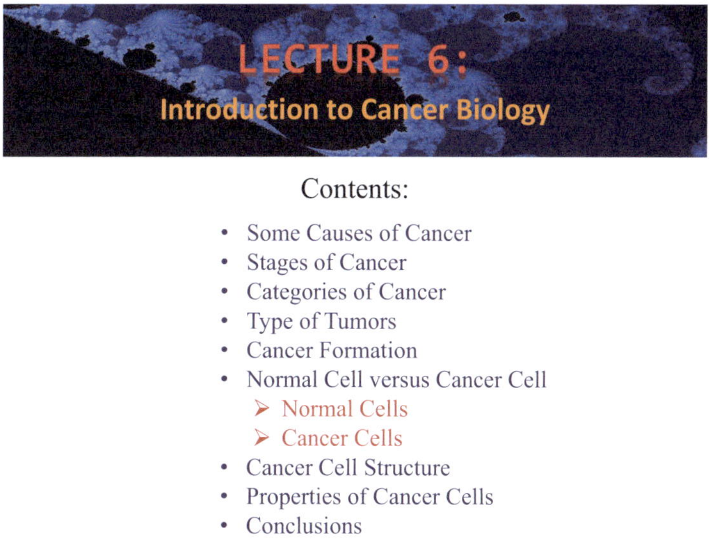

Slide 6.1 Contents of Lecture 6

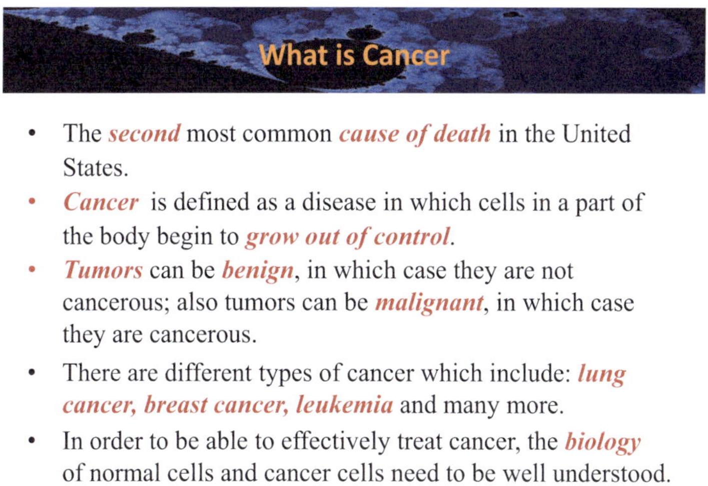

Slide 6.2 What is cancer

able to effectively treat cancer, the biology of normal cells and cancer cells needs to be well understood. Normal cells divide, grow, and die in an orderly fashion. They divide to replace worn-out and injured tissues, and they adhere to stop signals, thus not invading nearby tissues (Slide 6.3). On the other hand, cancer cells divide uncontrollably—they keep growing and never die; they invade other tissues and do not adhere to stop signals. Although there are different causes of cancer, the major

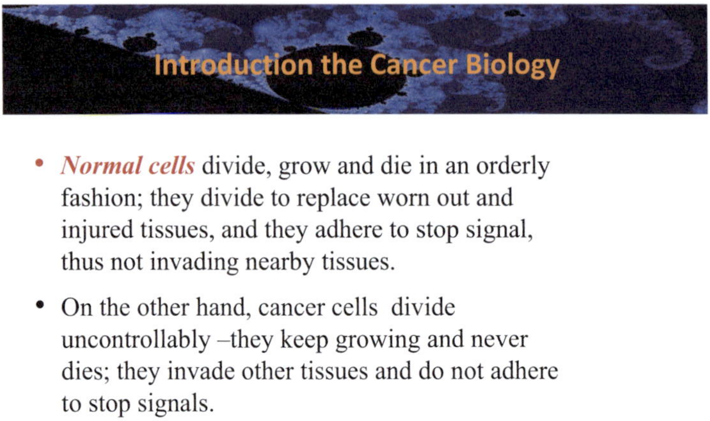

Slide 6.3 Introduction the cancer biology

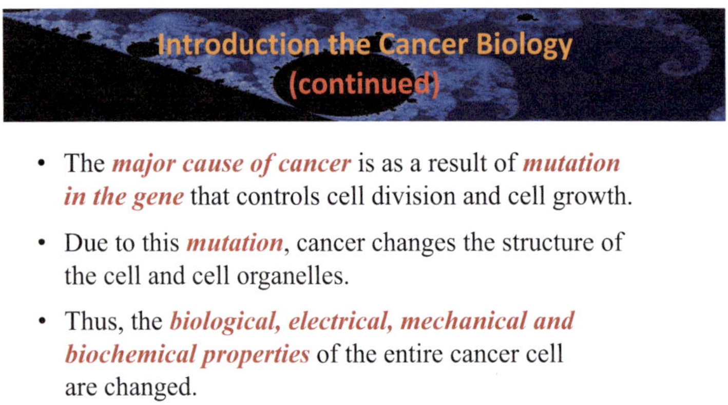

Slide 6.4 Introduction the cancer biology (continued)

cause is as a result of mutation in the gene that controls cell division and cell growth (Slide 6.4). Due to this mutation, cancer changes the structure of the cell and cell organelles, and thus the biological, electrical, mechanical, and biochemical properties of the entire cancer cell are changed. This change in the properties of the cancer cell is programmed by the cell in order to sustain the needs of these fast growing cells, for instance, the production of more enzymes to facilitate invasion of other tissues and the production of more blood vessels to sustain the blood, nutrient, and oxygen needs of these cells.

2.2.1 Some Causes of Cancer

Slides 6.5 and 6.6 describe some causes of cancer:

- *Chemicals/environment*: Chemicals such as asbestos, benzene, formaldehyde, and certain hair dyes are carcinogenic. Also, exposure to harmful rays like ultraviolet rays from the sun is a potential cause of cancer, especially skin cancer.
- *Food*: Acrylamide, artificial sweeteners, and fluoridated water are some of the food additives that are carcinogenic.
- *Genetics*: Some people develop cancer because they inherited it from their parents.
- *Hormones*: Some hormones like estrogen, which stimulates the proliferation of breast cells, increase the chances of cells being cancerous.
- *Infectious agents*: In most people, cancer occurs due to the patient suffering from diseases like human immunodeficiency virus infection, which lowers the immunity of the patient and predisposes them to any disease.
- *Radiation*: Exposure to certain types of radiation like those from X-rays, radioactive radon and exposure to pediatric computed tomography, and extreme UV light are thought to contribute to an individual developing cancer.
- *Tobacco*: Smoking is a well-known cause of lung cancer.
- *Weight*: Obesity is thought to contribute to cancer. In 2001, it was reported that cancer of the colon, breast cancer, and cancer of the kidney are associated with obesity.

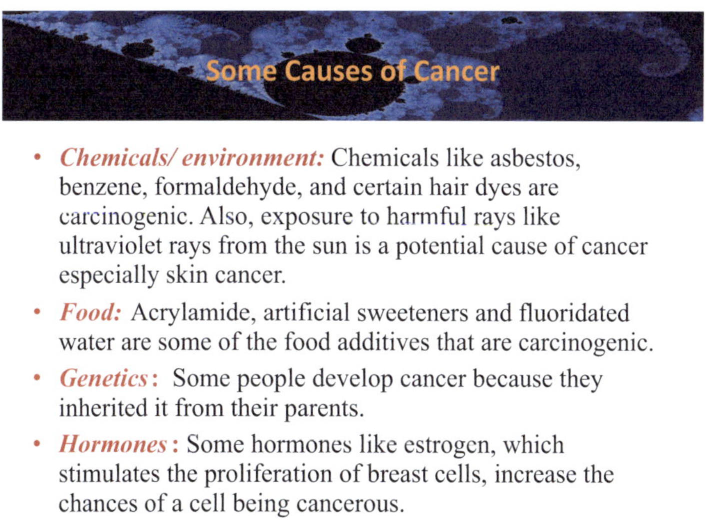

Slide 6.5 Some causes of cancer

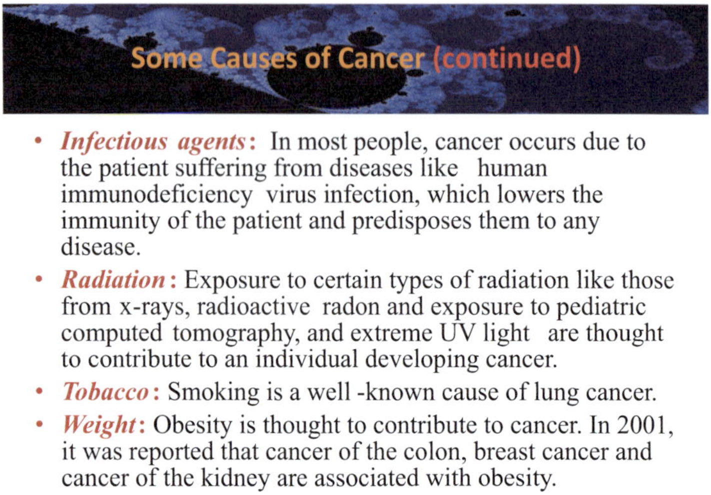

Slide 6.6 Some causes of cancer (continued)

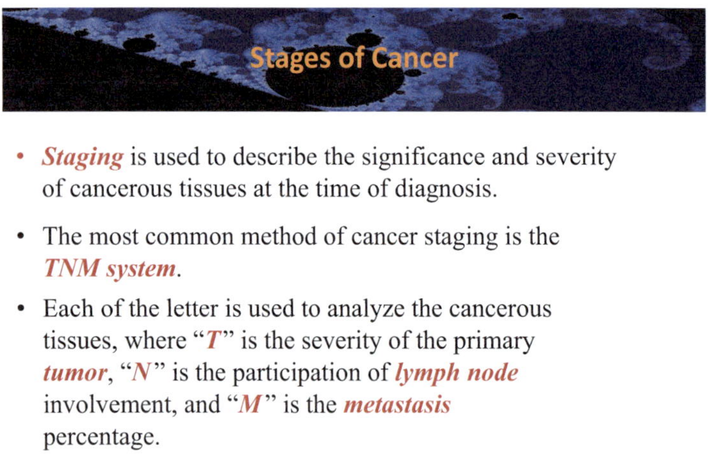

Slide 6.7 Stages of cancer

2.2.2 Stages of Cancer

Physicians have created a system which is used to describe the stages of cancers shown on Slide 6.7. *Staging* is used to describe the significance and severity of cancerous tissues at the time of diagnosis. The most common method of cancer staging is the *TNM* system. Each of the letters is used to analyze the cancerous tissues, where "*T*" is the severity of the primary tumor, "*N*" is the participation of the lymph node, and "*M*" is the metastasis percentage.

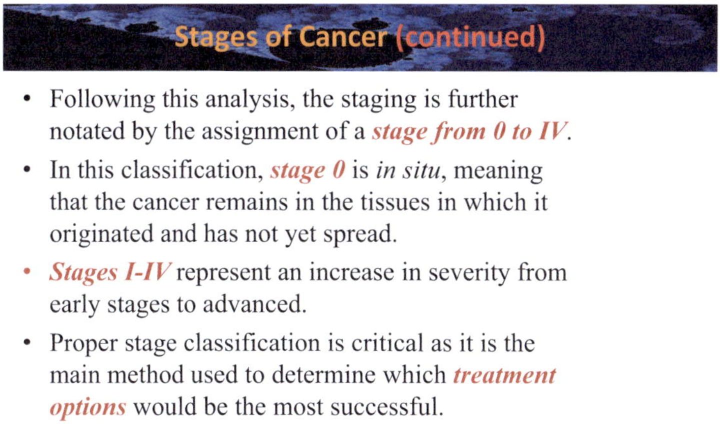

Slide 6.8 Stages of cancer (continued)

Following this analysis, the staging is further notated by the assignment of a stage from 0 to IV (Slide 6.8). In this classification, stage 0 is in situ, meaning that the cancer remains in the tissues in which it originated and has not yet spread. Stages I–IV represent an increase in severity from early to advanced. Proper stage classification is critical as it is the main method used to determine which treatment options would be the most successful.

2.2.3 Categories of Cancer

Categories of cancer are listed as follows (Slide 6.9):

- *Carcinoma*: These are cancers that develop in the skin or on tissues that line or cover internal organs [3].
- *Sarcoma*: These cancers start in the bone, cartilages, fats, muscles, blood vessels, and other connective and supportive tissues.
- *Leukemia*: These cancers start in blood-forming tissues, like bone marrow, and then lead to the production of large numbers of abnormal blood cells which enter the blood [4].
- *Lymphoma and myeloma*: These are cancers that develop in the cells of the immune system [4].
- *Cancer of the central nervous system*: These are cancers that begin in the tissue of the brain and spinal cord [4].

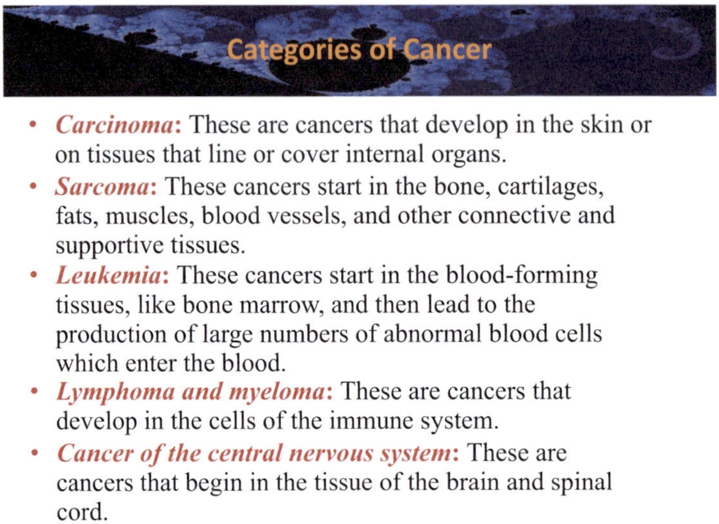

Slide 6.9 Categories of cancer

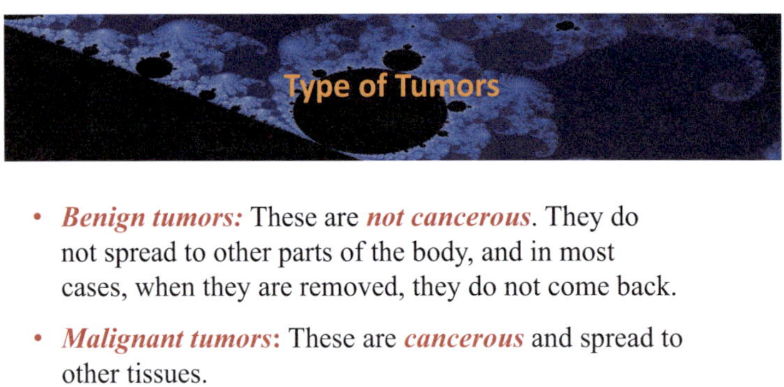

Slide 6.10 Types of tumors

2.2.4 *Types of Tumors*

Two types of tumors are as follows (Slide 6.10):

- *Benign tumors*: These are *not cancerous*. They do not spread to other parts of the body, and in most cases, when they are removed, they do not come back.
- *Malignant tumors*: These are *cancerous* and spread to other tissues.

2.2.5 Cancer Formation

Cancer cells are produced as a result of a series of mutations (permanent changes) in the genes discussed in Section 2.1 (Lecture 5) of this Chap. 2. These mistakes usually happen in mitosis during the cell division (Slide 6.11). As we discussed in the previous lecture, when cells are dividing, they go through checkpoints and specific regulatory genes (e.g., a tumor suppressor gene p53) to ensure that the replication process is correct. If there is a mutation or defect in any of these regulatory genes, then the checkpoints fail to detect problems with the cell division, and the cells divide despite being defective, leading to proliferated cell division and to cancer.

Thus, cancer is a genetic disease caused by multiple mutations of genes (Slide 6.12). Single mutation of DNA doesn't lead immediately to cancer. There are genes in control for DNA repair, correct cell division, differentiation of cells, contact recognition, and cell death, which will fix the first generation of mutations. For example, gene p53 is responsible for the actions that a cell has to take when it is mutated or damaged. This gene can either tell the cell to repair the DNA or it can tell the cell to start its death. Research shows that the cancer cells form in the third or even fourth generation of the mutated cell division.

Cancer cells do not repair DNA mutations because the genes that actually control DNA repair are mutated themselves (Slide 6.13). This leads to an increase in number of the mutations and to the phenomenon called *mutator phenotype*, where the cells are unable to transfer a copy of the nonmutated genetic information to daughter cells during replication. This alters the physical properties and functions of the daughter cells. For example, the cancer cells are

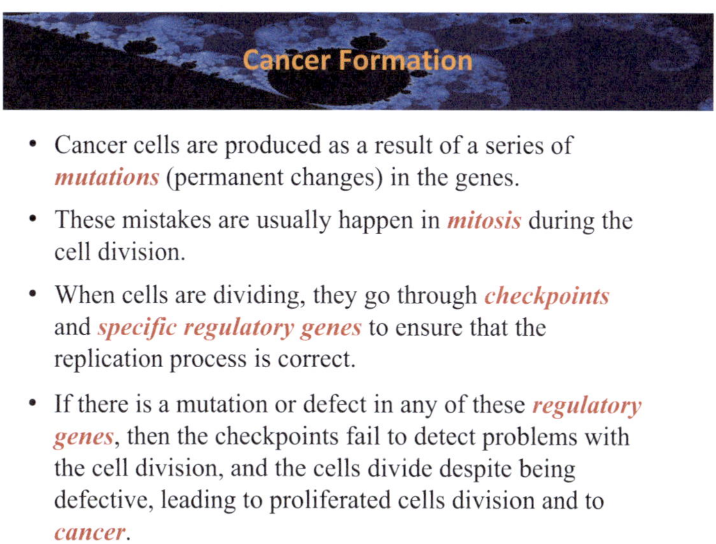

Slide 6.11 Cancer formation

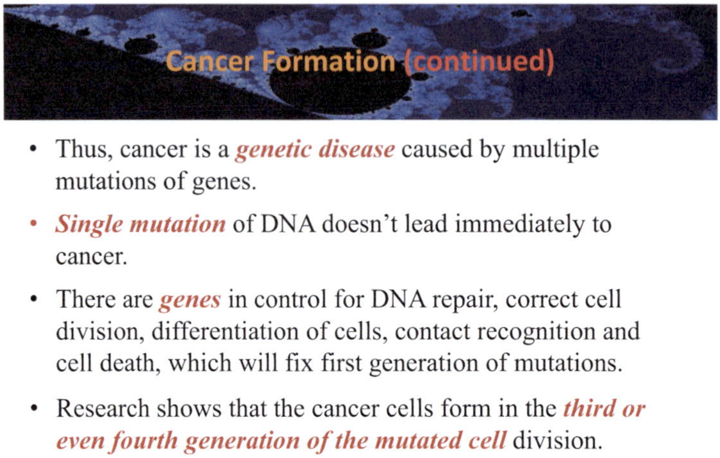

Slide 6.12 Cancer formation (continued)

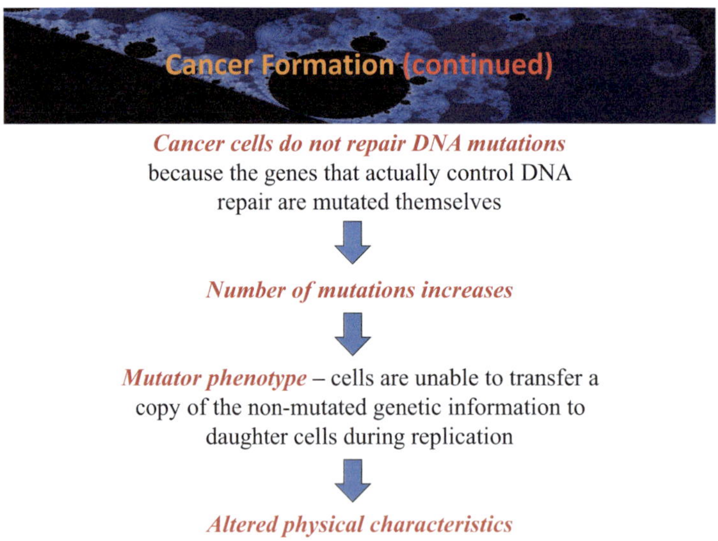

Slide 6.13 Cancer formation (continued)

able to avoid *apoptosis*, which is natural cell death controlled by genes that control the cell cycle (Slide 6.14). After the cancer cells are born, they initiate unregulated cell proliferation and are able to metastasize throughout the body via the bloodstream.

Genes controlling apoptosis are called *tumor-suppressor genes* (Slide 6.15). Gene p53 is such a gene, which makes a protein that inhibits or initiates the transcription of over 50 other genes. Production of p53 proteins increases when

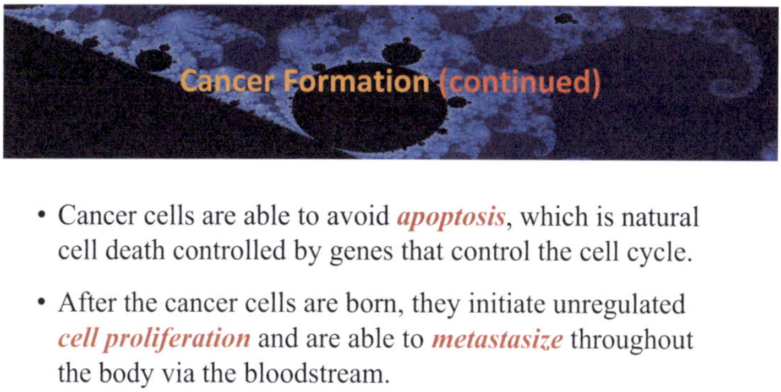

Slide 6.14 Cancer formation (continued)

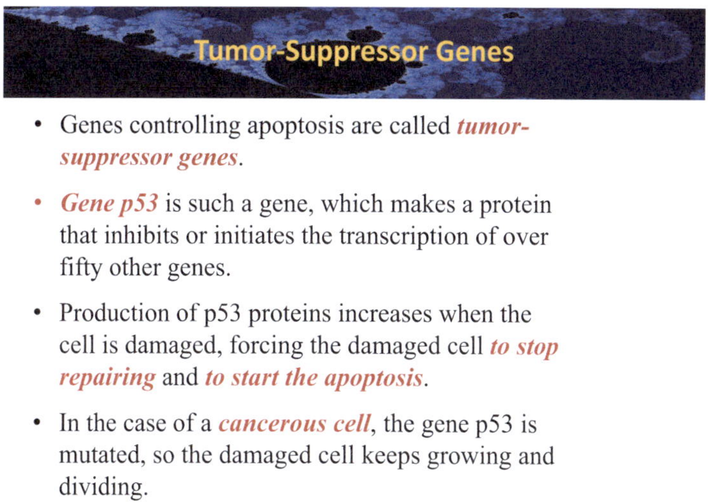

Slide 6.15 Tumor-suppressor genes

the cell is damaged, forcing the damaged cell to stop repairing and to start the apoptosis. In the case of a cancerous cell, gene p53 is mutated, so the damaged cell keeps growing and dividing.

2.2.6 *Normal Cells Versus Cancer Cells*

2.2.6.1 Normal Cells

A normal cell consists of *cell membranes, mitochondria, receptors, cytoskeleton, nucleus, ribosome, endoplasmic reticulum,* and many more. These organelles in

Normal Cells versus Cancer Cells
Normal Cells

Normal cellular activities are:

- ***Cell division*** – Grow, divide and die in an orderly fashion.
- ***Purpose*** – Cells divide to replace worn out tissues and repair injured cells.
- ***Contact inhibition*** – The division of cells stops when they are surrounded by neighboring cells.
- ***Cellular senescence*** – Cell division stops after a certain number of divisions.

Slide 6.16 Normal cells

Normal Cells versus Cancer Cells
Normal Cells (continued)

Normal cellular activities are:

- ***Apoptosis*** – Programmed cell death triggered by cell damage or by proteins – Poly(ADP-ribose) polymerase.
- ***Nature of invasion*** – Growing normal cells do not invade other tissues.

Normal cells are also typically characterized by their ***regular shape*** and ability to replicate in an ***orderly and consistent manner***.

Slide 6.17 Normal cells (continued)

normal form are involved in the maintenance of normal cellular activities shown in Slides 6.16 and 6.17:

- *Cell division*—Grow, divide, and die in an orderly fashion.
- *Purpose*—Cells divide to replace worn-out tissues and repair injured cells.
- *Contact inhibition*—The division of cells stops when they are surrounded by neighboring cells.
- *Cellular senescence*—Cell division stops after a certain number of divisions.
- *Apoptosis*—Programmed cell death triggered by cell damage or by proteins—Poly(ADP-ribose) polymerase.
- *Nature of invasion*—Growing normal cells do not invade other tissues.

Normal cells are also typically characterized by their regular shape and ability to replicate in an orderly and consistent manner.

2.2.6.2 Cancer Cells

Cancer cells are produced after multiple generations of DNA mutation (Slides 6.18 and 6.19). These mutations are caused by the radiation, reactive chemicals, reactions within cells, and errors in DNA transcription. Cancer cells are characterized by:

Normal Cells versus Cancer Cells
Cancer Cells

- *Cancer cells* are produced after multiple generations of DNA mutation.
- These *mutations* are caused by the radiation, reactive chemicals, reactions within cells, and errors in DNA transcription.
- *Cancer cells* are characterized by:
  - ➢ *Lifetime* – Cells do not die, avoiding apoptosis.
  - ➢ *Nature of invasion* – They can invade other tissues.

Slide 6.18 Cancer cells

Normal Cells versus Cancer Cells
Cancer Cells (continued)

Cancer cells are characterized by:

- *Growth* – they grow out of control.
- *Uncontrolled Division*
  - ➢ *Do not exhibit contact inhibition* – Cells keep dividing even when overcrowded, leading to tumor formation.
  - ➢ *Divide without an "all clear signal"* – cells do not stop dividing in the presence of DNA damage

Slide 6.19 Cancer cells (continued)

- *Lifetime*—Cells do not die, avoiding apoptosis.
- *Nature of invasion*—They can invade other tissues.
- *Growth*—They grow out of control.
- *Uncontrolled division*
  - *Do not exhibit contact inhibition*—Cells keep dividing even when overcrowded, leading to tumor formation.
  - *Divide without an "all clear signal"*—Cells do not stop dividing in the presence of DNA damage.

2.2.7 Cancer Cell Structure

As a result of multiple mutations, cancer cells have abnormal *nucleus, cytoplasm, cytoskeleton, cell membrane, ribosome, and mitochondria.*

Nuclear structure of cancer cells. There are several changes seen in the structure of a cancer cell nucleus (Slide 6.20), which include the size and shape of the nucleus, the number and size of nucleoli, the chromatin texture, the perinuclear compartment, and disruption of the promyelocytic leukemia body. For example, the nucleus of a lung carcinoma is malleable with frequent crush artifacts, the nucleus of papillary thyroid carcinoma has grooves and long clefts, the nuclei of some adenocarcinomas exhibit polylobulation (a nucleus divide into several lobes), while some other cancers have nuclei that exhibit indentation, undulations, and folds.

Changes in chromatin structure and texture. Chromatin coarsening, which is not common in normal cells, leads to large angular aggregates of heterochromatin. The degree of the coarsening correlates with the metastatic potential.

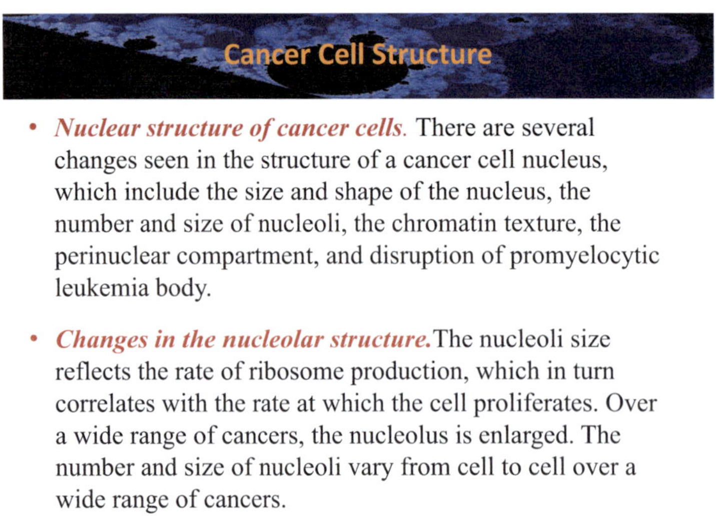

Slide 6.20 Cancer cell structure

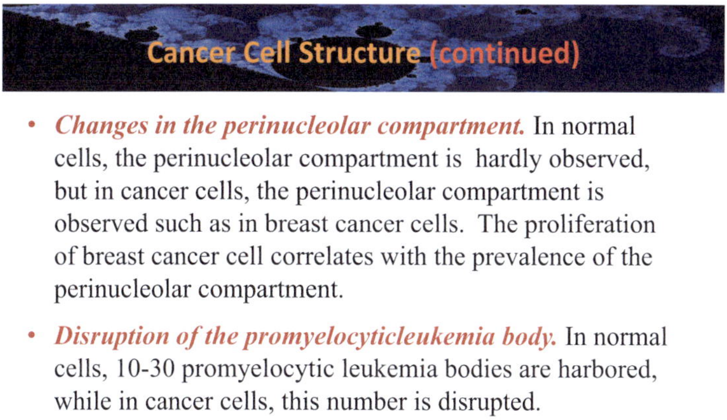

Slide 6.21 Cancer cell structure (continued)

Coarse aggregation of heterochromatin is seen on follicular thyroid adenoma, asymmetric aggregation of heterochromatin is seen over a wide range of cancers, dispersed heterochromatin is typically seen in lung cancer, while the loss of heterochromatin aggregates occurs in papillary thyroid carcinoma and in many different carcinomas.

Changes in nucleolar structure. The nucleoli size reflects the rate of ribosome production, which in turn correlates with the rate at which the cell proliferates (Slide 6.20). Over a wide range of cancers, the nucleolus is enlarged, while in the small-cell anaplastic lung carcinoma, the nucleolus is inconspicuous. The number and size of nucleoli vary from cell to cell over a wide range of cancers.

Changes in the perinucleolar compartment. In normal cells, the perinucleolar compartment is hardly observed, but in cancer cells, the perinucleolar compartment is observed such as in breast cancer cells (Slide 6.21). The proliferation of breast cancer cells correlates with the prevalence of the perinucleolar compartment.

Disruption of the promyelocytic leukemia body. In normal cells, 10–30 promyelocytic leukemia bodies are harbored, while in cancer cells, this number is disrupted (Slide 6.21).

Cytoplasmic structure of cancer cells. Many organelles in the cytoplasm of cancer cells are altered, which include: *mitochondria, cytoskeleton, cell membrane, and receptors* (Slide 6.22). In addition to morphological alterations in these organelles, the size of the cytoplasm is reduced in cancer cells.

The *cell membrane* is composed of phospholipids, sterols, and embedded proteins (Slide 6.22). The lipid and sterol content of the cell membrane are altered in the cancer cell. In some cancer cells, the cell membrane is smaller, while in other cancers cells, the cell membrane is enlarged. And because of this alteration, the structure and function of the cell membrane are also altered in cancer cells.

Mitochondria are the power house of the cell; it is where oxidation of food occurs in order to produce energy (Slide 6.23). Due to their high metabolic rate,

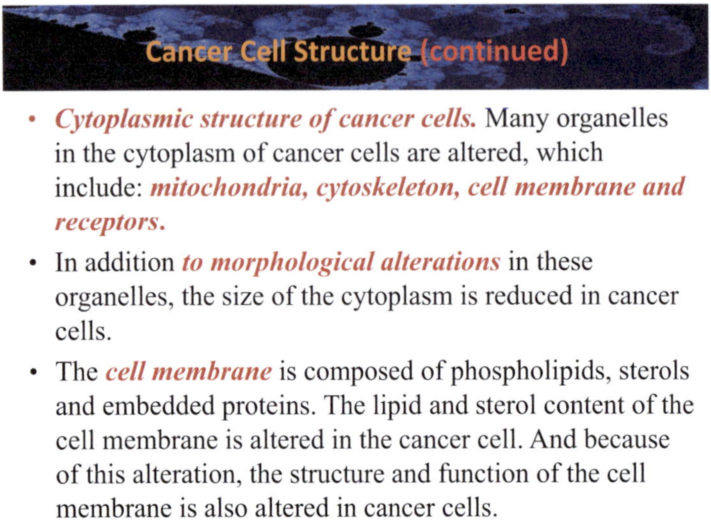

Slide 6.22 Cancer cell structure (continued)

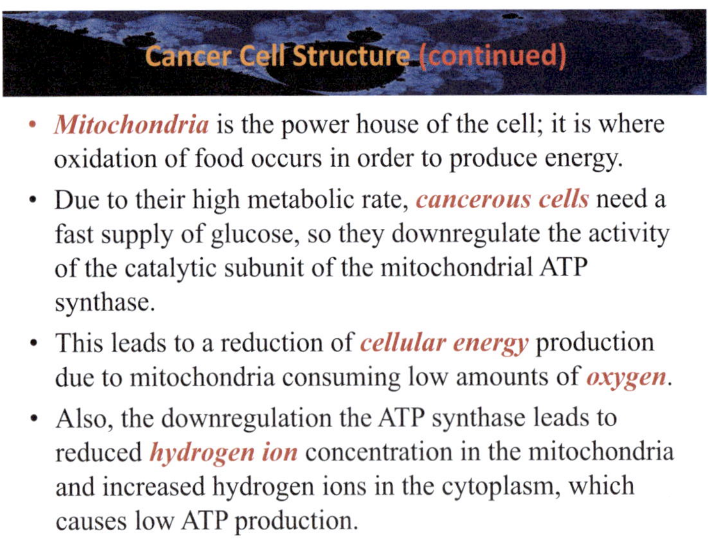

Slide 6.23 Cancer cell structure (continued)

cancer cells need a fast supply of glucose, so they downregulate the activity of the catalytic subunit of the mitochondrial ATP (adenosine triphosphate) synthase. This leads to a reduction of cellular energy production due to mitochondria consuming low amounts of oxygen. Also, the downregulation of the ATP synthase leads to reduced hydrogen ion concentration in the mitochondria and increased hydrogen ions in the cytoplasm, which causes low ATP production.

Cancer Cell Structure (continued)

- ***Cytoskeleton.*** The proteins that make up the cytoskeleton – ***microfilament, intermediate filament and microtubules*** – are reduced in cancer cells. Also, the structure of these proteins is altered, thereby altering their function.
- ***Receptors.*** Cancer cells alters the structure and function of some receptors in order to increase the survival and proliferation of the cells. For instance, i***ntegrins*** are ***cell surface receptors*** that are used by cancer cells to facilitate communication between the cytoplasm and the extracellular matrix. The Integrin is altered at the point of attachment to the extracellular matrix, so that it will function in favor of the cancer cell.
- Also, cancer cells use integrins to hinder ***apoptosis*** and facilitate cell motility.

Slide 6.24 Cancer cell structure (continued)

Cytoskeleton. The proteins that make up the cytoskeleton—microfilament, intermediate filament, and microtubules—are reduced in cancer cells (Slide 6.24). Also, the structure of these proteins is altered, thereby altering their function.

Receptors. Cancer cells alter the structure and function of some receptors in order to increase the survival and proliferation of the cells (Slide 6.24). For instance, integrins are cell surface receptors that are used by cancer cells to facilitate communication between the cytoplasm and the extracellular matrix. The integrin is altered at the point of attachment to the extracellular matrix, so that it will function in favor of the cancer cell. Also, cancer cells use integrins to hinder apoptosis and facilitate cell motility.

2.2.8 Properties of Cancer Cells

There are detectable changes in the properties of cancer cells as described as follows.

Biological properties of cancer cells (Slides 6.25, 6.26, and 6.27):

- Acquisition of self-sufficiency in growth signals: This property of cancer cells allows them to exhibit uncontrolled growth.
- Loss of sensitivity to antigrowth signals: This leads to uncontrolled growth in cancer cells.

- Loss of capacity for apoptosis: Cancer cells are able to grow even when there is genetic error, and when there is an antigrowth signal, they do not die.
- Loss of capacity for senescence: Because cancer cells lose the ability for senescence, they are able to replicate without limit, and they do not age nor die.
- Acquisition of sustained angiogenesis: This allows the tumor to grow beyond the limitations of passive nutrient diffusion [5], since cancer cells are highly metabolic and need a supply of oxygen.
- Acquisition of the ability to invade neighboring tissues: This is the major characteristic of invasive carcinoma.
- Acquisition of the ability to build metastases at distant sites: This is a major property of malignant tumors.
- Loss of the ability to repair genetic error: This results in an increase in mutation rates and leads to an increase in other adverse effects seen in cancer cells.
- Cancer cells are less adhesive than normal cells due to reduced cell-to-cell adhesion and reduced cell-to-extracellular matrix adhesion. This leads to metastasis and formation of large mass.
- Cancer cells do not differentiate, and because of this, they do not acquire a particular phenotype.
- Cancer cells secrete enzymes that facilitate their invasion to other cells. They also alter the shape and functions of some receptors, such as the shape and function of integrin so as to hinder apoptosis of cancer cells.

Electrical properties of cancer cells (Slide 6.28). The lipids, sterol content, and protein structure in the cell membrane of a cancer cell are altered, which results in reduction of the membrane potential. A low membrane potential is also as a result of increased intracellular sodium and decreased intracellular potassium. This causes the cancer cell membrane to be depolarized.

Properties of Cancer Cells

Biological properties of cancer cells:

- ***Acquisition of self-sufficiency in growth signals***: This property of cancer cell allows them to exhibit uncontrolled growth.
- ***Loss of sensitivity to anti-growth signals***: This also leads to uncontrolled growth in cancer cells.
- ***Loss of capacity for apoptosis***: Cancer cells are able to grow even when there is genetic error, and when there is an anti-growth signal, they do not die.

Slide 6.25 Properties of cancer cells

Properties of Cancer Cells (continued)

Biological properties of cancer cells (continued):

- ***Loss of capacity for senescence***: Because cancer cells lose the ability for senescence, they are able to replicate without limit, and they do not age nor die.
- ***Acquisition of sustained angiogenesis***: This allows the tumor to grow beyond the limitations of passive nutrient diffusion, since cancer cells are highly metabolic and need a supply of oxygen.
- ***Acquisition of the ability to invade neighboring tissues***: This is the major characteristic of invasive carcinoma.
- ***Acquisition of the ability to build metastases at distant sites***: This is a major property of malignant tumors.

Slide 6.26 Properties of cancer cells (continued)

Properties of Cancer Cells (continued)

Biological properties of cancer cells (continued):

- ***Loss of the ability to repair genetic error***: This results in an increase in mutation rates and leads to an increase in other adverse effects seen in cancer cells.
- ***Cancer cells are less adhesive*** than normal cells due to reduced cell-to-cell adhesion and reduced cell-to-extracellular matrix adhesion. This leads to metastasis and formation of large mass.
- ***Cancer cells do not differentiate***, and because of this, they do not acquire a particular phenotype.
- ***Cancer cells secrete enzymes*** that facilitate their invasion to other cells. They also alter the shape and functions of some receptors, such as the shape and function of integrin so as to hinder apoptosis of cancer cells.

Slide 6.27 Properties of cancer cells (continued)

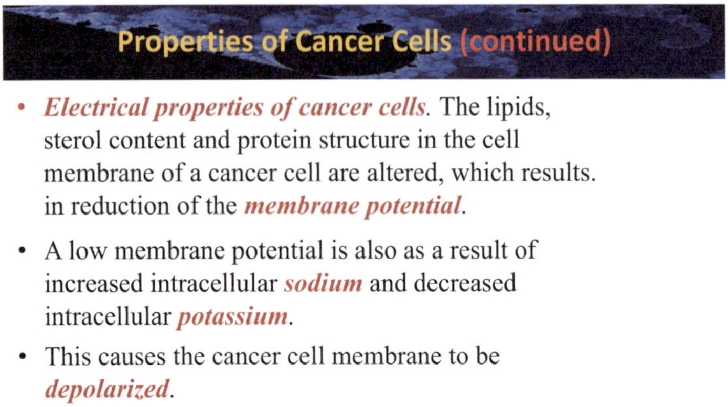

Slide 6.28 Properties of cancer cells (continued)

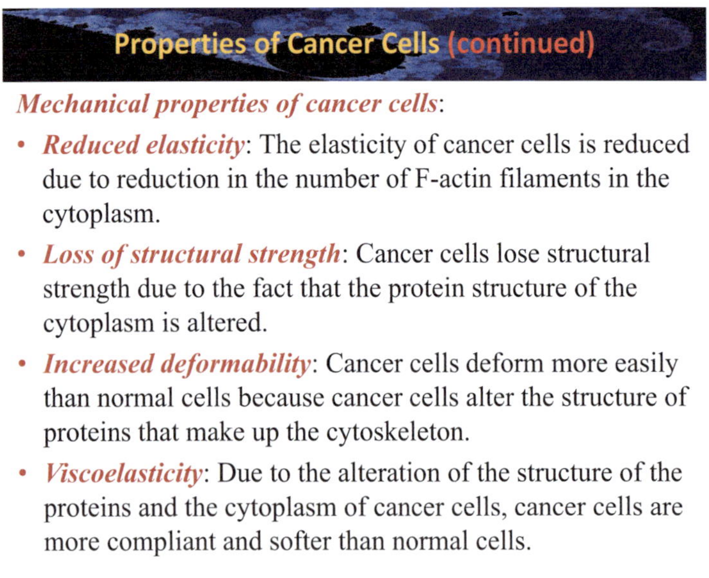

Slide 6.29 Properties of cancer cells (continued)

Mechanical properties of cancer cells (Slide 6.29):

- Reduced elasticity: The elasticity of cancer cells is reduced due to reduction in the number of F-actin filaments in the cytoplasm.
- Loss of structural strength: Cancer cells lose structural strength due to the fact that the protein structure of the cytoplasm is altered.

- Increased deformability: Cancer cells deform more easily than normal cells because cancer cells alter the structure of proteins that make up the cytoskeleton.
- Viscoelasticity: Due to the alteration of the structure of the proteins and the cytoplasm of cancer cells, cancer cells are more compliant and softer than normal cells.

Biochemical properties of cancer cells (Slides 6.30 and 6.31):

- Cancer cells have different lipids and sterol content: This affects the production of steroid hormones, thereby limiting the functions that these hormones perform.
- Cancer cells have different content of glycoprotein than normal cells: In cancer cells, there is a reduction in the number of normal growth receptors produced and an increase in the number of cancerous growth receptors produced.
- Gluconeogenesis: Cancer cells 'embrace' glucose. The amount of glucose made by a normal cell is not enough for the cancer cell, and so cancer cells make glucose from other sources like proteins and fats.
- Low pH: Cancer cells are highly metabolic because they are actively and continuously dividing. As a result of the high metabolic nature of cancer cells, there is decreased pH. Another reason for a decreased pH is lactic acid production due to low oxygen. The low pH in cancer cells affects the hydrogen ion gradient, which is a major path that drives the production of ATP.
- Glucose receptors: Cancer cells have large amounts of glucose receptors and glucose enzyme, because they need a large amount of glucose in order to keep up with the high metabolic activity they undergo.
- Low production of cellular energy: Cancer cells are highly metabolic. They need a faster supply of glucose and energy, so they downregulate the catalytic subunit of the mitochondrial ATP synthase and switch to anaerobic respiration (lactic acid pathway), which produces low hydrogen ion concentration and low amount of oxygen in mitochondria—this affects the mitochondrial function so that it produces low cellular energy.

In conclusion (Slide 6.32), the cause, structure, and properties of cancer cells need to be understood in order to effectively treat cancer. Although there are several causes of cancer, the major cause is mutation in the gene that controls cell division and cell growth. Due to this mutation, cancer cells divide and grow uncontrollably, they do not respond to stop signal, they do not die, and they invade other tissues. Because of their excessive growth, they change the structure and properties of cells in a way that ensures continual growth and survival of these cancer cells.

Properties of Cancer Cells (continued)

Biochemical properties of cancer cells:

- Cancer cells have different ***lipids and sterol*** content: This affects the production of steroid hormones, thereby limiting the functions that these hormones perform.
- Cancer cells have different content of ***glycoprotein*** than normal cells:. In cancer cells, there is a reduction in the number of normal growth receptors produced and an increase in the number of cancerous growth receptors produced.
- ***Gluconeogenesis***: Cancer cells 'embrace' glucose. The amount of glucose that the normal cell makes is not enough for the cancer cell, and so cancer cells make glucose from other sources like protein and fats.

Slide 6.30 Properties of cancer cells (continued)

Properties of Cancer Cells (continued)

Biochemical properties of cancer cells (continued):

- ***Low pH***: Cancer cells are highly metabolic because they are actively and continuously dividing. As a result of the high metabolic nature of cancer cells, there is decreased pH. Another reason for a decreased pH is lactic acid production due to low oxygen.
- ***Glucose receptors:*** Cancer cells have large amount of glucose receptors and glucose enzyme because they need a large amount of glucose in order to keep up with the high metabolic activity they undergo.
- ***Low production of cellular energy***: Low hydrogen ion concentration and low amount oxygen in the cell affects the mitochondrial function so that it produces low cellular energy.

Slide 6.31 Properties of cancer cells (continued)

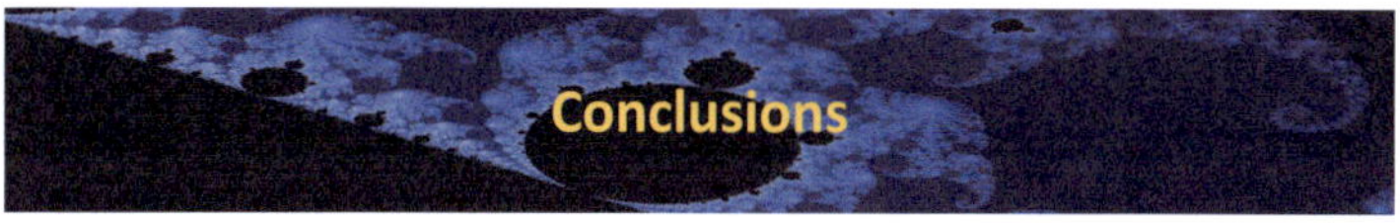

- The major cause of cancer is ***mutation*** in the gene that controls cell division and cell growth.
- Due to this mutation, cancer cells divide and grow ***uncontrollably***, they do not respond to stop signal, they do not die, and they also invade other tissues.
- Because of their excessive growth, they change the ***structure*** and ***properties*** of cells in a way that ensures continual growth and survival of these cancer cells.

Slide 6.32 Conclusions

2.3 Lecture 7: Selective Targeting of Cancer Cells (Slide 7.1)

- Passive Targeting
  - Angiogenesis
  - Leaky Vasculature
  - Tumor Microenvironment
  - Local Drug Application
  - Incubation Time
  - Adding Coatings
- Active Targeting
  - Antibodies
  - Overexpression of Epidermal Growth Factor Receptors
  - Conjugation of Monoclonal Antibodies
  - Carbohydrate-Directed Targeting
  - Receptor-Directed Targeting
  - Antigen-Directed Targeting
- Advances in Drug Delivery
- Conclusions

Current treatments of cancer are very invasive or intrusive through surgery, chemotherapy, and/or radiation therapy (Slide 7.2). The effectiveness of these treatment modalities is related to the ability to target and kill the cancer cells or tumor while minimizing the effects on the healthy cells. Side effects can change drastically the life of the patient. These changes are directly related to how effective the targeting of the tumor or cancer cells was. One important aspect of

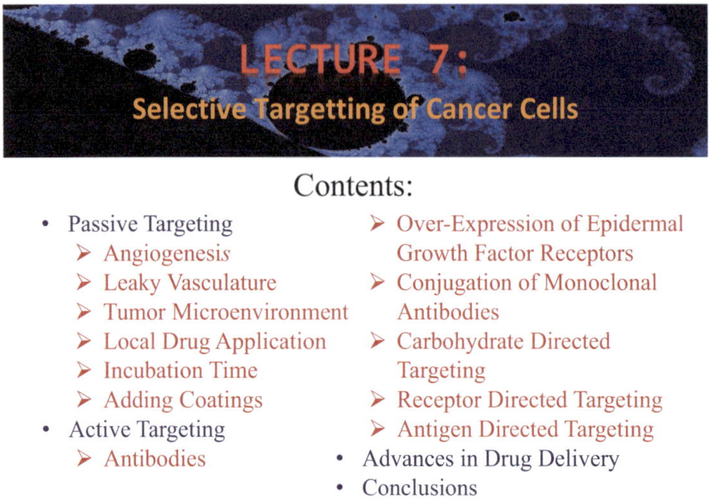

Slide 7.1 Contents of Lecture 7

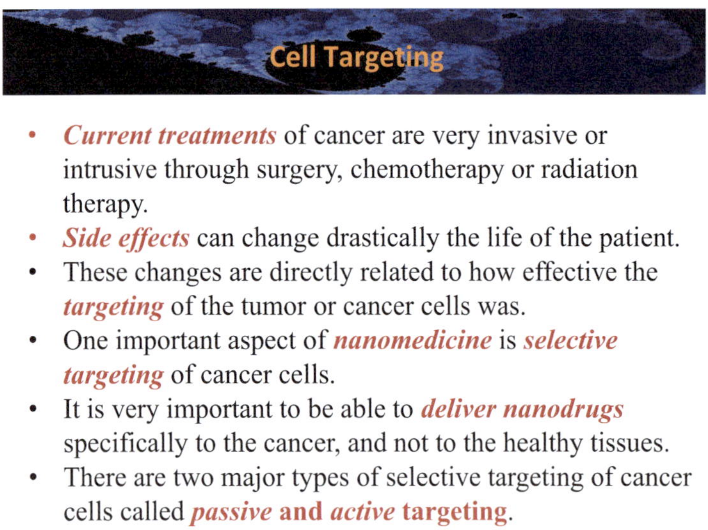

Slide 7.2 Cell targeting

nanomedicine is selective targeting of cancer cells. It is very important to be able to deliver nanodrugs specifically to the cancer and not to the healthy tissues. There are two major types of selective targeting of cancer cells called *passive* and *active* targeting (Slide 7.2).

2.3.1 *Passive Targeting*

In passive targeting, nanodrugs injected into the bloodstream circulate throughout the body to reach the tumor through its porous blood vessels (Slide 7.3). A phenomenon that can potentially be taken advantage of is *angiogenesis*. This is the process by which tumors signal blood vessels to supply nutrients to the tumor via *vascular epidermal growth factors* (VEGFs). The blood vessels have receptors, i.e., *VEGF receptors* that respond to the VEGFs by growing new vessels and capillaries to the tumor. The tumor then has a free supply of nutrients to feed from, stealing away from the host organism.

Passive targeting is divided into three submethods: *leaky vasculature, tumor microenvironment*, and *direct local application* (Slide 7.4). The basis is the same for all three, which is to take advantage of the permeability of the tumor tissue. Nanoparticles are used to deliver the drug which is manipulated to where it will activate upon exposure to the tumor environment. The *pH and redox potential* are some of the environmental factors which aid these processes.

The *leaky vasculature* method uses the ability of nanoparticles to penetrate the tumor (Slide 7.4). The new blood vessels grown for the tumor are hastily made and thus are very leaky. Nanoparticles that are injected into the arteries leading to the tumor site then can enter the tumor with great ease by passing through the porous blood vessels into the tumor cells. This phenomenon is known as *enhanced permeability and retention* (EPR). Since the tumor is uncontrollably growing, it is continually "hungry" and needs more and more nutrition. This "eating habit" of cancer cells can be used for the passive targeting by mixing the

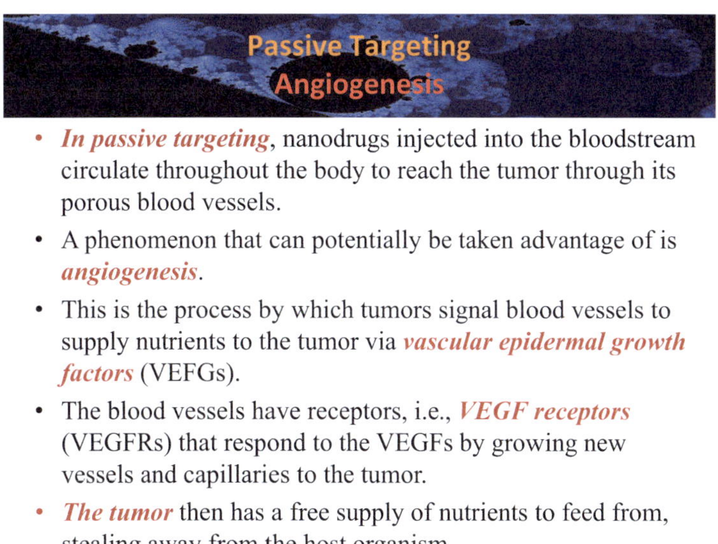

Slide 7.3 Passive targeting: angiogenesis

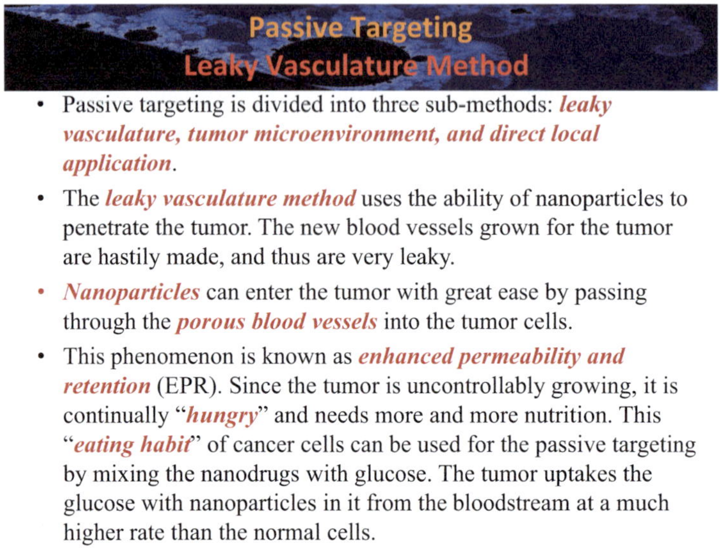

Slide 7.4 Passive targeting: leaky vasculature method

nanodrugs with glucose. The tumor uptakes the glucose with nanoparticles in it from the bloodstream at a much higher rate than the normal cells bringing the nanoparticles to the tumor site at higher concentration. Because the tumor lacks an effective drainage system, these nanoparticles carrying the drug will accumulate, therefore increasing the concentration of the chemotherapeutic drug in cancer in the tumor tissue.

The *effectiveness* of passive targeting depends on the *incubation time*, which is the time required to collect the nanoparticles at the tumor site (Slide 7.5). This time is a function of the size of the nanoparticle and the surface chemistry. Small particles (1–50 nm) pass through the cells rather quickly, while larger particles (50–100 nm) take more time to accumulate. For example, the incubation time for 1–50 nm nanoparticles is 1–5 h; and for 50–100 nm is 6–12 h. Using the glucose solution and surface chemistry can increase the uptake of the nanoparticles by the tumor cells and reduce the incubation time.

Adding coatings to the nanoparticles is another way to increase the effectiveness of the EPR phenomenon (Slide 7.6). Many types of coatings exist—organic and inorganic—that help increase the 'stickiness' of the nanoparticles. By adding '*arms*' to the nanoparticles, they are more likely to get caught within the cell wall. Examples of organic nanoparticles include proteins, carbohydrates, peptides, and nucleotides. The inorganic coatings consist of gold, silver, silica, polymers, and magnetic materials like iron oxide, cobalt oxide, and nickel oxide. These coatings can provide biocompatibility and stability of the nanoparticles and reduce the toxicity of the nanodrugs to the healthy tissue.

The body is fairly adept at recognizing foreign bodies introduced into the organism, so every nanoparticle must be biocompatible first and foremost, or risk

- The ***effectiveness*** of passive targeting depends on the ***incubation time***, which is the time required to collect the nanoparticles at the tumor site.
- The ***incubation time*** is a function of the size of the nanoparticle and the surface chemistry.
- ***Small particles*** (1-50 nm) pass through the cells rather quickly, while ***larger particles*** (50-100 nm) take more time to accumulate.
- For example, ***the incubation time*** for 1-50 nm nanoparticles is ***1-5 hours***; and for 50-100 nm is ***6-12 hours***.
- Using the ***glucose solution*** and ***surface chemistry*** can increase the uptake of the nanoparticles by the tumor cells and reduce the incubation time.

Slide 7.5 Passive targeting: incubation time

- ***Adding coatings*** to the nanoparticles is another way to increase the effectiveness of the ***EPR phenomenon***.
- Many types of coatings exist – ***organic and inorganic*** – that help increase the ***'stickiness'*** of the nanoparticles. By adding ***'arms*** to the nanoparticles, they are more likely to get caught within the cell wall.
- Examples of ***organic nanoparticles*** include proteins, carbohydrates, peptides and nucleotides.
- The ***inorganic coatings*** consist of gold, silver, silica, polymers, and magnetic materials like iron oxide, cobalt oxide, and nickel oxide.
- These coatings can provide ***biocompatibility*** and ***stability*** of the nanoparticles and ***reduce the toxicity*** of the nanodrugs to the healthy tissue.

Slide 7.6 Passive targeting: adding coatings

the body attacking the nanoparticles and spoiling the treatment (Slide 7.7). There are a few coatings that have been found to be biocompatible like silica and polyethylene glycol (PEG). Additionally, biotin and streptavidin coatings can act as masks or bonding agents for other materials.

- The body is fairly adept at recognizing foreign bodies introduced into the organism, so every nanoparticle must be ***biocompatible*** first and foremost, or risk the body attacking the nanoparticles and spoiling the treatment.
- There are a few coatings that have been found to be biocompatible like ***silica and polyethylene glycol*** (PEG). Additionally, ***biotin and streptavidin*** coatings can act as masks or bonding agents for other materials.
- An interesting application of these coatings is for those that are ***magnetic***. It has been shown that external magnetic fields can have advantageous effects on nanoparticles inside the body. It is possible under certain conditions to guide the nanoparticles using external fields.

Slide 7.7 Passive targeting: adding coatings (continued)

- Once the nanoparticle with its chemotherapeutic drug reaches its destination, the ***tumor microenvironment*** will make the drug active and volatile. Also called ***tumor-activated prodrug therapy***, this method takes advantage of the special tumor environment (like pH) as compared to the normal surrounding tissue.
- More invasive but possibly effective is ***local drug application***. Using the ***intra-tumoral administration technique***, the drug can be injected directly on the tumor. For example, p53 DNA loaded nanoparticles used to treat cancer cells have shown great results.

Slide 7.8 Passive targeting: tumor microenvironment and direct local application

An interesting application of these coatings is for those that are magnetic (Slide 7.7). It has been shown that external magnetic fields can have advantageous effects on nanoparticles inside the body. It is possible under certain conditions to guide the nanoparticles using external fields.

Once the nanoparticle with its chemotherapeutic drug reaches its destination, the *tumor microenvironment* will make the drug active and volatile (Slide 7.8). Also called *tumor-activated prodrug therapy*, this method takes advantage of the special tumor environment (like pH) as compared to the normal surrounding tissue.

More invasive but possibly very effective is *local drug application* (Slide 7.8). Using the *intratumoral administration technique*, the drug can be injected directly on the tumor. For example, p53 DNA-loaded nanoparticles used to treat cancer cells have shown great results.

2.3.2 Active Targeting

Active targeting presents an interesting and innovative way to selectively target cancer cells (Slide 7.9). Conversely to passive targeting, active targeting exploits the differences in receptors between the specific cancer cells and healthy cells. One way to do this is by taking advantage of the body's *own active targeting system*: *antibodies*. Antibodies are proteins that recognize foreign objects in the body and send signals to the white blood cells to come 'clean up' (Slide 7.9). Antibodies have the unique property in that they are specific to only one other antigen. The body's immune system contains hundreds of millions of different antibodies and is ready for just about anything. Using the body's own antibodies for the targeting of the tumor cells is ideal because it is not recognized by the immune system and will not be rejected.

The body's own antibodies can fight many diseases but not all of them (Slide 7.10). Because of this, the antibodies could be designed outside the body, grown in animals against the specific types of cancer, and then used for targeting

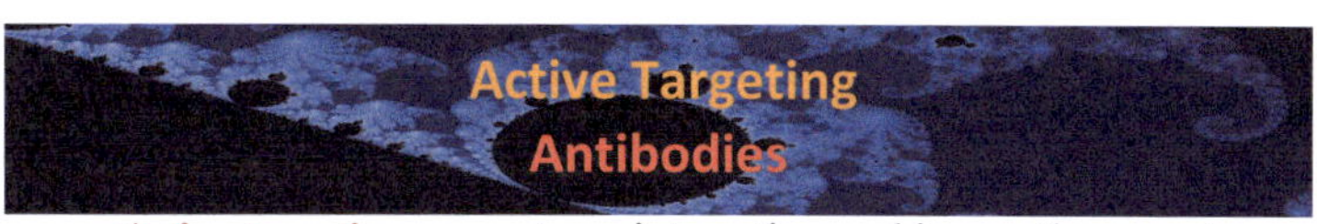

- ***Active targeting*** presents an interesting and innovative way to selectively target cancer cells.
- ***Active targeting*** exploits the differences in receptors between the specific cancer cells and healthy cells.
- One way to do this is by taking advantage of the body's ***own active targeting system***: ***antibodies***.
- ***Antibodies*** are proteins that recognize foreign objects in the body and send signals to the white blood cells to come 'clean up.'
- ***Antibodies*** have the unique property in that they are specific to only one ***antigen***.
- The body's ***immune system*** contains hundreds of millions of different antibodies and is ready for just about anything.
- Using the ***body's own antibody*** for the targeting of the tumor cells is ideal because it is not recognized by the immune system and will not be rejected.

Slide 7.9 Active targeting: antibodies

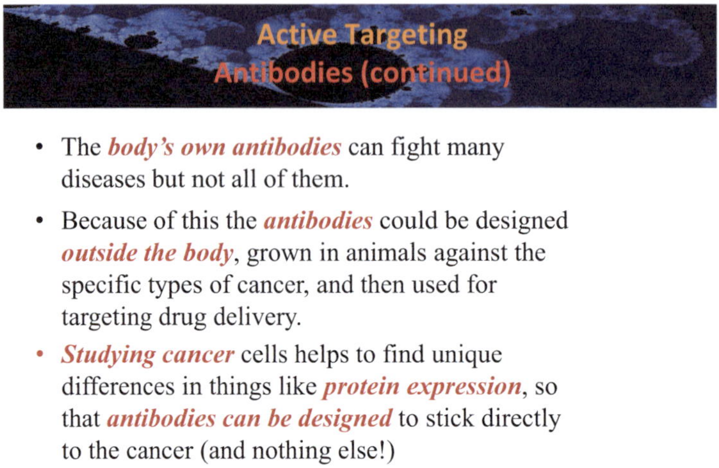

Slide 7.10 Active targeting: antibodies (continued)

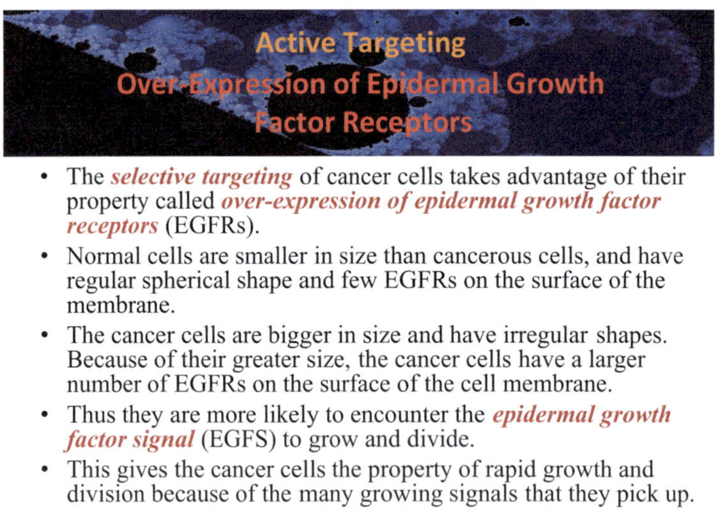

Slide 7.11 Active targeting: over-expression of epidermal growth factors receptors

drug delivery. Researchers want to exploit the idea that antibodies are specific to only one antigen, applying this to cancer cells. This means studying cancer cells can help find unique differences in things like protein expression, so that antibodies can be designed to stick directly to the cancer (and nothing else!).

The *selective targeting* of cancer cells takes advantage of their property called *overexpression of epidermal growth factor receptors* (EGFRs) presented in Slide 7.11. Normal cells are smaller in size than cancerous cells and have a regular spherical shape and few EGFRs on the surface of the membrane. The cancer cells

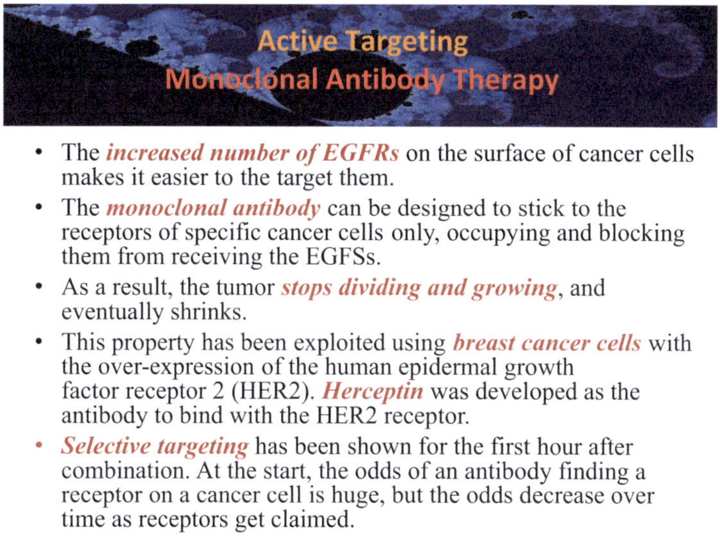

Slide 7.12 Active targeting: monoclonal antibody therapy

are larger and have irregular shapes. Because of their greater size, the cancer cells have a larger number of EGFRs on the surface of the cell membrane and thus are more likely to encounter the *epidermal growth factor signal* to grow and divide. Normal cells have only one or several EGFSs, whereas cancer cells have many EGFSs. This gives the cancer cells the property of rapid growth and division because of the many growing signals that they pick up.

The increased number of EGFRs on the surface of cancer cells makes it easier to target them (Slide 7.12). The monoclonal antibody can be designed to stick to the receptors of specific cancer cells only, occupying and blocking them from receiving the EGFSs, as schematically demonstrated in Slide 7.13. As a result, the tumor stops dividing and growing, and eventually shrinks. This property has been exploited using breast cancer cells with the overexpression of the *human epidermal growth factor receptor 2* (HER2). *Herceptin* was developed as the antibody to bind with the HER2 receptor. Selective targeting has been shown for the first hour after combination. This time dependency makes sense because of the sheer number of HER2 receptors on the cancer cells. At the start, the odds of an antibody finding a receptor on a cancer cell is huge, but the odds decrease over time as receptors get claimed.

The final step in this process is bringing in the nanomedicine component (Slide 7.14), which is done by *conjugation of monoclonal antibodies* to the nanoparticles. We can take the same antibodies described earlier to cover the nanoparticles. This will deliver and attach the nanoparticles to the cancer cell receptors only, producing the active selective targeting delivery system schematically shown in Slide 7.15. Then, for example, radiation can be applied to activate these nanoparticles to realize the nanodrug or heat them to a temperature high enough for biological ablation of the tumor. The study described earlier used

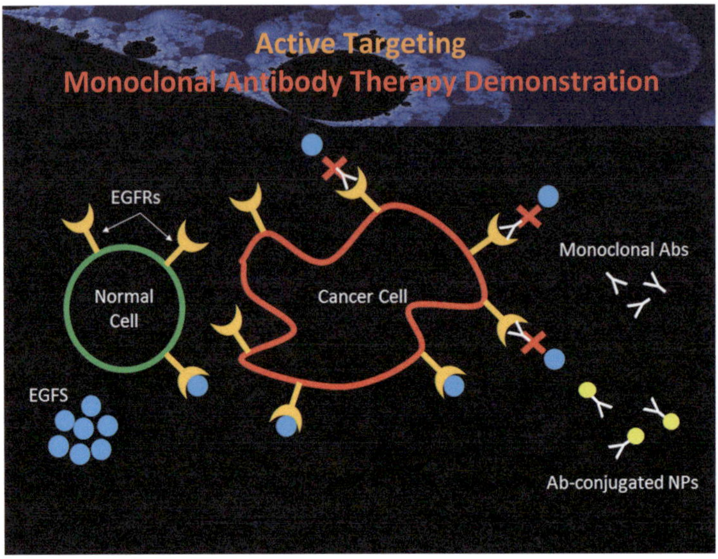

Slide 7.13 Active targeting: monoclonal antibody therapy demonstration

Active Targeting

Conjugation of Antibodies to Nanoparticles

- ***The final step*** in this process is bringing in the nanomedicine component.
- This is done by ***conjugation of monoclonal antibodies to the nanoparticles***. Researchers can take the same antibodies described above to cover the nanoparticles.
- This will deliver and attach the nanoparticles to the cancer cells receptors only, producing the ***active selective targeting delivery system***.
- Then, for example, the radiation can be applied ***to activate these nanoparticles*** to realize the nanodrug, or heat them to a temperature high enough for biological ablation of the tumor.

Slide 7.14 Active targeting: conjugation of antibodies to nanoparticles

Herceptin conjugated with PEGylated human serum albumin nanoparticles. The specificity of antibodies combined with the versatility of nanoparticles creates a highly efficient and effective tool which is very promising for the future of medicine.

Nanoparticles are configured such that they can attach to the specific type of proteins. There are two available techniques for bonding the nanoparticles to

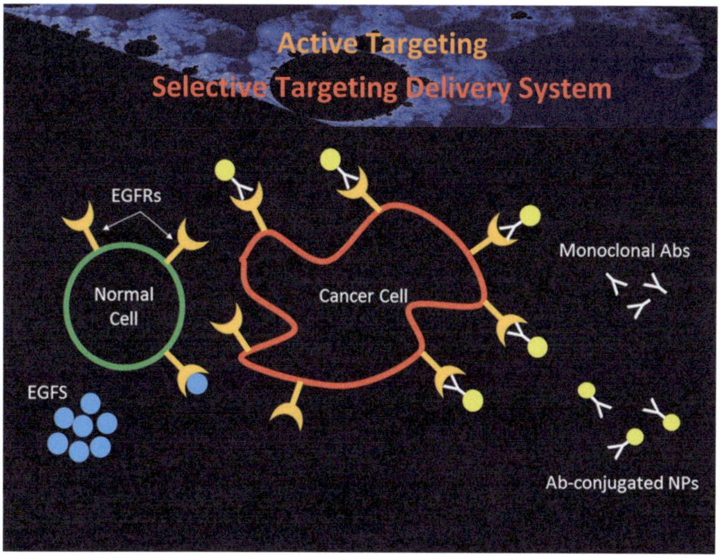

Slide 7.15 Active targeting: selective targeting delivery system

Antibody Conjugation

- ***There are two available techniques*** for bonding the nanoparticles to the antibody: ***direct bonding and indirect bonding***.
- In ***direct bonding***, the nanoparticles are mixed with the antibody, which sticks to the surface of the nanoparticle directly.
- However, the ***direct bonding*** is not an efficient technique for attaching the antibody to the surface of the nanodrug.
- In the ***indirect bonding*** technique, the nanoparticles are first covered by ***adaptor molecules*** like ***streptavidin, biotin, PEG***, etc., which are designed to stick effectively to the surface of the nanoparticle, such as metal, and covered secondly by the antibody.
- These ***adaptor molecules*** also have the ability to stick to the specific antibody used for the delivery system. The indirect bonding technique forms more stable drug delivery system.

Slide 7.16 Antibody conjugation

the antibody: *direct bonding* and *indirect bonding* (Slide 7.16). In the former, the nanoparticles are mixed with the antibody, which sticks to the surface of the nanoparticle directly. However, direct bonding is not an efficient technique for attaching antibody to the surface of the nanodrug. The antibody is designed to bind effectively to the receptors of the cancer cells, not to the surface of the nanoparticle,

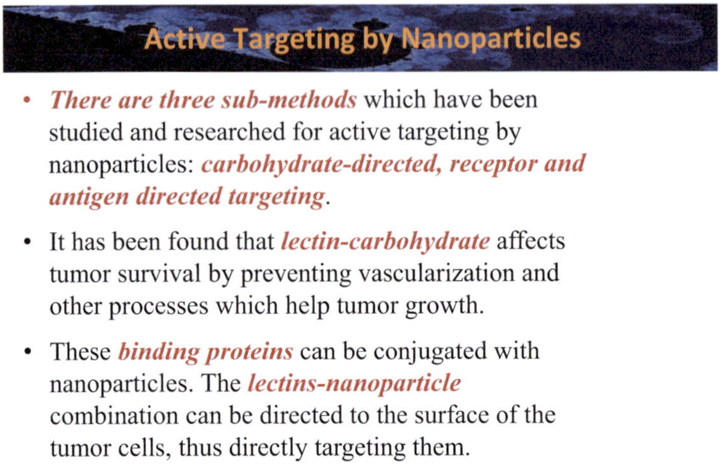

Slide 7.17 Active targeting by nanoparticles

which could be a metallic surface. Because of this, the direct bonding technique is unstable. In the indirect bonding technique, the nanoparticles are first covered by adaptor molecules such as streptavidin, biotin, PEG, etc., which are designed to stick effectively to the surface of the nanoparticle, such as metal, and covered secondly by the antibody (Slide 7.16). These adaptor molecules also have the ability to stick to the specific antibody used for the delivery system. The indirect bonding technique forms a more stable drug delivery system.

There are three submethods which have been studied and researched for active targeting by nanoparticles: *carbohydrate-directed, receptor, and antigen-directed targeting* (Slide 7.17). It has been found that *lectin-carbohydrate* affects tumor survival by preventing vascularization and other processes which help tumor growth. These binding proteins can be conjugated with nanoparticles. The lectin–nanoparticle combination can be directed to the surface of the tumor cells, thus directly targeting them.

Endocytosis is the process in which a cell takes in materials from the outside by engulfing and fusing them with its plasma membrane (Slide 7.18) [6]. The *receptors* and *antigens* of human cancer cells, as mentioned earlier, have great efficiency in uptake. A tumor cell needs to survive by providing more access to its internal environment. When *endocytosis* is used, the polymeric nanoparticles are taken into the microenvironment via interactions of receptors with ligands (molecules that binds to other molecules). A big advantage of this approach is that receptors or antigens can be recycled after delivery. The choice of a *ligand* or *antibody* requires great properties such as *immunogenicity, biodegradability, and physiochemical properties.*

The optimal and ideal antibody construction would be one that is incorporated with a nanoparticle which would result in increased affinity to the surface of the tumor cells receptors and have the ability to internalize the drug. The advantages of the antibody targeting are in specificity, accuracy, and biocompatibility, since they

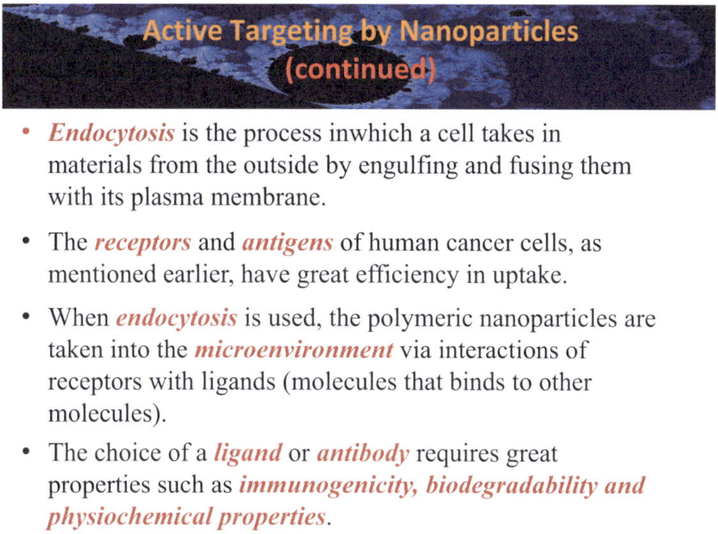

Slide 7.18 Active targeting by nanoparticles (continued)

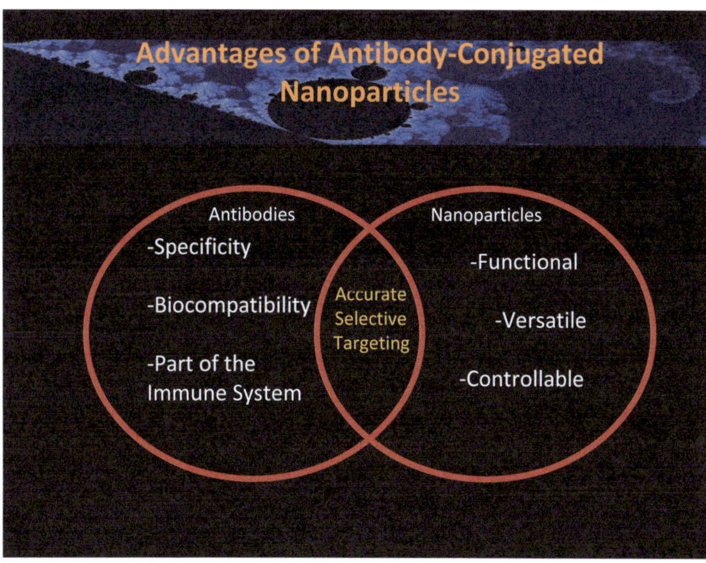

Slide 7.19 Advantages of antibody-conjugated nanoparticles

could be a part of immune system (Slide 7.19). The nanoparticles themselves have great benefits: they are functional, versatile, and controllable. All of this creates a very powerful and accurate selective targeting technique, which can be used for both the diagnosis and therapy of different diseases.

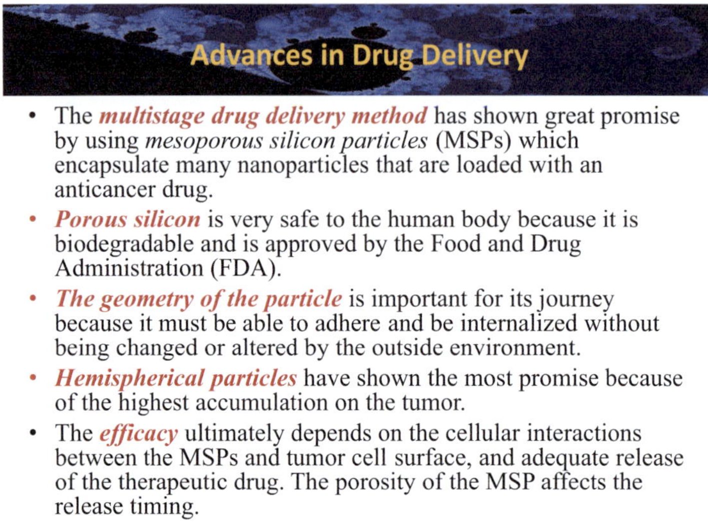

Slide 7.20 Advances in drug delivery

2.3.3 Advances in Drug Delivery

The *multistage drug delivery method* has shown great promise by using *mesoporous silicon particles* (MSPs), which encapsulate many nanoparticles that are loaded with an anticancer drug (Slide 7.20). Porous silicon is very safe to the human body because it is biodegradable and is approved by the Food and Drug Administration (FDA). The geometry of the particle is important for its journey because it must be able to adhere and be internalized without being changed or altered by the outside environment. Hemispherical particles have shown the most promise because of the highest accumulation on the tumor. The efficacy ultimately depends on the cellular interactions between the MSPs and tumor cell surface, and adequate release of the therapeutic drug. The porosity of the MSP affects the release timing. Encouraging results have shown great potential with this delivery method.

In conclusion, specifically designed nanoparticles can be selectively delivered to the tumor and then used in many diagnostic and treatment applications, including bioseparation, sensing and imaging, immunoassays, and purification, as demonstrated on the diagram in Slide 7.21. They can be used in MRIs as a diagnostic technique or as a therapeutic tool for drug delivery and tissue repair.

The general consensus of the scientific community on the best method for selectively targeting cancer cells appears to be antibody-conjugated nanoparticles. It is very accurate and has already been shown to work for in vivo studies.

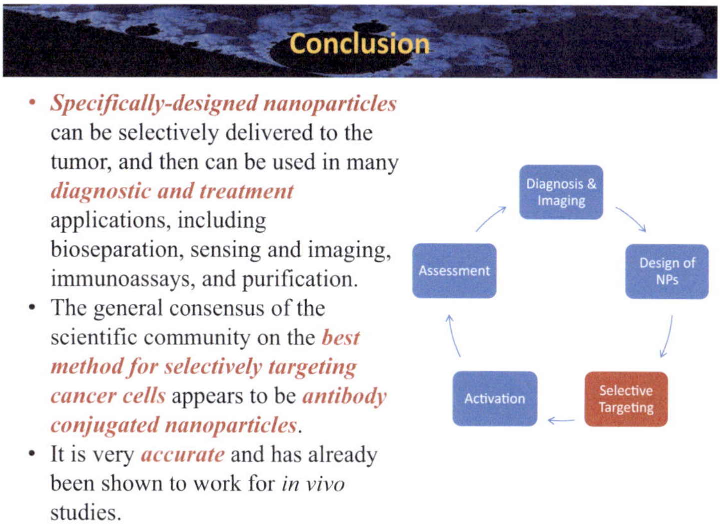

Slide 7.21 Conclusion

2.4 Lecture 8: Gene Therapy and Nanomedicine (Slide 8.1)

- Cancer and Gene Therapy
- Genes
- Mutations and Cancer
- Gene Therapy
- Nanomedicine for Gene Therapy
- Conclusions

2.4.1 Cancer and Gene Therapy

As discussed earlier in this chapter, deoxyribonucleic acid (DNA) contains all the genetic instructions for the body (Slide 8.2). Cells use these DNA instructions to create hormones, proteins, chemicals, and other macromolecules needed for the body to function. To be more efficient in its use of materials the body signals cells to differentiate into more specialized cell types, making different cells proficient at different tasks. For example, cardiac cells are streamlined to turn nutrients and oxygen into usable energy to keep pumping the heart again and again; they also support a large peak rate of energy usage—for situations when the heart needs to pump quickly and keep doing so for extended periods of time. On the other hand, pancreatic beta cells are fine-tuned to create insulin in relatively large volumes and store it, and can do their job with a much smaller peak rate of energy usage.

Slide 8.1 Contents of Lecture 8

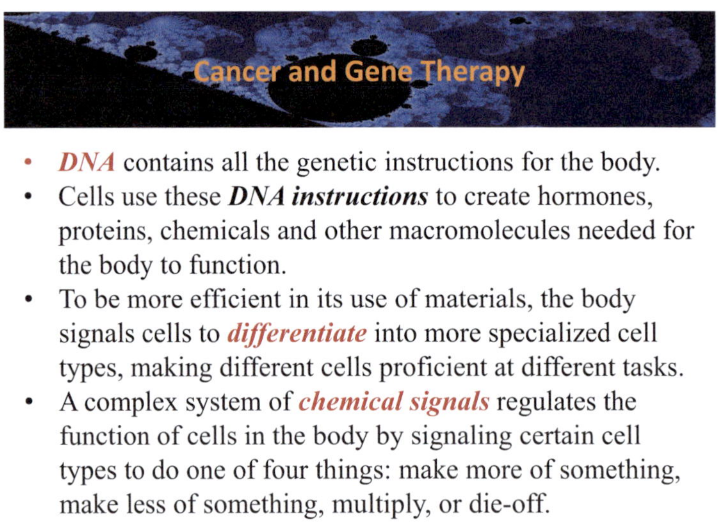

Slide 8.2 Cancer and gene therapy

Working together, certain differentiated cells come together to form tissues, which in turn form the different organs of the body.

A complex system of chemical signals regulates the function of cells in the body by signaling certain cell types to do one of four actions: make more of something, make less of something, multiply, or die-off (Slide 8.2). A lot of the time-differentiated cells will simply be tasked with producing more or less of the products they are specialized to produce. Sometimes a cell may be damaged more

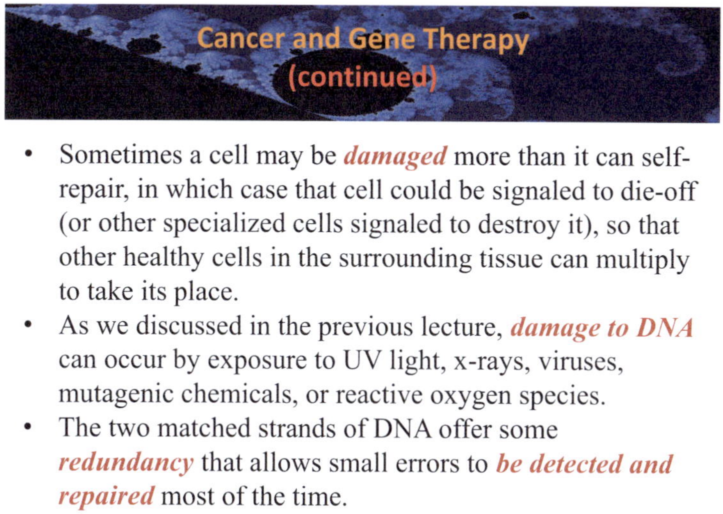

Slide 8.3 Cancer and gene therapy (continued)

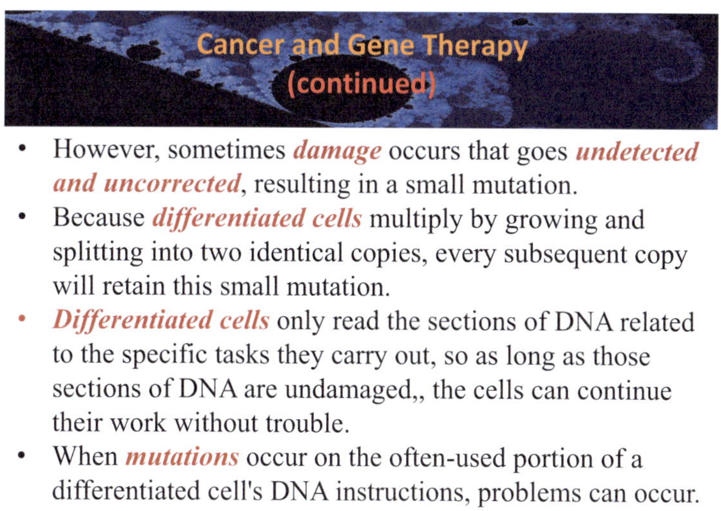

Slide 8.4 Cancer and gene therapy (continued)

than it can self-repair, in which case that cell could be signaled to die-off (or other specialized cells signaled to destroy it), so that other healthy cells in the surrounding tissue can multiply to take its place (Slide 8.3).

Damage to DNA can occur by exposure to ultraviolet (UV) light, X-rays, viruses, mutagenic chemicals, or reactive oxygen species (Slide 8.3). The two matched strands of DNA offer some redundancy that allows small errors to be detected and repaired most of the time. However, sometimes damage occurs that goes undetected and uncorrected, resulting in a small mutation (Slide 8.4). Because

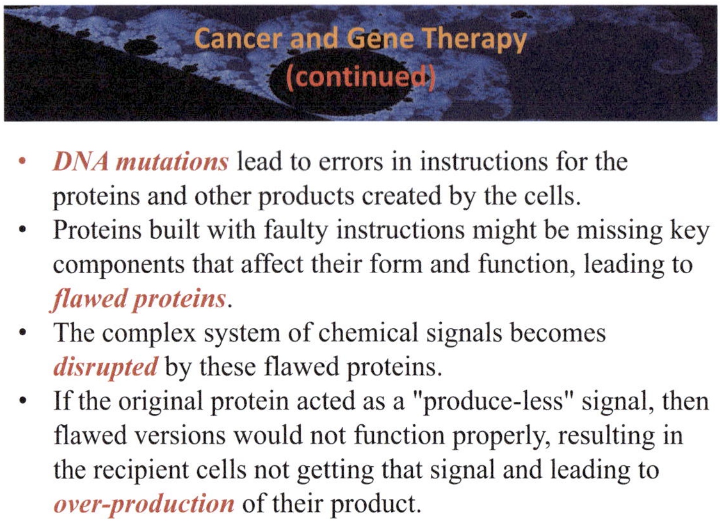

Slide 8.5 Cancer and gene therapy (continued)

differentiated cells multiply by growing and splitting into two identical copies, every subsequent copy will retain this small mutation. Differentiated cells only read the sections of DNA related to the specific tasks they carry out, so as long as those sections of DNA are undamaged, the cells can continue their work without trouble.

When mutations occur on the often-used portion of a differentiated cell's DNA instructions, problems can occur. Because DNA carries the instructions for the proteins and other products created by the cells, mutations lead to errors in those instructions (Slide 8.5). Proteins built with faulty instructions might be missing key components that affect their form and function (adding unnecessary components can also affect form and function), leading to flawed proteins. In the best case scenario, these flawed proteins are similar enough to their proper originals to carry out their intended functions without significant loss of efficiency. In the worst case, the flawed proteins not only fail to carry out their intended purpose but also interfere with the proper functioning of other proteins.

The complex system of chemical signals becomes disrupted by these flawed proteins (Slide 8.5). If the original protein acted as a "produce-less" signal, then flawed versions would not function properly, resulting in the recipient cells not getting that signal and leading to overproduction of their product. Similarly, a flawed "produce-more" signal could lead to underproduction by signal-recipient cells. Alternatively, if proteins involved in receiving incoming signals are flawed, the same sort of miscommunications can occur, even if all nearby cells are sending out all the correct signals.

Cancer is characterized by uncontrolled cell growth, which can be caused by a large number of possible signal disruptions (Slide 8.6). Most simply, if cancer cells have flawed "die-off" signal receptors, then the cells will multiply beyond their

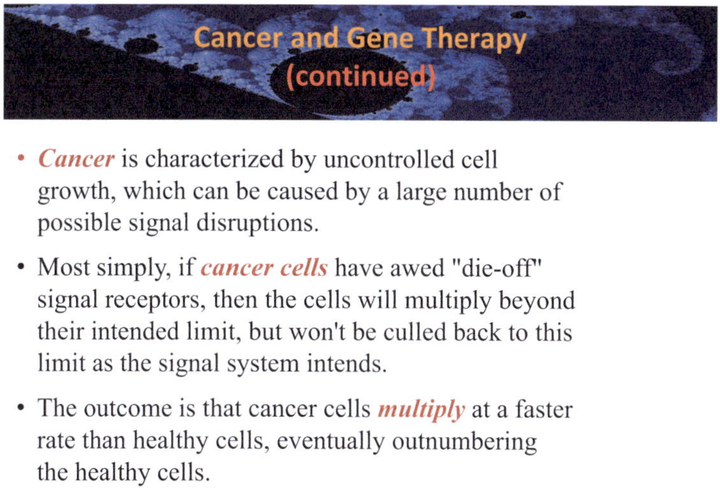

Slide 8.6 Cancer and gene therapy (continued)

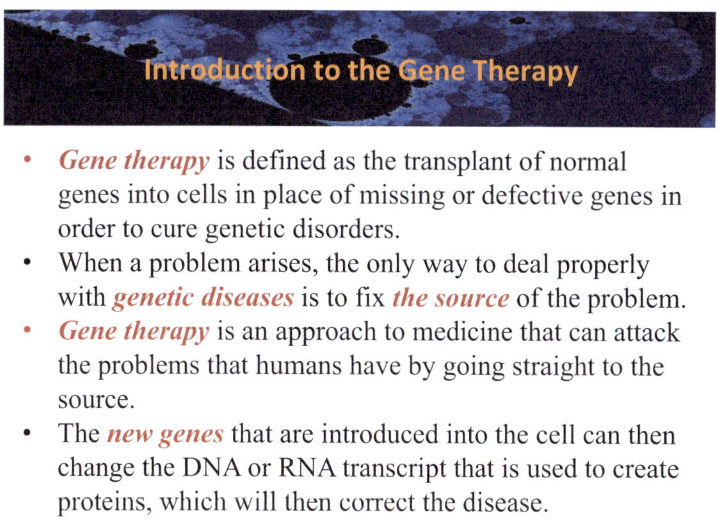

Slide 8.7 Introduction to the gene therapy

intended limit but will not be culled back to this limit as the signal system intends. Alternatively, cancer cells may overproduce signals for growth and multiplication to the extent that these signals overpower the contrary signals output from healthy cells. The outcome in any of these cases is that cancer cells multiply at a faster rate than healthy cells, eventually outnumbering the healthy cells.

Gene therapy is defined as the transplant of normal genes into cells in place of missing or defective genes in order to cure genetic disorders like cancer (Slide 8.7).

When a problem arises, the only way to deal properly with genetic diseases is to fix the source of the problem. Gene therapy is an approach to medicine that can attack the problems that humans have by going straight to the source. The new genes that are introduced into the cell can then change the DNA or RNA transcript that is used to create proteins, which will then correct the disease. Gene therapy is still in the experimental stage and is not yet widespread, but scientists are trying to figure out how to use gene therapy to correct *cystic fibrosis, stickle cell anemia, hemophilia, muscular dystrophy*, and even *cancer*.

As discussed earlier, nanomedicine allows for drugs to be delivered to a specific target cell. Gene therapy typically uses a virus to ensure that the new gene is delivered to the correct area in the body. Combining nanomedicine and gene therapy can allow for safer delivery of the genes to specific cells. Nanomedicine will be able to transport the genes while the gene therapy still processes the same way it normally does. In order to really understand gene therapy and how it ties into the nanomedicince, we have to first understand the gene itself.

2.4.2 Genes

Genes in a strict definition are working subunits of DNA that act as instructions to make proteins (Slide 8.8). There are about 100,000 of these genes in our body. It has been estimated by the *Human Genome* project that 20,000 to 25,000 of these genes are responsible for basic physical and functional units of heredity. Being units of heredity means that these genes are going to be responsible for things such as the

- ***Genes*** in a strict definition are working subunits of DNA that act as instructions to make proteins.
- There are about ***100,000 of*** these ***genes*** in our body.
- It has been estimated by the ***Human Genome*** project that ***20,000 to 25,000*** of these genes are responsible for basic ***physical and functional units of heredity***.
- Being ***units of heredity*** means that these genes are going to be responsible for things such as the color of our eyes and features that make individuals unique.
- These genes are responsible for more than just our ***physical features*** – they are responsible for how our ***bodies work***.

Slide 8.8 What are genes?

color of our eyes and features that make individuals unique. These genes are responsible for more than just our physical features—they are responsible for how our bodies work. This includes things in everyday life, such as production of insulin and how well the heart pumps blood into our body.

2.4.3 *Mutations and Cancer*

Genes are responsible for how healthy our body stays, and they are also responsible for defects or diseases that our bodies may have to deal with (Slide 8.9). Our bodies get these defects from gene mutations. Genes get their mutations during the process of cell replication—mitosis—which itself has several stages. During mitosis, all of the DNA is replicated so that the new cell also has a copy, but the DNA is not always replicated perfectly. As we defined earlier, the DNA mutation is a process where incorrect '*reading*' of RNA happens during the *transcription* event. In Section 2.1 (Lecture 5) of this Chap. 2, we discussed three major types of DNA mutations: nonsense, missense, and frameshift mutations. Mutations are the cause of some of the most problematic diseases, such as *Huntington's, type I diabetes*, most *heart diseases*, and *cancer*.

The question has been asked, "what types of mutations are actually responsible for cancer?" How things go wrong is actually a question we cannot answer because things seem to be random, and diseases like cancer may be caused by any number of these three mutations. When looking to do gene therapy, we want to know what in the DNA is exactly going wrong not necessarily how it goes wrong. The gene

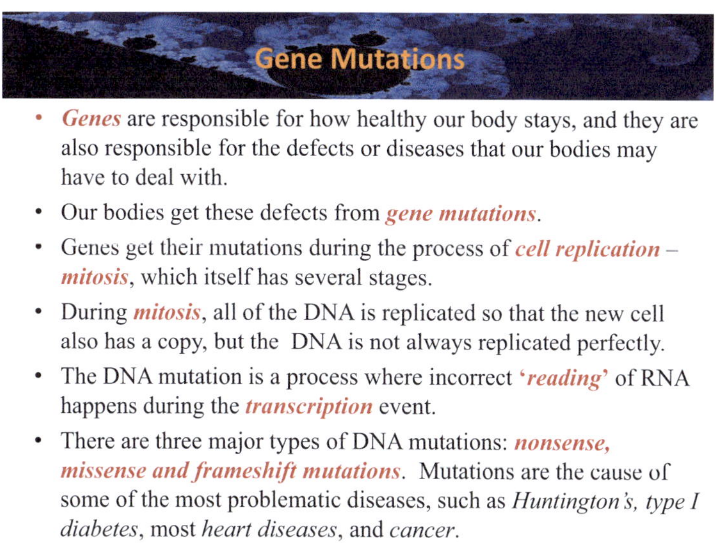

Slide 8.9 Gene mutations

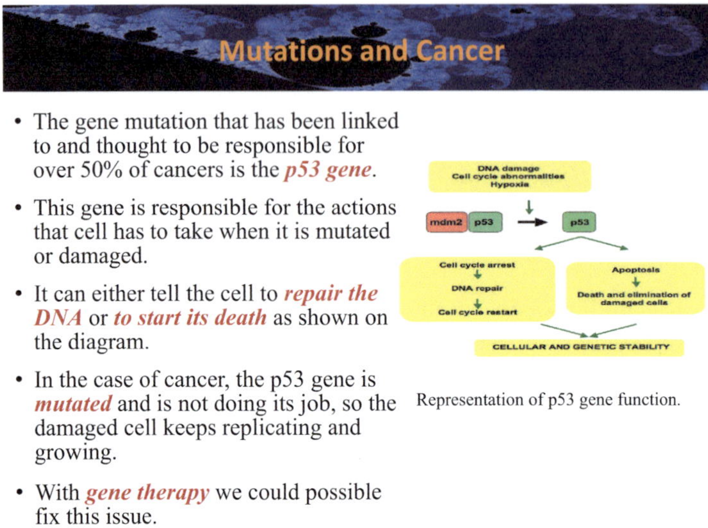

Slide 8.10 Mutations and cancer

mutation that has been linked to and thought to be responsible for over 50 % of cancers is the p53 gene (Slide 8.10). This gene is responsible for the actions that the cell has to take when it is mutated or damaged. It can either tell the cell to repair the DNA or to start its death, as shown on the diagram in Slide 8.10. In the case of cancer, the p53 gene is mutated and is not doing its job, so the damaged cell keeps replicating and keeps growing. With gene therapy we could possible fix this issue.

2.4.4 Gene Therapy

Gene therapy replaces the mutated genes by normal ones (Slide 8.11). These transplanted genes can do the following: (1) the DNA strand can be completely replaced, (2) the DNA can be used to knock out mutated DNA, or (3) it can be used to introduce a new gene into a cell in order to help it fight diseases.

The delivery of new genes is normally done through a carrier or virus (Slide 8.11). It would seem scary to use viruses, but they actually turn out to be perfect carriers because we know that a virus can inject its DNA into the hosting cell nucleus. If new DNA is stuck into the virus, it will inject the DNA into the cell nucleus so that it can go through mitosis with the new DNA. In doing so, there is one very important aspect: the virus must get through the endosomal process. This is the process that protects the cell nucleus from being hijacked by something like a virus. Normally this is not a problem for the virus, since it is coated with a protective outer layer that tricks the endosome. An *endosome* is a membrane-bound compartment inside eukaryotic cells.

Gene Therapy

- ***Gene therapy*** replaces the mutated genes by normal ones.
- These transplanted genes can do the following: ***(1) the DNA strand can be completely replaced; (2) the DNA can be used to knock out mutated DNA; or (3) it can be used to introduce a new gene into a cell in order to help it fight diseases***.
- The delivery of new genes is normally done through ***a carrier or virus***.
- The virus can ***inject its DNA*** into the hosting cell nucleus.
- In doing so, the virus must get through the ***endosomal process***.
- This is the process that ***protects*** the cell nucleus from being hi-jacked by something like a virus.
- Normally this is not a problem for the virus, since it is coated with a ***protective outer layer*** that tricks the endosome.
- ***An endosome*** is a membrane-bound compartment inside eukaryotic cells.

Slide 8.11 Gene therapy

Gene Therapy (continued)

- Gene therapy by using a virus as carrier of the DNA can be arranged in two modes***: inside or outside the body***.
- ***In the inside treatment***, the virus that carries the gene is injected directly into the part of the body that has the defective cells.
- ***In the outside treatment***, the blood of bone marrow is taken from the patient, and the immature cells are removed.
- Then the ***new genes are added*** to the cells and injccted back into the bloodstream. The cells can then travel to the bone marrow to mature and multiply rapidly, and then replace all of the defective cells.

Slide 8.12 Gene therapy (continued)

Gene therapy using a virus as the carrier of the DNA can be arranged in two modes: inside or outside the body, which gives doctors two different methods of utilization (Slide 8.12). In the inside treatment, the virus that carries the gene is injected directly into the part of the body that has the defective cells. In the outside

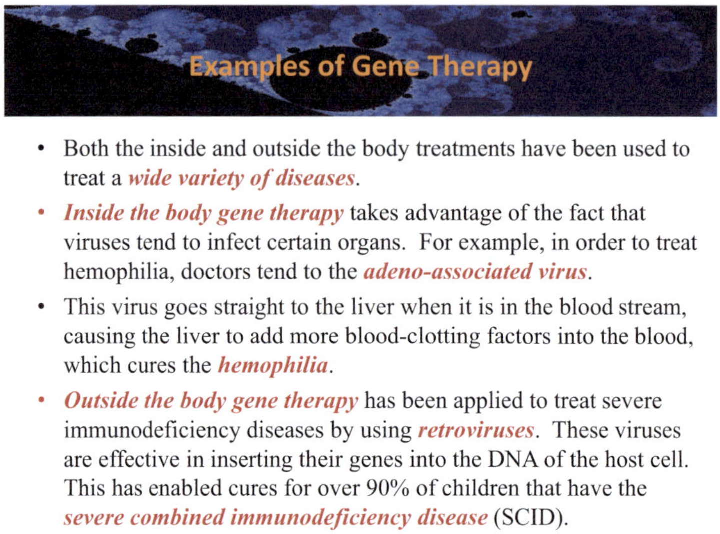

Slide 8.13 Examples of gene therapy

treatment, the blood of bone marrow is taken from the patient, and the immature cells are removed. Then the new genes are added to the cells and injected back into the bloodstream. The cells can then travel to the bone marrow to mature and multiply rapidly, and then replace all of the defective cells.

Both the inside and outside the body treatments have been used to treat a wide variety of diseases (Slide 8.13). Inside the body gene therapy takes advantage of the fact that viruses tend to infect certain organs. For example, in order to treat hemophilia, doctors tend to favor the adeno-associated virus. This virus goes straight to the liver when it is in the blood stream, causing the liver to add more blood-clotting factors into the blood, which cures the hemophilia. Outside the body gene therapy has been applied to treat severe immunodeficiency diseases by using retroviruses. These viruses are effective in inserting their genes into the DNA of the host cell. This has enabled cures for over 90 % of children that have the *severe combined immunodeficiency disease* (SCID).

There are some potential drawbacks with using a virus as the carrier for the new DNA (Slide 8.14). The main issue is that since it is a virus, the body could see it as a foreign object and attack it. Another issue is that the body could get infected with the virus.

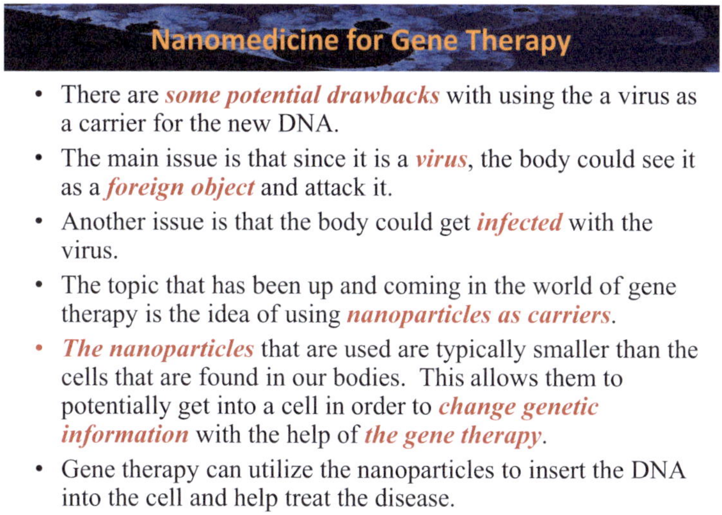

Slide 8.14 Nanomedicine for gene therapy

2.4.5 Nanomedicine for Gene Therapy

The topic that has been up and coming in the world of gene therapy is the idea of using nanoparticles as carriers (Slide 8.14). As mentioned before, nanomedicine uses nanoparticles in order to heal defective cells. The nanoparticles that are used are typically smaller than the cells that are found in our bodies. This allows them to potentially get into a cell in order to change genetic information with the help of gene therapy. Gene therapy can utilize the nanoparticles to insert the DNA into the cell and help treat the disease. For example, if a patient has cancer, gene therapy can possibly alter the genes causing the body to have a growth of cells in the tumor. This therapy is capable of keeping the patient alive and lowering the amount of side effects that come from the treatment.

Since there are some problems associated with viruses, research has taken a turn to try to replicate the natural process that viruses use (Slide 8.15). The three main characteristics of the virus that have to be replicated are its ability to (1) hold DNA, (2) make it through the cell's endosomal process, and most importantly, (3) place DNA inside the nucleus. A behavior that we want to modify is the virus' ability to replicate and stay inside the body. The idea is that whatever we put into the body, we want to be able to exit without harm to the patient.

A method for mimicking a virus is by encapsulating DNA into a nanoparticle itself (Slide 8.16). This allows direct DNA interaction if the nanocapsule gets through the endosomal process. Of course, that leads to the question as to how to protect that nanocapsule from the endosomal process? The answer is we coat it with the same type of protective layering composed of endosome-lytic components, such as lytic peptides and lytic lipids that a virus may use. The last hurdle to overcome is

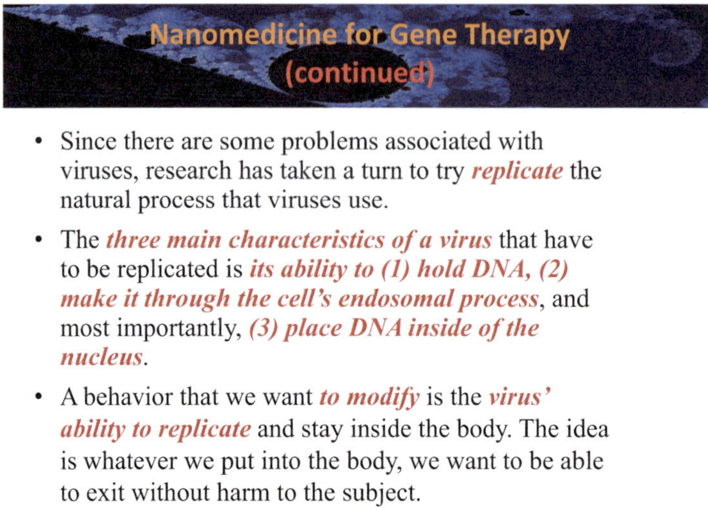

Slide 8.15 Nanomedicine for gene therapy (continued)

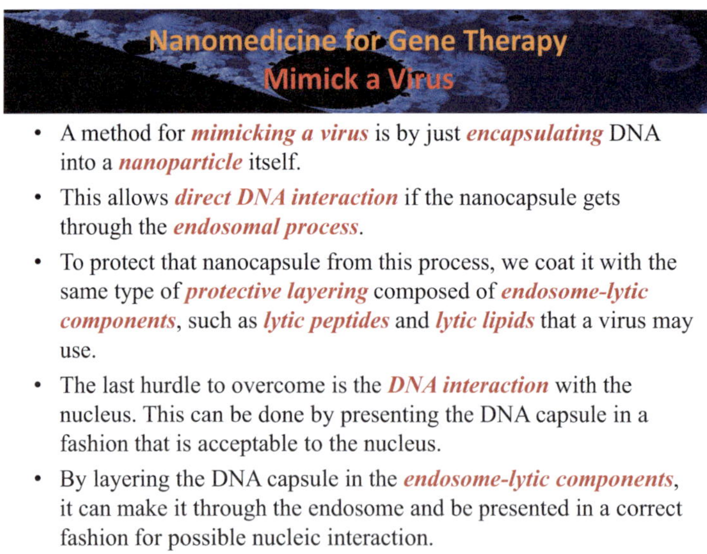

Slide 8.16 Nanomedicine for gene therapy: mimic a virus

the DNA interaction with the nucleus. This can be done by presenting the DNA capsule in a fashion that is acceptable to the nucleus. By layering the DNA capsule in the endosome-lytic components, it can make it through the endosome and be presented in a correct fashion for possible nucleic interaction. Examples of nonviral condensed DNA particles include cationic lipids, polylysine conjugates, liposomes, polyvinyl derivatives, and polycyanoacrylate nanoparticles.

Nanomedicine for Gene Therapy
Mimick a Virus (continued)

- Another possible way of mimicking the virus is by use of *dendrimers* discussed in Chapter 1.
- *Dendrimers* are compounds that are made up of branches around an inner core, where they serve as *a net-like material* to hold the DNA.
- This is achieved by *the electrostatic interaction* between the negativity-charged DNA and positive charge of the dendrimer's core.
- This method uses the same technique to get into the cell and escape from the *protective endosomal process* as the nanocapsules discussed above.
- Unlike the nanocapsule technique, the dendrimers method is heavily dependent on the *electrostatic interaction* between the particle and the DNA.

Slide 8.17 Nanomedicine for gene therapy: mimic a virus (continued)

Another possible way of mimicking the virus is by use of dendrimers discussed in Chap. 1. Dendrimers are compounds that are made up of branches around an inner core (Slide 8.17), where they serve as a net-like material to hold the DNA. This is achieved by the electrostatic interaction between the negatively charged DNA and positive charge of the dendrimer's core. This method uses the same technique to get into the cell and escape from the protective endosomal process as for the nanocapsules discussed earlier. Unlike the nanocapsule technique, the dendrimer method is heavily dependent on the electrostatic interaction between the particle and the DNA.

Once the nanoparticle that we have created is inside of the cell, how does the DNA get to the nucleus? There has been progress in this area of gene therapy (Slide 8.18). One way of targeting is to use a magnetic field to guide the magnetic nanoparticles into a targeted area. The same magnetic field would also be used to heat the nanoparticle so that it could release its DNA where needed.

Gene therapy has already enjoyed some success in the medical world (Slide 8.18). It has led to an increase in blood clotting in patients who have hemophilia. It has led to an increase in muscle control in patients who have Parkinson's disease, and of course it has led to regression of cancer. Hopefully these applications can be extended to cure diseases that are hereditary and reverse the mutations that cause unhealthy situations in the body such as cancer.

In conclusion, nanomedicine and gene therapy could open the door to many possibilities in the medical world (Slide 8.19). By combining these two fields, researchers should be able to create new methods that are safer for patients and

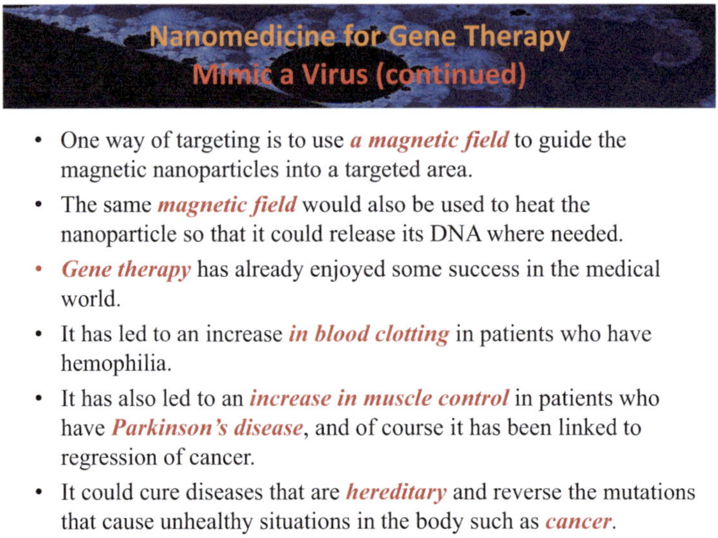

Slide 8.18 Nanomedicine for gene therapy: mimic a virus (continued)

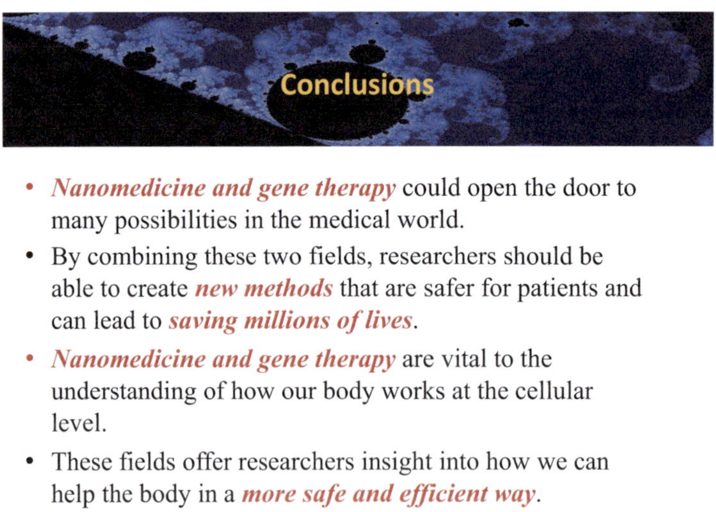

Slide 8.19 Conclusions

can lead to saving millions of lives. Nanomedicine and gene therapy are vital to the understanding of how our body works at the cellular level. These fields can offer researchers insight into how we can help the body in a more safe and efficient way.

Homework Exercises

Section 2.1: Introduction to Cell Biology

2.1. What functions do cells have in common?
2.2. What is mitosis?
2.3. Select one or more. A biological cell is made of:

(a) Carbohydrates
(b) Proteins
(c) Lysosomes
(d) Lipids
(e) Peroxisomes
(f) Nucleic acids
(g) All of the above
(h) None of the above

2.4. What are carbohydrates? Give some examples.
2.5. True or false? Carbohydrates control activity of the cells.
2.6. What are proteins?
2.7. List the functions of proteins in the cell.
2.8. True or false? Proteins are an excellent source of energy for cells.
2.9. What are lipids? Give some examples.
2.10. What functions do lipids have?
2.11. True or false? Lipids contain all the information needed to control and build cells.
2.12. What molecules are nucleic acids composed of?
2.13. What functions do nucleic acids have?
2.14. What is a nucleotide?
2.15. What are the differences between RNA and DNA?
2.16. True or false? Nucleic acids form the biological membrane.
2.17. True or false? DNA and RNA are made of nucleotides.
2.18. True or false? RNA is double stranded.
2.19. Select one or more. Cell organelles are:

(a) Nucleus
(b) Mitochondria
(c) Lysosome
(d) Ribosome
(e) Peroxisome
(f) Cytoskeleton
(g) Centrosome
(h) Endoplasmic reticulum
(i) All of the above
(j) None of the above

2.20. What is the main function of the nucleus in a cell?
2.21. What is a gene?
2.22. What is the main function of mitochondria?
2.23. What is a main function of a ribosome?
2.24. What is the main function of a cytoskeleton?
2.25. What is the main function of a plasma membrane?
2.26. What is a plasma membrane made of?
2.27. Select one or more. Stages of the cell cycle include:

(a) A stage
(b) C stage
(c) G1 and G2 stages
(d) T stage
(e) M stage
(f) All of the above
(g) None of the above

2.28. What are the checkpoints in a cell cycle?
2.29. When are genetic changes likely to happen?
2.30. What is a DNA mutation?
2.31. What happens if genes mutate?
2.32. Select one or more. Types of genetic changes are:

(a) Spontaneous mutation
(b) Induced mutation
(c) Gene amplification
(d) Missense mutation
(e) Nonsense mutation
(f) Viruses
(g) All of the above
(h) None of the above

2.33. What is a codon?
2.34. What is a nonsense mutation?
2.35. What is a missense mutation?
2.36. What is a gene amplification?
2.37. What is a frameshift mutation?
2.38. What is a spontaneous mutation?
2.39. What is an induced mutation?

Section 2.2: Introduction to Cancer Biology

2.40. What is a cancer?
2.41. What are the types of tumors?
2.42. What are the causes of cancer?

2.43. True or false? Some foods are carcinogenic.
2.44. True or false? Obesity contributes to cancer.
2.45. True or false? Tobacco is a well-known cause of lung cancer.
2.46. What is a common method of cancer staging? Describe the staging of cancer.
2.47. List and briefly describe the categories of cancer.
2.48. Describe a cancer formation.
2.49. What are regulatory genes?
2.50. True or false? Cancer forms in the third or fourth generation of mutated cells cycle.
2.51. True or false? Cancer cells are unable to avoid apoptosis.
2.52. What is a mechanism for cancer metastasizing?
2.53. What are tumor-suppressor genes?
2.54. How does the tumor-suppressor gene p53 work?
2.55. What is contact inhibition?
2.56. What is cellular senescence?
2.57. List and describe the common characteristics of cancer cells.
2.58. Describe the changes in a cancerous nucleus.
2.59. True or false? Cellular energy production increases in cancer cells.
2.60. What is the role of integrin?
2.61. List and describe the biological properties of cancer cells.
2.62. List and describe the electrical properties of cancer cells.
2.63. List and describe the mechanical properties of cancer cells.
2.64. List and describe the biochemical properties of cancer cells.
2.65. True or false? Cancer cells have a large amount of glucose receptors and glucose enzymes.

Section 2.3: Selective Targeting of Cancer Cells

2.66. Select one or more. Passive targeting include:

(a) Angiogenesis
(b) Antibodies
(c) Epidermal growth factors
(d) Adaptor molecules
(e) All of the above
(f) (a) and (c)
(g) (b) and (d)
(h) None of the above

2.67. What is angiogenesis?
2.68. What are three submethods of passive targeting?
2.69. What is enhanced permeability and retention?
2.70. What is the definition of incubation time?

2.71. Select one or more. What is the incubation time for nanoparticles of radii 1 to 50 nm?

(a) Less than 1 h
(b) 1–5 h
(c) 6–12 h
(d) 1–3 days
(e) All of the above
(f) (b) and (c)
(g) None of the above

2.72. Select one or more. Nanoparticle coatings are:

(a) Proteins
(b) Carbohydrates
(c) Peptides
(d) Nucleotides
(e) Polymers
(f) Iron oxide
(g) Cobalt oxide
(h) Nickel oxide
(i) All of the above
(j) (a) through (d)
(k) (e) through (h)
(l) None of the above

2.73. True or false? Nanoparticle coatings reduce biocompatibility.
2.74. What is a tumor microenvironment?
2.75. What is a local drug application?
2.76. True or false? Active targeting utilizes the differences in receptors of cancer and healthy cells.
2.77. True or false? Active targeting uses the body's own antibodies only.
2.78. True or false? Normal cells have a greater probability to receive epidermal growth factor signals than cancer cells.
2.79. True or false? Monoclonal antibodies block the receptors of cancer cells.
2.80. True or false? Monoclonal antibodies improve the ability of a cell to receive epidermal growth factor signals.
2.81. Describe the conjugation of antibodies to nanoparticles.
2.82. What are two techniques for bonding antibodies to nanoparticles?
2.83. True or false? Direct bonding is the most effective technique.
2.84. Select one or more. Adaptor molecules are used for:

(a) Attaching to antibodies
(b) Attaching to nanoparticles
(c) Increasing the toxicity of cancer cells
(d) Improving stability
(e) Reducing biocompatibility
(f) All of the above

(g) (c) through (e)
(h) (a), (b), and (d)
(i) None of the above

2.85. What is endocytosis?
2.86. List the advantages of the antibody treatments.
2.87. List the advantages of nanoparticle treatments.
2.88. What are the advantages of combining the antibody and nanoparticle techniques?

Section 2.4: Gene Therapy and Nanomedicine

2.89. True or false? Genes are subunits of DNA.
2.90. True or false? Genes contain the instructions to make proteins.
2.91. Select one. The total number of genes in a body is around:

(a) 10,000
(b) 20,000
(c) 25,000
(d) 50,000
(e) 100,000
(f) (b) and (c)
(g) All of the above
(h) None of the above

2.92. What is the definition of a gene therapy?
2.93. What is an endosomal process?
2.94. Select one or more. Types of DNA mutations are:

(a) Mitosis
(b) Nonsense
(c) Apoptosis
(d) Missense
(e) Frameshift
(f) Virus
(g) All of the above
(h) None of the above

2.95. True or false? Nonsense mutation is replacement of a single nucleotide.
2.96. True or false? Cancer cells are produced in the first mutation of DNA.
2.97. Select one or more. Gene therapy includes:

(j) Transplantation of normal genes into cells
(k) Replacing a mutated gene
(l) Inactivating or “knocking out” a mutated gene
(m) Introducing a new gene

(n) All of the above
(o) (a), (b), and (c)
(p) (b), (c), and (d)
(q) None of the above

2.98. Select one or more. Gene therapy uses:

(a) Carbon nanotubes
(b) Quantum dots
(c) Viruses
(d) Nanoparticles
(e) All of the above
(f) (c) and (d)
(g) (a) and (b)
(h) None of the above

2.99. Select one or more. Gene therapy could cause:

(a) Tumor
(b) Cancer
(c) Immune system reaction
(d) Infection
(e) Mitosis
(f) All of the above
(g) (c) and (d)
(h) None of the above

2.100. Select one or more. Nanogene therapy include:

(a) Conjugation
(b) Radiation therapy
(c) Plasmid DNA capsule
(d) All of the above
(e) Chemotherapy
(f) Dendrimers
(g) Surgery
(h) None of the above

Topics for Presentations and Writing Assignments

Presentation #1. Cell biology: Structure, functions and properties.
Presentation #2. Cancer cell structure and properties.
Presentation #3. Targeting cancer cells by nanoparticles: Mechanisms, methods, and future ideas.
Presentation #4. Gene therapy and nanotechnology.
Presentation #5. Biologically active food supplements for cancer prevention.

References

1. http://www.wikipedia.org
2. http://www.coursehero.com
3. https://cancerqld.org.au
4. http://nevadacancercoalition.org
5. A.G. Warrier et al., Biomolecular motor proteins as targets for cancer treatment—A computational study, in *2010 IEEE Students Technology Symposium* (2010), pp. 17–20
6. http://www.studystack.com

Chapter 3
Introduction to Cancer Therapy and Detection by Plasmonic Nanoparticles

Contents

Abstract The advancement of research in the areas of cellular biology and nanotechnology creates new opportunities for the diagnosis and treatment of diseases. The emergence of molecular targeting allows for the selective delivery of pharmaceutical or nanoparticle payloads to localized areas, while nanoparticle research offers more and more ways to engineer the physical properties of these payloads. This chapter introduces the surface plasmon resonance phenomenon and the use of plasmonic nanoparticles in diagnosis and cancer therapy. One of the therapies, called nanophotothermolysis, utilizes molecular targeting to deliver nanoparticles to the localized area, and then uses the plasmon absorption properties of the nanoparticles to generate heat in the targeted region. We discuss here the new dynamic modes in selective nanophotothermolysis of cancer, the design and methods of activation of the nanodrugs selectively delivered to the tumor site, and the plasmon resonance detection techniques using fiber optics.

Electronic supplementary material: The online version of this chapter (doi:10.1007/978-3-319-43577-0_3) contains supplementary material, which is available to authorized users.

R.R. Letfullin, T.F. George, *Computational Nanomedicine and Nanotechnology*,
DOI 10.1007/978-3-319-43577-0_3

3.1 Lecture 9: Introduction to Nanophototherapies

- Conventional Cancer Treatments
  - *Surgical Removal*
  - *Radiation Therapy*
  - *Chemotherapy*
- Nanophototherapies for Cancer Treatment
- Selective Nanophotothermolysis of Cancer Cells
  - *Definition*
  - *Methods of Nanoparticle Administration*
  - *Cell Ablation*
  - *Physical Concepts*
- New Dynamic Modes in Selective Nanophotothermolysis
  - *Cluster Aggregation Mode*
  - *Nanobubble Overlapping Mode*
  - *Thermal Explosion Mode*
- Conclusions
- References

This chapter focuses on targeted photothermal therapy of cancer, which combines molecular targeting mechanisms with the unique light absorption properties of nanoparticles to deliver heat and treat diseases at a subcellular level (Slide 9.1).

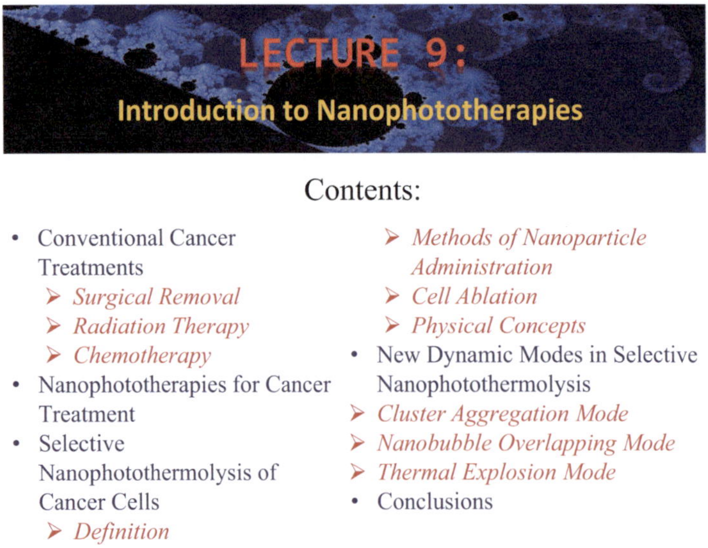

Slide 9.1 Contents of Lecture 9

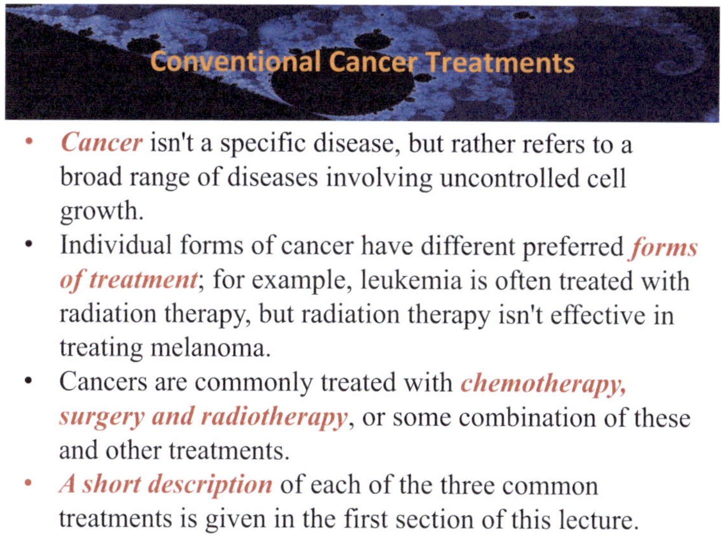

Slide 9.2 Conventional cancer treatments

To understand how photothermal therapies, like nanophotothermolysis and nanophotohyperthermia, could be useful, it is important to understand how current cancer treatments work.

Cancer isn't a specific disease, but rather refers to a broad range of diseases involving uncontrolled cell growth (Slide 9.2). Individual forms of cancer have different preferred forms of treatment; for example, leukemia is often treated with radiation therapy, but radiation therapy isn't effective in treating melanoma. Cancers are commonly treated with chemotherapy, surgery, and radiotherapy, or some combination of these and other treatments. A short description of each of the three common treatments is given in the first part of this lecture to create a framework for comparison to the newer photothermal therapy.

3.1.1 Conventional Cancer Treatments

Surgical Removal. Conceptually, the most basic approach of ridding the body of cancer is to physically remove tumors from the body (Slide 9.3). The main benefit of this approach is that large cancerous masses are removed from the body in a very short period of time. The downside is that this method is only viable for large isolated tumors that occur in operable parts of the body. If many small metastases are present, it may not be feasible to remove them all surgically, and if the metastases are too small, it may be impossible to remove them without also removing large amounts of healthy tissue. Also, in delicate areas like the brain, surgery may pose too many risks to justify this method of treatment.

- Conceptually, the most basic approach of ridding the body of cancer is to ***physically remove*** tumors from the body.
- The main benefit of this approach is that ***large cancerous masses*** are removed from the body in a very short period of time.
- The downside is that this method is only ***viable for large isolated tumors*** that occur in operable parts of the body.
- If many ***small metastases*** are present, it may not be feasible to remove them all surgically, and if the metastases are too small, it may be ***impossible to remove*** them without also removing large amounts of healthy tissue.
- Also, in delicate areas like ***the brain***, surgery may pose too many risks to justify this method of treatment.

Slide 9.3 Surgical removal

- ***X-rays*** can damage DNA with the extent of the damage ranging from small repairable changes, to small uncorrected changes (i.e., mutations), to large catastrophic changes (which trigger the destruction of the cell).
- ***Healthy cells*** have the ability to check for and repair small errors caused by x-ray exposure, using the redundancy of the two-strand DNA to detect whether one side is a proper match to the other.
- During ***cell division***, DNA is copied by separating the two strands then building matched copies directly onto each individual strand. While this is happening, the individual strands are more vulnerable to undetectable DNA damage, making them at greater risk of mutation.

Slide 9.4 Radiation therapy

Radiation Therapy. X-rays can damage DNA, with the extent of the damage ranging from small repairable changes to small uncorrected changes (i.e., mutations), to large catastrophic changes, which trigger the destruction of cells (Slides 9.4, 9.5, and 9.6).

Radiation Therapy (continued)

- ***Cancer cells*** divide the same way as healthy cells, so they too are more susceptible to x-ray -induced DNA mutations; as cancer cells multiply more often, they are more likely to be in this vulnerable state.
- Additionally, cancer cells may be less able to repair DNA damage if the proteins that perform this action have been ***mutated*** into less effective forms.
- ***Radiation therapy*** uses ***shaped radiation beams*** to limit exposure to non-tumor cells.
- ***Multiple beams*** can be used, all from different angles and intersecting on the tumor, to increase exposure to the target area (the tumor) while minimizing beam exposure to the surrounding tissues.

Slide 9.5 Radiation therapy (continued)

Radiation Therapy (continued)

- ***An alternate form*** places small amounts of ***radioactive material*** directly at the site of the tumor rather than using external radiation sources.
- In both cases, most of the radiation is ***focused on the tumor***, where it is hoped that enough damage has occurred to kill a large number of cancer cells.
- The ***downside to radiation therapy*** is the fact that beam radiation has to pass through healthy tissue to reach the cancerous tissues.
- The result of this is that a ***large number of healthy cells*** are exposed to radiation.
- Further complicating the problem is that the damage mechanism is potentially cancer-inducing if some fraction of healthy cells develop ***cancer-causing mutations***.

Slide 9.6 Radiation therapy (continued)

Healthy cells have the ability to check for and repair small errors caused by X-ray exposure, using the redundancy of the two-strand DNA to detect whether one side is a proper match to the other (Slide 9.4). During cell division, DNA is copied by separating the two strands and then building matched copies directly onto each individual strand. While this is happening, the individual strands are more vulnerable to undetectable DNA damage, making them at greater risk of mutation.

Cancer cells divide the same way as healthy cells, so they too are more susceptible to X-ray-induced DNA mutations; as cancer cells multiply more often, they are more likely to be in this vulnerable state (Slide 9.5). Additionally, cancer cells may be less able to repair DNA damage if the proteins that undergo this action have been mutated into less effective forms.

Radiation therapy uses shaped radiation beams to limit exposure to nontumor cells. Multiple beams can be used, all from different angles and intersecting on the tumor, to increase exposure to the target area (the tumor) while minimizing beam exposure to the surrounding tissues (Slide 9.5).

An alternate form places small amounts of radioactive material directly at the site of the tumor rather than using external radiation sources (Slide 9.6). In both cases, most of the radiation is focused on the tumor, where it is hoped that enough damage has occurred to kill a large number of cancer cells. Cancer cells that are not outright killed are likely to have suffered some DNA damage resulting in more new mutations. Radiation therapy often employs multiple treatments, which is beneficial for multiple reasons, particularly because cancer cells that weren't dividing during one treatment might be dividing during a later treatment (and would therefore be more vulnerable to radiation).

The downside to radiation therapy is that the beam radiation has to pass through healthy tissues to reach the cancerous tissues (Slide 9.6). The result of this is that a large number of healthy cells are exposed to radiation. Both beam shaping and multibeam convergence are used to maximize tumor damage while minimizing healthy cell damage, but some damage to healthy cells will always occur. Further complicating the problem is the fact that the damage mechanism is potentially cancer inducing if some fraction of healthy cells develop cancer-causing mutations.

Chemotherapy. Since the cancer cells are constantly growing and multiplying, most forms of chemotherapy are designed to impair mitosis (cell division) by introducing drugs (Slide 9.7). Many of these drugs directly target DNA during this process, preventing it from splitting and replicating, or by imitating DNA base pairs and thus preventing actual base pairs from forming new copies. Depending on the drugs used, cells may be stuck in this no division limbo until dying off from other factors, or the drug effects may be cytotoxic and actively kill off the cells.

One benefit of chemotherapy is that it is systemic (Slide 9.8). Because the drugs administered travel throughout the body, even metastatic cells found far from the tumor will receive treatment. This can destroy metastases before they have time to grow and develop into new tumors. For this reason, adjuvant chemotherapy is often administered in cases where cancers are surgically removed, in order to minimize the risk that any few remaining cancerous cells grow into new tumors.

The main drawback of chemotherapy is also due to the fact that it is systemic (Slide 9.8). Because the drugs target cells undergo cell division, even healthy cells undergoing cell division will be caught in the crossfire. This can have a large physical toll on the patient, depending on the choice of drugs and doses administered.

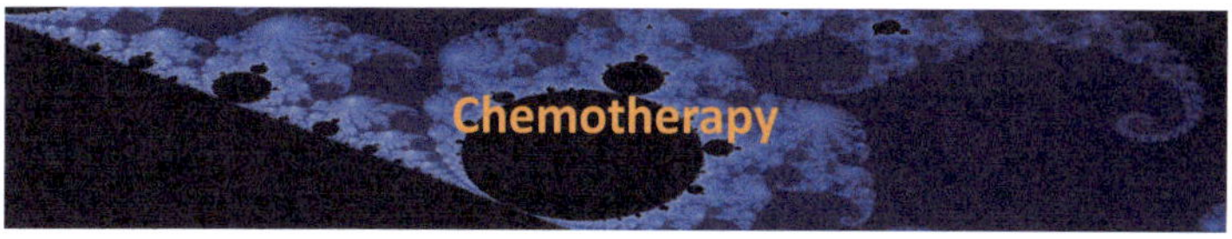

- Since the cancer cells are constantly growing and multiplying, most forms of ***chemotherapy*** are designed to ***impair mitosis*** (cell division) by introducing drugs.
- Many of these ***drugs*** directly target DNA during this process, ***preventing*** it from splitting and replicating, or by ***imitating*** DNA base pairs and thus preventing actual base pairs from forming new copies.
- Depending on the drugs used, cells may be stuck in this ***no-division limbo*** until dying-off from other factors, or they drugs effects may be cytotoxic and ***actively kill off*** the cells.

Slide 9.7 Chemotherapy

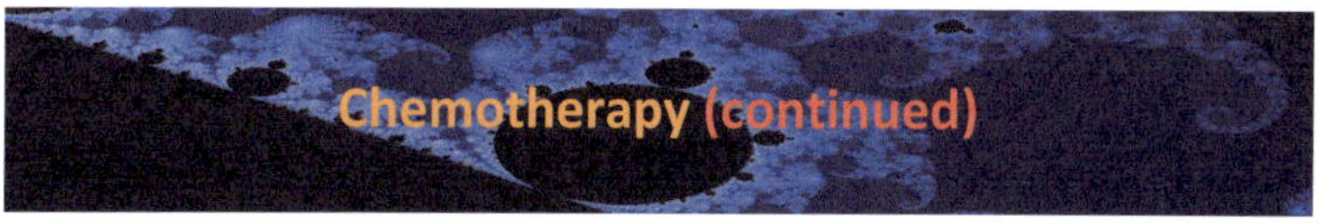

- ***One benefit*** of chemotherapy is the fact that it is ***systemic***.
- Because the drugs administered travel throughout the body, even ***metastatic cells*** found far from the tumor will receive treatment.
- For this reason, ***adjuvant chemotherapy*** is often administered in cases where cancers are ***surgically removed*** in order to minimize the risk that any few remaining cancerous cells grow into new tumors.
- ***The main drawback*** of chemotherapy is also due to the fact that it is systemic. Because the drugs target cells undergoing cell division, even healthy cells undergoing cell division will be caught in the crossfire.

Slide 9.8 Chemotherapy (continued)

3.1.2 Nanophototherapies for Cancer Treatment

From Chap. 2 it can be seen that there are many possible triggers leading to the creation of cancerous cells, and that individual cancer cells have the ability to develop into new tumors, making cancer a difficult disease to treat (Slides 9.9 and 9.10).

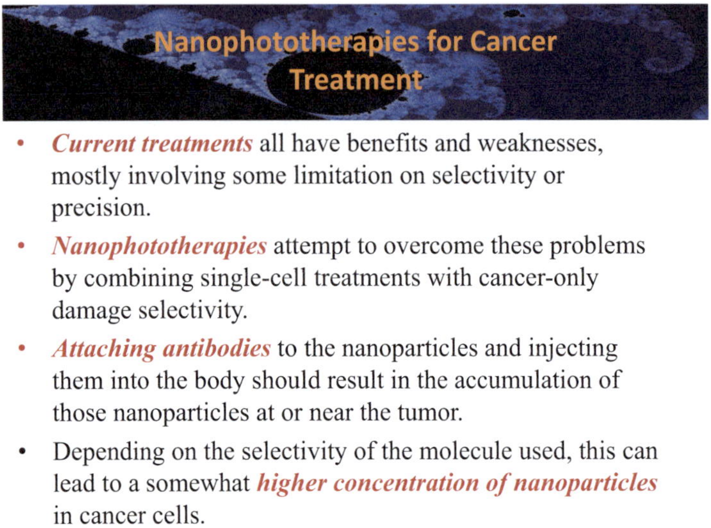

Slide 9.9 Nanophototherapies for cancer treatment

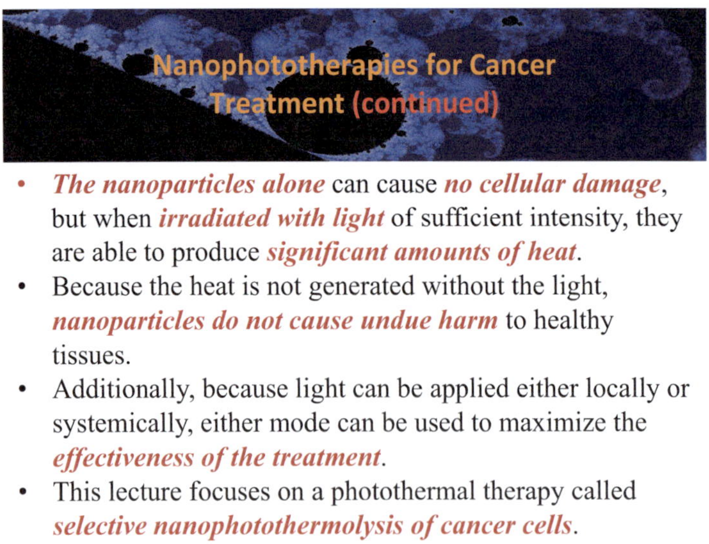

Slide 9.10 Nanophototherapies for cancer treatment (continued)

Current treatments all have benefits and weaknesses, mostly involving some limitation on selectivity or precision. Nanophototherapies attempt to overcome these problems by combining single-cell treatments with cancer-only damage selectivity (Slide 9.9).

As stated in Lecture 8, cancer cells have damaged DNA which causes them to produce flawed proteins, leading to an imbalance of other chemical signals. In some cases, these proteins can be recognized and bound to other proteins. In other cases, the imbalance of chemical signals results in strong demand for certain molecules in the cancerous region. In either case, attaching those proteins (antibodies) or molecules to small nanoparticles and injecting those nanoparticles into the body should result in the accumulation of those nanoparticles at or near the tumor (Slide 9.9). Depending on the selectivity of the molecule used, this can lead to a somewhat higher concentration of nanoparticles in cancer cells, or this could deposit nanoparticles only into cancer cells.

The nanoparticles alone can cause no cellular damage, but when irradiated with light of sufficient intensity, they are able to produce significant amounts of heat (Slide 9.10). Because the heat is not generated without the light, nanoparticles do not cause undue harm to healthy tissues. Additionally, because light can be applied either locally or systemically, either mode can be used to maximize the effectiveness of the treatment. This lecture focuses on a photothermal therapy called selective nanophotothermolysis of cancer cells.

3.1.3 *Selective Nanophotothermolysis of Cancer Cells*

Definition. Nanophotothermolysis is a selective thermal killing of cancer cells targeted by the absorbing nanoparticles heated by the radiation (Slide 9.11). The ultimate goal of this technology is to kill off cancer cells by means of the

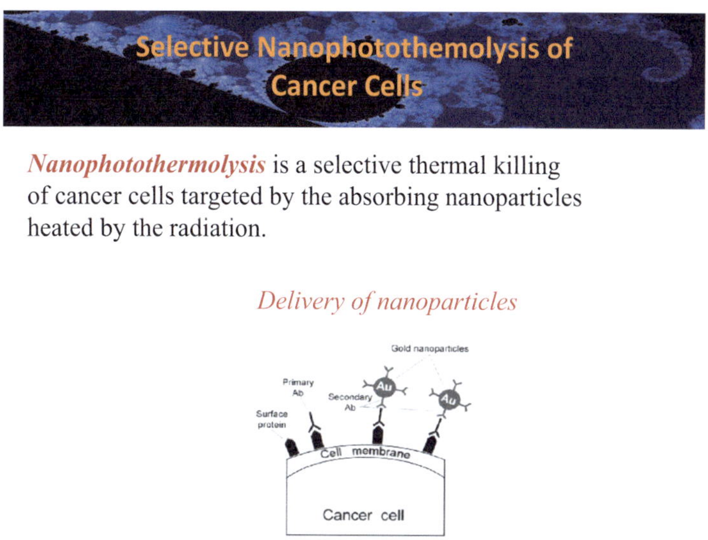

Slide 9.11 Selective nanophotothermolysis of cancer cells

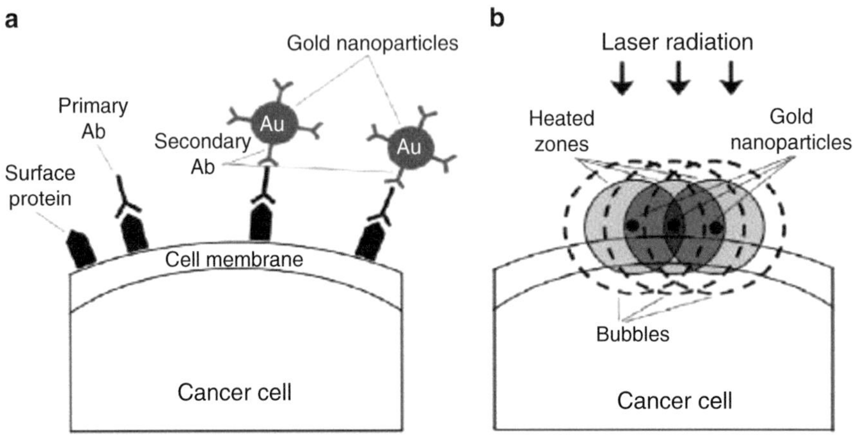

Fig. 3.1 Principle of selective nanophotothermolysis: (**a**) cancer cell targeted by Au nanoparticles using primary and secondary antibodies, (**b**) radiation heating of nanoparticles. Reprinted from [11]. © With permission from IOP Publishing

nanoparticles and radiation. How does it work? First, we have to deliver the nanoparticles and selectively attach them to the cancer cells only and accumulate them at in the tumor schematically shown on the Fig. 3.1a and diagram in Slide 9.11. As we discussed in Lecture 7, by coating the surface of nanoparticles with a specific protein (a *targeting agent*, normally an antibody), the nanoparticles will bind to a complementary protein such as found on a cancer cell [2], as shown in the slide.

Methods of nanoparticle administration. The administration of nanoparticles into the body can be done by using the different techniques such as direct microinjection, transdermal delivery, physiological transport, antibody conjugation, etc. (Slides 9.12, 9.13, 9.14, and 9.15).

In the *direct microinjection technique*, the nanoparticles are injected directly to the tumor through a fine needle (Slide 9.12). This technique is a simple and straightforward procedure. It works if the location of the tumor is known.

In the *transdermal delivery technique*, the nanoparticles are placed inside a film that has an internal teeth-like structure (Slide 9.13). Then this film with nanoparticles can be attached to the skin of cancer patients. The teeth penetrate deeply into the skin, delivering the nanoparticles. This technique can be applied to treating skin cancer.

If the cancer is hiding inside the body at unknown locations, we use another technique for administrating the nanoparticles called the *physiological transport technique*, whereby the nanoparticles are administrated intravenously (Slide 9.14). The size of the nanoparticles is small enough to swim through blood vessels and mirocapillaries, penetrating into any organ and even through the cell membranes to be collected inside the cells.

In order to find a tumor with selective attachment to the cancer cells only, the physiological transport technique must be supplied by the methods of *antibody*

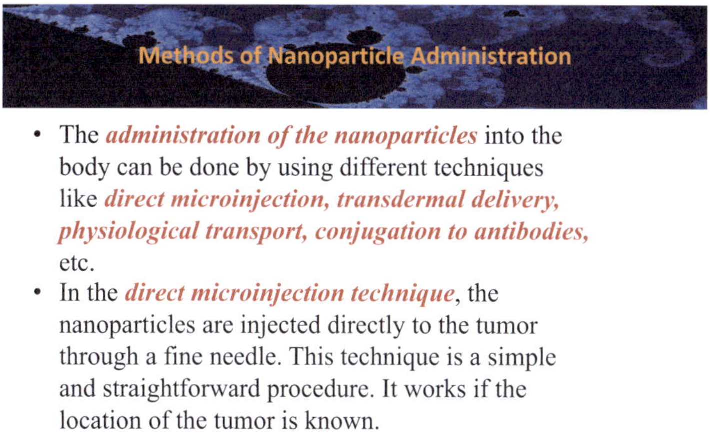

Slide 9.12 Methods of nanoparticle administration

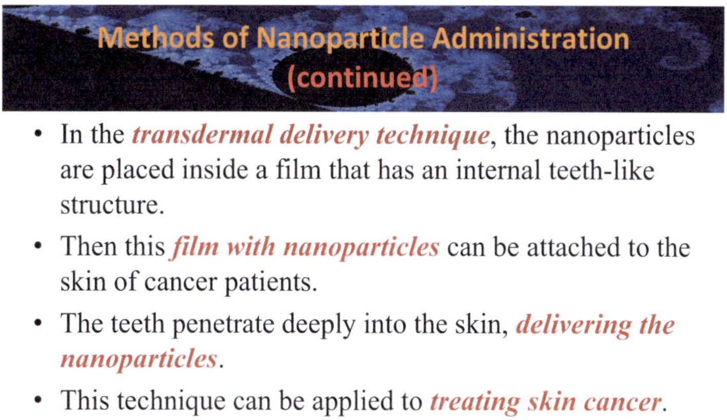

Slide 9.13 Methods of nanoparticle administration (continued)

conjugation to the nanoparticles, whereby the nanoparticles are covered by certain proteins (antibodies), which are designed to stick to the receptors of cancer cells only (Slide 9.15). This type of approach is discussed in detail in Lecture 7. The physiological transport technique supplied by the antibody conjugation method is a powerful and precise way for delivering nanodrugs to the tumor site, as proven in many experimental studies.

Cell Ablation. After the nanoparticles are bound to the cancerous cells, they can be heated with electromagnetic radiation (UV/visible/IR, RF, or X-ray pulses), inducing a variety of effects around the particles [1] as schematically shown in Fig. 3.1b and in Slide 9.16. The heated particle can cause the cell to experience hyperthermia, thermal ablation, bubble formation around the heated nanoparticle,

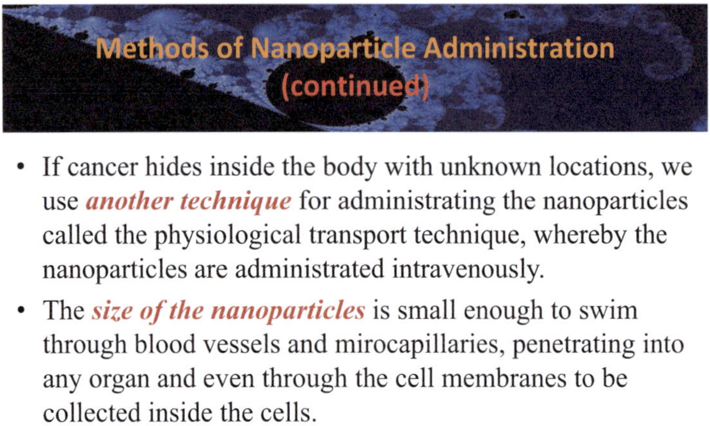

Slide 9.14 Methods of nanoparticle administration (continued)

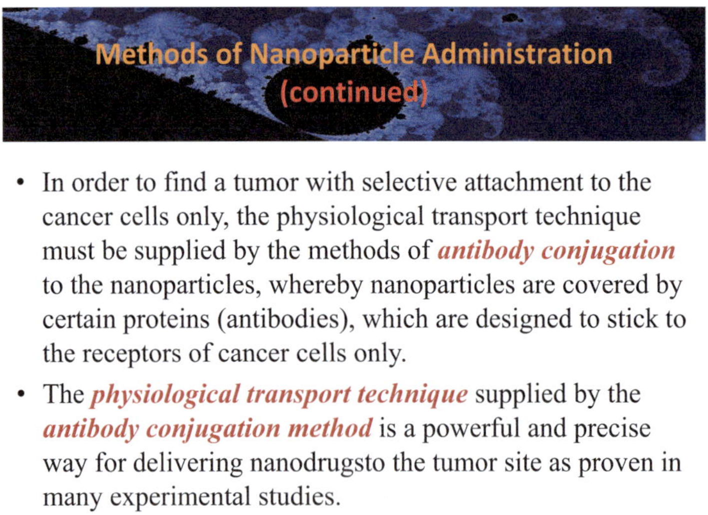

Slide 9.15 Methods of nanoparticle administration (continued)

evaporation of the surrounding medium, and sound generation. Alternatively, the nanoparticle itself can heat to the point of melting, evaporation, or explosion, causing further damage to cells [2]. These effects can be used to increase the sensitivity of photoacoustic diagnosis or aid in therapy by selective thermal killing of tumor cells into which absorbing nanoparticles have been incorporated [2].

Physical Concepts. Progress toward the development of the selective nanophotothermolysis technology requires the investigation of new physical concepts and new approaches to the study of radiation interactions with

Selective Nanophotothemolysis
Cell Ablation

Nanophotothermolysis is a selective thermal killing of cancer cells targeted by the absorbing nanoparticles heated by the radiation.

Delivery of nanoparticles

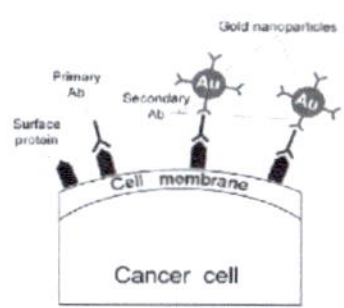

Radiation heating

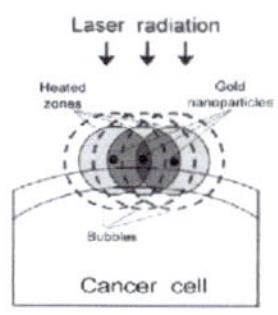

Cell killing effects:

- *thermal ablation*
- *bubbles formation*
- *sound generation*
- *shock waves*
- *explosive evaporation*

Slide 9.16 Selective nanophotothermolysis

Selective Nanophotothemolysis
Physical Concepts

- ***The theory*** of selective nanophotothermolysis includes many different sciences and topics like:
  - ***Nanoparticle optics***
  - ***Heat-mass transfer***
  - ***Cell ablation***
  - ***Microbubble dynamics***
  - ***Cluster aggregation***
  - ***Thermal explosion***
  - ***Shock wave generation***
  - ***Sound generation***
- Thus, the selective nanophotothermolysis is an ***interdisciplinary science*** which involves many topics from physics, biology and chemistry.

Slide 9.17 Selective nanophotothermolysis: physical concepts

nanostructures in biological systems [7] (Slide 9.17). The theory of selective nanophotothermolysis involves many different topics such as

- Nanoparticle optics
- Heat-mass transfer
- Cell ablation
- Microbubble dynamics

- Cluster aggregation
- Thermal explosion
- Shock wave generation
- Sound generation

We see that selective nanophotothermolysis is an interdisciplinary science which cuts across physics, biology, and chemistry. Some of the theoretical concepts listed earlier will be discussed in detail in later chapters.

3.1.4 New Dynamic Modes in Selective Nanophotothermolysis

Based on these concepts, we are able to predict the following new dynamic modes in selective nanophotothermolysis [1–4] (Slide 9.18):

- *Cluster Aggregation Mode*
- *Nanobubble Overlapping Mode*
- *Thermal Explosion Mode*

The *cluster aggregation mode* is realized when the incubation time is sufficiently long to allow a large number of nanoparticles to be collected at the tumor site (Slide 9.19). The nanoparticles collide with each other, stick together, and form a cluster shown in Fig. 3.2, which is a term used to describe an agglomeration of nanoparticles located closely together. The clusters are characterized by larger size and higher absorption efficiency than a single nanoparticle. The experimentally observed clusters have a size range 200–400 nm made of 10–30 nm nanoparticles. Because of this, the clusters provide a larger damage area in cell volume at a lower

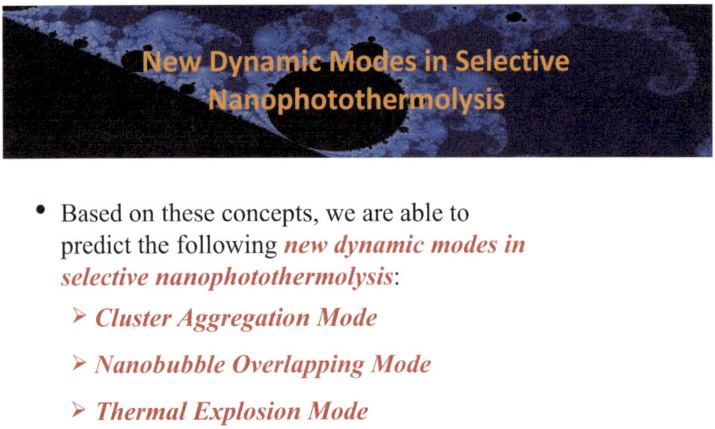

Slide 9.18 New dynamic modes in selective nanophotothermolysis

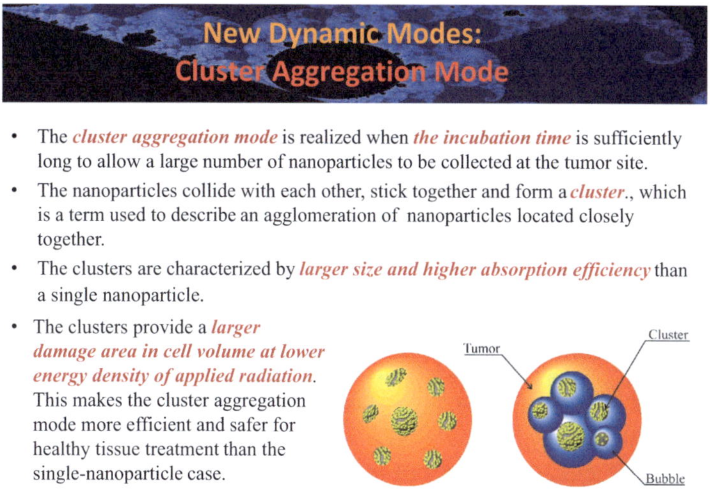

Slide 9.19 New dynamic modes: cluster aggregation mode

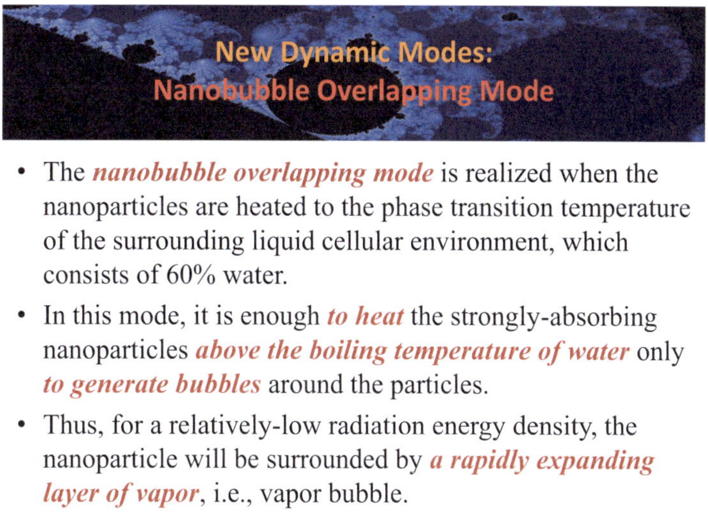

Slide 9.20 New dynamic modes: nanobubble overlapping mode

energy density of applied radiation. This makes the cluster aggregation mode more efficient and safer for healthy tissue treatment than the single-nanoparticle case.

The *nanobubble overlapping mode* is realized when the nanoparticles are heated to the phase transition temperature of the surrounding liquid cellular environment, which consists of 60 % water (Slides 9.20 and 9.21). In this mode, it is enough to heat the strongly absorbing nanoparticles above the boiling temperature of water only, or above the critical temperature of the phase transition in water, to generate

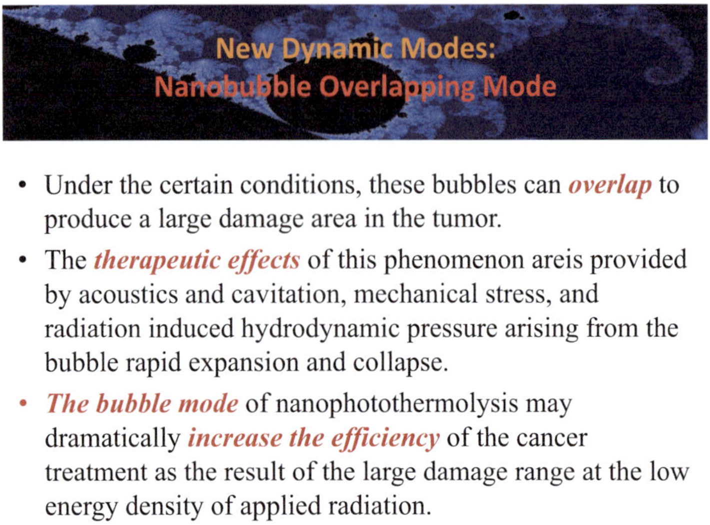

Slide 9.21 New dynamic modes: nanobubble overlapping mode (continued)

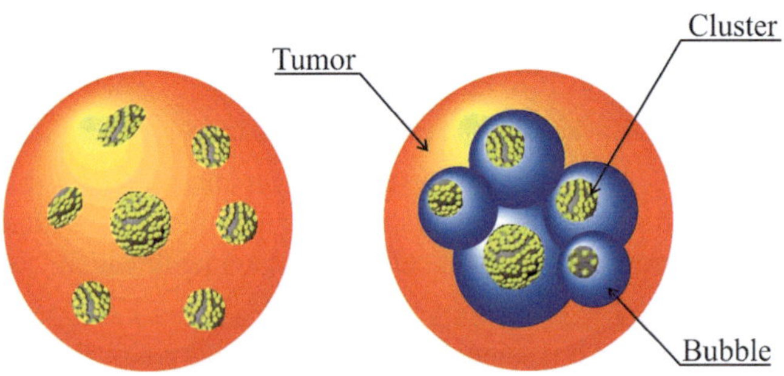

Fig. 3.2 Cluster aggregation and overlapping bubbles from nanoclusters inside a tumor. Reprinted from [3]. © With permission from World Scientific

bubbles around the particles or clusters [3] (Slide 9.20). Thus, for a relatively low radiation energy density, the nanoparticle will be surrounded by a rapidly expanding layer of vapor, i.e., vapor bubble. Under certain conditions to be discussed later in this book, these bubbles can overlap to produce a large damage area in the tumor schematically shown in Fig. 3.2 (Slide 9.21). The therapeutic effects of this phenomenon are provided by acoustics and cavitation, mechanical stress, and radiation-induced hydrodynamic pressure arising from the bubble rapid expansion and collapse. The bubble mode of nanophotothermolysis may dramatically increase the efficiency of the cancer treatment as the result of the large damage range at the low energy density of applied radiation [3].

The *thermal explosion mode* is realized for ultrashort pulse heating of the nanoparticles, when the heat is generated within a strongly absorbing target

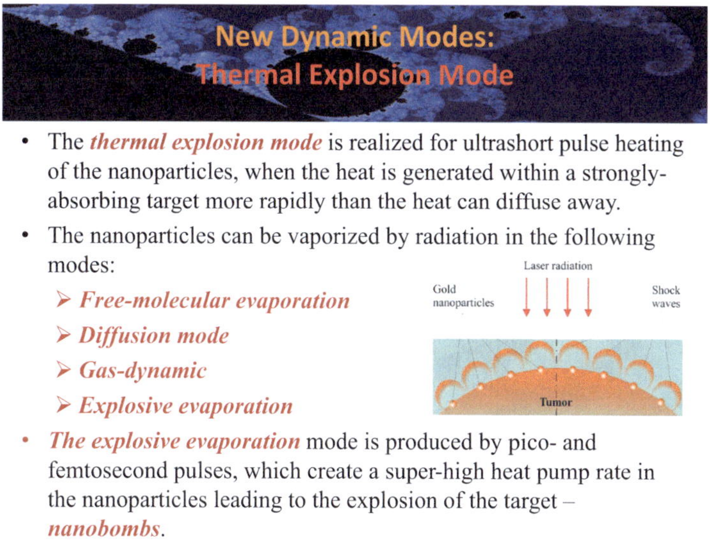

Slide 9.22 New dynamic modes: thermal explosion mode

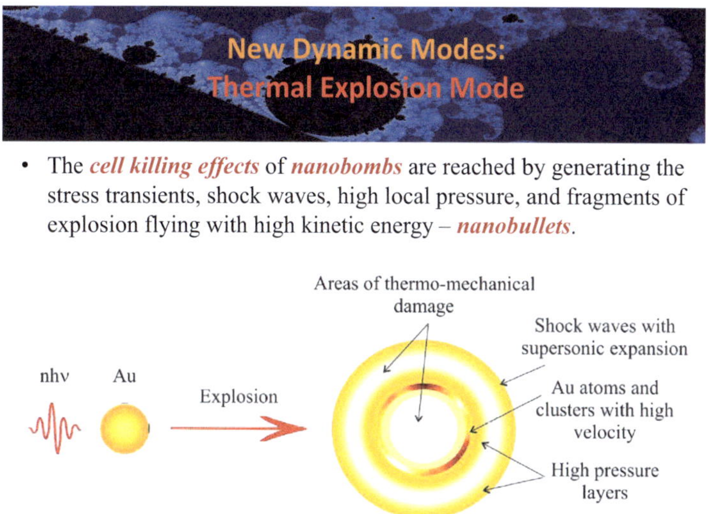

Slide 9.23 New dynamic modes: thermal explosion mode (continued)

more rapidly than the heat can diffuse away [3] (Slides 9.22, 9.23 and 9.24). The nanoparticles can be vaporized by radiation in the following modes:

- Free-molecular evaporation
- Diffusion
- Gas dynamic
- Explosive evaporation

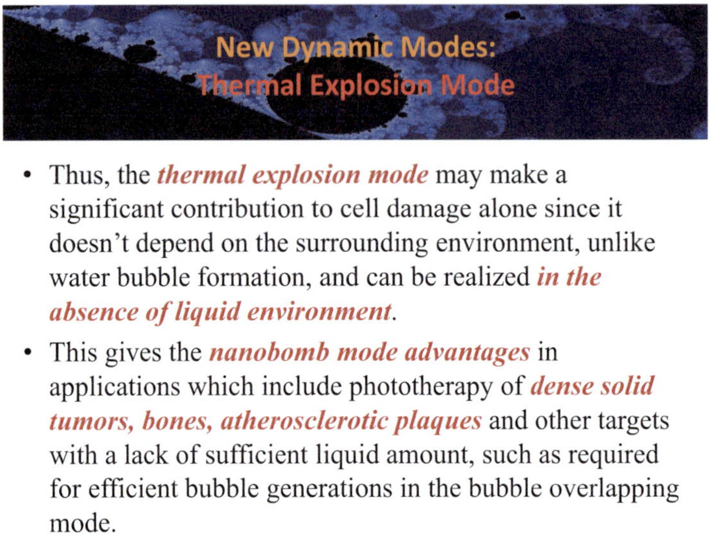

Slide 9.24 New dynamic modes: thermal explosion mode (continued)

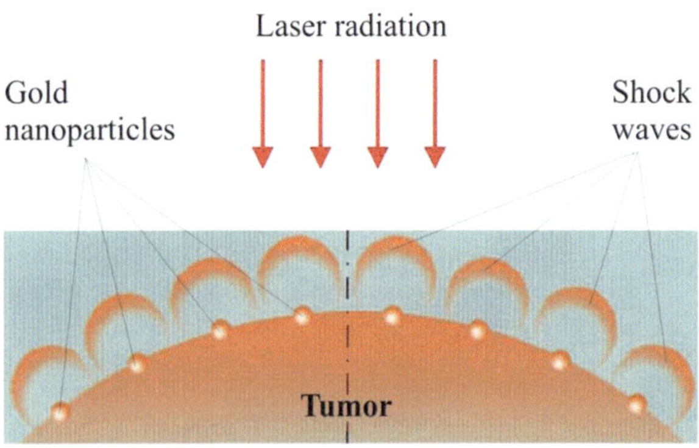

Fig. 3.3 Laser-induced explosion of gold nanoparticles. Reprinted from [3]. © With permission from World Scientific

The explosive evaporation mode is produced by pico- and femtosecond pulses, which create a super-high heat pump rate in the nanoparticles, leading to the explosion of the target—*nanobombs* shown in Fig. 3.3 (Slide 9.22). The cell killing effects of nanobombs are reached by generating the stress transients, shock waves, high local pressure, and fragments of explosion flying with high kinetic energy—*nanobullets*. A schematic picture of these complex physical effects is provided in Fig. 3.4 and Slide 9.23. Thus, the thermal explosion mode may make a significant contribution to cell damage alone since it doesn't depend on the surrounding environment, unlike water bubble formation, and can be realized in the absence of a liquid environment (Slide 9.24). This gives the nanobomb mode advantages in applications which include phototherapy of dense solid tumors, bones, atherosclerotic plaques, and other targets with a lack of sufficient liquid, such as required for efficient bubble generation in the bubble overlapping mode.

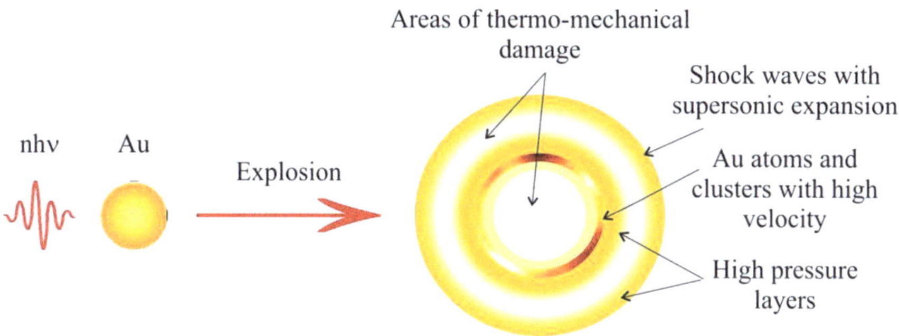

Fig. 3.4 A schematic picture of the complex physical effects accompanying the laser-induced explosion of Au nanoparticle. Reprinted from [3]. © With permission from World Scientific

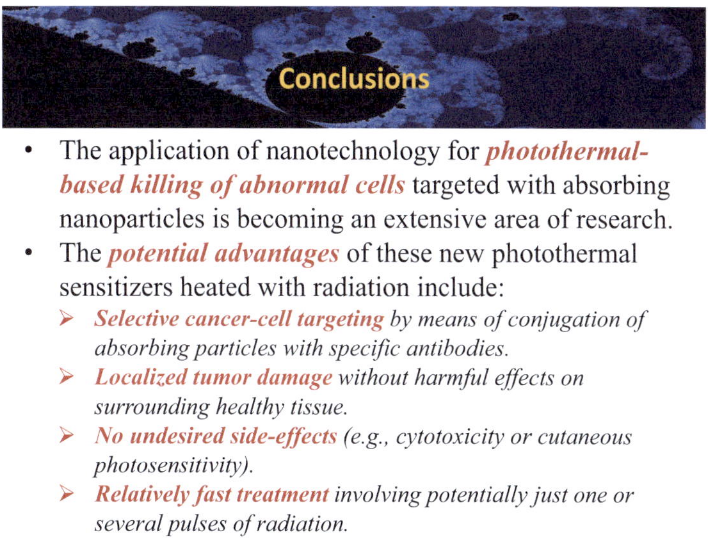

Slide 9.25 Conclusions

In conclusion, the application of nanotechnology for photothermal-based killing of abnormal cells (e.g., cancer cells) targeted with absorbing nanoparticles (e.g., gold solid nanospheres, nanoshells, or nanorods) is becoming an extensive area of research (Slide 9.25). The potential advantages of these new photothermal sensitizers heated with radiation include [3]

- Selective cancer cell targeting by means of conjugation of absorbing particles with specific antibodies.
- Localized tumor damage without harmful effects on surrounding healthy tissue.
- No undesired side effects (e.g., cytotoxicity or cutaneous photosensitivity).
- Relatively fast treatment involving potentially just one or several pulses of radiation.

3.2 Lecture 10: Nanodrug Design and Methods of Activation (Slide 10.1)

- Nanodrug Design
- Methods of Activation
  - *Ideal Radiation*
  - *X-Ray Activation*
  - *Activation in the Optical Range of the Spectrum*
  - *Microwave Activation*
  - *Activation by Magnetic Fields and Ultrasound Waves*
  - *RF Wave Activation*
- Conclusions
- References

3.2.1 Nanodrug Design

Nanotechnology has established its potential to significantly improve the performance characteristics of today's imaging and treatment modalities, and it offers a variety of new approaches for selective treatment and targeted nanodrug delivery [6] (Slides 10.2, 10.3, and 10.4). Nanodrugs/nanoparticles, selectively delivered to the tumor site, can be activated by radiation for drug release, or nanoparticles themselves can be used as a drug by producing biological damage through thermal,

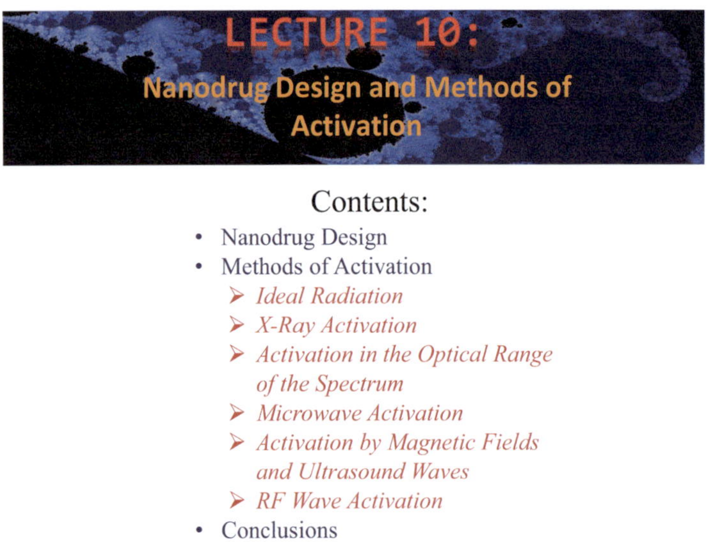

Slide 10.1 Contents of Lecture 10

Nanodrug Design

- ***Nanodrugs/nanoparticles***, selectively delivered to the tumor site, can ***be activated by radiation*** for drug release, or nanoparticles themselves can be used as a drug by producing biological damage through thermal, mechanical ablations or charged particle emission.
- ***By nanodrug activation***, we mean ***heating of nanoparticles, photoelectron emission, Compton scattering, explosion of nanoparticles, generation of shock waves, sound waves, mechanical waves of high pressure, nanobubble formations***, and so on.
- All these phenomena eventually ***kill the cancer cells*** and the tumor.

Slide 10.2 Nanodrug design

Nanodrug Design (continued)

- **Nanodrug design** includes the physical properties like material, optical and thermal, and morphological properties (shape, size and structure) of nanoparticles, which depend on the ***methods of nanodrug activation***.
- ***By methods of activation***, we mean using different types of ***radiation*** for a total body scan to activate the nanoparticles and reach any place inside the body.
- For example:
  - ➢ If activation is based on radiation ***in the optical range of the spectrum***, the best candidates for nanodrug design are ***plasmonic nanoparticles*** of different shapes, sizes and structures made of noble metals like ***Au, Ag and Pt***. Also, ***silica, glass and polymer nanoparticles*** can be used for the optical range of the spectrum activation because of their relatively high absorption in that range and the ability of silica shells to be tuned to the desired wavelength of radiation.

Slide 10.3 Nanodrug design (continued)

mechanical ablations, or charged particle emission (Slide 10.2). By *nanodrug activation*, we mean heating of nanoparticles, photoelectron emission, Compton scattering, explosion of nanoparticles, generation of shock waves, sound waves, mechanical waves of high pressure, nanobubble formation, and so on. All these phenomena eventually kill the cancer cells and the tumor [6].

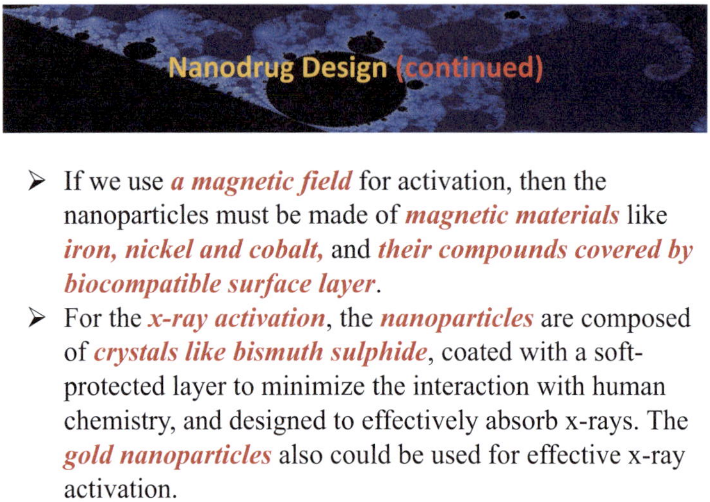

Slide 10.4 Nanodrug design (continued)

Nanodrug design includes the physical properties such as material, optical and thermal, and morphological properties (shape, size, and structure) of nanoparticles, which depend on the *methods of nanodrug activation* (Slide 10.3). By *methods of activation*, we mean using different types of radiation for a total body scan to activate the nanoparticles and reach any place inside the body. For example: [6]

- If activation is based on radiation in the optical range of the spectrum, the best candidates for nanodrug design are plasmonic nanoparticles of different shapes, sizes, and structures made of noble metals such as Au, Ag, and Pt [1]. Also, silica, glass, and polymer nanoparticles can be used for the optical range of the spectrum activation because of their relatively high absorption in that range and the ability of silica shells to be tuned to the desired wavelength of radiation (Slide 10.3).
- If we use the a magnetic field for activation, then the nanoparticles must be made of magnetic materials such as iron, nickel, and cobalt, and their compounds covered by a biocompatible surface layer (Slide 10.4).
- For X-ray activation, the nanoparticles are composed of crystals like bismuth sulfide, coated with a soft-protected layer to minimize the interaction with human chemistry, and designed to effectively absorb X-rays. The gold nanoparticles also could be used for effective X-ray activation [5] (Slide 10.4).

3.2.2 Methods of Activation

The radiation is characterized by a wide range of the electromagnetic spectrum, starting with the very short wavelengths of gamma and X-ray radiation, and ending with the very long wavelengths of radiofrequency (RF) radiation (Slide 10.5). The *ideal radiation* for nanoparticle activation would be one which has [6]

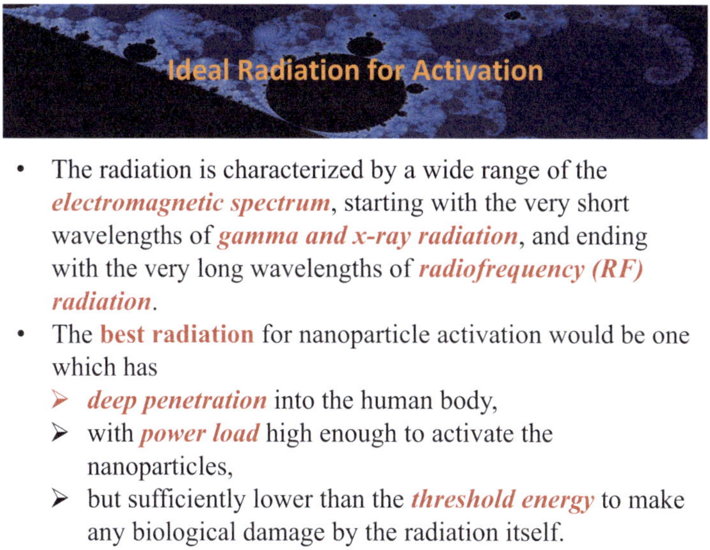

Slide 10.5 Ideal radiation for activation

- Deep penetration into the human body,
- With power load high enough to activate the nanoparticles,
- But sufficiently lower than the threshold energy to make any biological damage by the radiation itself.

Let us briefly examine the electromagnetic spectrum in order to find the optimal radiation for activation of nanoparticles inside the body.

X-ray Activation. Short electromagnetic waves like *X-rays* have good penetration depth, 10–25 cm depending on the exposure dose in both soft and bone tissues, and they have high-energy photons, enough to activate the nanoparticles [5] (Slide 10.6). The intravenously administrated nanoparticles into the body are selectively attached to the cancer cells. After a certain incubation time, depending on the size of nanoparticles, the X-rays can be applied for a total body scan by using a machine similar to a computed tomography (CT) scanner. The mass attenuation coefficient of soft tissue in the X-ray photon energies spectrum is three orders of magnitude less than the mass attenuation coefficient of gold in the same range of photon energies. This allows us to reduce the exposure rate to the level of diagnostic doses, just good enough to activate the nanoparticles without harming the patient.

The *therapeutic effect* of the localized activation of the nanoparticles by X-ray radiation is achieved by the emission of charged particles from the surface of the nanoparticles in photoelectric events, Compton scattering events, and thermionic emission, or heating the nanoparticles to temperatures high enough for cell ablation (Slide 10.7). The charged particles emitted from the surface of nanoparticles, selectively delivered to the cancer cells, destroy the DNA of the cancer cells through *direct* and *indirect* actions. DNA is the most sensitive element of the cell.

Methods of Activation
X-Ray Activation

- ***X-Ray Activation***. Short electromagnetic waves like *x-rays* have ***good penetration depth***, 10-25 cm depending on the exposure dose in both soft and bone tissues, and they have ***high-energy photons***, enough to activate the nanoparticles.
- The ***intravenously administrated*** nanoparticles into the body are ***selectively attached*** to the cancer cells.
- After a certain ***incubation time***, depending on the size of nanoparticles, the x-rays can be applied for ***a total body scan*** by using a machine similar to a computed tomography (CT) scanner.
- The ***mass attenuation coefficient*** of soft tissue in the x-ray photon energies spectrum is ***three orders of magnitude*** less than the mass attenuation coefficient of gold in the same range of photon energies.
- This allows us to ***reduce the exposure rate*** to the level of diagnostic doses, just good enough to activate the nanoparticles without harming the patient.

Slide 10.6 Methods of activation: X-ray activation

Methods of Activation
X-Ray Activation (continued)

- The ***therapeutic effect*** of the localized activation of the nanoparticles ***by x-ray radiation*** is achieved by ***emission of the charged particles from the surface of the nanoparticles in photoelectric events, Compton scattering events and thermionic emission, or heating the nanoparticles*** to the temperatures high-enough for cell ablation.
- The ***charged particles*** emitted from the surface of nanoparticles destroy the DNA of the cancer cells through ***direct and indirect actions***.
- In ***direct action***, the charged particles destroy the DNA directly.
- The ***indirect action*** means that the charged particles react with water contained in the cell medium, producing hydroxyl radicals which are toxic for DNA of cancer cells.
- ***X-ray would seem to be good*** for a total body scan to activate the nanoparticles inside the body.

Slide 10.7 Methods of activation: X-ray activation (continued)

In *direct action*, the charged particles destroy the DNA directly. The *indirect action* means that the charged particles react with water contained in the cell medium, producing hydroxyl radicals which are toxic for DNA of cancer cells. X-ray radiation would seem to be good for a total body scan to activate the nanoparticles inside the body.

Methods of Activation
Activation in the Optical Range of the Spectrum

- ***Activation in the Optical Range of the Spectrum***. The *optical range of the spectrum* includes ***short ultraviolet (UV) waves, visible light and infrared (IR) radiation***.
- A spectrum range from ***400 to 700 nm*** is called ***visible light***.
- Different types of ***lasers*** are reliable sources of coherent radiation in the optical range.
- Today, ***lasers*** are covering the ***whole optical spectrum*** from short UV up to long IR radiation.
- However, the ***penetration depth*** for most lasers in soft tissue is small and limited by the absorption of water.

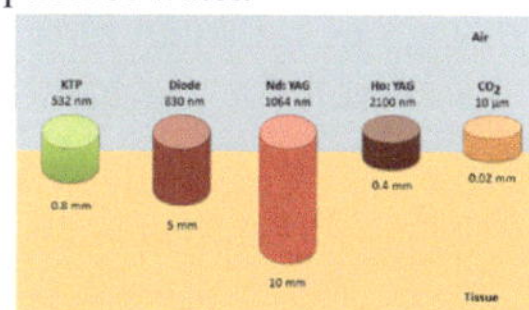

Slide 10.8 Methods of activation: activation in optical range of the spectrum

Methods of Activation
Activation in the Optical Range of the Spectrum (continued)

- As can be seen from the diagram, only the ***Nd:YAG laser*** with wavelength 1.06 μm has ***the deepest penetration of about 10 mm***, corresponding to the ***window of transparency of water***.
- Because of low penetration depth, researchers in laser physics use ***optical fibers*** to deliver optical light directly to the tumor.
- A ***single-mode optical fiber*** has a diameter of about 1 mm and is able to deliver light of high energy, enough to ablate the tumor, to almost any place inside the body.
- ***Laser surgery*** is widely used in clinics nowadays.
- However, because we have to know the exact location of the tumor in advance in order to deliver light to the right place, ***laser technology is not a universal technique*** of activating nanoparticles inside the body, since tumor or cancer cells can hide anywhere in the body.

Slide 10.9 Methods of activation: activation in optical range of the spectrum (continued)

Activation in the Optical Range of the Spectrum. The *optical range of spectrum* includes short ultraviolet (UV) waves, visible light, and infrared (IR) radiation (Slides 10.8 and 10.9). A spectrum range from 400 to 700 nm is called visible light. Different types of lasers are reliable sources of coherent radiation in the optical range. Today, lasers cover the whole optical spectrum from short UV up to long IR radiation.

However, the penetration depth for most lasers in soft tissue is small and limited by the absorption of water. Slide 10.8 demonstrates penetration depths for different lasers in the optical spectral range. As can be seen from this slide, only the Nd:YAG laser with wavelength 1.06 μm has the deepest penetration of about 10 mm, corresponding to the window of transparency of water (main component of soft tissue). Because of low penetration depth, researchers in laser physics use optical fibers to deliver optical light directly to the tumor inside the body (Slide 10.9) [6]. A single-mode optical fiber has a diameter of about 1 mm and is able to deliver light of high energy, enough to ablate the tumor, to almost any place inside the body. Laser surgery is widely used in clinics nowadays. However, because we have to know the exact location of the tumor in advance in order to deliver light to the right place, laser technology is not a universal technique of activating nanoparticles inside the body, since tumor or cancer cells can hide anywhere in the body.

Microwave Activation. If we search for the ideal radiation outside the optical spectral range, the longer waves in the *microwave range* also have good penetration depth and high power load that easily reaches high ablation temperatures in the tumor (Slide 10.10) [6]. But the problem with microwaves is water, the main content of the cells. The water absorbs a lot of microwave radiation, and a total body scan with high-energy microwaves will "cook" the patient. Because of this, researchers have to use antennas placed inside the tumor to deliver a controlled amount of microwave energy to limited places at the tumor site only. This technology works well for ablating large tumors with given locations, but not as a universal tool for a total body scan that is able to reach cancer at any unknown place [6].

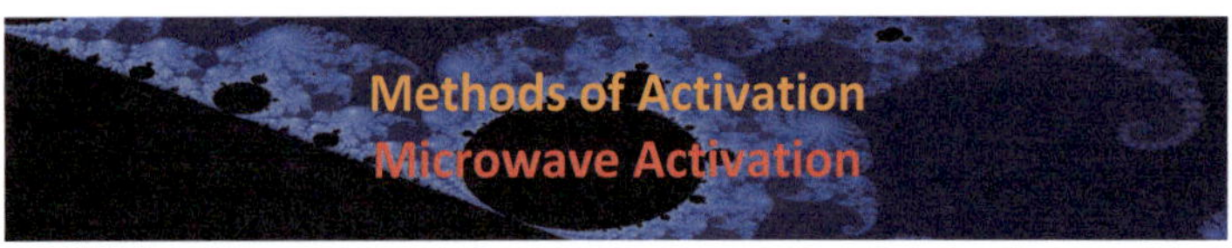

- If we search for the ideal or best radiation outside the optical spectral range, the longer waves in the ***microwave range*** also have ***good penetration depth*** and ***high power load*** that easily reaches high ablation temperatures in the tumor.
- But the problem with microwaves is ***water***, the main content of the cells.
- The water absorbs a lot of microwaves, and ***a total body scan*** with high-energy microwaves will "***cook***" the patient.
- Because of this, researchers have to use ***antennas*** placed inside the tumor to deliver a controlled amount of microwave energy to limited places at the tumor site only.
- This technology works well ***for ablating large tumors*** with given locations, but ***not as a universal tool*** for a total body scan that is able to reach cancer at any unknown place.

Slide 10.10 Methods of activation: microwave activation

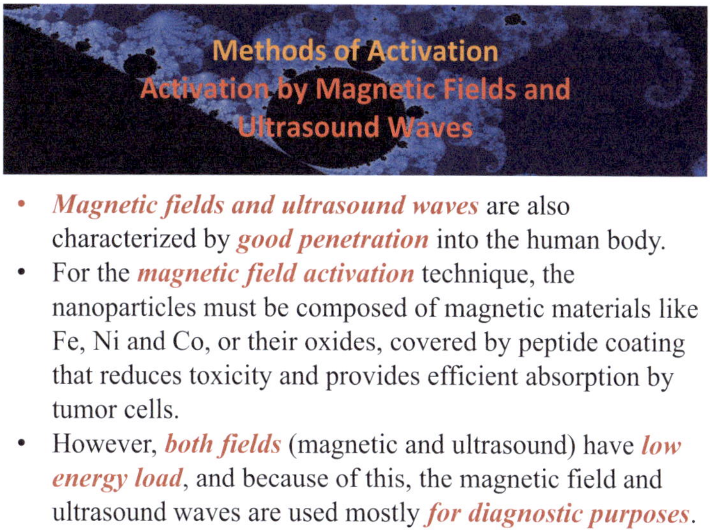

Slide 10.11 Methods of activation: activation by magnetic fields and ultrasound waves

Activation by Magnetic Fields and Ultrasound Waves. Magnetic fields and ultrasound waves are also characterized by good penetration into the human body (Slide 10.11). For the magnetic field activation technique, the nanoparticles must be composed of magnetic materials such as Fe, Ni, and Co, or their oxides, covered by peptide coating that reduces toxicity and provides efficient absorption by tumor cells [6]. However, both fields (magnetic and ultrasound) have low energy load, and because of this, the magnetic field and ultrasound waves are used mostly for diagnostic purposes.

At this point, we have briefly analyzed short X-rays, the optical spectral range, microwaves, magnetic fields, and ultrasound waves in search for a universal tool for nanoparticle activation inside the body (Slide 10.12). Each method has benefits as well as limitations [6]. For example, X-ray radiation is a good candidate for a total body scan to activate the nanoparticles inside the body. However, X-rays are toxic to the body and can cause a secondary cancer via direct and indirect DNA destruction in normal cells. The laser and microwave technologies work well at limited, well-known locations of tumor sites, so that these methods cannot be considered as a universal tool for activation of nanoparticles. Magnetic and ultrasound waves can easily penetrate the body but are too weak to cause any biological damage or heat the nanoparticles to the ablation temperatures.

RF Wave Activation. Let us continue searching for the best radiation for nanoparticle activation at longer wavelengths of the electromagnetic spectrum, i.e., *RF waves* (Slide 10.13). The human body, consisting of 60 % water, has negligible absorption of RF waves with penetration depth of 20–200 cm for 13.56 MHz wave frequency. The RF waves are weak and do not make any biological damage themselves. A therapeutic effect of RF waves can be reached by combining the

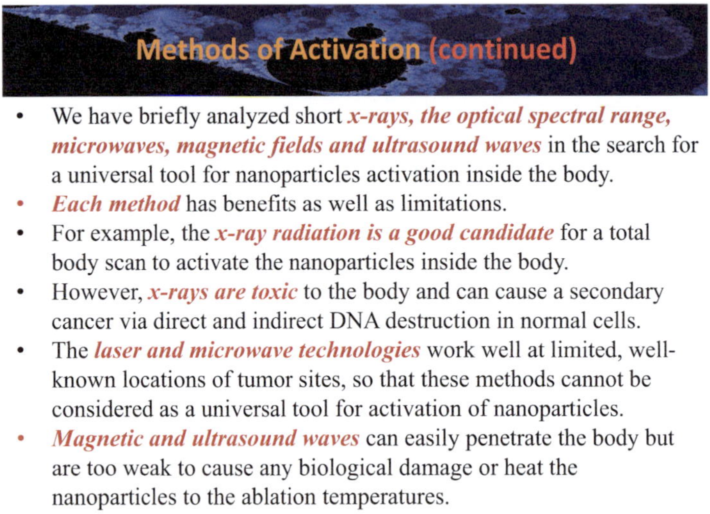

Slide 10.12 Methods of activation (continued)

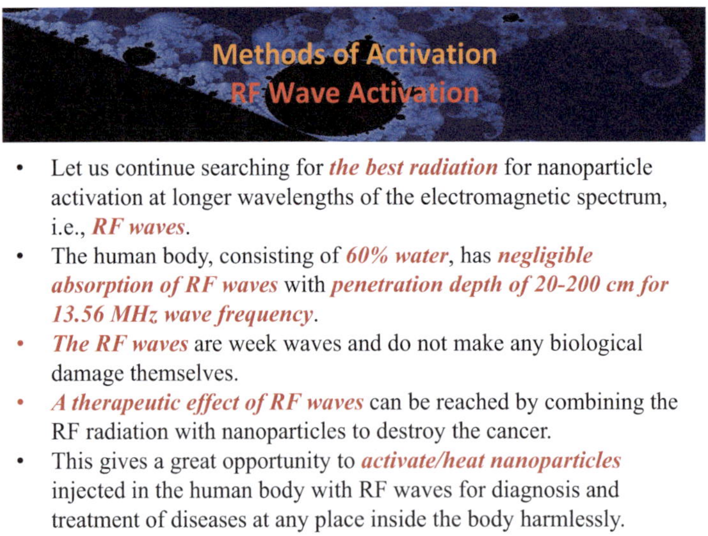

Slide 10.13 Methods of activation: RF wave activation

RF radiation with nanoparticles to destroy the cancer [6]. This gives a great opportunity to activate/heat nanoparticles injected in the human body with RF waves for diagnosis and treatment of diseases at any place inside the body harmlessly.

In conclusion, the use of nanoparticles can dramatically improve performance characteristics of today's imaging and treatment modalities such as CT scanners,

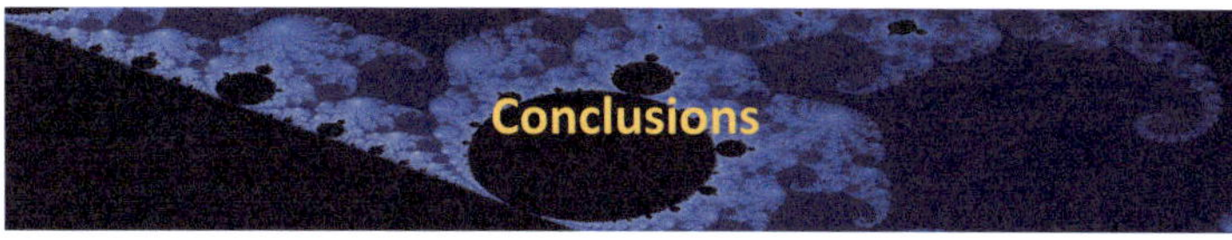

- The ***use of nanoparticles*** can dramatically improve performance characteristics of today's imaging and treatment modalities like CT scanners, positron emission tomography (PET) scanners, magnetic resonance imaging (MRI) and x-ray machines, and offers new approaches and unique technologies for selective treatment and targeted nanodrug delivery.
- ***Selective targeted delivery*** is reached through the conjugation of nanodrugs to antibodies designed to chemically react with the receptors of abnormal cells only.
- A nanodrug, selectively delivered to the tumor site, can then be ***activated by radiation*** for strong drug release, or nanoparticles can be used as a drug themselves by producing biological damage through thermal, mechanical ablations or charged particle emission.

Slide 10.14 Conclusions

- ***The nanodrug design*** depends on the ***methods of nanodrug activation***.
- ***The methods of activation*** mean using different types of radiation for a total body scan.
- ***Under activation of nanoparticles***, we mean heating them to initiate cell ablation, photoelectron emission, thermoionic emission, Compton scattering, generation of shock waves, sound waves, or mechanical waves of high pressure that are caused by the applied radiation.
- ***The ideal radiation*** for nanophotodynamic therapy involves deep penetration of the human body as well as enough power to activate the nanoparticles without harming the patient.
- The search for a universal tool for nanoparticles activation shows that very ends of the electromagnetic spectrum: ***x-rays*** on one side and ***RF waves*** on another end, ***are the best candidates*** for ideal radiation for a total body scan to activate the nanoparticles.

Slide 10.15 Conclusions (continued)

positron emission tomography (PET) scanners, magnetic resonance imaging (MRI), and X-ray machines, and offers new approaches and unique technologies for selective treatment and targeted nanodrug delivery (Slides 10.14 and 10.15). Selective targeted delivery is reached through the conjugation of nanodrugs to antibodies designed to chemically react with the receptors of abnormal cells only. A nanodrug, selectively delivered to the tumor site, can then be activated by radiation for strong drug release, or nanoparticles can be used as a drug themselves by producing biological damage through thermal, mechanical ablations, or charged particle emission (Slide 10.14) [6].

Nanodrug design depends on the *methods of nanodrug activation* (Slide 10.15). The *methods of activation* mean using different types of radiation for a total body scan. Under activation of nanoparticles, we mean heating them to initiate cell ablation, photoelectron emission, thermoionic emission, Compton scattering, generation of shock waves and sound waves, or mechanical waves of high pressure that are caused by the applied radiation. The ideal radiation for nanophotodynamic therapy involves deep penetration of the human body as well as enough power to activate the nanoparticles without harming the patient. The radiation is characterized by a wide range of the electromagnetic spectrum, from the very short wavelengths of gamma and X-rays radiation to the very long wavelengths of RF radiation. The search for a universal tool for nanoparticle activation shows that very ends of the electromagnetic spectrum—X-rays and RF waves—are the best candidates for a total body scan to activate the nanoparticles inside the body.

3.3 Lecture 11: Surface Plasmon Resonance Detection Techniques (Slide 11.1)

- Fiber Optics
  - Fundamentals of Fiber Optics
  - Evanescent Waves
  - Optical Fiber Biosensing
  - Medical Applications of Optical Fibers
- Optical Fiber Sensors Based on Surface Plasmon Resonances (SPRs)
  - Surface Plasmon Resonances and Polaritons
  - Fiber Optic SPR Sensors
  - Examples of Utilizing Optical Fibers as SPR Biosensors
- Conclusions

Previously we briefly introduced the application of plasmonic nanoparticles in the diagnosis and therapy of diseases. A potential area of nanomedicine involves the clinical application of fiber optics (Slide 11.2). Optical fibers are useful for communication, endoscopy, imaging, laser radiation delivery, and as biosensors. The ability to detect and observe areas of cancerous growth or even individual cancer cells will provide a great advantage for doctors or scientists in the fight against cancer.

Fiber optics can help in the war against cancer by closely monitoring various proteins or biomolecules that are present in higher concentrations when cancer

LECTURE 11:
Surface Plasmon Resonance Detection Techniques

Contents:

Slide 11.1 Contents of Lecture 11

Optical Fiber

- A potential area of **nanomedicine** involves ***clinical applications of fiber optics***.
- ***Optical fibers*** are useful for communication, endoscopy, imaging, laser radiation delivery, and as biosensors.
- ***Fiber optics*** can help in the war ***against cancer*** by closely monitoring various proteins or biomolecules that are present in higher concentrations when cancer grows.
- A promising detection technique involves ***surface plasmon resonances (SPRs).***
- ***SPR-coupled optical fibers*** have the capability to detect changes in the external refractive index, which will be a focus of this lecture.

Slide 11.2 Optical fiber

grows (Slide 11.2). Continuous development and miniaturization of applications will be beneficial in early detection of cancer. Different parameters of the fiber can be altered for specializing in direct or indirect monitoring of cancerous growth. Evanescent field sensing with fluorescence-based and label-free-based detection are beneficial to biomolecule detection. A promising detection technique involves surface plasmon resonances (SPRs), and this will be a focus of this section.

SPR-coupled optical fibers have the capability to detect changes in the external refractive index. With the ability to transmit optical signals over long distances and the mechanical flexibility for entering previously inaccessible locations, optical fibers present a very attractive method of sensing for SPR devices.

The following paragraphs will introduce the basics of optical fibers, fiber optic evanescent sensors, surface plasmon resonance biosensors, and protein or biomolecule concentration detection.

3.3.1 Fiber Optics

Fundamentals of Fiber Optics. Before discussing fiber optic biosensors, one should return to fiber optics themselves (Slide 11.3). An optical fiber is a narrow, flexible cylindrical silica glass rod core that is covered with a glass cladding. The fiber is then typically covered with a buffer and protective jacket. Its main purpose is to act as a dielectric waveguide or something that transmits light between its two ends. Since optical fibers can transmit light with low attenuation or signal loss, information and images can be sent rapidly through them. Digital information is encoded by having a laser switch on and off in order to represent bits. The optical fiber simply serves as a transmission device; one end has a transmitter while the other end has a receiver in order to decode the signal. Optical fibers offer several advantages such as low signal degradation, high speed and capacity for information transmission, and flexibility. Current applications include communication, sensing, power transmission, and imaging.

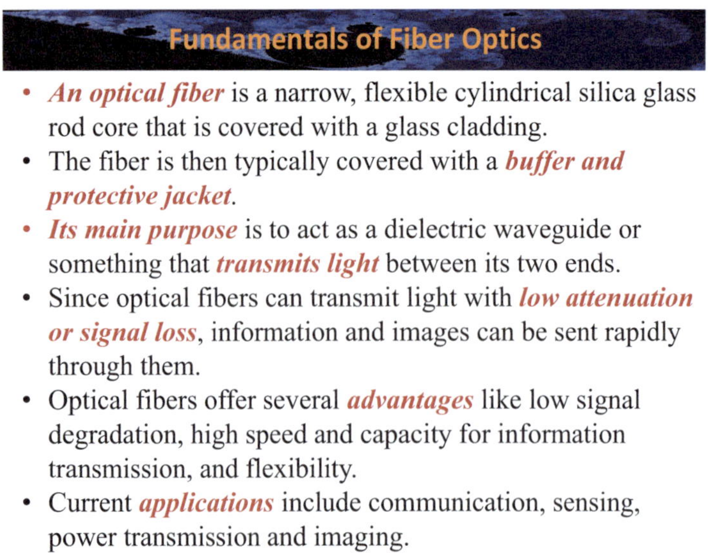

Slide 11.3 Fundamentals of fiber optics

- ***Two main principles*** that allow fiber optics to exist are ***Fermat's principle and Snell's law of refraction***.
- ***Fermat's principle states*** that light travels along a path of minimum time. The optical path will remain straight if light travels in a uniform medium.
- ***Snell's law of refraction*** governs the angle of incidence with which an incoming ray encounters a change in the optical index. It is described as

$$n_1 \sin(\theta_1) = n_2 \sin(\theta_2),$$

where $n_{1,2}$ are the respective indices of refraction (real component) of two media, θ_1 is the angle of incidence from the normal to the boundary of index change, and θ_2 is the angle of refraction of light in the second medium.

Slide 11.4 Fundamentals of fiber optics (continued)

Two main principles that allow fiber optics to exist are Fermat's principle and Snell's law of refraction (Slide 11.4). Fermat's principle states that light travels along a path of minimum time [8]. The optical path will remain straight if light travels in a uniform medium. Snell's law governs the angle of incidence with which an incoming ray encounters a change in the optical index. It is as expressed as

$$n_1 \sin(\theta_1) = n_2 \sin(\theta_2), \tag{3.1}$$

where $n_{1,2}$ are the respective indices of refraction (real component) of two media, θ_1 is an angle of incidence from normal to the boundary of index change, and θ_2 is an angle of refraction of light in the second medium.

Optical fibers guide light through *total internal reflection* (Slide 11.5), which occurs when the refractive index of the core is higher than that of the cladding as shown in Fig. 3.5. When a ray of light comes from a core medium with such a higher index of refraction and with angle of incidence high enough, the light will be reflected back into the first medium from the boundary of index change. The angle at which this occurs is called a critical angle and is defined as

$$\theta_{\mathrm{cr}} = \arcsin\left({}^{n_1}\!/_{n_2}\right). \tag{3.2}$$

Figure 3.5 and Slide 11.5 illustrate the effect of total internal reflection as well as the critical angle. If the conditions for this phenomenon are satisfied, the light remains inside the optical fiber and never leaves the core.

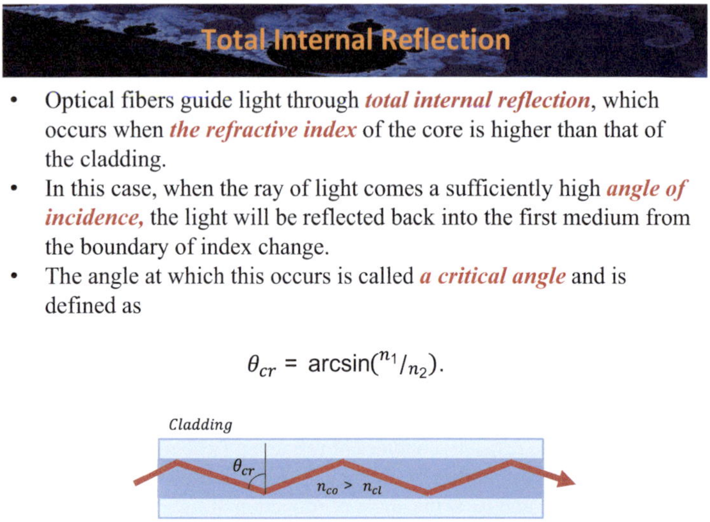

Slide 11.5 Total internal reflection

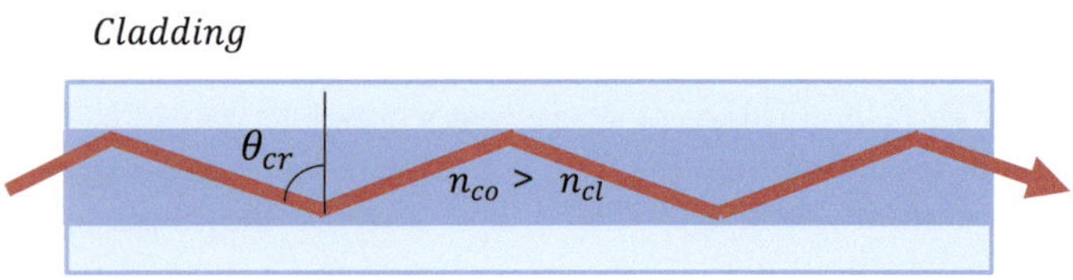

Fig. 3.5 Optical fiber guiding principle

Electromagnetic theory is used to describe how optical fibers transmit light. The guided waves must satisfy Maxwell's equations while imposing cylindrical boundary conditions; these solutions are called the modes.

Evanescent Waves. Optical fibers are useful in sensing applications which are arranged through the light interaction with a fiber's surrounding medium as shown in Fig. 3.6. One way to achieve this is to use the *evanescent field* of transmitted light (Slide 11.6) [9]. If the cladding of a particular fiber is removed, the evanescent waves can penetrate into the surrounding medium (Fig. 3.6). The penetration depth is determined by the distance x from the fiber core, where the amplitude of the evanescent light decreases to $1/e$ of the initial value E_0. The equation for the field amplitude $E(x)$ as a function of distance x outside core and the penetration depth d_p is given as

$$E(x) = E_0 \exp\left(-\frac{x}{d_p}\right). \tag{3.3}$$

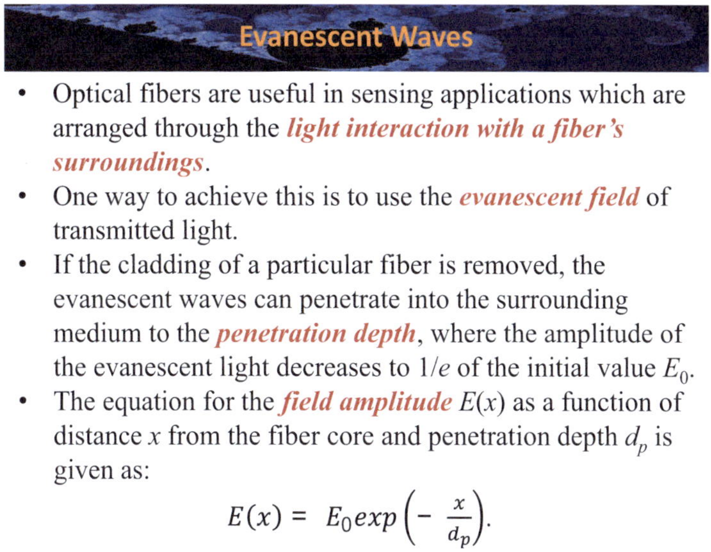

Slide 11.6 Evanescent waves

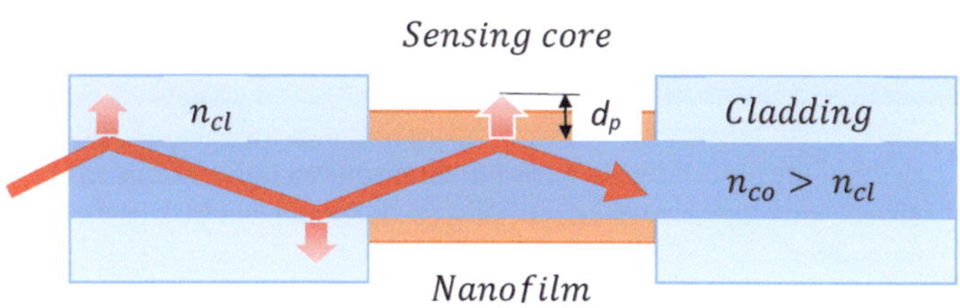

Fig. 3.6 Principle of optical fiber sensing

The penetration depth of the evanescent waves can be described by the angle of incidence (θ_1), wavelength of light (λ), and core (n_{co}) and cladding (n_{cl}) indices of refraction as (Slide 11.7)

$$d_p = \frac{\lambda}{2\pi\sqrt{n_{co}^2 \sin^2\theta_1 - n_{cl}^2}}. \tag{3.4}$$

As seen in Fig. 9.6, the light ray travels through the core; however, there is a small portion of light, called the evanescent waves, that decays outward from the core/cladding interface.

Optical Fiber Biosensing. The optical output of the fiber is a measure of the changes in magnitude of the evanescent field [9]. In sensor development, sensitivity is an important parameter to evaluate sensor performance. Sensitivity is determined by the strength of light–matter interaction on the sensor surface [10].

Evanescent Waves (continued)

- ***The penetration depth*** of the evanescent waves can be described by the angle of incidence (θ_1), wavelength of light (λ), and core (n_{co}) and cladding (n_{cl}) indices of refraction as:

$$d_p = \frac{\lambda}{2\pi\sqrt{n_{co}^2 \sin^2\theta_1 - n_{cl}^2}}.$$

- As seen on the figure below, the light ray travels through the core; however, there is a small portion of light, called ***the evanescent waves***, that decays outward from the core/cladding interface.

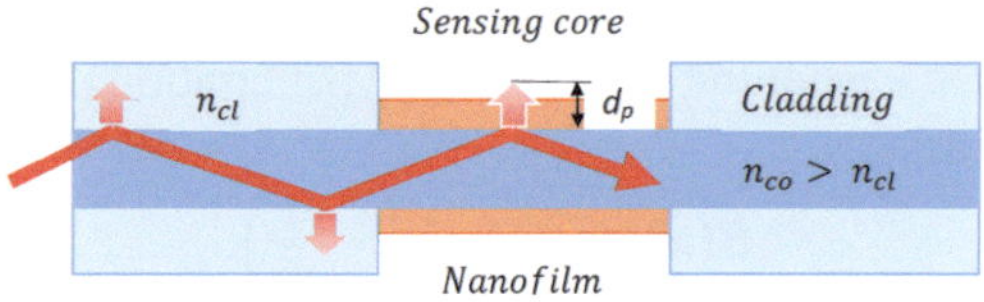

Slide 11.7 Evanescent waves (continued)

Fiber Optic Sensing

- ***Fiber optic sensing*** operates on the principle of modulation. One or more of the optical properties of the transmitted light, like intensity, phase, wavelength or polarization state, is modulated by a particular parameter.
- The ***interaction with a medium*** causes a change in the optical properties of light that can be detected by various means.
- ***Sources of light*** typically include lasers, lamps and light emitting diodes.
- The light is guided by ***the optical fiber*** where it exits through a probe. After it exits, it interacts with the parameter and is sent back to a ***detector***.
- Fiber optic sensors are ***advantageous*** because of their geometrical versatility, materials and safety. Their flexibility and size allows them to be inserted through needles and catheters. Optical fiber materials are inert and biocompatible, and they can be reused.

Slide 11.8 Fiber optic sensing

Fiber optic sensing operates on the principle of modulation (Slide 11.8). One or more of the optical properties of the transmitted light, such as intensity, phase, wavelength, or polarization state, is modulated. This interaction causes a change in the optical properties of light that can be detected by various means. The systems can be either intrinsic or extrinsic sensors. Extrinsic sensors utilize a transducer to

Fiber Optic Biosensor

- ***A fiber optic biosensor*** (FOB) is a specific sensor that uses optical properties in order to selectively detect biological species like DNA and proteins.
- Biosensors are extremely useful ***for diagnostic*** and quantitative purposes.
- Ideally, biosensors could detect analytes in ***low concentrations*** and ***various environmental conditions*** like ones that are turbid.
- Detection techniques are being developed ***in vitro and in vivo***.
- Since detection is highly selective, diseases and disorders could be more ***quickly and accurately diagnosed***.
- These new biosensing techniques will allow for ***real-time analytical measurements*** and eliminate the lapse of traditional sample preparations and measurements.

Slide 11.9 Fiber optic biosensor

modulate the light, while intrinsic sensors are directly modulated by the parameter. Sources of light typically include lasers, lamps, and light-emitting diodes. The light is guided by the optical fiber, where it exits through a probe. After that, it interacts with the parameter and is sent back to a detector. The detector analyzes the changes in the optical properties of transmitted light, like how intensity changed. This change must then be related back to the parameter in order to get useful information such as temperature. Fiber optic sensors are advantageous because of their geometrical versatility, materials, and safety. Their flexibility and size allows them to be inserted through needles and catheters. Optical fiber materials are inert and biocompatible, and they can be reused.

A fiber optic biosensor (FOB) uses optical properties in order to selectively detect biological species like DNA and proteins (Slide 11.9). Biosensors are extremely useful for diagnostic and quantitative purposes. Ideally, they could detect analytes in low concentrations and various environmental conditions like ones that are turbid. Detection techniques are being developed in vitro and in vivo. Since detection is highly selective, diseases and disorders could be more quickly and accurately diagnosed. These new biosensing techniques will allow for real-time analytical measurements and eliminate the lapse of traditional sample preparations and measurements.

In general, most biosensors use enzymes (Slide 11.10). Enzymes are specialized proteins that catalyze and facilitate reactions that take place in a living organism. Enzymes have high selectivity abilities for particular analytes because they are developed to only catalyze specific reactions. An enzyme is attached to the sensing core/probe of the FOB, shown in Slide 11.7. When the probe comes into contact with the analyte, the enzyme catalyzes a designated reaction. When the reaction

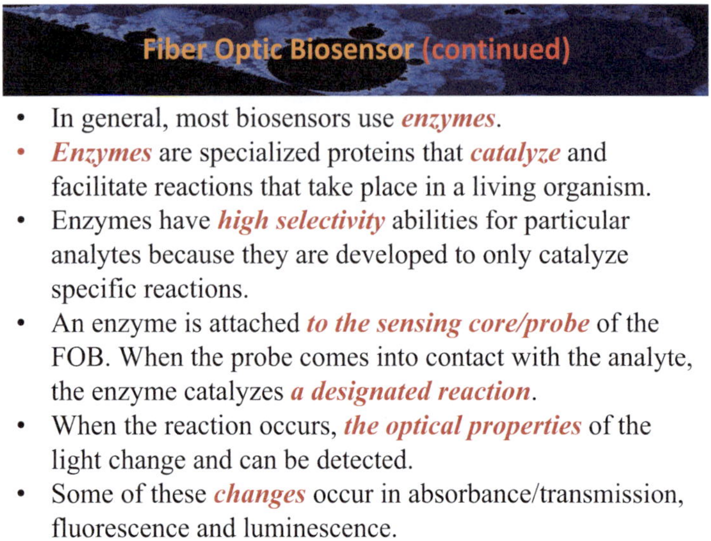

Slide 11.10 Fiber optic biosensor (continued)

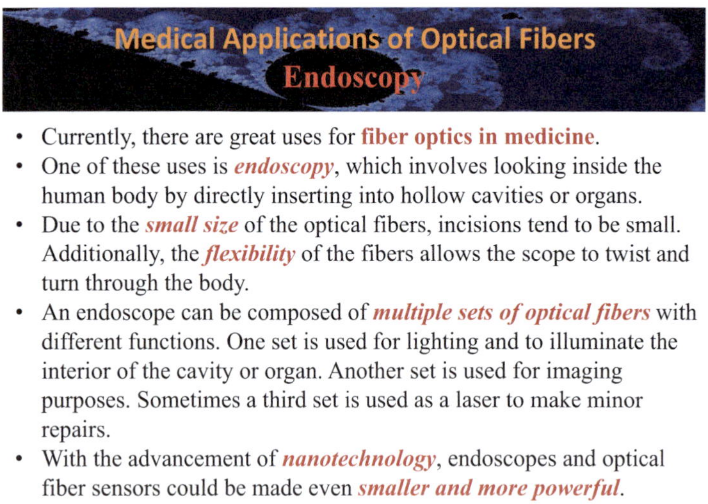

Slide 11.11 Medical applications of optical fibers: endoscopy

occurs, the optical properties of the light change and can be detected. Some of these changes occur in absorbance/transmission, fluorescence, and luminescence. There are many other forms of biosensing, but this lecture discusses only the ones most applicable to nanomedicine.

Medical Applications of Optical Fibers. Currently, there are great uses for fiber optics in medicine (Slide 11.11). One of these uses is endoscopy, which involves

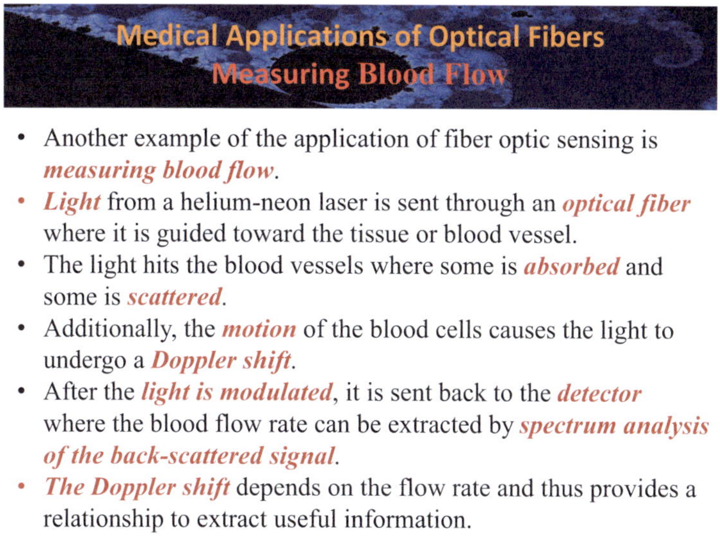

Slide 11.12 Medical applications of optical fibers: measuring blood flow

looking inside the human body by directly inserting into hollow cavities or organs. Due to the small size of the optical fibers, incisions tend to be small. Additionally, the flexibility of the fibers allows the scope to twist and turn through the body. An endoscope can be composed of multiple sets of optical fibers with different functions. One set is used for lighting and to illuminate the interior of the cavity or organ. Another set is used for imaging purposes. Sometimes a third set is used as a laser to make minor repairs. Even more so, some optical fibers are used for sensing things like temperature and pressure. With the advancement of nanotechnology, endoscopes and optical fiber sensors could be made even smaller and more powerful.

Another example of the application of fiber optic sensing is measuring blood flow (Slide 11.12). Light from a helium–neon laser is sent through an optical fiber where it is guided toward the tissue or blood vessel. The light hits the blood vessels, where some is absorbed and some is scattered. Additionally, the motion of the blood cells causes the light to undergo a Doppler shift. The blood cells acting as a parameter. After the light is modulated, it is sent back to the detector where the blood flow rate can be extracted by spectrum analysis of the back-scattered signal. The Doppler shift depends on the flow rate and thus provides a relationship to extract useful information. The goals of nanomedicine include making these fibers and probes smaller in order to be less invasive and to monitor the body at the molecular level. Research is being donc on detection techniques and relationships between modulation and the information wanted, like flow rate. The next horizon lies in fiber optic biosensors based on *surface plasmon resonances*.

3.3.2 *Optical Fiber Sensors Based on Surface Plasmon Resonances (SPRs)*

Surface Plasmon Resonances and Polaritons. One promising detection technique involves *surface plasmon resonances*, which occur on the surface of metal nanoparticles or nanofilms (Slide 11.13). The metal contains a large group of negatively-charged, free electrons. Collectively, the density of the free electrons can be treated as a plasma of particles. Additionally, the metal contains a lattice of positively charged ions that can be replaced by a constant positive background. Natural plasma oscillations occur when the free electrons oscillate around the stationary protons at a particular frequency. The conversion of the electrostatic potential energy to the kinetic energy of electrons causes these oscillations. A quantum of plasma oscillations is a plasmon. Surface plasmons occur along the direction of the metal–dielectric interface.

In order to excite surface plasmons, an external light radiation source with frequency in resonance with that of the plasmon oscillation must be applied (Slide 11.14). There are a few configurations of a metal–dielectric interface that support SPRs, like a prism with one side coated with a gold nanofilm and an optical fiber having a section covered by a micro- or nano-size thickness of metal film around the optical core. An optical fiber with the SPR sensing core is schematically shown on the figure in Slide 11.7. Adjusting the thickness of the nanofilm and frequency of the incident light allows for optimization of the SPR, which is the collective oscillations of electrons stimulated by light in resonance with the plasmon oscillation frequency. Evanescent waves with resonance frequency from

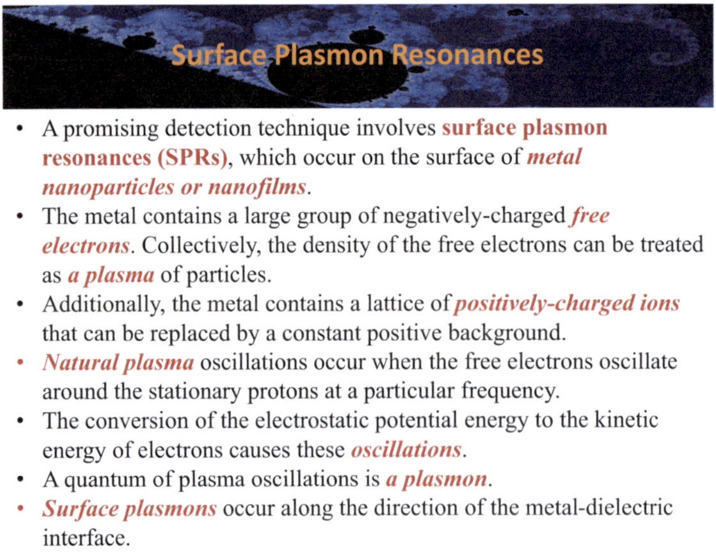

Slide 11.13 Surface plasmon resonances

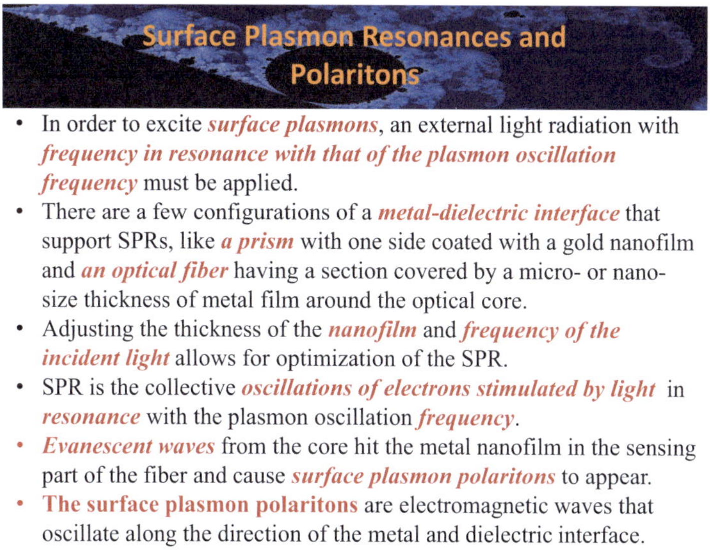

Slide 11.14 Surface plasmon resonances and polaritons

the core hit the metal nanofilm in the sensing part of the fiber and cause *surface plasmon polaritons* to appear. These polaritons are the electromagnetic waves that oscillate along the direction of the metal and dielectric or air interface. These oscillations become very sensitive to any change around the metal–dielectric boundary, and thus can be used as a detection technique and biosensor.

Fiber Optic SPR Sensors. SPR fibers have the capability to detect changes in the external refractive index (Slide 11.15). The optical fibers offer several advantages such as small size, flexibility, and remote sensing. Remote sensing could eventually occur within the human body itself.

A fiber optic SPR sensor is basically an optical fiber with one small adjustment. As mentioned earlier, in one section of the optical fiber, the cladding is removed and replaced with a metal nanofilm in order to promote SPRs. The nanofilm is then coated with the dielectric/sensing layer. This sensing layer can be coated with molecules that will bind to other target molecules. When a testing medium containing the analyte passes over the sensing core, molecular interactions occur between the analyte and the molecules on the sensing layer. These interactions change the refractive index of the medium.

Incident light with resonance frequency enters the optical fiber, strikes the metal–dielectric interface, and causes SPRs to occur (Slide 11.16). The reflected light passes to the end of the fiber and onto a photodetector, which measures the decrease in absorbance due to the change in the refractive index of the sensing layer. However, many new parameters have to be accounted and adjusted for, like the fiber geometry and numerical aperture. Many more parameters must be adjusted to construct calibration curves that analyze molecular interactions. Each sensor depends on the specificity of its function for detection of a particular analyte.

Fiber Optic SPR Sensors

- ***SPR fibers*** have the capability to detect changes in the external refractive index.
- The optical fibers offer several ***advantages*** like small size, flexibilit, and remote sensing. Remote sensing could eventually occur within the human body itself.
- ***A fiber optic SPR sensor*** is basically an optical fiber with one section of the cladding removed and replaced with the ***metal nanofilm*** in order to promote SPRs.
- ***The nanofilm*** is then coated with the dielectric/sensing layer.
- ***This sensing layer*** can be coated with molecules that will bind to other target molecules.

Slide 11.15 Fiber optic SPR sensors

Fiber Optic SPR Sensors (continued)

- When a testing medium containing the analyte passes over the sensing core, ***molecular interactions*** occur between the ***analyte*** and the ***molecules*** on the sensing layer.
- These interactions **change the refractive index** of the medium.
- ***Incident light*** with resonance frequency enters the optical fiber, strikes the metal-dielectric interface, and causes ***SPRs to occur.***
- ***A photodetector*** measures the decrease in absorbance due to the change in the refractive index of the sensing layer.
- ***Each sensor*** depends on the specificity of its function of detection of ***a particular analyte***.

Slide 11.16 Fiber optic SPR sensors (continued)

Examples of Utilizing Optical Fiber as SPR Biosensors. Examples of a current use of SPR fiber biosensors are presented in Slides 11.17, 11.18, and 11.19. For example, researchers from the Indiana Institute of Technology used an optical fiber biosensor to detect phenolic compounds (Slide 11.17). One of their goals was to measure the response of SPRs and relate this to the concentration of phenol. To construct the apparatus, they used an optical fiber with a section of exposed

Examples of SPR Fiber Biosensors
Phenol Concentration

- Researchers from the Indiana Institute of Technology used an optical fiber biosensor to detect ***phenolic compounds***.
- One of their goals was to measure the ***response of SPRs*** and relate this ***to the concentration of phenol***.
- To construct the apparatus, they used an optical fiber with a section of exposed cladding that was replaced ***with a 40 nm silver film***.
- ***The sensing layer*** was covered with ***immbolized tyrosinase in polyacrylamide gel***.
- ***Tyrosinase*** is an enzyme that catalyzes the breakdown of phenols.

Slide 11.17 Examples of SPR fiber biosensors: phenol concentration

Examples of SPR Fiber Biosensors
Phenol Concentration (continued)

- ***A flow cell*** was constructed around the fiber in order to deliver the solutions containing various concentrations of phenol.
- After the light passed through, it was analyzed by ***a spectrometer*** connected to a computer.
- ***The resonance wavelength*** was varied in order to find the new optimal wavelength for the SPRs. Each different concentration had a different optimal wavelength.
- ***Different concentrations*** caused varying indices of refraction.
- ***With the data collected***, the researchers were able to ***construct curves*** with polynomial fits of the resonance wavelength versus concentration.
- These curves could be used as ***calibrations*** to detect and quantify unknown ***concentrations of phenols***.
- Phenols are typically ***biohazards***, and this experiment offers a new and ***effective detection method***.

Slide 11.18 Examples of SPR fiber biosensors: phenol concentration (continued)

cladding that was replaced with a 40 nm silver film. The sensing layer was covered with immobilized tyrosinase in polyacrylamide gel. Tyrosinase is an enzyme that catalyzes the breakdown of phenols. A flow cell was constructed around the fiber in order to deliver the solutions containing various concentrations of phenol (Slide 11.18). After the light passed through, it was analyzed by a spectrometer connected to a computer.

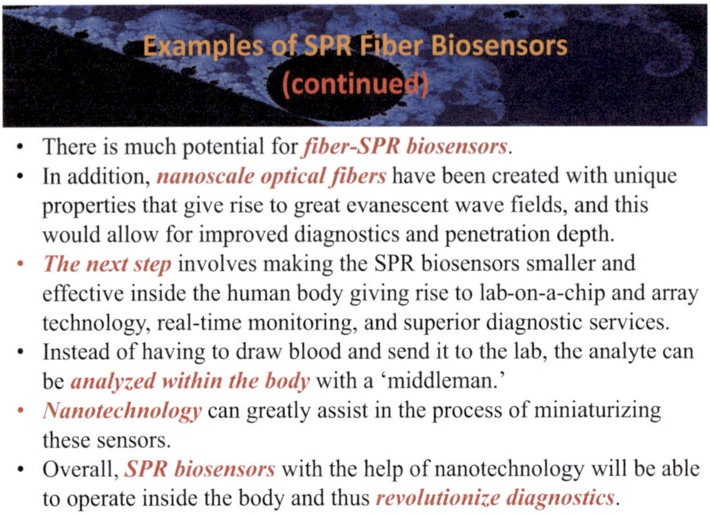

Slide 11.19 Examples of SPR fiber biosensors (continued)

The resonance wavelength was varied in order to find the new optimal wavelength for the SPRs. Each different concentration had a different optimal wavelength (Slide 11.18). Different concentrations caused varying indices of refraction. With the data collected, the researchers were able to construct curves with polynomial fits of the resonance wavelength versus concentration. These curves could be used as calibrations to detect and quantify unknown concentrations of phenols. Phenols are typically biohazards, and this experiment offers a new and effective detection method. This experiment has shown the SPR to be useful and effective in determining the concentration of a particular analyte. More research should be done to support different specificities with different molecules.

There is much potential for fiber-SPR biosensors (Slide 11.19). Studies like the one previously discussed have already demonstrated that optical biosensors in the lab are useful and effective. In particular, sensors utilizing SPRs will probably be the most useful. In addition, nanoscale optical fibers have been created with unique properties that give rise to great evanescent wave fields, and this would allow for improved diagnostics and penetration depth. Although much development is left, this could have many beneficial uses in nanomedicine.

The next step involves making the SPR biosensors smaller and effective inside the human body, giving rise to lab-on-a-chip and array technology, real-time monitoring, and superior diagnostic services. Instead of having to draw blood and send it to the lab, the analyte can be analyzed within the body with a 'middleman' (Slide 11.19). Nanotechnology can greatly assist in the process of miniaturizing these sensors. However, biochemical interactions must be developed and isolated for the various target analytes. Once the interactions are developed, calibration curves can be constructed that can determine unknowns like concentration. Overall,

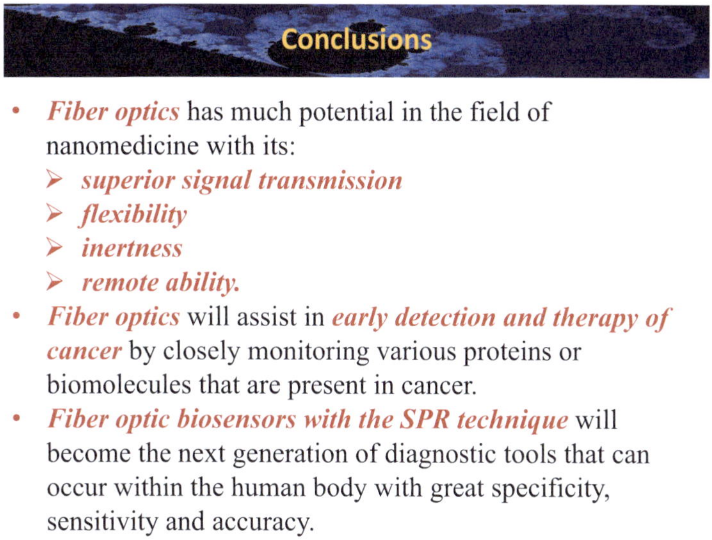

Slide 11.20 Conclusions

SPR biosensors with the help of nanotechnology will be able to operate inside the body and thus revolutionize diagnostics.

In conclusion (Slide 11.20), fiber optics has much potential in the field of nanomedicine with its superior signal transmission, flexibility, inertness, and remote ability. Fiber optics will assist in early detection and therapy of cancer by closely monitoring various proteins or biomolecules that are present in higher concentrations when cancerous growth is present. Additionally, fiber optic biosensors with the SPR technique will become the next generation of diagnostic tools that can occur within the human body with great specificity, sensitivity, and accuracy.

Homework Exercises

3.1. What are three conventional cancer treatments?
3.2. What are advantages and disadvantages of the surgical removal?
3.3. Describe a radiation therapy.
3.4. True or false? The radiation therapy is most effective in mitosis events.
3.5. Select one or more. The radiation therapy uses:

(a) Shaped radiation beams
(b) Multiple beams
(c) Photothermolysis
(d) Ultrasound

(e) Radioactive materials
(f) RF waves
(g) All of the above
(h) None of the above

3.6. What is a downside of radiation therapy?
3.7. How does chemotherapy work?
3.8. True or false? Chemotherapy prevents mitosis of the cells.
3.9. What are advantages and disadvantages of chemotherapy?
3.10. How does nanotherapy work?
3.11. True or false? Nanoparticles alone can cause cellular damage.
3.12. Select one or more. Nanophotothermolysis is selective thermal killing of cancer cells by:

(a) Surgery
(b) Nanoprticle activation
(c) Chemotherapy
(d) Radiation
(e) All of the above
(f) None of the above

3.13. List the nanoparticles administration techniques.
3.14. What is direct microinjection? What are the advantages and disadvantages of this technique?
3.15. What is transdermal delivery? What are the advantages and disadvantages of this technique?
3.16. What is the physiological transport technique? What are the advantages and disadvantages of this technique?
3.17. What is the conjugation to antibodies technique?
3.18. Select one or more. Mechanisms of cell killing include:

(a) Thermal ablation
(b) Bubbles formation
(c) Sound generation
(d) Shock waves
(e) Explosive evaporation
(f) All of the above
(g) None of the above

3.19. What is the definition of selective nanophotothermolysis?
3.20. Select one or more. The theory of selective nanophotothermolysis includes:

(a) Nanoparticle optics
(b) Heat-mass transfer
(c) Cell ablation
(d) Microbubble dynamics
(e) Cluster aggregation
(f) Thermal explosion

(g) Shock wave generation
(h) Sound generation
(i) All of the above
(j) None of the above

3.21. What are three main dynamic modes of selective nanophotothermolysis?
3.22. What is the cluster aggregation mode? What are the advantages and disadvantages of this mode?
3.23. True or false? The cluster aggregation mode provides small thermal damage area in a cell.
3.24. What is the nanobubble overlapping mode? What are the advantages and disadvantages of this mode?
3.25. True or false? The nanobubble overlapping mode can be used for therapy of dense solid tumors, bones and other targets with a lack of sufficient liquid.
3.26. What are the therapeutic effects of the nanobubble overlapping mode?
3.27. Select one or more. The particle evaporation modes are as follows:

(a) Free molecular
(b) Diffusion
(c) Gas dynamic
(d) Explosive
(e) All of the above
(f) None of the above

3.28. What is the thermal explosion mode? What are the advantages and disadvantages of this mode?
3.29. When is the thermal explosion mode realized?
3.30. Select one or more. "Nanobombs" are accompanied by:

(a) Stress transients waves
(b) Sound waves
(c) Shock waves
(d) Fragments of explosion
(e) Denaturation and coagulation
(f) All of the above
(g) None of the above

3.31. What are the therapeutic effects of "nanobombs?"
3.32. How can nanodrugs/nanoparticles be activated?
3.33. Select one or more. The nanoparticle activation include:

(a) Heating of nanoparticles
(b) Photoelectron emission
(c) Compton scattering,
(d) Explosion of nanoparticles
(e) Generation of shock waves
(f) Sound waves
(g) Mechanical waves of high pressure

(h) Nanobubble formation
(i) All of the above
(j) None of the above

3.34. True or false. Nanodrug design includes the morphological properties of nanoparticles.
3.35. True or false. Nanodrug design depends on the methods of activation.
3.36. What nanoparticles are used for the activation by radiation in the optical range of the spectrum?
3.37. What nanoparticles are used for activation by a magnetic field?
3.38. What nanoparticles are used for activation by X-ray radiation?
3.39. What are the characteristics of the ideal radiation for nanoparticle activation?
3.40. Select one or more. The radiation delivery techniques include:

(a) Microinjection
(b) Physiological transport
(c) Transdermal delivery
(d) Conjugation to antibodies
(e) All of the above
(f) None of the above

3.41. What are the therapeutic effects of nanoparticles activated by X-ray radiation?
3.42. What are the advantages and disadvantages of X-ray activation of nanoparticles?
3.43. True or false? In the direct action, charged particles react with water contained in the cell medium producing hydroxyl radicals which are toxic for DNA of cancer cells.
3.44. What is the optical range of the spectrum?
3.45. What is the penetration depth of different types of lasers into soft tissue?
3.46. What limits the penetration depth of optical radiation into biological tissue?
3.47. What is the window of transparency of biological tissue?
3.48. What are the advantages and disadvantages of using optical fibers for laser radiation delivery?
3.49. What are the advantages and disadvantages of using microwaves for tissue ablation?
3.50. What are the advantages and disadvantages of using a magnetic field and ultrasound for nanoparticle activation?
3.51. What are the advantages and disadvantages of using RF waves for nanoparticle activation?
3.52. What is an optical fiber?
3.53. What are the applications of the optical fibers?
3.54. What are the advantages of using optical fibers?
3.55. What are the fundamental principles of an optical fiber?
3.56. Describe Fermat's principle.
3.57. Describe Snell's law of refraction.

3.58. What is a total internal reflection?
3.59. What is a critical angle?
3.60. What is an evanescent field?
3.61. How is the penetration depth of evanescent waves defined?
3.62. How does the fiber optic sensing operate?
3.63. How is a sensing core of an fiber biosensor made?
3.64. What sources of light are typically used for fiber biosensors?
3.65. Why are enzymes are used in fiber optics biosensors?
3.66. What are the applications of optical fibers in medicine?
3.67. How are optical fibers used in endoscopy?
3.68. How does an optical fiber measure blood flow?
3.69. What is a surface plasmon resonance?
3.70. What is a plasmon?
3.71. What is a surface plasmon polariton?
3.72. True or false? A surface plasmon resonance can be initiated in any nanoparticle or nanofilm.
3.73. How are surface plasmons and polaritons induced?
3.74. True or false? A surface plasmon resonance is induced in a dielectric nanofilm by any radiation.
3.75. What main conditions must be satisfied to induce the surface plasmons?
3.76. How does a SPR fiber biosensor operate?
3.77. True or false? In a SPR fiber biosensor, a section of the cladding is removed and replaced with the dielectric nanofilm.
3.78. What is a sensing layer in a SPR fiber biosensor? How it is made?
3.79. How does an analyte bind to a sensing layer?
3.80. True or false? Each SPR fiber sensor depends on the specificity of its function of detection of a particular analyte.
3.81. Give examples of the usage of SPR fiber biosensors.

Topics for Presentations and Writing Assignments

Presentation #1. Methods and Tools for Cancer Treatment and Diagnosis in Conventional Medicine.
Presentation #2. Nanotechnology for Biophotonics: Bionanophotonics.
Presentation #3. Laser Ablation of Biological Tissues and Laser Surgery.
Presentation #4. Optics-Based Clinical Applications: Fiber Optics and Fiber Optics Biosensors.
Presentation #5. Nanoparticle-Enhanced X-Ray Therapy for Cancer.
Presentation #6. RF Activation of Nanoparticles for Cancer Treatment.

References

1. R.R. Letfullin, T.F. George, Plasmonic nanomaterials in nanomedicine, in *Springer Handbook of Nanomaterials*, ed. by R. Vajtai (Springer, Berlin, 2013), pp. 1063–1097
2. R.R. Letfullin, T.F. George, Nanomaterials in nanomedicine, in *Computational Studies of New Materials II: From Ultrafast Processes and Nanostructures to Optoelectronics, Energy Storage and Nanomedicine*, ed. by T.F. George, D. Jelski, R.R. Letfullin, G.P. Zhang (World Scientific, Singapore, 2011), pp. 103–129
3. R.R. Letfullin, C.B. Iversen, T.F. George, Modeling nanophotothermal therapy: Kinetics of thermal ablation of healthy and cancerous cell organelles and gold nanoparticles. Nanomed. Nanotechnol. Biol. Med. **7**, 137–145 (2011)
4. R.R. Letfullin, T.F. George, Laser ablation of biological tissue by short and ultrashort pulses, in *Computational Studies of New Materials II: From Ultrafast Processes and Nanostructures to Optoelectronics, Energy Storage and Nanomedicine*, ed. by T.F. George, D. Jelski, R.R. Letfullin, G.P. Zhang (World Scientific, Singapore, 2011), pp. 191–218
5. R.R. Letfullin, C.E.W. Rice, T.F. George, X-ray optics of gold nanoparticles. Appl. Opt. **53**, 7208–7214 (2014)
6. R.R. Letfullin, A.R. Letfullin, T.F. George, Absorption efficiency and heating kinetics of nanoparticles in the RF range for selective nanotherapy of cancer. Nanomed. Nanotechnol. Biol. Med. **11**, 413–420 (2015)
7. R.R. Letfullin, C.B. Iversen, T.F. George, Modeling nanophotothermal therapy: Kinetics of thermal ablation of healthy and cancerous cell organelles and gold nanoparticles. Nanomed. Nanotechnol. Biol. Med. **7**, 137–145 (2011)
8. http://web.mit.edu/ytc/www/HLMA/report.pdf
9. A. Leung, P.M. Shankar, R. Mutharasan, A review of fiber-optic biosensors. Sens. Actuator. B **125**, 688–703 (2007)
10. T. Dar, J. Homola, B.M.A. Rahman, M. Rajarajan, Label-free slot-waveguide biosensor for the detection of DNA hybridization. Appl. Opt. **51**, 8195–8202 (2012)
11. V.P. Zharov, R.R. Letfullin, E.N. Galitovskaya, Microbubbles-overlapping mode for laser killing of cancer cells with absorbing nanoparticle clusters. J. Phys. D: Appl. Phys. **38**, 2571–2581 (2005)

Chapter 4
Introduction to Light–Particle Interactions

Contents

Abstract In the previous chapter, we introduced targeted photothermal therapy of cancer, which combines molecular targeting mechanisms with the unique light absorption properties of nanoparticles to deliver heat and treat diseases at a subcellular level. To understand how photothermal therapies like nanophotothermolysis and nanophotohyperthermia could be useful, it is important to understand how radiation interacts with nanoparticles. This chapter models the light absorption properties of nanoparticles in the *X-ray, optical, and RF ranges of the spectrum*. To find the optimal wavelength of radiation and optimal size range of nanoparticles for effective heating/activating of the nanoparticles, we use an extended *Lorentz–Mie* diffraction theory, taking into account the *plasmon-resonance absorption effect* observed in metal nanoparticles. This chapter

---

Electronic supplementary material: The online version of this chapter (doi:10.1007/978-3-319-43577-0_4) contains supplementary material, which is available to authorized users.

R.R. Letfullin, T.F. George, *Computational Nanomedicine and Nanotechnology*,
DOI 10.1007/978-3-319-43577-0_4

contains material adapted from our publications listed at the end of the chapter. The detailed list of references and reviews on a given topic of this chapter can be found in those original papers.

4.1 Lecture 12: Nanoparticle Optics—Lorenz–Mie Solutions (Slide 12.1)

- Introduction to Waves
- Interaction of Light with a Medium
  - *Index of Refraction of a Medium*
  - *Propagation of Light*
- Maxwell's Equations
- Lorenz–Mie Formalism
- Conclusions
- References

4.1.1 Introduction to Waves

Light can be produced by oscillations/transitions of electrons inside atoms/molecules (Slide 12.2). Light has a wave-particle duality nature. In optical phenomena such as scattering, diffraction, and interference, many photons of light are involved, not just a single photon. A beam of light with a large number of photons behaves as a wave.

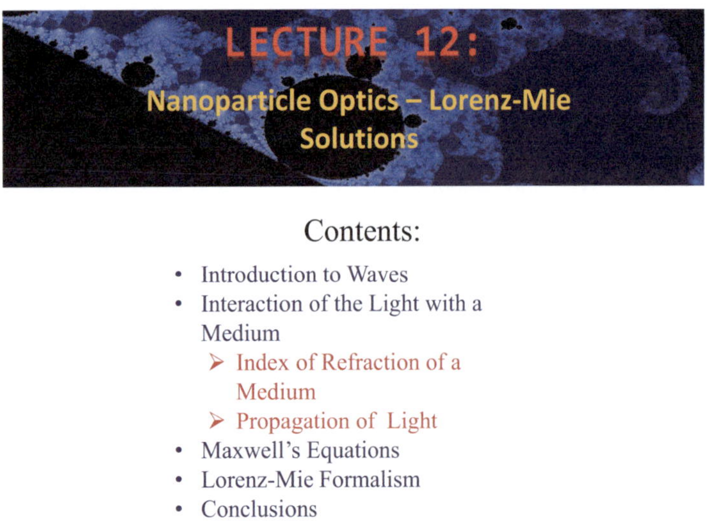

Slide 12.1 Contents of Lecture 12

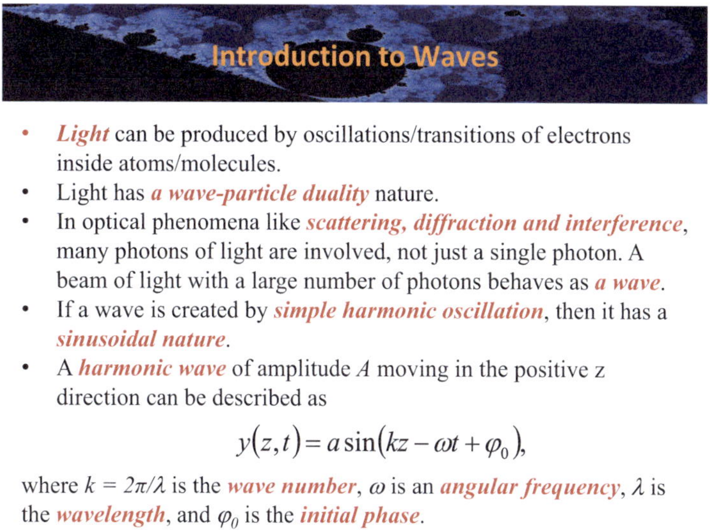

Slide 12.2 Introduction to waves

Fig. 4.1 Sinusoidal wave

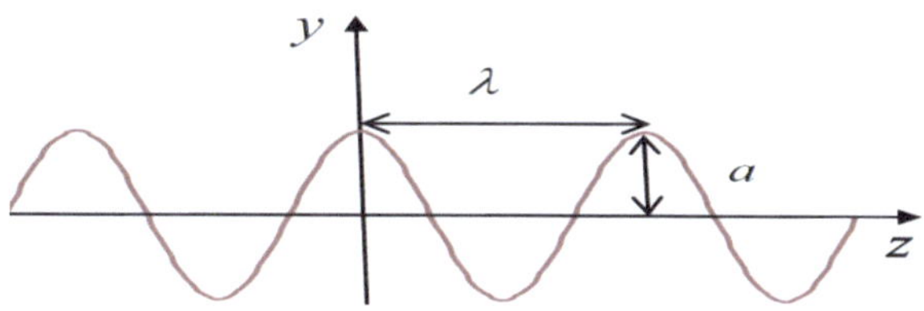

If a wave is created by simple harmonic oscillation, it has a sinusoidal nature as shown in Fig. 4.1 and in Slide 12.3. A harmonic wave of amplitude A moving in the positive z direction can be described as (Slide 12.2)

$$y(z,t) = A\sin(kz - \omega t + \varphi_0), \tag{4.1}$$

where $k = \frac{2\pi}{\lambda}$ is the wave number, ω is the angular frequency, λ is the wavelength, and φ_0 is the initial phase. The *amplitude* A is the maximum displacement of a point on the wave from the equilibrium position, and λ is the distance between identical points on the wave (Slide 12.3).

The wave is also described by the *period* $T = \frac{1}{f}$ and *frequency* f of oscillation (Slide 12.3), where T is the time for a point on the wave to undergo one complete oscillation, and f is the number of oscillations per unit time. A regular frequency f relates to the *angular frequency* of the wave as $\omega = 2\pi f$.

The *speed of the wave* v can be defined as the rate at which the wave moves one wavelength λ in one period (Slide 12.4),

$$v = \lambda/T = \lambda f, \tag{4.2}$$

which is constant and depends only on the medium of propagation. The speed of light does not depend on the amplitude, wavelength, or period of the wave.

Introduction to Waves
Amplitude, Frequency, Period

- The ***amplitude a*** is the maximum displacement of a point on the wave from the equilibrium position.
- The ***wavelength λ*** is the distance between identical points on the wave.
- The ***period T*** is the time for a point on the wave to undergo one complete oscillation.
- The ***frequency f*** of the wave is the number of oscillations per unit time.
- The ***regular frequency*** *f* relates to ***the angular frequency*** of the wave as $\omega = 2\pi f$.
- The wave is also described by the ***period*** $T = \frac{1}{f}$.

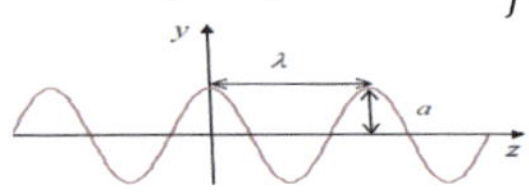

Slide 12.3 Introduction to waves: amplitude, frequency, period

Introduction to the Waves
Speed of the Wave

- The ***speed of the wave*** *v* can be defined as the rate at which the wave moves one wavelength λ in one period *T*:

$$v = \lambda / T = \lambda f.$$

- The ***speed of the wave*** is constant and depends only on the medium of propagation. The speed of light does not depend on the amplitude, wavelength or period of the wave.
- By using the ***relationships among T, f and ω***,

$$T = \frac{1}{f} = \frac{2\pi}{\omega},$$

a few useful formulas describing the wave can then be derived:

$$\lambda = \frac{v}{f}, \quad \lambda = vT, \quad \lambda = \frac{2\pi v}{\omega}, \quad \lambda = \frac{2\pi}{k}, \quad v = \frac{\omega}{k}.$$

Slide 12.4 Introduction to waves: speed of the wave

By using the relationships among $T, f,$ and ω (Slide 12.4),

$$T = \frac{1}{f} = \frac{2\pi}{\omega}, \tag{4.3}$$

Introduction to Waves
Electromagnetic Wave

- Light is an ***electromagnetic wave*** made of electric $\boldsymbol{E}$ and magnetic $\boldsymbol{B}$ fields oscillating perpendicularly to each other.
- An electromagnetic wave is a ***transverse wave*** where both vectors $\boldsymbol{E}$ and $\boldsymbol{B}$ are oscillating in the perpendicular direction to the wave propagation.
- A traveling electromagnetic wave ***transports energy***.
- The direction of the energy transport as well as the direction of wave propagation is described by the ***Poynting vector S***,
$$\boldsymbol{S} = \frac{\boldsymbol{E}\times\boldsymbol{B}}{\mu_0},$$
where μ_0 is the ***vacuum permeability***.
- The ***magnitude of S*** is directly related to the ***energy*** being transported by the wave,
$$S = \frac{EB}{\mu_0} = \frac{E^2}{\mu_0 c},$$ where we used the relation $E = cB$.

Slide 12.5 Introduction to waves: electromagnetic wave

a few useful formulas describing a wave can then be derived (Slide 12.4):

$$\lambda = \frac{v}{f}, \quad \lambda = vT, \quad \lambda = \frac{2\pi v}{\omega}, \tag{4.4}$$

$$\lambda = \frac{2\pi}{k}, \quad v = \frac{\omega}{k}. \tag{4.5}$$

Light is an *electromagnetic wave* made of electric $\boldsymbol{E}$ and magnetic $\boldsymbol{B}$ fields oscillating perpendicularly to each other (Slide 12.5). An electromagnetic wave is a transverse wave where both vectors $\boldsymbol{E}$ and $\boldsymbol{B}$ are oscillating in the perpendicular direction to the wave propagation. A traveling electromagnetic wave transports energy. The direction of the energy transport as well as the direction of wave propagation is described by the *Poynting vector* $\boldsymbol{S}$ (Slide 12.5),

$$\boldsymbol{S} = \frac{\boldsymbol{E}\times\boldsymbol{B}}{\mu_0}, \tag{4.6}$$

where μ_0 is the vacuum permeability. The magnitude of $\boldsymbol{S}$ is directly related to the energy being transported by the wave (Slide 12.5),

$$S = \frac{EB}{\mu_0} = \frac{E^2}{\mu_0 c}, \tag{4.7}$$

where we have used the relation $E = cB$.

The energy delivered by the electromagnetic wave can be described by three main quantities: *absolute amount of energy U*, *power P*, and *intensity I* (Slide 12.6).

Introduction to Waves
Energy of Electromagnetic Wave

- The ***energy*** delivered by the electromagnetic wave can be described by three main quantitites: ***absolute amount of energy U, power P*** and ***intensity I***.
- The ***power P*** is an absolute amount of energy delivered per unit time, $P = U/t$.
- The ***intensity*** of the wave is the power delivered per unit area, $I = P/a$, where a is a cross-section area of the light.
- The ***intensity*** of the harmonic waves can be found by ***averaging the Poynting vector S*** over a cycle as
$$I = \langle S \rangle = \frac{\langle E^2 \rangle}{\mu_0 c} = \frac{E_0^2}{\mu_0 c} \langle sin^2(kz - \omega t) \rangle = \frac{1}{2}\frac{E_0^2}{\mu_0 c} = \frac{c}{2}\frac{B_0^2}{\mu_0}.$$
- The ***light pressure*** also can be found by the average value of the Poynting vector as
$$p = \langle S \rangle / c \quad \textit{for absorbing surface}$$
$$p = 2\langle S \rangle / c \quad \textit{for reflecting surface}$$

Slide 12.6 Introduction to waves: energy of electromagnetic wave

The power is the absolute amount of energy delivered per unit time, $P = U/t$. The intensity of the wave is the power delivered per unit area, $I = P/a$, where a is the cross-section area of the light. The intensity of the harmonic waves can be found by averaging the Poynting vector $\boldsymbol{S}$ over a cycle as (Slide 12.6)

$$I = \langle S \rangle = \frac{\langle E^2 \rangle}{\mu_0 c} = \frac{E_0^2}{\mu_0 c} \left\langle \sin^2(kz - \omega t) \right\rangle = \frac{1}{2}\frac{E_0^2}{\mu_0 c} = \frac{c}{2}\frac{B_0^2}{\mu_0}. \tag{4.8}$$

The *light pressure* also can be found by the average value of the Poynting vector as (Slide 12.6)

$$p = \langle S \rangle / c \quad \text{for absorbing surface,} \tag{4.9}$$

$$p = 2\langle S \rangle / c \quad \text{for reflecting surface.} \tag{4.10}$$

4.1.2 Interaction of Light with a Medium

Index of Refraction of a Medium (Slide 12.7). The speed of light in vacuum is $c = 3 \times 10^8$ m/s. When light enters a medium, it interacts with atoms and molecules of the medium. These interactions slow down light to a speed $v < c$ in the medium. Since different media have different densities of atoms and molecules, light has different speeds in the different media. Because the wavelength of the light is related to the velocity of the wave as $\lambda = \frac{v}{f}$, the wavelength decreases as well inside

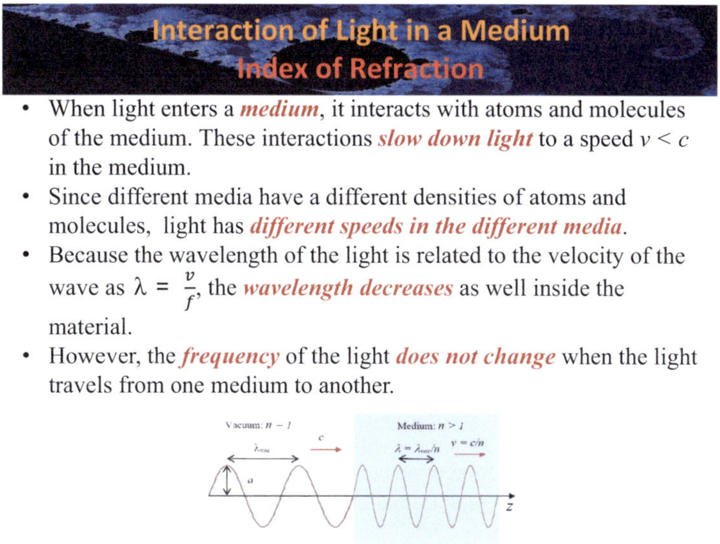

Slide 12.7 Interaction of light in a medium: index of refraction

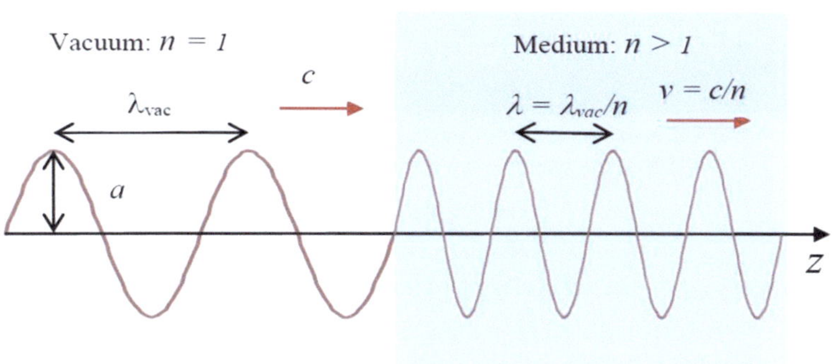

Fig. 4.2 Propagation of the wave in the medium

the material as shown in Fig. 4.2 and Slide 12.7. However, the frequency of the light does not change when the light travels from one medium to another.

Thus, it is natural to introduce the optical characteristic of the medium through the speed of light as a ratio of the speed in vacuum to the speed in the medium: $n = c/v$ (Slide 12.8). This number is called *index of refraction* of the medium, which is a main optical characteristic of any medium. Since $c \geq v$, the refractive index is equal to or greater than one. For example, the index of refraction of the air is approximately equal to 1 since the air molecules are far apart, and light can propagate through air almost without disturbance. The index of refraction of water is 1.33. Glass has an index of refraction 1.5. Each medium has its own index of refraction.

Interaction of Light in a Medium Index of Refraction (continued)

- Thus, it is natural to introduce the optical characteristic of the medium through the speed of light as ***a ratio of the speed*** of light in vacuum to the speed of light in the medium: $n = c/v$.
- This number is called ***index of refraction of the medium***, which is a main optical characteristic of any medium.
- The index of refraction shows ***how many times the speed of light*** in the vacuum is greater than the speed of light in the particular medium.
- ***Since*** $c \geq v$, the refractive index is equal to or ***greater than one***.
- For example, the index of refraction of ***the air*** is approximately equal to 1 since the air molecules are far apart and light can propagate through air almost without disturbance.
- The index of refraction of ***the water*** is 1.33.
- ***A glass*** has the index of refraction 1.5. Each medium has its own index of refraction.

Slide 12.8 Interaction of light in a medium: index of refraction (continued)

Propagation of Light

- If the ***index of refraction*** of the medium is known, then a ***speed and wavelength*** of the wave ***in medium*** can easily be found:

$$v = \frac{c}{n}, \quad \lambda = \frac{c}{fn} = \frac{\lambda_{vac}}{n},$$

where $\lambda_{vac} = \frac{c}{f}$ is a wavelength of the wave in a vacuum.

- ***Propagation of the Light***. Since there are different optical phenomena happen with the light in the medium like ***scattering, refraction, absorption, dispersion***, etc., in general the index of refraction is ***a complex number***:

$$m = n - ik,$$

where n is a ***real part***, which describes the optical phenomena like ***scattering and refraction*** of the light; k is an ***imaginary part***, which is responsible for ***absorption*** of the light in the medium.

Slide 12.9 Propagation of light

If the index of refraction of the medium is known, then the speed and wavelength of the wave in a medium can easily be found as (Slide 12.9)

$$v = \frac{c}{n}, \quad \lambda = \frac{c}{fn} = \frac{\lambda_{\text{vac}}}{n}, \tag{4.11}$$

where $\lambda_{\text{vac}} = \frac{c}{f}$ is the wavelength of the wave in vacuum.

Propagation of Light. Since different optical phenomena can happen with the light in a medium, such as scattering, refraction, absorption, and dispersion, in general the index of refraction is a complex number (Slide 12.9),

$$m = n - ik. \tag{4.12}$$

Here, n is the real part, which describes the optical phenomena like scattering and refraction of the light, and k is the imaginary part, which is responsible for absorption of the light in the medium.

Let us consider the Beer–Lambert experimental arrangement shown in Fig. 4.3 and in Slide 12.10 to describe monochromatic light propagation through two media

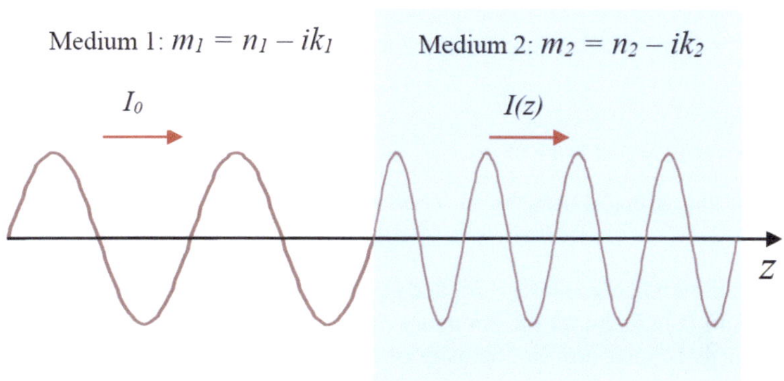

Fig. 4.3 The Beer–Lambert experiment

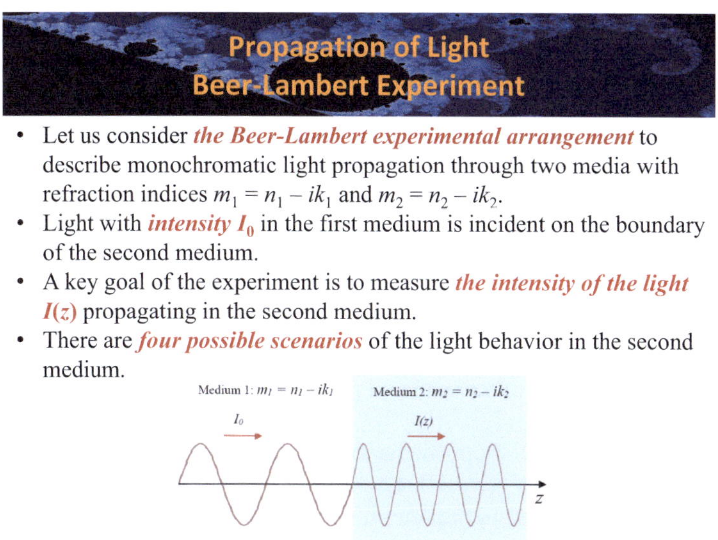

Slide 12.10 Propagation of light: Beer–Lambert experiment

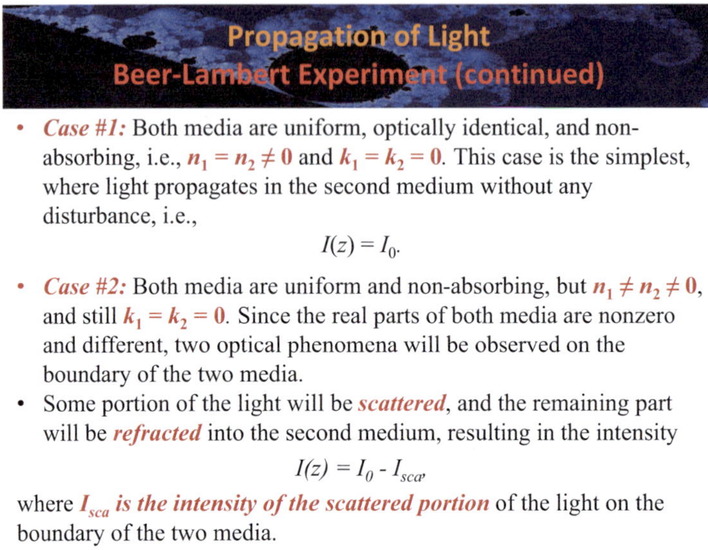

Slide 12.11 Propagation of light: Beer–Lambert experiment (continued)

with refraction indices $m_1 = n_1 - ik_1$ and $m_2 = n_2 - ik_2$. Light with intensity I_0 in the first medium is incident on the boundary of the second medium. A key goal of the experiment is to measure the intensity of the light $I(z)$ propagating in the second medium. There are four possible scenarios of the light behavior in the second medium.

Case #1 (Slide 12.11): both media are uniform, optically identical, and nonabsorbing, i.e., $n_1 = n_2 \neq 0$ and $k_1 = k_2 = 0$. This case is the simplest, where light propagates in the second medium without any disturbance, i.e.,

$$I(z) = I_0. \tag{4.13}$$

Case #2 (Slide 12.11): Both media are uniform and nonabsorbing, but $n_1 \neq n_2 \neq 0$, and still $k_1 = k_2 = 0$. Since the real parts of both media are nonzero and different, two optical phenomena will be observed on the boundary of two media. Some portion of the light will be scattered, and the remaining part will be refracted into the second medium, resulting in the intensity

$$I(z) = I_0 - I_{\text{sca}}, \tag{4.14}$$

where I_{sca} is the intensity of the scattered portion of the light on the boundary of the two media.

Case #3 (Slide 12.12): Both media are uniform, but a second medium is an absorbing medium with $n_1 = n_2 \neq 0$, $k_1 = 0$ and $k_2 \neq 0$. In this case, the intensity of the light in the second medium is determined by the absorption phenomenon only.

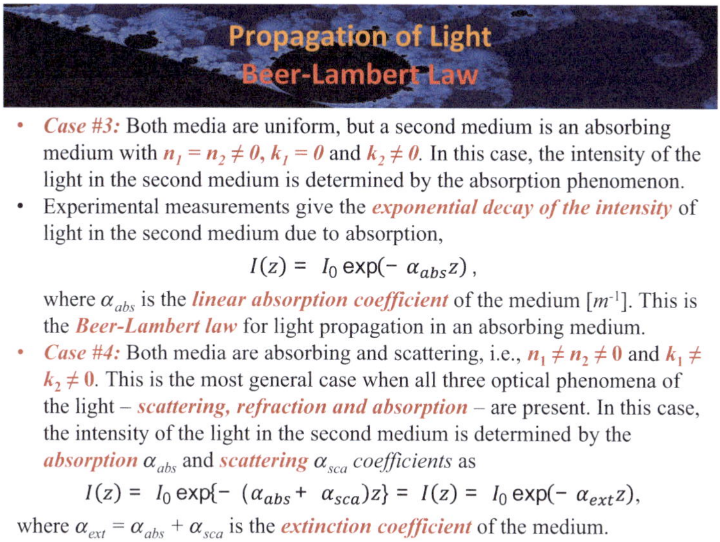

Slide 12.12 Propagation of light: Beer–Lambert law

Experimental measurements give the exponential decay of the intensity of light in the second medium due to absorption:

$$I(z) = I_0 \exp(-\alpha_{\text{abs}} z), \tag{4.15}$$

where α_{abs} is the *linear absorption coefficient* of the medium [m^{-1}]. This is the *Beer–Lambert law* for light propagation in an absorbing medium.

Case #4 (Slide 12.12): Both media are absorbing and scattering, i.e., $n_1 \neq n_2 \neq 0$ and $k_1 \neq k_2 \neq 0$. This is the most general case when all three optical phenomena of the light—scattering, refraction, and absorption—are present. In this case, the intensity of the light in the second medium is determined by the *absorption* α_{abs} and *scattering* α_{sca} *coefficients* as

$$I(z) = I_0 \exp\{-(\alpha_{\text{abs}} + \alpha_{\text{sca}})z\} = I(z) = I_0 \exp(-\alpha_{\text{ext}} z), \tag{4.16}$$

where $\alpha_{\text{ext}} = \alpha_{\text{abs}} + \alpha_{\text{sca}}$ is the *extinction coefficient* of the medium.

4.1.3 Maxwell's Equations

In the case of scattering and absorption caused by *nanoparticles*, the coefficients α_{ext}, α_{abs}, and α_{sca} can be determined from Maxwell's equations by taking into account the morphological and optical properties of the particles as well as the

Maxwell's Equations

- In the case of scattering and absorption caused ***by nanoparticles***, the coefficients α_{ext}, α_{abs} and α_{sca} can be determined from ***Maxwell's equations*** by taking into account the morphological and optical properties of the particles as well as the medium surrounding them.
- ***Maxwell's equations*** in differential form are given by

$$\nabla \times \boldsymbol{B} = \mu_0 \boldsymbol{J} + \mu_0 \varepsilon_0 \frac{\partial \boldsymbol{E}}{\partial t},$$
$$\nabla \times \boldsymbol{E} = -\frac{\partial \boldsymbol{B}}{\partial t},$$
$$\nabla \cdot \boldsymbol{E} = \frac{\rho}{\varepsilon_0},$$
$$\nabla \cdot \boldsymbol{B} = 0,$$

where $\boldsymbol{J}$ is the ***current density***, ρ is the ***charge density***, ε_0 is the ***vacuum permittivity***, and μ_0 is the ***vacuum permeability***.

- ***The first equation*** is Ampere's law.
- ***The second equation*** is a Faraday's law of induction.
- ***The third equation*** is a Gauss' law for electricity, and the ***fourth equation*** is a Gauss' law for magnetism.

Slide 12.13 Maxwell's equations

medium surrounding them (Slide 12.13). These equations unite the electric and magnetic phenomena and are given in differential form as (Slide 12.13)

$$\begin{aligned}\nabla \times \boldsymbol{B} &= \mu_0 \boldsymbol{J} + \mu_0 \varepsilon_0 \frac{\partial \boldsymbol{E}}{\partial t},\\ \nabla \times \boldsymbol{E} &= -\frac{\partial \boldsymbol{B}}{\partial t},\\ \nabla \cdot \boldsymbol{E} &= \frac{\rho}{\varepsilon_0},\\ \nabla \cdot \boldsymbol{B} &= 0,\end{aligned} \tag{4.17}$$

where $\boldsymbol{J}$ is the current density, ρ is the charge density, ε_0 is the vacuum permittivity, and μ_0 is the vacuum permeability. The first equation in the system (Eq. (4.17)) is known as Ampere's law. The second equation is Faraday's law of induction. The third equation is Gauss' law for electricity, and the fourth is Gauss' Law for magnetism (Slide 12.13).

Making a reasonable assumption that there is no inherent free charge on the spheres and no current flow (i.e., $\rho_j = 0$ and $J_f = 0$), the system of equations (4.17) can be simplified to the following form (Slide 12.14):

$$\begin{aligned}\nabla \times \boldsymbol{B} &= \mu_0 \varepsilon_0 \frac{\partial \boldsymbol{E}}{\partial t}, \quad \nabla \cdot \boldsymbol{B} = 0,\\ \nabla \times \boldsymbol{E} &= -\frac{\partial \boldsymbol{B}}{\partial t}, \quad \nabla \cdot \boldsymbol{E} = 0.\end{aligned} \tag{4.18}$$

Wave Equations

- If there is ***no inherent free charge*** on the spheres and ***no current flow*** (i.e., $\rho_j = 0$ and $J_f = 0$), Maxwell's equations take the following form:

$$\nabla \times \boldsymbol{B} = \mu_0 \varepsilon_0 \frac{\partial \boldsymbol{E}}{\partial t}, \qquad \nabla \cdot \boldsymbol{B} = 0,$$

$$\nabla \times \boldsymbol{E} = -\frac{\partial \boldsymbol{B}}{\partial t}, \qquad \nabla \cdot \boldsymbol{E} = 0.$$

- ***This system*** can be reduced to two differential equations (wave equations) for the electric and magnetic fields:

$$\nabla^2 \boldsymbol{B} = \mu_0 \varepsilon_0 \frac{\partial^2 \boldsymbol{B}}{\partial t^2},$$

$$\nabla^2 \boldsymbol{E} = \mu_0 \varepsilon_0 \frac{\partial^2 \boldsymbol{E}}{\partial t^2}.$$

- These are the ***wave equations*** that describe the propagation of the electromagnetic wave and all phenomena associated with wave behavior, like scattering, diffraction, interference, etc.

Slide 12.14 Wave equations

By applying the transformations described in any standard physics or optics textbooks, this system can be reduced to the following two differential equations (*wave equations*) for the electric and magnetic fields (Slide 12.14):

$$\nabla^2 \boldsymbol{B} = \mu_0 \varepsilon_0 \frac{\partial^2 \boldsymbol{B}}{\partial t^2}, \qquad \nabla^2 \boldsymbol{E} = \mu_0 \varepsilon_0 \frac{\partial^2 \boldsymbol{E}}{\partial t^2}. \tag{4.19}$$

These are the *wave equations* that describe the propagation of the electromagnetic wave and all phenomena associated with wave behavior, such as scattering, diffraction, interference, etc.

Expressing the time dependence of the travelling plane waves as $\boldsymbol{E} \to \boldsymbol{E}e^{-i\omega t}$ and $\to \boldsymbol{B}e^{-i\omega t}$ the wave equations (4.19) can be transformed to the space vector Helmholtz equations (Slide 12.15),

$$\nabla^2 \boldsymbol{B} + k^2 \boldsymbol{B} = 0, \qquad \nabla^2 \boldsymbol{E} + k^2 \boldsymbol{E} = 0, \tag{4.20}$$

where $k^2 = \mu_0 \varepsilon_0 \omega^2$.

Helmholtz Equations

- Expressing the time dependence of the travelling plane-waves as $\boldsymbol{E} \rightarrow \boldsymbol{E}e^{-i\omega t}$ and $\boldsymbol{B} \rightarrow \boldsymbol{B}e^{-i\omega t}$, the ***wave equations*** can be transformed to the ***space vector Helmholtz equations***,

$$\nabla^2\boldsymbol{B} + k^2\boldsymbol{B} = 0,$$
$$\nabla^2\boldsymbol{E} + k^2\boldsymbol{E} = 0,$$

where $k^2 = \mu_0\varepsilon_0\omega^2$.

- The scattering of the wave on a spherical particle can be described by the ***scalar wave equation*** for the electrical field ***in the spherical system of coordinates***:

$$\frac{1}{r}\cdot\frac{\partial^2}{\partial r^2}\cdot(r\cdot E)+\frac{1}{r^2\cdot\sin\theta}\cdot\frac{\partial}{\partial\theta}\cdot\left(\sin\theta\cdot\frac{\partial E}{\partial\theta}\right)+\frac{1}{r^2\cdot\sin^2\theta}\cdot\frac{\partial^2 E}{\partial\phi^2}+k^2E=0.$$

Slide 12.15 Helmholtz equations

The scattering of the wave on a spherical particle can be described by the scalar wave equation for the electrical field in the spherical system of coordinates (Slide 12.15):

$$\frac{1}{r}\cdot\frac{\partial^2}{\partial r^2}\cdot(r\cdot E)+\frac{1}{r^2\cdot\sin\theta}\cdot\frac{\partial}{\partial\theta}\cdot\left(\sin\theta\cdot\frac{\partial E}{\partial\theta}\right)+\frac{1}{r^2\cdot\sin^2\theta}\cdot\frac{\partial^2 E}{\partial\varphi^2}+k^2E=0. \tag{4.21}$$

4.1.4 Lorenz–Mie Formalism

An analytical solution of the wave equation (4.21) for the spherical particles was found by the Danish physicist Ludwig Lorenz in 1890, and also independently by the German physicist Gustav Mie in 1908, who was working on the problem of explaining the color of colloidal gold particles in water (Slide 12.16). Thus, the optimal wavelength of radiation and optimal size range of nanoparticles for effective heating/activating of the nanoparticles can be found by using an extended *Lorentz–Mie* diffraction theory, taking into account *the plasmon-resonance absorption effect* observed in metal nanoparticles [1].

In the most general case, the Lorentz–Mie solutions give the *scattering matrix* of particle j, $S^j\ (\theta,\varphi)$ [1]. This consists of four complex functions, $S_i^j(\theta,\varphi)$ ($i=1, 2, 3, 4$), describing the amplitude and phase of a scattered scalar wave in any direction (Slide 12.16). Forward scattering ($\theta=0°$) contains the attenuation process of the electromagnetic wave, and for the case of spherical particles,

Lorenz-Mie Formalism

- An ***analytical solution*** of the wave equation for the spherical particles was found by the Danish physicist ***Ludwig Lorenz*** in 1890 in Danish, and also independently by the German physicist ***Gustav Mie*** in 1908, who was working on the problem of explaining the color of colloidal gold particles in water.
- Thus, optimal wavelength of a radiation and optimal size range of nanoparticles for effective heating/activating the nanoparticles can be found by using an ***extended Lorentz-Mie diffraction theory***, taking into account the ***plasmon-resonance absorption effect*** observed in metal nanoparticles.
- In the most general case, the Lorentz-Mie solutions give the ***scattering matrix of particle j***, $S^j(\theta,\varphi)$. This consists of four complex functions, $S_i^j(\theta,\varphi)$ $(i = 1,2,3,4)$, describing the amplitude and phase of a scattered scalar wave in any direction. Forward scattering ($\theta = 0°$) contains the attenuation process of an electromagnetic wave, and for the case of spherical particles, $S_3^j = S_4^j = 0$.

Slide 12.16 Lorenz–Mie formalism

Lorenz-Mie Formalism (continued)

- We can limit the description to a single ***scattering amplitude function***:

$$S^j(0) = S_1^j(0) = S_2^j(0) = \frac{1}{2}\Sigma_{l=1}^{\infty}(2l+1)\left(a_l^j + b_l^j\right).$$

- ***The Mie coefficients a_l and b_l*** contain the characteristics of the dispersal medium and are calculated through the ***cylindrical Bessel function*** of the first kind, $\psi_l(y)$, and the ***Hankel function*** of the second kind, $\xi_l(\rho)$, both with half-integral indices:

$$a_l = \frac{\psi_l'(y)\psi_l(\rho) - \widetilde{m}\psi_l(y)\psi_l'(\rho)}{\psi_l'(y)\xi_l(\rho) - \widetilde{m}\psi_l(y)\xi_l'(\rho)},\quad b_l = \frac{\widetilde{m}\psi_l'(y)\psi_l(\rho) - \psi_l(y)\psi_l'(\rho)}{\widetilde{m}\psi_l'(y)\xi_l(\rho) - \psi_l(y)\xi_l'(\rho)}.$$

- Here, $\widetilde{m} = m_0/m_1$ is the relative value of the refractive index of thc medium.

Slide 12.17 Lorenz–Mie formalism (continued)

$S_3^j = S_4^j = 0$. Thus, we can limit the description to a single *scattering amplitude function* [1] (Slide 12.17):

$$S^j(0) = S_1^j(0) = S_2^j(0) = \frac{1}{2}\sum_{l=1}^{\infty}(2l+1)\left(a_l^j + b_l^j\right) \tag{4.22}$$

The *Mie coefficients* a_l and b_l contain the characteristics of the dispersal medium and are calculated through the cylindrical Bessel function of the first kind, $\psi_l(y)$, and the Hankel function of the second kind, $\xi_l(\rho)$, both with half-integral indices (Slide 12.17):

$$a_l = \frac{\psi_l'(y)\psi_l(\rho) - \widetilde{m}\psi_l(y)\psi_l'(\rho)}{\psi_l'(y)\xi_l(\rho) - \widetilde{m}\psi_l(y)\xi_l'(\rho)}, \quad b_l = \frac{\widetilde{m}\psi_l'(y)\psi_l(\rho) - \psi_l(y)\psi_l'(\rho)}{\widetilde{m}\psi_l'(y)\xi_l(\rho) - \psi_l(y)\xi_l'(\rho)}. \tag{4.23}$$

Here, $\widetilde{m} = m_0/m_1$ is the relative value of the refractive index of the medium; $m_0 = n_0 - ik_0$ and $m_1 = n_1 - ik_1$ are the complex refractive indices of the particle material and the aqueous suspension, respectively; $\rho = 2\pi r_0/\lambda$ is the *Mie parameter*; and $y = 2\pi r_0 n_0/\lambda$, $\psi_l(u) = (\pi u/2)^{1/2} J^{(1)}_{l+1/2}$, $\xi_l(u) = (\pi u/2)^{1/2} H^{(2)}_{l+1/2}$, and $\psi_l' = d\psi_l(u)/du$. With knowledge of $S^j(0)$, it is possible to calculate the optical properties of the particles (i.e., the dimensionless *efficiencies of scattering*, $Q^j_{\text{sca}}(\rho,\widetilde{m}) = \sigma^j_{\text{sca}}(\rho,\widetilde{m})/\sigma_0$ *absorption*, $Q^j_{\text{abs}}(\rho,\widetilde{m}) = \sigma^j_{\text{abs}}(\rho,\widetilde{m})/\sigma_0$ and *extinction*, $Q^j_{\text{ext}}(\rho,\widetilde{m}) = \sigma^j_{\text{ext}}(\rho,\widetilde{m})/\sigma_0$ of the radiation at a given wavelength) as (Slide 12.18)

$$\begin{aligned} Q^j_{\text{ext}}(\rho,\widetilde{m}) &= \frac{4\pi}{k^2}\,\text{Re}\{S^j(0)\}, \\ Q^j_{\text{sca}}(\rho,\widetilde{m}) &= \frac{2}{\rho^2}\sum_{l=1}^{\infty}(2l+1)\left\{\left|a_l^j\right|^2 + \left|b_l^j\right|^2\right\}, \\ Q^j_{\text{abs}}(\rho,\widetilde{m}) &= Q_{\text{ext}}(\rho,\widetilde{m}) - Q_{\text{sca}}(\rho,\widetilde{m}), \end{aligned} \tag{4.24}$$

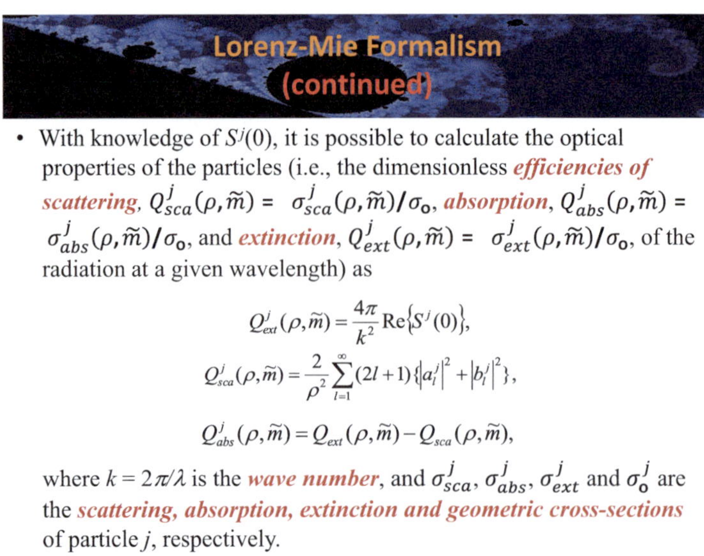

Slide 12.18 Lorenz–Mie formalism (continued)

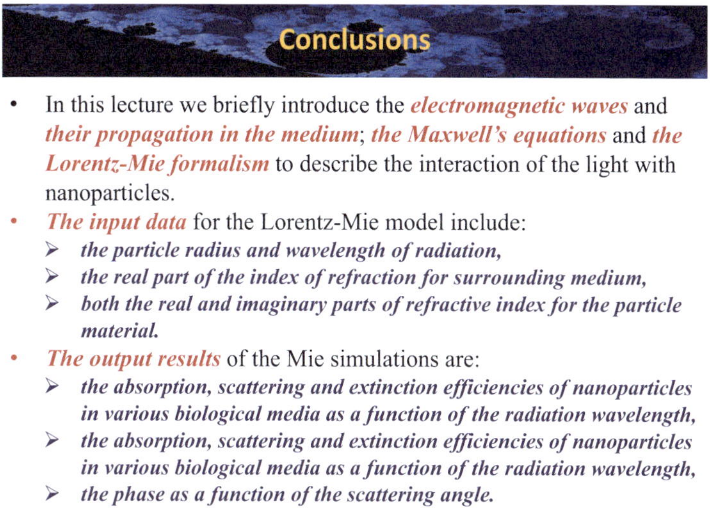

Slide 12.19 Conclusions

where $k = 2\pi/\lambda$ is the wave number, and $\sigma^{j}_{\text{sca}}(\rho, \widetilde{m})$, $\sigma^{j}_{\text{abs}}(\rho, \widetilde{m})$, $\sigma^{j}_{\text{ext}}(\rho, \widetilde{m})$, and σ_0 are the *scattering, absorption, extinction, and geometric cross-sections* of particle j, respectively.

Thus, the Lorentz–Mie formalism (Eqs. (4.22)–(4.24)) requires the use of two dimensionless input parameters, $\rho = 2\pi r_0/\lambda$ and $\delta = \rho\widetilde{m}$, where $\widetilde{m}$ is the relative value of the complex refractive index of the nanoparticles in the surrounding medium at the wavelength λ [1]. It allows us to simulate the absorption, scattering, and extinction efficiencies of nanoparticles in various biological media as a function of the wavelength of the radiation and radius of nanoparticles, as well as the phase functions as a function of the scattering angle.

In conclusion, in this lecture we briefly introduce the electromagnetic waves and their propagation in the medium; the Maxwell's equations and the Lorentz–Mie formalism to describe the interaction of the light with nanoparticles (Slide 12.19). The input data for the Lorentz–Mie model include:

- The particle radius and wavelength of radiation,
- The real part of the index of refraction for the surrounding medium,
- Both the real and imaginary parts of refractive index for the particle material.

The output results of the Mie simulations are as follows:

- The absorption, scattering, and extinction efficiencies of nanoparticles in various biological media as a function of the radiation wavelength,
- The absorption, scattering, and extinction efficiencies of nanoparticles in various biological media as a function of the nanoparticle radius,
- The phase as a function of the scattering angle.

4.2 Lecture 13: X-ray Optics of Nanoparticles (Slide 13.1)

- Introduction
- Wave Optics vs. X-Ray Optics
- Optical Constants for X-Ray Photons
  - *Simulation Example*
- Nanoparticle Optics in the X-Ray Range of the Spectrum
  - *X-Ray Optics of Gold Nanoparticles*
- Effective Intensity-Absorption Efficiency
- Conclusions

4.2.1 Introduction

In Sect. 3.2 (Lecture 10) of Chap. 3, we discussed methods *of nanodrug activation*. By m*ethods of activation*, we mean using different types of radiation for a total body scan (Slide 13.2). For the activation of nanoparticles, we heat them to initiate cell ablation, photoelectron emission, thermoionic emission, Compton scattering, generation of shock waves and sound waves, or mechanical waves of high pressure that are caused by the applied radiation. Unfortunately, biological tissues are very good at scattering visible light, which leads to lower light intensities

Contents:

- Introduction
- Wave Optics vs X-Ray Optics
- Optical Constants for X-Ray Photons
- Nanoparticle Optics in the X-Ray Range of the Spectrum
- Effective Intensity-Absorption Efficiency
- Conclusions

Slide 13.1 Contents of Lecture 13

Introduction

- By ***methods of activation***, we mean using different types of radiation for a total body scan.
- ***For the activation of nanoparticles***, we heat them to initiate cell ablation, photoelectron emission, thermoionic emission, Compton scattering, generation of shock waves and sound waves, or mechanical waves of high pressure that are caused by the applied radiation.
- Unfortunately, ***biological tissues*** are very good at ***scattering visible light***, which leads to lower light intensities reaching the nanoparticles if they are under too much tissue, and also leads to greater light absorption in the surrounding tissues.
- Researchers are trying to find wavelengths that make a good compromise between ***maximizing heat generation in the nanoparticle and minimizing the scattering*** and absorption by surrounding tissues.

Slide 13.2 Introduction

reaching the nanoparticles if they are under too much tissue, and also leads to greater light absorption (and therefore heat generation) in the surrounding tissues [2] (Slide 13.2).

Researchers are trying to find wavelengths that make a good compromise between maximizing heat generation in the nanoparticle and minimizing the scattering and absorption by surrounding tissues [2]. The search for a universal tool for nanoparticle activation presented in Sect. 3.2 showed that the very ends of the electromagnetic spectrum—X-rays and radiofrequency (RF) waves—are the best candidates for a total body scan to activate the nanoparticles inside the body (Slide 13.3). For example, short X-rays have excellent ability to penetrate into the human body and have high photon energies, giving a great opportunity to activate/heat nanoparticles inside the body for diagnostic and treatment purposes [2].

In this Lecture 13, we focus on X-ray optics of nanoparticles. For example, gold nanoparticles have been investigated as contrast agents for traditional X-ray medical procedures, utilizing the strong absorption characteristics of the nanoparticles to enhance the contrast of the detected X-ray image [20]. For that reason, here we use X-ray radiation to investigate the optical properties of gold nanoparticles as contrast agents in the X-ray range of the spectrum.

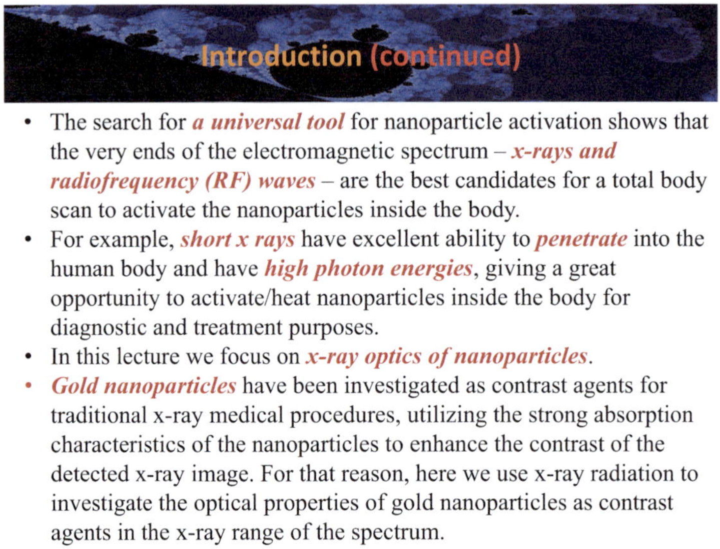

Slide 13.3 Introduction (continued)

4.2.2 *Wave Optics vs. X-ray Optics*

This section contains material reproduced from our publication [2]. Both X-rays and visible light are forms of electromagnetic radiation, but for a number of reasons are often treated differently in practice (Slide 13.4). Optics traditionally describes electromagnetic radiation by its wavelength using a linear scale, with visible light comprising the spectrum from violet to red (380–750 nm). Even now, the optics commonly uses the linear wavelength convention, though this area of study has expanded beyond visible light into both ultraviolet (UV) radiation (10–400 nm—extreme UV is classified down to 10 nm) and infrared (IR) radiation (700–1000 μm). X-rays, however, were originally discovered in early cathode ray experiments where high velocity electrons collide with the anode, producing bremsstrahlung radiation of X-rays. The maximum velocity of the electrons hitting the anode was dependent upon the voltages used to accelerate them from cathode to anode, and these voltages ranged from hundreds to hundreds of thousands of volts. The best way to describe this radiation was therefore using a logarithmic scale with the units of electron volts, eV, which describes the energy of the radiation.

The electromagnetic spectrum defines X-rays by the energy range of 124 eV to 1 keV [2] (Slide 13.4). The modern wave-particle understanding of electromagnetic radiation shows that the wavelength (λ) of light and the energy (E) of its photons are related by the equation

$$E = hc/\lambda, \tag{4.25}$$

where h is Planck's constant and c is the speed of light. Thus, conversion from photon wavelength to photon energy is relatively simple. However, one single

Wave Optics vs X-Ray Optics

Wave Optics	X-Ray Optics
Wavelength Frequency	Photon Energy: 124 eV to 124 keV.
Intensity	Dose, Exposure
Index of Refraction	KERMA, Stopping Power

- The electromagnetic spectrum defines ***x-rays*** by the energy range of ***124 eV to 1 keV***.
- The modern ***wave-particle*** understanding of electromagnetic radiation shows that the wavelength (λ) of light and energy (E) of its photons are related by the equation

$$E = hc/\lambda,$$

where h is ***Planck's constant*** and c is the speed of light.

Slide 13.4 Wave optics vs. X-ray optics

convention does not win over the other because of the convenience of the historical units in each respective field. In optics, phenomena like interference typically display observable and measurable qualities, like fringe spacing, proportionally to λ. Furthermore, over most of the optical spectrum, from far-UV (122 nm) to near-IR (1.4 μm), the wavelength changes by only a single order of magnitude, making the linear nanometer scale very convenient. The X-ray spectrum, however, covers energies spanning three orders of magnitude, and thus a logarithmic scale is generally much more suitable.

In this section, we simulate the optical constants for pure gold over a wide range of X-ray photon energies, 0.1 keV to 10 MeV. Then, we use these optical constants to calculate the absorption efficiency of various size gold nanoparticles in the X-ray range of the spectrum. Utilizing the tungsten anode spectral model interpolating polynomials (TASMIP) model [3], we introduce the *effective intensity-absorption efficiency* values for gold nanoparticles of different radii, which take into account the spectral distribution of energies generated by typical tungsten anode X-ray tubes with kilovolt potentials ranging from 50 to 150 kVp.

4.2.3 Optical Constants for X-ray Photons

This section contains material reproduced from our publication [2]. As we discussed in Sect. 4.1 of this chapter, in the optical sciences, materials are described by their complex refractive index, m (Slide 13.5). The real part is the refractive index, n, and describes how a material diffracts light with λ. The imaginary part, k, is related to the propagation loss associated with light absorption by the material,

Optical Constants for X-Ray Photons

- ***In the optical sciences***, materials are described by their ***complex refractive index, m***.
- ***The real part is the refractive index, n***, and describes how a material diffracts light with λ.
- ***The imaginary part, k,*** is related to the propagation loss associated with light absorption by the material, where $k \neq 0$ denotes attenuation of the incoming light: $m = n - ik$.
- ***X-rays*** are not absorbed by matter as strongly as visible light; therefore, most materials are relatively ***transparent to x-rays***.
- ***X rays*** also ***do not diffract much*** when passing through media, so that n is almost always very nearly equal to 1.
- ***In x-ray optics***, it is therefore more common to use the following definition for ***the complex index***:

$$m = 1 - \delta - i\beta,$$

where δ is the ***refractive index decrement***, and β is the ***absorption index***.

Slide 13.5 Optical constants for X-ray photons

where $k \neq 0$ denotes attenuation of the incoming light: $m = n - ik$. X-rays are not absorbed by matter as strongly as visible light, so that most materials are relatively transparent to X-rays. Medical X-ray imaging devices are able to produce images of bones and tissues utilizing that fact. X-rays also do not diffract much when passing through media, so that n is almost always very nearly equal to 1. In X-ray optics, it is therefore more common to use the following definition for the complex index [2] (Slide 13.5):

$$m = 1 - \delta - i\beta, \tag{4.26}$$

where δ is the refractive index decrement, and β is the absorption index. For photon energies above 30 eV, these optical properties of a material can be described in terms of the atomic scattering factors of the atoms by the equation (Slide 13.6) [4]

$$m = 1 - \frac{r_e}{2\pi}\lambda^2\sum_j N_j \cdot \widetilde{f}_j, \tag{4.27}$$

where r_e is the classical electron radius, the subscript j refers to different atomic elements in the unit volume, N_j refers to the number of atoms of element j per unit volume, and $\widetilde{f}_j$ is the complex atomic scattering factor $\widetilde{f}_j$ Slide 13.6:

$$\begin{aligned} \widetilde{f}_j &= f_1 + if_2 \\ f_1 &= \frac{\sigma_a}{2r_e a} \\ f_2 &= Z^* + \frac{1}{\pi r_e hc}\int_0^\infty \frac{\varepsilon^2\sigma_a(\varepsilon)}{E^2 - \varepsilon^2}d\varepsilon \end{aligned} \tag{4.28}$$

Optical Constants for X-Ray Photons (continued)

- For photon energies above 30 eV, these optical properties of a material can be described in terms of the ***atomic scattering factors*** of the atoms by the equation

$$m = 1 - \frac{r_e}{2\pi}\lambda^2 \sum_j N_j \cdot \tilde{f}_j,$$

where r_e is the classical electron radius, the subscript j refers to different atomic elements in the unit volume, N_j refers to the number of atoms of element j per unit volume, and $\tilde{f}_j$ is the complex atomic scattering factor.

- The ***atomic scattering factors*** are the real (f_1) and imaginary (f_2) components of the complex atomic scattering factor:

$$\tilde{f}_j = f_1 + if_2.$$

Slide 13.6 Optical constants for X-ray photons (continued)

Here, the real part, f_1, is proportional to the atomic photoabsorption cross-section, σ_a, the imaginary part, f_2, can be found from f_1 through the Kramers–Kronig relation [4], and Z^* is the atomic number of the element modified by a small relativistic correction. The complex index of refraction can therefore be calculated using the photoabsorption cross-section. However, rather than calculating the atomic scattering factors by this method we will choose to use tabulated values for f_1 and f_2 from available data [5, 6]. Using these relations, we can find expressions for δ and β in terms of f_1 and f_2 (Slide 13.7):

$$\delta = \frac{r_e}{2\pi}\lambda^2 \sum_j N_j \cdot f_{j,1}, \quad \beta = \frac{r_e}{2\pi}\lambda^2 \sum_j N_j \cdot f_{j,2}. \tag{4.29}$$

Therefore, in the situation where the medium is composed of a single element (in our case, gold), we have (Slide 13.8)

$$m(\lambda) = \left(1 - \frac{r_e}{2\pi}\lambda^2 \cdot N \cdot f_2(\lambda)\right) - i \cdot \left(\frac{r_e}{2\pi}\lambda^2 \cdot N \cdot f_1(\lambda)\right), \tag{4.30}$$

and in the case that f_1 and f_2 are tabulated in terms of energy rather than wavelength (Slide 13.8),

$$m(E) = \left(1 - \frac{r_e}{2\pi}\left(\frac{E}{hc}\right)^2 \cdot N \cdot f_2(E)\right) - i \cdot \left(\frac{r_e}{2\pi}\left(\frac{E}{hc}\right)^2 \cdot N \cdot f_1(E)\right) \tag{4.31}$$

The summation technique, however, is still useful for determining the optical properties of compounds and mixtures of elements. Because the effective distance

Optical Constants for X-Ray Photons (continued)

- The real part, f_1, is proportion to the atomic photoabsorption cross-section, σ_a. The imaginary part, f_2, can be found from f_1 through the Kramers-Kronig relation as

$$f_1 = \frac{\sigma_a}{2r_e a}, \qquad f_2 = Z^* + \frac{1}{\pi r_e hc}\int_0^\infty \frac{\varepsilon^2\sigma_a(\varepsilon)}{E^2-\varepsilon^2}d\varepsilon,$$

where Z^* is the atomic number of the element modified by a small relativistic correction.

- Rather than calculating the atomic scattering factors by this method, we will choose to use tabulated values for f_1 and f_2 from available data. Using these relations. we can find expressions for δ and b in terms of f_1 and f_2:

$$\beta = \frac{r_e}{2\pi}\lambda^2\sum_j N_j \cdot f_{j,2} \qquad \delta = \frac{r_e}{2\pi}\lambda^2\sum_j N_j \cdot f_{j,1}$$

Slide 13.7 Optical constants for X-ray photons (continued)

Optical Constants for X-Ray Photons (Index of Refraction)

- Therefore, in the situation where the medium is composed of a single element (in our case, gold), we have

$$m(\lambda) = \left(1 - \frac{r_e}{2\pi}\lambda^2 \cdot N \cdot f_2(\lambda)\right) - i \cdot \left(\frac{r_e}{2\pi}\lambda^2 \cdot N \cdot f_1(\lambda)\right)$$

and in the case that f_1 and f_2 are tabulated in terms of energy rather than wavelength,

$$m(E) = \left(1 - \frac{r_e}{2\pi}\left(\frac{E}{hc}\right)^2 \cdot N \cdot f_2(E)\right) - i \cdot \left(\frac{r_e}{2\pi}\left(\frac{E}{hc}\right)^2 \cdot N \cdot f_1(E)\right)$$

- The summation technique, however, is still useful for determining the *optical properties of compounds and mixtures of elements*. Because the effective distance between chemically bonded atoms is much larger than an x-ray wavelength, no significant resonance effects are observed.

Slide 13.8 Optical constants for X-ray photons: index of refraction

between chemically bonded atoms is much larger than an X-ray wavelength, no significant resonance effects are observed, and optical constants of mixtures can be calculated based on the weighted summation of the individual constituent elements. The mass fractions of individual elements in different biological tissues can be found, with the optical properties of these tissues calculated in a very straightforward manner. This is very different from the visual, UV, and IR optical properties of biological tissues, which can show widely different behavior because of various resonance effects caused by certain pigments being predominant in specific tissues.

Simulation Example. The optical constants are simulated for pure gold over a wide range of photon energies, 0.1 keV to 10 MeV, utilizing the methods described earlier, and presented in Figs. 4.4, 4.5, and 4.6. Figure 4.4 shows the real, f_1, and imaginary, f_2, parts of the complex atomic scattering factor for gold computed by using the available tabulated values from [4] (Slide 13.9). The refractive index decrement, δ, and the absorption index, β, are computed from Eq. (4.30) and shown in Fig. 4.5 (Slide 13.10). The complex index of refraction for pure gold in the X-ray range of the spectrum is calculated from Eqs. (4.30) and (4.31); it is shown as a function of energy in Fig. 4.6a and as a function of wavelength in Fig. 4.6b (Slide 13.11). It is evident from both figures that X-rays do not diffract much when passing through media, so that n is nearly equal to 1 for almost all photon energies staring from 1 keV and higher. The noticeable absorption of X-rays is observed only for low-energy photons (long wavelengths) with energies less than 100 keV (for a wavelength range of 1–100 nm), corresponding to the photoelectric absorption range (Slide 13.11). The absorption part of the refractive index, k, drops rapidly for higher energy photons (for shorter wavelengths), substantially increasing the penetration depth of the high-energy photons into the material accompanied by the Compton scattering effect.

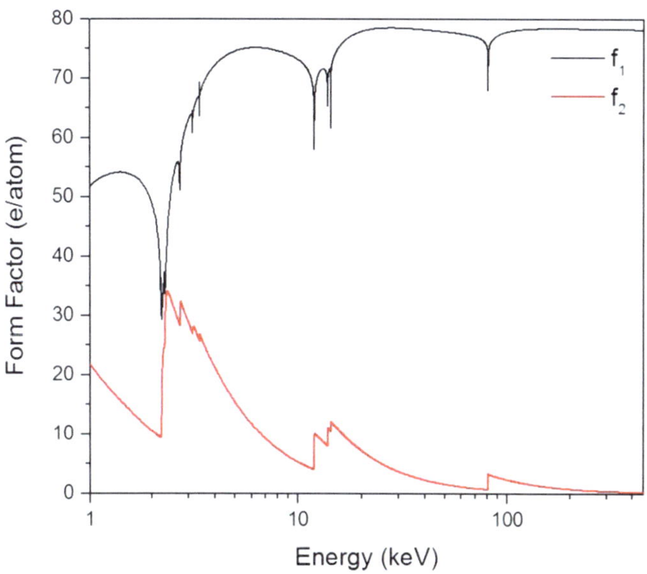

Fig. 4.4 Real, f_1, and imaginary, f_2, parts of the complex atomic scattering factor for pure gold computed by using the tabulated values from [5] for photon energies over the 2–433 keV range. Reprinted from [2]. © With permission from OSA

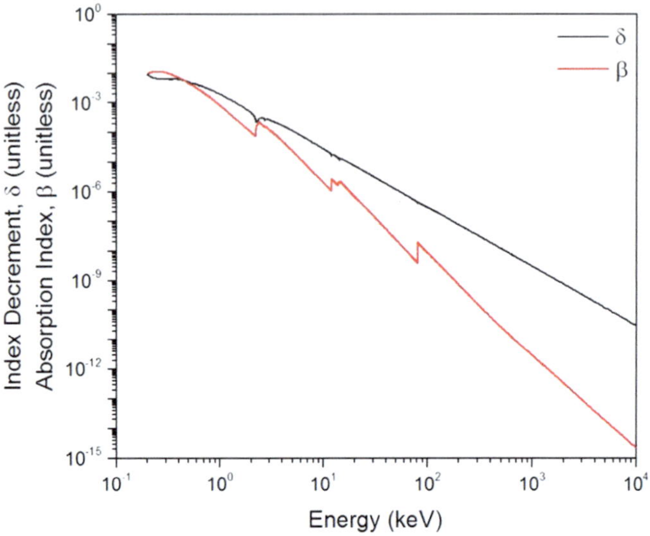

Fig. 4.5 Refractive index decrement, δ, and the absorption index, β, for pure gold for photon energies over the 0.1 keV to 10 MeV range. Reprinted from [2]. © With permission from OSA

4.2.4 Nanoparticle Optics in the X-ray Range of the Spectrum

This section contains material adopted from our publication [2]. In Sect. 4.1 (Lecture 12) of this chapter, we introduced the Lorenz–Mie theory describing the scattering of electromagnetic radiation by a small sphere (Slide 13.12). This theory helps to explain the mechanism of scattering by small particles as well as nanoparticle excitation. By using the indices of refraction for the nanoparticle and surrounding medium, the Lorenz–Mie theory allows us to calculate the optical properties of nanoparticles, such as absorption, scattering, and extinction efficiencies. Since we have found the real and imaginary parts of index of refraction given by Eqs. (4.30) and (4.31), now we can use Lorenz–Mie formalism to simulate the nanoparticle optics in the X-ray range of the spectrum (Slide 13.12).

As we saw in Sect. 4.1, the Lorenz–Mie solutions are analytic but take the form of an analytic infinite series given by Eq. (4.24). Because of this series, exact solutions cannot technically be achieved, but computed solutions can be brought arbitrarily close to the exact solution simply by computing more and more terms in the series. Essentially, Lorenz–Mie scattering models a homogeneous sphere in an infinite surrounding medium (also homogeneous) being struck by incoming light. The solution requires us to define the size of the sphere, the wavelength of the incoming light, and the complex index of refraction of the sphere relative to the surrounding medium for a given wavelength of the radiation. From this, the Lorenz–Mie theory produces many results such as Q_{ext}, the extinction efficiency;

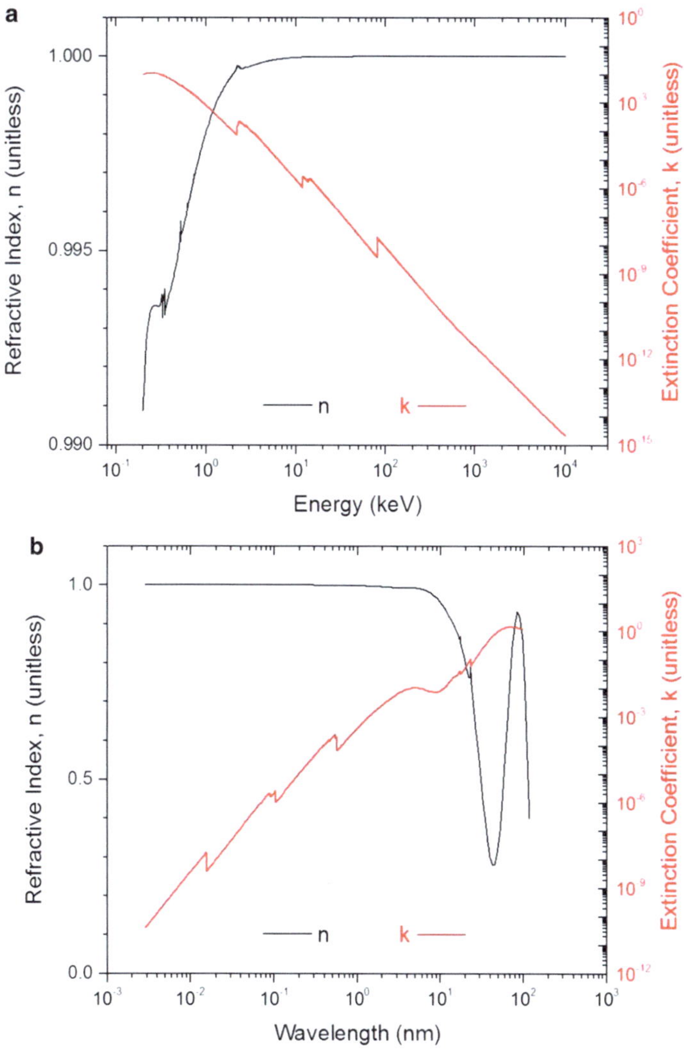

Fig. 4.6 Complex index of refraction real, n, and imaginary, k, parts for pure gold for (**a**) X-ray photon energies over the range of 0.1 keV to 10 MeV range, and (**b**) as a function of wavelength of X-ray radiation valid for 0.00286 nm $\leq \lambda \leq$ 116 nm [5]. Reprinted from [2]. © With permission from OSA

Q_{sca}, the scattering efficiency; Q_{abs}, the absorption efficiency; and Q_{back}, the backscattering efficiency.

X-ray Optics of Gold Nanoparticles. The complex index of refraction for pure gold was computed using the methods described in the previous section and, taking that result, the Lorenz–Mie scattering theory was used to compute absorption efficiencies for gold nanoparticles with radii ranging from 5 to 50 nm (see Fig. 4.7). The results give the absorption efficiency as a function of wavelength [Fig. 4.7a] or energy [Fig. 4.7b] for a given nanoparticle size (Slide 13.13).

Optical Constants for X-Ray Photons (Simulation Example)

- ***The optical constants*** are simulated ***for pure gold*** over a wide range of photon energies, 0.1 keV to 10 MeV, utilizing the methods described above.
- The figure below shows the real, f_1, and imaginary, f_2, parts of the ***complex atomic scattering factor*** for gold computed by using available tabulated values.

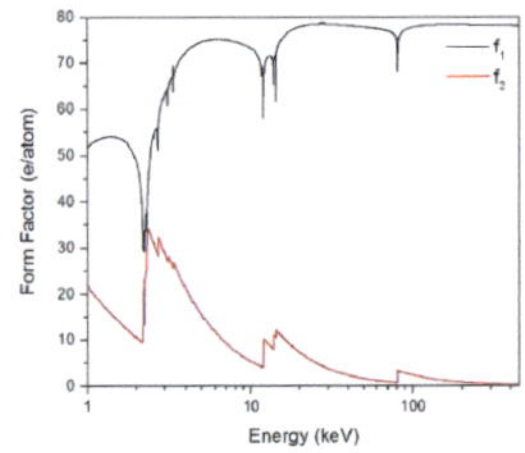

Slide 13.9 Optical constants for X-ray photons: simulation example

Optical Constants for X-Ray Photons (Simulation Example continued)

- The refractive index ***decrement***, δ, and the ***absorption index***, β, are computed from Eq. (4.30) and shown in the figure below.

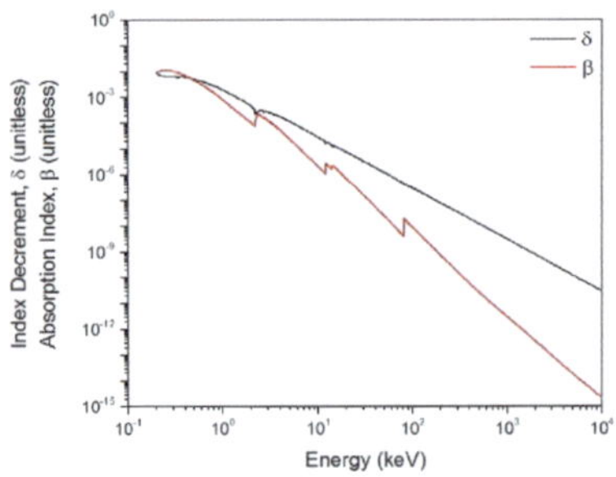

Slide 13.10 Optical constants for X-ray photons: simulation example continued

A large extinction coefficient means that little energy from an incoming wave will continue past the particle; that is, few photons will encounter the sphere and continue in the same direction afterward. According to the Lorenz–Mie formalism, extinguished light must have either been absorbed or scattered, and so the scattering

Optical Constants for X-Ray Photons (Simulation Example continued)

- The complex index of refraction for pure gold in the x-ray range of the spectrum is calculated from Eqs. (4.30) and (4.31); it is shown as a function of energy (a) and as a function of wavelength (b).

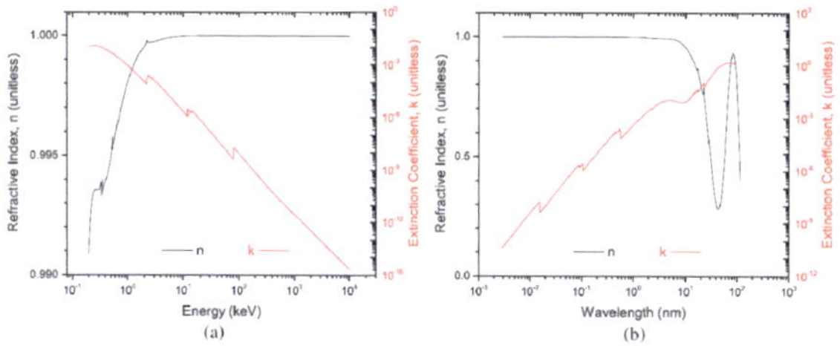

- It is evident from both figures that x-rays do not diffract much when passing through media, so that n is nearly equal to 1. The noticeable absorption of x-rays is observed only for low-energy photons with energies less than 100 keV, corresponding to the photoelectric absorption range.

Slide 13.11 Optical constants for X-ray photons: simulation example continued

Nanoparticle Optics in the X-Ray Range of the Spectrum

- Earlier we introduced the ***Lorenz-Mie theory*** describing the scattering of electromagnetic radiation by a small sphere.
- This theory helps to explain the ***mechanism of scattering*** by small particles as well as ***nanoparticle excitation***.
- By using ***the indices of refraction*** for the nanoparticle and surrounding medium, the Lorenz-Mie theory allows us to calculate the optical properties of nanoparticles, like ***absorption, scattering and extinction efficiencies***.
- Since we have found the real and imaginary parts of the index of refraction given by Eqs. (4.30) and (4.31), now we can use the Lorenz–Mie formalism to simulate the ***nanoparticle optics in x-ray range of the spectrum***.

Slide 13.12 Nanoparticle optics in the X-ray range of the spectrum

and absorption efficiencies must add to equal the extinction coefficient. Additionally, the backscattering efficiency is specifically in the reverse direction. Additional information regarding the far-field angular scattering patterns can be obtained from the Lorenz–Mie theory, but for our purposes, only the absorption efficiency Q_{abs} is needed to describe the energy absorption by the nanoparticles.

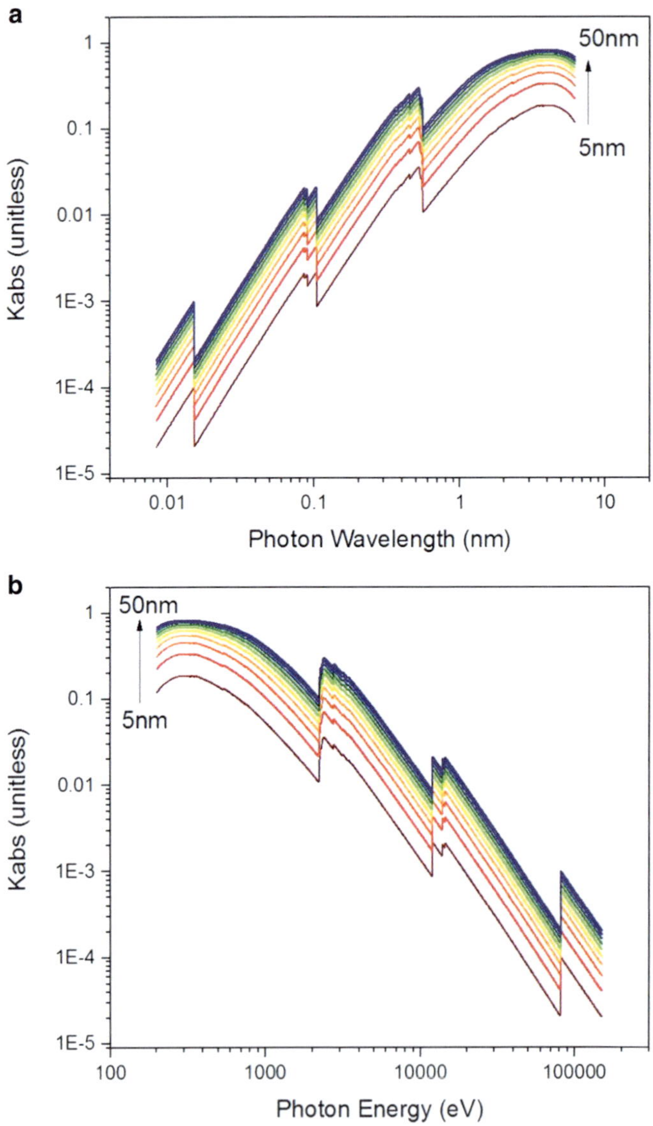

Fig. 4.7 Absorption efficiency values for gold nanoparticles in a vacuum calculated using the Mie theory. The same absorption efficiency data are presented (**a**) in relation to the wavelength, which is more common for the optical sciences, and (**b**) in relation to the photon energy, as is the norm in X-ray research. Reprinted from [2]. © With permission from OSA

4.2.5 Effective Intensity-Absorption Efficiency

This section contains material adopted from our publication [2]. The optical properties of the nanoparticles are required for many applications, such as nanophotothermolysis and nanophotohyperthermia (Slide 13.14), which exploit

X-Ray Optics of Gold Nanoparticles

- The results give the ***absorption efficiency*** as a function of wavelength (a) or energy (b) for a given gold nanoparticle size.

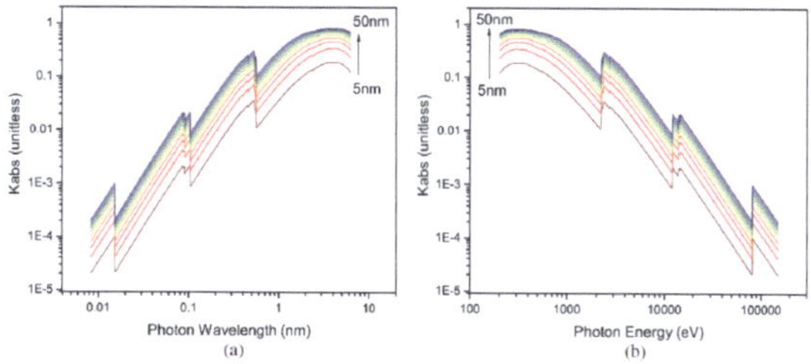

- A ***large extinction coefficient*** means that little energy from an incoming wave will continue past the particle; that is, few photons will encounter the sphere and continue in the same direction afterward.

Slide 13.13 X-ray optics of gold nanoparticles

Effective Intensity-Absorption Efficiency

- The ***optical properties of the nanoparticles*** are required for many applications, such as ***nanophotothermolysis*** and ***nanophotohyperthermia***, which exploit the strong light absorption properties of nanoparticles to generate heat in local targeted areas of the body, as discussed earlier.
- The ***time-dependent thermal model of heating the nanoparticles*** by radiation contains the product $Q_{abs}I_0$ describing the amount of radiation energy of intensity I_0 absorbed by the nanoparticle.
- The product in this form assumes a ***monochromatic light source***, which is sufficient for describing laser light.
- However, the output from most ***medical diagnostic x-ray devices*** follows a ***wide spectrum of photon energies***, i.e. $I_0(\varepsilon)$.
- Because of this, it is better ***to integrate over all spectral energies*** when trying to determine the amount of heat generated in the nanoparticle, and define an "***effective value***" for the combined terms $Q_{eff} \cdot I_{eff}$ as:

$$Q_{eff}I_{eff} = \int_0^\infty Q_{abs}(\varepsilon)I_0(\varepsilon)d\varepsilon.$$

Slide 13.14 Effective intensity-absorption efficiency

the strong light absorption properties of nanoparticles to generate heat in local targeted areas of the body, as discussed in Sect. 3.1 (Lecture 9) of Chap. 3. The time-dependent thermal model of heating the nanoparticles by radiation using the one temperature model (OTM) [7] is introduced in Chap. 6. This model contains the product $Q_{abs}I_0$ describing the amount of radiation energy of intensity I_0 absorbed by the nanoparticle. The product in this form assumes a monochromatic light source, which is sufficient for describing laser light. However, the output from most medical diagnostic X-ray devices follows a wide spectrum of photon energies, i.e., $I_0(\varepsilon)$. Because of this, it is better to integrate over all spectral energies when trying to determine the amount of heat generated in the nanoparticle, and define an "effective value" for the combined terms $Q_{\mathrm{eff}} \cdot I_{\mathrm{eff}}$ as (Slide 13.14)

$$Q_{\mathrm{eff}} I_{\mathrm{eff}} = \int_0^\infty Q_{\mathrm{abs}}(\varepsilon) I_0(\varepsilon) d\varepsilon. \tag{4.32}$$

To compute the effective intensity-absorption efficiency, it is therefore necessary to model the spectrum output by a typical X-ray source. The true spectra will vary from one machine to another and depend greatly on the materials used in the X-ray tube construction and filters used to harden the beam.

X-rays produced by X-ray tubes are generated by bremsstrahlung and X-ray fluorescence (Slide 13.15). Most commonly, X-ray tubes use tungsten anodes, giving characteristic peaks at 59.3 and 67.2 keV above the general bremsstrahlung output. Many numerical and semiempirical models are available from the literature. Here we use the program Spektr [8] to produce realistic X-ray spectra based on the TASMIP (*Tungsten Anode Spectral Model Using Interpolating Polynomials*) model of Boone and Seibert [3]. The results of simulations for spectra generated by Spektr are presented in Fig. 4.8 (Slide 13.15). Then, the values for the effective intensity-absorption efficiency, $Q_{\mathrm{eff}} \cdot I_{\mathrm{eff}}$, are computed by multiplying the spectral values from Fig. 4.8 by the absorption efficiency of gold nanoparticles of different radii from Fig. 4.7. Table 4.1 lists the effective intensity-absorption efficiency values calculated for nanoparticles of radii varying from 5 to 50 nm absorbing

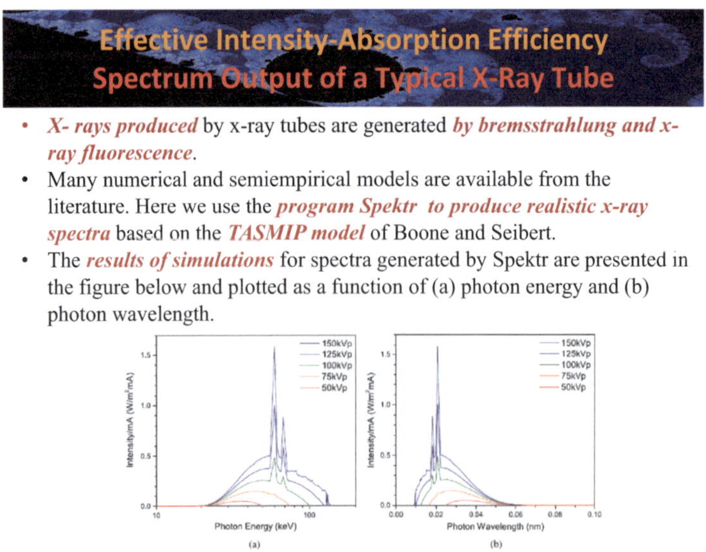

Slide 13.15 Effective intensity-absorption efficiency: spectrum output of a typical X-ray tube

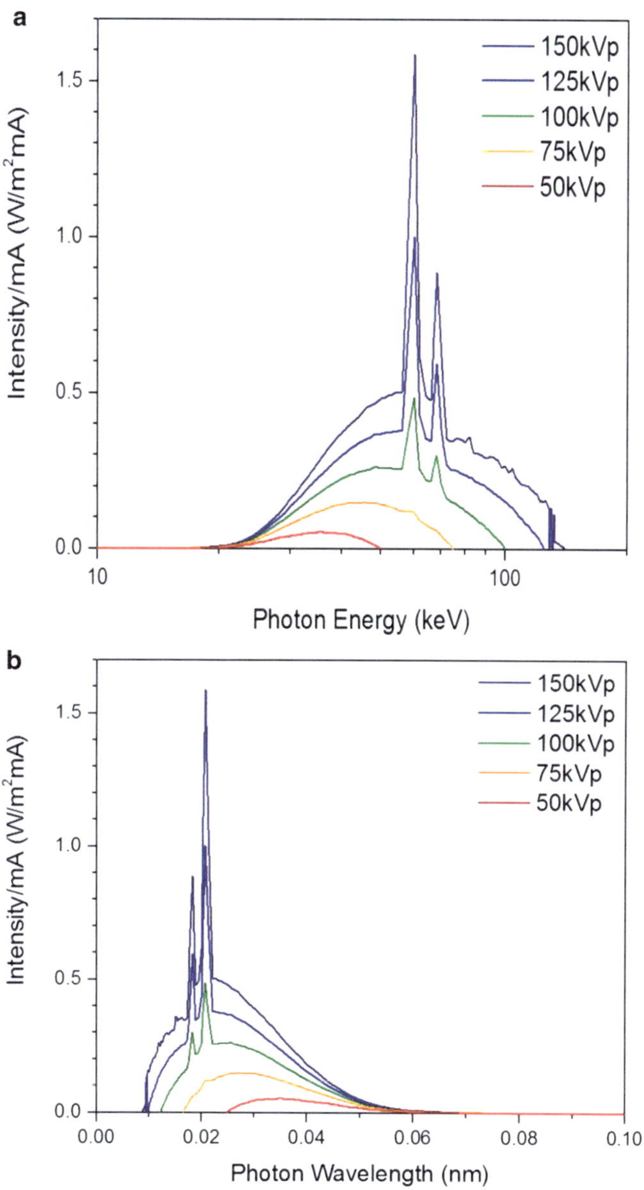

Fig. 4.8 X-ray spectra exported from Spektr [8], then converted to intensity units per unit of current (mA), plotted as a function of (**a**) photon energy and (**b**) photon wavelength. All five spectra assume an X-ray tube employing a tungsten anode with inherent filtration by 2.5 mm of aluminum, but varying the potential difference across the tube from 50 to 150 kV in 25 kV increments. Reprinted from [2]. © With permission from OSA

Table 4.1 Values (per mA) of the effective intensity-absorption efficiency, $Q_{\mathrm{eff}} \cdot I_{eff}$, computed for nanoparticles of radii varying from 5 to 50 nm absorbing radiation generated by tungsten anode X-ray tubes with kilovolt potentials ranging from 50 to 150 kVp. Reprinted from [2]. © With permission from OSA

	$Q_{\mathrm{eff}} \cdot I_{eff}$ ($10^{-6} \times \mathrm{W/m^2}$)										Exposure
kVp	5 nm	10 nm	15 nm	20 nm	25 nm	30 nm	35 nm	40 nm	45 nm	50 nm	mR
50	0.229	0.459	0.688	0.917	1.146	1.375	1.604	1.833	2.062	2.290	1.3314
75	0.666	1.331	1.996	2.661	3.326	3.991	4.656	5.320	5.984	6.648	3.9734
100	1.257	2.514	3.771	5.027	6.283	7.539	8.795	10.050	11.306	12.561	7.5995
125	2.014	4.027	6.040	8.053	10.065	12.077	14.089	16.100	18.111	20.122	12.0014
150	2.794	5.588	8.381	11.174	13.966	16.758	19.550	22.341	25.132	27.922	16.8100

Effective Intensity-Absorption Efficiency (continued)

- The table below lists the *effective intensity-absorption efficiency values* (per mA) calculated *for nanoparticles* of radii varying *from 5 to 50 nm* absorbing radiation generated by *tungsten anode x-ray tubes* with kilovolt potentials ranging *from 50 to 150 kVp*.

kVp	$Q_{eff} \cdot I_{eff}$ $(10^{-6} \times W/m^2)$										Exposure
	5 nm	10nm	15nm	20nm	25nm	30nm	35nm	40nm	45nm	50nm	mR
50	0.229	0.459	0.688	0.917	1.146	1.375	1.604	1.833	2.062	2.290	1.3314
75	0.666	1.331	1.996	2.661	3.326	3.991	4.656	5.320	5.984	6.648	3.9734
100	1.257	2.514	3.771	5.027	6.283	7.539	8.795	10.050	11.306	12.561	7.5995
125	2.014	4.027	6.040	8.053	10.065	12.077	14.089	16.100	18.111	20.122	12.0014
150	2.794	5.588	8.381	11.174	13.966	16.758	19.550	22.341	25.132	27.922	16.8100

- The 50 nm Au particles have $Q_{eff} \cdot I_{eff}$ values ranging from *2.290 to 27.922* $\cdot 10^{-6}$ *W/m²*, with the corresponding *exposures* ranging from *1.33 to 16.81 mR/mA*.
- Typically, *a single chest x-ray* will give an *exposure of 6-25 mrem*, and standard *x-ray computed tomography scans* can range from *300 to 2700 mrem*.

Slide 13.16 Effective intensity-absorption efficiency (continued)

radiation generated by tungsten anode X-ray tubes with kilovolt potentials ranging from 50 to 150 kVp (Slide 13.16). Figure 4.9 demonstrates the examples of such simulations of the intensity-absorption efficiency values for the 50 nm gold sphere in the 1–150 keV X-ray spectrum (Slide 13.17). The integrals of these curves are tabulated in Table 4.1.

The calculated effective intensity-absorption efficiency values for a 50 nm particle are on the order of 10^{-6} W/m^2 per amp, as shown in the two rightmost columns of Table 4.1 (Slide 13.16). The 50 nm Au particles have $Q_{\text{eff}} \cdot I_{\text{eff}}$ ranging from 2.290 to 27.922×10^{-6} W/m^2, with the corresponding exposures ranging from 1.33 to 16.81 mR/mA. Typically, a single chest X-ray will give an exposure of 6–25 mrem, and standard X-ray computed tomography scans can range from 300 to 2700 mrem.

The output from Spektr assumes a standardized exposure time of 1 s at 100 cm from the focal spot. By doubling the amperage while cutting the exposure time in half, the overall exposure should be unchanged, while the light intensity has doubled. Decreasing the exposure time, the effective intensity could be increased to the extent that nanophotothermolysis of cancer cells could be done at radiation therapy equivalent exposures. The research on laser-generated X-rays hints at a promising future for X-ray light sources [9, 10]. Output from these new X-ray emitters already produces pulses on the nanosecond timescale, and additionally, the output is much more directional and focused, limiting unnecessary exposure to healthy tissues.

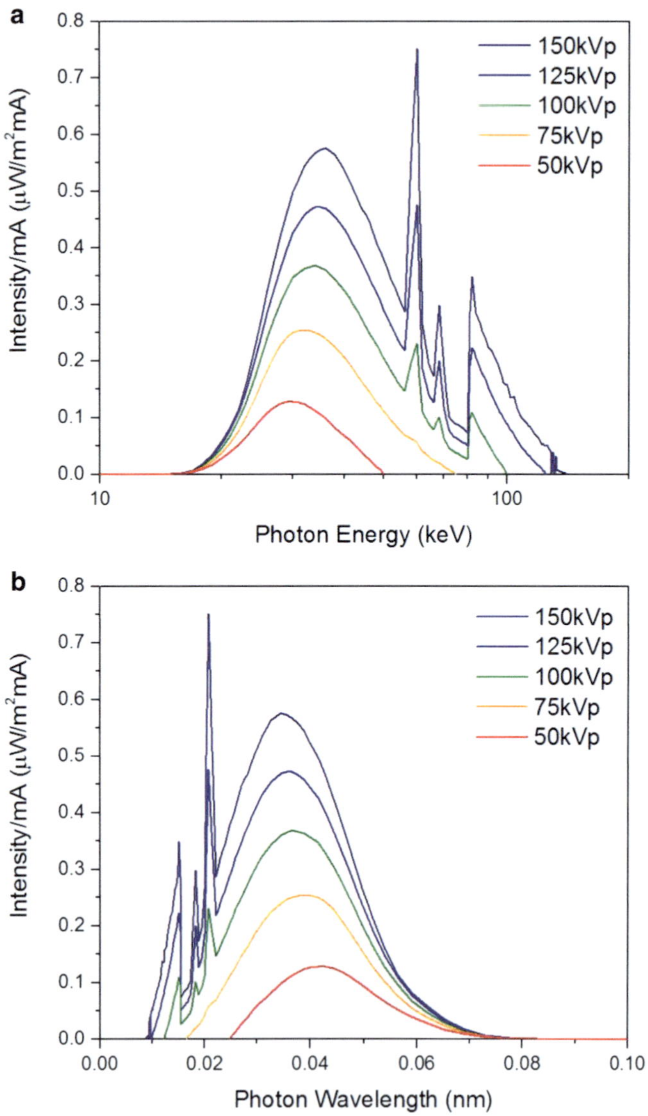

Fig. 4.9 Intensity-absorption efficiency values for the 50 nm gold sphere over the 1–150 keV X-ray spectrum, plotted as a function of (**a**) photon energy and (**b**) photon wavelength. Reprinted from [2]. © With permission from OSA

In conclusion (Slide 13.18), by utilizing the Kramers–Kronig relation for complex atomic scattering factors and the Lorenz–Mie theory, we have found the optical properties of nanoparticles in the X-ray spectrum: real and imaginary parts of the index of refraction for a nanoparticle material in a low-absorbing medium, and the absorption efficiency of various size gold nanoparticles for photon energies over the range of 1–150 keV. We have used a TASMIP model to produce

Effective Intensity -Absorption Efficiency (continued)

- ***Intensity-absorption efficiency*** values for the ***50 nm gold sphere*** over the ***1-150 keV x-ray spectrum***, plotted as a function of (a) photon energy and (b) photon wavelength.

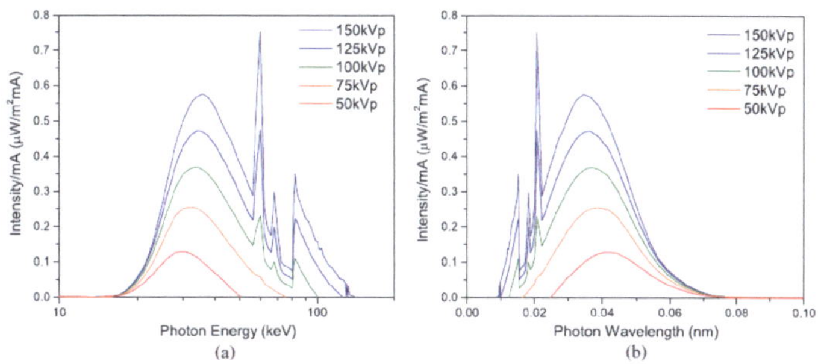

Slide 13.17 Effective intensity-absorption efficiency (continued).

Conclusions

- By utilizing the ***Kramers–Kronig relation*** for complex atomic scattering factors and the ***Lorenz–Mie theory***, we have found the ***optical properties of nanoparticles in the x-ray spectrum***:
  - ***the real and imaginary parts of the index of refraction*** for a nanoparticle material in a low-absorbing medium, and
  - ***the absorption efficiency*** of various size gold nanoparticles for photon energies over the range of 1-150 keV.
- We have used a ***TASMIP model*** to produce ***realistic x-ray spectra*** generated by typical tungsten anode x-ray tubes.
- We have introduced the new optical values for the ***effective intensity-absorption efficiency*** of nanoparticles, which take into account a wide and filtered output spectrum of photon energies from most medical diagnostic x-ray devices.

Slide 13.18 Conclusions

realistic X-ray spectra generated by typical tungsten anode X-ray tubes and introduced the new optical values for the effective intensity-absorption efficiency of nanoparticles, which take into account a wide and filtered output spectrum of photon energies from most medical diagnostic X-ray devices.

4.3 Lecture 14: RF Optics of Nanoparticles (Slide 14.1)

- Introduction
- Method

  One-Temperature Model

- Absorption Efficiency of Nanoparticles in the RF Range of the Spectrum
  - *RF Absorption Efficiency of Metal Nanoparticles*
  - *RF Absorption Efficiency of Dielectric Nanoparticles*
- Conclusions
- References

4.3.1 Introduction

This section contains material adopted from our publication [11]. In the previous Sect. 4.2 (Lecture 13) of this chapter, we discussed X-rays as a universal tool for nanoparticle activation (Slide 14.2). X-rays have the ability to penetrate into the human body and have high photon energies, giving a great opportunity to activate/heat nanoparticles inside the body for diagnostic and treatment purposes. However, the X-ray itself is toxic to the body and can cause a secondary cancer at the therapeutic doses.

Contents:

- Introduction
- Method

  One-Temperature Model

- Absorption Efficiency of Nanoparticles in the RF Range of the Spectrum
  - ➢ *RF Absorption Efficiency of Metal Nanoparticles*
  - ➢ *RF Absorption Efficiency of Dielectric Nanoparticles*
- Conclusions

Slide 14.1 Contents of Lecture 14

Introduction

- In Section 4.2 we discussed ***x-rays as a universal tool*** for nanoparticle activation.
- ***X-rays*** have the ability to ***penetrate*** into the human body and have ***high photon energies***, giving a great opportunity to ***activate/heat nanoparticles*** inside the body for diagnostic and treatment purposes.
- However, the ***x-ray*** itself is ***toxic*** to the body and can cause a ***secondary cancer*** at the therapeutic doses.
- Another candidate for ***an ideal radiation*** for nanoparticles activation is ***radiofrequency (RF) waves***.
- ***RF waves*** have an excellent ability to ***penetrate*** into the human body, giving a great opportunity to ***activate/heat nanoparticles*** delivered inside the body as a contrast agent for diagnosis and treatment purposes.

Slide 14.2 Introduction

Another candidate for an ideal radiation for nanoparticle activation is radiofrequency (RF) waves [11] (Slide 14.2) as presented in Sect. 3.2 (Lecture 10) of Chap. 3. RF waves have an excellent ability to penetrate into the human body, giving a great opportunity to activate/heat nanoparticles delivered inside the body as a contrast agent for diagnosis and treatment purposes. However, the heating of NPs in the RF spectral range is controversial in the research community because of the low power load of RF waves and low absorption efficiency of nanoparticles in the RF range (Slide 14.3). For example, a theoretical paper by Hanson et al. [12] analyzed several absorption mechanisms, including both classical and quantum effects, by which nonmagnetic spherical nanoparticles can absorb RF and far-infrared radiation, which can be subsequently released to their host medium as heat. They concluded that none of these absorption mechanisms can increase temperatures by tens of degrees as reported in many RF-GHz experiments.

On the other hand, there is recent experimental evidence (Slide 14.4) of using metal NPs to deposit heat in a local confined space for RF ablation of cancer [13–16], summarized in Table 4.2. For example, Raoof et al. [16] reported the death of liver cancer cells after targeting those cells with Au nanoparticles and exposing them to an RF field. Moran et al. [13] examined the capacitive RF heating properties of Au nanoparticles with respect to their volume fraction and diameter. According to Raoof et al. [16], the heating rate is 0.4 °C/s for a nanoparticle concentration of 1000 μg/mL in water solution. In these experiments it has been shown that using RF fields along with conjugated gold nanoparticles significantly decreases the volumes of pancreatic tumor cells. Ninety-nine percent of tumor death in human gastrointestinal cancer cells has been demonstrated using a 13.56 MHz RF field and 1–5 min exposures for a concentration of 67 μM/L Au

- However, the heating of NPs in the RF spectral range is controversial in the research community because of the ***low power load of RF waves*** and ***low absorption efficiency*** of nanoparticles in the RF range.
- For example, a theoretical paper by Hanson *et al.* analyzed several absorption mechanisms, including both classical and quantum effects, by which nonmagnetic spherical nanoparticles can absorb RF and far-infrared radiation, which can be subsequently released to their host medium as heat.
- They concluded that none of these absorption mechanisms can increase temperatures by tens of degrees as reported in many RF-GHz experiments.

Slide 14.3 Introduction (continued)

- On the other hand, there is recent ***experimental evidence*** of using metal NPs to deposit heat in a local confined space for RF ablation of cancer.
- For example, ***Raoof et al.*** reported the death of liver cancer cells after targeting those cells with Au nanoparticles and exposing them to an RF field.
- ***Moran et al.*** examined the capacitive RF heating properties of Au nanoparticles with respect to their volume fraction and diameter.
- ***In these experiments*** it has been shown that using RF fields along with conjugated gold nanoparticles significantly decreases the volumes of pancreatic tumor cells.
- 99% tumor death in human gastrointestinal cancer cells has been demonstrated using a 13.56 MHz RF field and 1-5 min exposures for a concentration of 67 μM/L Au nanoparticles.

Slide 14.4 Introduction (continued)

nanoparticles [17]. Overall, all of the experiments listed in Table 4.2 have reached significant temperature changes that confirm the possibility of cell ablation (Slide 14.5).

In order to resolve the crisis between theoretical conclusions and experimental evidence, and to prove that RF radiation is capable of heating the nanoparticles, we

Table 4.2 Experimental data for RF wave heating of NPs. Reprinted from [11]. © With permission from Elsevier

Reference		Li et al. [18]	Raoof et al. [16]	Moran et al. [13]	Gannon et al. [17]	Glazer et al. [15]
Radiation	Frequency	13.56 MHz	13.56 MHz	13.56 MHz	13.56 MHz	13.56 MHz
	Power and exposure time	25 W and 1600 s	600 W and 120 s	600 W and 120 s	600 W and 300 s	600 W and 60 s
		26 W and 285 s			800 W and 300 s	
		27 W and 205 s				
		28 W and 1600 s				
		29 W and 250 s				
Particle	Material	Gold	Gold	Gold	Gold	Gold
	Structure	Sphere	Sphere	Sphere	Sphere	sphere
	Radius and concentration	10 nm and 5.7×10^{12} cm^{-3}	10 nm and 50 μg/mL	5 nm and 36 ppm	2.5 nm and 67 μM/L	5 nm and 100 μg/L
		20 nm and 7×10^{11} cm^{-3}				
		50 nm and 4.5×10^{10} cm^{-3}				
		100 nm and 5.6×10^{9} cm^{-3}				
		200 nm and 7×10^{8} cm^{-3}				
Sample	Medium	Deionized water	Aqueous solution at pH 7.4	Deionized water	Deionized water	Deionized water
	Size	0.5 mL	1.3 mL	1.5 mL		1.5 mL
Maximum temperature		360 K	348 K	345 K	373 K	324 K

Introduction (continued)

- Overall, ***all of the experiments*** listed in the table below have ***reached significant temperature changes*** that confirm the possibility of cell ablation.

Experimental data for RF wave heating of NPs.

Reference		Li et al[5]	Raoof et al[6]	Moran et al[1]	Gannon et al[7]	Glazer et al[3]
Radiation	Frequency	13.56 MHz	13.56 MHz	13.56 MHz	13.56 MHz	13.56 MHz
	Power & exposure Time	25 W & 1600 s 26 W & 285 s 27 W & 205 s 28 W & 1600 s 29 W & 250 s	600 W & 120 s	600 W & 120 s	600 W & 300 s 800 W & 300 s	600 W & 60 s
Particle	Material	Gold	Gold	Gold	Gold	Gold
	Structure	Sphere	Sphere	Sphere	Sphere	sphere
	Radius & concentration	10 nm & 5.7×10^{12} cm^{-3} 20 nm & 7×10^{11} cm^{-3} 50 nm & 4.5×10^{10} cm^{-3} 100 nm & 5.6×10^{9} cm^{-3} 200 nm & 7×10^{8} cm^{-3}	5 nm & 50 μg/mL	5 nm & 36 ppm	2.5 nm & 67 μM/L	5 nm & 100 μg/L
Sample	Medium	Deionized water	Aqueous solution at pH 7.4	Deionized water	Deionized water	Deionized water
	Size	0.5 mL	1.3 mL	1.5 mL		1.5 mL
Maximum Temperature		360 K	348 K	345 K	373 K	324 K

- In this lecture, we use ***a phenomenological approach*** to find ***the absorption efficiency*** for different types of nanoparticles.

Slide 14.5 Introduction (continued)

have to determine the actual absorption efficiency of nanoparticles in the RF spectral range [11]. In this lecture, we use a phenomenological approach to find the absorption efficiency for different types of nanoparticles in the RF range through the study of heating and cooling kinetics of those particles in a biological environment. The time-dependent simulations are carried out for different nonstationary RF–nanoparticle interactions in single-pulse and multipulse modes of heating. The detailed analysis includes the effect of the radiation field structure (pulse shape, pulse duration, repetition rate, and time interval between pulses) on the kinetics of heating and cooling of nanoparticles, as well as the conditions and limitations for the heat accumulation effect in the biological tissue containing nanostructures.

4.3.2 *Method*

This section contains material adopted from our publication [11]. We use a reverse phenomenological approach to estimate the absorption efficiency for different types of nanoparticles in the radiofrequency range of the spectrum (Slide 14.6). The reverse method consists of three steps:

1. Plugging the available RF experimental data into a well-proven thermal model for heating the nanoparticles, known as the one-temperature model (OTM) developed by Letfullin et al. [1];
2. Solving the OTM numerically to compute temperature curves based on the nanoparticle's shape, its composition material, the type and intensity of radiation it is exposed to, and its surrounding medium; and

3. Determining the absorption efficiency of NPs in the RF spectral range by adjusting the OTM results until the theoretical temperature change meets that observed experimentally under the same heating conditions.

The one-temperature model [1] is derived in Chap. 6 of this book with provided computer code for simulation of heating of nanoparticles by radiation (Slide 14.7).

Method

- We use ***a reverse phenomenological approach*** to estimate the absorption efficiency for different types of nanoparticles in the radiofrequency range of the spectrum.
- The reverse method consists of ***three steps***:
  (1) plugging the ***available RF experimental data*** into a well-proven thermal model for heating the nanoparticles, known as the ***one-temperature model (OTM)*** developed by Letfullin *et al.*;
  (2) solving the OTM numerically to ***compute temperature curves*** based on the nanoparticle's shape, its composition material, the type and intensity of radiation it is exposed to, and its surrounding medium; and
  (3) determining ***the absorption efficiency*** of NPs in the RF spectral range ***by adjusting the OTM*** results until the ***theoretical temperature*** change meets that ***observed experimentally*** under the same heating conditions.

Slide 14.6 Method

One-Temperature Model

- ***The one-temperature model*** is derived in Chapter 6 with provided computer code for simulation of heating of nanoparticles by radiation.
- Here, the OTM is briefly demonstrated below, with a table that explains each variable:

Table. OTM variables and their units.

Q_{abs}	Absorption efficiency of NP
I_0 [W/cm^2]	Intensity of RF radiation
$f(t)$	Pulse shape
r_0	Radius of NP
$C(T)$	Specific heat of NP
ρ_0	Density of NP
μ_0	Heat conductivity of surrounding medium
T	Final temperature of NP
s	Exponential power
T_0	Initial temperature of NP
L	Latent heat
dt	Time step

$$\frac{dT}{dt} = \frac{3Q_{abs}I_0 f(t)}{4r_0 C(T)\rho_0} - \frac{\mu_0 T}{(s+1)r_0^2 C(T)\rho_0}\left[\left(\frac{T}{T_0}\right)^{s+1} - 1\right] + \frac{3L}{r_0 C(T)}\frac{dr_0}{dt}.$$

Slide 14.7 One-temperature model

Table 4.3 OTM variables. Reprinted from [11]. © With permission from Elsevier

Q_{abs}	Absorption efficiency of NP
I_0 [W/cm^2]	Intensity of RF radiation
$f(t)$	Pulse shape
r_0	Radius of NP
$C(T)$	Specific heat of NP
ρ_0	Density of NP
μ_0	Heat conductivity of surrounding medium
T	Final temperature of NP
s	Exponential power
T_0	Initial temperature of NP
L	Latent heat
dt	Time step

One-Temperature Model (continued)

- The ***first term*** on the right side of the equation expresses the amount of energy absorbed by the spherical nanoparticle. It depends on the size, material properties of the nanoparticle, absorption efficiency, and the intensity of the radio wave.
- The ***second term*** describes nanoparticle's energy loss due to thermal diffusion into the surrounding medium. This term is governed by the size and material properties of the nanoparticle, as well as the thermal properties of the medium.
- The ***final term*** deals with evaporation of the nanoparticle, if the phase transition temperature is reached.

$$\frac{dT}{dt} = \frac{3Q_{abs}I_0 f(t)}{4r_0 C(T)\rho_0} - \frac{\mu_0 T}{(s+1)r_0^2 C(T)\rho_0}\left[\left(\frac{T}{T_0}\right)^{s+1} - 1\right] + \frac{3L}{r_0 C(T)}\frac{dr_0}{dt}.$$

Slide 14.8 One-temperature model (continued)

Here, the OTM is briefly demonstrated later, followed by Table 4.3 that explains each variable:

$$\frac{dT}{dt} = \frac{3Q_{\text{abs}}I_0 f(t)}{4r_0 C(T)\rho_0} - \frac{\mu_0 T}{(s+1)r_0^2 C(T)\rho_0}\left[\left(\frac{T}{T_0}\right)^{s+1} - 1\right] + \frac{3L}{r_0 C(T)}\frac{dr_0}{dt}. \tag{4.33}$$

The first term on the right side of the equation expresses the amount of energy absorbed by the spherical nanoparticle (Slide 14.8). It depends on the size, material properties of the nanoparticle, absorption efficiency, and the intensity of the radio wave. The second term describes nanoparticle's energy loss due to thermal diffusion into the surrounding medium. This term is governed by the size and material properties of the nanoparticle, as well as the thermal properties of the medium. The final term deals with evaporation of the nanoparticle, if the phase transition temperature is reached.

4.3.3 Absorption Efficiency of Nanoparticles in the RF Range of the Spectrum

This section contains material adopted from our publication [11]. Various types of NPs have been examined and are described as follows.

RF Absorption Efficiency of Metal Nanoparticles (Slide 14.9). Gold is nontoxic to living tissue, and Au nanoparticles have been well studied. In an experimental paper by Moran et al. [13], the authors used a colloidal solution of gold nanoparticles with radius 5 nm in deionized water to expose it to a radiofrequency field. The results of this research are summarized in Table 4.2, where the achieved temperature change is $\Delta T = 45$ K above ambience. This temperature change is attained then in the OTM by varying the absorption efficiency value until the maximum temperature matches the maximum temperature change achieved in the experimental study. The computer simulation technique for OTM is described in Chap. 6. Here we present the results of the simulations for heating a 5-nm radius gold nanoparticle by a single RF pulse performed under the same experimental conditions and shown in Fig. 4.10 (Slide 14.10). Our simulations show that the absorption efficiency of Au nanoparticles in the RF range corresponding to this experimental change in temperature is 1.2×10^{-9}.

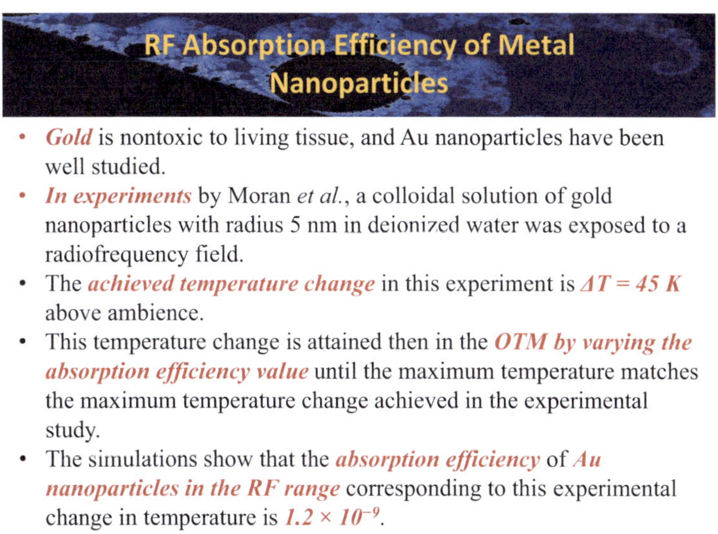

Slide 14.9 RF absorption efficiency of metal nanoparticles

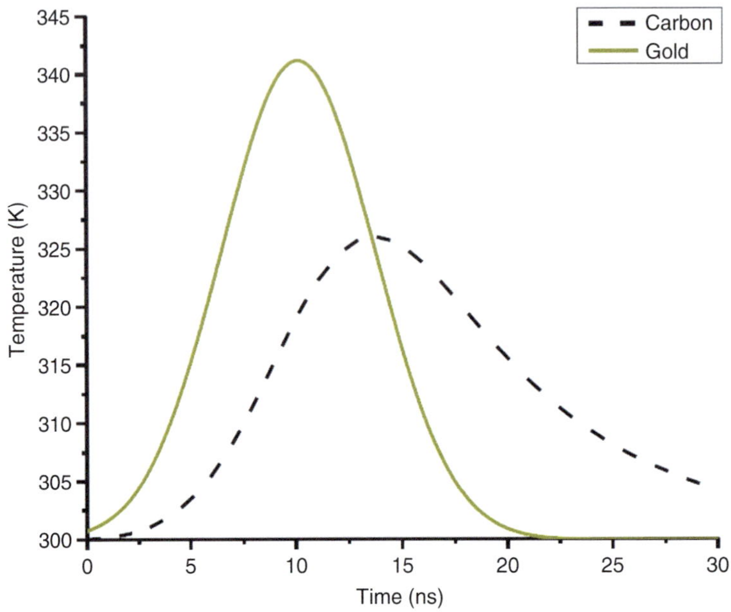

Fig. 4.10 Heating of metal and dielectric nanoparticles by RF radiation. Reprinted from [11]. © With permission from Elsevier

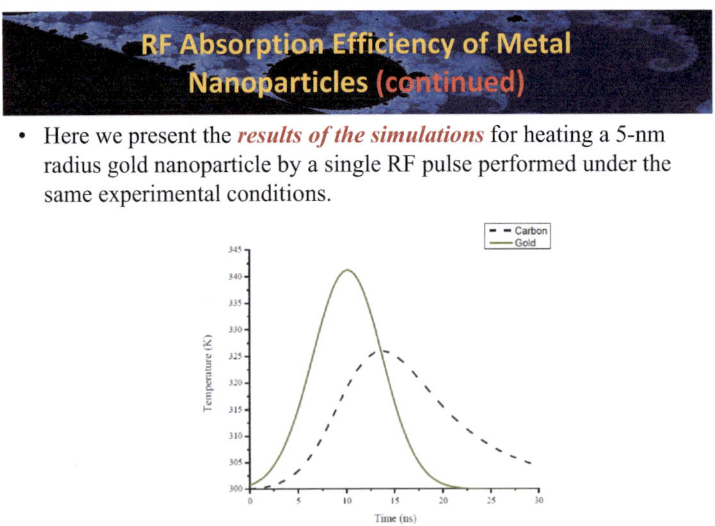

Slide 14.10 RF absorption efficiency of metal nanoparticles (continued)

Similar experiments were conducted by Gannon et al. [17], who measured the heat generation in a 2.5-nm Au nanoparticle colloidal solution by RF waves (13.56 MHz; 600 and 800 W). Table 4.2 shows the data collected from these experiments. Applying the technique described before, we find that the average

RF Absorption Efficiency of Metal Nanoparticles (continued)

- Similar ***experiments*** were conducted by Gannon *et al.*, who measured the ***heat generation in a 2.5-nm Au nanoparticle*** colloidal solution ***by RF waves (13.56 MHz, 600 & 800 W)***.
- Applying the technique described above, we find that the ***average absorption efficiency*** for 2.5-nm gold nanoparticles with concentration 11.1 μM/L is $\boldsymbol{0.8 \times 10^{-9}}$ for 13.56 MHz radiation and 200-800 W power range.
- ***In the experiments*** by Raoof *et al.* and Moran *et al.*, the highest temperature reached was 348 K.
- The ***absorption efficiencies*** that we calculated using experimental data from Moran *et al.* and Raoof *et al.* were averaged to give $\boldsymbol{1.15 \times 10^{-9}}$ for identical conditions used in the experiments: both used 5-nm radius Au nanoparticles, 600 W of power, 13.56 MHz, and 120 s heating time.

Slide 14.11 RF absorption efficiency of metal nanoparticles (continued)

absorption efficiency for 2.5-nm gold nanoparticles with concentration 11.1 μM/L is 0.8×10^{-9} for 13.56 MHz radiation and 200–800 W power range (Slide 14.11).

In the paper by Raoof et al. [16], the authors investigate the effects of the surrounding medium, specifically the pH levels, on heating gold nanoparticles by RF radiation (Slide 14.11). The highest temperature on the experimental curves is 348 K, which occurs at a pH of 7.4. Since this is only slightly differs from water's pH level of 7, in OTM we use the same thermal conductivity as water. The highest temperature of 48 °C is close to what Moran et al. [13] reached as well. The absorption efficiencies that we calculated using experimental data from Moran et al. and Raoof et al. were averaged to give 1.15×10^{-9} for identical conditions used in the experiments: both used 5-nm radius Au nanoparticles, 600 W of power, 13.56 MHz, and 120 s heating time (Slide 14.11).

There are also theoretical papers that have examined the absorption of gold nanoparticles in the RF range. For example, a paper by Hanson et al. [12] describes different theoretical mechanisms to analyze the heating of metallic nanoparticles by RF radiation. This paper calculates the absorption efficiency for Au nanoparticles of 20-nm radius in water exposed to radiation with frequencies ranging from 1 MHz to 1 THz using two different methods: (1) Mie diffraction theory and (2) thermodynamical analysis of electronic absorption due to surface discrepancies. Both coated and uncoated nanoparticles are considered. The coating is assumed to be the most common types used in combination with nanoparticles, such as an oxide layer, sodium citrate, or layer of proteins that enable selective binding to cells. The results of classic Mie theory simulations from this paper give a range of absorption efficiencies equaling 10^{-8} to 10^{-7} for coated Au nanoparticles.

RF Absorption Efficiency of Dielectric Nanoparticles

- Lots of research has been done with ***carbon nanotubes*** in recent years.
- For example, ***Gannon et al.*** achieved a rise in temperature of 25-45°C above ambience using single-wall carbon nanotubes that range from 300 nm to 1 μm in length at a concentration of 50 mg/L exposed to an RF field with a peak field strength of approximately 15 kV/m.
- The ***OTM simulations*** were performed using the same conditions as Gannon's experiment to compute the absorption efficiency for a wide range of carbon nanoparticles sizes of 5 nm-500 nm.
- All of the obtained ***absorption efficiencies for carbon nanoparticles*** are in the range of ***1.01×10^{-8} to 8.70×10^{-6}***.
- The reason for ***a large range of absorption efficiencies*** is because of ***various possibilities between dimensions of length***.

Slide 14.12 RF absorption efficiency of dielectric nanoparticles

Our simulations based on a reverse phenomenological approach are close to their results for the absorption efficiency for coated nanoparticles, which is in the range of 10^{-8} to 10^{-7} for larger Au particles of 20 nm in radius. Our simulations were based on the experimental conditions and performed for much smaller nanoparticles with radii of 2.5 and 5 nm, giving the absorption efficiency in the range $0.8–1.2 \times 10^{-9}$. For the purpose of cancer therapy, we use Au nanoparticles coated with antibodies for binding to the cancer cells; and some coating also comes from fabrication of nanoparticles, such as an oxide layer. However, we do not support the statement made in [12] that the RF radiation doesn't produce any heat in metallic nanoparticles.

RF Absorption Efficiency of Dielectric Nanoparticles. Lots of research has been done with carbon nanotubes in recent years (Slide 14.12). For example, Gannon et al. [19] achieved a rise in temperature of 25–45 °C above ambience using single-wall carbon nanotubes that range from 300 nm to 1 μm in length at a concentration of 50 mg/L exposed to an RF field with a peak field strength of approximately 15 kV/m. We performed OTM simulations using the same conditions as Gannon's paper to compute the absorption efficiency for a wide range of carbon nanoparticles sizes of 5–500 nm. The curves achieved through the OTM simulations can be viewed in Fig. 4.11 and summarized in Table 4.4. All of the obtained absorption efficiencies for carbon nanoparticles are in the range of 1.01×10^{-8} to 8.70×10^{-6}. The reason for a large range of absorption efficiencies is because of various possibilities between dimensions of length. By extrapolating absorption efficiencies from the simulation results, we plot the curve shown in Fig. 4.11, which can be used to predict the absorption efficiency for carbon nanoparticles of various sizes (Slide 14.13).

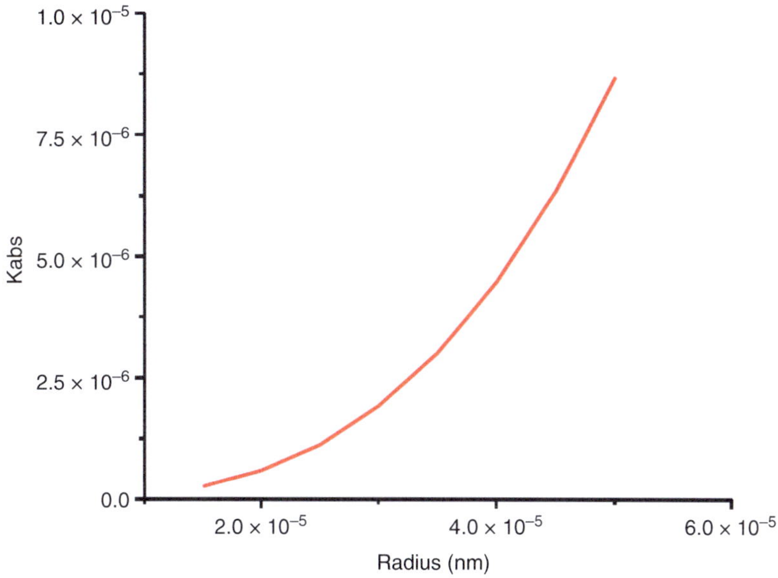

Fig. 4.11 Absorption efficiency of the carbon nanoparticles in the RF range of the spectrum as a function of particle size. Reprinted from [11]. © With permission from Elsevier

Table 4.4 Absorption efficiencies of different materials in the RF range of the spectrum calculated for radiation power 600 W and for fixed maximum temperature observed in experiments. Reprinted from [11]. © With permission from Elsevier

Material	Absorption efficiency	Maximum temperature (K)
Metal (aluminum oxide, gold, magnesium oxide, nickel oxide, silver) sphere, 2.5-nm radius	0.8×10^{-9}	366.5
Metal sphere, 5-nm radius	1.15×10^{-9}	349.1
Metal sphere, 20-nm radius	6.10×10^{-9}	360.1
Metal sphere, 50-nm radius	2.73×10^{-8}	360.2
Metal sphere, 100-nm radius	1.47×10^{-7}	360.5
Dielectric (carbon, fullerene, glass, polystyrene) sphere, 5-nm radius	1.21×10^{-11}	300.6
Dielectric sphere, 50-nm radius	1.01×10^{-8}	326.0
Dielectric sphere, 100-nm radius	7.62×10^{-8}	339.0
Dielectric sphere, 150-nm radius	26×10^{-8}	344.6
Dielectric sphere, 250-nm radius	112×10^{-8}	344.8
Dielectric sphere, 500-nm radius	870×10^{-8}	345.2

The results for the absorption efficiency in the RF spectral range along with the highest temperature achieved for metal and dielectric nanoparticles studied are summarized in Table 4.4 (Slide 14.14). Our simulations showed [11] that all of the metals have the same heating kinetics and almost identical time–temperature profiles for the same conditions of RF heating. Single-pulse heating of different

RF Absorption Efficiency of Dielectric Nanoparticles (continued)

- ***By extrapolating absorption efficiencies*** from the simulation results, the curve shown in the figure below is plotted, which can be used to predict the absorption efficiency for carbon nanoparticles of various sizes.

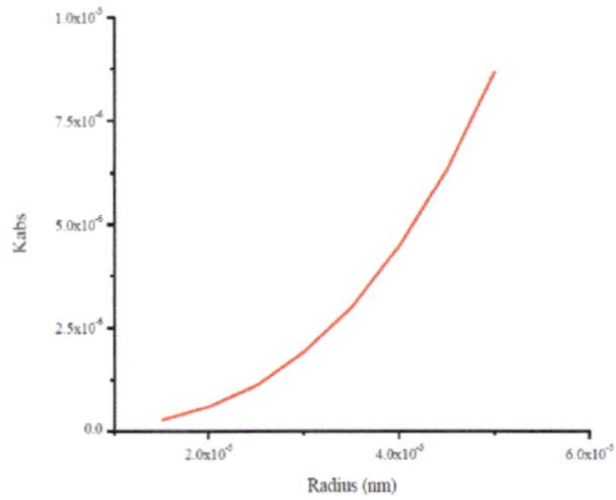

Slide 14.13 RF absorption efficiency of dielectric nanoparticles (continued)

RF Absorption Efficiency of Nanoparticles

- The results for the ***absorption efficiency in the RF spectral range*** along with the highest temperature achieved ***for metal and dielectric nanoparticles*** studied are summarized in Table below.

Material	Absorption efficiency	Maximum temperature (K)
Metal (aluminum oxide, gold, magnesium oxide, nickel oxide, silver) sphere, 2.5-nm radius	0.8×10^{-9}	366.5
Metal sphere, 5-nm radius	1.15×10^{-9}	349.1
Metal sphere, 20-nm radius	6.10×10^{-9}	360.1
Metal sphere, 50-nm radius	2.73×10^{-8}	360.2
Metal sphere, 100-nm radius	1.47×10^{-7}	360.5
Dielectric (carbon, fullerene, glass, polystyrene) sphere, 5-nm radius	1.21×10^{-11}	300.6
Dielectric sphere, 50-nm radius	1.01×10^{-8}	326.0
Dielectric sphere, 100-nm radius	7.62×10^{-8}	339.0
Dielectric sphere, 150-nm radius	26×10^{-8}	344.6
Dielectric sphere, 250-nm radius	112×10^{-8}	344.8
Dielectric sphere, 500-nm radius	870×10^{-8}	345.2

Slide 14.14 RF absorption efficiency of nanoparticles

dielectric NPs, like in the metal case, has similar curves as well for the same conditions of RF heating (Slide 14.15).

In conclusion (Slide 14.16), we have applied a reverse phenomenological approach to determine the absorption efficiency for different types of nanoparticles

RF Absorption Efficiency of Nanoparticles (continued)

- The curves achieved through the ***OTM simulations*** can be viewed in the figure below.

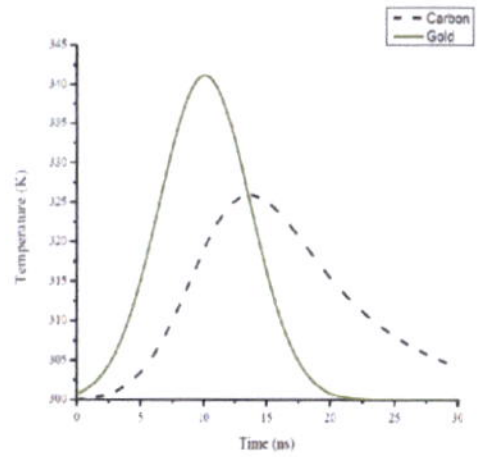

- The simulations show that ***all of the metals*** have the ***same heating kinetics*** and almost identical time-temperature profiles ***for the same conditions of RF heating***.
- The heating of different ***dielectric nanoparticles***, like in the metal case, has ***similar curves*** as well for the same conditions of RF heating.

Slide 14.15 RF absorption efficiency of nanoparticles (continued)

Conclusions

- We have applied ***a reverse phenomenological approach*** to determine the ***absorption efficiency*** for different types of ***nanoparticles*** in the ***RF range of the spectrum*** by matching the theoretically achieved maximum temperature of the nanoparticles to the experimentally observed temperature change at the same heating conditions.
- ***This method*** based on the study of ***heating kinetics*** gives the following estimations for ***the absorption efficiency of nanoparticles*** in the RF range of the spectrum:
  - The ***absorption efficiency*** of the researched ***metal nanoparticles*** with radius 5 nm is ~ 1.15×10^{-9}.
  - The ***absorption efficiencies of carbon spherical nanoparticles*** with radii of 5 nm and 500 nm are 1.21×10^{-11} and 8.70×10^{-6}, respectively.
  - The ***absorption efficiency of dielectric nanoparticles*** such as glass, polystyrene and fullerene with radius of 5 nm is estimated to be 1.21×10^{-11}.

Slide 14.16 Conclusions

in the RF range of the spectrum by matching the theoretically achieved maximum temperature of the nanoparticles to the experimentally observed temperature change at the same heating conditions. This method based on the study of heating kinetics gives the following estimations for the absorption efficiency of nanoparticles in the RF range of the spectrum:

- The absorption efficiency of the researched metal nanoparticles with radius 5 nm is ~1.15×10^{-9}.
- The absorption efficiencies of carbon spherical nanoparticles with radii of 5 and 500 nm are 1.21×10^{-11} and 8.70×10^{-6}, respectively.
- The absorption efficiency of dielectric nanoparticles such as glass, polystyrene, and fullerene with radius of 5 nm is estimated to be 1.21×10^{-11}.

Homework Exercises

4.1. What is a source of the light?
4.2. When is light treated as a photon?
4.3. When is light considered as a wave?
4.4. What is the definition of a traveling wave?
4.5. What is the amplitude of a wave?
4.6. What is the wavelength of a wave?
4.7. What is a period of the wave?
4.8. What is the frequency of a wave?
4.9. What is the speed of a wave?
4.10. What is the wavelength for 13.5 MHz radiation? What is this type of radiation?
4.11. What is the general solution for a traveling wave?
4.12. What is the frequency range for the visible light spectrum?
4.13. What is the period for 600 nm light?
4.14. For a radiowave with wavelength 30.0 cm, amplitude 5.0 cm, and initial phase π rad, find the following:

(a) Wave number
(b) Period
(c) Regular and angular frequencies
(d) General solution for this wave, $y(z,t)$, traveling to the positive z-direction.

4.15. If at the initial moment of the time $t = 0$ s an electromagnetic wave with frequency 1.0 THz has the displacements $z = 5.0$ cm, $y(z,0) = 1.0$ cm and amplitude 3.0 cm, find the following:

(a) Initial phase of the wave
(b) General solution for this wave, $y(z,t)$, traveling to the positive z-direction.

4.16. A sinusoidal wave with an amplitude of 1.0 cm and a frequency of 100 Hz travels at 200 m/s in the positive z-direction. At $t = 0$ s, the point $x = 1.0$ cm is on a crest of the wave.

(a) Determine the amplitude a, frequency f, wavelength λ, wavenumber k, angular frequency ω, period T, and initial phase φ_0 of this wave.
(b) Write the equation for the wave's displacement $y(z,t)$ as it travels.

4.17. A wave on a string is described by the displacement $y(z,t) = (4.0\ \text{cm})\ \sin[(25.0\ \text{rad/m})z—(550\ \text{rad/s})t]$, where z is in m and t is in s.

(a) What is the maximum displacement of a point on the string?
(b) What is the displacement $y(z,t)$ at time 5.0 s?
(c) What is the maximum speed of a point on the string?

4.18. True or false? An electromagnetic wave is a transverse wave.
4.19. True or false? A traveling wave transports matter.
4.20. What is a Poynting vector?
4.21. What is the power of a wave?
4.22. What is the intensity of a wave?
4.23. A 5.0 mW laser beam has a diameter 5.0 mm.

(a) What is the intensity of the beam?
(b) What is an electric field E?
(c) What is a magnetic field B?

4.24. A high-power laser beam with intensity $I = 3 \times 10^6\ \text{W/m}^2$ is reflected from the surface of a mirror. What light pressure has been delivered to the mirror?
4.25. Why does light have a different speeds in different media?
4.26. True or false? Light has different frequencies in different media.
4.27. True or false? Light has same wavelength in different media.
4.28. What is the index of refraction of a medium? What does it show?
4.29. True or false? The index of refraction is always equal or greater than 1.
4.30. Calculate the speed of light in:

(a) Water
(b) Glass
(c) Diamond

4.31. A He–Ne laser has the wavelength 632.8 nm in vacuum. What is the wavelength of this light in:

(a) Water
(b) Glass
(c) Diamond

4.32. What optical phenomena are described by the real part of the index of refraction?
4.33. What optical phenomena are described by the imaginary part of the index of refraction?
4.34. Sketch the Beer–Lambert experimental arrangement.
4.35. What is the goal of the Beer–Lambert experiment?
4.36. Describe four main cases of the behavior of light in the second medium from the Beer–Lambert experiment.
4.37. If light lost a half of its initial intensity due to absorption in the second medium at 1.0 cm depth, what is the linear absorption coefficient of that medium?

4.38. Laser light of power 10 mW with diameter 5.0 mm is incident on the boundary of two media, air and glass. If 25 % of the light is scattered on the boundary and 35 % is absorbed in a 10 cm thick glass plate,

(a) What is intensity of the light in the first $\boldsymbol{E} \rightarrow \boldsymbol{E}e^{-i\omega t}$ medium?
(b) What is intensity of the light after it passed a glass plate?
(c) What is the extinction coefficient?

4.39. Describe each equation in the system of Maxwell's equations.
4.40. True or false? The system of Maxwell's equations can be reduced to two differential equations for the electric and magnetic fields
4.41. What does a wave equation describe?
4.42. By using the transformations $\boldsymbol{E} \rightarrow \boldsymbol{E}e^{-i\omega t}$ and $\rightarrow \boldsymbol{B}e^{-i\omega t}$ derive the Helmholtz equations (4.20) from the wave equations (4.19).
4.43. What is the Lorentz–Mie theory able to calculate?
4.44. What is the relative index of refraction of a medium?
4.45. What are the absorption, scattering, and extinction efficiencies?
4.46. What input data are required for the Lorentz–Mie simulations?
4.47. What output data can be obtained from the Lorentz–Mie simulations?
4.48. What types of radiation can be considered as a universal tool for nanoparticle activation?
4.49. What is the photon energy range of X-ray radiation in the electromagnetic spectrum?
4.50. What is the wavelength range of X-ray radiation in the electromagnetic spectrum?
4.51. According to what properties does wave optics describe electromagnetic radiation?
4.52. According to what properties does X-ray optics describe electromagnetic radiation?
4.53. What is the photon energy of the He–Ne laser light ($\lambda = 633$ nm)?
4.54. True or false? The photon energy is different in different media?
4.55. What is the photon energy range of visible light?
4.56. True or false? X-rays are absorbed by matter as strongly as visible light.
4.57. True or false? Most materials are relatively transparent to X-rays.
4.58. True or false? X-rays diffract a lot when passing through media.
4.59. True or false? The real part of the index of refraction n is almost always nearly equal to 1 for most materials in the X-ray range of the spectrum.
4.60. For what X-ray photon energies can the optical properties of a material be described in terms of the atomic scattering factors?
4.61. True or false? The atomic scattering factor has real and imaginary parts?
4.62. True or false? The real part of the atomic scattering factor, f_1, is proportion to the atomic photoabsorption cross-section, σ_a.
4.63. In what range of photon energies is the absorption of X-rays maximum?
4.64. What does a large extinction coefficient of a nanoparticle mean?
4.65. What does the product $Q_{abs}I_0$ describe?

4.66. True or false? The output from most medical diagnostic X-ray devices has a wide spectrum of photon energies.
4.67. Why do we integrate over all spectral energies when trying to determine the amount of heat generated in a nanoparticle?
4.68. Select one or more. The X-rays produced by X-ray tubes are generated by:

(a) X-ray fluorescence
(b) Photoelectron emission
(c) Compton scattering
(d) Bremsstrahlung
(e) All of the above
(f) None of the above

4.69. What does the program Spektr produce?
4.70. What material is commonly used for an X-ray tube anode?
4.71. What is the effective intensity-absorption efficiency?
4.72. What is a typical exposure in mrem of a single chest X-ray?
4.73. What is an exposure in mrem of standard X-ray computed tomography scans?
4.74. Why can radiofrequency waves be considered as a universal tool for nanoparticle activation?
4.75. What is the frequency range of RF waves?
4.76. What is the wavelength of 13.5 MHz RF waves?
4.77. Describe a few experiments on nanoparticle heating by RF waves.
4.78. What are three steps in a reverse phenomenological approach to find the absorption efficiency for different types of nanoparticles in the RF range of spectrum?
4.79. Describe three terms of the one-temperature model.
4.80. Select one or more. The amount of energy absorbed by a spherical nanoparticle depends on the:

(a) Size of the nanoparticle
(b) Material properties of the nanoparticle
(c) Absorption efficiency
(d) Intensity of the radio wave.
(e) All of the above
(f) None of the above

4.81. Select one or more. The energy lost from the surface of a nanoparticle due to thermal diffusion into the surrounding medium is governed by the:

(a) Size of the nanoparticle
(b) Material properties of the nanoparticle
(c) Thermal properties of the medium
(d) Absorption efficiency
(e) Intensity of the radiowave
(f) All of the above
(g) None of the above

4.82. What is a typical temperature change observed in nanoparticle heating experiments by RF radiation?
4.83. What are the dimensions of single-wall carbon nanotubes?
4.84. What is the average value of the absorption efficiency in the RF spectral range for metal nanoparticles?
4.85. What is the average value of the absorption efficiency in the RF spectral range for dielectric nanoparticles?

Topics for Presentations and Writing Assignments

Presentation #1. Nanophotonics: Definition, Methods, and Applications.
Presentation #2. X-ray Optics: Definition, Methods, and Applications.
Presentation #3. RF Optics: Definition, Methods, and Applications.

References

1. R.R. Letfullin, T.F. George, Plasmonic nanomaterials in nanomedicine, in *Springer Handbook of Nanomaterials*, ed. by R. Vajtai (Springer, Berlin, 2013), pp. 1063–1097
2. R.R. Letfullin, C.E.W. Rice, T.F. George, X-ray optics of gold nanoparticles. Appl. Opt. **53**, 7208–7214 (2014)
3. J.M. Boone, J.A. Seibert, An accurate method for computer-generating tungsten anode x-ray spectra from 30 to 140 kV. Med. Phys. **24**, 1661–1670 (1997)
4. E.M. Gullikson, "Atomic scattering factors", Sect. 1.7, in *X-Ray Data Booklet*, ed. by A.C. Thompson (Lawrence Berkeley National Laboratory, Berkeley, 2009), pp. 44–52
5. C.T. Chantler, Theoretical form factor, attenuation, and scattering tabulation for Z = 1-92 from E = 1-10 eV to E = 0.4-1.0 MeV. J. Phys. Chem. Ref. Data **24**, 71 (1995)
6. P.P. Kane, L. Kissel, R.H. Pratt, S.C. Roy, Elastic scattering of γ-rays and x-rays by atoms. Phys. Rep. **140**, 75–159 (1986)
7. R.R. Letfullin, T.F. George, G.C. Duree, B.M. Bollinger, Ultrashort laser pulse heating of nanoparticles: comparison of theoretical approaches, *Adv. Opt. Technol.* **2008**, ID 251718/1-8 (2008).
8. J. Siewerdsen, A. Waese, D. Moseley, S. Richard, D. Jaffray, Spektr: a computational tool for x-ray spectral analysis and imaging system optimization. Med. Phys. **31**, 3057–3067 (2004)
9. B.J. MacGowan, L.B. Da Silva, D.J. Fields, C.J. Keane, J.A. Koch, R.A. London, D.L. Matthews, S. Maxon, S. Mrowka, A.L. Osterheld, J.H. Scofield, G. Shimkaveg, J.E. Trebes, R.S. Walling, Short wavelength x-ray laser research at the Lawrence Livermore National Laboratory. Phys. Fluids B **4**, 2326–2337 (1992)
10. J. Nilsen, B.J. MacGowan, L.B. Da Silva, J.C. Moreno, Prepulse technique for producing low-Z Ne-like x-ray lasers. Phys. Rev. A **48**, 4682–4685 (1993)
11. R.R. Letfullin, A.R. Letfullin, T.F. George, Absorption efficiency and heating kinetics of nanoparticles in the RF range for selective nanotherapy of cancer. Nanomed.: Nanotechnol. Biol. Med. **11**, 413–420 (2015)
12. G.W. Hanson, R.C. Monreal, S.P. Apell, Electromagnetic absorption mechanisms in metal nanospheres: bulk and surface effects in radiofrequency-terahertz heating of nanoparticles. J. Appl. Phys. **109**, 124306/1-6 (2011)

13. C.H. Moran, S.M. Wainerdi, T.K. Cherukuri, C. Kittrell, B.J. Wiley, N.W. Nicholas, Size-dependent Joule heating of gold nanoparticles using capacitively coupled radiofrequency fields. Nano Res. **2**, 400–405 (2009)
14. H.M. Gach, A. Balachandrasekarana, T. Naira, RF hyperthermia using conductive nanoparticles. Proc. Soc. Photo Opt. Instrum. Eng. **7548**, 41/1–41/11 (2010)
15. E.S. Glazer, C. Zhu, K.L. Massey, C.S. Thompson, W.D. Kaluarachchi, A.N. Hamir, S.A. Curley, Noninvasive radiofrequency field destruction of pancreatic adenocarcinoma xenografts treated with targeted gold nanoparticles. Clin. Cancer Res. **16**, 5712–5721 (2010)
16. M. Raoof, S.J. Corr, W.D. Kaluarachchi, K.L. Massey, K. Briggs, C. Zhu, M.A. Cheney, L.J. Wilson, S.A. Curley, Stability of antibody-conjugated gold nanoparticles in the endolysosomal nanoenvironment: implications for noninvasive radiofrequency-based cancer therapy. Nanomedicine **8**, 1096–1105 (2012)
17. C.J. Gannon, C.R. Patra, R. Bhattacharya, Intracellular gold nanoparticles enhance non-invasive radiofrequency thermal destruction of human gastrointestinal cancer cells. J. Nanobiotechnol. **6**, 1–9 (2008)
18. D. Li, Y.S. Jung, S. Tan, H.K. Kim, E. Chory, D.A. Geller, Negligible absorption of radiofrequency radiation by colloidal gold nanoparticles. J. Colloid Interface Sci. **358**, 47–53 (2011)
19. C.J. Gannon, P. Cherukuri, B.I. Yakobson, L. Cognet, J.S. Kanzius, C. Kittrell, R.B. Weisman, M. Pasquali, H.K. Schmidt, R.E. Smalley, S.A. Curley, Carbon nanotube-enhanced thermal destruction of cancer cells in a noninvasive radiofrequency field. Cancer **110**, 2654–2665 (2007)
20. http://www.science.gov/

Chapter 5
Modeling Nanoparticle Optics

Contents

Abstract As we discussed in the preceding Chap. 4, the optimal range of nanoparticle sizes and optimal wavelength of radiation for effective activating/heating of nanoparticles in tumor cells can be calculated on the basis of the Lorenz–Mie diffraction theory at the single-scattering approximation. The Mie formalism requires the use of two dimensionless input parameters: the complex refractive

---

Electronic supplementary material: The online version of this chapter (doi:10.1007/978-3-319-43577-0_5) contains supplementary material, which is available to authorized users.

R.R. Letfullin, T.F. George, *Computational Nanomedicine and Nanotechnology*,
DOI 10.1007/978-3-319-43577-0_5

index of the nanoparticle and the surrounding medium at the given wavelength. In this chapter we introduce a computer software and perform Mie simulations of absorption, scattering, and extinction efficiencies of different types of nanoparticles as a function of particle size and the wavelength of radiation in different surrounding biological media. Also, we calculate and then use the differences in optical properties of normal and cancer cell organelles in order to design new cancer detection and treatment methods.

5.1 Computer Practicum 1 (Lecture 15): Lorenz–Mie Simulations of Nanoparticle Optics (Slide 15.1)

- Statement of the Research Task
- Input Data
- Mie Simulations of the Absorption and Scattering Spectrum of Nanoparticles
- Practice Example 1: Optical Properties of Gold Nanoparticles in Low-Absorbing Biological Media
  - *Absorption and Scattering Spectrum*
  - *Absorption and Scattering vs. Particle Size*
  - *Optical Properties of Gold Nanoparticles in Low-Absorbing Media*
  - *Conclusions*

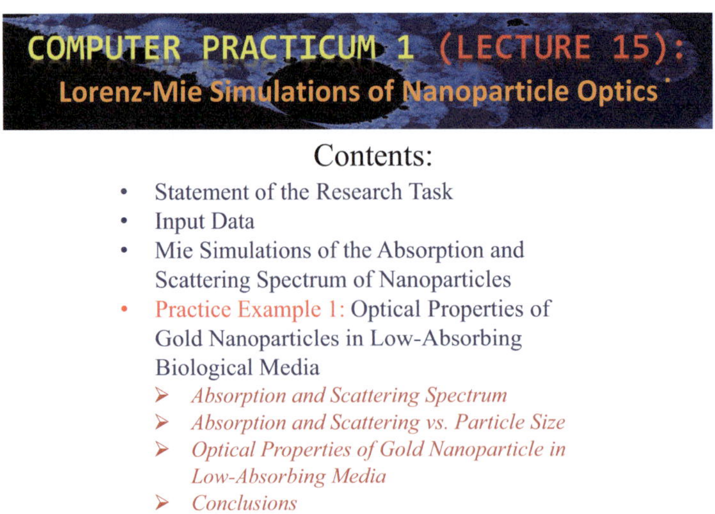

Slide 15.1 Contents of Lecture 15

5.1.1 *Statement of the Research Task*

The *statement of the research task* is schematically demonstrated in Slide 15.2, showing the Au nanoparticles selectively delivered to the cancer cells. Since the size of the nanoparticles is smaller than the pores of the cell membrane, they can be collected inside the cancer cells in the cytoplasm or even inside the cell organelles such as nucleus, mitochondria, ribosome, etc. After delivery of the nanoparticles, the radiation is applied to activate or heat the nanoparticles to initiate the cancer cell death. The research task is to model the radiation interaction with nanoparticles to (Slide 15.2):

- Find the absorption and scattering efficiencies of the nanoparticles in biological media;
- Define the optimal wavelength of radiation for nanotherapy and diagnosis;
- Determine the optimal size of nanoparticles for therapeutic and diagnostic applications;
- Simulate the nanoparticle heating kinetics;
- Evaluate the biological ablation and thermal damage area;
- Make the recommendations for experimental realization of this technology.

A first step in the modeling of the radiation interaction with nanoparticle is to simulate the nanoparticle optics and find the absorption and scattering efficiencies of the nanoparticles in the surrounding biological medium.

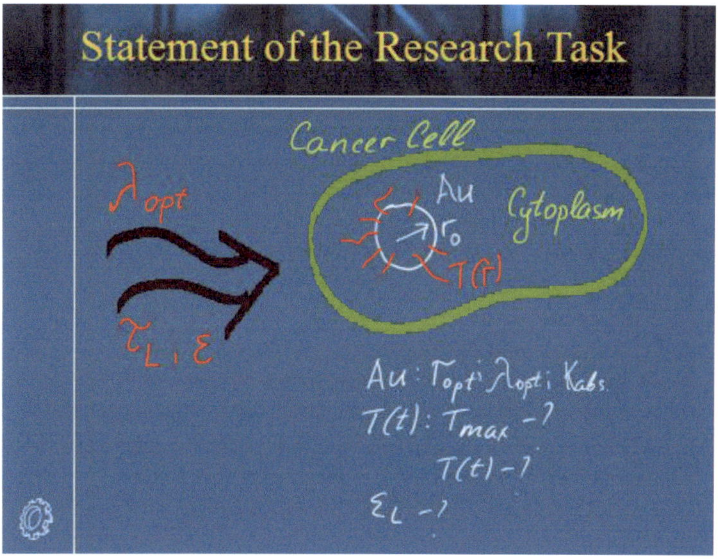

Slide 15.2 Statement of research task

5.1.2 Input Data

In Sect. 4.1 of Chapter 4 (Lecture 12), we introduced the Maxwell's equations and Lorenz–Mie solutions to find the optical properties of a spherically symmetric particle. These final solutions characterize absorption and scattering for incoming light of a given wavelength in terms of the nanoparticle size and the electromagnetic properties of the nanoparticle and surrounding media. Starting in general terms with the electric permittivity ε and the magnetic permeability μ the solutions utilize an assumption that the material is nonmagnetic, and thus $\mu \approx 1$ and $\varepsilon = m^2$. Therefore, absorption and scattering efficiencies of the nanoparticle exposed to incoming light of wavelength λ can be calculated with only three input parameters (Slide 15.3):

- Nanoparticle radius r_0;
- Complex refractive index m_0 of the sphere;
- Complex refractive index m_1 of the surrounding medium.

The nanoparticle radius is a parameter that can be engineered to modify the light absorption properties of a particle (Slide 15.3). The complex refractive indices of the nanoparticle material and surrounding medium, however, are dependent upon the physical situation under investigation. The complex refractive index of the surrounding medium corresponds to the type of biological tissue being treated, and therefore is treated as a nonalterable parameter. The complex refractive index of the spherical nanoparticle could technically be engineered by making the nanoparticles from different materials. However, because this application relies on the biological inertness of the nanoparticles, the choice of materials becomes

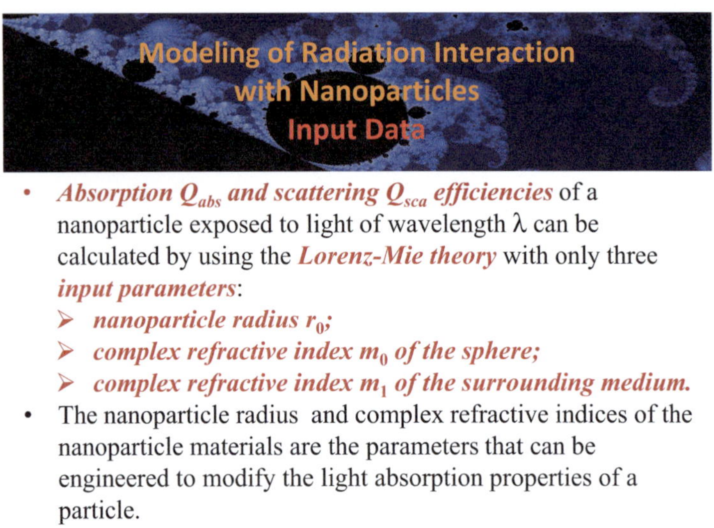

Slide 15.3 Input data

more limited. Gold nanoparticles have been well studied and shown to be biologically inert, making them a good material to work with. Nanoparticles made from other materials (e.g., silver, copper, iron, polystyrene, silica, carbon, etc.) have shown varying degrees of cytotoxicity or genotoxicity. The examples of the complex refractive indices of some nanoparticle materials and surrounding media are given in Tables 5.1 and 5.2 at room temperature for ordinary rays in optical range of the spectrum (Slide 15.4).

Table 5.1 Complex refractive indices for some nanoparticle materials at room temperature for ordinary rays

Substance	Wavelength, λ (nm)	Refractive index, $m = n_0 - ik_0$		Reference
		n_0	k_0	
Gold (Au) bulk	100	1.1549	0.7857	Generated from RefractiveIndex.info
	200	1.2430	1.1510	
	300	1.5253	1.8080	
	400	1.5689	1.8389	
	500	0.7862	1.6734	
	600	0.16714	2.9854	
	700	0.082853	3.9736	
	800	0.07949	4.8339	
	900	0.088565	5.6272	
	1000	0.098721	6.3854	
Gold (Au) nanosphere 13.5 ± 1 nm diameter	300	1.4480397	0.716125	Generated from Luxpop.com
	350	1.533208	0.746771	
	400	1.584586	0.743689	
	450	1.531719	0.771014	
	500	1.388243	0.949755	
	550	1.426692	1.380822	
	600	1.828947	1.724205	
	650	2.362782	1.8	
	700	2.809680	1.564931	
	750	3.018045	1.246967	
	800	3.073308	0.965949	
	850	3.058271	0.769416	
	900	3.019086	0.628180	
	950	2.981756	0.530989	
SiAu nanosphere Au core 15.2 ± 2 nm diameter Si shell thickness 8.5 ± 1 nm	300	1.541772	0.201164	Generated from Luxpop.com
	350	1.545569	0.174389	
	400	1.545569	0.169219	
	450	1.523734	0.194775	
	500	1.527848	0.283392	
	550	1.714992	0.349912	
	600	1.815823	0.170988	

(continued)

Table 5.1 (continued)

Substance	Wavelength, λ (nm)	Refractive index, $m = n_0 - ik_0$		Reference
		n_0	k_0	
	650	1.766983	0.088608	
	700	1.729219	0.061829	
	750	1.709922	0.051786	
	800	1.692905	0.047411	
	850	1.688098	0.044954	
	900	1.679024	0.042290	
	950	1.676487	0.039916	
Silver (Ag) bulk	50	0.87682	0.55759	Generated from RefractiveIndex.info
	100	1.2928	0.58146	
	200	1.0790	1.2470	
	300	1.4157	0.95567	
	400	0.05397	2.1391	
	500	0.0520	3.1412	
	600	0.049204	3.9981	
	700	0.050996	4.8143	
	800	0.053831	5.6056	
	900	0.060326	6.3872	
	1000	0.069764	7.1582	
Silver (Ag) nanocubes 30 ± 1 nm	350	0.603232	0.846639	Generated from Luxpop.com
	400	0.763798	1.297153	
	450	0.920250	1.603809	
	500	1.046559	2.019048	
	550	1.411849	2.546369	
	600	2.182226	3.001190	
	650	3.205745	2.910615	
	700	3.874566	2.209014	
	750	3.988270	1.535516	
	800	3.875404	1.093527	
	850	3.741472	0.838542	
SiAg nanosphere Ag core 50 ± 5 nm diameter Si shell thickness 25 ± 5 nm	350	1.231461	0.089412	Generated from Luxpop.com
	400	1.065202	0.332778	
	450	1.477034	0.701176	
	500	1.730557	0.358092	
	550	1.674007	0.220176	
	600	1.647940	0.181176	
	650	1.643564	0.151031	
	700	1.635955	0.123415	
	750	1.630992	0.100047	
	800	1.623999	0.080966	
	850	1.610986	0.068627	

(continued)

Table 5.1 (continued)

Substance	Wavelength, λ (nm)	Refractive index, $m = n_0 - ik_0$		Reference
		n_0	k_0	
Aluminum (Al) bulk	100	0.0350	0.76571	Generated from RefractiveIndex.info
	200	0.12285	2.2879	
	300	0.24757	3.4765	
	400	0.46555	4.7121	
	500	0.78833	5.8655	
	600	1.1414	6.9256	
	700	1.6620	8.0439	
	800	2.7075	8.2713	
	900	2.0362	8.2147	
	1000	1.5148	9.2640	
Nickel (Ni) bulk	200	1.2241	1.6768	Generated from RefractiveIndex.info
	300	1.7320	2.0053	
	400	1.5892	2.3709	
	500	1.7072	2.9974	
	600	1.9378	3.5260	
	700	2.1782	3.9723	
	800	2.4020	4.3650	
	900	2.6030	4.7227	
	1000	2.7810	5.0575	
Carbon (C) graphite bulk	50	0.39598	0.64089	Generated from RefractiveIndex.info
	100	2.2734	0.72470	
	200	0.82868	1.3102	
	300	2.8120	1.8407	
	400	2.6506	1.2236	
	500	2.6518	1.3987	
	600	2.7894	1.5552	
	700	2.9236	1.6645	
	800	3.0304	1.7609	
	900	3.1187	1.8607	
	1000	3.1987	1.9659	
Fullerene* (C_{60}) nanosphere	52.8	0.2	1.3	Calculated from the ε_1 and ε_2 values
	66.3	0.9	2	
	101.6	1.8	2.4	
	123.98	2.2	1.7	
	139.3	2.4	1.6	
	161	2	1	
	213.8	1	3	
	269.5	1.3	4.5	
	344.4	3.2	4	
	400	2.51	0.63	

(continued)

Table 5.1 (continued)

Substance	Wavelength, λ (nm)	Refractive index, $m = n_0 - ik_0$		Reference
		n_0	k_0	
	500	2.24	0.45	
	633	2.27	0.42	
	700	2.25	0.22	
	800	2.18	0.1	
	900	2.11	0.05	
	1000	2.01	0.01	
Polystyrene bulk	400	1.605	0.0005	Ma et al. [1]
	500	1.590	0.0003	
	600	1.584	0.0006	
	700	1.580	0.0006	
	800	1.578	0.0007	
	900	1.5763	0.0009	
	1000	1.5758	0.0012	
Glass (Si) bulk	400	1.47018	0.015	Generated from Luxpop.com
	500	1.46239	0.018	
	600	1.4581	0.021	
	700	1.45536	0.025	
	800	1.45338	0.031	
	900	1.45182	0.037	
	1000	1.45048	0.042	
Aluminum oxide (Al_2O_3) bulk	300	1.8147	1.8056	Generated from Luxpop.com
	400	1.786631	1.778098	
	500	1.774359	1.766152	
	600	1.767568	1.759480	
	700	1.763228	1.755202	
	800	1.760346	1.752368	
	900	1.757843	1.749846	
	1000	1.755761	1.748174	
Chromium (Cr) bulk	200	1.396849	1.7	Generated from Luxpop.com
	300	1.524596	2.333515	
	400	2.013515	2.846894	
	500	2.782013	3.305055	
	600	3.193535	3.3	
	700	3.053757	3.385303	
	800	3.163203	3.461602	
	900	3.310445	3.524748	
	1000	3.438488	3.571676	

*The n_0 and k_0 components of the complex refractive index for fullerene are calculated from the available electric permittivity ε_1 and ε_2 values by using the following equations (Slide 15.4):

$$\varepsilon = \varepsilon_1 + i\varepsilon_2 = (n_0 + ik_0)^2, \tag{5.1}$$

(continued)

Table 5.1 (continued)

$$\varepsilon_1 = n_0^2 + k_0^2, \tag{5.2}$$

$$\varepsilon_2 = 2n_0k_0, \tag{5.3}$$

$$n_0 = \sqrt{\frac{\varepsilon_1 + \sqrt{\varepsilon_1^2 + \varepsilon_2^2}}{2}}, \tag{5.4}$$

$$k_0 = \sqrt{\frac{-\varepsilon_1 + \sqrt{\varepsilon_1^2 + \varepsilon_2^2}}{2}}. \tag{5.5}$$

Table 5.2 Real parts of the refractive indices for surrounding media at room temperature for ordinary rays (Slide 15.5)

Wavelength	100 nm	200 nm	300 nm	400 nm	500 nm	600 nm	700 nm
Medium	n_1	n_1	n_1	n_1	n_1	n_1	n_1
Water	1.346	1343	1.341	1.339	1.334	1.332	1.331
Blood	1.2	1.25	1.30	1.354	1.4	1.39	1.383
Cytoplasm	1.32	1.33	1.34	1.35	1.36	1.365	1.367
Collagen	1.75	1.70	1.65	1.6	1.55	1.5	1.45
Cell membrane	1.66	1.62	1.58	1.54	1.5	1.46	1.40

5.1.3 *Mie Simulations of the Absorption and Scattering Spectrum of Nanoparticles*

In this section we introduce a computer software and perform Lorenz–Mie simulations of absorption, scattering, and extinction efficiencies of nanoparticles in surrounding biological media (Slide 15.6). We have developed an effective algorithm to perform the simulations [2], where the cylindrical functions of real or imaginary arguments and their derivatives, which occur in the expressions for the Mie coefficients a_l and b_l (see Eqs. (4.2)), are calculated as the ratio of the function and its derivative by using the recurrent relationships for direct and inverse recursions. Such an approach allows us to make an effective and accurate determination of the optical properties of a dispersive medium over a broad range of Mie parameters, $\rho = 2\pi r_0/\lambda = 0.001 - 1500$, below the diffraction limit simultaneously with small and large values of the real n_0 and the imaginary k_0 parts of the refractive index of the particle's and surrounding medium substance.

Examples of the Complex Refractive Indices of Some Nanoparticle Materials

Substance	Wavelength, λ, nm	Refractive Index, $m = n_0 - ik_0$		Reference
		n_0	k_0	
Gold (Au) Bulk	100	1.1549	0.7857	Generated from RefractiveIndex.info
	200	1.2430	1.1510	
	300	1.5253	1.8080	
	400	1.5689	1.8389	
	500	0.7862	1.6734	
	600	0.16714	2.9854	
	700	0.082853	3.9736	
	800	0.07949	4.8339	
	900	0.088565	5.6272	
	1000	0.098721	6.3854	
Gold (Au) Nanosphere 13.5 ±1 nm diameter	300	1.4480397	0.716125	Generated from Luxpop.com
	350	1.533208	0.746771	
	400	1.584586	0.743689	
	450	1.531719	0.771014	
	500	1.388243	0.949755	
	550	1.426692	1.380822	
	600	1.828947	1.724205	
	650	2.362782	1.8	
	700	2.809680	1.564931	
	750	3.018045	1.246967	
	800	3.073308	0.965949	
	850	3.058271	0.769416	
	900	3.019086	0.628180	
	950	2.981756	0.530989	
SiAu Nanosphere Au core 15.2 ±2 nm diameter Si shell thickness 8.5±1 nm	300	1.541772	0.201164	Generated from Luxpop.com
	350	1.545569	0.174389	
	400	1.545569	0.169219	
	450	1.523734	0.194775	
	500	1.527848	0.283392	
	550	1.714992	0.349912	
	600	1.815823	0.170988	
	650	1.766983	0.088608	
	700	1.729219	0.061829	
	750	1.709922	0.051786	
	800	1.692905	0.047411	
	850	1.688098	0.044954	
	900	1.679024	0.042290	
	950	1.676487	0.039916	

The n_0 and k_0 components of the complex refractive index for fullerene are calculated from the available electric permittivity ε_1 and ε_2 values by using the following equations:

$$\varepsilon = \varepsilon_1 + i\varepsilon_2 = (n_0 + ik_0)^2$$

$$\varepsilon_1 = n_0^2 + k_0^2$$

$$\varepsilon_2 = 2n_0k_0$$

$$n_0 = \sqrt{\frac{\varepsilon_1 + \sqrt{\varepsilon_1^2 + \varepsilon_2^2}}{2}}$$

$$k_0 = \sqrt{\frac{-\varepsilon_1 + \sqrt{\varepsilon_1^2 + \varepsilon_2^2}}{2}}$$

Slide 15.4 Examples of the complex refractive indices of some nanoparticle materials

Real Refractive Indices for Surrounding Media at Room Temperature for Ordinary Ray

Wavelength	100 nm	200 nm	300 nm	400 nm	500 nm	600 nm	700 nm
Medium	n_1	n_1	n_1	n_1	n_1	n_1	n_1
Water	1.346	1343	1.341	1.339	1.334	1.332	1.331
Blood	1.2	1.25	1.30	1.354	1.4	1.39	1.383
Cytoplasm	1.32	1.33	1.34	1.35	1.36	1.365	1.367
Collagen	1.75	1.70	1.65	1.6	1.55	1.5	1.45
Cell membrane	1.66	1.62	1.58	1.54	1.5	1.46	1.40

Slide 15.5 Real refractive indices for surrounding media at room temperature for ordinary rays

Similar algorithms are used in many available software designed to run Lorenz–Mie simulations. Mie-type codes such as *MiePlot, BHMIE, Mie_Single, DMIES, MieSolid, nLayerMieCode***,** etc., can be easily found on the Internet—some examples are listed on Wikipedia [3]. For instance, the *MiePlot* software offers an easy-to-use interface providing graphs of intensity vs. scattering angle, wavelength, radius, and refractive index [4]. It is free for users and can be downloaded from

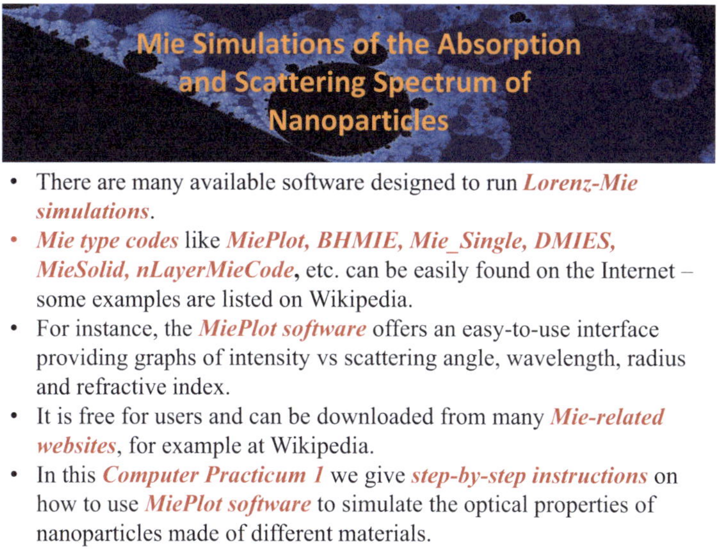

Slide 15.6 Mie simulations of the absorption and scattering spectrum of nanoparticles

many Mie-related websites, for example, at Wikipedia [3]. Following we give step-by-step instructions on how to use *MiePlot* software to simulate the optical properties of nanoparticles made of different materials over a wide range of particle sizes and radiation wavelengths. We also provide our own Mie code designed on MATLAB in Appendix A, which is a simple but efficient computer code.

Step 1: Click on MiePlot v3501.exe file which opens the interface shown in Fig. 5.1 and Slide 15.7.

Step 2: In the right top window, select the "*Mie*" program, and in the window below select "*Qext/Qsca/Qabs vs. radius*," as shown in the red ellipse in Fig. 5.1 and Slide 15.8, to perform Mie simulations of the particle's optical properties vs. radius and to find an optimal radius.

Step 3 selects the index of refraction for a nanoparticle/sphere. In the menu bar, click "*Advanced*," which opens an interactive small window shown in Fig. 5.2 and Slide 15.9. Select "*Refractive Index*" in that window highlighted by blue background color and marked by the first/largest red ellipse in the figure. It opens another small interactive window with two options: "*Sphere*" and "*Surrounding medium*." Select "*Sphere*" highlighted by blue background color and marked by the middle-sized red circle in the figure, which opens a window with material options for the sphere. The index of refraction of the listed materials in that window is built-in to the *MiePlot* code. For example, the simulations select "*Gold*" highlighted by blue background color and circled by the smallest red ellipse in Fig. 5.2 (Slide 15.9).

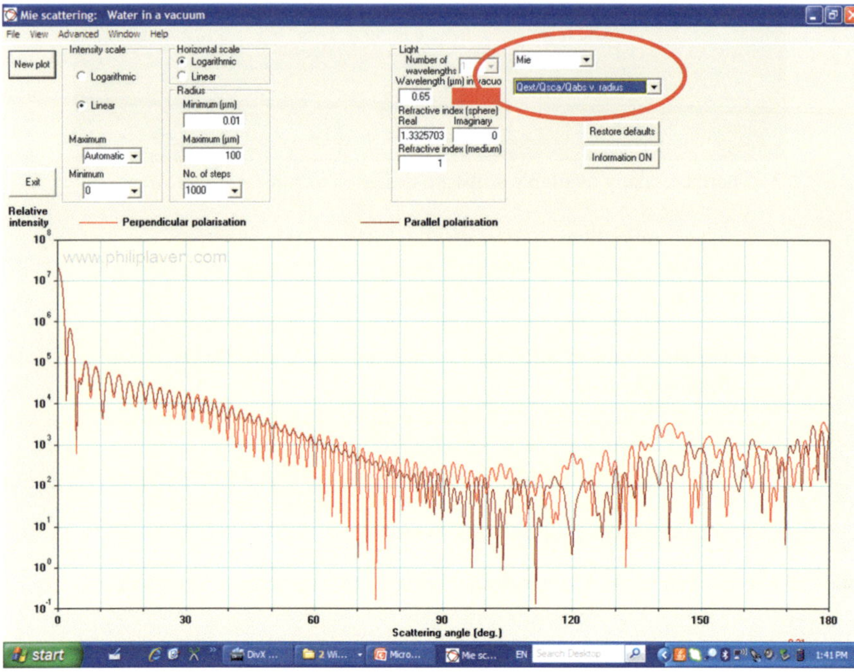

Fig. 5.1 Interface of MiePlot v3501

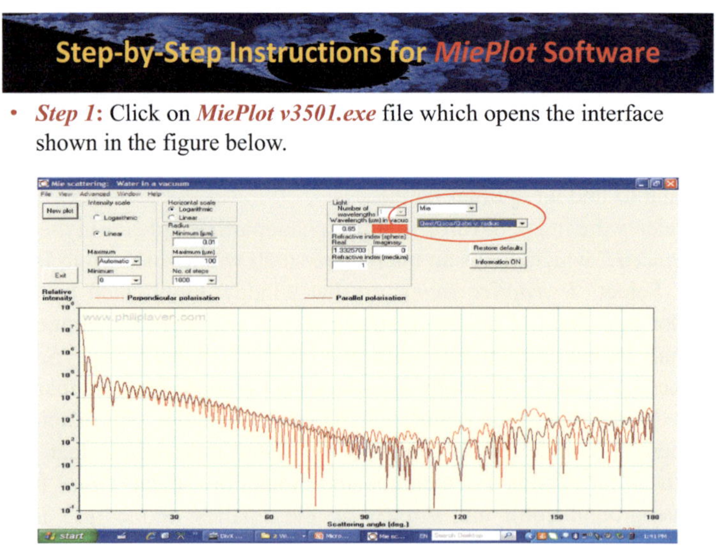

Slide 15.7 Step-by-step instructions for *MiePlot* software

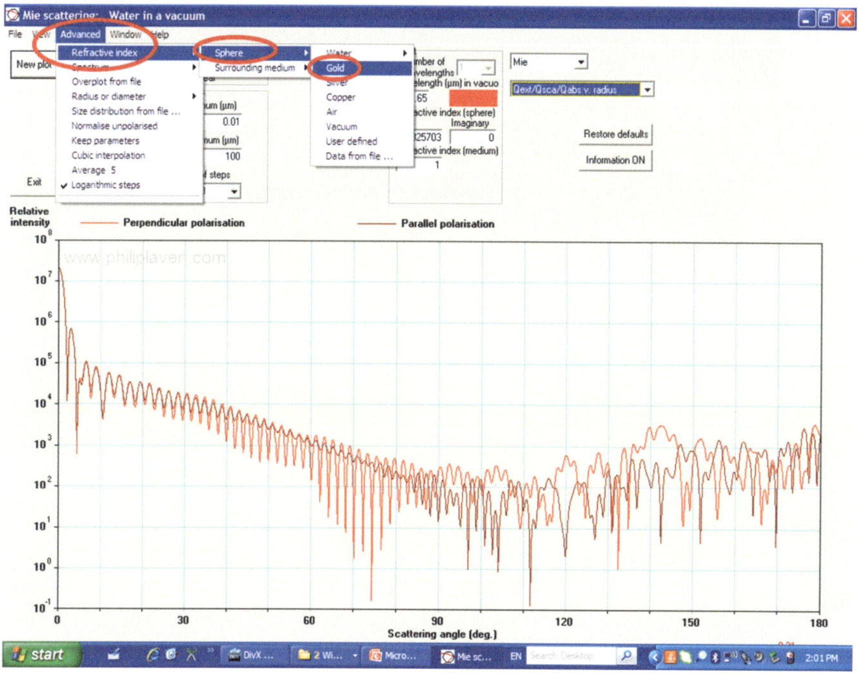

Fig. 5.2 Selecting the index of refraction for a nanoparticle/sphere

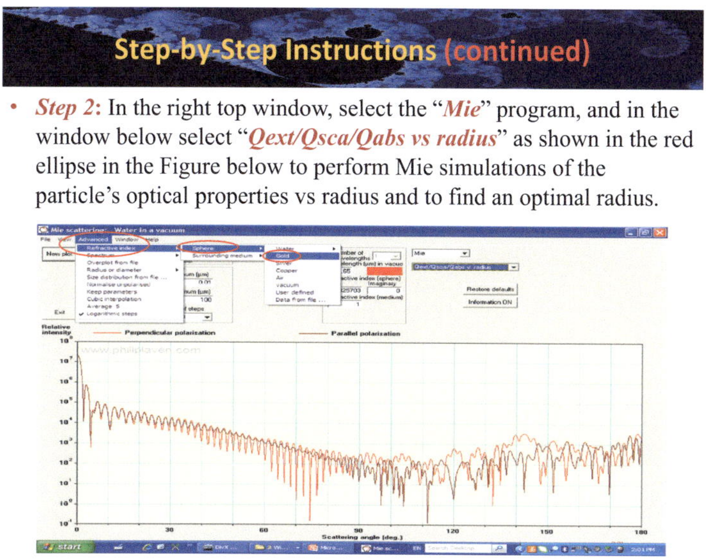

Slide 15.8 Step-by-step instructions (continued)

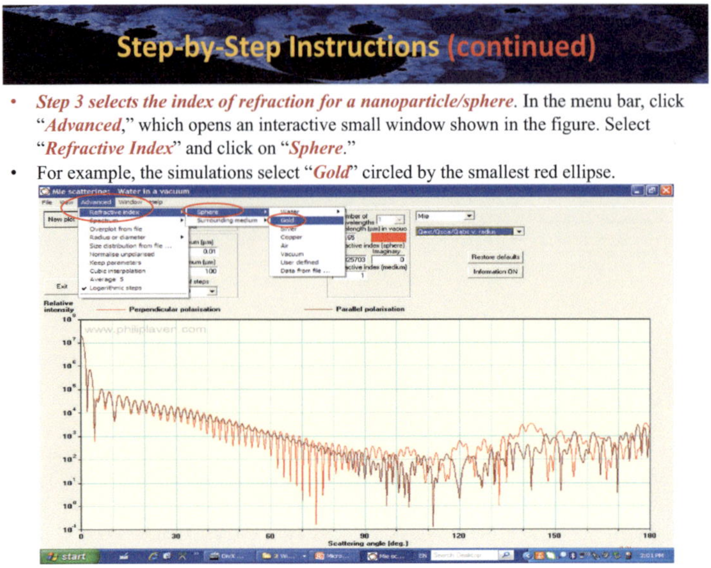

Slide 15.9 Step-by-step instructions (continued)

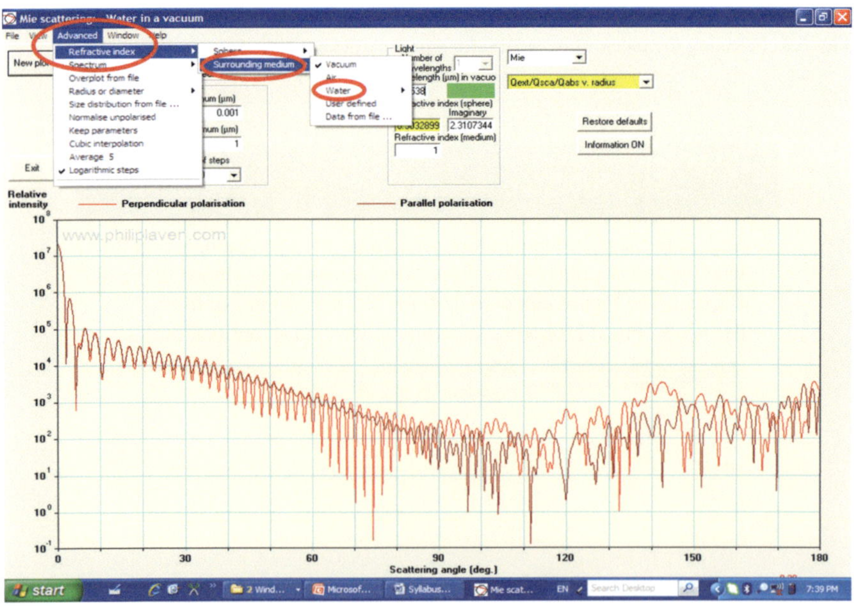

Fig. 5.3 Selecting the index of refraction for the surrounding medium

Step 4 selects the index of refraction for the surrounding medium. This procedure is very similar to Step 3. In the menu bar, click "*Advanced*," which opens an interactive small window shown in Fig. 5.3 and Slide 15.10. Select "*Refractive Index*" in that window highlighted by blue background color and marked by the

Step-by-Step Instructions (continued)

- ***Step 4 selects the index of refraction for the surrounding medium.*** In the menu bar click "***Advanced***," which opens an interactive small window shown in the figure.
- Select "***Refractive Index***" marked by first/largest red ellipse and click on "***Surrounding medium***."
- For example, the simulations select "***Water***" circled by smallest red ellipse.

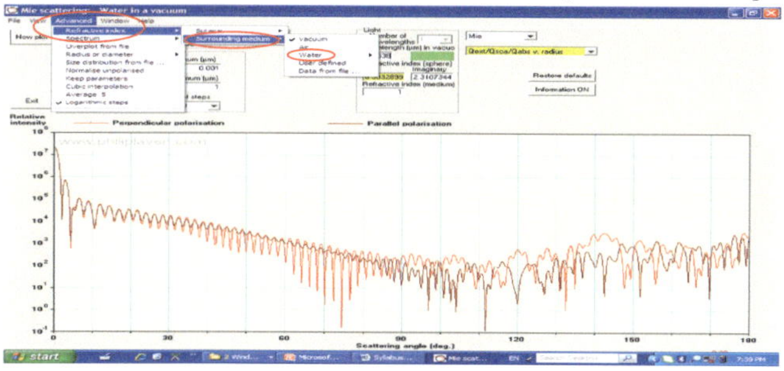

Slide 15.10 Step-by-step instructions (continued)

first/largest red ellipse in the figure. It opens another small interactive window with two options: "*Sphere*" and "*Surrounding medium*." This time select "*Surrounding medium*" highlighted by blue background color and marked by the middle-sized red ellipse on the figure, which opens a window with media options. The index of refraction of the listed media in that window is built-in to the *MiePlot* code. The biological soft tissues and cells mostly consist of water up to 60–70 % of the total content. For example, simulations select "*Water*" highlighted by blue background color and circled by the smallest red ellipse in Fig. 5.3.

Step 5 chooses a radius range for the sphere/particle. In the two windows titled by the common name "*Radius*," introduce the minimum and maximum radii for the particle. Choose a wide range of the particle radii from 1 to 1000 nm. Note that the radius units in *MiePlot* software are given in μm. Thus, input 0.001 μm in the window "*Minimum*" under the "*Radius*," and input 1 μm in the window "*Maximum*" as shown in the red ellipse in Fig. 5.4 and Slide 15.11.

Step 6 selects a wavelength for the radiation. In the group titled "*Light*," input the wavelength of the radiation in μm. For the example in Fig. 5.4, type "0.530" μm (green light) in the window "*Wavelength*," as shown in the blue ellipse and in Slide 15.12.

Step 7 runs the simulations for absorption, scattering, and extinction efficiencies vs. the particle radius for a given wavelength of the radiation. Check the "*Linear*" scale in the group titled "*Intensity scale*" and check the "*Logarithmic*" scale in the group titled "*Horizontal scale*" as shown in the red ellipses in Fig. 5.5 and Slide 15.13. Select "*1000*" in the "*No. of steps*" window marked by the blue ellipse.

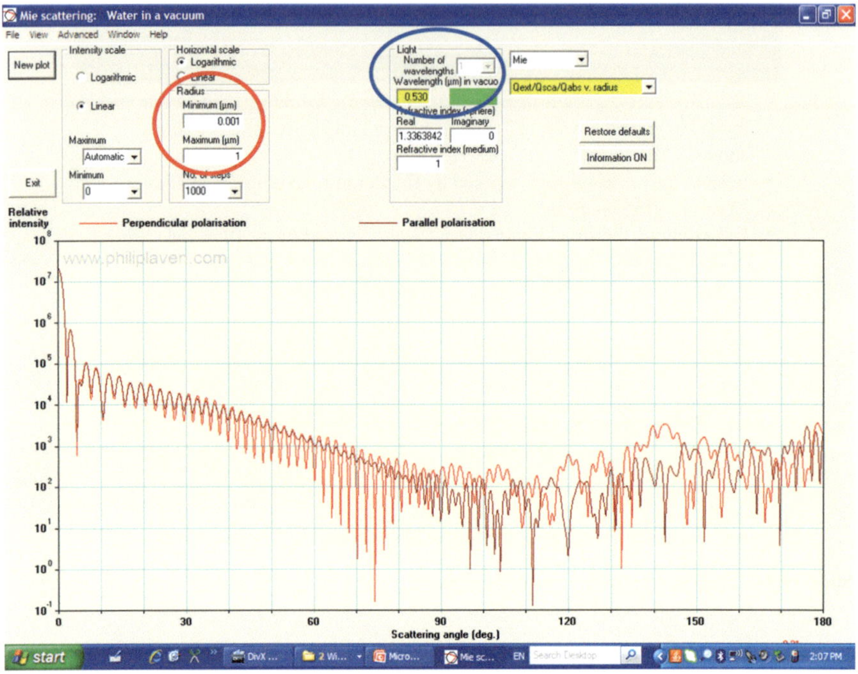

Fig. 5.4 Choosing a radius range for the particle and selecting a wavelength for the radiation

Step-by-Step Instructions (continued)

- ***Step 5 chooses a radius range for the sphere/particle.*** In the two windows titled by common name "*Radius*," introduce the minimum and maximum radii for the particle.
- Choose a wide range of the particle radius from ***1 to 1000 nm***. Note that the radius units in *MiePlot* software are given in µm.
- Thus, input ***0.001 µm*** in the window "***Minimum***" under the "***Radius***", and input ***1 µm*** in the window "***Maximum***" as shown in the red ellipse in the figure below.

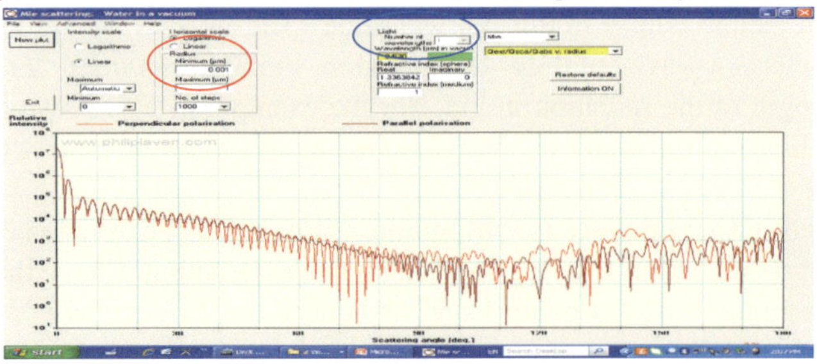

Slide 15.11 Step-by-step instructions (continued)

Step-by-Step Instructions (continued)

- ***Step 6 selects a wavelength for the radiation***. In the group titled "***Light***," input the wavelength of the radiation in μm.
- For the example in the figure below, type "***0.530***" μm (green light) in the window "***Wavelength***" as shown in the blue ellipse.

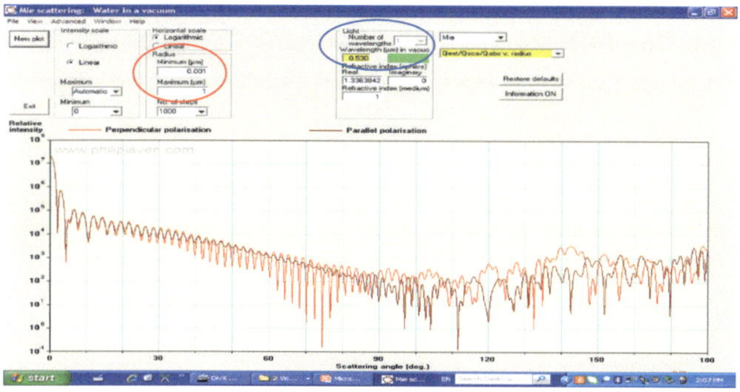

Slide 15.12 Step-by-step instructions (continued)

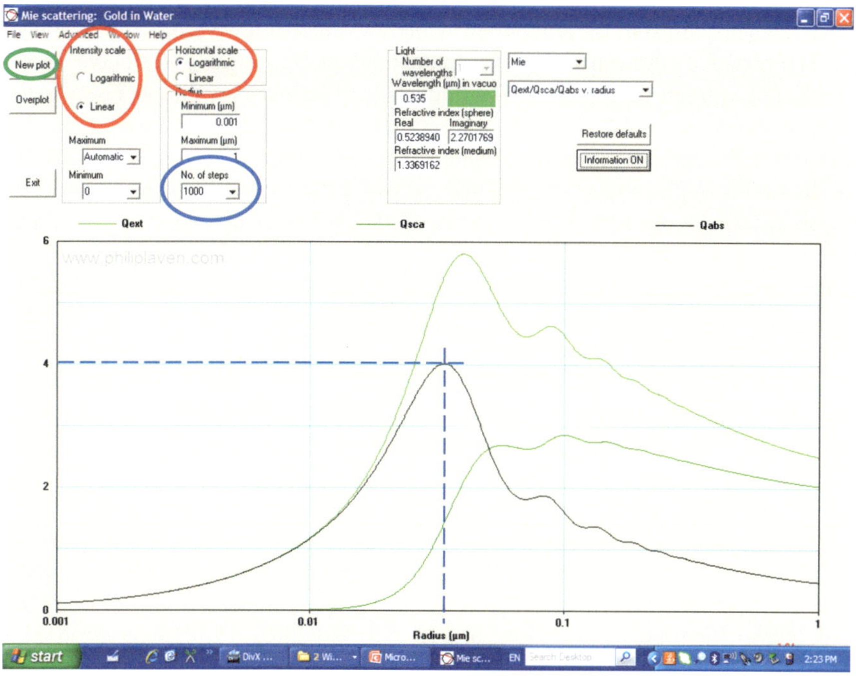

Fig. 5.5 Running the simulations for the extinction Q_{ext}, scattering Q_{sca} and absorption Q_{abs} efficiencies vs. the particle radius

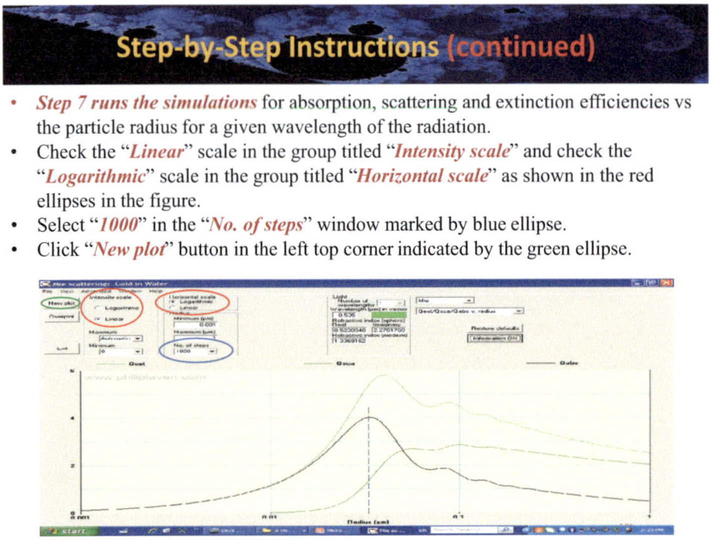

Slide 15.13 Step-by-step instructions (continued)

Click "*New plot*" button in the left top corner indicated by the green ellipse in Fig. 5.5. This will start the simulations which will take a few seconds to complete.

The results of the simulations for the extinction Q_{ext} (light green curve), scattering Q_{sca} (green curve), and absorption Q_{abs} (black curve) efficiencies vs. the particle radius are shown in Fig. 5.5 and Slide 15.14. From these simulations we can determine the optimal radius of the nanoparticle of a given material for therapeutic or diagnostic applications. This optimal radius for the therapeutic uses corresponds to the maximum absorption point shown by the dashed blue lines in Fig. 5.5. The optimal radius range for diagnostic applications is determined by the region where the scattering efficiency Q_{sca} (green curve) dominates over the absorption efficiency Q_{abs} (black curve). *Practice Example 1* presents the output results of these simulations and conclusions in more detail.

Step 8. The results of the simulations can be saved in numerical format as a text file, which can be opened by, for example, the *Notepad* program. The numerical format allows us to find the exact values for efficiencies and the radius of the nanoparticle. To save the results numerically, go to the main menu bar, and under "*File*" menu click on "*Save results as text file*" as shown in Fig. 5.6 and Slide 15.15.

Step 9 simulates the absorption and scattering spectrum of the nanoparticle vs. the wavelength of radiation. In the top right window select the "*Mie*" program, and in the window below select "*Qext/Qsca/Qabs vs. wavelength*" as shown by the red ellipse highlighted in yellow in Fig. 5.7 and Slide 15.16 to perform Mie simulations of the particle's optical properties vs. wavelength of radiation and to find an optimal wavelength. For example, in the simulations we use a gold nanoparticle in a surrounding medium of water as introduced in *Steps 3 and 4*.

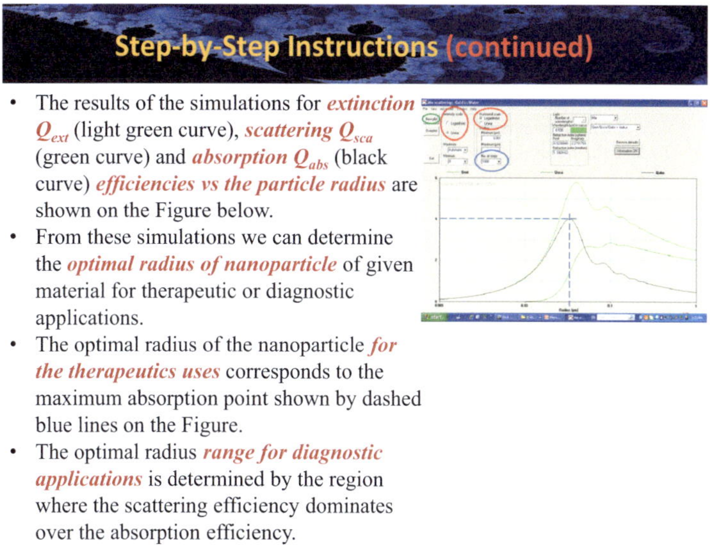

Slide 15.14 Step-by-step instructions (continued)

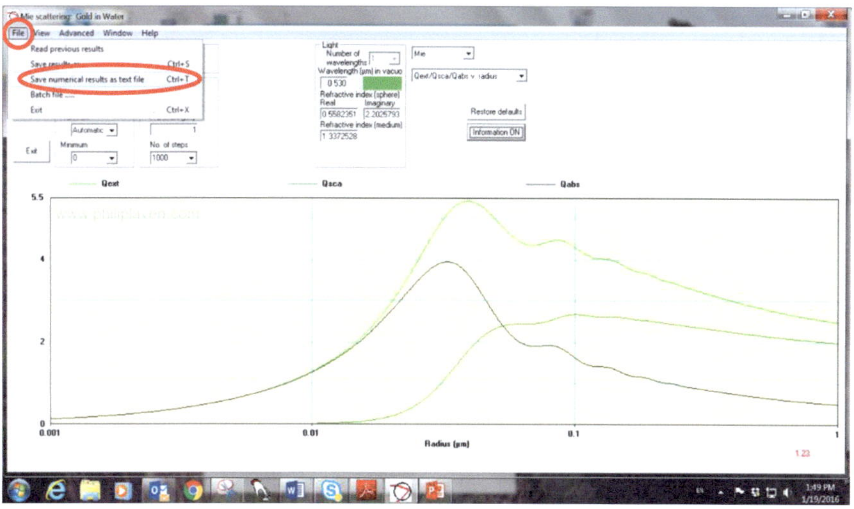

Fig. 5.6 Saving the results of simulations in numerical format as a text file

Step 10 selects the radius of the particle. In the window "*Radius*" of the group titled "*Drop size*," input the optimal radius of the nanoparticle in µm. The simulations performed in Step 6 show that the maximum absorption of the gold nanoparticle for 530 nm light is observed for a nanoparticle radius of 35 nm. For example, simulations enter "0.035" µm in the window "*Radius*" as shown in the blue ellipse in Fig. 5.7 and Slide 15.17.

Step-by-Step Instructions (continued)

- ***Step 8. The results of the simulations can be saved in numerical format as a text file***, which can be opened by, for example, the *Notepad* program.
- To save the results numerically, go to the main menu bar, and under "***File***" menu click on "***Save results as text file***" as shown in the figure below.

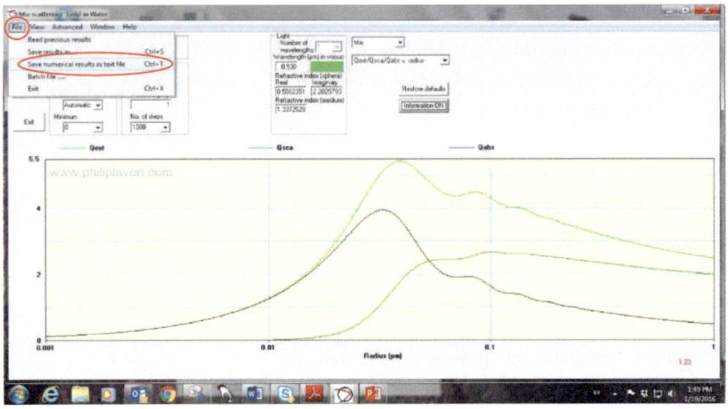

Slide 15.15 Step-by-step instructions (continued)

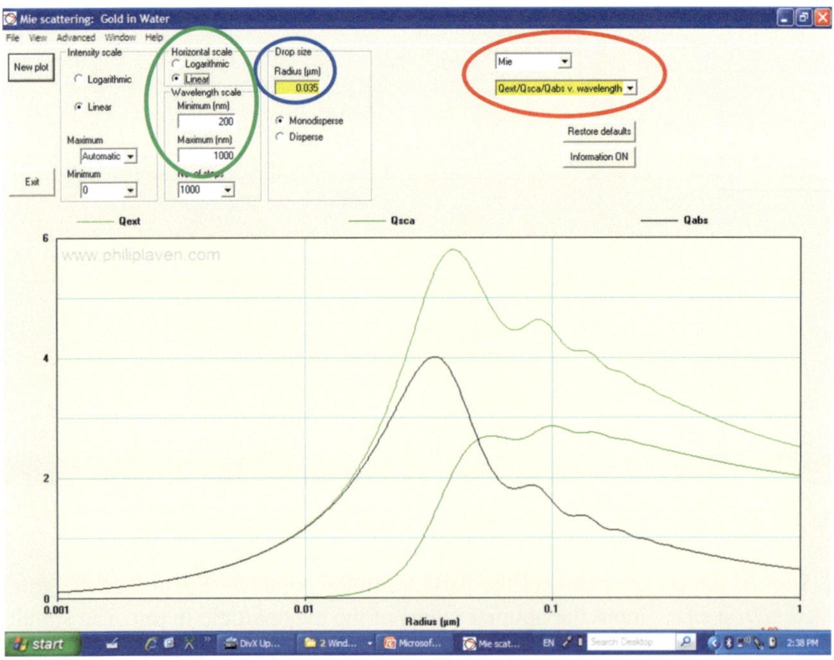

Fig. 5.7 Simulating the absorption and scattering spectrum of a nanoparticle vs. the wavelength of radiation

Step-by-Step Instructions (continued)

- ***Step 9 simulates the absorption and scattering spectrum of the nanoparticle vs the wavelength of radiation.***
- In the top right window select the "*Mie*" program, and in the window below select "*Qext/Qsca/Qabs vs wavelength*" as shown in the red ellipse in the figure.
- For example, in the simulations we use a ***gold nanoparticle in a surrounding medium of water*** as introduced in ***Steps 3 and 4***.

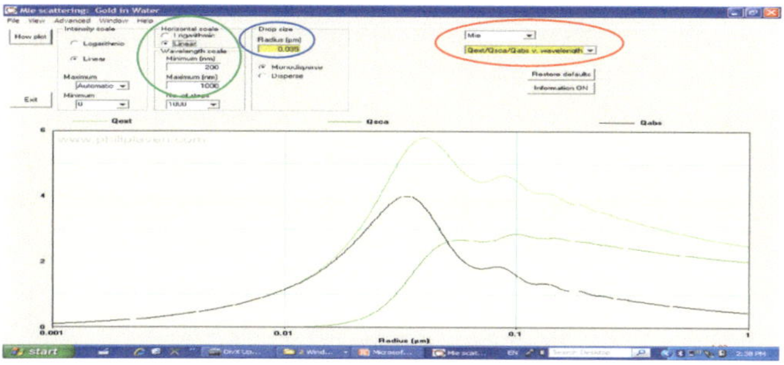

Slide 15.16 Step-by-step instructions (continued)

Step-by-Step Instructions (continued)

- ***Step 10 selects the radius of the particle.*** In the window "***Radius***" of the group titled "***Drop size***," input the optimal radius of the nanoparticle in μm. The simulations performed in ***Step 6*** show that the maximum absorption of the gold nanoparticle for 530 nm light is observed for the nanoparticle radius of ***35 nm***. For the example in the figure below, enter "***0.035***" μm in the window "***Radius***" as shown in a blue ellipse.

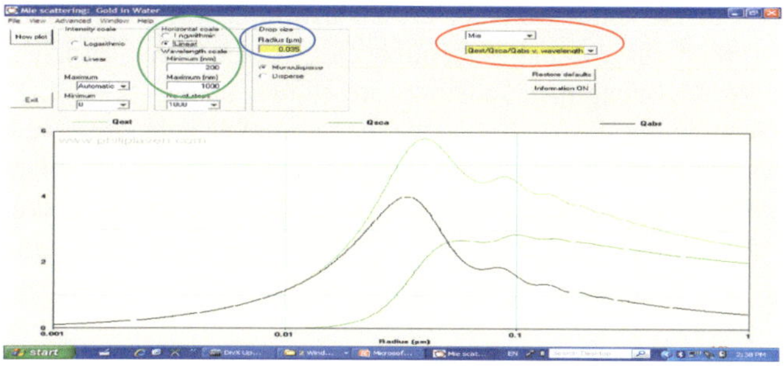

Slide 15.17 Step-by-step instructions (continued)

Step 11 chooses the range for the wavelength of radiation. We have run most simulations in the optical range of spectrum from 200 up to 1000 nm. Thus, in the two windows under "*Wavelength scale*," enter the minimum (200 nm) and maximum (1000 nm) wavelength of the radiation as shown in the green ellipse in Fig. 5.7 and Slide 15.18.

Step-by-Step Instructions (continued)

- ***Step 11 chooses a range for the wavelength of radiation.***
- We have run most simulations in the optical range of the spectrum ***from 200 nm up to 1000 nm***.
- Thus, in the two windows under "***Wavelength scale***," enter the minimum (**200 nm**) and maximum (**1000 nm**) wavelength of the radiation as shown in the green ellipse in the figure below.

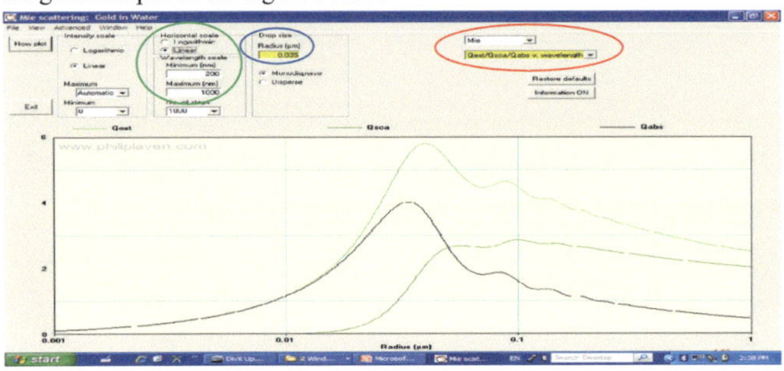

Slide 15.18 Step-by-step instructions (continued)

Step 12 runs the simulations for absorption, scattering, and extinction efficiencies of the nanoparticle vs. the wavelength of the radiation. Leave the "*Linear*" scale in the group titled "*Intensity scale*" and check the "*Linear*" scale in the group titled "*Horizontal scale*" as shown in the red ellipse in Fig. 5.8. Leave "*1000*" in the "*No. of steps*" window marked by the blue ellipse. Click the "*New plot*" button in the left top corner circled by the green ellipse in Fig. 5.8 and Slide 15.19. This will start the simulations, which will take a few seconds to complete.

The results of the simulations for the extinction Q_{ext} (light green curve), scattering Q_{sca} (green curve), and absorption Q_{abs} (black curve) efficiencies of the particle vs. the wavelength of radiation are shown in Fig. 5.8 and Slide 15.20. From these simulations we can determine the optimal wavelength of radiation for a given nanoparticle which best suits therapeutic or diagnostic applications. The optimal wavelength of light for therapeutics uses corresponds to the maximum absorption point shown by the dashed blue lines in Fig. 5.8. The optimal wavelength range for diagnostic applications is determined by the region where the scattering efficiency Q_{sca} (green curve) dominates over the absorption efficiency Q_{abs} (black curve). *Practice Example 1* presents the output results of these simulations and conclusions in more detail. The results of the simulations can be saved in numerical format as a text file as shown in Fig. 5.6.

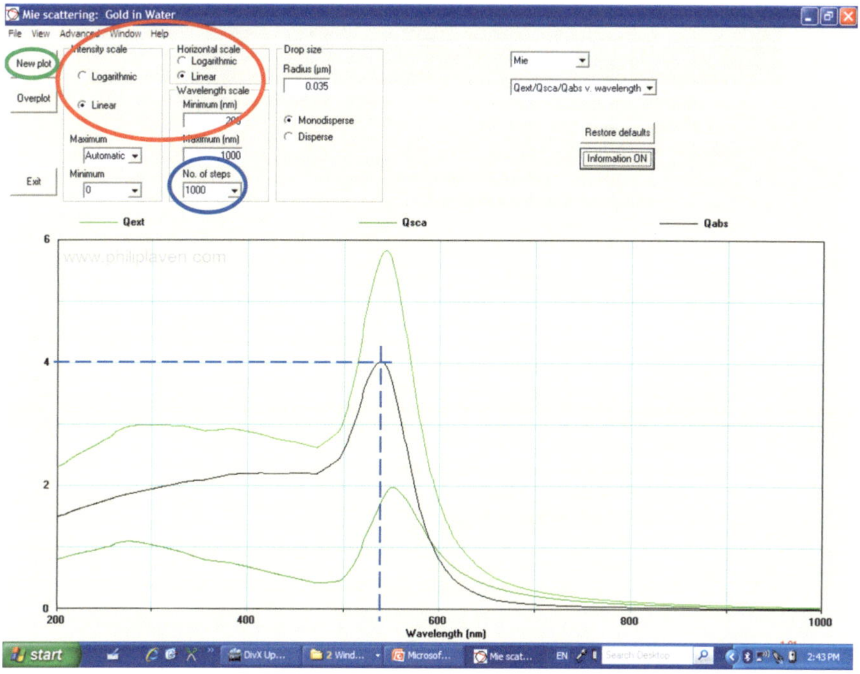

Fig. 5.8 Running the simulations for the extinction Q_{ext}, scattering Q_{sca}, and absorption Q_{abs} efficiencies of the nanoparticle vs. wavelength of radiation

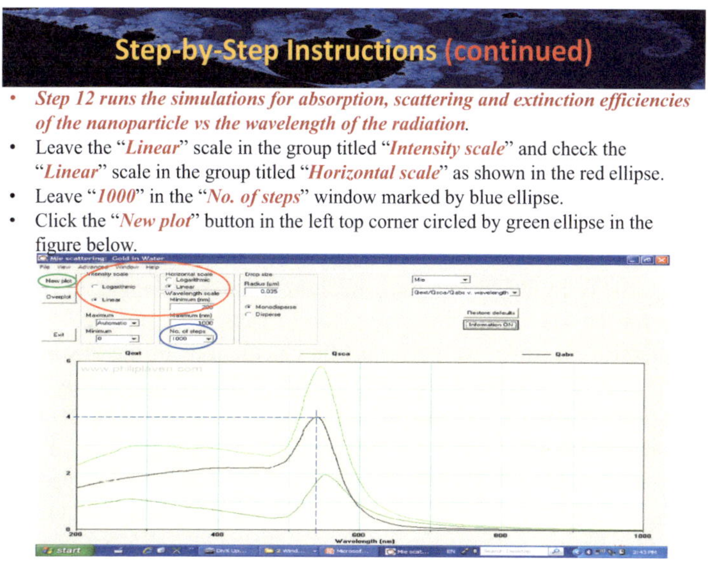

Slide 15.19 Step-by-step instructions (continued)

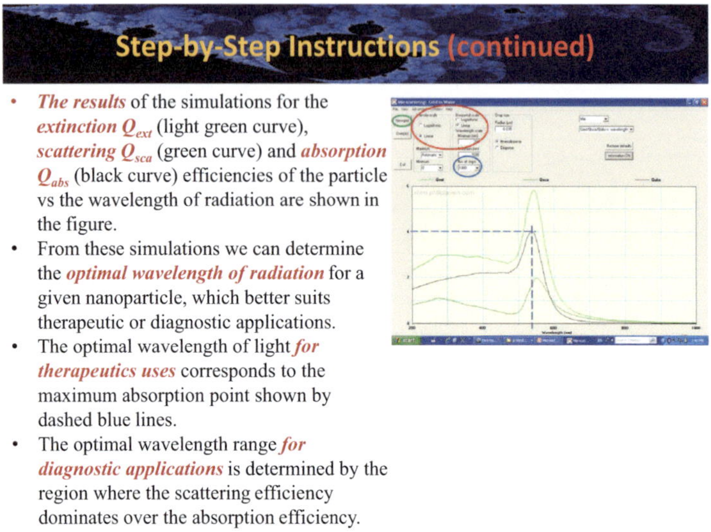

Slide 15.20 Step-by-step instructions (continued)

5.1.4 *Practice Example 1: Optical Properties of Gold Nanoparticles in Low-Absorbing Biological Media*

5.1.4.1 Absorption and Scattering Spectrum (Slide 15.21)

The optimal wavelength of laser radiation and the optimal size range of nanoparticles for effective laser killing of cancer cell can be found by using Lorenz–Mie diffraction theory at the single-scattering approximation described in Sect. 4.1 (*Lecture 12*) of Chap. 4. The computer code for Mie simulations with step-by-step instructions are provided in the *Computer Practicum 1* (*Lecture 15*) of this Chap. 5.

The Mie formalism requires the use of two dimensionless input parameters, $\rho = 2\pi r_0/\lambda$ and $\delta = \rho\widetilde{m}$, where $\widetilde{m}$ is the relative value of the complex refractive index of the nanoparticles in the surrounding medium at the wavelength λ [7]. The input optical data for the Mie simulations are given in Tables 5.1 and 5.2.

In this practice example, we perform Mie calculations of the absorption, Q_{abs}, and scattering, Q_{sca}, efficiencies for gold nanoparticles in an aqueous suspension. The results of these simulations are plotted in Fig. 5.9 and Slide 15.21. Figure 5.9a shows the absorption and scattering spectrum of the gold particle in the visible range of the spectrum, $\lambda = 400$–700 nm, and Fig. 5.9b illustrates the dependences of Q_{abs} and Q_{sca} on particle size for the optimal wavelength. It is evident from these figures that Q_{abs} has a strong maximum at the wavelength $\lambda = 538.3$ nm for a gold-particle radius of 35 nm. Gold nanoparticles are effective absorbers of the radiation in the spectral range of 400–580 nm, where $Q_{abs} > 1$, Q_{sca}. In this range, the

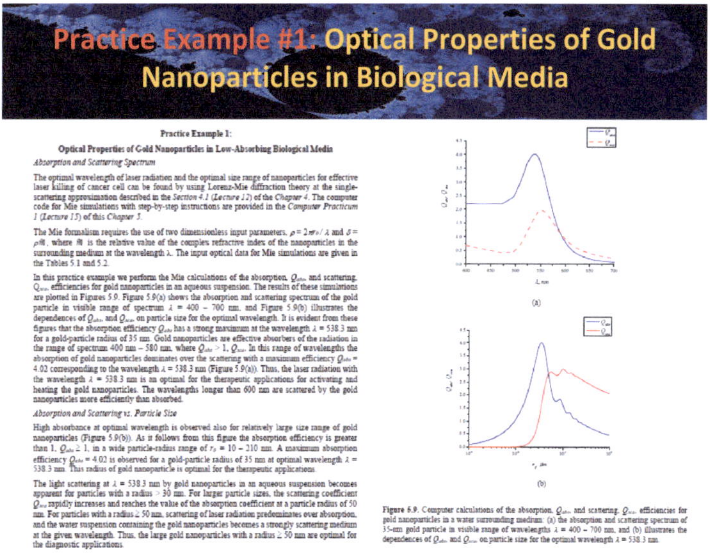

Slide 15.21 Practice Example #1: Optical properties of gold nanoparticles in biological media

absorption of gold nanoparticles dominates over the scattering with a maximum efficiency $Q_{\text{abs}} = 4.02$ corresponding to the wavelength $\lambda = 538.3$ nm (Fig. 5.9a). Thus, the laser radiation with the wavelength $\lambda = 538.3$ nm is optimal for therapeutic applications for activating and heating the gold nanoparticles. The wavelengths longer than 600 nm are scattered by the gold nanoparticles more efficiently than absorbed.

5.1.4.2 Absorption and Scattering vs. Particle Size (Slide 15.22)

High absorbance at the optimal wavelength is observed also for a relatively large size range of gold nanoparticles (Fig. 5.9b). As follows from this figure, the absorption efficiency is greater than 1, $Q_{\text{abs}} \geq 1$, over a wide particle radius range of $r_0 = 10$-210 nm. A maximum absorption efficiency $Q_{\text{abs}} = 4.02$ is observed for a gold particle radius of 35 nm at the optimal wavelength $\lambda = 538.3$ nm. This radius is thus optimal for therapeutic applications.

The light scattering at $\lambda = 538.3$ nm by gold nanoparticles in an aqueous suspension becomes apparent for particles of radii > 30 nm [7]. For larger particle sizes, the scattering coefficient Q_{sca} rapidly increases and reaches the value of the absorption coefficient at the particle radius of 50 nm. For particles with radii ≥ 50 nm, scattering of laser radiation predominates over absorption, and the water suspension containing the gold nanoparticles becomes a strongly scattering medium at the given wavelength. Thus, the large gold nanoparticles with a radii ≥ 50 nm are optimal for diagnostic applications.

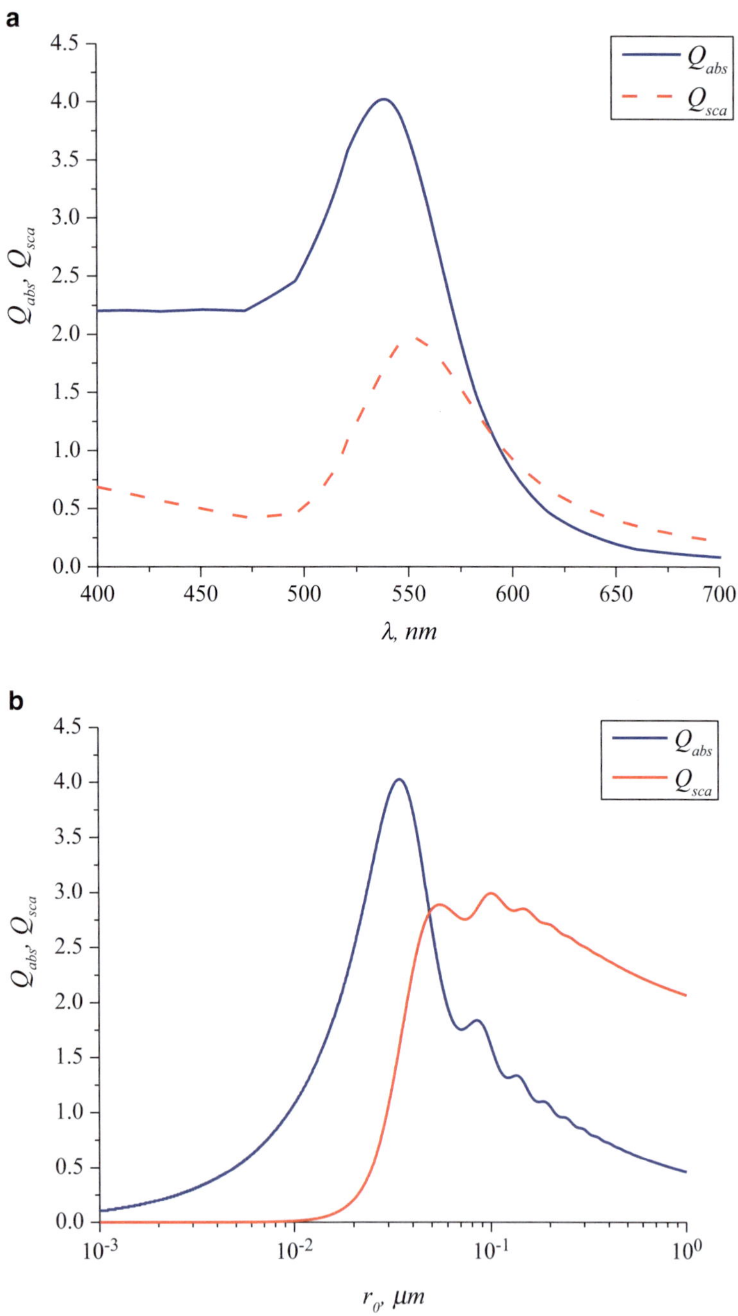

Fig. 5.9 Computer calculations of the absorption, Q_{abs}, and scattering, Q_{sca}, efficiencies for gold nanoparticles in a water surrounding medium: (**a**) the absorption and scattering spectrum of 35-nm gold particle in the visible range of wavelengths $\lambda = 400$–700 nm; and (**b**) illustrates the dependences of Q_{abs} and Q_{sca} on particle size for the optimal wavelength $\lambda = 538.3$ nm. Reprinted from [23]. © With permission from World Scientific

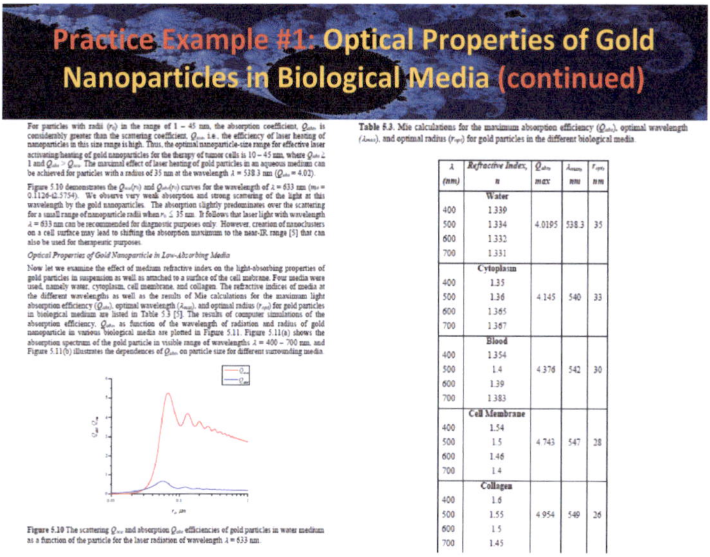

Slide 15.22 Practice Example #1: Optical properties of gold nanoparticles in biological media (continued)

For particles with radii (r_0) in the range of 1–45 nm, Q_{abs} is considerably greater than Q_{sca}, i.e., the efficiency of laser heating of nanoparticles in this size range is high. Thus, the optimal nanoparticle size range for effective laser activating/heating of gold nanoparticles for therapy of tumor cells is 10–45 nm, where $Q_{abs} \geq 1$ and $Q_{abs} > Q_{sca}$. The maximal effect of laser heating of gold particles in an aqueous medium can be achieved for particles with radius 35 nm at the wavelength $\lambda = 538.3$ nm ($Q_{abs} = 4.02$) [7].

Figure 5.10 demonstrates the $Q_{sca}(r_0)$ and $Q_{abs}(r_0)$ curves for the wavelength $\lambda = 633$ nm ($m_0 = 0.1126 - i2.5754$). We observe very weak absorption and strong scattering of the light at this wavelength by the gold nanoparticles. The absorption slightly predominates over the scattering for a small range of nanoparticle radii when $r_0 \leq 35$ nm. It follows that laser light with wavelength $\lambda = 633$ nm can be recommended for diagnostic purposes only. However, creation of nanoclusters on a cell surface may lead to shifting the absorption maximum to the near-IR range [5] that can also be used for therapeutic purposes.

5.1.4.3 Optical Properties of Gold Nanoparticles in Low-Absorbing Media (Slide 15.22)

Now let we examine the effect of the medium refractive index on the light-absorbing properties of gold particles in suspension as well as attached to the surface of the cell membrane. Four media are used: water, cytoplasm, cell

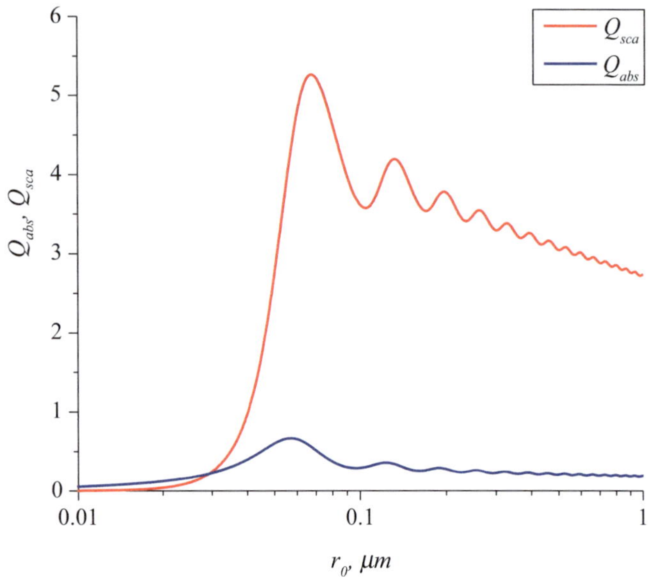

Fig. 5.10 Scattering Q_{sca} and absorption Q_{abs} efficiencies of gold particles in water medium as a function of the particle for the laser radiation of wavelength $\lambda = 633$ nm

membrane, and collagen. The refractive indices of media at the different wavelengths, as well as the results of Mie calculations for the maximum light absorption efficiency Q_{abs}, optimal wavelength λ_{max}, and optimal radius r_{opt} for gold particles in a biological medium, are listed in Table 5.3 [5]. The results of computer simulations of Q_{abs} as a function of the wavelength of radiation and radius of gold nanoparticle in various biological media are plotted in Fig. 5.11. Figure 5.11a shows the absorption spectrum of the gold particle in visible range of wavelengths $\lambda = 400$–700 nm, and Fig. 5.11b illustrates the dependences of Q_{abs} on particle size for different surrounding media.

As follows from the calculations, the optical properties of the gold nanoparticles in low-absorbing biological media are different from the case of a surrounding water medium. The maximal peak of the absorption efficiency increases continuously with increasing refractive index of the surrounding media. At the same time, the optimal wavelength of the incident light corresponding to the maximum absorption changes from green to red light. This is an effect of a red shift of the optimal wavelength for absorption maximum with a refractive index change (see Fig. 5.11a). The maximum absorption by gold nanoparticles is observed for surrounding collagen medium, where the absorption efficiency Q_{abs} reaches 4.954 for the optimal wavelength 549 nm. The presence of low-absorbing biological surrounding also shifts the optimal radius of the gold nanoparticle for effective absorption in a smaller size region from 35 nm for the water medium to 26 nm for the collagen case (see Table 5.3). Thus, the absorption

Table 5.3 Mie calculations of the maximum absorption efficiency Q_{abs}, optimal wavelength λ_{max}, and optimal radius r_{opt} for gold particles in different biological media. Reprinted from [23].

λ (nm)	Refractive index, n	Q_{abs} max	λ_{max} (nm)	r_{opt} (nm)
Water				
400	1.339	4.0195	538.3	35
500	1.334			
600	1.332			
700	1.331			
Cytoplasm				
400	1.35	4.145	540	33
500	1.36			
600	1.365			
700	1.367			
Blood				
400	1.354	4.376	542	30
500	1.4			
600	1.39			
700	1.383			
Cell membrane				
400	1.54	4.743	547	28
500	1.5			
600	1.46			
700	1.4			
Collagen				
400	1.6	4.954	549	26
500	1.55			
600	1.5			
700	1.45			

efficiency of gold nanoparticles is higher in the biological medium as compared to the water medium [7].

As can be seen from Table 5.3 and Slide 15.22, the maximum absorption efficiency of gold nanoparticles Q_{abs} for the optimal wavelength of radiation at the optimal size of particle is in the range of 4–5 in a biological medium. This means that the absorption cross-section of gold nanoparticles exceeds the geometric cross-section by a factor of 4–5(!) for the optimal wavelength. This happens due to the plasmon resonance absorption effect. The metal nanoparticles have a free electron subsystem called a plasma, which oscillates at a certain frequency called a plasmon frequency. If the external light frequency incident on the nanoparticle is in resonance with plasmon frequency of electrons, it creates this plasmon resonance effect when a particle has a huge absorption efficiency observed in Figure 5.11 and shown in Slide 15.23. The plasmon resonance effect is observed for conducting nanoparticles in the optical range of the spectrum of incident radiation.

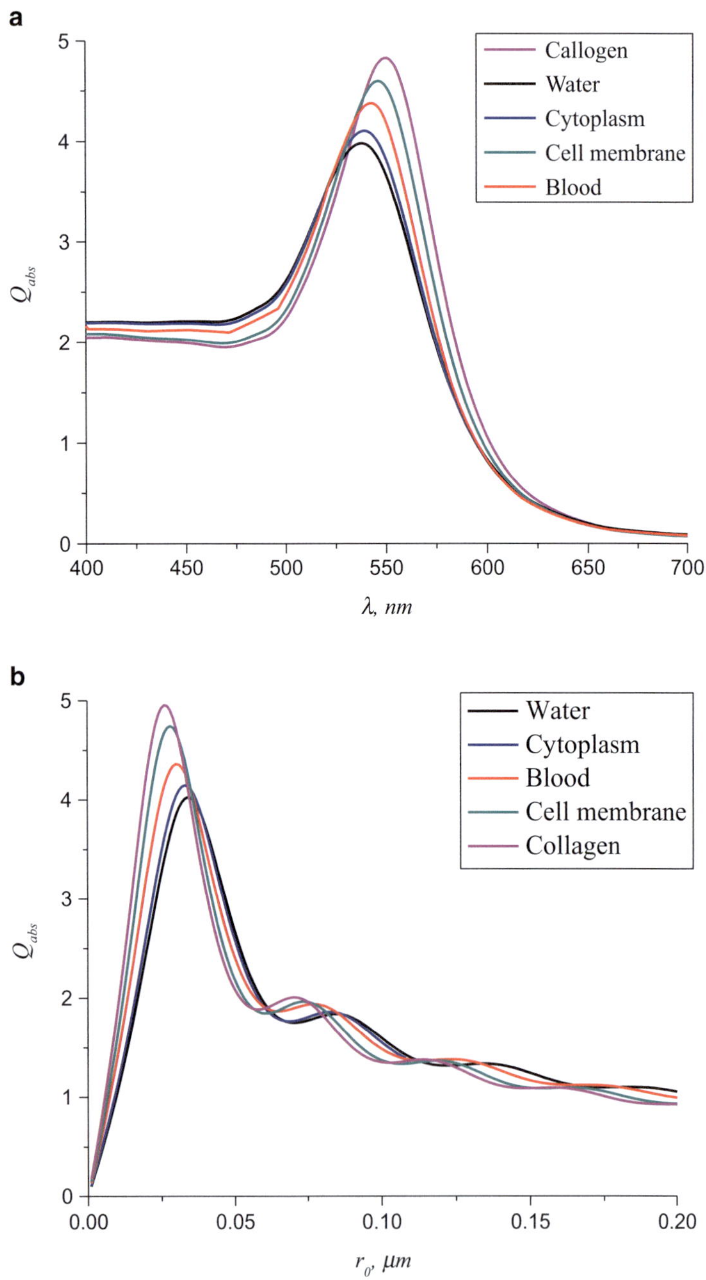

Fig. 5.11 Absorption spectrum of the gold nanoparticle in the visible range of wavelengths $\lambda = 400–700$ nm (**a**), and the absorption efficiency Q_{abs} as a function of particle size in different surrounding media (**b**). Reprinted from [23]. © With permission from World Scientific

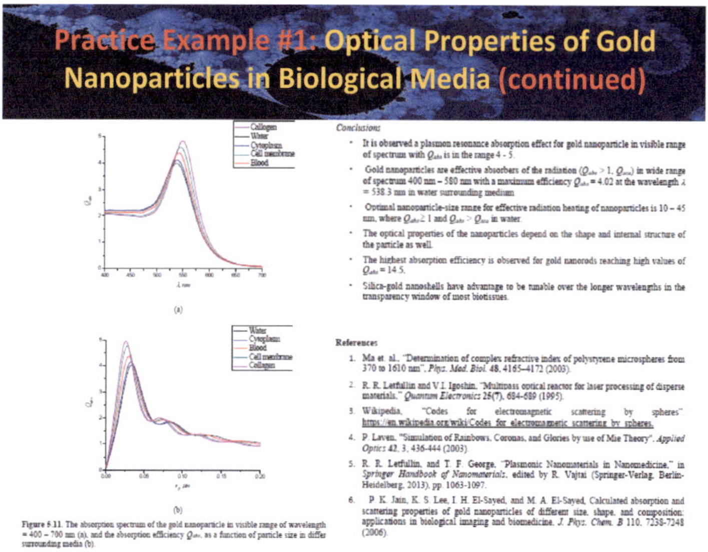

Slide 15.23 Practice Example #1: Optical properties of gold nanoparticles in biological media (continued)

Table 5.4 Examples of absorption efficiency values for different nanoparticle shapes and structures at various wavelengths. Reprinted from [24]. © With permission from Elsevier

Nanoparticle type	Optimal radius	Optimal wavelength	Absorption efficiency	Reference
Gold nanospheres	35 nm	532 nm	4.02	[5]
	40 nm	550 nm	3.5	
Silica–gold	120 nm	1100 nm	3.5	[6]
Nanoshells	120 nm	750 nm	3.75	
Gold nanorods	11.4 nm	800 nm	14	[6]
	17.9 nm	825 nm	14.5	

The optical properties of the nanoparticles depend on the shape and internal structure of the particle as well. Table 5.4 contains results of the calculations for the absorption efficiency of gold nanospheres, silica–gold nanoshells, and gold nanorods. For gold nanoparticles, the highest absorption efficiency is observed for nanorods; however, silica–gold nanoshells are tunable over the longer wavelengths in the transparency window of most biotissues. We should note that other materials can exhibit strong absorption efficiency, such as silver nanospheres, which show strong absorption around 394 nm in the UV range of the spectrum [7].

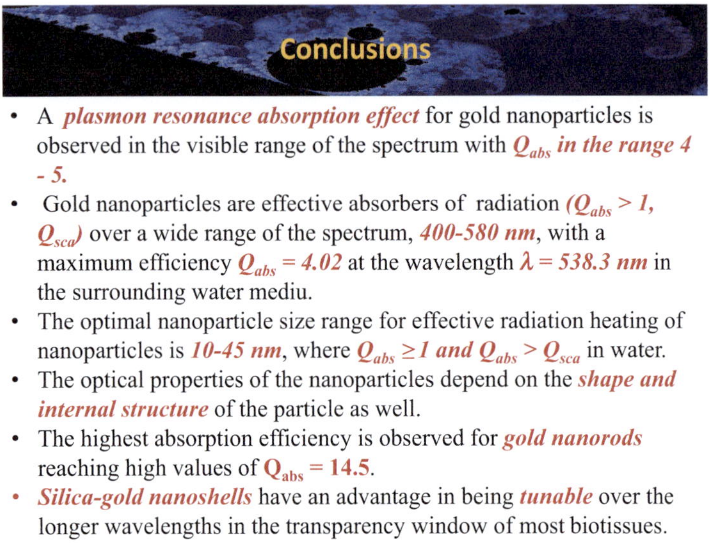

Slide 15.24 Conclusions

5.1.4.4 **Conclusions** (Slide 15.24)

- A plasmon resonance absorption effect for gold nanoparticles is observed in the visible range of the spectrum with Q_{abs} in the range 4–5.
- Gold nanoparticles are effective absorbers of radiation ($Q_{\mathrm{abs}} > 1$, Q_{sca}) over a wide range of the spectrum, 400–580 nm, with a maximum efficiency $Q_{\mathrm{abs}} = 4.02$ at the wavelength $\lambda = 538.3$ nm in the surrounding water medium.
- The optimal nanoparticle size range for effective radiation heating of nanoparticles is 10–45 nm, where $Q_{\mathrm{abs}} \geq 1$ and $Q_{\mathrm{abs}} > Q_{\mathrm{sca}}$ in water.
- The optical properties of the nanoparticles depend on the shape and internal structure of the particle as well.
- The highest absorption efficiency is observed for gold nanorods reaching high values of $Q_{\mathrm{abs}} = 14.5$.
- Silica–gold nanoshells have an advantage in being tunable over the longer wavelengths in the transparency window of most biotissues.

5.2 Computer Practicum 2 (Lecture 16): Simulations of Optical Properties of Biological Particles (Slide 16.1)

- Statement of the Research Task
- Input Data
- Mie Simulations for User-Defined Nanoparticles and Surrounding Media

  - *Mie Simulations for Any Surrounding Media*
  - *Mie Simulations for Any Particle Material*
  - *Creating a Rix File for Any Particle Material*
- Optical Properties of Normal and Cancer Cell Organelles
- Practice Example 2: Simulation of Optical Properties of Normal and Cancerous Ribosomes in Cytoplasm
- Practice Example 3: Simulation of Optical Properties of Bone Particles
- References

5.2.1 Statement of the Research Task

In cancer cells, it is observed that there are distinct changes (Slide 16.2) in the nucleus, mitochondria, ribosome, and cytoskeleton, which are the organelles in the cells, and hence the structure and function of these organelles in normal and cancer cells must be examined, as discussed in Chap. 2. Normal cell organelles are typically characterized by their regular spherical shape and smaller size. As a result of multiple mutations, cancer cells have abnormal cell organelles with irregular shape and much larger size. These distinct differences in the morphological properties of normal and cancer cells can be used to design new diagnostic and therapy techniques without involving the nanoparticles as contrast agents.

A biological cell has a nonuniform structure made of cell organelles. The *Statement of the Research Task* for this case is demonstrated in Slide 16.3,

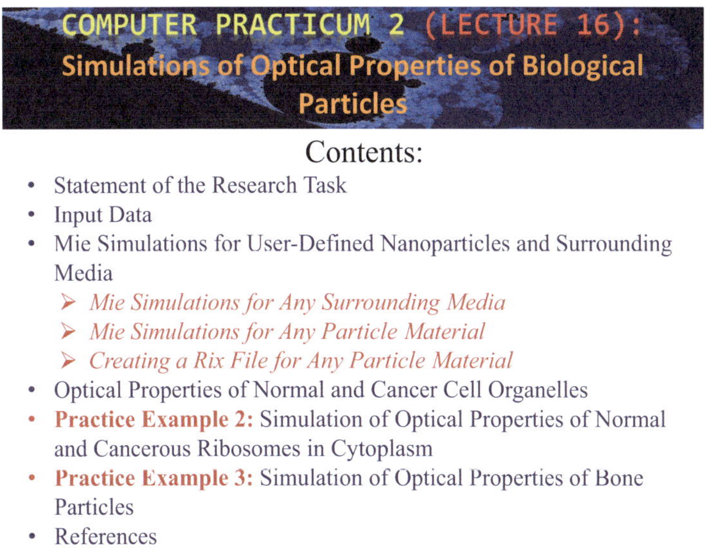

Slide 16.1 Contents of Lecture 16

Differences in Morphological Properties of Normal and Cancer Cells

- ***In cancer cells***, it is observed that there are ***distinct changes*** in the nucleus, mitochondria, ribosome and cytoskeleton, which are the organelles in the cells, and hence the ***structure and properties*** of these organelles in normal and cancer cells must be examined.
- Cell organelles in ***normal cells*** are typically characterized by their ***regular spherical shape and smaller size***.
- As a result of multiple mutations, ***cancer cells*** have abnormal cell organelles with ***irregular shape and much larger size***.
- ***These distinct differences*** in morphological properties of normal and cancer cells can be used to ***design new diagnostic and therapy techniques*** without involving the nanoparticles as contrast agents.

Slide 16.2 Differences in morphological properties of normal and cancer cells

Statement of the Research Task

The research task is to model the radiation interaction with biological particles to

- ***find the absorption and scattering efficiencies*** of the normal and cancerous cell organelles;
- ***define the optimal wavelength of radiation*** for the therapy and diagnosis of cancer cells;
- ***simulate the heating*** of biological particles;
- ***evaluate the biological ablation*** and thermal damage area;
- and ***make recommendations*** for experimental realization of this technology.

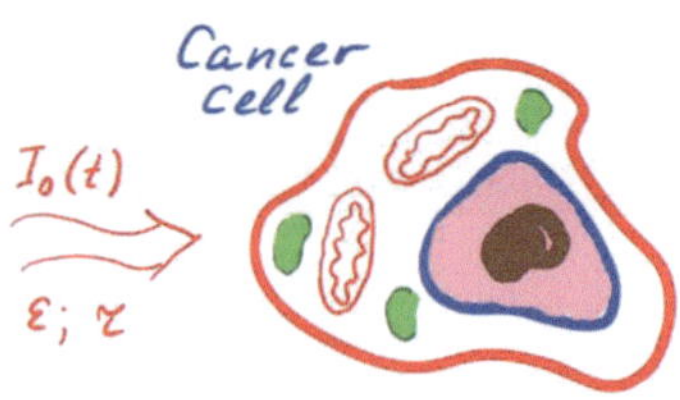

Slide 16.3 Statement of the research task

schematically showing the cell organelles. These organelles can serve a role as absorbing centers for upcoming radiation, which can be used for localized heating and biological ablation of the cell. Since the cancer cell organelles are irregular in size and shape, the scattering and absorption efficiencies, as well as the heating kinetics of cancer cells, will be different than normal ones, giving the ability to selectively destroy cancer cells. Our research task is to model the radiation interaction with biological particles to (Slide 16.3):

- Find the absorption and scattering efficiencies of the normal and cancerous cell organelles;
- Define the optimal wavelength of radiation for therapy and diagnosis of cancer cells;
- Simulate the heating of biological particles;
- Evaluate the biological ablation and thermal damage area;
- Make recommendations for experimental realization of this technology.

In this section, we simulate the optical properties of the biological particles and find the absorption and scattering efficiencies of the normal and cancerous cell organelles.

5.2.1.1 Input Data

Acquiring data for organelles is difficult due to the wide variety of situations under which they must act; the added complexity of the large number of cancers makes finding very specific data difficult [7]. Table 5.5 summarizes the experimental data for the diameters of various types of normal and cancerous cell organelles [7] (Slide 16.4). Using a variety of cited dimensions and organelle properties, we have compiled Table 5.6 for the 'average' organelle size and complex indices of refraction to perform Mie simulations of the absorption and scattering efficiencies of the normal and cancerous cell organelles (Slide 16.5). The real and imaginary parts of the refractive index for cell organelles in Table 5.6 are collected from

Table 5.5 Experimental data for healthy and tumor cells. Reprinted from [24]. © With permission from Elsevier

Source	Cancer	Organelle	Healthy diameter (μm)	Well differentiated diameter (μm)	Moderately differentiated diameter (μm)	Poorly differentiated diameter (μm)
[15]	Hepatocellular carcinoma	Nucleus	7.5	8.5	8.8	10.3
[15]	Hepatocellular carcinoma	Mitochondria	0.5	0.9	0.9	0.66
[16]	Colorectal adenocarcinoma	Nucleus	5.2	7.2	–	–
Source	Cancer	Organelle	Benign diameter (μm)		In situ diameter (μm)	Invasive diameter (μm)
[17]	Ductal breast carcinoma	Nucleus	7		11.6	8.33
[18]	Intraductal breast cancer	Nucleus	8.58		–	8.97
[19]	Breast cancer	Nucleus	5.6		–	5.8

Table 5.5 Experimental data for healthy and tumor cells.

Cancer	Organelle	Healthy diameter (μm)	Well differentiated diameter (μm)	Moderately differentiated diameter (μm)	Poorly differentiated diameter (μm)
Hepatocellular carcinoma	Nucleus	7.5	8.5	8.8	10.3
Hepatocellular carcinoma	Mitochondria	0.5	0.9	0.9	0.66
Colorectal adenocarcinoma	Nucleus	5.2	7.2	-	-
Cancer	Organelle	Benign diameter (μm)	*In situ* diameter (μm)	Invasive diameter (μm)	
Ductal breast carcinoma	Nucleus	7	11.6	8.33	
Intraductal breast cancer	Nucleus	8.58	-	8.97	
Breast cancer	Nucleus	5.6	-	5.8	

Slide 16.4 Simulating optical properties of biological particles: Input data

Table 5.6 Optical data and average size for normal and cancerous intracellular organelles: mitochondria, nucleus, ribosome, and cytoskeleton in the visible range of the spectrum at room temperature

Organelles	Average healthy radius Nr_0 (μm)	Average cancerous radius Cr_0 (μm)	Wavelength λ (nm)	Refractive index $m = n_0 - ik_0$ n_0	k_0
Lysosomes	0.1	0.25	400	1.578	0
			500	1.582	0
			600	1.596	0.01
			700	1.6	0.015
Mitochondria	1.5 (length)	3 (length)	400	1.38	0
			500	1.393	0
			600	1.408	0
			700	1.442	0.01
			800	1.466	0.016
			900	1.478	0.02
			1000	1.491	0.024
Nucleus	3	5	400	1.1	0
			500	1.21	0.01
			600	1.32	0.015
			700	1.41	0.031
			800	1.46	0.043
			900	1.5	0.049
			1000	1.54	0.051

(continued)

Table 5.6 (continued)

Organelles	Average healthy radius Nr_0 (μm)	Average cancerous radius Cr_0 (μm)	Wavelength λ (nm)	Refractive index $m = n_0 - ik_0$	
				n_0	k_0
Ribosome	0.015	0.03	400	1.15	0
			500	1.1625	0
			600	1.175	0.01
			700	1.202	0.015
			800	1.21	0.022
			900	1.212	0.0235
			1000	1.213	0.0241
Cytoskeleton	0.004	0.012	400	1.36	0
			500	1.36	0
			600	1.365	0.01
			700	1.367	0.015

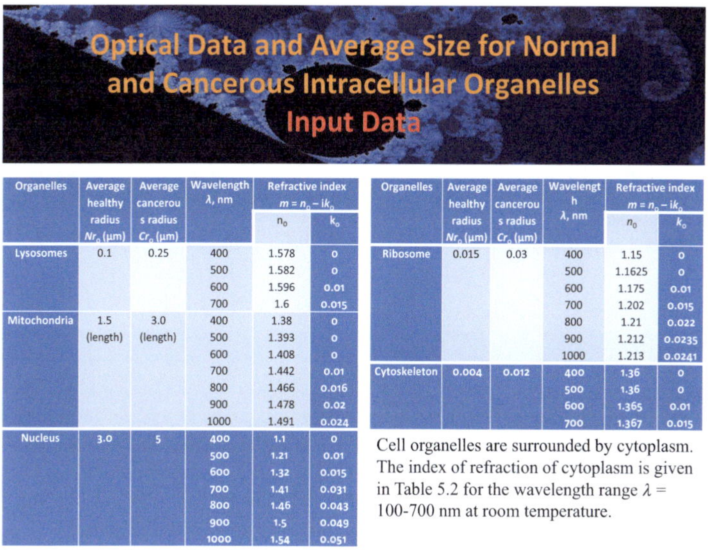

Slide 16.5 Optical data and average size for normal and cancerous intracellular organelles: input data

available data [8–14]. Cell organelles are surrounded by the cytoplasm, whose index of refraction is given in Table 5.2 for the wavelength range $\lambda = 100$–700 nm at room temperature.

5.2.2 *Mie Simulations for User-Defined Nanoparticles and Surrounding Media*

In the Computer Practicum in Sect. 5.1 of this Chap. 5, we introduced step-by-step instructions (steps 1–12) as to how to simulate the optical properties of the particles as a function of their size and wavelength of radiation by using *MiePlot* software (Slide 16.6). We have completed simulations incorporated into the *MiePlot* code for the particle material and surrounding media. Let us customize the *MiePlot* software to perform the simulations for the user-defined particles and surrounding media.

Mie Simulations for Any Surrounding Media. The first two steps, *Steps 1* and *2*, are the same to start the simulations with the *MiePlot* code. The initial simulations are run for a gold nanoparticle vs. the radius for a wavelength of radiation 538 nm. We repeat *Steps 3, 5, and 6* to set up the particle material (gold), radius range, and wavelength of incident radiation (Slide 16.6).

Step 13 introduces the index of refraction for the user-defined surrounding medium (Slide 16.7). In the menu bar click "*Advanced*," which opens an interactive small window shown in Fig. 5.12 and Slide 16.8. Select "*Refractive Index*" in that window highlighted by the blue background color and marked by the first largest red ellipse in the figure. It opens another small interactive window with two options: "*Sphere*" and "*Surrounding medium*." Select "*Surrounding medium*" highlighted by the blue background color and marked by the middle-sized red ellipse, which opens a window with media options. To create a user-defined surrounding medium, click on "*User defined*" circled by the smallest red ellipse in Fig. 5.12 (Slide 16.8). It opens a small window with a table inside shown in Fig. 5.13 (Slide 16.9).

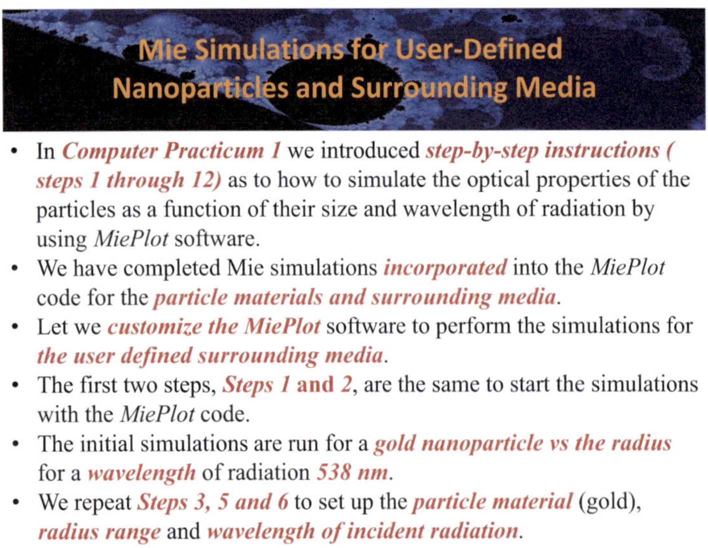

Slide 16.6 Mie simulations for user-defined nanoparticles and surrounding media

Mie Simulations for Any Surrounding Media

- ***Step 13*** is introducing the index of refraction for the ***user-defined surrounding medium***.
- In the menu bar click "***Advanced***," which opens an interactive small window shown in the figure on the next slide.
- Select "***Refractive Index***" in that window marked by the first largest red ellipse in the figure.
- It opens another small interactive window with two options: "*Sphere*" and "*Surrounding medium*." Select "***Surrounding medium***" marked by the middle-sized red ellipse, which opens a window with media options.
- To create a user-defined surrounding medium, click on "***User defined***" circled by smallest red ellipse.
- It opens a small window with a table inside shown in the figure.

Slide 16.7 Mie simulations for any surrounding media

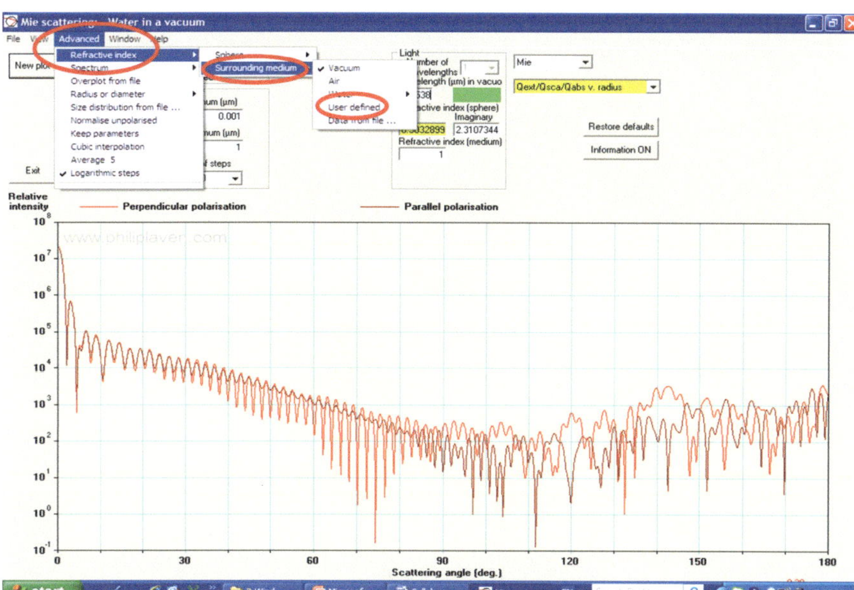

Fig. 5.12 Selecting the index of refraction for the user-defined surrounding medium

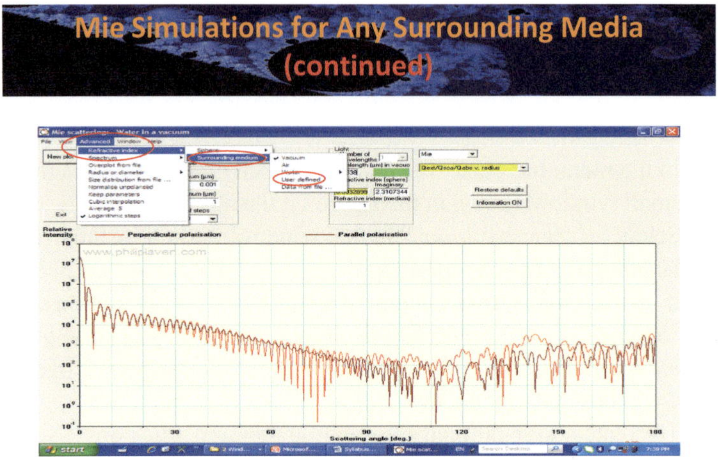

Figure: Selecting the index of refraction for the user-defined surrounding medium.

Slide 16.8 Mie simulations for any surrounding media (continued)

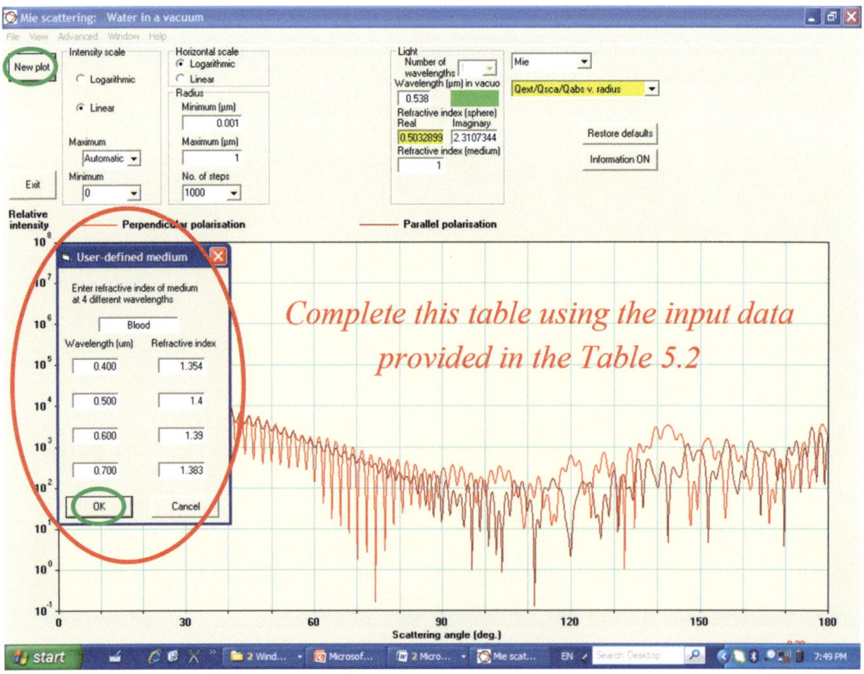

Fig. 5.13 Entering the index of refraction for the user-defined medium

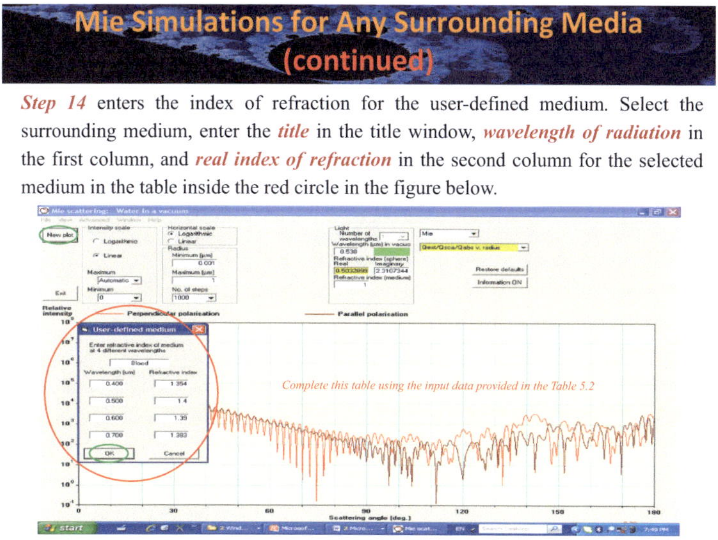

Slide 16.9 Mie simulations for any surrounding media (continued)

Step 14 enters the index of refraction for the user-defined medium (Slide 16.9). Select the surrounding medium, for example, from Table 5.2, and enter the title in the title window, wavelength of radiation in the first column, and real index of refraction in the second column for the selected medium in the table inside the red circle in Fig. 5.13. The sample simulations are performed for gold particles in surrounding blood medium, so that the optical data for blood should be entered in that table. Then click "*OK*" as shown in the small green ellipse inside the red circle in Fig. 5.13. Use *Step 7* to run the simulations for absorption, scattering, and extinction efficiencies of the nanoparticle vs. the particle radius in the user-defined surrounding medium.

Figure 5.14 shows the results of these simulations for gold nanoparticle in a blood surrounding medium (Slide 16.10). This way we can perform simulations for any surrounding medium.

Mie Simulations for Any Particle Material. Step 15 (Slide 16.11) introduces the user-defined nanoparticle/sphere material. In the menu bar click "*Advanced*," which opens an interactive small window shown in Fig. 5.15. Select "*Refractive Index*" in that window highlighted in blue background color and marked by the first largest red ellipse in the figure (Slide 16.12). It opens another small interactive window with two options: "*Sphere*" and "*Surrounding medium*." Select "*Sphere*" highlighted by the blue background color and marked by the middle-sized red ellipse in the figure (Slide 16.12), which opens a window with material options for the sphere. Select "*Data from file...*" circled by the blue ellipse in Fig. 5.15 (Slide 16.13). It opens a folder with "*rix*" files as shown in Fig. 5.16.

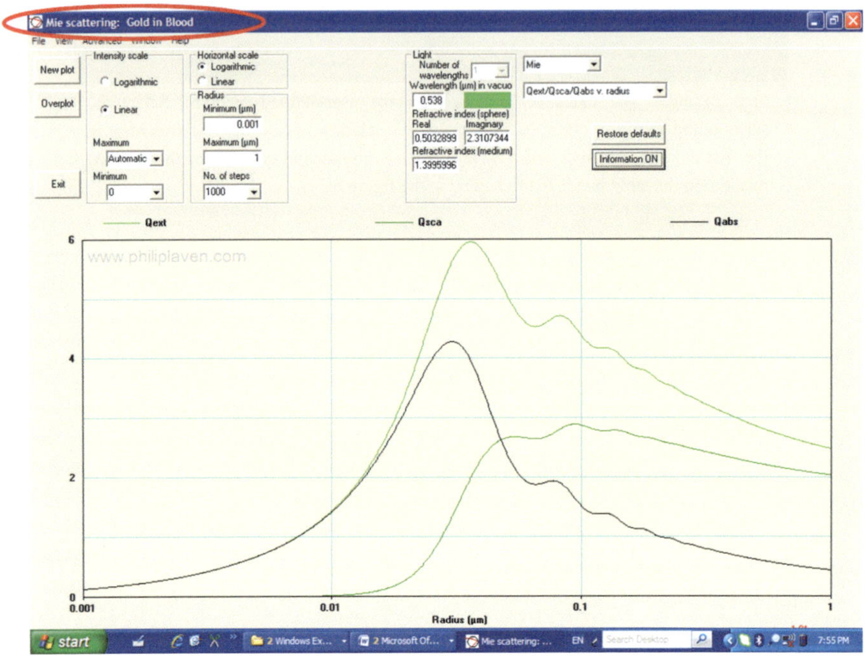

Fig. 5.14 Results of simulations for the extinction Q_{ext}, scattering Q_{sca} and absorption Q_{abs} efficiencies vs. the particle radius for a gold nanoparticle in a surrounding blood medium

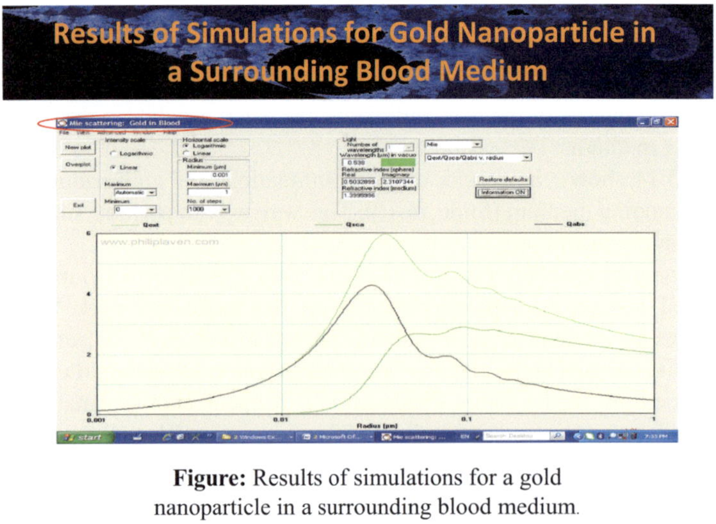

Slide 16.10 Results of simulations for gold nanoparticle in a surrounding blood medium

Mie Simulations for Any Particle Material

- ***Step 15*** introduces the user-defined ***nanoparticle/sphere material***.
- In the menu bar click "***Advanced***," which opens an interactive small window shown in the figure on the next slide.
- Select "***Refractive Index***" in that window marked by the first largest red ellipse. It opens another small interactive window with two options: "*Sphere*" and "*Surrounding medium*."
- Select "***Sphere***" marked by the middle-sized red ellipse in the figure, which opens a window with material options for the sphere.
- Select "***Data from file...***" circled by the blue ellipse in the figure.
- It opens a folder with "***rix***" files.

Slide 16.11 Mie simulations for any particle material

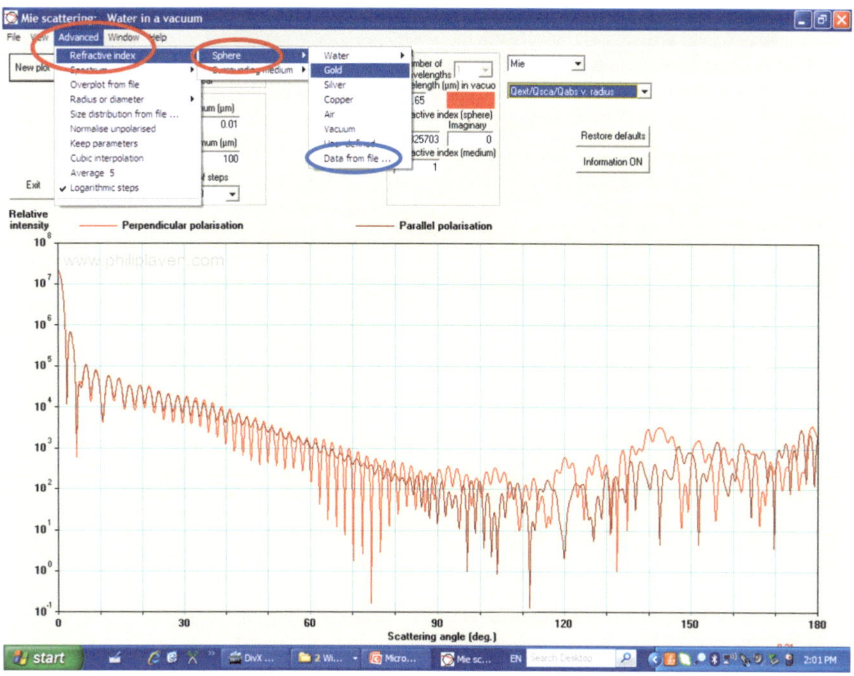

Fig. 5.15 Introducing the user-defined nanoparticle/sphere material

Mie Simulations for Any Particle Material (continued)

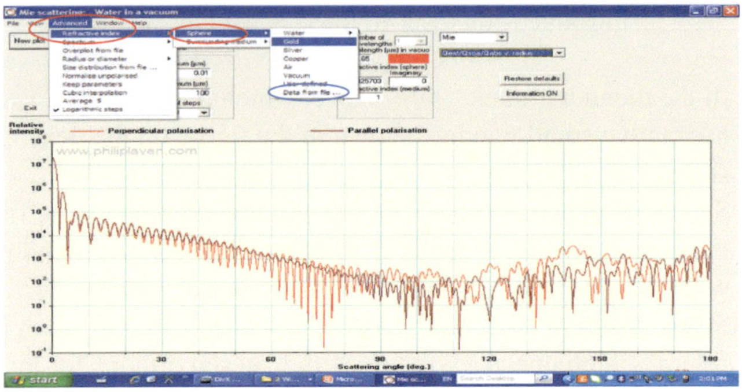

Figure: Introducing the user-defined nanoparticle/sphere material.

Slide 16.12 Mie simulations for any particle material (continued)

Mie Simulations for Any Particle Material (continued)

- ***Step 16*** selects a ***rix*** file. The ***rix*** files are *MiePlot* files, which contain the complex refractive index for a specific material. Select a ***rix*** file from the *MiePlot* host folder for specific material, for example "*Fullerene.rix*," as shown in the figure.

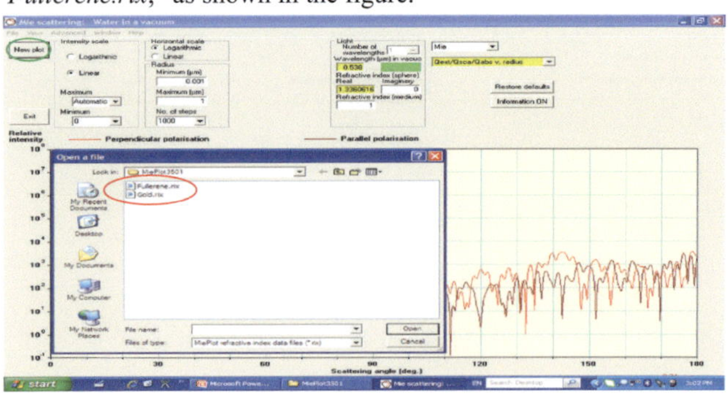

Slide 16.13 Mie simulations for any particle material (continued)

Step 16 selects a *rix* file (Slide 16.13). The *rix* files are *MiePlot* files, which contain the complex refractive index for a specific material. Select a *rix* file from the *MiePlot* host folder for a specific material, for example "*Fullerene.rix*," as shown in Fig. 5.16.

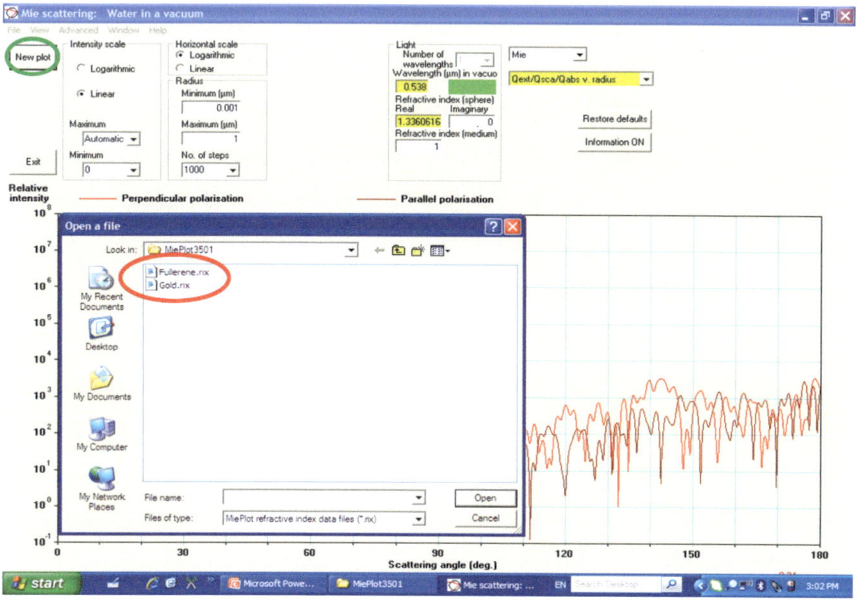

Fig. 5.16 Selecting a *rix* file

Step 17. Run the simulations by clicking on the "*New plot*" button in the top left corner circled by a green ellipse (Slide 16.13). Results of simulations for a fullerene nanoparticle vs. radius for the radiation wavelength of 538 nm are presented in Fig. 5.17 and Slide 16.14.

Creating a Rix File for Any Particle Material. Open any *rix* file, for example, the "*Gold.rix*" file, provided in the *MiePlot* host folder by using the *Notepad* editor as shown in Fig. 5.18 and Slides 16.15 and 16.16. There are instructions for the *rix* file at the top of the file circled by an orange ellipse on the figure. Following is a copy of these instructions from the *rix* file (Slide 16.15):

- *Line 2: Sample refractive index data as a function of wavelength.*
- *Line 4: Do not modify lines 1–12.*
- *Line 5: Data starts at line 13.*
- *Line 6: Line 13 specifies the name (e.g., "Gold") of the material.*
- *Line 7: Line 14 specifies the units of wavelength (m, mm, μm, or nm).*
- *Line 8: Line 15 indicates whether the data is complex or real.*
- *Line 9: Complex data must be entered in the format (wavelength + real + imaginary).*
- *Line 10: Real data must be entered in the format (wavelength + real).*
- *Line 11: Line 16 specifies the number of data points (count carefully!).*
- *Line 12: Line 17 is the first line of the refractive index data.*

Step 18. Change the title of the data shown in the small blue ellipse in Fig. 5.18 and Slide 16.16.

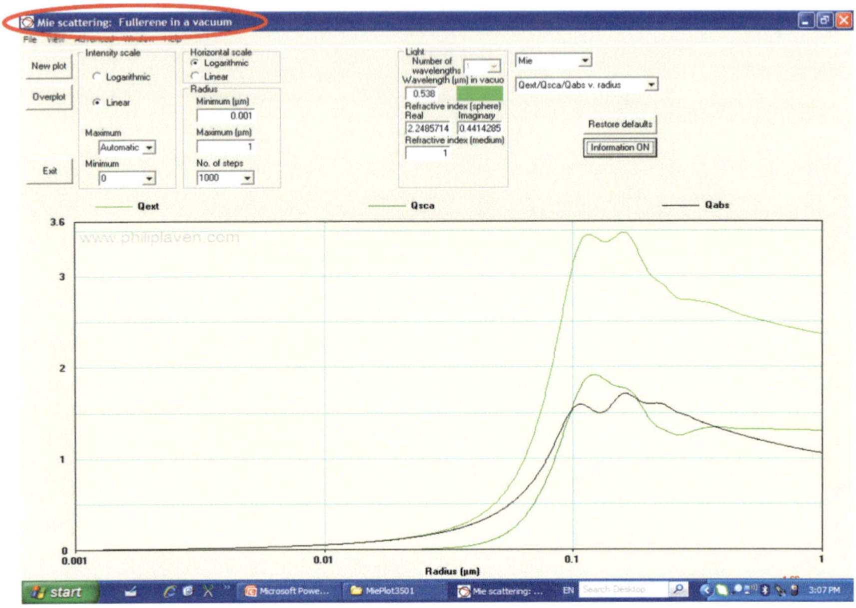

Fig. 5.17 Results of simulations for a fullerene nanoparticle vs. radius for the radiation wavelength of 538 nm

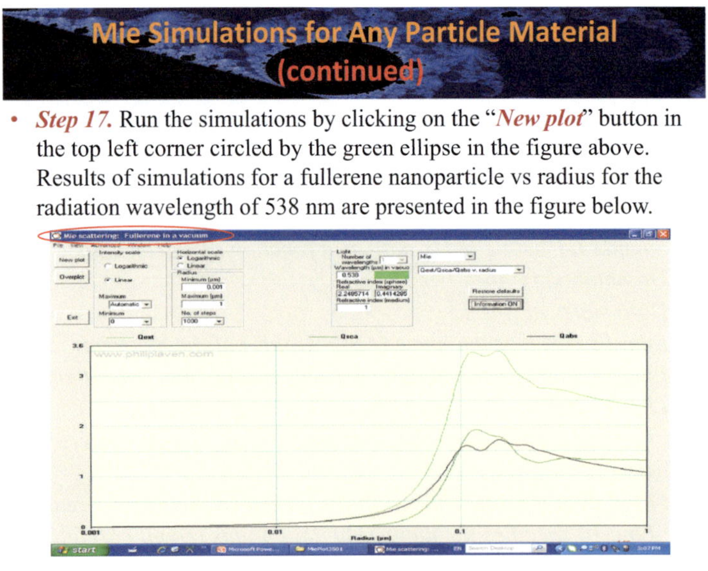

Slide 16.14 Mie simulations for any particle material (continued)

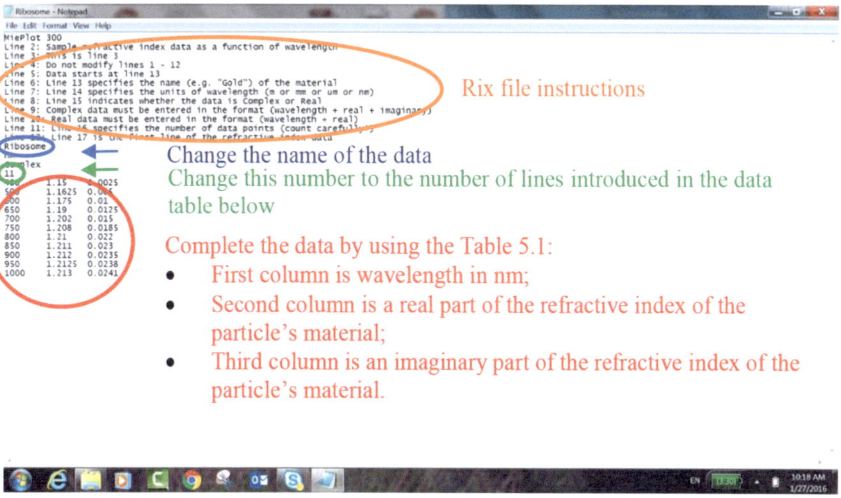

Fig. 5.18 Creating a rix file for any particle material

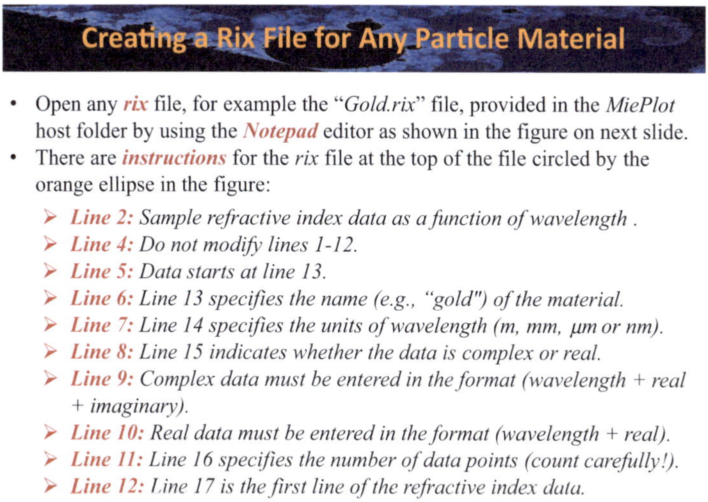

Slide 16.15 Creating a *rix* file for any particle material

Step 19. Complete the data in the table within the big red circle in Fig. 5.18 (Slide 16.16) using optical data, for example, given in Table 5.1. The first column in that rix file table is the wavelength, second column is the real part of the refractive index of the selected particle's material, and the third column is the imaginary part of the refractive index of the particle's material. The unit for the wavelength in Fig. 5.18 is shown in nm. However, it can be changed to mm, μm, or nm in line 14 of the *rix* file.

Step 20. Change the number in the green circle in Fig. 5.18 to the number of lines introduced in the data table (Slide 16.17). Note: no extra line is allowed between that number in the green circle and data table for the index of refraction.

Creating a Rix File for Any Particle Material (continued)

- *Step 18.* Change the title of the data shown in the small blue ellipse in the figure.
- *Step 19.* Complete the data in the table within the big red circle using optical data, for example, given in Table 5.1.
- The ***first column*** in that ***rix*** file table is ***wavelength; second column*** is a ***real part of the refractive index*** of selected particle's material; and ***third column is the imaginary part of the refractive index*** of the particle's material. The unit for the wavelength in the figure is shown in nm. However, it can be changed to mm, μm or nm in the line 14 of the *rix* file.

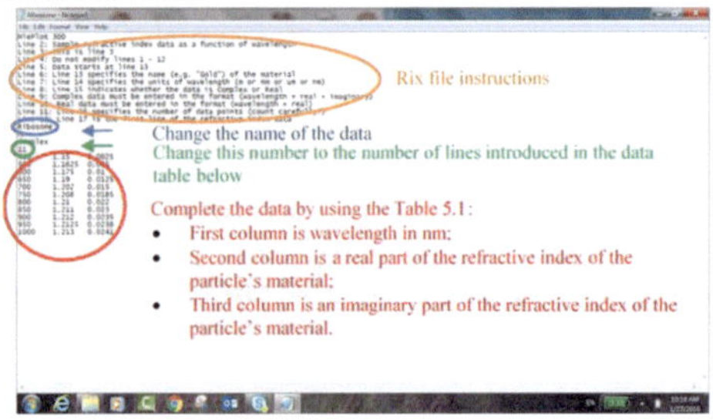

Slide 16.16 Creating a *rix* file for any particle material (continued)

Creating a Rix File for Any Particle Material (continued)

- *Step 20.* Change the number in the green circle in Figure 5.18 to the number of lines introduced in the data table. *Note:* no extra line is allowed between that number in the green circle and data table for the index of refraction.
- *Step 21.* Save the changes as a *rix* file in the *MiePlot* host folder by giving a new file name corresponding to the material's name and keeping a rix extension of the file.
- *Step 22.* Run the simulations by using ***Steps 15-17***. Thus, by creating the new *rix* files, we can simulate any particle material.

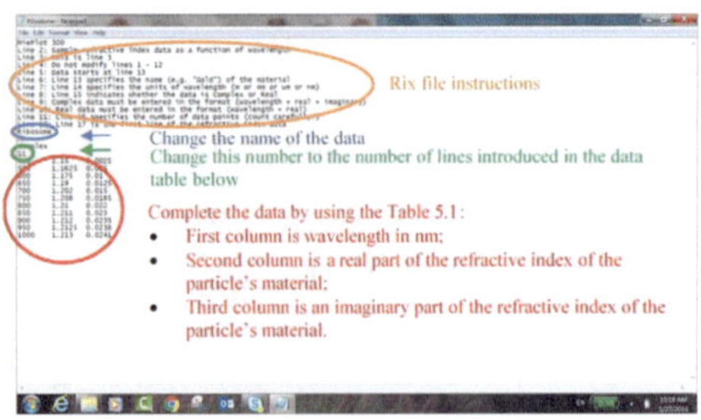

Slide 16.17 Creating a *rix* file for any particle material (continued)

Step 21. Save the changes as a *rix* file in the *MiePlot* host folder by giving a new file name corresponding to the material's name and keeping a rix extension of the file (Slide 16.17).

Step 22. Run the simulations by using *Steps 15–17*. Thus, by creating the new *rix* files, we can simulate any particle material (Slide 16.17).

Using a *Step 8*, the results of the simulations can be saved in numerical format as a text file, which can be open by, for example, the *Notepad* application.

5.2.3 *Optical Properties of Normal and Cancer Cell Organelles*

By creating the *rix* files for the cell organelles and user-defined surrounding medium (cytoplasm) and using data from Tables 5.2, 5.5, and 5.6, we have performed the Mie simulations for the optical properties of normal and tumor cell organelles (Slide 16.18). The results of these calculations for the absorption efficiency of normal and cancerous nuclei and mitochondria are listed in Table 5.7 [7]. The *Practice Examples 2 and 3* present the output results of these simulations and conclusions in more detail.

Optical Properties of Normal and Cancer Cell Organelles

- By creating the ***rix*** files for the cell organelles and ***user-defined surrounding medium*** (cytoplasm) and using data from Tables 5.2, 5.5 and 5.6, we have performed the Mie simulations for the ***optical properties of normal and tumor cell organelles***.

Cancer	Organelle	Organelle State	Diameter (μm)	Absorption Efficiency (at λ=700 nm)
Hepatocellular carcinoma	Nucleus	Healthy	7.5	0.944183362
		Well differentiated	8.5	0.964865712
		Moderately differentiated	8.8	0.969903343
		Poorly differentiated	10.3	0.989209877
Hepatocellular carcinoma	Mitochondria	Healthy	1.1	0.134234082
		Well differentiated	0.9	0.110949860
		Moderately differentiated	0.9	0.110949860
		Poorly differentiated	0.66	0.082130836
Ductal breast carcinoma	Nucleus	Benign	7.0	0.930908259
		In situ	11.6	1.000194442
		Invasive	8.33	0.961677234
Intraductal breast cancer	Nucleus	Benign	8.58	0.966255375
		Invasive	8.97	0.972557278
Breast cancer	Nucleus	Benign	5.6	0.880181004
		Invasive	5.8	0.889019082

Slide 16.18 Optical properties of normal and cancer cell organelles

Table 5.7 Computational input/output data for normal and cancer cell organelles. Reprinted from [24].

Cancer	Organelle	Organelle state	Diameter (μm)	Absorption efficiency (at $\lambda = 700$ nm)[a]
Hepatocellular carcinoma	Nucleus	Healthy	7.5	0.944183362
		Well differentiated	8.5	0.964865712
		Moderately differentiated	8.8	0.969903343
		Poorly differentiated	10.3	0.989209877
Hepatocellular carcinoma	Mitochondria	Healthy	1.1	0.134234082
		Well differentiated	0.9	0.110949860
		Moderately differentiated	0.9	0.110949860
		Poorly differentiated	0.66	0.082130836
Ductal breast carcinoma	Nucleus	Benign	7.0	0.930908259
		In situ	11.6	1.000194442
		Invasive	8.33	0.961677234
Intraductal breast cancer	Nucleus	Benign	8.58	0.966255375
		Invasive	8.97	0.972557278
Breast cancer	Nucleus	Benign	5.6	0.880181004
		Invasive	5.8	0.889019082

[a]700 nm was chosen because it is the wavelength with the greatest absorption efficiency difference between cancerous and healthy organelles at a wavelength in the range 400–1000 nm.

5.2.4 *Practice Example 2: Simulation of Optical Properties of Normal and Cancerous Ribosomes in Cytoplasm*

5.2.4.1 **Abstract** (Slide 16.19)

The optical properties of ribosomes are examined in order to find the differences in absorption and scattering properties of normal and cancerous ribosomes to design new diagnostic and therapeutic techniques. Additionally, the scattering properties have been analyzed to determine the spectrum of radiation where ribosomes are more proficient at scattering or absorbing light. By understanding how the optical properties vary in normal cells compared to cancerous cells, conclusions can be made about whether or not ribosomes are optimal organelles for absorbing light for therapeutic purposes or for scattering light for diagnostic imaging applications.

5.2.4.2 Input Parameters

The function of a ribosome is to translate RNA into amino acids that are used to make proteins. Ribosomes themselves are made up of two components, RNA and proteins, and they are located in the cytosol of cells or attached to the rough

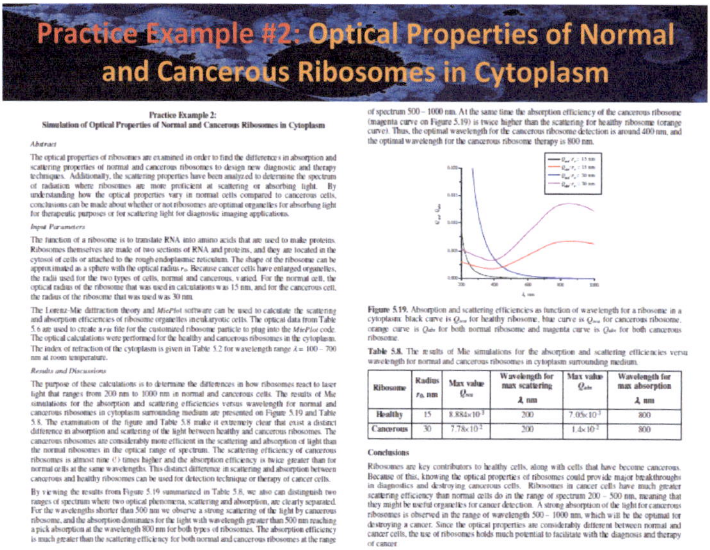

Slide 16.19 Practice Example #2: Optical properties of normal and cancerous ribosomes in cytoplasm

endoplasmic reticulum. The shape of the ribosome can be approximated as a sphere with the optical radius r_0. Because cancer cells have enlarged organelles, the radii used for the two types of cells, normal and cancerous, are varied. For the normal cell, the optical radius of the ribosome that was used in calculations was 15 nm, and for the cancerous cell, the radius of the ribosome that was used was 30 nm.

The Lorenz–Mie diffraction theory and *MiePlo*t software can be used to calculate the scattering and absorption efficiencies of ribosome organelles in eukaryotic cells. The optical data from Table 5.6 are used to create a *rix* file for the customized ribosome particle to plug into the *MiePlot* code. Optical calculations have been performed for healthy and cancerous ribosomes in the cytoplasm. The index of refraction of the cytoplasm is given in Table 5.2 for the wavelength range $\lambda = 100$–700 nm at room temperature.

5.2.4.3 Results and Discussion

The purpose of these calculations is to determine the differences between normal and cancerous cells in how ribosomes react to laser light ranging from 200 to 1000 nm. The results of Mie simulations for the absorption and scattering efficiencies vs. wavelength for normal and cancerous ribosomes in the cytoplasm surrounding medium are presented on Fig. 5.19 and Table 5.8 (Slide 16.19). An examination of the figure and Table 5.8 makes it clear that there exists a distinct difference in absorption and scattering of the light between healthy and cancerous ribosomes. The cancerous ribosomes are considerably more efficient in the scattering and

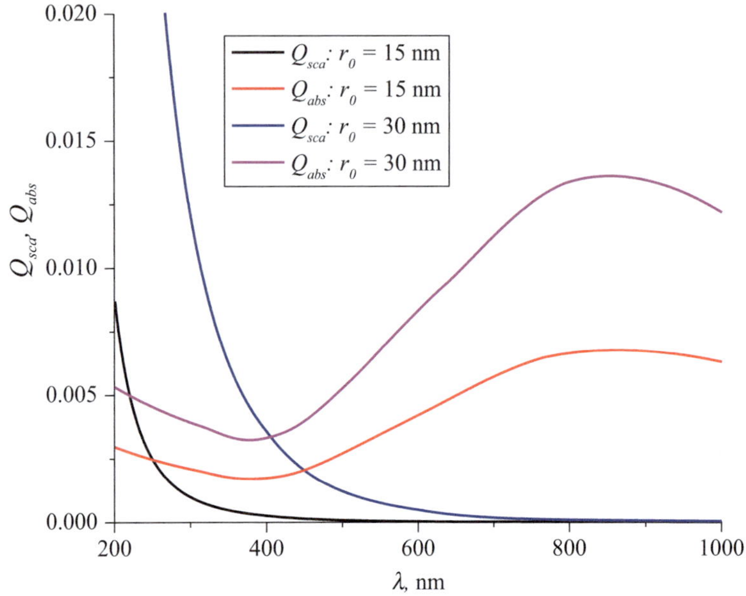

Fig. 5.19 Absorption and scattering efficiencies as functions of the wavelength for a ribosome in a cytoplasm: the *black curve* is Q_{sca} for healthy ribosome, the *blue curve* is Q_{sca} for cancerous ribosome, the *orange curve* is Q_{abs} for normal ribosome, and the *magenta curve* is Q_{abs} for cancerous ribosome

Table 5.8 Results of Mie simulations for the absorption and scattering efficiencies vs. wavelength for normal and cancerous ribosomes in a cytoplasm surrounding medium

Ribosome	Radius r_0 (nm)	Max value Q_{sca}	Wavelength for max scattering λ (nm)	Max value Q_{abs}	Wavelength for max absorption λ (nm)
Healthy	15	8.884×10^{-3}	200	7.05×10^{-3}	800
Cancerous	30	7.78×10^{-2}	200	1.4×10^{-2}	800

absorption of light than the normal ribosomes in the optical range of the spectrum. The scattering efficiency of cancerous ribosomes is almost nine (!) times higher, and the absorption efficiency is twice greater than for normal cells at the same wavelengths. This distinct difference in scattering and absorption between cancerous and healthy ribosomes can be used for detection techniques or therapy of cancer cells.

By viewing the results from Fig. 5.19 summarized in Table 5.8 (Slide 16.19), we also can distinguish two ranges of the spectrum where two optical phenomena—scattering and absorption—are clearly separated. For wavelengths shorter than 500 nm, we observe strong scattering of the light by cancerous ribosome, and the absorption dominates for the light with wavelengths greater than 500 nm, reaching a pick absorption at 800 nm for both types of ribosomes. The absorption efficiency is much greater than the scattering efficiency for both normal and cancerous ribosomes in the spectral range of 500–1000 nm. At the same time, the absorption efficiency of the cancerous ribosome (magenta curve on Fig. 5.19) is twice higher than the scattering for healthy ribosomes (orange curve). Thus, the optimal wavelength for the cancerous ribosome detection is around 400 nm, and the optimal wavelength for the cancerous ribosome therapy is 800 nm (Slide 16.19).

5.2.4.4 Conclusions

Ribosomes are key contributors to healthy cells, along with cells that have become cancerous. Because of this, knowing the optical properties of ribosomes could provide major breakthroughs in diagnostics and destroying cancerous cells. Ribosomes in cancer cells have much greater scattering efficiency than normal cells in the spectral range of 200–500 nm, meaning that they might be useful organelles for cancer detection. Strong absorption of light for cancerous ribosomes is observed in the wavelength range 500–1000 nm, which will be optimal for destroying cancer. Since the optical properties are considerably different between normal and cancer cells, the use of ribosomes holds much potential to facilitate the diagnosis and therapy of cancer.

5.2.5 Practice Example 3: Simulation of Optical Properties of Bone Particles

5.2.5.1 **Abstract** (Slides 16.20 and 16.21)

Successful implementation of lasers in soft-tissue surgery has fueled interest in the implementation of similar hard-tissue applications in laser dentistry and laser medicine. However, unlike most soft tissues, which are composed mostly of water and some organic matter, hard tissues contain a large mineral component which affects the tissue's optical and thermal properties. This can make soft-tissue lasers unsuitable for hard-tissue applications depending on the laser wavelength. For these reasons, Mie simulations for bone tissue particles are performed to optimize hard-tissue laser surgery and laser dentistry.

5.2.5.2 Input Parameters

The difficulties in modeling laser heating and ablation of bone tissue arise from its complex structure which combines mineral lattice with an organic component. The inorganic mineral portion, which composes about 70 % of total bone material, is formed mostly from calcium phosphate in a crystalline structure known as

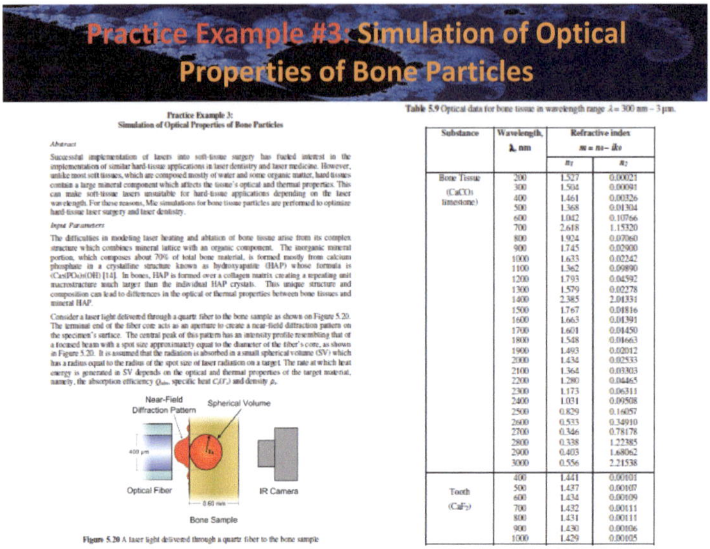

Practice Example #3: Simulation of Optical Properties of Bone Particles

Practice Example 3:
Simulation of Optical Properties of Bone Particles

Abstract

Successful implementation of lasers into soft-tissue surgery has fueled interest in the implementation of similar hard-tissue applications in laser dentistry and laser medicine. However, unlike most soft tissues, which are composed mostly of water and some organic matter, hard tissues contain a large mineral component which affects the tissue's optical and thermal properties. This can make soft-tissue lasers unsuitable for hard-tissue applications depending on the laser wavelength. For these reasons, Mie simulations for bone tissue particles are performed to optimize hard-tissue laser surgery and laser dentistry.

Input Parameters

The difficulties in modeling laser heating and ablation of bone tissue arise from its complex structure which combines mineral lattice with an organic component. The inorganic mineral portion, which composes about 70% of total bone material, is formed mostly from calcium phosphate in a crystalline structure known as hydroxyapatite (HAP) whose formula is $(Ca_5(PO_4)_3(OH)$ [14]. In bones, HAP is formed over a collagen matrix creating a repeating unit macrostructure much larger than the individual HAP crystals. This unique structure and composition can lead to differences in the optical or thermal properties between bone tissues and mineral HAP.

Consider a laser light delivered through a quartz fiber to the bone sample as shown on Figure 5.20. The terminal end of the fiber core acts as an aperture to create a near-field diffraction pattern on the specimen's surface. The central peak of this pattern has an intensity profile resembling that of a focused beam with a spot size approximately equal to the diameter of the fiber's core, as shown in Figure 5.20. It is assumed that the radiation is absorbed in a small spherical volume (SV) which has a radius equal to the radius of the spot size of laser radiation on a target. The rate at which heat energy is generated in SV depends on the optical and thermal properties of the target material, namely, the absorption efficiency Q_{abs}, specific heat $C_i(T_i)$ and density ρ_i.

Figure 5.20 A laser light delivered through a quartz fiber to the bone sample

Table 5.9 Optical data for bone tissue in wavelength range λ = 300 nm – 3 μm.

Substance	Wavelength, λ, nm	Refractive index $m = n_1 - ik_1$	
		n_1	k_1
Bone Tissue ($CaCO_3$ limestone)	200	1.527	0.00021
	300	1.504	0.00091
	400	1.461	0.00326
	500	1.368	0.01304
	600	1.042	0.10766
	700	2.618	1.15320
	800	1.924	0.07060
	900	1.745	0.02900
	1000	1.633	0.02242
	1100	1.362	0.09890
	1200	1.793	0.04592
	1300	1.579	0.02278
	1400	2.385	2.01331
	1500	1.767	0.01816
	1600	1.663	0.01391
	1700	1.601	0.01450
	1800	1.548	0.01663
	1900	1.493	0.02012
	2000	1.434	0.02533
	2100	1.364	0.03303
	2200	1.280	0.04465
	2300	1.173	0.06311
	2400	1.031	0.09508
	2500	0.829	0.16057
	2600	0.533	0.34910
	2700	0.346	0.78178
	2800	0.338	1.22385
	2900	0.403	1.68062
	3000	0.556	2.21538
Tooth (CaF_2)	400	1.441	0.00101
	500	1.437	0.00107
	600	1.434	0.00109
	700	1.432	0.00111
	800	1.431	0.00111
	900	1.430	0.00106
	1000	1.429	0.00105

Slide 16.20 Practice Example #3: Simulations of optical properties of bone particles

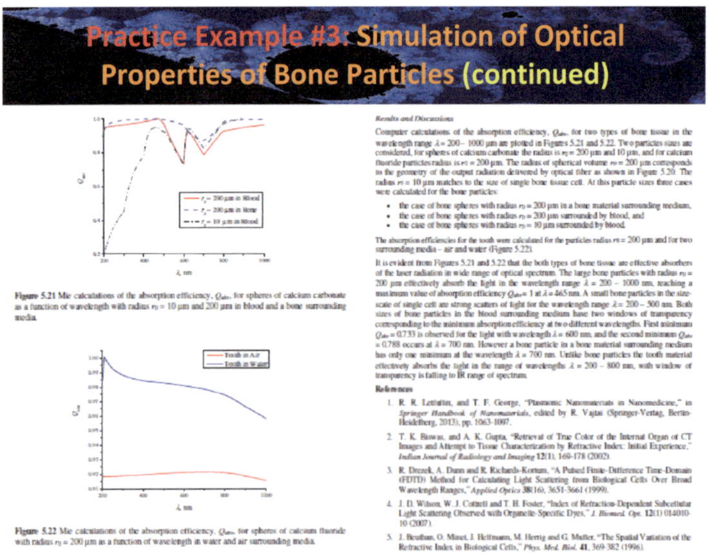

Practice Example #3: Simulation of Optical Properties of Bone Particles (continued)

Figure 5.21 Mie calculations of the absorption efficiency, Q_{abs}, for spheres of calcium carbonate as a function of wavelength with radius r_0 = 10 μm and 200 μm in blood and a bone surrounding media

Figure 5.22 Mie calculations of the absorption efficiency, Q_{abs}, for spheres of calcium fluoride with radius r_0 = 200 μm as a function of wavelength in water and air surrounding media

Results and Discussions

Computer calculations of the absorption efficiency, Q_{abs}, for two types of bone tissue in the wavelength range λ = 200 – 1000 μm are plotted in Figures 5.21 and 5.22. Two particles sizes are considered, for spheres of calcium carbonate the radius is r_0 = 200 μm and 10 μm, and for calcium fluoride particles radius is r_0 = 200 μm. The radius of spherical volume r_0 = 200 μm corresponds to the geometry of the output radiation delivered by optical fiber as shown in Figure 5.20. The radius r_0 = 10 μm matches to the size of single bone tissue cell. At this particle sizes three cases were calculated for the bone particles:

- the case of bone spheres with radius r_0 = 200 μm in a bone material surrounding medium,
- the case of bone spheres with radius r_0 = 200 μm surrounded by blood, and
- the case of bone spheres with radius r_0 = 10 μm surrounded by blood.

The absorption efficiencies for the tooth were calculated for the particles radius r_0 = 200 μm and for two surrounding media – air and water (Figure 5.22).

It is evident from Figures 5.21 and 5.22 that the both types of bone tissue are effective absorbers of the laser radiation in wide range of optical spectrum. The large bone particles with radius r_0 = 200 μm effectively absorb the light in the wavelength range λ = 200 – 1000 nm, reaching a maximum value of absorption efficiency Q_{abs} = 1 at λ = 465 nm. A small bone particles in the size-scale of single cell are strong scatters of light for the wavelength range λ = 200 – 500 nm. Both sizes of bone particles in the blood surrounding medium have two windows of transparency corresponding to the minimum absorption efficiency at two different wavelengths. First minimum Q_{abs} = 0.733 is observed for the light with wavelength λ = 600 nm, and the second minimum Q_{abs} = 0.788 occurs at λ = 700 nm. However a bone particle in a bone material surrounding medium has only one minimum at the wavelength λ = 700 nm. Unlike bone particles the tooth material effectively absorbs the light in the range of wavelengths λ = 200 – 800 nm, with window of transparency is falling to IR range of spectrum.

References

1. R. R. Letfullin, and T. F. George, "Plasmonic Nanomaterials in Nanomedicine," in *Springer Handbook of Nanomaterials*, edited by R. Vajtai (Springer-Verlag, Berlin-Heidelberg, 2013), pp. 1063-1097.
2. T. K. Biswas, and A. K. Gupta, "Retrieval of True Color of the Internal Organ of CT Images and Attempt to Tissue Characterization by Refractive Index: Initial Experience," *Indian Journal of Radiology and Imaging* **12**(1), 169-178 (2002).
3. R. Drezek, A. Dunn and R. Richards-Kortum, "A Pulsed Finite-Difference Time-Domain (FDTD) Method for Calculating Light Scattering from Biological Cells Over Broad Wavelength Ranges," *Applied Optics* **38**(16), 3651-3661 (1999).
4. J. D. Wilson, W. J. Cottrell and T. H. Foster, "Index of Refraction-Dependent Subcellular Light Scattering Observed with Organelle-Specific Dyes," *J. Biomed. Opt.* **12**(1) 014010-10 (2007).
5. J. Beuthan, O. Minet, J. Helfmann, M. Herrig and G. Muller, "The Spatial Variation of the Refractive Index in Biological Cells," *Phys. Med. Biol.* **41**, 369-382 (1996).

Slide 16.21 Practice Example #3: Simulations of optical properties of bone particles (continued)

hydroxyapatite (HAP) whose formula is $(Ca_5(PO_4)_3(OH)$ [20]. In bones, HAP is formed over a collagen matrix creating a repeating unit macrostructure much larger than the individual HAP crystals. This unique structure and composition can lead to differences in the optical or thermal properties between bone tissues and mineral HAP.

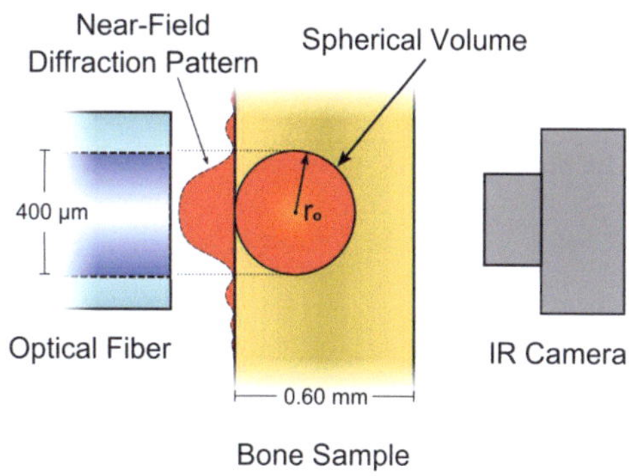

Fig. 5.20 Laser light delivered through a quartz fiber to the bone sample. Reprinted from [20]. © With permission from World Scientific

Consider laser light delivered through a quartz fiber to a bone sample as shown in Fig. 5.20 and Slide 16.20. The terminal end of the fiber core acts as an aperture to create a near-field diffraction pattern on the specimen's surface. The central peak of this pattern has an intensity profile resembling that of a focused beam with a spot size approximately equal to the diameter of the fiber's core, as shown in the figure. It is assumed that the radiation is absorbed in a small spherical volume (SV) which has a radius equal to that of the spot size of laser radiation on the target. The rate at which heat energy is generated in SV depends on the optical and thermal properties of the target material, namely, the absorption efficiency Q_{abs}, specific heat $C_s(T_s)$, and density ρ_s.

The absorption efficiency of bone tissue for laser radiation of wavelength λ and radius r_0 of the SV can be estimated by using Mie diffraction theory at the single-scattering approximation. As stated previously, the primary mineral constituent of bone is hydroxyapatite, $Ca_5(PO_4)_3(OH)$. The hydroxyl group is commonly replaced by carbonate, chloride, or fluoride. For the bulk of bone tissues, carbonate replaces the OH^- ion [20], so that bones are modeled as calcium carbonate ($CaCO_3$). Fluorinated dental treatments replace the hydroxyl group with a fluorine ion, so that tooth enamel properties are modeled as calcium fluoride (CaF_2) for these purposes.

Table 5.9 contains the complex refractive indices for bone ($CaCO_3$) [21] and tooth [22] at the body temperature of 35–55 °C (Slide 16.20). Tooth properties are modeled as calcium fluoride (CaF_2) [22], since calcium ions and fluorine ions both play important roles in tooth structure.

5.2.5.3 Results and Discussions

Computer calculations of Q_{abs} for two types of bone tissue in the wavelength range $\lambda = 200$–1000 μm are plotted in Figs. 5.21 and 5.22 (Slide 16.21). Two particles sizes are considered: for spheres of calcium carbonate, the radius is $r_0 = 200$ and 10 μm, and for calcium fluoride particles the radius is $r_0 = 200$ μm. The radius of

Table 5.9 Optical data for bone tissue in the wavelength range $\lambda = 300$ nm to 3 μm

Substance	Wavelength, λ, nm	Refractive index $m = n_0 - ik_0$	
		n_1	n_2
Bone tissue ($CaCO_3$, limestone)	200	1.527	0.00021
	300	1.504	0.00091
	400	1.461	0.00326
	500	1.368	0.01304
	600	1.042	0.10766
	700	2.618	1.1532
	800	1.924	0.0706
	900	1.745	0.029
	1000	1.633	0.02242
	1100	1.362	0.0989
	1200	1.793	0.04592
	1300	1.579	0.02278
	1400	2.385	2.01331
	1500	1.767	0.01816
	1600	1.663	0.01391
	1700	1.601	0.0145
	1800	1.548	0.01663
	1900	1.493	0.02012
	2000	1.434	0.02533
	2100	1.364	0.03303
	2200	1.28	0.04465
	2300	1.173	0.06311
	2400	1.031	0.09508
	2500	0.829	0.16057
	2600	0.533	0.3491
	2700	0.346	0.78178
	2800	0.338	1.22385
	2900	0.403	1.68062
	3000	0.556	2.21538
Tooth (CaF_2)	400	1.441	0.00101
	500	1.437	0.00107
	600	1.434	0.00109
	700	1.432	0.00111
	800	1.431	0.00111
	900	1.43	0.00106
	1000	1.429	0.00105

spherical volume $r_0 = 200$ μm corresponds to the geometry of the output radiation delivered by optical fiber as shown in Fig. 5.20. The radius $r_0 = 10$ μm matches the size of a single bone tissue cell. At this particle size, three cases have been calculated for the bone particles (Slide 16.21):

- The case of bone spheres with radius $r_0 = 200$ μm in a surrounding bone medium,
- The case of bone spheres with radius $r_0 = 200$ μm surrounded by blood,
- The case of bone spheres with radius $r_0 = 10$ μm surrounded by blood.

The absorption efficiencies for tooth have been calculated for the particle radius $r_0 = 200$ μm and for two surrounding media—air and water (Fig. 5.22).

It is evident from Figs. 5.21 and 5.22 (Slide 16.21) that both types of bone tissue are effective absorbers of the laser radiation over a wide range of the optical spectrum. The large bone particles with radius $r_0 = 200$ μm effectively absorb the light in the wavelength range $\lambda = 200$–1000 nm, reaching a maximum value of the absorption efficiency $Q_{abs} = 1$ at $\lambda = 465$ nm. The small bone particles at the size scale of a single cell are strong scatters of light for the wavelength range $\lambda = 200$–500 nm. Both sizes of bone particles in the surrounding blood medium have two windows of transparency corresponding to the minimum absorption efficiency at two different wavelengths. The first minimum, $Q_{abs} = 0.733$, is observed for the light with wavelength $\lambda = 600$ nm, and the second minimum, $Q_{abs} = 0.788$, occurs at $\lambda = 700$ nm. However, a bone particle in a surrounding bone medium has only one minimum at 700 nm. Unlike bone particles, tooth material effectively absorbs light in the wavelength range $\lambda = 200$–800 nm, with the window of transparency falling to IR range of the spectrum (Slide 16.21).

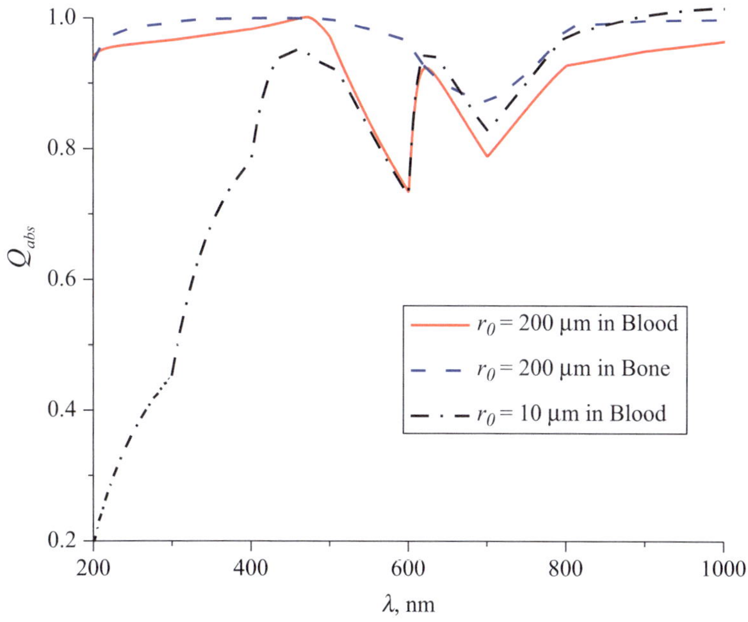

Fig. 5.21 Mie calculations of Q_{abs} for spheres of calcium carbonate as a function of wavelength with radius $r_0 = 10$ and 200 μm in blood and surrounding bone medium

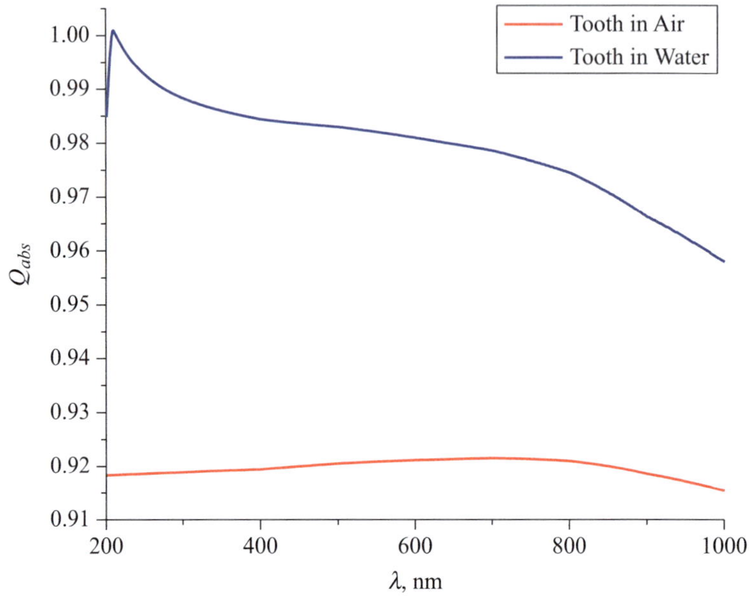

Fig. 5.22 Mie calculations of the absorption efficiency Q_{abs} for spheres of calcium fluoride with radius $r_0 = 200$ μm as a function of wavelength in water and surrounding medium

Appendix A: Mie Code

This appendix contains the MATLAB code designed to compute the Mie coefficients. The comments within the MATLAB code should explain the used variables, procedures, and functions.

Note: This code comes directly from the MATLAB file and is therefore formatted to word-wrap according to MATLAB standard character-per-line limits.

```
function result = Mie.abcd (m, x)
% Computes the Mie coefficients: a_n, b_n, c_n, d_n.
%
% A criterion for approximating the minimum number of terms required to
% reach convergence (rather than determining convergence itera-
tively) is
% approximated by "nmax = 2 + x + 4*x^(1/3) " as suggested on p. 477 of
% Bohren & Huffman, Absorption and Scattering of Light by Small Parti-
cles.
%
% Mie coefficients a_1 through a_nmax, b_1 through b_nmax, etc. are
% calculated using the Eqs. (4.2).
%
```

```
% The only inputs are the size parameter x and the index parameter m.
% x=k*r, where k is the wavenumber and r is the sphere radius.
% Because k=2pi/wavelength, x results in a dimensionless parameter
% provided r and lambda are in the same units (e.g. nm).
% m is the relative complex refractive index m sphere/m medium.
%
% Colin E.W. Rice and Renat R. Letfullin - Rose-Hulman Institute of Technology

nmax=round(2+x+4*x^(1/3));   % convergence criteria to determine
                              number of terms

n=(1:nmax);                  % list of order, n
n_half=(n+0.5);              % list of order, n+1/2 (needed in Bessel
                              functions)

mx=m.*x;                     % m*x, defined for easier typing/reading
                              below
mm=m.*m;                     % m^2, again defined for easier typing.

jx=besselj(n_half,x).*sqrt(0.5*pi./x);      % spherical Bessel func-
                                              tion j_n(x)
jmx=besselj(n_half,mx).*sqrt(0.5*pi./mx);   % spherical Bessel func-
                                              tion j_n(mx)
yx=bessely(n_half,x).*sqrt(0.5*pi./x);      % spherical Bessel func-
                                              tion y_n(x)
hx=jx+1i*yx;                                % spherical Hankel func-
                                              tion h_n(x)

% The formula for above coefficients j_n(x), j_n(mx) and h-n(x)
% use derivatives [x.j_n(x)]' and [mx.j_n(mx)]' and
% [x.h_n(x)]' which can be calculated from recurrence relations using
% the above in addition to j_[n-1](x), j_[n-1](mx), and h_[n-1](x). The
% above jx, jmx, hx functions range from n=1..nmax so the (n-1) versions
% will instead range from n=0..(nmax-1). These can be made simply by
% inserting the 0th order term to the front (and dropping the last term).
% The above naming convention is kept, with suffix 0 to denote the addition
% of the above 0^th order in the [n-1]th order terms.

jx0=[sin(x)/x, jx(1:nmax-1)];      % spherical Bessel function j_[n-1]
                                     (x)
jmx0=[sin(mx)/mx, jmx(1:nmax-1];   % spherical Bessel function j_[n-1]
                                     (mx)
yx0=[-cos(x)/x, yx(1:nmax-1];      % spherical Bessel function y_[n-1]
                                     (x)
```

```
hx0=jx0+1i*yx0;                  % spherical Hankel function h_[n-1](x)

% With both the (n)th and (n-1)th order terms defined, the recurrence
% relations for [x. j_n(x)]' and [mx. j_n(mx)]' and [x. h_n(x)]' can be used.
% The above naming convention is used with suffix p to denote the prime'
% status when taking the derivative.

jxp=x.* jx0-n.*jx;          % [x.j_n(x)]' = x.j_[n-1](x) - n.j_n(x)
jmxp=mx.*jmx0-n.*jmx;       % [mx.j_n(mx)]' = mx.j_[n-1](mx) - n.j_n(mx)
hxp=x.*hx0-n.*hx;           % [x.h_n(x)]' = x.h_[n-1](x) - n.h_n(x)

% Finally the coefficients can be calculated from all of the above
% Bessel and Hankel functions.

an = (mm.*jmx.*jxp-jx.*jmxp)./(mm.*jmx.*hxp-hx.*jmxp);
bn = (jmx.*jxp-jx.*jmxp)./(jmx.*hxp-hx.*jmxp);
cn = (jx.*hxp-hx.*jxp)./(jmx.*hxp-hx.*jmxp);
dn = (m.*jx.*hxp-m.*hx.*jxp)./(mm.*jmx.*hxp-hx.*jmxp);

% Return the resulting lists of coefficients.
result=[an; bn; cn; dn];
end
```

Homework Exercises

Section 5.1: Lorenz–Mie Simulations of Nanoparticle Optics

5.1. Describe the research task in modeling the radiation interaction with nanoparticles.
5.2. List three input parameters required for Lorenz–Mie calculations.
5.3. What are the output results from Lorenz–Mie simulations?
5.4. True or false? The complex refractive index of the nanoparticle could be engineered.
5.5. Why is a gold nanoparticle good for biological applications?
5.6. True or False? The index of refraction of materials doesn't depend on the wavelength of the radiation.
5.7. How is the optimal range of the wavelengths determined for therapeutic applications of gold nanoparticles?

5.8. How is the optimal range of the wavelengths defined for the diagnostic applications of gold nanoparticles?
5.9. At what wavelength is a plasmon resonance absorption observed for gold nanoparticles?
5.10. True or false? The surrounding medium doesn't affect the absorption properties of the nanoparticles.
5.11. True or false? The optical properties of the nanoparticles depend on the shape and internal structure of the particle.
5.12. Perform Mie simulations of absorption and scattering efficiencies for a silver nanosphere vs. the particle radius ($r_0 = 1–1000$ nm) and radiation wavelength ($\lambda = 200–1000$ nm) in cytoplasm, blood, and collagen by using optical data from Tables 5.1 and 5.2. Find the optimal wavelength of radiation and optimal particle radius for therapeutic and diagnostic applications of silver nanospheres. Make conclusions about the scattering and absorption properties of silver nanospheres similar to the Practice Example #1.
5.13. Perform Mie simulations of absorption and scattering efficiencies for a silica–gold nanoparticle vs. the particle radius ($r_0 = 1–1000$ nm) and radiation wavelength ($\lambda = 200–1000$ nm) in cytoplasm, blood, and collagen by using optical data from Tables 5.1 and 5.2. Find the optimal wavelength of radiation and optimal particle radius for therapeutic and diagnostic applications of silica–gold nanoparticles. Make conclusions about the scattering and absorption properties of silica–gold nanoparticles similar to the Practice Example #1.
5.14. Perform Mie simulations of absorption and scattering efficiencies for a silver nanocube vs. the particle radius ($r_0 = 1–1000$ nm) and radiation wavelength ($\lambda = 200–1000$ nm) in cytoplasm, blood, and collagen by using optical data from Tables 5.1 and 5.2. Find the optimal wavelength of radiation and optimal radius of particle for therapeutic and diagnostic applications of silver nanocubes. Make the conclusions about the scattering and absorption properties of silver nanocubes similar to the Practice Example #1.
5.15. Perform Mie simulations of absorption and scattering efficiencies for a silica–silver nanoparticle vs. the particle radius ($r_0 = 1–1000$ nm) and radiation wavelength ($\lambda = 200–1000$ nm) in cytoplasm, blood, and collagen by using optical data from Tables 5.1 and 5.2. Find the optimal wavelength of radiation and optimal particle radius for therapeutic and diagnostic applications of silica–silver nanoparticles. Make conclusions about the scattering and absorption properties of silica–silver nanoparticles similar to the Practice Example #1.
5.16. Perform Mie simulations of absorption and scattering efficiencies for an aluminum nanoparticle vs. the particle radius ($r_0 = 1–1000$ nm) and radiation wavelength ($\lambda = 200–1000$ nm) in cytoplasm, blood, and collagen by using optical data from Tables 5.1 and 5.2. Find the optimal wavelength of radiation and optimal particle radius for therapeutic and diagnostic applications of aluminum nanoparticles. Make conclusions about the scattering and absorption properties of aluminum nanoparticles similar to the Practice Example #1.

5.17. Perform Mie simulations of absorption and scattering efficiencies for a nickel nanoparticle vs. the particle radius ($r_0 = 1$–1000 nm) and radiation wavelength ($\lambda = 200$–1000 nm) in cytoplasm, blood, and collagen by using optical data from Tables 5.1 and 5.2. Find the optimal wavelength of radiation and optimal particle radius for therapeutic and diagnostic applications of nickel nanoparticles. Make conclusions about the scattering and absorption properties of nickel nanoparticles similar to the Practice Example #1.

5.18. Perform Mie simulations of absorption and scattering efficiencies for a carbon nanoparticle vs. the particle radius ($r_0 = 1$–1000 nm) and radiation wavelength ($\lambda = 200$–1000 nm) in cytoplasm, blood, and collagen by using optical data from Tables 5.1 and 5.2. Find the optimal wavelength of radiation and optimal particle radius for therapeutic and diagnostic applications of carbon nanoparticles. Make conclusions about the scattering and absorption properties of carbon nanoparticles similar to the Practice Example #1.

5.19. Perform Mie simulations of absorption and scattering efficiencies for a fullerene nanosphere vs. the particle radius ($r_0 = 1$–1000 nm) and radiation wavelength ($\lambda = 200$–1000 nm) in cytoplasm, blood, and collagen by using optical data from Tables 5.1 and 5.2. Find the optimal wavelength of radiation and optimal particle radius for therapeutic and diagnostic applications of fullerene nanospheres. Make conclusions about the scattering and absorption properties of fullerene nanospheres similar to the Practice Example #1.

5.20. Perform Mie simulations of absorption and scattering efficiencies for a polystyrene nanoparticle vs. the particle radius ($r_0 = 1$–1000 nm) and radiation wavelength ($\lambda = 200$–1000 nm) in cytoplasm, blood, and collagen by using optical data from Tables 5.1 and 5.2. Find the optimal wavelength of radiation and optimal particle radius for therapeutic and diagnostic applications of polystyrene nanoparticles. Make conclusions about the scattering and absorption properties of polystyrene nanoparticles similar to the Practice Example #1.

5.21. Perform Mie simulations of absorption and scattering efficiencies for a glass nanoparticle vs. the particle radius ($r_0 = 1$–1000 nm) and radiation wavelength ($\lambda = 200$–1000 nm) in cytoplasm, blood, and collagen by using optical data from Tables 5.1 and 5.2. Find the optimal wavelength of radiation and optimal particle radius for therapeutic and diagnostic applications of glass nanoparticles. Make conclusions about the scattering and absorption properties of glass nanoparticles similar to the Practice Example #1.

5.22. Perform Mie simulations of absorption and scattering efficiencies for an aluminum oxide nanoparticle vs. the particle radius ($r_0 = 1$–1000 nm) and radiation wavelength ($\lambda = 200$–1000 nm) in cytoplasm, blood, and collagen by using optical data from Tables 5.1 and 5.2. Find the optimal wavelength of radiation and optimal particle radius for therapeutic and diagnostic applications of aluminum oxide nanoparticles. Make conclusions about the scattering and absorption properties of aluminum oxide nanoparticles similar to the Practice Example #1.

5.23. Perform Mie simulations of absorption and scattering efficiencies for a chromium nanoparticle vs. the particle radius ($r_0 = 1$–1000 nm) and radiation wavelength ($\lambda = 200$–1000 nm) in cytoplasm, blood, and collagen by using optical data from Tables 5.1 and 5.2. Find the optimal wavelength of radiation and optimal particle radius for therapeutic and diagnostic applications of chromium nanoparticles. Make conclusions about the scattering and absorption properties of chromium nanoparticles similar to the Practice Example #1.
5.24. Perform Mie simulations for gold nanoparticles and silica–gold nanospheres in water, blood, and cell membrane by using optical data from Tables 5.1 and 5.2. Make a comparative analysis of the scattering and absorption efficiencies for those particles similar to the Practice Example #1.
5.25. Perform Mie simulations for silver nanocubes and silica–silver nanospheres in water, blood, and cell membrane by using optical data from Tables 5.1 and 5.2. Make a comparative analysis of the scattering and absorption efficiencies for those particles similar to the Practice Example #1.
5.26. Perform Mie simulations for aluminum nanoparticles and nickel nanoparticles vs. the particle radius and radiation wavelength in water, blood, and cell membrane by using optical data from Tables 5.1 and 5.2. Make a comparative analysis of the scattering and absorption efficiencies for those particles similar to the Practice Example #1.
5.27. Perform Mie simulations for carbon nanoparticles and fullerene nanospheres vs. the particle radius and radiation wavelength in water, blood, and cell membrane by using optical data from Tables 5.1 and 5.2. Make a comparative analysis of the scattering and absorption efficiencies for those particles similar to the Practice Example #1.
5.28. Perform Mie simulations for polystyrene and glass nanoparticles vs. the particle radius and radiation wavelength in water, blood, and cell membrane by using optical data from Tables 5.1 and 5.2. Make a comparative analysis of the scattering and absorption efficiencies for those particles similar to the Practice Example #1.
5.29. Perform Mie simulations for aluminum oxide and chromium nanoparticles vs. the particle radius and radiation wavelength in water, blood, and cell membrane by using optical data from Tables 5.1 and 5.2. Make a comparative analysis of the scattering and absorption efficiencies for those particles similar to the Practice Example #1.

Section 5.2: Simulations of Optical Properties of Biological Particles

5.30. What are the differences in the morphological properties of normal and cancer cells?
5.31. How can the differences in the morphological properties of normal and cancer cells be used for diagnostic and therapeutic applications?

5.32. Describe the research task in modeling the radiation interaction with biological particles.
5.33. Describe how to create a *rix* file for the *MiePlot* software.
5.34. True or false? Cancerous ribosomes are considerably more efficient in the scattering and absorption of light than normal ribosomes in the optical range of spectrum.
5.35. In what wavelength range of the spectrum are cancer ribosomes strong scatters of light?
5.36. In what wavelength range of the spectrum are cancer ribosomes strong absorbers of light?
5.37. What is the optimal wavelength for cancerous ribosome detection?
5.38. What is the optimal wavelength for cancerous ribosome therapy?
5.39. True or false? About 70 % of total bone material is formed mostly from calcium phosphate in a crystalline structure known as hydroxyapatite.
5.40. Describe the experimental setup shown in Fig. 5.20.
5.41. True or false? Bone tissue is an effective absorber of laser radiation over a wide range of the optical spectrum.
5.42. What are the wavelengths of two windows of transparency of bone tissue corresponding to the minimum absorption efficiency?
5.43. What are the differences in optical properties (absorption and scattering) of bone particles and tooth material?
5.44. Perform Mie calculations and a comparative analysis of the absorption and scattering efficiencies for normal and cancerous cell mitochondria surrounded by cytoplasm in the optical range of spectrum ($\lambda = 400$ nm to 1 μm) by using optical data from Tables 5.2 and 5.6.
5.45. Perform Mie calculations and comparative analysis of the absorption and scattering efficiencies for normal and cancerous cell nuclei surrounded by cytoplasm in the optical range of the spectrum ($\lambda = 400$ nm to 1 μm) by using optical data from Tables 5.2 and 5.6.
5.46. Perform Mie calculations and comparative analysis of the absorption and scattering efficiencies for normal and cancerous cell lysosome surrounded by cytoplasm in the optical range of the spectrum ($\lambda = 400$ nm to 1 μm) by using optical data from Tables 5.2 and 5.6.
5.47. Perform Mie calculations and comparative analysis of the absorption and scattering efficiencies for normal and cancerous cell cytoskeleton surrounded by cytoplasm in the optical range of the spectrum ($\lambda = 400$ nm to 1 μm) by using optical data from Tables 5.2 and 5.6.
5.48. Perform Mie simulations of the scattering efficiency for a bone particle vs. the particle radius ($r_0 = 10$–1000 μm) and radiation wavelength ($\lambda = 200$–1000 nm) in water, blood, and air by using optical data from Tables 5.2 and 5.9. Find the optimal wavelength of radiation and optimal particle radius for diagnostic applications. Make conclusions about the scattering properties of bone particles similar to the Practice Example #3.
5.49. Perform Mie simulations of the scattering efficiency for a tooth particle vs. the particle radius ($r_0 = 10$–1000 μm) and radiation wavelength

($\lambda = 200$–1000 nm) in water, blood, and air by using optical data from Tables 5.2 and 5.9. Find the optimal wavelength of radiation and optimal particle radius for diagnostic applications. Make conclusions about the scattering properties of tooth particles similar to the Practice Example #3.

References

1. X. Ma et al., Determination of complex refractive index of polystyrene microspheres from 370 to 1610 nm. Phys. Med. Biol. **48**, 4165–4172 (2003)
2. R.R. Letfullin, V.I. Igoshin, Multipass optical reactor for laser processing of disperse materials. Quantum Electron. **25**, 684–689 (1995)
3. Wikipedia, "Codes for Electromagnetic Scattering by Spheres," https://en.wikipedia.org/wiki/Codes_for_electromagnetic_scattering_by_spheres
4. P. Laven, Simulation of rainbows, coronas, and glories by use of Mie theory. Appl. Optics **42** (3), 436–444 (2003)
5. R.R. Letfullin, T.F. George, New dynamic modes for selective laser cancer nanotherapy, in *Computational Studies of New Materials II: From Ultrafast Processes and Nanostructures to Optoelectronics, Energy Storage and Nanomedicine*, ed. by T.F. George, D. Jelski, R.R. Letfullin, G.P. Zhang (World Scientific, Singapore, 2011), pp. 131–172
6. P.K. Jain, K.S. Lee, I.H. El-Sayed, M.A. El-Sayed, Calculated absorption and scattering properties of gold nanoparticles of different size, shape, and composition: applications in biological imaging and biomedicine. J. Phys. Chem. **110**, 7238–7248 (2006)
7. R.R. Letfullin, T.F. George, Plasmonic nanomaterials in nanomedicine, in *Springer Handbook of Nanomaterials*, ed. by R. Vajtai (Springer, Berlin, 2013), pp. 1063–1097
8. T.K. Biswas, A.K. Gupta, Retrieval of true color of the internal organ of CT images and attempt to tissue characterization by refractive index: initial experience. Indian J. Radiol. Imaging **12**, 169–178 (2002)
9. R. Drezek, A. Dunn, R. Richards-Kortum, A pulsed finite-difference time-domain (FDTD) method for calculating light scattering from biological cells over broad wavelength ranges. Appl. Optics **38**, 3651–3661 (1999)
10. J.D. Wilson, W.J. Cottrell, T.H. Foster, Index of refraction-dependent subcellular light scattering observed with organelle-specific dyes. J. Biomed. Opt. **12**(1), 014010 (2007)
11. J. Beuthan, O. Minet, J. Helfmann, M. Herrig, G. Muller, The spatial variation of the refractive index in biological cells. Phys. Med. Biol. **41**, 369–382 (1996)
12. S. Johnsen, E.A. Widder, The physical basis of transparency in biological tissue: ultrastructure and the minimization of light scattering. J. Theor. Biol. **199**(2), 181–198 (1999)
13. B. Chance, M. Cope, E. Gratton, N. Ramanujam, B. Tromberg, Phase measurement of light absorption and scatter in human tissue. Rev. Sci. Instrum. **69**, 3457–3481 (1998)
14. S. Bassenett, On the mechanism of organelle degradation in the vertebrate lens. Exp. Eye Res. **88**, 133–139 (2009)
15. T. Torimura, T. Ueno, S. Inuzuka, Y. Kimura, P. Ko, M. Kin, T. Minetoma, T. Majima, M. Sata, H. Abe, K. Tanikawa, Ultrastructural observation on hepatocellular carcinoma: correlation of tumor grade and degree of atypia of cell organelles by morphometry. Med. Electron Microsc. **26**, 19–28 (1993)
16. G.T. Deans, P.W. Hamilton, P.C.H. Watt, M. Heatly, K. Williamson, C.C. Patterson, B.J. Rowlands, G. Parks, R. Spence, Morphometric analysis of colorectal cancer. Diseases of the Colon and Rectum **36**, 450–456 (1993)
17. M.M. Radwan, K.A. Amer, N.M. Mokhtar, M.A. Kandil, A.M. El-Barbary, H.A. Aiad, Nuclear morphometry in ductal breast carcinoma with correlation to cell proliferative activity and prognosis. J. Egypt. Natl. Canc. Inst. **15**, 169–182 (2003)

18. D. Ozaki, Y. Kondo, Comparative morphometric studies of benign and malignant intraductal proliferative lesions of the breast by computerized image analysis. Hum. Pathol. **26**, 1109–1113 (1995)
19. Y. Cui, E.A. Koop, P.J. van Diest, R.A. Kandel, T.E. Rohan, Nuclear morphometric features in benign breast tissue and risk of subsequent breast cancer. Breast Cancer Res. Treat. **104**, 103–107 (2007)
20. R.R. Letfullin, T.F. George, Laser ablation of biological tissue by short and ultrashort pulses, in *Computational Studies of New Materials II: From Ultrafast Processes and Nanostructures to Optoelectronics, Energy Storage and Nanomedicine*, ed. by T.F. George, D. Jelski, R.R. Letfullin, G.P. Zhang (World Scientific, Singapore, 2011), pp. 191–218
21. M.R. Querry, G. Osborne, K. Lies, R. Jordon, R.M. Coveney Jr., Complex refractive index of limestone in the visible and infrared. Appl. Optics **17**, 353–356 (1978)
22. I.H. Malitson, A redetermination of some optical properties of calcium fluoride. Appl. Optics **2**, 1103–1107 (1963)
23. R.R. Letfullin, T.F. George, Nanomaterials in nanomedicine, in *Computational Studies of New Materials II: From Ultrafast Processes and Nanostructures to Optoelectronics, Energy Storage and Nanomedicine*, ed. by T.F. George, D. Jelski, R.R. Letfullin, G.P. Zhang (World Scientific, Singapore, 2011), pp. 103–129
24. R.R. Letfullin, C.B. Iversen, T.F. George, Modeling nanophotothermal therapy: Kinetics of thermal ablation of healthy and cancerous cell organelles and gold nanoparticles. Nanomed.: Nanotechnol. Biol. Med. **7**, 137–145 (2011)

Chapter 6
Thermal Interaction of Particle and Radiation

Contents

Electronic supplementary material: The online version of this chapter (doi:10.1007/978-3-319-43577-0_6) contains supplementary material, which is available to authorized users.

R.R. Letfullin, T.F. George, *Computational Nanomedicine and Nanotechnology*,
DOI 10.1007/978-3-319-43577-0_6

Abstract The interaction between nanoparticles and radiation holds great interest in nanomedicine, introducing such possibilities as selective cell targeting by nanoparticles to create highly localized thermal cell damage. The evolution of the thermal fields in the particle depends on an interplay of heat generation and thermal diffusion governed by material properties. This chapter first describes the thermodynamic model used and then focuses on the heating and cooling kinetics of nanoparticles by short and ultrashort laser pulses in the femtosecond, picosecond, and nanosecond regimes. The time-dependent simulations and detailed analyses are carried out for heating of nanoparticles, including biological particles. This chapter contains material adapted from our publications listed at the end of the chapter. A detailed list of references and reviews on a given topic of this chapter can be found in those original papers.

6.1 Lecture 17: Kinetics of Heating and Cooling of Nanoparticles: Time-Dynamic Models (Slide 17.1)

- Introduction
- Thermodynamics of Nanoparticles
- One-Temperature Model
- Two-Temperature Model

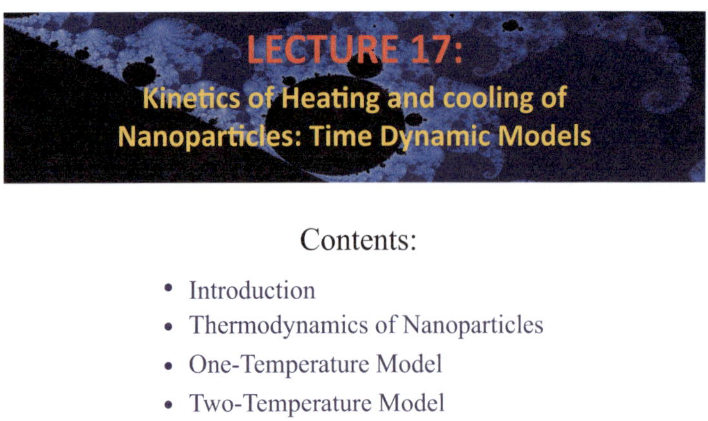

Slide 17.1 Contents of Lecture 17

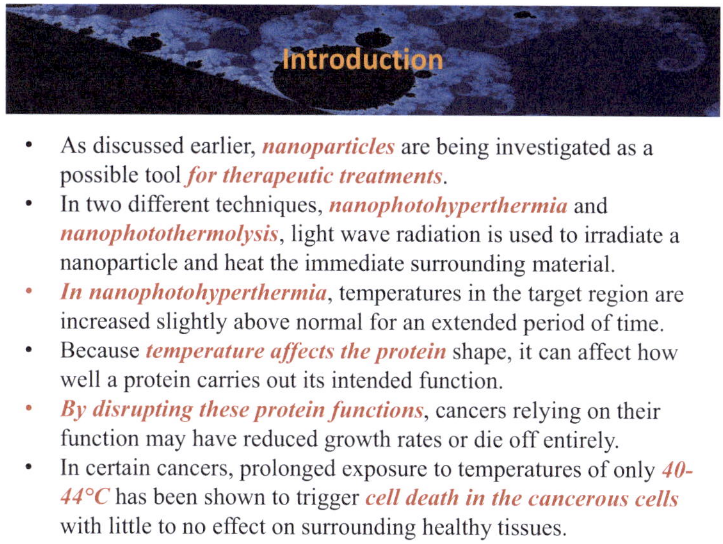

Slide 17.2 Introduction

6.1.1 Introduction

This section contains material reproduced from our publications [1–3]. As discussed in Chap. 3 (Lecture 9), nanoparticles are being investigated as a possible tool for therapeutic treatments (Slide 17.2). In two different techniques, nanophotohyperthermia and nanophotothermolysis, light wave radiation is used to irradiate a nanoparticle and heat the immediate surrounding material. In nanophotohyperthermia, temperatures in the target region are increased slightly above normal for an extended period of time. Because temperature affects the protein shape, it can affect how well a protein carries out its intended function. By disrupting these protein functions, cancers relying on their function may have reduced growth rates or die off entirely. In certain cancers, prolonged exposure to temperatures of only 40–44 °C has been shown to trigger cell death in cancerous cells with little to no effect on the surrounding healthy tissues [1]. In nanophotothermolysis, however (Slide 17.3), high-intensity pulses are used to heat the nanoparticles very quickly, reaching thresholds for various nonlinear synergistic effects in the targeted region, including thermal coagulation, vaporization, sound production, bubble formation, and shock-wave generation, which have shown great potential for selective damage to cancer cells, bacteria, viruses, and DNA.

More importantly, these therapies are poised to remove many of the detrimental effects seen in current oncological treatments [1]. Since the nanoparticles are not cytotoxic in and of themselves, they can be delivered into the body without harming or hindering healthy cell functions; this is in contrast to chemotherapy,

Introduction (continued)

- ***In nanophotothermolysis***, however, high-intensity pulses are used to heat the nanoparticles very quickly, reaching thresholds for various nonlinear synergistic effects in the targeted region, including ***thermal coagulation, vaporization, sound production, bubble formation and shock-wave generation***, which have shown great potential for selective damage to cancer cells, bacteria, viruses and DNA.
- The extent of the ***particle heating*** depends on many factors, divided primarily into three categories:
  - ***(1) particle material, size and shape;***
  - ***(2) radiation wavelength, energy density, duration, pulse shape and generation frequency;***
  - ***(3) properties of the surrounding medium.***

Slide 17.3 Introduction (continued)

which may have detrimental effects that are pervasive to many types of healthy tissues. The damage mechanism, heat, does not kill cells by creating DNA damage in the way that radiation therapy does, and therefore removes the risk that healthy cells get DNA defects, resulting in reduced viability or mutation. Like chemotherapy or radiation therapy, this can be used in cases where the pervasiveness of metastases or the delicate nature of local tissues would make traditional surgery infeasible. However, the successful use of such therapies in cancer treatment relies on two main factors: targeted delivery of the nanoparticles to the cancer cells, which we studied in Chap. 2 (Lecture 7), and controlled heating of the nanoparticles to induce the desired thermal damage mechanisms, which we will study in this chapter.

The extent of the particle heating depends on many factors, divided primarily into three categories [3] (Slide 17.3): (1) particle material, size, and shape; (2) radiation wavelength, energy density, duration, pulse shape, and generation frequency; and (3) properties of the surrounding medium. Even with the many factors involved in an experimental setup that contribute to the individual heating characteristics of a nanoparticle, several factors stay relatively constant across many experimental situations. Specifically, spherical gold nanoparticles are commonly used because they are relatively easy to fabricate, nontoxic, easily conjugated to antibodies, and strong absorbers. Nanosecond pulse-width lasers with pulse-firing frequencies of 10 Hz are often used because they are widely available and cheaper than pico- and femtosecond lasers or lasers with higher firing frequencies. Additionally, water or a phosphate buffered saline solution is employed as a surrounding medium due to large similarities to a biological cell.

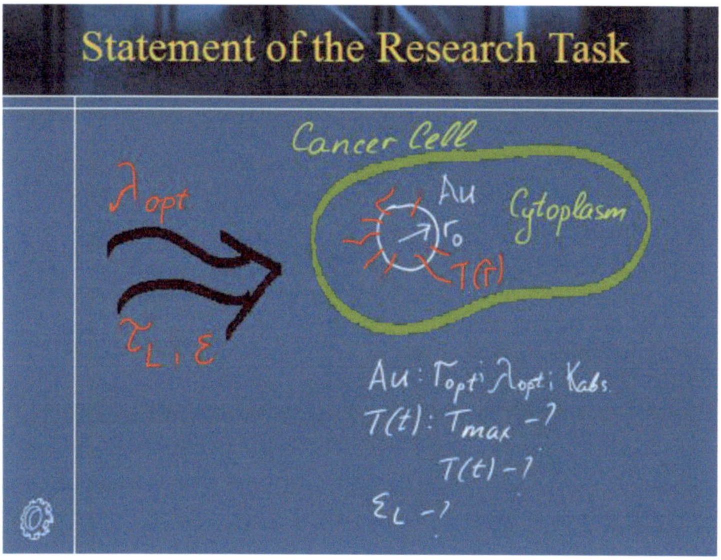

Slide 17.4 Statement of research task

6.1.2 *Thermodynamics of Nanoparticles*

The *research task* schematically shown in Slide 17.4 involves the heating of gold nanoparticles selectively delivered to the cancer cells by pulsed radiation. In Chap. 5, we used the generalized Lorenz–Mie diffraction theory to determine the absorption characteristics of spherical particles in various biological media. Using the complex indices of refraction for the medium and particle, as well as the particle size, we determined the absorption efficiency of the spherical particle at a given wavelength. In this chapter, we use the heat-mass transfer equation to model the time evolution of the thermal fields in the nanoparticle.

The changing temperature of a particle and its surroundings can be described by a set of continuity equations which track heat generation, heat diffusion, and mass transfer across a boundary (Slide 17.5). Fluid heat transfer models often have to treat heat diffusion and mass transfer, which can be due to local motion or a velocity vector representing fluid flow passing through the boundary in question. One such heat transfer equation, which will be used as a starting point, is

$$\rho_{\mathrm{f}} C_{\mathrm{f}}(T_{\mathrm{f}}) \frac{DT_{\mathrm{f}}}{Dt} - \beta_{\mathrm{f}} T_{\mathrm{f}} \frac{DP_{\mathrm{f}}}{Dt} = \rho_{\mathrm{f}} \Phi_{\mathrm{f}} + \nabla \cdot (\mu_{\mathrm{f}}(T_{\mathrm{f}}) \nabla T_{\mathrm{f}}), \tag{6.1}$$

where $C_{\mathrm{f}}(T_{\mathrm{f}})$, ρ_{f}, and μ_{f} represent the specific heat (at constant pressure P_{f} in the case of a fluid), density, and thermal conductivity of the fluid. The term β_{f} is the thermal expansion coefficient, Φ_{f} is the thermal dissipation function (viscous dissipation function in some other sources), and the operator $\frac{D}{Dt}$ is the material derivative defined by $\frac{D}{Dt} = \frac{\partial}{\partial t} + u \cdot \nabla$.

Thermodynamics of Nanoparticles

- The changing temperature of a particle and its surroundings can be described by a set of ***continuity equations*** which track ***heat generation, heat diffusion and mass transfer across a boundary***.
- One such heat transfer equation, which will be used as a starting point, starts off as

$$\rho_f C_f(T_f)\frac{DT_f}{Dt} - \beta_f T_f\frac{DP_f}{Dt} = \rho_f \Phi_f + \nabla\cdot(\mu_f(T_f)\nabla T_f),$$

where the terms $C_f(T_f)$, ρ_f and μ_f represent the specific heat (at constant pressure P_f in the case of a fluid), density and thermal conductivity of the fluid. The term β_f is the thermal expansion coefficient, Φ_f is the thermal dissipation function (viscous dissipation function in some other sources), and the operator $\frac{D}{Dt}$ is the material derivative defined by $\frac{D}{Dt} = \frac{\partial}{\partial t} + u\cdot\nabla$.

Slide 17.5 Thermodynamics of nanoparticles

Setting Up Limitations

- If we assume that the particle undergoes little to ***no thermal expansion***, then the second term can be eliminated as $\beta_f \to 0$.
- Further assuming that the particle is a ***rigid solid***, and will not break up while being heated, pretty much excludes any other forms of mass transfer that might be possible in other fluid models.
- Assuming the ***fluid velocity*** around the particle is near zero, the thermal dissipation function tends to zero, $\Phi_f \to 0$.
- Within the boundary, where the particle is solid metal, there is ***no fluid motion***, and therefore the viscous dissipation function is identically equal to zero.

Slide 17.6 Setting up limitations

In the case of a small spherical nanoparticle, it makes sense to define the boundary at the surface of the sphere at distance r_0 from the center. If we assume that the particle undergoes little to no thermal expansion, then the second term can be eliminated as $\beta_f \to 0$ (Slide 17.6). Further assumption that the particle is a rigid

Developing a Thermodynamic Model

Thus the original heat transfer equation for fluids has been simplified by a number of assumptions that remove terms, but one term has to be added to account for the heat generated in the nanoparticle as light is absorbed. Following all the reductions plus a single addition, the heat transfer equation takes the form

$$\rho_f C_f \frac{\partial T_f}{\partial t} = \nabla \cdot \left(\mu_f \nabla T_f\right) + q_{gen},$$

where q_{gen} is the heat generated in the nanoparticle due to the absorption of the radiation. Converting to spherical coordinates, while assuming $T_f(r,\theta,\varphi,t) \approx T_f(r,t)$ (i.e., the temperature function is radially symmetric) gives

$$\rho_f C_f \frac{\partial T_f}{\partial t} = \frac{1}{r^2}\frac{\partial}{\partial r}\left(r^2 \mu_f \frac{\partial T_f}{\partial r}\right) + q_{gen}.$$

Slide 17.7 Developing a thermodynamic model

solid, and will not break up while being heated, pretty much excludes any other forms of mass transfer that might be possible in other fluid models.

The thermal dissipation function Φ_{f} describes the heat generated by viscous forces and is proportional to the squares of the spatial derivatives of the fluid viscosity's vector components. Assuming the fluid velocity around the particle is near zero, the square of the spatial derivative of this velocity would be exceedingly small, and thus $\Phi_{\mathrm{f}} \rightarrow 0$ (Slide 17.6). Within the boundary, where the particle is solid metal, there is no fluid motion, and therefore the viscous dissipation function is identically equal to zero.

Lastly, buried in $\frac{DT_{\mathrm{f}}}{Dt}$ is a convective term concerning the fluid velocity and temperature gradient. Once again, because the nanoparticle is solid, there will be no fluid velocity within the boundary; furthermore, outside the boundary it is assumed that the relative fluid velocity of the nanoparticle and medium is essentially zero as well.

Thus, the original heat transfer equation for fluids (6.1) has been simplified by a number of the earlier assumptions that remove terms, but one term has to be added to account for the heat generated in the nanoparticle as light is absorbed (Slide 17.7). Following all the reductions, and a single addition (6.1) takes the form

$$\rho_{\mathrm{f}} C_{\mathrm{f}} \frac{\partial T_{\mathrm{f}}}{\partial t} = \nabla \cdot (\mu_{\mathrm{f}} \nabla T_{\mathrm{f}}) + q_{\mathrm{gen}}, \tag{6.2}$$

where q_{gen} is the heat generated in the nanoparticle due to the absorption of the radiation. If the earlier is converted from the generalized differential form to

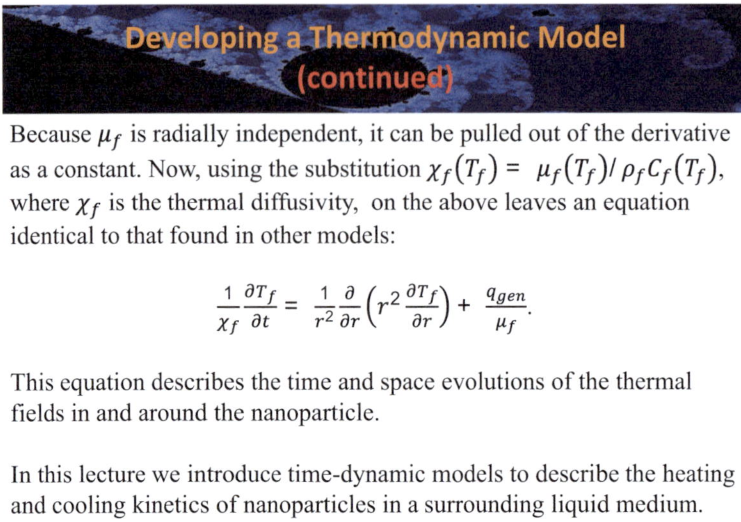

Slide 17.8 Developing a thermodynamic model (continued)

spherical coordinates, while assuming $T_f(r, \theta, \varphi, t) \approx T_f(r, t)$ (i.e., the temperature function is radially symmetric), then we have (Slide 17.7)

$$\rho_f C_f \frac{\partial T_f}{\partial t} = \frac{1}{r^2}\frac{\partial}{\partial r}\left(r^2 \mu_f \frac{\partial T_f}{\partial r}\right) + q_{gen}. \tag{6.3}$$

Because μ_f is radially independent, it can be pulled out of the derivative as a constant. Now, using the substitution $\chi_f(T_f) = \mu_f(T_f)/\rho_f C_f(T_f)$, where χ_f is the thermal diffusivity, on the earlier leaves an equation identical to that found in other models (Slides 17.8):

$$\frac{1}{\chi_f}\frac{\partial T_f}{\partial t} = \frac{1}{r^2}\frac{\partial}{\partial r}\left(r^2 \frac{\partial T_f}{\partial r}\right) + \frac{q_{gen}}{\mu_f}. \tag{6.4}$$

This equation describes the time and space evolutions of the thermal fields in and around the nanoparticle (Slide 17.8). In this chapter, we introduce the time dynamic models to describe the heating and cooling kinetics of nanoparticles in a surrounding liquid medium.

A main mechanism of heating a nanoparticle by radiation is the inverse bremsstrahlung absorption of photons by free electrons, followed by heat transferred from the electron gas into the material lattice and further to the surrounding medium (Slide 17.9). These processes do not occur instantaneously: time must be allowed for the cooling of the electrons and the heating of the lattice producing a difference between electron (T_e) and phonon (T_s) subsystem temperatures. The electron–phonon coupling process has several characteristic time scales [2]:

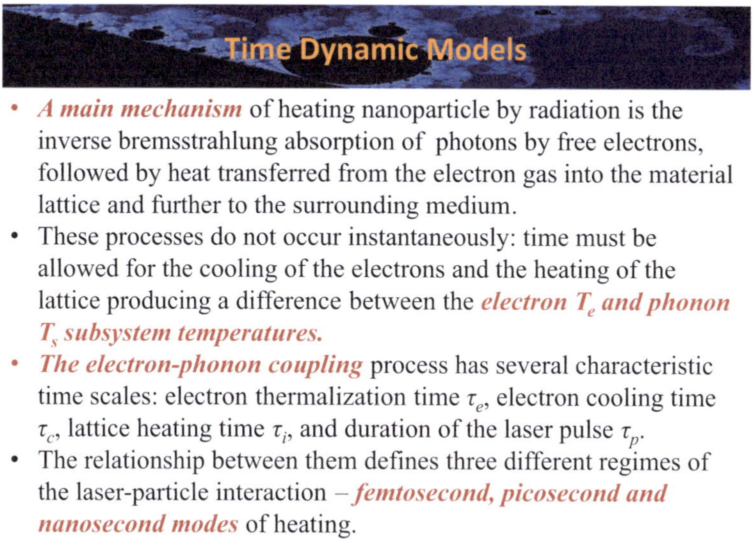

Slide 17.9 Time dynamic model

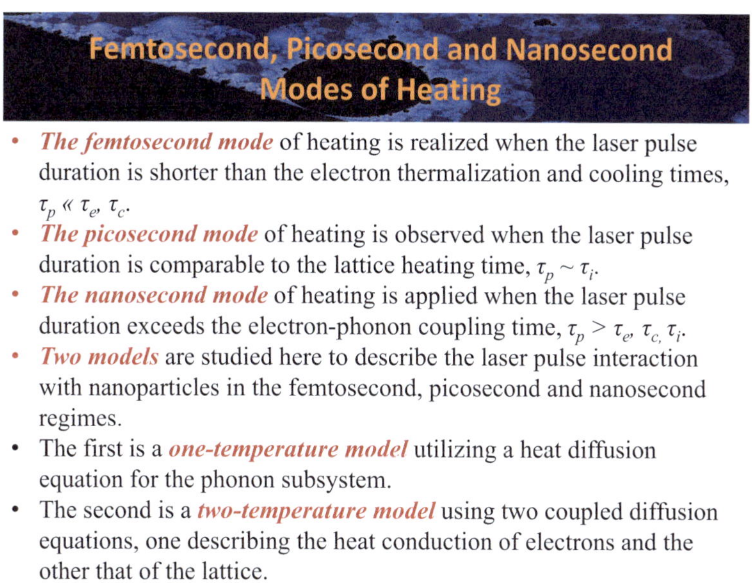

Slide 17.10 Femtosecond, picosecond, and nanosecond modes of heating

electron thermalization time τ_e, electron cooling time τ_c, lattice heating time τ_i, and duration of the laser pulse τ_p. The relationship between them defines three different regimes of the laser–particle interaction—femtosecond, picosecond, and nanosecond modes of heating (Slide 17.10):

- The femtosecond mode of heating is realized when the laser pulse duration is shorter than the electron thermalization and cooling times, $\tau_p \ll \tau_e$, τ_c.
- The picosecond mode of heating is observed when the laser pulse duration is comparable to the lattice heating time, $\tau_p \sim \tau_i$.
- The nanosecond mode of heating is applied when the laser pulse duration exceeds the electron–phonon coupling time, $\tau_p > \tau_e$, τ_c, τ_i.

Two models are studied here to describe the laser pulse interaction with nanoparticles in the femtosecond, picosecond, and nanosecond regimes [2] (Slide 17.10). The first is a one-temperature model utilizing a heat diffusion equation for the phonon subsystem and applying a uniform heating approximation throughout the particle volume. The second is a two-temperature model using two coupled diffusion equations, one describing the heat conduction of electrons and the other that of the lattice.

6.1.3 One-Temperature Model

During the interaction of a laser pulse of intensity I_0 and pulse duration τ_p with a metal nanoparticle of radius r_0, the laser energy is absorbed by free electrons and then transferred from the electron gas into the material lattice. In the one-temperature model (OTM), it is assumed that the electron heat transfer into the phonon subsystem is very fast, i.e., the electron and lattice temperatures are equal, $T_e = T_s$, at any instant in time [2] (Slide 17.11).

One-Temperature Model

- In the ***one-temperature model*** (OTM), it is assumed that the electron heat transfer into the phonon subsystem is very fast, i.e., ***the electron and lattice temperatures are equal***, $T_e = T_s$, at any instant in time.
- To investigate the time dependence of the temperature within the nanoparticle, we begin by assuming that the ***temperature within the nanoparticle is uniform***, T_0.
- The ***high thermal conductivity of gold*** (being three orders of magnitude better than that of the surrounding material) allows for heat within the particle to diffuse very quickly in relative terms.
- To study the ***time dependence of the temperature*** within the nanoparticle $T_s(t)$, we start by integrating the heat diffusion equation over the spherical volume of the nanoparticle,

$$\iiint_V \left(\rho_s C_s \frac{\partial T_s}{\partial t}\right) dV = \iiint_V \left(\frac{1}{r^2}\frac{\partial}{\partial r}\left(r^2 \mu_s \frac{\partial T_s}{\partial r}\right) + q_{gen}\right) dV$$

Slide 17.11 One-temperature model

One-Temperature Model (continued)

$$\frac{4\pi r_0^3}{3}\left(\rho_s C_s \frac{\partial T_s}{\partial t}\right) = 4\pi \int_{r=0}^{r_0} \left(\frac{1}{r^2}\frac{\partial}{\partial r}\left(r^2 \mu_s \frac{\partial T_f}{\partial r}\right)\right) r^2 dr + 4\pi \int_{r=0}^{r_0} q_{gen}\, r^2 dr$$

$$V_0\left(\rho_s C_s \frac{\partial T_s}{\partial t}\right) = 4\pi \left[r^2 \mu_s \frac{\partial T_s}{\partial r}\right]_{r=0}^{r_0} + 4\pi \int_{r=0}^{r_0} q_{gen}\, r^2 dr$$

$$V_0 \rho_s C_s \frac{\partial T_s}{\partial t} = 4\pi r_0^2 \left(\mu_s \frac{\partial T_s}{\partial r}\right) + 4\pi\left(\frac{1}{4} I_0 f(t) Q_{abs} r_0^2\right)$$

$$\frac{\partial T_s}{\partial t} = \frac{S_0}{V_0 \rho_s C_s}\left(\mu_s \frac{\partial T_s}{\partial r}\right) + \frac{S_0}{V_0 \rho_s C_s}\left(\frac{1}{4} I_0 f(t) Q_{abs}\right)$$

Slide 17.12 One-temperature model (continued)

To investigate the time dependence of the temperature within the nanoparticle we begin by assuming that the temperature is uniform, T_0 (Slide 17.11). This will be investigated further in the next model, but for now we say that the high thermal conductivity of gold (being three orders of magnitude better than that of the surrounding material) allows for heat within the particle to diffuse very quickly in relative terms. To study the time dependence of the temperature within the nanoparticle $T_s(t)$, we start at Eq. (6.3), replacing the subscript "f" for fluid by the subscript "s" for solid nanoparticle, and integrate over the spherical volume of the nanoparticle (Slides 17.11, 17.12, 17.13),

$$\iiint_V \left(\rho_s C_s \frac{\partial T_s}{\partial t}\right) dV = \iiint_V \left(\frac{1}{r^2}\frac{\partial}{\partial r}\left(r^2 \mu_s \frac{\partial T_s}{\partial r}\right) + q_{gen}\right) dV$$

$$\frac{4\pi r_0^3}{3}\left(\rho_s C_s \frac{\partial T_s}{\partial t}\right) = 4\pi \int_{r=0}^{r_0} \left(\frac{1}{r^2}\frac{\partial}{\partial r}\left(r^2 \mu_s \frac{\partial T_f}{\partial r}\right)\right) r^2 dr + 4\pi \int_{r=0}^{r_0} q_{gen} r^2 dr$$

$$V_0\left(\rho_s C_s \frac{\partial T_s}{\partial t}\right) = 4\pi \left[r^2 \mu_s \frac{\partial T_s}{\partial r}\right]_{r=0}^{r_0} + 4\pi \int_{r=0}^{r_0} q_{gen} r^2 dr$$

$$V_0 \rho_s C_s \frac{\partial T_s}{\partial t} = 4\pi r_0^2 \left(\mu_s \frac{\partial T_s}{\partial r}\right) + 4\pi\left(\frac{1}{4} I_0 f(t) Q_{abs} r_0^2\right)$$

One-Temperature Model (continued)

$$\frac{\partial T_s}{\partial t} = \frac{3}{r_0 \rho_s C_s}\left(\mu_s \frac{\partial T_s}{\partial r}\right) + \frac{3 I_0 f(t) Q_{abs}}{4 r_0 \rho_s C_s}.$$

Here, S_0, V_0 and Q_{abs} are the surface area, volume and absorption efficiency of the nanoparticle with radius r_0; μ_s is the thermal conductivity of a surrounding medium; I_0 is the radiation intensity; and $f(t)$ is the time profile of a radiation pulse.

Provided the thermal conductivity of the nanoparticle is large and the particle very small, the particle will diffuse heat within the boundary quickly enough to create a quasi-stationary heat exchange within a diffusion time frame on the order of $\tau_D \sim r_0^2/4\chi_s$. If the heating occurs using pulses of radiation with duration $\tau_p > \tau_D$, then the ***quasi-stationary approximation*** should suffice.

Slide 17.13 One-temperature model (continued)

$$\frac{\partial T_s}{\partial t} = \frac{S_0}{V_0 \rho_s C_s}\left(\mu_s \frac{\partial T_s}{\partial r}\right) + \frac{S_0}{V_0 \rho_s C_s}\left(\frac{1}{4} I_0 f(t) Q_{abs}\right)$$
$$\frac{\partial T_s}{\partial t} = \frac{3}{r_0 \rho_s C_s}\left(\mu_s \frac{\partial T_s}{\partial r}\right) + \frac{3 I_0 f(t) Q_{abs}}{4 r_0 \rho_s C_s} \tag{6.5}$$

Here, S_0, V_0 and Q_{abs} are the surface area, volume, and absorption efficiency of the nanoparticle with radius r_0; μ_s is the thermal conductivity of a surrounding medium; I_0 is the radiation intensity; and $f(t)$ is the time profile of a radiation pulse (Slide 17.13).

Provided the thermal conductivity of the nanoparticle is large and the particle very small, the particle will diffuse heat within the boundary quickly enough to create a quasi-stationary heat exchange within a diffusion time frame on the order of $\tau_D \sim r_0^2/4\chi_s$. If the heating occurs using pulses of radiation with duration $\tau_p > \tau_D$, then the quasi-stationary approximation should suffice (Slide 17.13).

As the thermal conductivity of the immediate surrounding material depends on the temperature difference between the sphere and medium, the relationship $\mu_s = \mu_\infty \left(\frac{T_s}{T_\infty}\right)^\alpha$ can be used [2] to find the following analytical expression for the temperature profile (Slide 17.14),

$$T_s(r) = T_\infty \left[1 + \frac{r_0}{r}\left(\left(\frac{T_s}{T_\infty}\right)^{\alpha+1} - 1\right)\right]^{1/(\alpha+1)}, \tag{6.6}$$

where μ_∞ is the heat conductivity of the surrounding medium at normal temperature T_∞, and the power exponent α is constant and is a property of the surrounding

One-Temperature Model (continued)

As the *thermal conductivity* of the immediate surrounding material depends on the temperature difference between the sphere and medium, the relationship $\mu_s = \mu_\infty \left(\frac{T_s}{T_\infty}\right)^\alpha$ can be used to find the following analytical expression for the temperature profile

$$T_s(r) = T_\infty \left[1 + \frac{r_0}{r}\left(\left(\frac{T_s}{T_\infty}\right)^{\alpha+1} - 1\right)\right]^{1\ (\alpha+1)},$$

where μ_∞ is the heat conductivity of the surrounding medium at normal temperature T_∞, and the power exponent α is constant and is a property of the surrounding medium. For a surrounding liquid medium similar to a biological environment, the power exponent is $\alpha \approx 1$.

Slide 17.14 One-temperature model (continued)

One-Temperature Model (continued)

From the above equation, an expression for $\mu_s \frac{\partial T_s}{\partial r}$ can be found:

$$\mu_s \frac{\partial T_s}{\partial r} = -\frac{\mu_\infty T_\infty r_0}{(\alpha+1)r^2}\left[\left(\frac{T_s}{T_\infty}\right)^{\alpha+1} - 1\right].$$

Substitution of this equation into the first term of the heat diffusion equation give

$$\frac{\partial T_s}{\partial t} = \frac{3I_0 f(t) Q_{abs}}{4r_0 \rho_s C_s} - \frac{3\mu_\infty T_\infty r_0}{(\alpha+1)r_0^2 \rho_s C_s}\left[\left(\frac{T_s}{T_\infty}\right)^{\alpha+1} - 1\right],$$

which is the OTM.

Here, the first term on the right side of the equation describes the heat generation into the nanoparticle due to radiation energy absorption by the nanoparticle. The second term describes the energy losses from the surface of the particle into the surrounding medium due to the heat diffusion process.

Slide 17.15 One-temperature model (continued)

medium. For a surrounding liquid medium similar to a biological environment, the power exponent is $\alpha \approx 1$. From Eq. (6.6), an expression for $\mu_s \frac{\partial T_s}{\partial r}$ can be found [2] (Slide 17.15):

$$\mu_s \frac{\partial T_s}{\partial r} = -\frac{\mu_\infty T_\infty r_0}{(\alpha + 1)r^2}\left[\left(\frac{T_s}{T_\infty}\right)^{\alpha+1} - 1\right]. \tag{6.7}$$

Substitution of Eq. (6.7) into the first term of the right side of Eq. (6.5) gives

$$\frac{\partial T_\mathrm{s}}{\partial t} = \frac{3I_0 f(t) Q_\mathrm{abs}}{4r_0 \rho_\mathrm{s} C_\mathrm{s}} - \frac{3\mu_\infty T_\infty r_0}{(\alpha + 1) r_0^2 \rho_\mathrm{s} C_\mathrm{s}}\left[\left(\frac{T_\mathrm{s}}{T_\infty}\right)^{\alpha+1} - 1\right], \tag{6.8}$$

which is called the one-temperature model (OTM) [2]. This model will be used to evaluate the temperatures reached inside the nanoparticle as a function of time. Here, the first term on the right side of the equation describes the heat generation into the nanoparticle due to radiation energy absorption by the nanoparticle. The second term describes the energy losses from the surface of the particle into the surrounding medium due to the heat diffusion process.

6.1.4 Two-Temperature Model

Free electrons with minimal capacity for heat inside the nanoparticles are the first to absorb the energy of incident radiation, rapidly attaining high temperatures T_e and transferring thermal energy to the material lattice, increasing the temperature T_s of the nanoparticle (Slide 17.16). These processes do not occur instantaneously: time must be allowed for the cooling of the electrons and the heating of the lattice. The electron cooling and lattice heating have time delays on the order of femtoseconds and picoseconds, respectively [2]. For the heating of nanoparticles by ultrashort

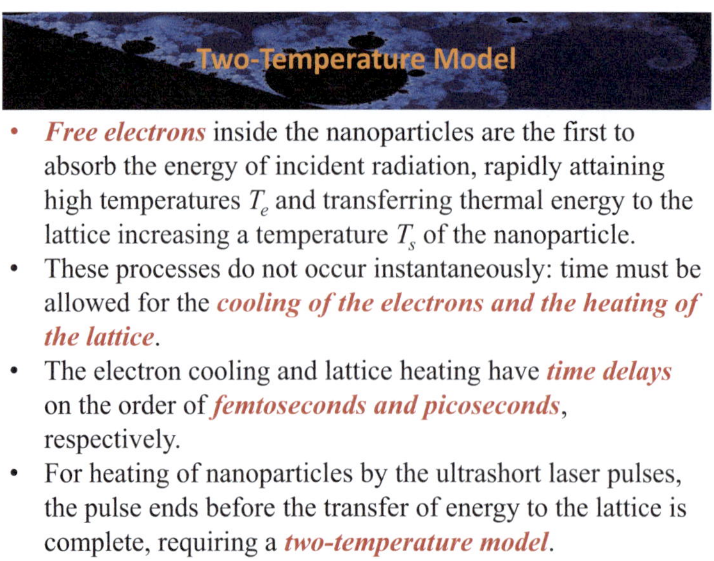

Slide 17.16 Two-temperature model

Two-Temperature Model (continued)

- In the two-temperature model (TTM), the temperature relaxation time and heat transfer depth into the nanoparticle can be modeled by two coupled diffusion equations, one describing the heat conduction of electrons and the other that in the lattice.
- Both equations are connected by a term that is proportional to the electron-phonon coupling constant γ and to the temperature difference between electrons and lattice.
- By adding a heat exchange term between the surface of the particle and the surrounding medium, the set of equations describing the heating of the electron and lattice subsystems and the energy transfer between the particles and surrounding medium has the form shown on the next slide.

Slide 17.17 Two-temperature model (continued)

laser pulses, the pulse ends before the transfer of energy to the lattice is complete, requiring a two-temperature model in order to describe the further conversion of energy from electron excitation to heat within the lattice system.

In the two-temperature model (TTM), the temperature relaxation time and heat transfer depth into the nanoparticle can be modeled by two coupled diffusion equations, one describing the heat conduction of electrons and the other that in the lattice (Slide 17.17). Both equations are connected by a term that is proportional to the electron–phonon coupling constant γ and to the temperature difference between electrons and lattice [2]. By adding a heat exchange term between the surface of the particle and the surrounding medium, the set of equations describing the heating of electron and lattice subsystems and the energy transfer between the particles and surrounding medium has the form (Slide 17.18)

$$C_{\mathrm{e}}(T_{\mathrm{e}})\frac{\partial T_{\mathrm{e}}}{\partial t} = -\frac{\partial Q(z)}{\partial z} - \gamma(T_{\mathrm{e}} - T_{\mathrm{s}}) + S, \tag{6.9}$$

$$C_{\mathrm{s}}(T_{\mathrm{s}})\frac{\partial T_{\mathrm{s}}}{\partial t} = \gamma(T_{\mathrm{e}} - T_{\mathrm{s}}) - \frac{3\mu_{\infty}T_{\infty}r_0}{(\alpha+1)r_0^2\rho_{\mathrm{s}}}\left[\left(\frac{T_{\mathrm{s}}}{T_{\infty}}\right)^{\alpha+1} - 1\right].$$

Here, $Q(z) = -k_{\mathrm{e}}\frac{\partial T_{\mathrm{e}}}{\partial z}$ is the heat flux, z is the direction perpendicular to the target surface, k_{e} is the electron thermal conductivity, $S = I_0 f(t)\,\alpha\exp(-\delta z)$ is the laser heating source term, δ is the material absorption coefficient, and $C_{\mathrm{e}}(T_{\mathrm{e}})$ and $C_{\mathrm{s}}(T_{\mathrm{s}})$ are the temperature-dependent heat capacities (per unit volume) of the electron and lattice subsystems. Comparative simulations of the two modeling approaches are performed in the next Lecture 18 for nanoparticle heating in the femtosecond, picosecond, and nanosecond regimes, thus providing an effective modeling method for further nanomedicine research to explore.

Two Temperature Model (continued)

$$C_e(T_e)\frac{\partial T_e}{\partial t} = -\frac{\partial Q(z)}{\partial z} - \gamma(T_e - T_s) + S,$$

$$C_s(T_s)\frac{\partial T_s}{\partial t} = \gamma(T_e - T_s) - \frac{3\mu_\infty T_\infty r_0}{(\alpha + 1)r_0^2 \rho_s}\left[\left(\frac{T_s}{T_\infty}\right)^{\alpha+1} - 1\right].$$

Here, $Q(z) = -k_e\frac{\partial T_e}{\partial z}$ is the heat flux; z is the direction perpendicular to the target surface; k_e is the electron thermal conductivity; $S = I_0 f(t)\,\alpha\,\exp(-\delta z)$ is the laser heating source term; δ is the material absorption coefficient; and $C_e(T_e)$ and $C_s(T_s)$ are the temperature-dependent heat capacities (per unit volume) of the electron and lattice subsystems.

Slide 17.18 Two-temperature model (continued)

Conclusions

- ***In conclusion***, we have introduced the ***heat-mass transfer equation*** to model the time evolution of the thermal fields in the nanoparticle.
- ***Photons*** emitted from the laser pulse ***are absorbed by free electrons*** within the nanoparticle via the inverse bremsstrahlung process and transferred to the lattice subsystem, and then to the surrounding medium.
- These processes do not occur instantaneously: time must be allowed for the cooling of the electrons and the heating of the lattice producing a difference between the ***electron T_e and phonon T_s subsystem*** temperatures.

Slide 17.19 Conclusions

In conclusion, we have introduced the heat-mass transfer equation to model the time evolution of the thermal fields in the nanoparticle (Slide 17.19). Photons emitted from the laser pulse are absorbed by free electrons within the nanoparticle via the inverse bremsstrahlung process and transferred to the lattice subsystem, and then to the surrounding medium. These processes do not occur instantaneously [2]: time must be allowed for the cooling of the electrons and the heating of the lattice, producing a difference between electron T_e and phonon T_s subsystem temperatures.

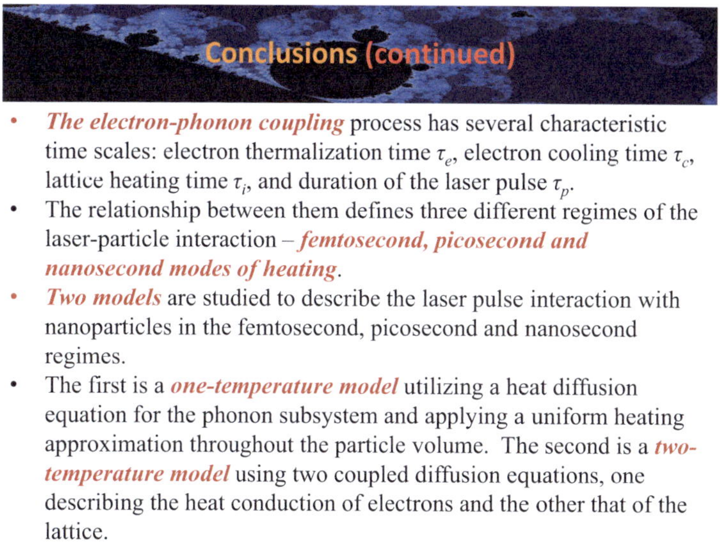

Slide 17.20 Conclusions (continued)

The electron–phonon coupling process has several characteristic time scales (Slide 17.20): electron thermalization time τ_{e}, electron cooling time τ_{c}, lattice heating time τ_{i}, and duration of the laser pulse τ_{p}. The relationship between them defines three different regimes of the laser–particle interaction—femtosecond, picosecond, and nanosecond modes of heating.

Two models are studied to describe the laser pulse interaction with nanoparticles in the femtosecond, picosecond, and nanosecond regimes [2] (Slide 17.20). The first is a one-temperature model utilizing a heat diffusion equation for the phonon subsystem and applying a uniform heating approximation throughout the particle volume. The second is a two-temperature model using two coupled diffusion equations, one describing the heat conduction of electrons and the other that of the lattice.

6.2 Lecture 18: Ultrashort Laser Pulse Heating of Nanoparticles in Femtosecond, Picosecond, and Nanosecond Modes (Slide 18.1)

- Introduction
- Comparison of Theoretical Approaches
  - *One-Temperature Model*
  - *Two-Temperature Model*

- Femtosecond Pulses
- Picosecond Pulses
- Nanosecond Pulses
- Conclusions and Summary

6.2.1 Introduction

In the previous lecture we introduced two models to describe the heating of nanoparticles in time (Slide 18.2). One is a one-temperature model (OTM) utilizing a heat diffusion equation for the phonon subsystem and applying a uniform heating approximation throughout the particle volume. The other is a two-temperature model (TTM) describing heat transfer from the electron subsystem to the lattice. In this lecture we perform the comparative simulations of the two modeling approaches to confirm that OTM is an appropriate approximation for nanoparticle heating in the femtosecond, picosecond, and nanosecond regimes, thus providing an effective modeling method for further nanomedicine research to explore. This section contains material reproduced from our publications [2–4].

The application of ultrashort laser pulse thermal-based killing of abnormal cells (e.g., cancer cells) targeted with absorbing nanoparticles (e.g., solid gold nanospheres, nanoshells, or nanorods) has become a major area of research (see, for example, Ref. [3] and references therein). High laser energies in ultrashort pulses can be concentrated on an extremely small biological mass infused with metallic nanoparticles, and the heat gained by the particles can be used as a protein

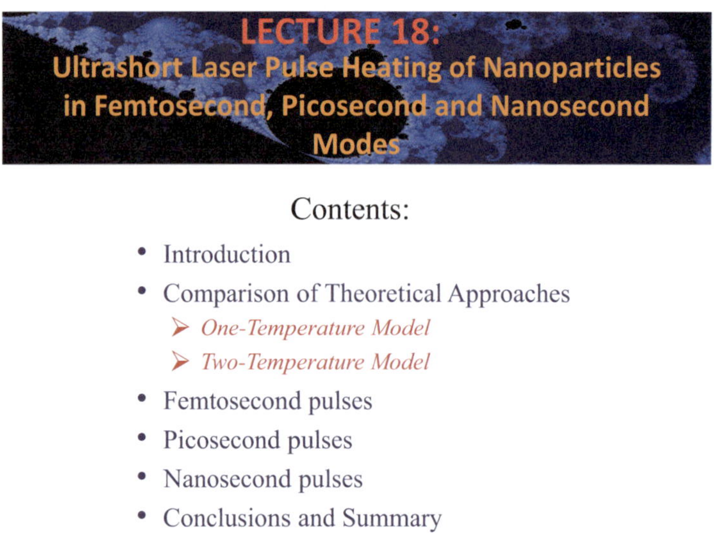

Slide 18.1 Contents

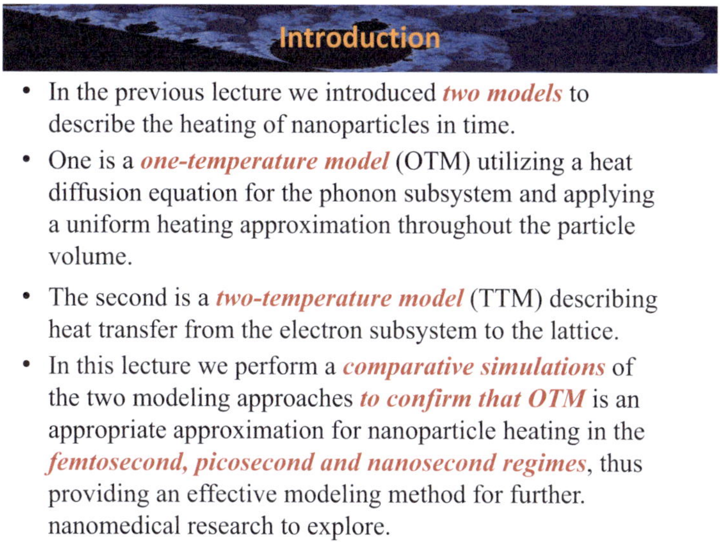

Slide 18.2 Introduction

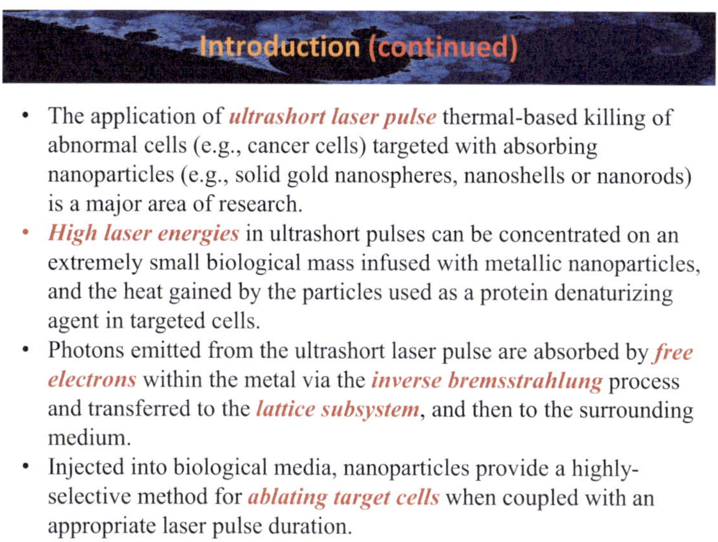

Slide 18.3 Introduction (continued)

denaturizing agent in targeted cells (Slide 18.3). Photons emitted from thc ultrashort laser pulse are absorbed by free electrons within the metal via the inverse bremsstrahlung process and transferred to the lattice subsystem, and then to the surrounding medium. Injected into biological media, nanoparticles provide a highly selective method for ablating target cells when coupled with the appropriate laser pulse duration.

Introduction (continued)

- ***Ultrashort pulses***, specifically those in the ***femtosecond and picosecond ranges***, impose several challenges in modeling material response.
- ***Free electrons*** with minimal capacity for heat are the first to absorb energy, rapidly attaining high temperatures and transferring thermal energy to the material lattice.
- These processes do not occur instantaneously: ***time must be allowed for the cooling of the electrons and the heating of the lattice***.
- The electron cooling and lattice heating have time delays ***on the order of femtoseconds and picoseconds***, respectively. Ultrashort laser pulses end before the transfer of energy to the lattice is complete, requiring ***two-temperature model*** in order to describe the further conversion of energy from electron excitation to heat within the lattice system.
- In this lecture we demonstrate that a simpler, ***one-temperature model*** utilizing the uniform heating approximation is appropriate for understanding ***ultrashort pulse interactions*** with nanoparticles.

Slide 18.4 Introduction (continued)

Traditional knowledge of laser–nanoparticle interactions has necessitated specialized models for each case, dependent upon the laser pulse duration [2]. In the previous lecture we introduced a dual-temperature model for the ultrashort laser pulse mode, calculating electron and lattice subsystem temperatures. Pulses of longer duration are modeled using a uniform heating one-temperature model (OTM) derived in the previous lecture. The OTM is an appropriate approximation for pulse durations greatly exceeding the electron–phonon coupling time. This approximation is reasonable for particles sizes not much larger than the laser wavelength, which is completely applicable for the heating of nanoparticles.

Ultrashort pulses, specifically those in the femtosecond and picosecond ranges, impose several challenges in modeling material response (Slide 18.4). Free electrons with minimal capacity for heat are the first to absorb energy, rapidly attaining high temperatures and transferring thermal energy to the material lattice. These processes do not occur instantaneously: time must be allowed for the cooling of the electrons and the heating of the lattice. As shown in [4], the electron cooling and lattice heating have time delays on the order of femtoseconds and picoseconds, respectively. Ultrashort laser pulses end before the transfer of energy to the lattice is complete, requiring the two-temperature model in order to describe the further conversion of energy from electron excitation to heat within the lattice system.

In this lecture we demonstrate that a simpler, one-temperature model utilizing the uniform heating approximation is appropriate for understanding ultrashort laser pulse interactions with metal nanoparticles [2] (Slide 18.4). The approximation may be used in this situation due to the extremely small size of the nanoparticles in comparison to the wavelength of laser radiation. Using this idea, the time delay between the electron and lattice interactions will be bypassed, and the simpler model will ultimately yield results similar to the two-temperature model.

Comparison of Theoretical Approaches

- ***One-Temperature Model.*** After ***integration of the heat diffusion equation*** over the volume of a nanoparticle performed in the previous lecture in a uniform heating approximation and ***adding a particle evaporation term***, the OTM takes the following form

$$\frac{dT_s}{dt} = \frac{3Q_{abs}I_0 f(t)}{4r_0 C_s(T_s)\rho_s} - \frac{3\mu_\infty T_s}{(\alpha+1)r_0^2 C_s(T_s)\rho_s}\left[\left(\frac{T_s}{T_\infty}\right)^{\alpha+1} - 1\right] + \frac{3L}{r_0 C_s(T_s)}\frac{dr_0}{dt}.$$

- The ***first term*** on the right side of the equation describes the heat generation into the spherical volume due to laser energy absorption by the nanoparticle.
- The ***second term*** describes the energy losses from the surface of the particle into the surrounding medium due to the heat diffusion process.
- The ***last term*** describes the energy losses due to the evaporation of the particle.

Slide 18.5 Comparison of theoretical approaches

6.2.2 Comparison of Theoretical Approaches

One-Temperature Model (Slide 18.5). After integration of the heat diffusion equation over the volume of a nanoparticle performed in the previous lecture with a uniform heating approximation and adding a particle evaporation term, the OTM takes the following form [4]:

$$\frac{dT_{\mathrm{s}}}{dt} = \frac{3Q_{\mathrm{abs}}I_0 f(t)}{4r_0 C_{\mathrm{s}}(T_{\mathrm{s}})\rho_{\mathrm{s}}} - \frac{3\mu_\infty T_{\mathrm{s}}}{(\alpha+1)r_0^2 C_{\mathrm{s}}(T_{\mathrm{s}})\rho_{\mathrm{s}}}\left[\left(\frac{T_{\mathrm{s}}}{T_\infty}\right)^{\alpha+1} - 1\right] + \frac{3L}{r_0 C_{\mathrm{s}}(T_{\mathrm{s}})}\frac{dr_0}{dt}. \tag{6.10}$$

Here once again, the first term on the right side of the equation describes the heat generation into the spherical volume due to laser energy absorption by the nanoparticle. The second term describes the energy losses from the surface of the particle into the surrounding medium due to the heat diffusion process. The last term describes the energy losses due to the evaporation of the particle. This evaporation depends on the laser pulse characteristics and particle properties, and it can be realized in five different regimes: free molecular, convective, diffusive, gas dynamic, and explosive [5] modes of evaporation. For example, within the approximation of free molecular flow, the evaporation term in Eq. (6.10) can be written as [6]

Comparison of Theoretical Approaches (continued)

- ***Two-Temperature Model***. For TTM we introduced in the previous lecture ***two coupled diffusion equations***, one describing the heat conduction of electrons and the other that in the lattice.
- Both equations are connected by a term that is proportional to the ***electron-phonon coupling*** constant γ and to the ***temperature difference*** between the electrons and lattice.
- The expressions for the temperature dependence of the heat capacities, as well as the values for the electron-phonon coupling constant γ and material data, are listed in the ***table*** on the next slide.

$$C_e(T_e)\frac{\partial T_e}{\partial t} = -\frac{\partial Q(z)}{\partial z} - \gamma(T_e - T_s) + S,$$

$$C_s(T_s)\frac{\partial T_s}{\partial t} = \gamma(T_e - T_s) - \frac{3\mu_\infty T_\infty r_0}{(\alpha+1)r_0^2\rho_s}\left[\left(\frac{T_s}{T_\infty}\right)^{\alpha+1} - 1\right].$$

Slide 18.6 Comparison of theoretical approaches (continued)

$$4\pi r_0^2 \frac{dr_0}{dt}\rho_s = -\eta 4\pi r_0^2 V_s(T_s)\rho_v(T_s)\,, \tag{6.11}$$

where η is the accommodation coefficient, and $V_s(T_s)$ and $\rho_s(T_s)$ are, respectively, the average velocity and density of the vapor at temperature T_s. If the heating of the nanoparticle occurs below the temperature of the phase transition in the particle material, the third term on the right side of Eq. (6.10) can be neglected.

Two-Temperature Model (Slide 18.6). For TTM, in the previous lecture we introduced two coupled diffusion equations (6.9), one describing the heat conduction of electrons and the other that in the lattice [2]. Both equations are connected by a term that is proportional to the electron–phonon coupling constant γ and to the temperature difference between electrons and lattice. The expressions for the temperature dependence of the heat capacities, as well as the values for the electron–phonon coupling constant γ and material data, are listed in Table 6.1 (Slide 18.7).

Comparative simulations using the models described earlier have been performed for laser heating of a gold nanoparticle with radius $r_0 = 20$ nm in a surrounding medium of water [2]. The same set of input data presented in Table 6.1 was used for the calculations in both models. The temperature dependences of the electron heat capacity for gold and specific heat and thermal conductivity for water were obtained by interpolating the experimental data available in the literature (references are listed in Table 6.1). As can be seen from the table, the electron heat capacity is much less than the lattice heat capacity, and therefore electrons can be heated to very high transient temperatures. Then the evolution of the electron

Table 6.1 Input parameters used for simulations in both models. Reproduced from the following open access publication: [2]

Parameters	Magnitude and units	Reference
Laser pulse shape	$f(t) = \exp\left[-(at-b)^2 - (at-c)^2\right], \quad a,\ b,\ c = \text{const}]$	
Energy density	$E = 10^{-3}\ \text{J/cm}^2$	
Volume density of the gold	$\rho_0 = 0.0193\ \text{kg/cm}^3$	
Electron heat capacity (gold)	$C_e(T_e) = aT_e^4 - bT_e^3 + cT_e^2 - dT_e + e, \left(\frac{\text{J}}{\text{K}^2\ \text{cm}^3}\right)$ $a = 6.0 \times 10^{-17}$ $b = 3.0 \times 10^{-12}$ $c = 5.07 \times 10^{-8}$ $d = 6.7589 \times 10^{-5}$ $e = 7.207 \times 10^{-2}$	[7]
Electron–phonon coupling constant for gold	$\gamma = 2.27 \times 10^{10} \frac{\text{W}}{\text{K cm}^3}$	[7]
Gold specific heat	$C = 129 \frac{\text{J}}{\text{K kg}}$	
Specific heat of water	$C_w(T_s) = a(1 + b(T_s - 293\ \text{K})) \left(\frac{\text{J}}{\text{K kg}}\right)$ $a = 4182.0$ $b = 1.016 \times 10^{-4}$	
Radius of gold nanoparticle	$r_0 = 20$ nm	
Absorption efficiency of gold nanoparticle	$k_{abs} = 4.0195$ ($r_0 = 20$ nm and $\lambda = 528$ nm)	[3]
Absorption coefficient of gold nanoparticle	$\alpha = 8.736 \times 10^5\ \text{cm}^{-1}$ ($r_0 = 20$ nm and $= 528$ nm)	[4]
Power exponent	$s = 1.0$	
Thermal conductivity of water	$\mu_0(T_s) = a(1 + b(T_s - \text{T}_\infty)) \left(\frac{\text{W}}{\text{K cm}}\right)$ $a = 0.00597$ $b = 1.78 \times 10^{-3}$	

temperature involves energy transfer to the lattice and energy losses duc to the electron heat transport into the target. The electron–phonon coupling process has several characteristic time scales: electron thermalization time τ_e, electron cooling time τ_c, lattice heating time τ_i, and duration of the laser pulse τ_p. The relationship between them defines three different regimes of the laser–metal interaction—femtosecond, picosecond, and nanosecond modes of heating.

6.2.3 Femtosecond Pulses

Here we reproduce our results published in [2–4]. TTM and OTM have been solved numerically to predict the time dependence of the electron and lattice temperatures

Input Parameters Used for Simulations in Both Models

Parameters	Magnitude and Units	Reference
Laser pulse shape	$f(t) = \exp[-(at-b)^2 - (at-c)^2]$, $a, b, c = const$	
Energy density	$E = 10^{-3}$ J/cm^2	
Volume density of the gold	$\rho_0 = 0.0193\ kg/cm^3$	
Electron heat capacity (gold)	$C_e(T_e) = aT_e^4 - bT_e^3 + cT_e^2 - dT_e + e, \left(\frac{J}{K^2\,cm^3}\right)$ $a = 6.0 \times 10^{-17}$ $b = 3.0 \times 10^{-12}$ $c = 5.07 \times 10^{-8}$ $d = 6.7589 \times 10^{-5}$ $e = 7.207 \times 10^{-2}$	[5]
Electron-phonon coupling constant for gold	$\gamma = 2.27 \times 10^{10} \frac{W}{K\,cm^3}$	[5]
Gold specific heat	$C = 129 \frac{J}{K\,kg}$	
Specific heat of water	$C_w(T_s) = a(1 + b(T_s - 293\,K)), \quad \left(\frac{J}{K\,kg}\right)$ $a = 4182.0$ $b = 1.016 \times 10^{-4}$	
Radius of gold nanoparticle	$r_0 = 20$ nm	
Absorption efficiency of gold nanoparticle	$k_{abs} = 4.0195$ ($r_0 = 20$ nm *and* $\lambda = 528$ nm)	[1]
Absorption coefficient of gold nanoparticle	$\alpha = 8.736 \times 10^5\ cm^{-1}$ ($r_0 = 20$ nm *and* = 528 nm)	[2]
Power exponent	$s = 1.0$	
Thermal conductivity of water	$\mu_0(T_s) = a(1 + b(T_s - T_\infty)), \quad \left(\frac{W}{K\,cm}\right)$ $a = 0.00597$ $b = 1.78 \times 10^{-3}$	

Slide 18.7 Input parameters used for simulations in both models

Femtosecond Pulses

- ***TTM and OTM*** have been solved numerically to predict the time dependence of the electron and lattice temperatures ***in the femtosecond mode*** when the laser pulse duration is shorter than the electron thermalization and cooling times, $\tau_p \ll \tau_e, \tau_c$.
- The calculations were performed for a ***laser pulse energy density*** of $E = 1.0$ mJ/cm^2 and pulse duration of $\tau_p = 60$ fs.
- The ***time profile*** of the femtosecond laser pulse given in ***Table 6.1*** and shown in ***Figure 6.1(b)*** (solid curve) corresponds to the experimentally-observed output from an amplified Ti:sapphire laser (Legend-HE from Coherent Inc.).
- The ***laser flux*** is chosen at the level 1.0 mJ/cm^2 to provide cell lethality due to thermal ablation by a heated nanoparticle during a single laser pulse.
- The ***results of the simulations*** for the heating of a gold nanoparticle by a femtosecond laser pulse are shown in ***Figures 6.1 and 6.2*** on the next slide.

Slide 18.8 Femtosecond pulses

in the femtosecond mode when the laser pulse duration is shorter than the electron thermalization and cooling times, $\tau_p \ll \tau_e$, τ_c (Slide 18.8). The calculations were performed for a laser pulse energy density of $E = 1.0$ mJ/cm^2 and pulse duration of $\tau_p = 60$ fs. The time profile of the femtosecond laser pulse given in Table 6.1 and shown in Fig. 6.1b (solid curve) corresponds to the experimentally observed output from an amplified Ti:sapphire laser (Legend-HE from Coherent Inc.). The laser flux

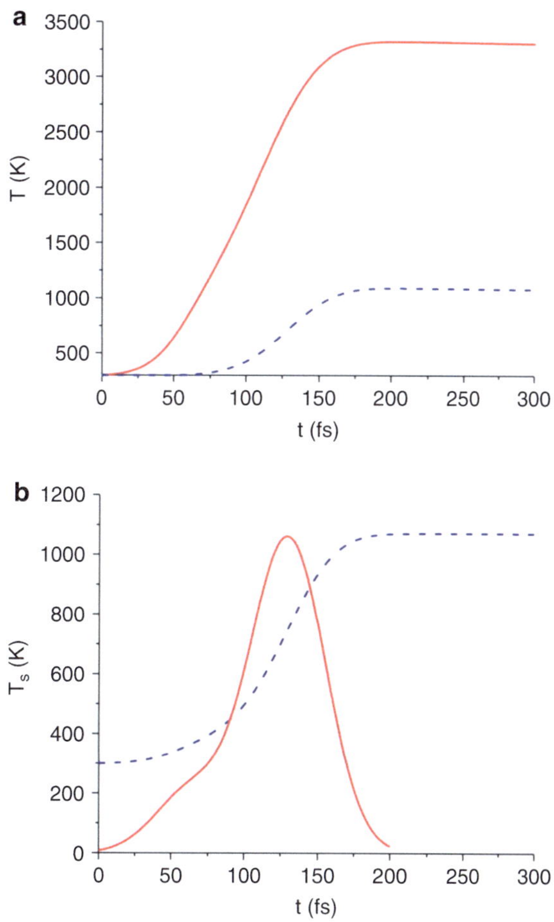

Fig. 6.1 (**a**) Electron (*solid curve*) and lattice (*dashed curve*) temperature evolutions on the femtosecond time scale for a gold nanoparticle predicted by TTM. (**b**) Evolution of the nanoparticle temperature (*dashed curve*) after the femtosecond laser pulse, predicted by OTM, and laser pulse shape (*solid curve*). Reproduced from the following open access publication: [2]

is chosen at the level 1.0 mJ/cm^2 to provide the cell lethality due to thermal ablation by a heated nanoparticle during a single laser pulse. The results of the simulations for the heating of a gold nanoparticle by a femtosecond laser pulse are shown in Figs. 6.1 and 6.2. Figure 6.1a displays the results obtained by TTM, and Fig. 6.1b presents the results of OTM simulations.

As follows from Fig. 6.1a, thermal equilibrium among the excited electrons with equilibrium temperature $T_e \approx 3300$ K for a given laser fluence is established within 175 fs (Slide 18.9). We should note that the equilibrium temperature for electrons is reached long after the end of a laser pulse, which had a duration of 60 fs. The electrons remain in the thermal equilibrium state from several hundred femtoseconds up to 1 ps (see Fig. 6.3a). Then the electrons cool exclusively by coupling to the lattice, resulting in a linear decay of the electron temperature during the first 10 ps

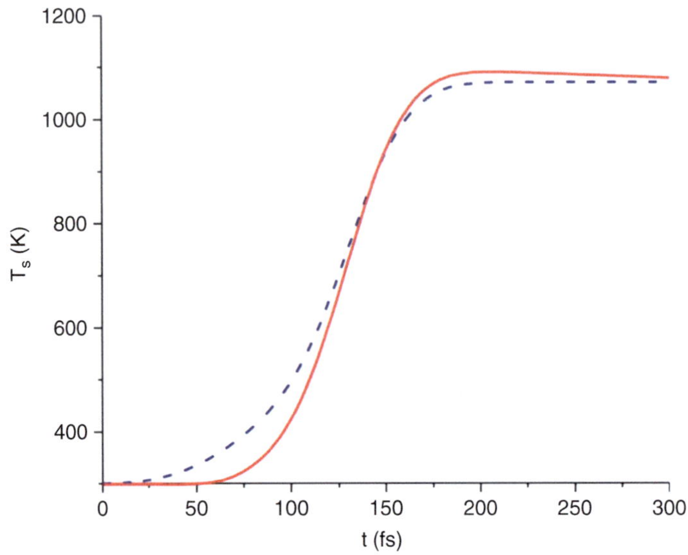

Fig. 6.2 Comparison of nanoparticle temperature evolutions after the 60-fs laser pulse as predicted by TTM (*solid curve*) and OTM (*dashed curve*). Reproduced from the following open access publication: [2]

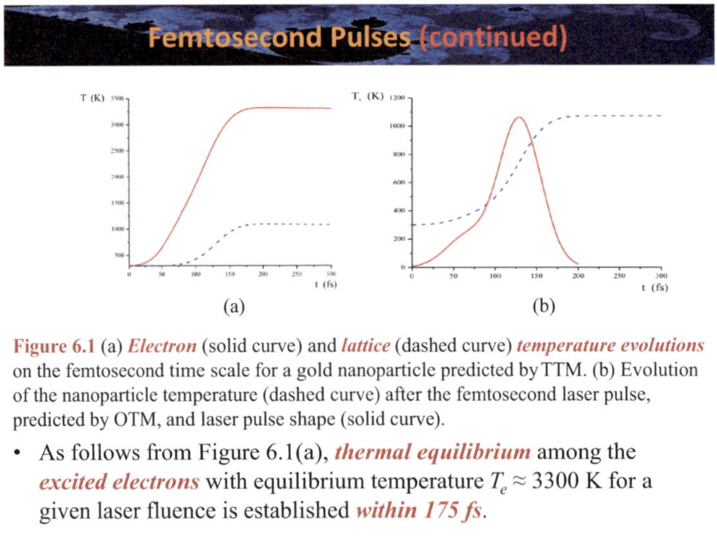

Slide 18.9 Femtosecond pulses (continued)

(Fig. 6.3a). Our simulations agreed with the electron relaxation time measured in [8] for femtosecond pulse excitation of a DNA-modified gold nanoparticle.

The slow rate of electron heat diffusion into the phonon subsystem on the femtosecond time scale results in a delay of about 100 fs in the heating of the

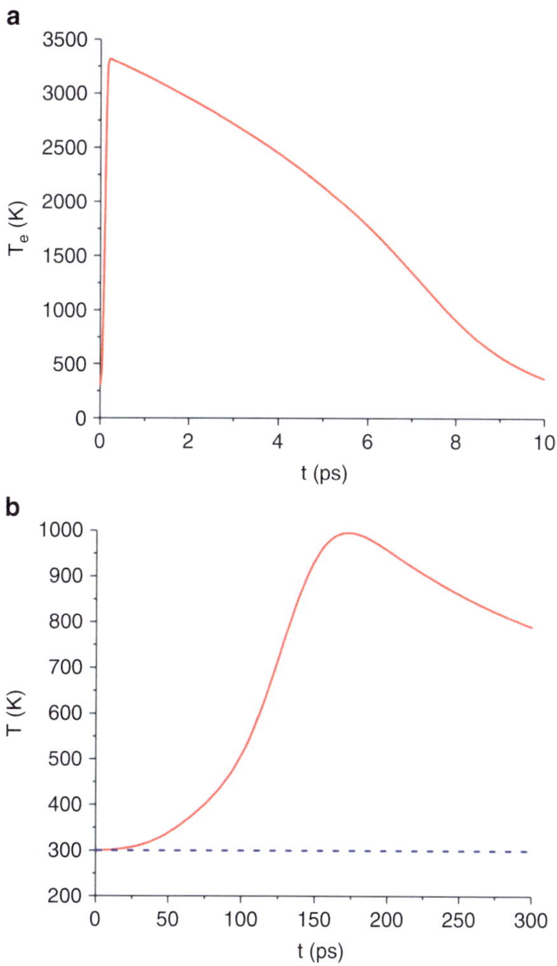

Fig. 6.3 (**a**) Electron temperature relaxation on the picosecond time scale. (**b**) Temperature–time distributions for a gold nanoparticle predicted by OTM (*solid curve*) and TTM (*dashed curve*) after the 60-ps laser pulse. Reproduced from the following open access publication: [2]

bulk sample (dashed curve in Fig. 6.1a). Once the electron thermal equilibrium is established, a hot electron bath raises the temperature of the cold lattice up to 1090 K for a given laser energy density $E = 1$ mJ/cm^2.

The results of heating the gold nanoparticle obtained by OTM are demonstrated in Fig. 6.1b. This figure also shows the femtosecond laser pulse time profile used in the calculations. Comparative simulations for the evolution of the nanoparticle temperature using both models under the same conditions are presented in Fig. 6.2 (Slide 18.10). It follows from these simulations that both models demonstrate the same scenario in the heating kinetics of a metal nanoparticle by a femtosecond laser pulse. Both models reveal approximately a 100 fs time delay in the heating of the particle, followed by a maximum lattice temperature of around

Femtosecond Pulses (continued)

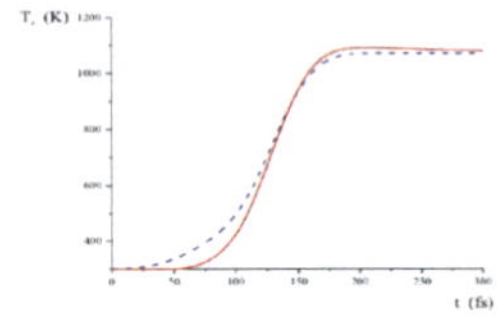

Figure 6.2 Comparison of nanoparticle temperature evolutions after the 60-fs laser pulse as predicted by TTM (solid curve) and OTM (dashed curve).

- It follows from these simulations that ***both models demonstrate the same scenario in the heating kinetics*** of a metal nanoparticle by a femtosecond laser pulse. Both models reveal approximately a 100 fs time delay in the heating of the particle, followed by a maximum lattice temperature of around 1090 K within 175 fs after the end of a laser pulse. Even the maximum values of the particle temperature predicted by both models are the same.
- The ***saturation parts*** in the lattice temperature curves are explained by negligibly-small heat diffusion from the surface of the nanoparticle into the surrounding medium on the femtosecond time scale. A comparison of these two models shows that the simpler OTM gives the same results as the more precise TTM.

Slide 18.10 Femtosecond pulses (continued)

1090 K within 175 fs after the end of a laser pulse. Even the maximum values of the particle temperature predicted by both models are the same (see Fig. 6.2).

The saturation parts in the lattice temperature curves are explained by negligibly small heat diffusion from the surface of the nanoparticle into the surrounding medium on the femtosecond time scale (Slide 18.10). A slight difference in the slopes of the temperature curves within the first 100 fs of heating occurs due to the assumption made in OTM that the electron heat transfer into the lattice subsystem is very fast. Because of this, the particle temperature in OTM promptly follows electron thermal behavior. A comparison of these two models shows that the simpler OTM gives the same results as the more precise TTM. Thus, OTM provides an adequate description of the laser heating of nanoparticles in the femtosecond regime.

6.2.4 *Picosecond Pulses*

In this mode, the constants *a*, *b*, and *c* in the time profile for the laser pulse have been chosen to provide the laser pulse width of 60 ps at FWHM with the same pulse shape listed in Table 6.1 and shown in Fig. 6.1b. Here we reproduce our results published in [2–4]. As can be seen from Figs. 6.1a and 6.3a, in the picosecond regime the electron thermalization time τ_e and electron cooling time τ_c are much less than the duration τ_p of the 60 ps laser pulse (Slide 18.11). Hence, during the first 10 ps the electron subsystem has already been completely cooled (Fig. 6.3a), and the electron temperature above the ambient no longer exists for the

Picosecond Pulses

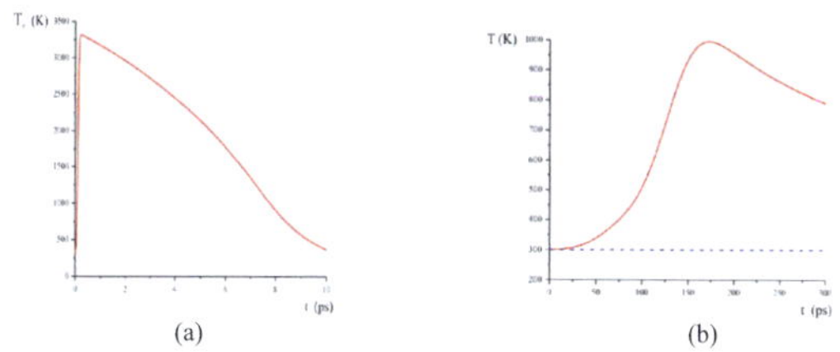

Figure 6.3 (a) Electron temperature relaxation on the picosecond time scale. (b) Temperature-time distributions for a gold nanoparticle predicted by OTM (solid curve) and TTM (dashed curve) after the 60-ps laser pulse.

- As can be seen from Figures 6.1(a) and 6.3(a), in the ***picosecond regime*** the electron thermalization time τ_e and electron cooling time τ_c are much less than the duration τ_p of the 60-ps laser pulse.
- Hence, during the first 10 ps, the ***electron subsystem*** has already been completely ***cooled*** (Figure 6.3(a)), and the electron temperature above the ambient temperature no longer exists for the considered picosecond mode of heating. This is confirmed by our calculations presented in Figure 6.3(b) (dashed curve).

Slide 18.11 Picosecond pulses

Picosecond Pulses (continued)

- Thus, ***TTM*** provides a very good ***approximation for the femtosecond mode*** as soon as the electron temperature exists, but it fails to describe the laser heating of nanoparticles for longer pulse durations in the picosecond and nanosecond regimes.
- Opposite to TTM, ***OTM*** describes the ***picosecond heating kinetics*** very well.
- A typical ***time evolution*** of the particle temperature predicted by OTM on the picosecond time scale is displayed in Figure 6.3(b) (solid curve).
- The ***main feature*** is the appearance of ***heat lost*** from the surface of the nanoparticle into the surrounding medium ***on the picosecond time scale***. After about 200 ps, cooling of the nanoparticle begins due to heat diffusion into the water. The maximum temperature reached by a 20-nm gold particle for the given laser pulse is 995 K. This temperature is sufficient to initiate any thermal killing mechanisms in cancer cells.

Slide 18.12 Picosecond pulses (continued)

considered picosecond mode of heating. This is confirmed by our calculations presented in Fig. 6.3b (dashed curve). Thus, TTM provides a very good approximation for the femtosecond mode as soon as the electron temperature exists, but it fails to describe the laser heating of nanoparticles for longer pulse durations in the picosecond and nanosecond regimes (Slide 18.12).

Opposite to TTM, OTM describes the picosecond heating kinetics very well. A typical time evolution of the particle temperature predicted by OTM on the picosecond time scale is displayed in Fig. 6.3b (solid curve) for the same material constants listed in Table 6.1. The main feature is the appearance of heat lost from the surface of the nanoparticle into the surrounding medium on the picosecond time scale (Slide 18.12). After about 200 ps, cooling of the nanoparticle begins due to heat diffusion into the water. The maximum temperature reached by a 20-nm gold particle for the given laser pulse is 995 K. This temperature is sufficient to initiate any thermal killing mechanisms in cancer cells.

6.2.5 Nanosecond Pulses

For laser heating of metal nanoparticles in the nanosecond regime, the characteristic lattice heating time τ_i is much smaller than the laser pulse duration: $\tau_p \gg \tau_i$ (Slide 18.13). This means that the temperature inside the nanoparticle is nearly uniform over the whole particle on the time scale of the laser pulse duration τ_p. In this case, the electron and lattice temperatures are equal ($T_e = T_s$), so that the homogeneous heating of the particle and quasi-steady heat exchange with the surrounding medium can be described by just OTM. The characteristic lattice heating time τ_i required for the formation of a quasi-stationary temperature profile across the nanoparticle can be estimated from the formula $\tau_i = r_0^2/4\chi$, where r_0 is the particle radius and χ the thermal diffusivity of the particle material. For a gold nanoparticle ($\chi = 1.18 \times 10^{-4}$ m^2/s) with radius $r_0 = 20$–30 nm, the lattice heat

Nanosecond Pulses

- For laser heating of metal nanoparticles in the ***nanosecond regime***, the characteristic lattice heating time τ_i is much smaller than the laser pulse duration: $\tau_p \gg \tau_i$.
- This means that the ***temperature inside*** the nanoparticle is nearly ***uniform*** over the whole particle on the time scale of the laser pulse duration τ_p.
- In this case, the ***electron and lattice temperatures are equal***, $T_e = T_s$, so that the homogeneous heating of the particle and quasi-steady heat exchange with the surrounding medium can be described by just OTM.
- The ***characteristic lattice heating time*** τ_i required for the formation of a quasi-stationary temperature profile across the nanoparticle can be estimated from the formula

$$\tau_i = r_0^2/4\chi,$$

where r_0 is the particle radius and χ the thermal diffusivity of the particle material.

Slide 18.13 Nanosecond pulses

Nanosecond Pulses (continued)

- ***Sample calculations*** have been carried out using OTM for gold nanoparticles with radii r_0 = 30-35 nm in different surrounding biomedia for an incident laser pulse of energy density E = 10 mJ/cm^2 and pulse duration τ_p = 8 ns.
- The ***laser pulse profile*** and duration 8 ns have been chosen to be close to those used in the experiments. We should note that 10 mJ/cm^2 is comparable to the laser fluence currently used in the photothermolysis of cancer cells.
- The ***kinetics of heating and cooling*** the gold nanoparticle are demonstrated in Figure 6.4, where (a) illustrates the time dynamics of laser heating of a 30-nm gold particle in different biological media: blood, human prostate, tumor and fat.
- Figure 6.4(b) shows results of ***thermal calculations*** for a 35-nm gold particle, which is heated and cooled in water at different heat transfer rates α.

Slide 18.14 Nanosecond pulses (continued)

diffusion time is $\tau_i \sim 2 \times 10^{-12}$ $s \ll \tau_p \sim 10^{-8}$ s. Here we reproduce our results published in [2–4].

Sample calculations have been carried out using OTM for gold nanoparticles with radii $r_0 = 30$–35 nm in different surrounding biomedia for an incident laser pulse of energy density $E = 10$ mJ/cm^2 and pulse duration $\tau_p = 8$ ns with the time profile shown in Fig. 6.1b (Slide 18.14). The laser pulse profile and duration 8 ns have been chosen to be close to those used in the experiments. The laser flux chosen is ten times higher than in the regimes considered earlier to provide approximately the same maximum nanoparticle temperature as observed for femtosecond and picosecond laser heating. We should note that 10 mJ/cm^2 is comparable to the laser fluence currently used in the photothermolysis of cancer cells [3]. The kinetics of heating and cooling the gold nanoparticle are demonstrated in Fig. 6.4, where (a) illustrates the time dynamics of laser heating of a 30-nm gold particle in different biological media: blood, human prostate, tumor, and fat. The thermophysical characteristics of gold and biological surrounding media for different temperatures are listed in Table 6.2 (Slide 18.15). Figure 6.4b shows results of thermal calculations for a 35-nm gold particle, which is heated and cooled in water at different heat transfer rates α.

It follows from these calculations that during the laser pulse duration, the transfer of heat from the nanoparticle into the surrounding medium is slight, and the particle rapidly reaches a high temperature (Slide 18.16). The heating rate is about 10^{12} Ks^{-1}. The temperature of the particle continues to rise even after the end of the laser pulse. The highest temperature, 770 K, for a given laser pulse fluence is observed for the heating time of 13.5 ns, when the laser pulse has already degraded (see Fig. 6.4a). After that time, the transfer of heat from the particle to the

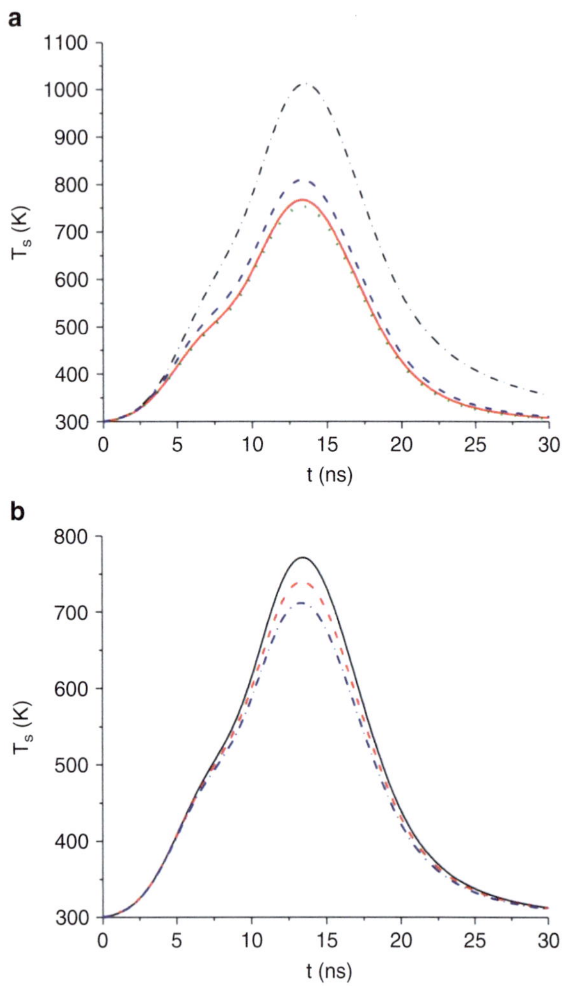

Fig. 6.4 Kinetics of heating and cooling of a gold nanoparticle by a nanosecond laser pulse of energy density 10 mJ/cm^2 and duration 8 ns. Calculations have been made by using OTM. (**a**) Illustration of the time dynamics of laser heating of a 30-nm gold particle in different biological media: fat (*dashed-dotted curve*), blood (*dashed curve*), tumor (*solid curve*), and prostate (*dotted curve*). (**b**) Results of thermal calculations for a 35-nm gold particle: heating and cooling in the water surrounding medium at different heat transfer rates α: $\alpha = 1.0$ (*solid curve*), $\alpha = 1.25$ (*dashed curve*), and $\alpha = 1.5$ (*dashed-dotted curve*). Reproduced from the following open access publication: [2]

surrounding medium becomes increasingly important, since the energy source is no longer present in the system. The temperature of the particle and surrounding medium remains high (~400 K) up to 20 ns, exceeding the laser pulse duration by 2.5 times. The total time for one cycle (heating from the initial temperature 300 K to maximum temperature, followed by cooling back to the initial temperature) is about 30 ns (Slide 18.16).

We have also examined the effect of different biological surroundings on the laser heating dynamics of 30-nm gold particles (Slide 18.17). Four biomedia were used: blood, human prostate, tumor, and fat. Results of computer simulations of the time–temperature profiles of gold nanoparticles in various biological media, performed by using OTM, are plotted in Fig. 6.4a. As followed from our calculations, the laser heating and temperature behavior of the gold nanoparticles in blood, prostate, and tumor are comparable to the surrounding water medium case, since the

Table 6.2 Thermophysical properties of the gold particle and surrounding biological tissue. Reproduced from the following open access publication: [2]

Material	Specific heat C (J/K kg)	Interval of T (K)	Thermal conductivity μ_0 (W/m K)	Thermal diffusivity χ (m^2/s)
Gold	129	273–373	318	1.18×10^{-4}
Water	4181.6–4215.6 $C(T) = 4182\ (1 + 1.016 \times 10^{-4}$ $(T - T_\infty)$	273–373	0.597–0.682 $\mu\ (T) = 0.597\ (1 + 1.78 \times 10^{-3})\ (T - T_\infty)$	1.43×10^{-7}
Human prostate	3740	310	0.529	
Blood	3645–3897	273–373	0.48–0.6	1.6×10^{-7}
Cytoplasm	4181.6–4215.6 $C(T) = 4182(1 + 1.016 \times 10^{-4})$ $(T - T_\infty)$	273–373	0.59	0.597–0.682 $\mu\ (T) = 0.597(1 + 1.78 \times 10^{-3})$ $(T - T_\infty)$
Fat	2975	273–373	0.185–0.233	
Tumor	3160	310	0.561	
Skin		273–373	0.210–0.410	

Thermophysical Properties of a Gold Particle and Surrounding Biological Tissue

Material	Specific heat C (J/K kg)	Interval of T (K)	Thermal conductivity $_0$ (W/m K)	Thermal diffusivity χ(m²/s)
Gold	129	273 – 373	318	1.18×10^{-4}
Water	4181.6 –4215.6 $C(T) = 4182\ (1{+}1.016 \times10^{-4}\ (T{-}T_{\infty})$	273–373	0.597–0.682 $\mu\ (T) = 0.597\ (1{+}1.78\times10^{-3}\ (T{-}T_{\infty})$	1.43×10^{-7}
Human prostate	3740	310	0.529	
Blood	3645 – 3897	273–373	0.48–0.6	1.6×10^{-7}
Cytoplasm	4181.6 – 4215.6 $C(T) = 4182(1{+}1.016\times10^{-4}\ (T{-}T_{\infty})$	273 – 373	373	0.597 – 0.682 $\mu\ (T) = 0.597(1{+}1.78\times10^{-3}\ (T{-}T_{\infty})$
Fat	2975	273–373	0.185–0.233	
Tumor	3160	310	0.561	
Skin		273 – 373	0.210–0.410	

Slide 18.15 Thermophysical properties of a gold particle and surrounding biological tissue

Nanosecond Pulses (continued)

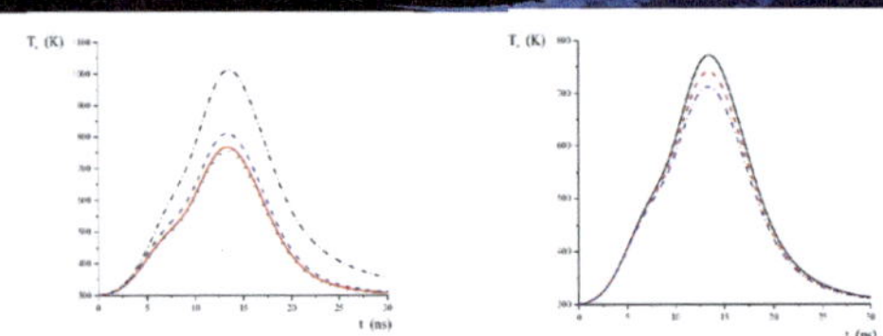

(a) (b)

Figure 6.4 (a) Illustration of the time dynamics of laser heating of a 30-nm gold particle in different biological media: fat (dashed-dotted curve), blood (dashed curve), tumor (solid curve), and prostate (dotted curve). (b) Results of thermal calculations for a 35-nm gold particle: heating and cooling in the surrounding water medium at different heat transfer rates α: $\alpha = 1.0$ (solid curve), $\alpha = 1.25$ (dashed curve), and $\alpha = 1.5$ (dashed-dotted curve).

- It follows from these calculations that the highest temperature, 770 K, for a given laser pulse fluenceis observed for the heating time of 13.5 ns, when the laser pulse has already degraded.
- After that time, the ***transfer of heat*** from the particle to the surrounding medium becomes ***increasingly important***, since the energy source is no longer present.
- The temperature of the particle and surrounding medium remains high (~ 400 K) upto 20 ns, exceeding the laser pulse duration by 2.5 times. The total time for one heating cycle is about 30 ns.

Slide 18.16 Nanosecond pulses (continued)

thermodynamic properties of those media are very close to each other (see Table 6.2). This means that for thermal calculations of laser heating of biological media, the thermal properties of water can be used if the water content in the media is high. But the heating of a gold nanoparticle in fat is substantially different from the water case, since fat has low thermal characteristics (Slide 18.18). Here we

Nanosecond Pulses (continued)

- We have also examined the ***effect of different biological surroundings*** on the laser heating dynamics of 30 nm-gold particles.
- ***Four biomedia*** were used: blood, human prostate, tumor and fat.
- Results of ***computer simulations*** in various biological media, performed by using OTM, are plotted in Figure 6.4(a).
- As follows from our calculations, the laser heating and temperature behavior of the gold nanoparticles in blood, prostate and tumor are ***comparable to the surrounding water medium*** case, since the thermodynamic properties of those media are very close to each other.
- This means that for thermal calculations of laser heating of ***biological media***, the thermal properties of ***water*** can be used if the water content in the media is high.

Slide 18.17 Nanosecond pulses (continued)

Nanosecond Pulses (continued)

- But the ***heating*** of a gold nanoparticle ***in fat*** is substantially different from the water case, since fat has low thermal characteristics.
- Here we observe ***higher overheating*** of the particle at the same energy level and duration of the laser pulse due to the relatively ***low thermal conductivity of fat*** as compared to other biomedia.
- The ***temperature dynamics*** of the particle is sensitive to the ***power exponent*** α used in the temperature dependence of the heat lost from the surface of the nanoparticle into the surrounding medium, shown in Figure 6.4(b)).
- The value of ***α = 1 better corresponds to the real biological surrounding***. The power $\alpha > 1$ describes the medium with high thermophysical characteristics, like the special cooling liquids and metals. The medium with $\alpha < 1$ has a low thermal conductivity and can be used as a thermal isolator.

Slide 18.18 Nanosecond pulses (continued)

observe higher overheating of the particle at the same energy level and duration of the laser pulse due to the relatively low thermal conductivity of fat as compared to other biomedia.

The temperature dynamics of the particle is sensitive to the power exponent α used in the temperature dependence of the heat lost from the surface of the nanoparticle into the surrounding medium, shown in Fig. 6.4b (Slide 18.18).

Nanosecond Pulses (continued)

- It is interesting to investigate the ***effect of the particle's radius*** on the temperature dynamics of the nanoparticle heated by the nanosecond laser radiation in the biological surroundings.
- There are ***two competitive factors*** here. On one hand, according to the Mie diffraction theory, ***the absorption efficiency*** of the gold nanoparticle ***drops with decreasing size*** of the particle.
- On the other hand, the ***heating rate increases for smaller particles***, as follows from OTM.
- To find which factor has a stronger effect on the effective laser heating of a gold nanoparticle, we have calculated the maximal ***temperature profile*** for different nanoparticle radii in blood and ***compared*** it to the gold nanoparticle ***absorption curve***.
- ***The results of these simulations*** are performed by using OTM and presented in Figure 6.5 on the next slide.

Slide 18.19 Nanosecond pulses (continued)

The value of $\alpha = 1$ better corresponds to the real biological surrounding. The power $\alpha > 1$ describes the medium with high thermophysical characteristics, like the special cooling liquids and metals. The medium with $\alpha < 1$ has a low thermal conductivity and can be used as a thermal isolator (Slide 18.18).

It is interesting to investigate the effect of the particle's radius on the temperature dynamics of the nanoparticle heated by the nanosecond laser radiation in the biological surroundings (Slide 18.19). There are two competitive factors here. On one hand, according to the Mie diffraction theory, the absorption efficiency of the gold nanoparticle drops with decreasing size of the particle. On the other hand, the heating rate increases for smaller particles, as follows from Eq. (6.10). To find which factor has a stronger effect on the effective laser heating of a gold nanoparticle, we have calculated the maximal temperature profile for different nanoparticle radii in blood and compared it to the gold nanoparticle absorption curve.

The results of these simulations are performed by using OTM and presented in Fig. 6.5. It follows from this figure that the optical effect is much stronger than the thermal effect when the radius of nanoparticle is less than 35 nm. In the radii range 1–35 nm, the overheating effect of the particle behaves according to the absorption efficiency (Slide 18.20). For a radius of 35 nm, the thermal processes dominate over the optical properties. For large radii ($\geq$35 nm), the maximal temperature profile is saturated with oscillations, repeating the maxima and minima of the absorption curve. The saturation of the maximal temperature curve for large particle radii is explained by the balance between heating of the particle due to absorption of laser energy and energy losses from the surface of the particle because of heat diffusion into the surrounding biological medium.

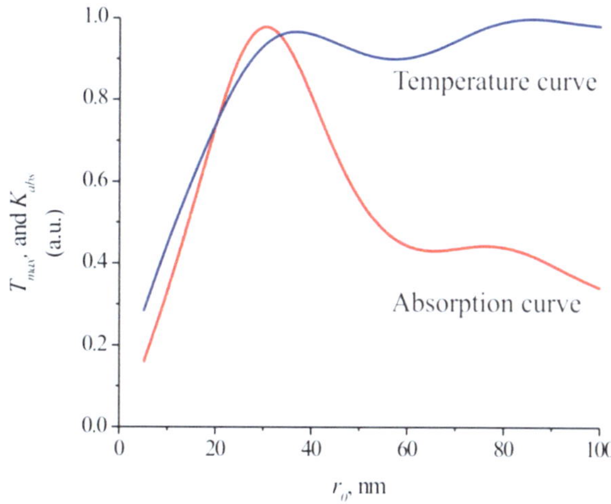

Fig. 6.5 Nondimensional (a.u.) absorption efficiency K_{abs} and maximal temperature T_{max} curves as a function of the particle's radius for gold nanoparticles in blood heated by a laser pulse of energy density 10 mJ/cm^2 and duration 8 ns. Reproduced from the following open access publication: [2]

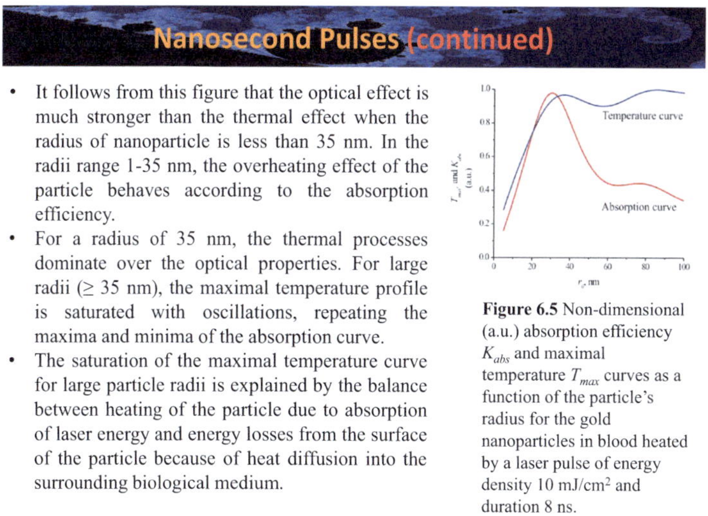

Slide 18.20 Nanosecond pulses (continued)

6.2.6 Conclusions and Summary

The comparative analysis of OTM and TTM for heating of a metal nanoparticle in the femtosecond, picosecond, and nanosecond regimes has shown [2–4] that (Slides 18.21 and 18.22):

- In the femtosecond mode, the thermal equilibrium among the excited electrons is established within the first 175 fs, long after the end of the laser pulse duration.
- The electrons remain in the thermal equilibrium state up to 1 ps.

- Both models demonstrate the same scenario in the heating kinetics of a metal nanoparticle by a femtosecond laser pulse: about a 100 fs time delay in the heating of the particle is observed, until reaching a maximum lattice temperature and saturation in temperature curves after 175 fs.
- The electron cooling time due to coupling to the lattice is about 10 ps, which imposes an upper time limit for TTM application.
- TTM gives a very good approximation for the femtosecond mode while an electron temperature exists, but it fails to describe the laser heating of nanoparticles for longer pulse durations in the picosecond and nanosecond regimes.
- OTM shows that the heat lost from the surface of the nanoparticle into the surrounding medium becomes noticeable after 200 ps.
- The heating of a metal nanoparticle by a nanosecond laser pulse in fat provides higher particle overheating than in blood, prostate, and water as surrounding media due to the thermally isolating property of the fat.
- The optical properties of the nanoparticle have a much stronger effect on the heating dynamics in the nanosecond mode than the thermal effects when the radius of the particle is less than 35 nm. For larger particles, the thermal processes dominate the optical properties, and the temperature curve is determined by the balance between heating of the nanoparticle and energy losses from the surface of the particle due to heat diffusion into the surrounding biological medium.

Thus, the comparison of the two models shows that OTM provides an adequate description of the laser heating of nanoparticles in the femtosecond, picosecond, and nanosecond regimes (Slide 18.22).

Conclusions and Summary

The ***comparative analysis of OTM and TTM*** for heating of a metal nanoparticle in the ***femtosecond, picosecond and nanosecond regimes*** has shown that:

- In the ***femtosecond mode***, the thermal equilibrium among the excited electrons is established within the first 175 fs, long after the end of the laser pulse duration.
- The ***electrons*** remain in the thermal equilibrium state up to 1 ps.
- ***Both models demonstrate the same scenario*** in the heating kinetics of a metal nanoparticle by a femtosecond laser pulse: about a 100 fs time delay in the heating of the particle is observed, until reaching a maximum lattice temperature and saturation in temperature curves after 175 fs.
- The ***electron cooling time*** due to coupling to the lattice is about 10 ps, which imposes an upper time limit for applying TTM.

Slide 18.21 Conclusions and summary

Conclusions and Summary

- ***TTM*** gives a very good approximation for the ***femtosecond mode*** while an electron temperature exists, but it fails to describe the laser heating of nanoparticles for longer pulse durations in the picosecond and nanosecond regimes.
- ***OTM*** shows that the ***heat lost*** from the surface of the nanoparticle into the surrounding medium becomes noticeable after 200 ps.
- The ***optical properties*** of the nanoparticle have a much stronger effect on the heating dynamics in the nanosecond mode than the thermal effects when the radius of the particle is ***less than 35 nm***.
- For ***larger particles***, the ***thermal processes dominate*** the optical properties, and the temperature curve is determined by the balance between heating of the nanoparticle and energy losses from the surface of the particle.
- Thus, the comparison of the two models shows that ***OTM provides an adequate description*** of the laser heating of nanoparticles ***in the femtosecond, picosecond and nanosecond regimes.***

Slide 18.22 Conclusions and summary (continued)

6.3 Computer Practicum 3 (Lecture 19): Single-Pulse Heating of Nano/Bioparticles (Slide 19.1)

- Introduction
- Practice Problem 4: Time Dynamics of the Gold Nanoparticle Temperature in a Single-Pulse Heating Mode
  - *Abstract*
  - *Model*
  - *Time Profile of the Laser Pulse*
  - *Heat Transfer to the Surrounding Medium*
  - *Effect of the Energy Density on the Time Dynamics of Nanoparticle Heating in the Single-Pulse Mode*
  - *Effect of the Particle Size and Shape on the Time Dynamics of Nanoparticle Heating*
  - *Effect of the Pulse Duration in the Single-Pulse Mode of Nanoparticle Heating*
  - *Conclusion*
- Practice Problem 5: Thermal Dynamics of Normal and Cancerous Ribosome in Cytoplasm Heated by a Single Pulse of Laser Radiation
  - *Abstract*
  - *Procedure and Results*

Contents:

- Introduction
- **Practice Problem 4:** Time Dynamics of the Gold Nanoparticle Temperature in a Single Pulse Heating Mode
  - *Abstract*
  - *Model*
  - *Time Profile of the Laser Pulse*
  - *Heat Transfer to the Surrounding Medium*
  - *Effect of the Energy Density on the Time Dynamics of Nanoparticle Heating in the Single-Pulse Mode*
  - *Effect of the Particle Size and Shape on the Time Dynamics of Nanoparticle Heating*
  - *Effect of the Pulse Duration in the Single-Pulse Mode of Nanoparticle Heating*
  - *Conclusion*
- **Practice Problem 5:** Thermal Dynamics of Normal and Cancerous Ribosome in Cytoplasm Heated by a Single Pulse of Laser Radiation
  - *Abstract*
  - *Procedure and Results*

Slide 19.1 Contents

6.3.1 Introduction

This section contains material adopted from our publication [9]. In the previous lecture, we showed that the one-temperature model (OTM) is an appropriate approximation for nanoparticle heating in the femtosecond, picosecond, and nanosecond regimes, thus providing an effective modeling tool for further nanomedicine research to explore. *The research task* for this computer practicum (schematically shown in Slide 19.2) is to numerically solve the OTM to determine the nanoparticle temperature as a function of time for a real laser pulse profile and make the recommendations on energy density of the radiation required to reach a desired maximum temperature by nanoparticles in the surrounding biological medium. The OTM (Slide 19.3), described in Sects. 6.1 and 6.2 of this chapter, is given by Eq. (6.10), repeated here as (6.12) as follows:

$$\frac{dT_s}{dt} = \frac{3Q_{abs}I_0 f(t)}{4r_0 C_s(T_s)\rho_s} - \frac{3\mu_\infty T_s}{(\alpha+1)r_0^2 C_s(T_s)\rho_s}\left[\left(\frac{T_s}{T_\infty}\right)^{\alpha+1} - 1\right] + \frac{3L}{r_0 C_s(T_s)}\frac{dr_0}{dt}. \tag{6.12}$$

The heat generation in the nanoparticle volume is due to laser energy absorption by the nanoparticle. In the *Computer Practicum 1* (Sect. 5.1), we simulated the absorption characteristics of the nanoparticles in various biomedia by using a generalized Lorenz–Mie diffraction theory, which takes into account the plasmon resonance absorption effect observed for metal nanoparticles in the optical range of

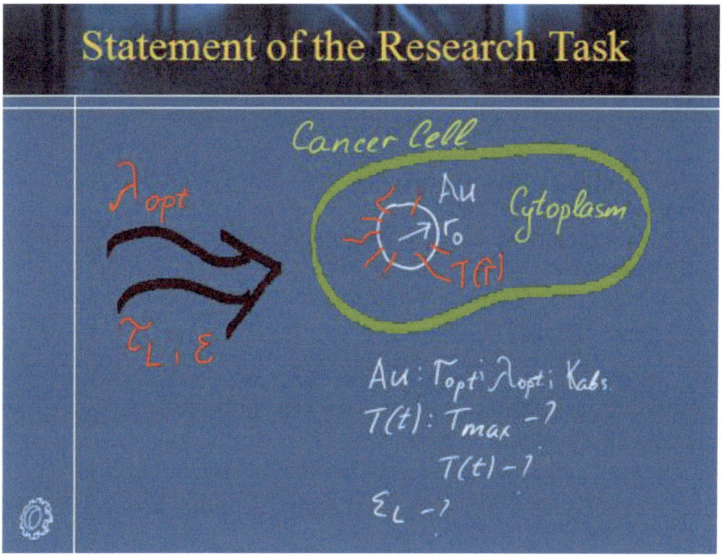

Slide 19.2 Statement of the research task

One-Temperature Model

$$\frac{dT_s}{dt} = \frac{3Q_{abs}I_0 f(t)}{4r_0 C_s(T_s)\rho_s} - \frac{3\mu_\infty T_s}{(\alpha+1)r_0^2 C_s(T_s)\rho_s}\left[\left(\frac{T_s}{T_\infty}\right)^{\alpha+1} - 1\right] + \frac{3L}{r_0 C_s(T_s)}\frac{dr_0}{dt}$$

- The ***absorption efficiency*** Q_{abs} of the nanoparticles in various biomedia have been simulated in the ***Computer Practicum 1*** (Section 5.1) by using a generalized Lorenz-Mie diffraction theory.
- Using the complex indices of refraction for the medium and particle (which depend on wavelength as well as the particle size), we have found the ***absorption efficiency*** of spherical particles at a given wavelength within the bounds of the refractive index values provided.
- The ***results*** of these calculations for absorption efficiency of gold nanospheres, silica-gold nanoshells and gold nanorods are listed in ***Table 5.4.***

Slide 19.3 One-temperature model

the spectrum (Slide 19.3). Using the complex indices of refraction for the medium and particle (which depend on wavelength, as well as particle size), we found the absorption efficiency of spherical particles at the given wavelength within the bounds of the refractive index values provided. The results of these calculations

Assumptions and Limitations

- ***We assume*** here that the possible ***vapor bubble formation*** around the nanotarget in the surrounding aqueous medium due to heating of the nanoparticles ***does not change the absorption efficiency of the nanoparticles***, since the bubble is transparent for the laser wavelength that is used, and the bubble could not change considerably the plasmon resonance absorption of the metal nanoparticles. This is a phenomenon which is related to electron subsystem oscillations in the particle materials rather than in the surrounding medium.
- ***We also limit the maximum temperature*** of the theoretical calculations ***to the melting point of the material*** (for example, $T_M \sim$ 1336 K for bulk gold material and remaining above 1100 K for gold particles larger than 5 nm in diameter), so that the particles will not undergo advanced phenomena of heating (evaporation, melting or explosion), ***but require that the nanoparticles surpass 433 K as required for protein denaturing.***

Slide 19.4 Assumptions and limitations

for the absorption efficiency of gold nanospheres, silica–gold nanoshells, and gold nanorods are listed in Table 5.4. The highest absorption efficiency is observed for gold nanorods; however, silica–gold nanoshells are tunable to the longer wavelengths in the transparency window of most biotissues.

We assume here (Slide 19.4) that the possible vapor bubble formation around the nanotarget in the aqueous surrounding medium due to heating of the nanoparticles does not change the absorption efficiency of the nanoparticles, since the bubble is transparent for the laser wavelength that is used, and the bubble could not change considerably the plasmon resonance absorption of the metal nanoparticles. This is a phenomenon which is related to electron subsystem oscillations in the particle materials rather than in the surrounding medium.

We also limit (Slide 19.4) the maximum temperature of the theoretical calculations to the melting point of the material (e.g., $T_M \sim 1336$ K for bulk gold material and remaining above 1100 K for gold particles larger than 5 nm in diameter), so that the particles will not undergo an advanced phenomena of heating (evaporation, melting, or explosion), but require that the nanoparticles surpass 433 K as required for protein denaturing. Thus, the last term in Eq. (6.12), which deals with evaporation of the nanoparticle, can be neglected.

The heating kinetics of the nanoparticles is sensitive to the time structure of the radiation delivery described by the function $f(t)$ in the first term on the right side of (6.12), where $f(t)$ determines the pulse shape, pulse duration, and time interval between the pulses for the case of a multipulse heating mode. A few examples of laser pulse profiles as shown in Slide 19.5 include rectangular, triangle, and Gaussian shapes. In OTM simulations, we use a Gaussian profile, which is a most

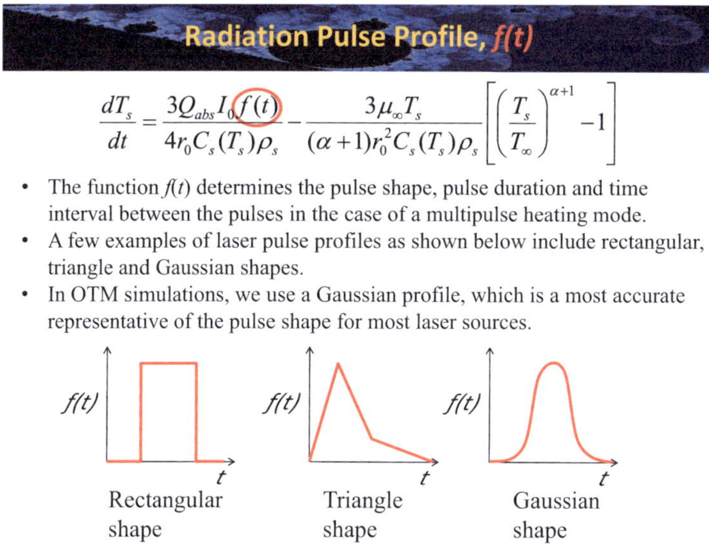

Slide 19.5 Radiation pulse profile

accurate representation of the pulse shape for most laser sources. Thus, a single pulse can be described by a Gaussian-like curve,

$$f(t) = \mathrm{e}^{-(bt-d)^2}, \tag{6.13}$$

where b and d are constants defining the pulse duration. The input data for OTM simulations are listed in Tables 6.1, 6.2, and 6.3.

We have developed an effective computer Maple, shown on the slides and Appendix B, to perform simulations of the nanoparticle heating kinetics by using the OTM model. The first step is to compute a Gaussian pulse profile by using Eq. (6.13), shown in Fig. 6.6 and Slide 19.6. The values for the constants a and b in $f(t)$ are chosen in such a way as to form a pulse of 8–10 ns duration as used in many experimental studies. The duration is determined as a full time interval corresponding to half of the maximum amplitude of $f(t)$.

A Maple code begins with the command *restart*, which clears the internal memory of the Maple kernel, making it ready to accept the new data. We *restart* the Maple *with plot* and *DEtools* packages providing the graphical and differential equation solving tools. Then we define the Gaussian function $f(t)$ and *plot* the pulse shape in a defined time interval as shown in Fig. 6.6 and Slide 19.6.

The next step in developing a computer code is to introduce input data from Tables 6.1, 6.2, and 6.3 as required for the simulations, listed in Figs. 6.7, 6.8, and 6.9 and Slide 19.7 with names and units. We input the specific heat C of the surrounding medium, nanoparticle radius r_0, maximum absorption efficiency Q_{abs}, initial temperature T_0, pulse duration τ, pulse energy density E, power exponent α, density of the nanoparticle ρ, and heat conductivity of the surrounding medium μ,

Table 6.3 Thermal properties of some nanoparticle materials

Material	Specific heat C (J/Kkg)	Interval of T (K)	Phase transition point T_{ph} (K)	Thermal conductivity μ_0 (W/mK)	Thermal diffusivity[a] χ (m^2/s)	Density ρ (kg/m^3)
Silver (Ag)	187	100	1234.93	444	0.000226	10500
	225	200		430	0.000182	
	235	300		429	0.000174	
	239	400		425	0.000169	
	250	600		412	0.000157	
	262	800		396	0.000144	
	277	1000		379	0.00013	
Fullerene (C_{60})	1600	300	800–1073 (sublimes)	670	1.45e-7	1720
Polystyrene	1170	300	463–533	0.13	1.06e-7	1050
Glass (Si)	259	100	773–1923	884	0.001465	2300
	556	200		264	0.000204	
	712	300		148	8.92e-5	
	790	400		98.9	5.37e-5	
	867	600		61.9	3.06e-5	
	913	800		42.4	1.99e-5	
	946	1000		31.2	1.42e-5	
Carbon (C)	136	100	3800	4970	0.0165	2210
	411	200		3230	0.00356	
	709	300		1950	0.00125	
	992	400		1390	0.000634	
	1406	600		892	0.000287	
	1650	800		667	0.000183	
	1793	1000		534	0.000135	
Aluminum oxide (Al_2O_3)	765	300	2327	46	1.51e-5	3979
	940	400		32.4	8.68e-6	
	1110	600		18.9	4.29e-6	
	1180	800		13	2.78e-6	
	1225	1000		10.5	2.16e-6	
Gold (Au)	109	100	1337.33	327	0.000155	19300
	124	200		323	0.000135	
	129	300		317	0.000127	
	131	400		311	0.000123	
	135	600		298	0.000114	
	140	800		284	0.000105	
	145	1000		270	9.65e-5	
Magnesium Oxide (MgO)	907.8	298	3125–3873	49.286	1.51E-05	3.6
Nickel Oxide (NiO)	478.6	298	2228–2257	82	2.45E-05	7

[a]The thermal diffusivity in Table 6.3 is calculated from the specific heat, thermal conductivity and density of each material using the relation $\chi = \mu / C\rho$

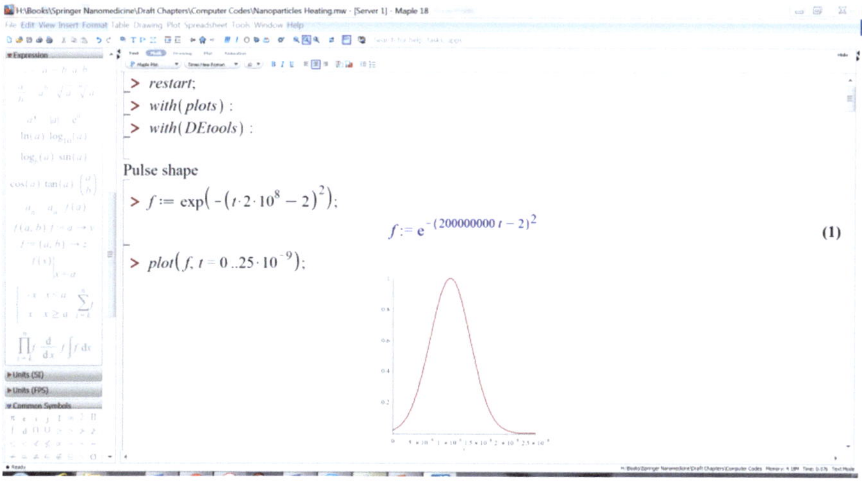

Fig. 6.6 Modeling the radiation pulse profile

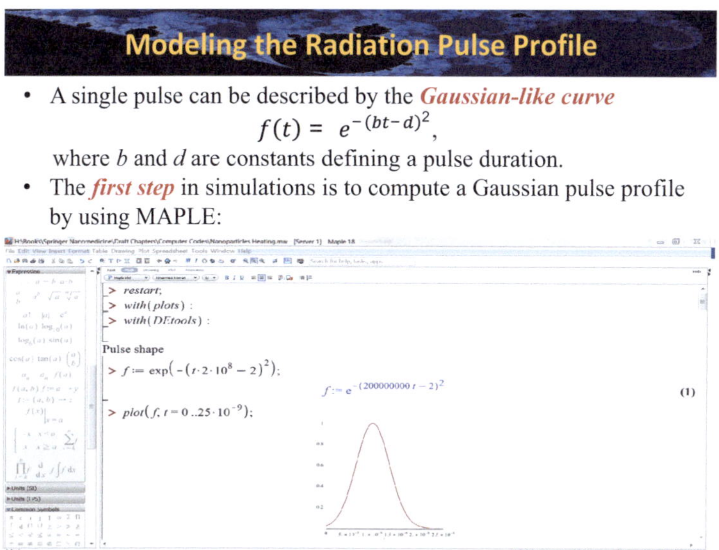

Slide 19.6 Modeling the radiation pulse profile

and calculate the initial incident intensity of the radiation by using the formula $I_0 = E/\tau$. E is a variable which can be adjusted to achicve a desirable maximum temperature of the nanoparticle. The maximum absorption efficiency Q_{abs} values are obtained from Mie simulations performed for the optimal radiation wavelength and given radius of the nanoparticle.

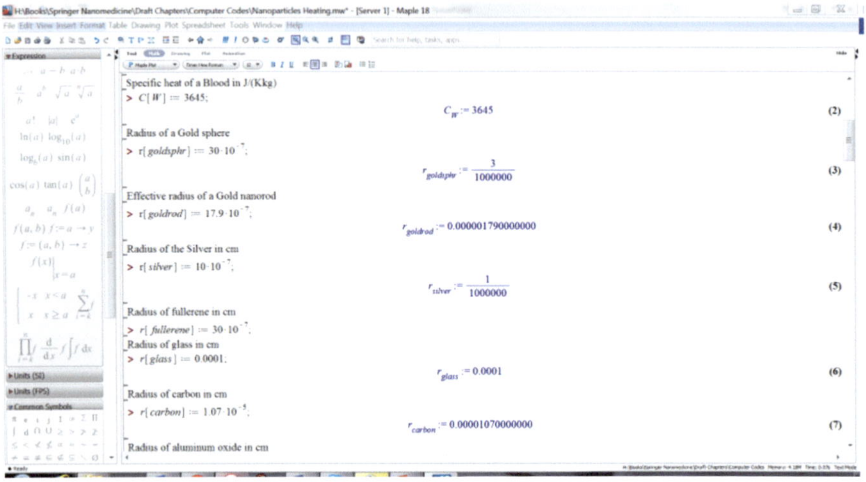

Fig. 6.7 Input data for OTM

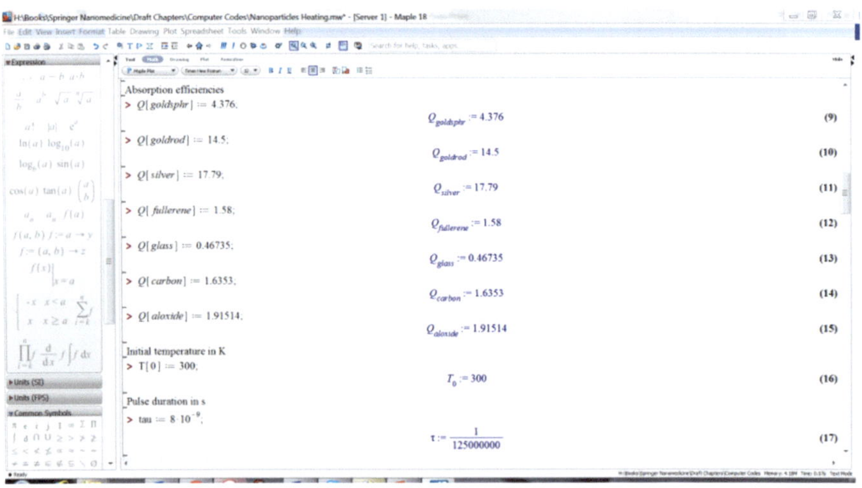

Fig. 6.8 Input data for OTM (continued)

Then we type the OTM differential equation (6.12) without a last evaporation term for each nanoparticle, shown in Fig. 6.10 and Slide 19.8. To solve these equations, we use the *dsolve* command, which finds a numerical solution for an ordinary differential equation using a *Fehlberg fourth-fifth order Runge–Kutta* (rkf45) method as shown in Fig. 6.11 and Slide 19.9.

The output results of these simulations can be arranged in numerical or graphical formats. To record the results numerically, we create a matrix zz which contains the

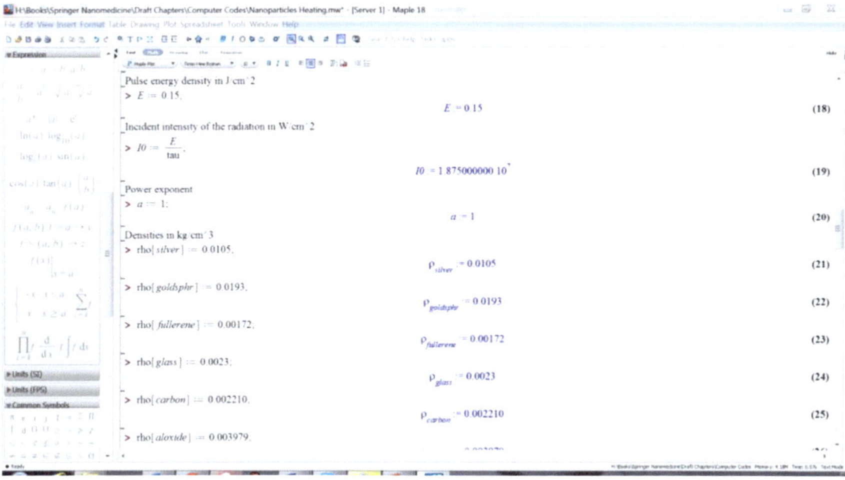

Fig. 6.9 Input data for OTM (continued)

Input Data for OTM

- The *next step* in developing a computer code is to introduce *input data* from Tables 6.1, 6.2 and 6.3, as required for the simulations, listed below with names and units.

Material	Specific heat C (J/Kkg)	Phase transition point T_{ph} (K)	Thermal conductivity μ_0 (W/mK)	Thermal diffusivity χ (m²/s)	Density ρ (kg/m³)
Silver (Ag)	187 225 235 239 250 262 277	1234.93	444 430 429 425 412 396 379	.000226 .000182 .000174 .000169 .000157 .000144 .00013	10500
Fullerene (C_{60})	1600	800-1073 (sublimes)	670	1.45e-7	1720
Polystyrene	1170	463-533	.13	1.06e-7	1050

Slide 19.7 Input data for OTM

time and temperature values, shown in Fig. 6.12 and Slide 19.10. For graphical presentation of the output results, we use the *odeplot* command specifying a time interval and temperature frame for the output graph, as shown in Fig. 6.12 and Slide 19.10. To see the actual graph, we use the *display* command, which plots the graph in a defined time interval and temperature frame, as shown in Fig. 6.13 and Slide 19.11.

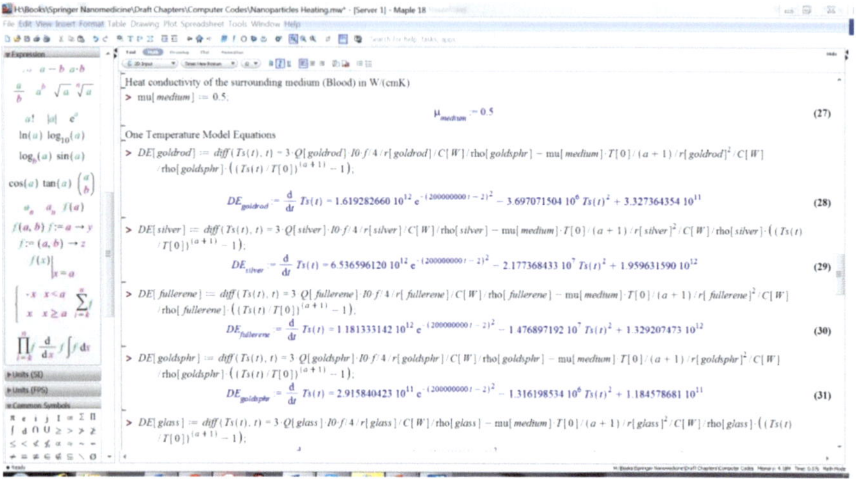

Fig. 6.10 Modeling OTM differential equations

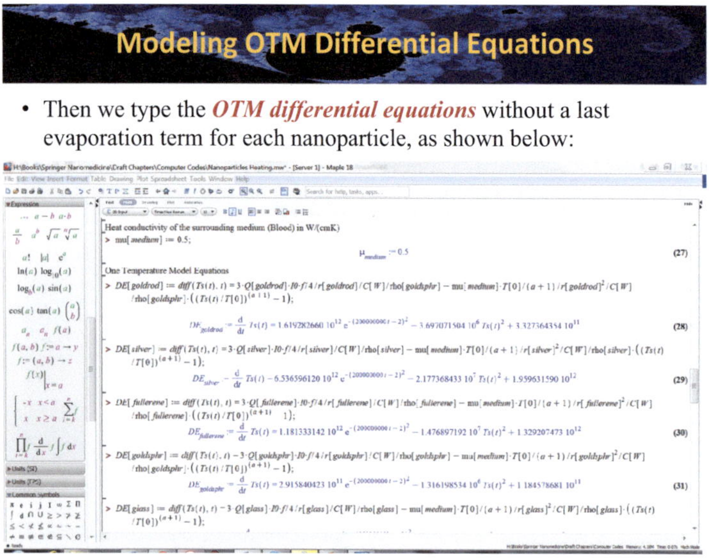

Slide 19.8 Modeling OTM differential equations

The results of these simulations as shown in Fig. 6.13 and Slide 19.11 provide the time dynamics of heating and cooling of nanoparticles in different surrounding biological media, allowing us to make recommendations on the radiation energy density, pulse duration, particle size, and shape. The following Practice Examples present the output results of these simulations in detail for specific nano/bioparticles.

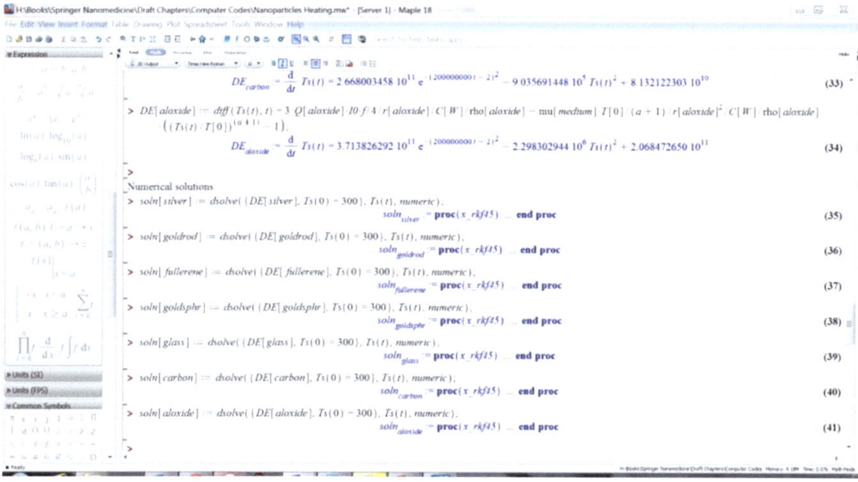

Fig. 6.11 Numerical solution

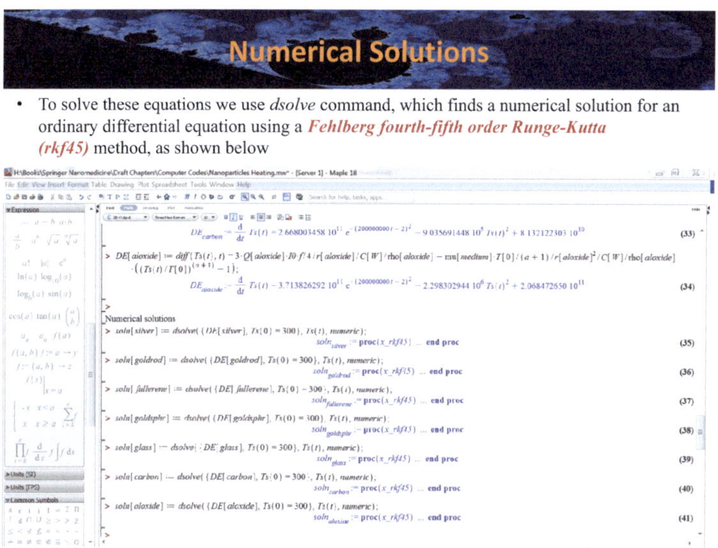

Slide 19.9 Numerical solutions

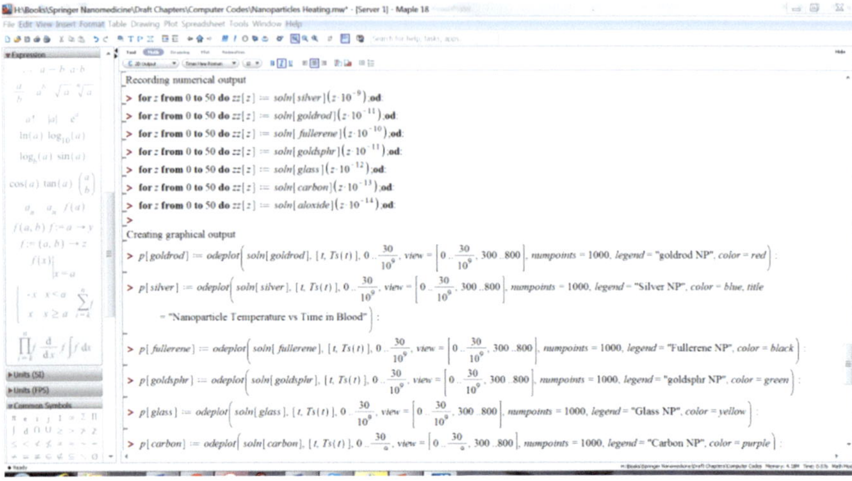

Fig. 6.12 Arranging the output results in numerical and graphical formats

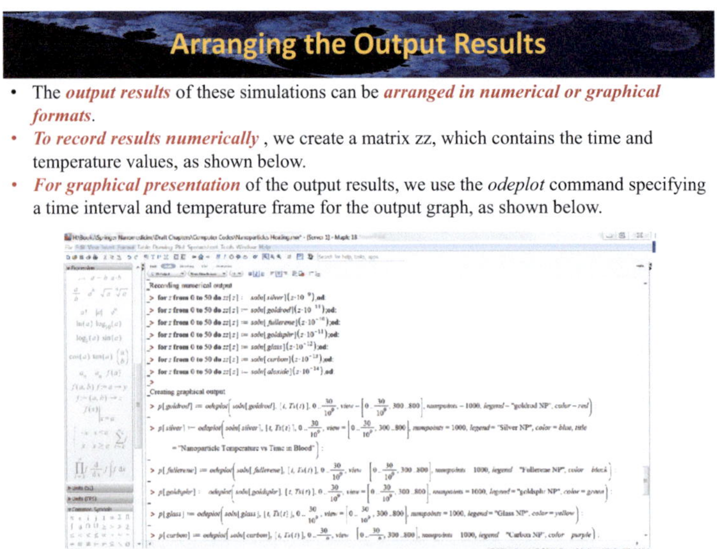

Slide 19.10 Arranging the output results

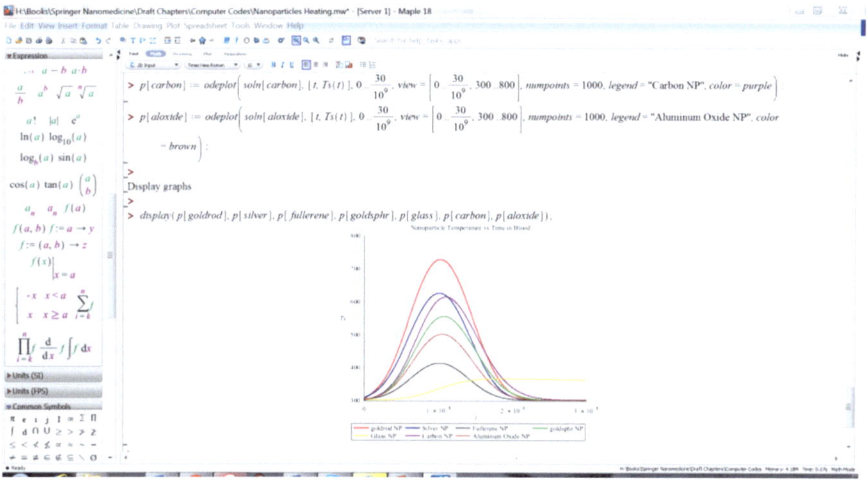

Fig. 6.13 Plots of the temperature profiles of nanoparticles in different surrounding biological media

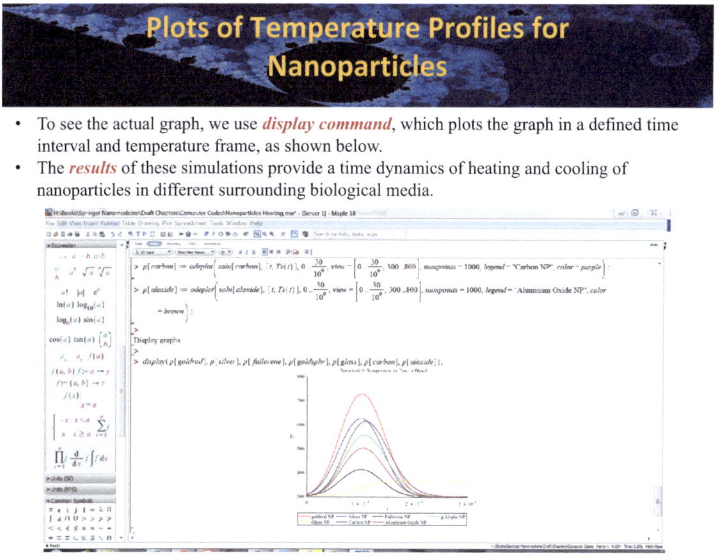

Slide 19.11 Plots of temperature profiles for nanoparticles

6.3.2 *Practice Problem 4: Time Dynamics of the Gold Nanoparticle Temperature in a Single-Pulse Heating Mode*

6.3.2.1 **Abstract** (Slides 19.12 and 19.13)

The impulse of electromagnetic radiation to a particle results in the heating of the particle as well as the surrounding medium over time. This computer practicum studies the time dependence of the temperature of a gold nanoparticle when irradiated by a single short laser pulse. For this procedure, the energy density of a single pulse is altered in order to manipulate the time dynamic heating curve to reach the maximum temperature values sufficient to initiate one or a combination of the cancer cell killing mechanisms, such as thermal ablation, bubble formation, sound generation, high pressure and shock waves generation, to cause biological damage to the tumor. This section contains material adapted from our publication [9].

6.3.2.2 Model

Recent advances in nanotechnology allow for the destruction of cancer cells via the laser heating of nanoparticles. It is important to look at the time dependence of these tiny systems to be able to engineer the correct dosage of nanoparticles and amount

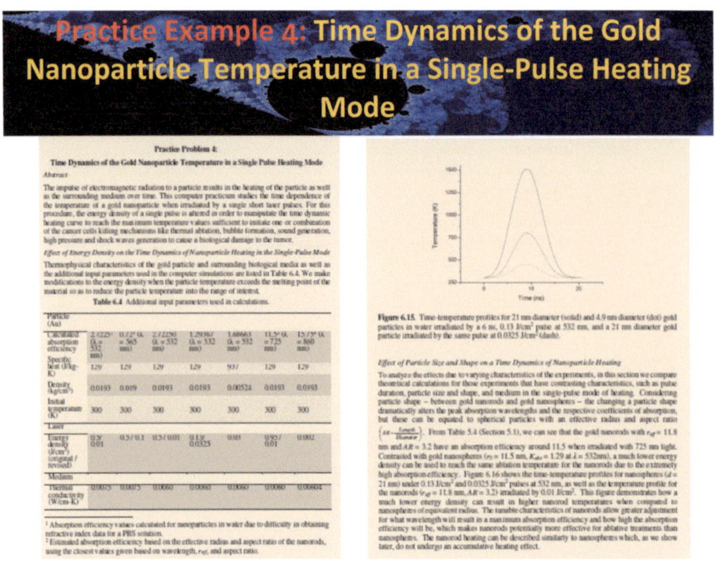

Slide 19.12 Practice Example 4: Time dynamics of the gold nanoparticle temperature in a single-pulse heating mode

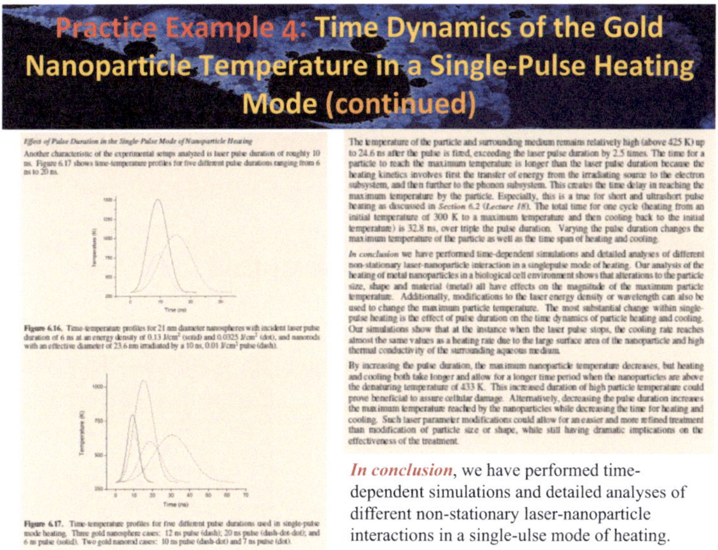

Slide 19.13 Practice Example 4: time dynamics of the gold nanoparticle temperature in a single-pulse heating mode (continued)

of radiation in order to supply the maximum amount of damage while containing the damage to only the cancer cells. The simulations are performed for a gold nanoparticle heated by a single pulse of 8.0 ns. The absorbed laser energy is quasi-instantaneously converted into the heat of the nanoparticle. The characteristic heat diffusion time is determined mainly by the size of the nanoparticle and can be found as

$$\tau_D = r_0^2/4\chi, \tag{6.14}$$

where r_0 is the particle radius, and χ is the thermal diffusivity of the particle material. For a gold nanoparticle ($\chi = 1.18 \times 10^{-4}$ m^2/s) of radius $r_0 = 20$–30 nm, the heat diffusion time is $\tau_D \sim 10^{-12}$s $\ll \tau_p \sim 10^{-8}$ s, where τ_p is the laser pulse duration. Comparing the heat diffusion length δ in the gold to the particle size, $\delta = 2\sqrt{\chi\tau_D}$, we deduce that the temperature is nearly uniform over the whole particle on the time scale of the laser pulse duration τ_p.

In this homogeneous heating approximation and quasi-steady heat exchange with the surrounding biological medium, the time dynamics of the nanoparticle temperature can be described by OTM given by Eq. (6.12). The focus of this computer practicum is to simulate a single-pulse mode of laser heating of nanoparticles. We limit the maximum temperature of the theoretical calculations to the melting point of the material ($T_M \sim 1336$ K for bulk gold material and remaining above 1100 K for gold particles larger than 5 nm in diameter), so that the particles will not undergo an advanced phenomena of heating (evaporation, melting, or explosion), but require that the nanoparticles surpass 433 K as required for protein denaturing [9].

6.3.2.3 Time Profile of the Laser Pulse

Most experiments utilize a Gaussian beam pulse profile given by Eq. (6.13). The constants a and b in the function $f(t)$ are chosen to provide the pulse width in the range 6–20 ns at FWHM. Figure 6.14 shows the pulse profile for laser pulse duration 8 ns (6.14a) and 10 ns (6.14b).

6.3.2.4 Heat Transfer to the Surrounding Medium

It is assumed here that the main mechanism of energy losses from the surface of a particle is heat diffusion into the surrounding medium. The temperature

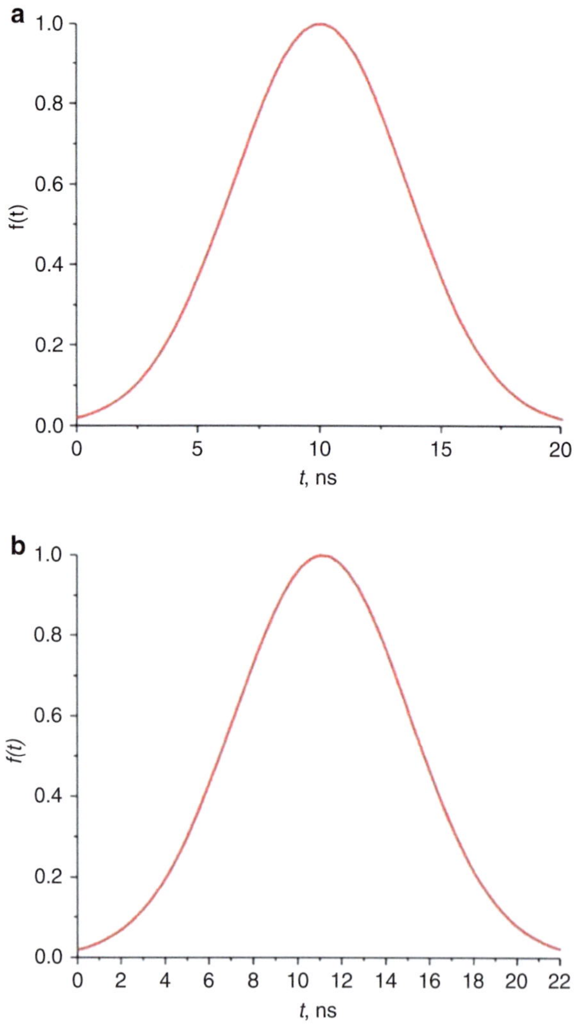

Fig. 6.14 Laser pulse time profiles for laser pulse durations 8 ns (**a**) and 10 ns (**b**)

dependence of the thermal conductivity of the biological tissue can be approximated by a power law as

$$\mu(T) = \mu_0 \left(\frac{T}{T_0} \right)^{\alpha}, \tag{6.15}$$

where the power exponent s describes the heat transfer rate in the medium. Appendix B contains the MAPLE code designed to simulate the nanoparticle heating kinetics by using OTM.

6.3.2.5 Effect of the Energy Density on the Time Dynamics of Nanoparticle Heating in the Single-Pulse Mode

Thermophysical characteristics of the gold particle and surrounding biological media, as well as the additional input parameters used in the computer simulations, are listed in Table 6.4. We make modifications to the energy density when the particle temperature exceeds the melting point of the material so as to reduce the particle temperature into the range of interest.

The purpose of this computer practicum is to heat the nanoparticles in a single-pulse mode below the melting temperature of the nanoparticle. Here we reproduce our results published in [9].

As a case for reducing the incident energy density, we have performed calculations to determine how changing the energy density would change the time required to reach the nanoparticle maximum temperature. Our calculations show [9] that changing the energy density in the chosen range has a very small or no effect on the time it takes for the particle to heat to the maximum temperature. Figure 6.15 demonstrates this with three different time–temperature profiles: a 21-nm diameter gold sphere irradiated by a 0.13 J/cm^2 pulse (solid) and 0.0325 J/cm^2 pulse (dash), and a 4.9-nm diameter gold sphere irradiated by a 0.13 J/cm^2 pulse (dot). Changing the energy density from a 0.13 J/cm^2 pulse to 0.0325 J/cm^2 pulse has no effect on the time required for the particle to reach its maximum temperature, which occurs 9 ns after the incident pulse in both cases. This is due to the fast transfer of energy from the irradiating source to the electron lattice, and further to the phonon system which has a picosecond time scale, as shown in Sect. 6.2 (Lecture 18), and is much shorter than the nanosecond duration of the laser pulse. The time for electron–phonon thermalization is more dependent on the specific heat of the electron and particle structures, rather than the energy density.

While the energy density of the laser pulse on the nanosecond time scale does not affect the time required to reach the maximum particle temperature (though it *does* affect the magnitude of temperature reached), the cooling (heat dissipation) time for the particles is shown to be longer for particles that underwent heating with higher energy density lasers [9]. The cooling time for the particles is slightly affected by the magnitude of the maximum temperature and is thus linked to the

Table 6.4 Additional input parameters used in the calculations. Reprinted from [9]. © With permission from Elsevier

Reference	[9]	[10]	[11]	[12]	[13]	[14]	[15]
Particle (Au)							
Calculated absorption efficiency	2.7225 ($\lambda = 532$ nm)	0.72 ($\lambda = 565$ nm)	2.72250 ($\lambda = 532$ nm)	1.29367 ($\lambda = 532$ nm)	1.68663 ($\lambda = 532$ nm)	11.5 ($\lambda = 725$ nm)	15.75 ($\lambda = 860$ nm)
Specific heat (J/kg K)	129	129	129	129	937	129	129
Density (kg/cm^3)	0.0193	0.019	0.0193	0.0193	0.00524	0.0193	0.0193
Initial temperature (K)	300	300	300	300	300	300	300
Laser							
Energy density (J/cm^2) (original/revised)	0.5/0.01	0.5/0.1	0.5/0.01	0.13/0.0325	0.03	0.95/0.01	0.002
Medium							
Thermal conductivity (W/cm K)	0.0075	0.0075	0.0060	0.0060	0.0060	0.0060	0.00604

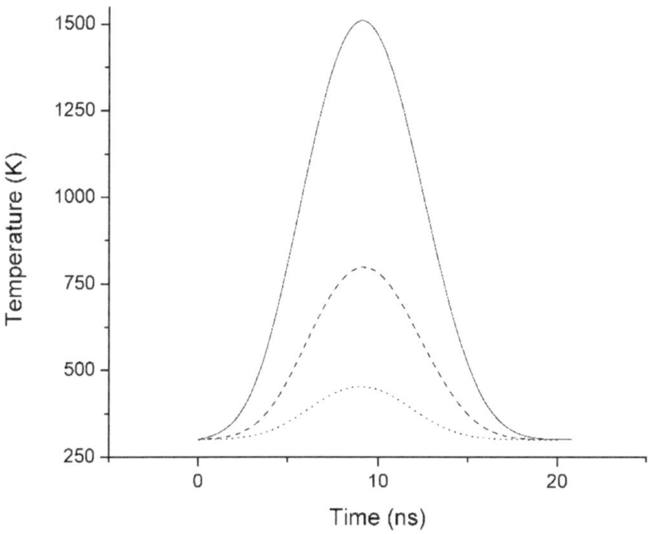

Fig. 6.15 Time–temperature profiles for 21-nm diameter (*solid*) and 4.9-nm diameter (*dot*) gold particles in water irradiated by a 6-ns, 0.13 J/cm^2 pulse at 532 nm, and a 21-nm diameter gold particle irradiated by the same pulse at 0.0325 J/cm^2 (*dash*). Reprinted from [9]. © With permission from Elsevier

energy density. Between a 0.13 J/cm^2 pulse and a 0.0325 J/cm^2 pulse and heating the particles to 1500 K and 800 K, respectively, the cooling time is only slightly longer for the higher energy density pulse.

For the 0.13 J/cm^2 pulse, the particle cools back to the ambient temperature in 11.4 ns after the peak, while after the 0.0325 J/cm^2 pulse the particle takes 10.6 ns after the maximum to return to the ambient temperature [9].

The minimal effect of changing the energy density on the time required to heat and cool the nanoparticles is a fundamental limitation of multipulse heating kinetics of metal nanoparticles [9]. This is not to be confused with the *maximum temperature* reached by nanoparticles irradiated by sources of different energy density; the magnitude of the temperature is directly related to and greatly dependent on the energy density of the laser. Given the minimal effect that the energy density has on the time dynamics of heating and cooling, we alter the energy densities provided in the analyzed papers to maintain a valid calculation of particle temperatures (which requires the particles to not undergo high-temperature phenomena). This is deemed appropriate due to the minimal effect of the energy density on the time dynamics of nanoparticle heating.

6.3.2.6 Effect of the Particle Size and Shape on the Time Dynamics of Nanoparticle Heating

To analyze the effects due to varying the characteristics of the experiments, in this section we compare theoretical calculations for those experiments that have

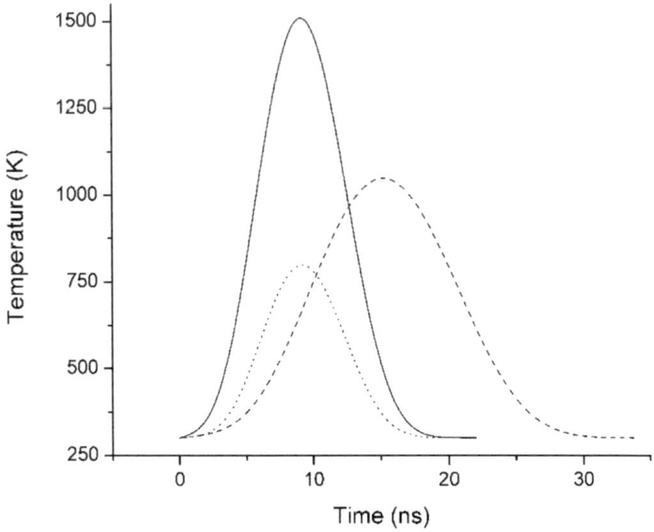

Fig. 6.16 Time–temperature profiles for 21-nm diameter nanospheres with incident laser pulse duration of 6 ns at an energy density of 0.13 J/cm^2 (*solid*) and 0.0325 J/cm^2 (*dot*), and nanorods with an effective diameter of 23.6 nm irradiated by a 10-ns, 0.01 J/cm^2 pulse (*dash*). Reprinted from [9]. © With permission from Elsevier

contrasting characteristics, such as pulse duration, particle size and shape, and medium in the single-pulse mode of heating. Here we reproduce our results published in Ref. [9]. Considering particle shape—between gold nanorods and gold nanospheres—changing a particle shape dramatically alters the peak absorption wavelengths and the respective coefficients of absorption, but these can be equated to spherical particles with an effective radius and aspect ratio $\left(\mathrm{AR} = \frac{\mathrm{Length}}{\mathrm{Diameter}}\right)$. From Table 5.4 (Sect. 5.1), we can see that the gold nanorods with $r_{\mathrm{eff}} = 11.8$ nm and AR = 3.2 have an absorption efficiency around 11.5 when irradiated with 725-nm light. Contrasted with gold nanospheres ($r_0 = 11.5$ nm, $K_{\mathrm{abs}} = 1.29$ at $\lambda = 532$ nm), a much lower energy density can be used to reach the same ablation temperature for the nanorods due to the extremely high absorption efficiency. Figure 6.16 shows the time–temperature profiles for nanospheres ($d = 21$ nm) under 0.13 J/cm^2 and 0.0325 J/cm^2 pulses at 532 nm, as well as the temperature profile for the nanorods ($r_{\mathrm{eff}} = 11.8$ nm, AR = 3.2) irradiated by 0.01 J/cm^2. This figure demonstrates how a much lower energy density can result in higher nanorod temperatures when compared to nanospheres of equivalent radius. The tunable characteristics of nanorods allow greater adjustment for what wavelength will result in a maximum absorption efficiency and how high the absorption efficiency will be, which makes nanorods potentially more effective for ablative treatments than nanospheres. The nanorod heating can be described similarly to nanospheres which, as we show later, do not undergo an accumulative heating effect.

6.3.2.7 Effect of the Pulse Duration in the Single-Pulse Mode of Nanoparticle Heating

Another characteristic of the experimental setups analyzed is a laser pulse duration of roughly 10 ns [9]. Figure 6.17 shows time–temperature profiles for five different pulse durations ranging from 6 to 20 ns. It follows from our calculations that during the laser pulse, the transfer of heat from the nanoparticles into the surrounding media is slight, and the particles rapidly reach their maximum temperature. The heating rate is about 10^{12} Ks^{-1}, depending on the pulse duration and incident energy density. After the laser pulse stops, the transfer of heat from the particle to the surrounding medium becomes critical, since the energy source is no longer present in the system. At this instant, the cooling rate reaches almost the same values as a heating rate due to large surface area of the nanoparticle and high thermal conductivity of the surrounding aqueous medium.

The temperature of the particle and surrounding medium remains relatively high (above 425 K) up to 24.6 ns after the pulse is fired, exceeding the laser pulse duration by 2.5 times [9]. The time for a particle to reach the maximum temperature is longer than the laser pulse duration because the heating kinetics involves first the transfer of energy from the irradiating source to the electron subsystem, and then further to the phonon subsystem. This creates a time delay in reaching the maximum temperature by the particle. This is especially true for short and ultrashort pulse heating as discussed in Sect. 6.2 (Lecture 18). The total time for one cycle (heating from an initial temperature of 300 K to a maximum temperature and then

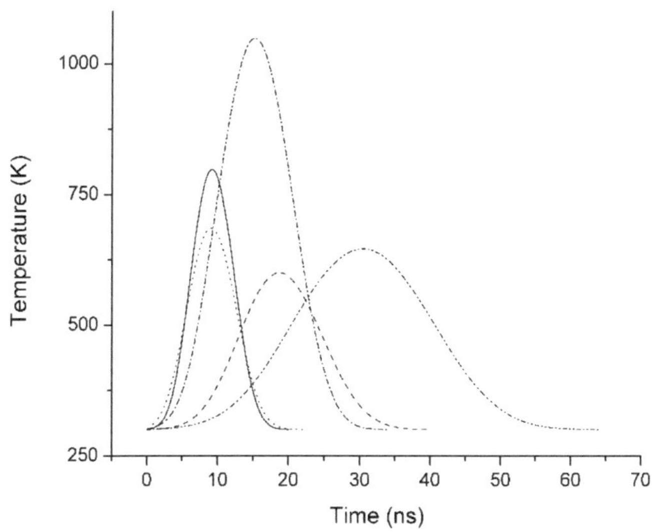

Fig. 6.17 Time–temperature profiles for five different pulse durations used in single-pulse mode heating. Three gold nanosphere cases: 12-ns pulse (*dash*), 20-ns pulse (*dash-dot-dot*), and 6 ns pulse (*solid*). Two gold nanorod cases: 10-ns pulse (*dash-dot*) and 7-ns pulse (*dot*). Reprinted from [9]. © With permission from Elsevier

cooling back to the initial temperature) is 32.8 ns, over triple the pulse duration. Varying the pulse duration changes the maximum temperature of the particle as well as the time span of heating and cooling.

In conclusion, we have performed time-dependent simulations and detailed analyses of different nonstationary laser–nanoparticle interactions in a single-pulse mode of heating [9]. Our analysis of the heating of metal nanoparticles in a biological cell environment shows that alterations to the particle size, shape, and material (metal) all have effects on the magnitude of the maximum particle temperature. Additionally, modifications to the laser energy density or wavelength can be used to change the maximum particle temperature. The most substantial change within single-pulse heating is the effect of pulse duration on the time dynamics of particle heating and cooling. Our simulations show that at the instance when the laser pulse stops, the cooling rate reaches almost the same values as the heating rate due to the large surface area of the nanoparticle and high thermal conductivity of the surrounding aqueous medium.

By increasing the pulse duration, the maximum nanoparticle temperature decreases, but heating and cooling both take longer and allow for a longer time period when the nanoparticles are above the denaturing temperature of 433 K. This increased duration of high particle temperature could prove beneficial to assure cellular damage. Alternatively, decreasing the pulse duration increases the maximum temperature reached by the nanoparticles while decreasing the time for heating and cooling. Such laser parameter modifications could allow for an easier and more refined treatment than modification of particle size or shape, while still having dramatic implications for the effectiveness of the treatment [9].

6.3.3 Practice Problem 5: Thermal Dynamics of Normal and Cancerous Ribosome in Cytoplasm Heated by a Single Pulse of Laser Radiation

6.3.3.1 Abstract (Slide 19.14)

In this computer practicum, we perform time-dependent simulations of soft tissue–laser interactions to determine the potential for cancer treatment via organelle-specific ablation. Given the effects of cancer on cell organelles, here we investigate if it is possible to selectively heat cancer organelles with lasers as a method of cancer treatment without using nanoparticles. By utilizing the physical changes experienced by ribosome organelles in abnormal cells, we can determine how laser irradiation would affect cancerous cells in comparison to healthy cells. The critical temperature for organelle heating is 425 K as that results in protein denaturing and thermal ablation. This section contains material adapted from our publication [16].

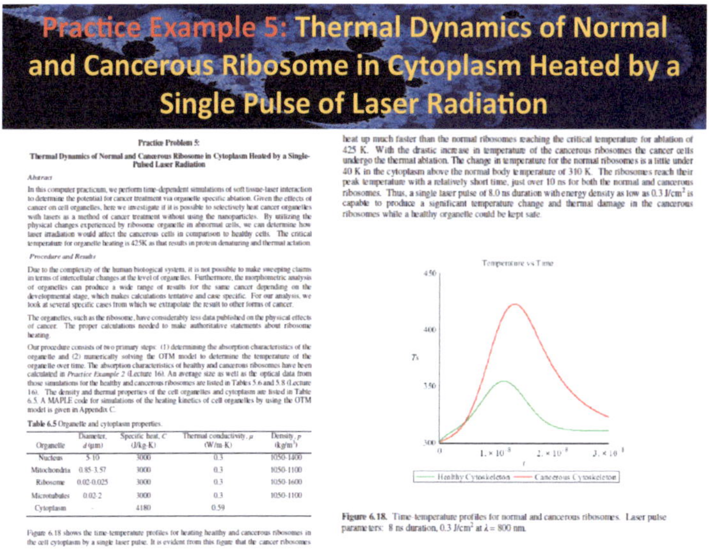

Practice Example 5: Thermal Dynamics of Normal and Cancerous Ribosome in Cytoplasm Heated by a Single Pulse of Laser Radiation

Practice Problem 5:

Thermal Dynamics of Normal and Cancerous Ribosome in Cytoplasm Heated by a Single-Pulsed Laser Radiation

Abstract

In this computer practicum, we perform time-dependent simulations of soft tissue-laser interaction to determine the potential for cancer treatment via organelle specific ablation. Given the effects of cancer on cell organelles, here we investigate if it is possible to selectively heat cancer organelles with lasers as a method of cancer treatment without using the nanoparticles. By utilizing the physical changes experienced by ribosome organelle in abnormal cells, we can determine how laser irradiation would affect the cancerous cells in comparison to healthy cells. The critical temperature for organelle heating is 425K as that results in protein denaturing and thermal ablation.

Procedure and Results

Due to the complexity of the human biological system, it is not possible to make sweeping claims in terms of intercellular changes at the level of organelles. Furthermore, the morphometric analysis of organelles can produce a wide range of results for the same cancer depending on the developmental stage, which makes calculations tentative and case specific. For our analysis, we look at several specific cases from which we extrapolate the result to other forms of cancer.

The organelles, such as the ribosome, have considerably less data published on the physical effects of cancer. The proper calculations needed to make authoritative statements about ribosome heating.

Our procedure consists of two primary steps: (1) determining the absorption characteristics of the organelle and (2) numerically solving the OTM model to determine the temperature of the organelle over time. The absorption characteristics of healthy and cancerous ribosomes have been calculated in *Practice Example 2* (Lecture 16). An average size as well as the optical data from those simulations for the healthy and cancerous ribosomes are listed in Tables 5.6 and 5.8 (Lecture 16). The density and thermal properties of the cell organelles and cytoplasm are listed in Table 6.5. A MAPLE code for simulations of the heating kinetics of cell organelles by using the OTM model is given in Appendix C.

Table 6.5 Organelle and cytoplasm properties.

Organelle	Diameter, d (μm)	Specific heat, C (J/kg-K)	Thermal conductivity, μ (W/m-K)	Density, ρ (kg/m^3)
Nucleus	5-10	3000	0.3	1050-1400
Mitochondria	0.85-3.57	3000	0.3	1050-1100
Ribosome	0.02-0.025	3000	0.3	1050-1600
Microtubules	0.02-2	3000	0.3	1050-1100
Cytoplasm	-	4180	0.59	

Figure 6.18 shows the time-temperature profiles for heating healthy and cancerous ribosomes in the cell cytoplasm by a single laser pulse. It is evident from this figure that the cancer ribosomes heat up much faster than the normal ribosomes reaching the critical temperature for ablation of 425 K. With the drastic increase in temperature of the cancerous ribosomes the cancer cells undergo the thermal ablation. The change in temperature for the normal ribosomes is a little under 40 K in the cytoplasm above the normal body temperature of 310 K. The ribosomes reach their peak temperature with a relatively short time, just over 10 ns for both the normal and cancerous ribosomes. Thus, a single laser pulse of 8.0 ns duration with energy density as low as 0.3 J/cm^2 is capable to produce a significant temperature change and thermal damage in the cancerous ribosomes while a healthy organelle could be kept safe.

Figure 6.18. Time-temperature profiles for normal and cancerous ribosomes. Laser pulse parameters: 8 ns duration, 0.3 J/cm^2 at λ = 800 nm.

Slide 19.14 Practice Example 5: thermal dynamics of normal and cancerous ribosome in cytoplasm heated by a single pulse of laser radiation

6.3.3.2 Procedure and Results

Due to the complexity of the human biological system, it is not possible to make sweeping claims in terms of intercellular changes at the level of organelles. Furthermore, the morphometric analysis of organelles can produce a wide range of results for the same cancer depending on the developmental stage, which makes calculations tentative and case specific. For our analysis, we look at several specific cases from which we extrapolate the result to other forms of cancer [16].

Organelles, such as ribosome, have considerably less data published on the physical effects of cancer. Proper calculations are needed to make authoritative statements about ribosome heating.

Our procedure consists of two primary steps [16]: (1) determining the absorption characteristics of the organelle and (2) numerically solving the OTM model to determine the temperature of the organelle over time. The absorption characteristics of healthy and cancerous ribosomes have been calculated in *Practice Example 2* (Lecture 16). An average size as well as the optical data from those simulations for the healthy and cancerous ribosomes is listed in Tables 5.6 and 5.8 (Lecture 16). The density and thermal properties of the cell organelles and cytoplasm are listed in Table 6.5. A MAPLE code for simulations of the heating kinetics of cell organelles by using the OTM model is given in Appendix C.

Figure 6.18 shows the time–temperature profiles for heating healthy and cancerous ribosomes in the cell cytoplasm by a single laser pulse. It is evident from this figure that the cancer ribosomes heat up much faster than the normal ribosomes,

Table 6.5 Organelle and cytoplasm properties

Organelle	Diameter, d (μm)	Specific heat, C (J/kg K)	Thermal conductivity, μ (W/m K)	Density, p (kg/m^3)
Nucleus	5–10	3000	0.3	1050–1400
Mitochondria	0.85–3.57	3000	0.3	1050–1100
Ribosome	0.02–0.025	3000	0.3	1050–1600
Microtubules	0.02–2	3000	0.3	1050–1100
Lysosomes	0.1–0.25	3000	0.3	1050–1100
Cytoskeleton	0.004–0.012	3000	0.3	1050–1100
Cytoplasm	–	4180	0.59	

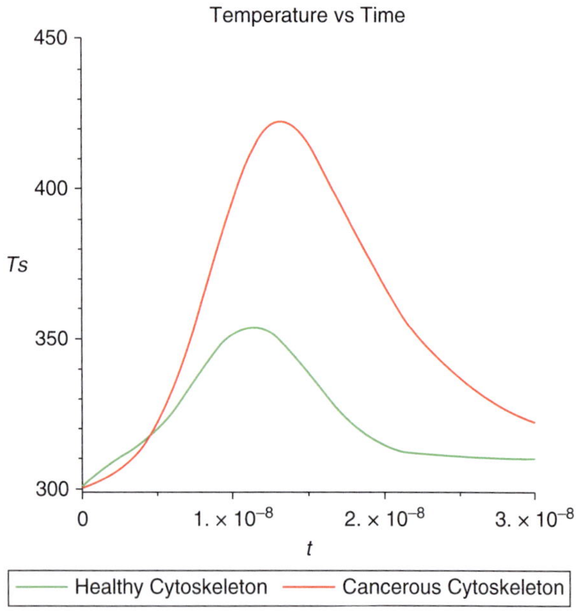

Fig. 6.18 Time–temperature profiles for normal and cancerous ribosomes. Laser pulse parameters: 8 ns duration, 0.3 J/cm^2 at $\lambda = 800$ nm

reaching the critical temperature for ablation of 425 K. With the dramatic increase in temperature of the cancerous ribosomes, the cancer cells undergo thermal ablation. The change in temperature for the normal ribosomes is a little under 40 K in the cytoplasm above the normal body temperature of 310 K. The ribosomes reach their peak temperature with a relatively short time, just over 10 ns for both the normal and cancerous ribosomes. Thus, a single laser pulse of 8.0 ns duration with energy density as low as 0.3 J/cm^2 is capable of producing a significant temperature change and thermal damage in the cancerous ribosomes, while a healthy organelle would be kept safe.

Appendix B: Maple Code for Nanoparticle Heating in a Single-Pulse Mode

This appendix contains a MAPLE code designed for using the OTM model to find the heating kinetics of different nanoparticles in a surrounding blood medium heated by a single radiation pulse (Slide 19.15). The comments within the MAPLE code should explain the variables, procedures, and functions that are used.

Note: This code comes directly from the MAPLE file and is therefore formatted to word-wrap according to MAPLE standard character-per-line limits.

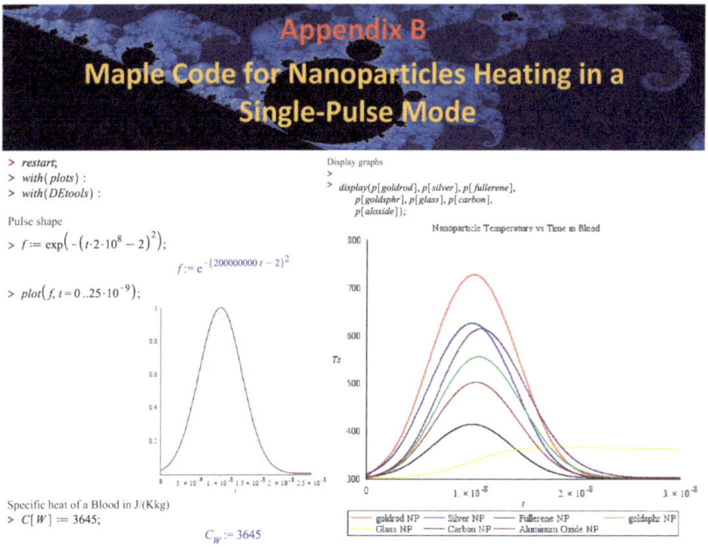

Slide 19.15 Appendix B: maple code for nanoparticles heating in a single-pulse mode

> $restart;$
> $with(plots):$
> $with(DEtools):$

Pulse shape

> $f := \exp\left(-\left(t \cdot 2 \cdot 10^8 - 2\right)^2\right);$

$$f := \mathrm{e}^{-(200000000\,t - 2)^2}$$

> $plot\left(f, t = 0..25 \cdot 10^{-9}\right);$

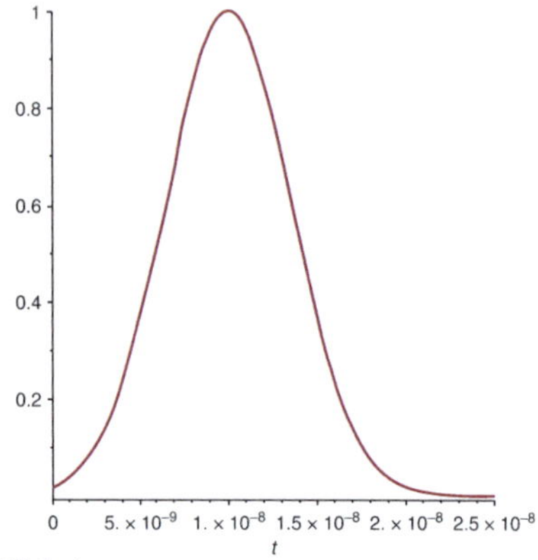

Specific heat of a blood in J/(K kg)

> $C[W] := 3645;$

$$C_W := 3645$$

Radius of a gold sphere

> $\mathrm{r}[goldsphr] := 30 \cdot 10^{-7};$

$$r_{goldsphr} := \frac{3}{1000000}$$

Effective radius of a gold nanorod

> $\mathrm{r}[goldrod] := 17.9 \cdot 10^{-7};$

$$r_{goldrod} := 0.000001790000000$$

Radius of silver in cm

> $\mathrm{r}[silver] := 10 \cdot 10^{-7};$

$$r_{silver} := \frac{1}{1000000}$$

Radius of fullerene in cm

> $r[fullerene] := 30 \cdot 10^{-7};$

$$r_{fullerene} := \frac{3}{1000000}$$

Radius of glass in cm

```
> r[glass] := 0.0001;
```

$$r_{glass} := 0.0001$$

Radius of carbon in cm

```
> r[carbon] := 1.07·10^(-5);
```

$$r_{carbon} := 0.00001070000000$$

Radius of aluminum oxide in cm

```
> r[aloxide] := 5.0·10^(-6);
```

$$r_{aloxide} := 0.000005000000000$$

Absorption efficiencies

```
> Q[goldsphr] := 4.376;
```

$$Q_{goldsphr} := 4.376$$

```
> Q[goldrod] := 14.5;
```

$$Q_{goldrod} := 14.5$$

```
> Q[silver] := 17.79;
```

$$Q_{silver} := 17.79$$

```
> Q[fullerene] := 1.58;
```

$$Q_{fullerene} := 1.58$$

```
> Q[glass] := 0.46735;
```

$$Q_{glass} := 0.46735$$

```
> Q[carbon] := 1.6353;
```

$$Q_{carbon} := 1.6353$$

```
> Q[aloxide] := 1.91514;
```

$$Q_{aloxide} := 1.91514$$

Initial temperature in K

```
> T[0] := 300;
```

$$T_0 := 300$$

Pulse duration in s

```
> tau := 8·10^(-9);
```

$$\tau := \frac{1}{125000000}$$

Pulse energy density in J/cm^2

> $E := 0.15;$

$$E := 0.15$$

Incident intensity of the radiation in W/cm^2

> $I0 := \frac{E}{\text{tau}};$

$$I0 := 1.875000000\ 10^7$$

Power exponent

> $a := 1;$

$$a := 1$$

Densities in kg/cm^3

> $\text{rho}[silver] := 0.0105;$

$$\rho_{silver} := 0.0105$$

> $\text{rho}[goldsphr] := 0.0193;$

$$\rho_{goldsphr} := 0.0193$$

> $\text{rho}[fullerene] := 0.00172;$

$$\rho_{fullerene} := 0.00172$$

> $\text{rho}[glass] := 0.0023;$

$$\rho_{glass} := 0.0023$$

> $\text{rho}[carbon] := 0.002210;$

$$\rho_{carbon} := 0.002210$$

> $\text{rho}[aloxide] := 0.003979;$

$$\rho_{aloxide} := 0.003979$$

Heat conductivity of the surrounding medium (blood) in W/(cm K)

> $\text{mu}[medium] := 0.5;$

$$\mu_{medium} := 0.5$$

One-temperature model (OTM) equations

>
$$DE[goldrod] := diff(Ts(t), t) = 3 \cdot Q[goldrod] \cdot I0 \cdot f / 4 / r[goldrod] / C[W] / \text{rho}[goldsphr] - \text{mu}[medium] \cdot T[0] / (a+1) / r[goldrod]^2 / C[W] / \text{rho}[goldsphr] \cdot \big((Ts(t) / T[0])^{(a+1)} - 1 \big);$$

$$DE_{goldrod} := \frac{d}{dt} Ts(t) = 1.619282660\ 10^{12}\ e^{-(200000000\,t-2)^2} - 3.697071504\ 10^{6}\ Ts(t)^2 + 3.327364354\ 10^{11}$$

> $DE[silver] := diff(Ts(t), t) = 3 \cdot Q[silver] \cdot I0 \cdot f / 4 / r[silver] / C[W] / \mathrm{rho}[silver] - \mathrm{mu}[medium] \cdot T[0] / (a+1) / r[silver]^2 / C[W] / \mathrm{rho}[silver] \cdot \left((Ts(t)/T[0])^{(a+1)} - 1 \right);$

$$DE_{silver} := \frac{d}{dt} Ts(t) = 6.536596120\ 10^{12}\ e^{-(200000000\,t-2)^2} - 2.177368433\ 10^{7}\ Ts(t)^2 + 1.959631590\ 10^{12}$$

> $DE[fullerene] := diff(Ts(t), t) = 3 \cdot Q[fullerene] \cdot I0 \cdot f / 4 / r[fullerene] / C[W] / \mathrm{rho}[fullerene] - \mathrm{mu}[medium] \cdot T[0] / (a+1) / r[fullerene]^2 / C[W] / \mathrm{rho}[fullerene] \cdot \left((Ts(t)/T[0])^{(a+1)} - 1 \right);$

$$DE_{fullerene} := \frac{d}{dt} Ts(t) = 1.181333142\ 10^{12}\ e^{-(200000000\,t-2)^2} - 1.476897192\ 10^{7}\ Ts(t)^2 + 1.329207473\ 10^{12}$$

> $DE[goldsphr] := diff(Ts(t), t) = 3 \cdot Q[goldsphr] \cdot I0 \cdot f / 4 / r[goldsphr] / C[W] / \mathrm{rho}[goldsphr] - \mathrm{mu}[medium] \cdot T[0] / (a+1) / r[goldsphr]^2 / C[W] / \mathrm{rho}[goldsphr] \cdot \left((Ts(t)/T[0])^{(a+1)} - 1 \right);$

$$DE_{aloxide} := \frac{d}{dt} Ts(t) = 3.713826292\ 10^{11}\ e^{-(200000000\,t-2)^2} - 2.298302944\ 10^{6}\ Ts(t)^2 + 2.068472650\ 10^{11}$$

>

Numerical solutions

```
> soln[silver] := dsolve({DE[silver], Ts(0) = 300}, Ts(t), numeric);
```

$$soln_{silver} := \mathbf{proc}(x\_rkf45)\ \dots\ \mathbf{end\ proc}$$

```
> soln[goldrod] := dsolve({DE[goldrod], Ts(0) = 300}, Ts(t), numeric);
```

$$soln_{goldrod} := \mathbf{proc}(x\_rkf45)\ \dots\ \mathbf{end\ proc}$$

```
> soln[fullerene] := dsolve({DE[fullerene], Ts(0) = 300}, Ts(t), numeric);
```

$$soln_{fullerene} := \mathbf{proc}(x\_rkf45)\ \dots\ \mathbf{end\ proc}$$

```
> soln[goldsphr] := dsolve({DE[goldsphr], Ts(0) = 300}, Ts(t), numeric);
```

$$soln_{goldsphr} := \mathbf{proc}(x\_rkf45)\ \dots\ \mathbf{end\ proc}$$

```
> soln[glass] := dsolve({DE[glass], Ts(0) = 300}, Ts(t), numeric);
```

$$soln_{glass} := \mathbf{proc}(x\_rkf45)\ \dots\ \mathbf{end\ proc}$$

```
> soln[carbon] := dsolve({DE[carbon], Ts(0) = 300}, Ts(t), numeric);
```

$$soln_{carbon} := \mathbf{proc}(x\_rkf45)\ \dots\ \mathbf{end\ proc}$$

```
> soln[aloxide] := dsolve({DE[aloxide], Ts(0) = 300}, Ts(t), numeric);
```

$$soln_{aloxide} := \mathbf{proc}(x\_rkf45)\ \dots\ \mathbf{end\ proc}$$

>

Recording numerical output

> **for** z **from** 0 **to** 50 **do** $zz[z] := soln[silver](z \cdot 10^{-9})$;**od**:

> **for** z **from** 0 **to** 50 **do** $zz[z] := soln[goldrod](z \cdot 10^{-11})$;**od**:

> **for** z **from** 0 **to** 50 **do** $zz[z] := soln[fullerene](z \cdot 10^{-10})$;**od**:

> **for** z **from** 0 **to** 50 **do** $zz[z] := soln[goldsphr](z \cdot 10^{-11})$;**od**:

> **for** z **from** 0 **to** 50 **do** $zz[z] := soln[glass](z \cdot 10^{-12})$;**od**:

> **for** z **from** 0 **to** 50 **do** $zz[z] := soln[carbon](z \cdot 10^{-13})$;**od**:

> **for** z **from** 0 **to** 50 **do** $zz[z] := soln[aloxide](z \cdot 10^{-14})$;**od**:

Creating graphical output

>
$$p[goldrod] := odeplot\left(soln[goldrod], [t, Ts(t)], 0\,..\frac{30}{10^9}, view = \left[0\,..\frac{30}{10^9}, 300\,..800\right], numpoints = 1000, legend = \text{"goldrod NP"}, color = red\right):$$

> $p[silver] := odeplot\left(soln[silver], [t, Ts(t)], 0 .. \frac{30}{10^9}, view = \left[0 .. \frac{30}{10^9}, 300 .. 800\right], numpoints = 1000, legend = \text{"Silver NP"}, color = blue, title = \text{"Nanoparticle Temperature vs Time in Blood"}\right) :$

> $p[fullerene] := odeplot\left(soln[fullerene], [t, Ts(t)], 0 .. \frac{30}{10^9}, view = \left[0 .. \frac{30}{10^9}, 300 .. 800\right], numpoints = 1000, legend = \text{"Fullerene NP"}, color = black\right) :$

> $p[goldsphr] := odeplot\left(soln[goldsphr], [t, Ts(t)], 0 .. \frac{30}{10^9}, view = \left[0 .. \frac{30}{10^9}, 300 .. 800\right], numpoints = 1000, legend = \text{"goldsphr NP"}, color = green\right) :$

> $p[glass] := odeplot\left(soln[glass], [t, Ts(t)], 0 .. \frac{30}{10^9}, view = \left[0 .. \frac{30}{10^9}, 300 .. 800\right], numpoints = 1000, legend = \text{"Glass NP"}, color = yellow\right) :$

> $p[carbon] := odeplot\left(soln[carbon], [t, Ts(t)], 0 .. \frac{30}{10^9}, view = \left[0 .. \frac{30}{10^9}, 300 .. 800\right], numpoints = 1000, legend = \text{"Carbon NP"}, color = purple\right) :$

$$> p[aloxide] := odeplot\left(soln[aloxide], [t, Ts(t)], 0..\frac{30}{10^9}, view = \left[0..\frac{30}{10^9}, 300..800\right], numpoints = 1000, legend = \text{"Aluminum Oxide NP"}, color = brown\right):$$

\>

Display graphs

\>

$$> display(p[goldrod], p[silver], p[fullerene], p[goldsphr], p[glass], p[carbon], p[aloxide]);$$

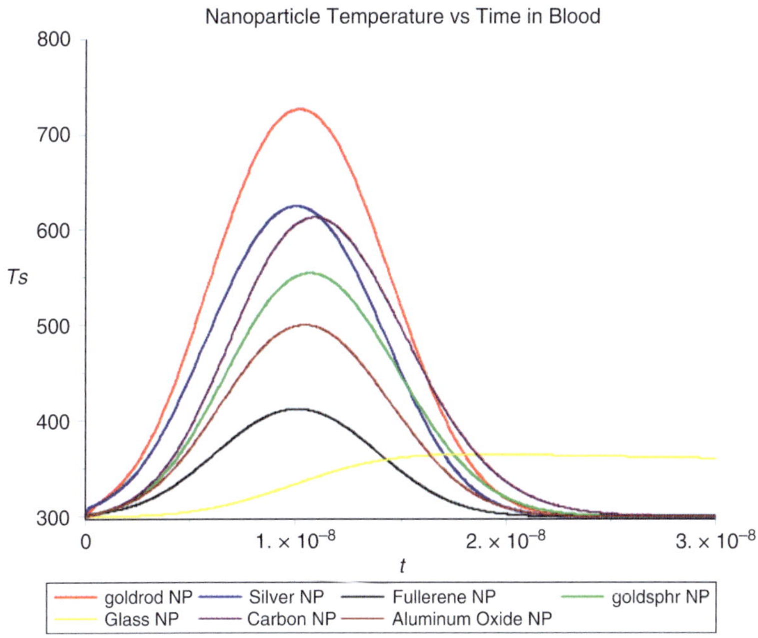

Appendix C: Maple Code for Cell Organelle Heating in a Single-Pulse Mode

This appendix contains the MAPLE code designed to use the OTM model to find the heating kinetics of healthy and cancerous ribosomes in a cytoplasm heated by single pulse of radiation (Slide 19.16). The comments within the MAPLE code should explain the variables, procedures, and functions that are used.

Note: This code comes directly from the MAPLE file and is therefore formatted to word-wrap according to MAPLE standard character-per-line limits.

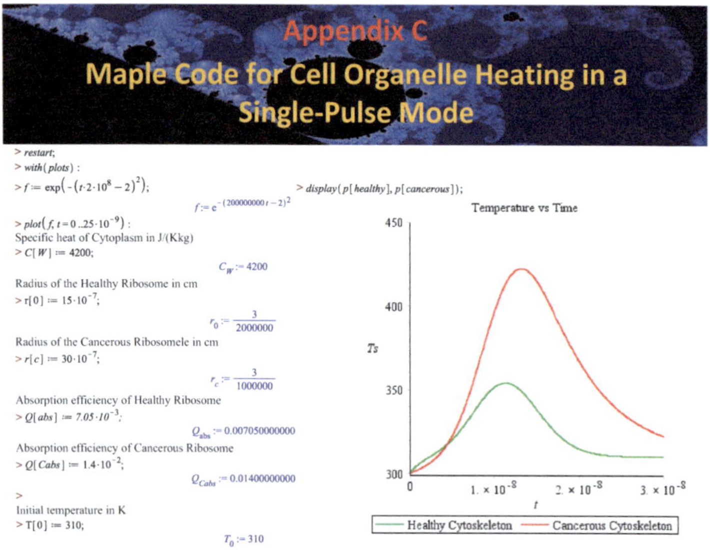

Slide 19.16 Appendix C: maple code for cell organelle heating in a single-pulse mode

> *restart*;

> *with*(*plots*) :

> $f := \exp\left(-\left(t \cdot 2 \cdot 10^8 - 2\right)^2\right);$

$$f := e^{-(200000000\,t-2)^2}$$

> $plot\left(f, t = 0 ..25 \cdot 10^{-9}\right)$:

Specific heat of cytoplasm in J/(K kg)

> $C[W] := 4200;$

$$C_W := 4200$$

Radius of healthy ribosome in cm

> $r[0] := 15 \cdot 10^{-7};$

$$r_0 := \frac{3}{2000000}$$

Radius of cancerous ribosomele in cm

> $r[c] := 30 \cdot 10^{-7};$

$$r_c := \frac{3}{1000000}$$

Absorption efficiency of healthy ribosome

> $Q[abs] := 7.05 \cdot 10^{-3};$

$$Q_{abs} := 0.007050000000$$

Absorption efficiency of cancerous ribosome

> $Q[Cabs] := 1.4 \cdot 10^{-2};$

$$Q_{Cabs} := 0.01400000000$$

>

Initial temperature in K

> $T[0] := 310;$

$$T_0 := 310$$

Laser pulse duration in s

> $tau := 8 \cdot 10^{-9};$

$$\tau := \frac{1}{125000000}$$

Laser pulse energy density in J/cm^2

> $E := 0.3;$

$$E := 0.3$$

Incident intensity of the laser light in W/cm^2

> $I0 := \frac{E}{tau};$

$$I0 := 3.75000000\ 10^7$$

Power exponent

> $a := 1;$

$$a := 1$$

>

Density of rganelle in kg/cm^3

```
> rho[0] := 0.001050;
```

$$\rho_0 := 0.001050$$

Heat conductivity of cytoplasm in W/(cm K)

```
> mu[medium] := 0.0059;
```

$$\mu_{medium} := 0.0059$$

One-temperature model (OTM) equations

```
>
> DE[healthy] := diff(Ts(t), t) = 3·Q[abs]·I0·f/4/r[0]
    /C[W]/rho[0] - mu[medium]·T[0]/(a + 1)
    /r[0]^2/C[W]/rho[0]·((Ts(t)/T[0])^(a+1) - 1);
```

$$DE_{healthy} := \frac{d}{dt} Ts(t) = 2.997448979\ 10^{10}\ e^{-(200000000\,t - 2)^2} - 9.590455057\ 10^{5}\ Ts(t)^2 + 9.216427310\ 10^{10}$$

```
> DE[cancerous] := diff(Ts(t), t) = 3·Q[Cabs]·I0·f/4
    /r[c]/C[W]/rho[0] - mu[medium]·T[0]/(a + 1)
    /r[c]^2/C[W]/rho[0]·((Ts(t)/T[0])^(a+1) - 1);
```

$$DE_{cancerous} := \frac{d}{dt} Ts(t) = 2.976190477\ 10^{10}\ e^{-(200000000\,t - 2)^2} - 2.397613765\ 10^{5}\ Ts(t)^2 + 2.304106828\ 10^{10}$$

Numerical solutions

```
> soln[healthy] := dsolve({DE[healthy], Ts(0) = 300},
    Ts(t), numeric);
```

$soln_{healthy}$:= **proc**(*x_rkf45*) ... **end proc**

```
> soln[cancerous] := dsolve({DE[cancerous], Ts(0)
    = 300}, Ts(t), numeric);
```

$soln_{cancerous}$:= **proc**(*x_rkf45*) ... **end proc**

Arranging the numerical and graphical output

```
> for z from 0 to 50 do zz[z] := soln[healthy](z·10^(-9));od:

> for z from 0 to 50 do zz[z] := soln[cancerous](z·10^(-9));
    od:
```

$$> p[healthy] := odeplot\left(soln[healthy], [t, Ts(t)], 0..\frac{30}{10^9}, view = \left[0..\frac{30}{10^9}, 300..450\right], numpoints = 1000, legend = \text{"Healthy Cytoskeleton"}, color = green, title = \text{"Temperature vs Time"}\right):$$

$$> p[cancerous] := odeplot\left(soln[cancerous], [t, Ts(t)], 0..\frac{30}{10^9}, view = \left[0..\frac{30}{10^9}, 300..450\right], numpoints = 1000, legend = \text{"Cancerous Cytoskeleton"}, color = red\right):$$

$$> display(p[healthy], p[cancerous]);$$

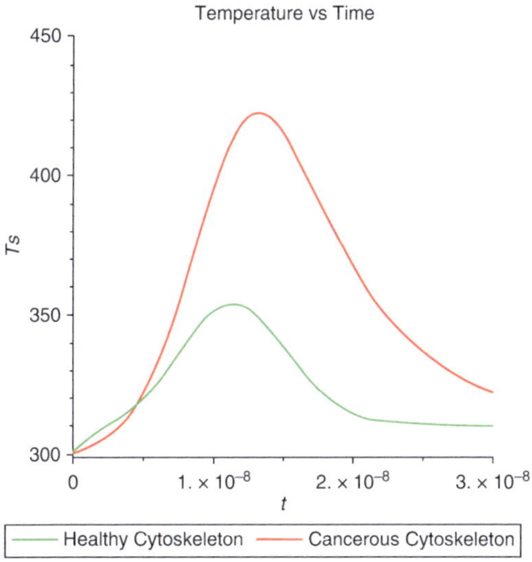

Homework Exercises

Section 6.1: Kinetics of Heating and Cooling of Nanoparticles: Time-Dynamic Models

6.1. What is nanophotohyperthermia? How is it different from nanophotothermolysis?
6.2. What is nanophotothermolysis? How is it different from nanophotohyperthermia?

6.3. What temperature range can trigger cell death in cancerous cells?
6.4. List possible cancer cells killing mechanisms caused by heat.
6.5. List three main factors affecting the heating of nanoparticles.
6.6. Select one or more. The main factors affecting the heating of nanoparticles are as follows:

(a). Particle material, size, and shape
(b). Radiation wavelength, energy density, duration, pulse shape, and generation frequency
(c). Properties of the surrounding medium
(d). Thermal coagulation
(e). Shock-wave generation
(f). Sound production
(g). All of the above
(h). None of the above

6.7. Describe the research task in modeling the thermal interaction of particles and radiation.
6.8. Describe the terms in the heat transfer equation.
6.9. True or false? The heat transfer equation describes the following phenomena: heat generation, heat diffusion, and mass transfer across a boundary of the system.
6.10. List the limitations and assumptions used in a thermodynamic model for heating the nanoparticles.
6.11. What is a main mechanism of heating nanoparticle by radiation?
6.12. True or false? In the event of heating, the nanoparticle has two temperatures: electron T_e and phonon T_s subsystem temperatures.
6.13. What is the electron–phonon coupling time?
6.14. List three main time dynamic heating modes?
6.15. True or false? The femtosecond mode of heating is realized when the laser pulse duration is longer than the electron thermalization and cooling times.
6.16. True or false? The picosecond mode of heating is observed when the laser pulse duration is comparable to the lattice heating time.
6.17. True or false? The nanosecond mode of heating is applied when the laser pulse duration is shorter than the electron–phonon coupling time.
6.18. What is a main assumption made in one-temperature model?
6.19. True or false? Heat diffuses very slowly within a gold nanoparticle.
6.20. Describe the terms in the one-temperature model.
6.21. True or false? Electron cooling has a time delay on the order of picoseconds.
6.22. True or false? Lattice heating has a time delay on the order of femtoseconds.
6.23. In which case is the two-temperature model applicable?
6.24. True or false? The ultrashort laser pulse ends before the transfer of energy to the lattice is complete.
6.25. Describe the terms and equations in the two-temperature model.
6.26. List the characteristic time scales for the electron–phonon coupling process.

Section 6.2: Ultrashort Laser Pulse Heating of Nanoparticles in Femtosecond, Picosecond, and Nanosecond Modes

6.27. True or false? Free electrons are the first to absorb radiation energy, rapidly attaining high temperatures and transferring thermal energy to the material lattice.
6.28. What are the output results from the two-temperature model?
6.29. What is the time range for establishing thermal equilibrium among the excited electrons in a nanoparticle?
6.30. Is the one-temperature model appropriate for understanding ultrashort pulse interactions with nanoparticles?
6.31. What is the uniform heating approximation?
6.32. What is the time range for the electrons cooling within a metal nanoparticle?
6.33. In what time range is the two-temperature model applicable?
6.34. True or false? Heat lost from the surface of the nanoparticle into the surrounding medium is observed on the femtosecond time scale.
6.35. Calculate the characteristic lattice heating time for a silver nanoparticle of 10-nm radius.
6.36. Calculate the characteristic lattice heating time for a fullerene nanoparticle of 10-nm radius.
6.37. Calculate the characteristic lattice heating time for an aluminum oxide nanoparticle of 10-nm radius.
6.38. Calculate the characteristic lattice heating time for a magnesium oxide nanoparticle of 10-nm radius.
6.39. Calculate the characteristic lattice heating time for a nickel oxide nanoparticle of 10-nm radius.
6.40. True or false? The highest temperature of a nanoparticle heated by an 8.0-ns laser pulse is observed when the pulse has already degraded.
6.41. True or false? The cooling kinetics of a nanoparticle in a surrounding biological medium has the same rate as the heating kinetics of the nanoparticle.
6.42. Describe the effect of different biological surroundings on the laser heating dynamics of nanoparticles.
6.43. Why is higher overheating observed in a surrounding fat medium?
6.44. Describe the effect of the particle's radius on the temperature dynamics of the nanoparticle.
6.45. Name two competitive factors affecting laser heating of a nanoparticle.
6.46. True or false? The absorption efficiency of a gold nanoparticle increases with decreasing size of the particle.
6.47. True or false? The heating rate increases for smaller particles.
6.48. In what radii range is the optical effect on heating nanoparticle stronger than thermal effects?
6.49. For what size range do thermal processes dominate over the particle's optical properties?

6.50. Explain the effect of saturation of the maximal temperature curve observed for large particle radii shown on Fig. 6.5.
6.51. How long do the electrons remain in the thermal equilibrium state?
6.52. On what time scale does heat lost from the surface of the nanoparticle into the surrounding medium become noticeable?

Section 6.3: Computer Practicum 3: Single-Pulse Heating of Nano/Bioparticles

6.53. List the assumptions made in one-temperature model simulations.
6.54. True or false? Possible vapor bubble formation around a nanotarget in a surrounding aqueous medium due to heating of the nanoparticle does change the absorption efficiency of the nanoparticles.
6.55. Name three main laser pulse profiles.
6.56. How is the pulse duration defined?
6.57. How is the intensity of the radiation calculated?
6.58. Perform time dynamic simulations of a silver nanoparticle heated by a 8.0-ns single laser pulse for an optimal radius of a particle in cytoplasm, fat, and tumor surrounding media by using thermal data from Tables 6.1, 6.2, 6.3, and 6.4 and the Maple code in Appendix B. Find the optimal radius of the nanoparticle, optimal wavelength of the radiation, absorption efficiency, and energy density of the radiation required to achieve a maximum temperature of nanoparticle in each surrounding medium. Write a report of your simulation results according to the Practice Example 4. Make conclusions about the heating and cooling kinetics of the nanoparticle in different surrounding media.
6.59. Perform time dynamic simulations of a fullerene nanoparticle heated by a 8.0-ns single laser pulse for an optimal radius of a particle in cytoplasm, fat, and tumor surrounding media by using thermal data from Tables 6.1, 6.2, 6.3, and 6.4 and the Maple code in Appendix B. Find the optimal radius of the nanoparticle, optimal wavelength of the radiation, absorption efficiency, and energy density of the radiation required to achieve a maximum temperature of nanoparticle in each surrounding medium. Write a report of your simulation results according to the Practice Example 4. Make conclusions about the heating and cooling kinetics of the nanoparticle in different surrounding media.
6.60. Perform time dynamic simulations of a polystyrene nanoparticle heated by a 8.0-ns single laser pulse for an optimal radius of a particle in cytoplasm, fat, and tumor surrounding media by using thermal data from Tables 6.1, 6.2, 6.3, and 6.4 and the Maple code in Appendix B. Find the optimal radius of the nanoparticle, optimal wavelength of the radiation, absorption efficiency, and energy density of the radiation required to achieve a maximum temperature of nanoparticle in each surrounding medium. Write a report of your

simulation results according to the Practice Example 4. Make conclusions about the heating and cooling kinetics of the nanoparticle in different surrounding media.

6.61. Perform time dynamic simulations of a glass nanoparticle heated by a 8.0-ns single laser pulse for an optimal radius of a particle in cytoplasm, fat, and tumor surrounding media by using thermal data from Tables 6.1, 6.2, 6.3, and 6.4 and the Maple code in Appendix B. Find the optimal radius of the nanoparticle, optimal wavelength of the radiation, absorption efficiency, and energy density of the radiation required to achieve a maximum temperature of nanoparticle in each surrounding medium. Write a report of your simulation results according to the Practice Example 4. Make conclusions about the heating and cooling kinetics of the nanoparticle in different surrounding media.

6.62. Perform time dynamic simulations of a carbon nanoparticle heated by a 8.0-ns single laser pulse for an optimal radius of a particle in cytoplasm, fat, and tumor surrounding media by using thermal data from Tables 6.1, 6.2, 6.3, and 6.4 and the Maple code in Appendix B. Find the optimal radius of the nanoparticle, optimal wavelength of the radiation, absorption efficiency, and energy density of the radiation required to achieve a maximum temperature of nanoparticle in each surrounding medium. Write a report of your simulation results according to the Practice Example 4. Make conclusions about the heating and cooling kinetics of the nanoparticle in different surrounding media.

6.63. Perform time dynamic simulations of an aluminum oxide nanoparticle heated by a 8.0-ns single laser pulse for an optimal radius of a particle in cytoplasm, fat, and tumor surrounding media by using thermal data from Tables 6.1, 6.2, 6.3, and 6.4 and the Maple code in Appendix B. Find the optimal radius of the nanoparticle, optimal wavelength of the radiation, absorption efficiency, and energy density of the radiation required to achieve a maximum temperature of nanoparticle in each surrounding medium. Write a report of your simulation results according to the Practice Example 4. Make conclusions about the heating and cooling kinetics of the nanoparticle in different surrounding media.

6.64. Perform time dynamic simulations of a magnesium oxide nanoparticle heated by a 8.0-ns single laser pulse for an optimal radius of a particle in cytoplasm, fat, and tumor surrounding media by using thermal data from Tables 6.1, 6.2, 6.3, and 6.4 and the Maple code in Appendix B. Find the optimal radius of the nanoparticle, optimal wavelength of the radiation, absorption efficiency, and energy density of the radiation required to achieve a maximum temperature of nanoparticle in each surrounding medium. Write a report of your simulation results according to the Practice Example 4. Make conclusions about the heating and cooling kinetics of the nanoparticle in different surrounding media.

6.65. Perform time dynamic simulations of a nickel oxide nanoparticle heated by a 8.0-ns single laser pulse for an optimal radius of a particle in cytoplasm, fat, and tumor surrounding media by using thermal data from Tables 6.1, 6.2, 6.3, and 6.4 and the Maple code in Appendix B. Find the optimal radius of the nanoparticle, optimal wavelength of the radiation, absorption efficiency, and energy density of the radiation required to achieve a maximum temperature of nanoparticle in each surrounding medium. Write a report of your simulation results according to the Practice Example 4. Make conclusions about the heating and cooling kinetics of the nanoparticle in different surrounding media.

6.66. Perform time dynamic simulations of normal and cancerous cell nucleus heating by a 10.0-ns single laser pulse in cytoplasm by using thermal data from Tables 6.1, 6.2, 6.3, 6.4, and 6.5 and the Maple code in Appendix C. Write a report of your simulation results according to the Practice Example 5. Make conclusions about how laser heating would affect the cancerous organelles in comparison to healthy organelles.

6.67. Perform time dynamic simulations of normal and cancerous cell mitochondria heating by a 10.0-ns single laser pulse in cytoplasm by using thermal data from Tables 6.1, 6.2, 6.3, 6.4, and 6.5 and the Maple code in Appendix C. Write a report of your simulation results according to the Practice Example 5. Make conclusions about how laser heating would affect the cancerous organelles in comparison to healthy organelles.

6.68. Perform time dynamic simulations of normal and cancerous cell microtubules heating by a 10.0-ns single laser pulse in cytoplasm by using thermal data from Tables 6.1, 6.2, 6.3, 6.4, and 6.5 and the Maple code in Appendix C. Write a report of your simulation results according to the Practice Example 5. Make conclusions about how laser heating would affect the cancerous organelles in comparison to healthy organelles.

6.69. Perform time dynamic simulations of normal and cancerous cell lysosomes heating by a 10.0-ns single laser pulse in cytoplasm by using thermal data from Tables 6.1, 6.2, 6.3, 6.4, and 6.5 and the Maple code in Appendix C. Write a report of your simulation results according to the Practice Example 5. Make conclusions about how laser heating would affect the cancerous organelles in comparison to healthy organelles.

6.70. Perform time dynamic simulations of normal and cancerous cell cytoskeleton heating by a 10.0-ns single laser pulse in cytoplasm by using thermal data from Tables 6.1, 6.2, 6.3, 6.4, and 6.5 and the Maple code in Appendix C. Write a report of your simulation results according to the Practice Example 5. Make conclusions about how laser heating would affect the cancerous organelles in comparison to healthy organelles.

References

1. R.R. Letfullin, C.E.W. Rice, T.F. George, Bone cancer therapy by plasmonic nanoparticles. Ther. Deliv. **2**, 1259–1273 (2011)

2. R.R. Letfullin, T.F. George, G.C. Duree, B.M. Bollinger, Ultrashort laser pulse heating of nanoparticles: comparison of theoretical approaches. Adv. Opt. Technol. **2008**, ID 251718/1-8 (2008)
3. R.R. Letfullin, T.F. George, Plasmonic nanomaterials in nanomedicine, in *Springer Handbook of Nanomaterials*, ed. by R. Vajtai (Springer, Berlin, 2013), pp. 1063–1097
4. R.R. Letfullin, T.F. George, Nanoscale materials in strong ultrashort laser pulses, in *Computational Studies of New Materials II: From Ultrafast Processes and Nanostructures to Optoelectronics, Energy Storage and Nanomedicine*, ed. by T.F. George, D. Jelski, R.R. Letfullin, G.P. Zhang (World Scientific, Singapore, 2011), pp. 37–64
5. R.R. Letfullin, C. Joenathan, T.F. George, V.P. Zharov, Laser-induced explosion of gold nanoparticles: potential role for nanophotothermolysis of cancer. Nanomedicine **1**, 473–480 (2006)
6. R.R. Letfullin, V.I. Igoshin, T.F. George, Chemical laser based on a photon-branched chain reaction, in *Modern Topics in Chemical Physics*, ed. by T.F. George, X. Sun, G.P. Zhang (Research Signpost, Trivandrum, 2002), pp. 183–286
7. Z. Lin, L.V. Zhigilei, V. Celli, Electron-phonon coupling and electron heat capacity of metals under conditions of strong electron-phonon nonequilibrium. Phys. Rev. B **77**, 075133/1–17 (2008)
8. S.I. Anisimov, B.L. Kapeliovich, T.L. Perel'man, Electron emission from metal surfaces exposed to ultrashort laser pulses. Sov. Phys. JETP **39**, 375–377 (1974)
9. R.R. Letfullin, C.B. Iversen, T.F. George, Modeling nanophotothermal therapy: kinetics of thermal ablation of healthy and cancerous cell organelles and gold nanoparticles. Nanomed: Nanotechnol. Biol. Med. **7**, 137–145 (2011)
10. C.M. Pitsillides, E.K. Joe, X. Wei, R.R. Anderson, C.P. Lin, Selective cell targeting with light-absorbing microparticles and nanoparticles. Biophys. J. **84**, 4023–4032 (2003)
11. V.P. Zharov, R.R. Letfullin, E.N. Galitovskaya, Microbubbles-overlapping mode for laser killing of cancer cells with absorbing nanoparticle clusters. J. Phys. D. Appl. Phys. **38**, 2571–2581 (2005)
12. Z. Peng, T. Walther, K. Kleinermanns, Influence of intense pulsed laser irradiation on optical and morphological properties of gold nanoparticle aggregates produced by surface acid-base reactions. Langmuir **21**, 4249–4253 (2005)
13. V.S. Kalambur, E.K. Longmire, J.C. Bischof, Cellular level loading and heating of superparamagnetic iron oxide nanoparticles. Langmuir **23**, 12329–12336 (2007)
14. E.Y. Hleb, D.O. Lapotko, Photothermal properties of gold nanoparticles under exposure to high optical energies. Nanotechnology **19**, 1–10 (2008)
15. H. Takahashi, T. Niidome, A. Nariai, Y. Niidome, S. Yamada, Gold nanorod-sensitized cell death: microscopic observation of single living cells irradiated by pulsed near-infrared laser light in the presence of gold nanorods. Chem. Lett. **35**, 500–501 (2006)
16. R.R. Letfullin, T.F. George, Laser ablation of biological tissue by short and ultrashort pulses, in *Computational Studies of New Materials II: From Ultrafast Processes and Nanostructures to Optoelectronics, Energy Storage and Nanomedicine*, ed. by T.F. George, D. Jelski, R.R. Letfullin, G.P. Zhang (World Scientific, Singapore, 2011), pp. 191–218

Chapter 7
Time Dynamic Modes of Nano/Bioparticle Heating

Contents

Abstract Nanoparticles are being researched as a noninvasive method for selectively killing cancer cells. With particular antibody coatings on nanoparticles, they attach to abnormal cells of interest (cancer or otherwise). Once attached, nanoparticles can be heated with UV/visible/IR, RF, or X-ray pulses, heating the surrounding area of the cell to the point of death. Researchers often use single-pulse or multipulse modes of laser heating when conducting nanoparticle ablation

Electronic supplementary material: The online version of this chapter (doi:10.1007/978-3-319-43577-0_7) contains supplementary material, which is available to authorized users.

R.R. Letfullin, T.F. George, *Computational Nanomedicine and Nanotechnology*,
DOI 10.1007/978-3-319-43577-0_7

research. In this chapter, time-dependent simulations and detailed analyses are carried out for different nonstationary pulsed laser–nanoparticle interaction modes, and the advantages and disadvantages of single-pulse and multipulse (set of short pulses) heating of nanoparticles are shown. Simulations are performed for metal nanoparticles in a surrounding biological medium, as well as for healthy and cancerous cell organelles heated by optical, RF, and X-ray pulses. This chapter contains material adapted from our publications [1–3]. A detailed list of references and reviews on a given topic of this chapter can be found in those original papers.

7.1 Lecture 20: Modeling Photothermal Heating in a Multipulse Radiation Mode (Slide 20.1)

- Introduction
- Experiments
- Multipulse Mode of Heating
- Conclusions and Summary

7.1.1 Introduction

The application of nanotechnology for laser thermal-based killing of abnormal cells (e.g., cancer cells) targeted with absorbing nanoparticles (e.g., gold solid nanospheres, nanoshells, or nanorods) is becoming an extensive area of research [1]. As discussed in Sect. 2.3 (Lecture 7), by coating the surface of nanoparticles with a specific protein (a "targeting agent," normally an antibody), the nanoparticles will bind to a complementary protein such as found on a cancer

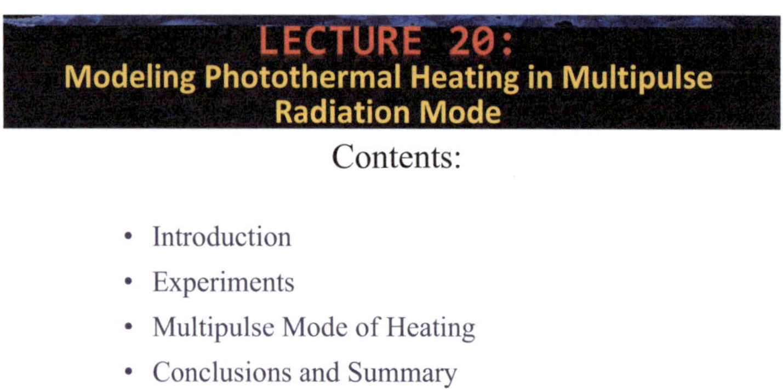

Slide 20.1 Contents

Introduction

- After the ***nanoparticles*** are bound to cancerous cells, they ***can be heated*** with electromagnetic radiation, ***inducing a variety of effects*** around the particles.
- ***These effects*** can be used to increase the sensitivity of photoacoustic ***diagnosis or aid in therapy***, such as ***selective photothermolysis***.
- ***The potential advantages*** of these photothermal sensitizers heated with short laser pulses may include:
    - ➢ ***selective cancer-cell targeting*** by means of conjugation of absorbing particles (e.g., gold nanospheres, nanoshells or nanorods) with specific antibodies;
    - ➢ ***localized tumor damage*** without harmful effects on surrounding healthy tissue;
    - ➢ ***absorption at X-rays or RF range*** wavelengths in the transparency window of most biotissues;
    - ➢ ***no undesired side effects*** (e.g., cytotoxicity or cutaneous photosensitivity); and
    - ➢ relatively ***fast treatment*** involving potentially just one or several pulses.

Slide 20.2 Introduction

cell. After the nanoparticles are bound to the cancerous cells, they can be heated with electromagnetic radiation, inducing a variety of effects around the particles [2] (Slide 20.2). The heated particle can cause the cell to experience hyperthermia, resulting in surface-protein denaturing and changing membrane permeability. Alternatively, the nanoparticles themselves can be heated to the point of melting, evaporation, or explosion [3], causing further damage to cells. These effects can be used to increase the sensitivity of photoacoustic diagnosis or aid in therapy, such as selective photothermolysis, by selective thermal killing of tumor cells into which absorbing nanoparticles have been incorporated. The potential advantages of these new photothermal sensitizers heated with short laser pulses include [2] (Slide 20.2):

- Selective cancer cell targeting by means of conjugation of absorbing particles (e.g., gold nanospheres, nanoshells, or nanorods) with specific antibodies.
- Localized tumor damage without harmful effects on surrounding healthy tissue.
- Absorption at X-ray or RF wavelengths in the transparency window of most biotissues.
- No undesired side effects (e.g., cytotoxicity or cutaneous photosensitivity).
- Relatively fast treatment involving potentially just one or several pulses.

As discussed in Chap. 6, the extent of the particle heating depends on many factors, divided primarily into three categories (Slide 20.3): (1) particle material, size, and shape; (2) laser pulse wavelength, energy density, duration, pulse shape, and generation frequency; and (3) properties of the surrounding medium.

Even with the many factors involved in an experimental setup that contribute to the individual heating characteristics of a nanoparticle, several factors stay relatively constant across many experimental situations [1] (Slide 20.3). Specifically,

Introduction (continued)

- The extent of the ***particle heating*** depends on many factors, divided primarily into ***three categories***:
  (1) ***particle material, size and shape***;
  (2) ***laser pulse wavelength, energy density, duration, pulse shape and generation frequency***; and
  (3) ***properties of the surrounding medium***.
- Even with the many factors involved in an experimental setup that contribute to the individual heating characteristics of a nanoparticle, ***several factors stay relatively constant*** across many experimental situations:
  - ➢ Specifically, ***spherical gold nanoparticles*** are commonly used because they are relatively easy to fabricate, nontoxic, easily conjugated to antibodies, and strong absorbers.
  - ➢ ***Nanosecond pulse-width*** lasers with pulse-firing frequencies of 10 Hz are often used because they are widely available and cheaper than pico- and femtosecond lasers or lasers with higher firing frequencies.
  - ➢ Additionally, ***water or a phosphate buffered saline solution*** is employed as a surrounding medium.

Slide 20.3 Introduction (continued)

spherical gold nanoparticles are commonly used because they are relatively easy to fabricate, nontoxic, easily conjugated to antibodies, and strong absorbers. Nanosecond pulse-width lasers with pulse-firing frequencies of 10 Hz are often used because they are widely available and cheaper than pico- and femtosecond lasers or lasers with higher firing frequencies. Additionally, water or a phosphate buffered saline solution is employed as a surrounding medium due to large similarities with a biological cell. Several experiments that use these factors are studied in our analysis later [4–10].

7.1.2 Experiments

Nanoparticle heating has been used experimentally for bacterial killing (Slide 20.4). Spherical gold nanoparticles of diameters 10, 20, and 40 nm are heated with laser light, where it has been determined that at 3–5 J/cm^2, only 1–3 pulses are required for harmful effects on bacteria, while at 0.5–1 J/cm^2, at least 100 pulses are required to produce harmful effects [4]. Micro- and nanoparticles have been explored for selective cell therapy, finding that for 20-nm diameter gold nanospheres irradiated by 20 ns, 532 nm pulses at 0.5 J/cm^2, cellular damage is far more influenced by the number of particles attached to the cell than by the number of pulses [5]. Bubble formation around nanoparticles has been investigated [6]. Using a smooth distribution of 40-nm diameter gold nanospheres on a cell, it is found that 30 pulses of 0.5 J/cm^2 or one pulse of ~2 J/cm^2 results in complete cell death. Alternatively, with cluster formations, cell damage occurs after 100 laser pulses at 80 mJ/cm^2. Other

Experiments

Reference	[4]	[5]	[6]	[7]	[8]	[9]	[10]
Laser	Nd:YAG	Nd:YAG	Nd:YAG	Nd:YAG	Nd:YAG	Ti:sapphire	Nd:YAG
Wavelength (nm)	532	565	532	532	532	750	1064
Pulse energy (J/cm^2)	0.5	0.5	0.5	0.13	0.03	0.95	.002
Pulse duration (ns)	12	20	12	6	7	10	7
Frequency of generation (Hz)	10	10	10	10	20	11.8 MHz / 50 MHz	10
Particle							
Material	Gold	Gold	Gold	Gold	Iron Oxide	Gold	Gold
Structure	Sphere	Sphere	Sphere	Sphere	Sphere	Rod	Rod
Diameter (nm)	40	30	40	21/4.9	10	23.6	22.76
Aspect ratio	-	-	-	-	-	3.2	5.9
Other							
Number of pulses	100	100	30	3000	600	2	10
Surroun. Medium	PBS	PBS	Water	Water	Water	Water	Water

Slide 20.4 Experiments

experiments have looked at the effects of laser irradiation on particle size and peak absorption wavelength [7]. It is shown that using a Nd:YAG laser and 21-nm spherical gold nanoparticles, the irradiation causes the particles to fragment to an average diameter of 4.9 nm, changing the peak absorption wavelengths. Iron oxide nanoparticles have been used to compare RF heating with multipulse laser heating, where it is determined that multipulse lasers produce a higher cellular uptake of particles, enabling greater cellular damage [8]. Gold nanorods have been used to determine high-energy effects of lasers on nanorods and nanospheres, such as how long the particles maintain photothermal properties (before deterioration) and the effects of multiple pulses in the form of a "pulse train," determining that very rapid pulse repetition leads to increased bubble effects [9]. Lastly, gold nanorods irradiated with multipulse near-IR lasers have been applied for successful selective cell damage [10].

One of the greatest potential benefits from multipulse lasers is accumulative heating in the target. If such lasers could be used to quickly accumulate heat in nanoparticles with many consecutive low-energy pulses, then the laser energy density would not be a prohibitive aspect of the treatment. Multiple low-energy density pulses could be used rather than a single high-energy density pulse to achieve the same nanoparticle temperature, sparing healthy cells from excessive heating as well as assuring that the energy density required for treatment is below the medical standard of 100 mJ/cm^2 [6].

The focus of this chapter is to compare and contrast the effects of multipulse (set of short pulses) versus single-pulse laser heating of nano- and bioparticles. In this section we reproduce our results published in [1–3]. We limit the maximum

Table 7.1 Experimental data for multipulse laser nanoparticle heating (Slide 20.4). Reprinted from [2]. © With permission from Elsevier.

Reference	[4]	[5]	[6]	[7]	[8]	[9]	[10]
Laser	Nd: YAG	Nd: YAG	Nd: YAG	Nd: YAG	Nd: YAG	Ti:sapphire	Nd: YAG
Wavelength (nm)	532	565	532	532	532	750	1064
Pulse energy (J/cm^2)	0.5	0.5	0.5	0.13	0.03	0.95	0.002
Pulse duration (ns)	12	20	12	6	7	10	7
Frequency of generation (Hz)	10	10	10	10	20	11.8 MHz/ 50 MHz	10
Particle							
Material	Gold	Gold	Gold	Gold	Iron oxide	Gold	Gold
Structure	Sphere	Sphere	Sphere	Sphere	Sphere	Rod	Rod
Diameter (nm)	40	30	40	21/4.9	10	23.6	22.76
Aspect ratio	–	–	–	–	–	3.2	5.9
Other							
Number of pulses	100	100	30	3000	600	2	10
Surrounding Medium	PBS	PBS	Water	Water	Water	Water	Water

temperature of the theoretical calculations to the melting point of the material ($T_M \sim 1336$ K for bulk gold material and remaining above 1100 K for gold particles larger than 5 nm in diameter), so that the particles will not undergo advanced phenomena of heating (evaporation, melting, or explosion), but require that the nanoparticles surpass 433 K as required for protein denaturing. Based on theoretical calculations of the time dynamics of nanoparticle heating in various experimental circumstances, we will demonstrate that the "standard" experimental setup does not benefit from a multipulse over single-pulse laser heating mode of metal nanoparticles with nanosecond pulses. However, changing the target material to, for example, biological particles like cell organelles, our calculations show that the organelles undergo accumulative heating, whereas metal nanoparticles do not experience accumulative heating, even at high-frequency generation rates. The summary data obtained from experimental studies can be found in Table 7.1 (Slide 20.4).

7.1.3 Multipulse Mode of Heating

Using OTM and a function to simulate multiple pulses, we calculate the time profile of the nanoparticle temperature in a multipulse heating mode to determine what would enable accumulative heating of nanoparticles (Slide 20.5). To simulate multiple pulses, we use the same function as in the single-pulse simulations, but with the addition of several values of b to shift each consecutive pulse [1]:

$$f(t) = e^{-(a\times t-b_1)^2} + e^{-(a\times t-b_2)^2} + e^{-(a\times t-b_3)^2}. \tag{7.1}$$

Multipulse Mode of Heating

- The *focus* of this lecture is to compare and contrast the *effects of multipulse* (set of short pulses) versus single-pulse laser heating of nano- and bioparticles.
- *Using OTM and a function to simulate multiple pulses*, we calculate the time profile of the nanoparticle temperature in a multipulse heating mode to determine what would enable accumulative heating of nanoparticles.
- *To simulate multiple pulses*, we use the same function as in the single-pulse simulations, but with the addition of several values of b to shift each consecutive pulse:

$$f(t) = e^{-(a\times t-b_1)^2} + e^{-(a\times t-b_2)^2} + e^{-(a\times t-b_3)^2}$$

- Appendix D contains a Maple code for particle heating in a multipulse mode.

Slide 20.5 Multipulse mode of heating

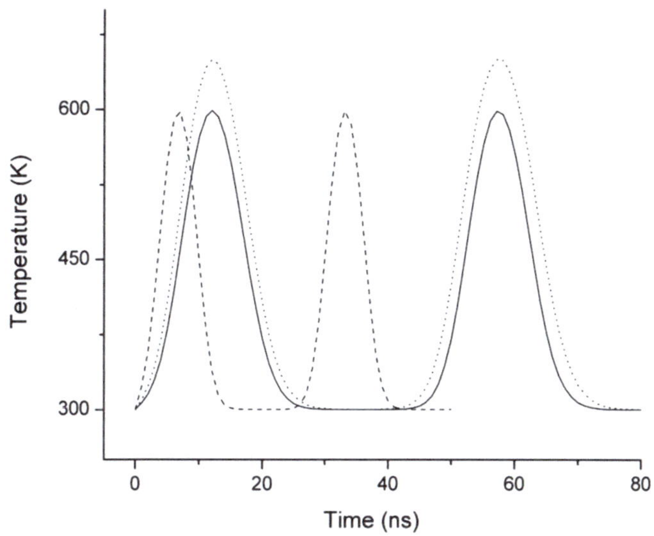

Fig. 7.1 Multipulse time–temperature profiles of a gold nanoparticle for 10 MHz (*solid*), 37 MHz (*dot*), and 62.5 MHz (*dash*) pulse generation frequencies. Reprinted from [18]. © With permission from World Scientific.

Appendix D contains a Maple code for particle heating in a multipulse mode. Three multipulse cases are shown in Fig. 7.1 and Slide 20.6, demonstrating the Au nanoparticle heating and cooling over time for the experimental conditions, listed in Table 7.1. These simulations utilize 37 and 62.5 MHz generation frequencies due

Multipulse Time-Temperature Profiles of Gold Nanoparticles

- The ***simulations*** utilize 37 MHz, 62.5 MHz and 10 MHz pulse generation frequencies.
- ***These profiles*** demonstrate visually that even at relatively high generation frequencies, ***multiple pulses do not produce any accumulative heating effect in metal nanoparticles***.
- The ***particle reaches*** the peak temperature and falls to equilibrium (300 K) in roughly three times the pulse duration, but the time delay between the pulses in experiments exceeds the pulse duration by five to ten million times.

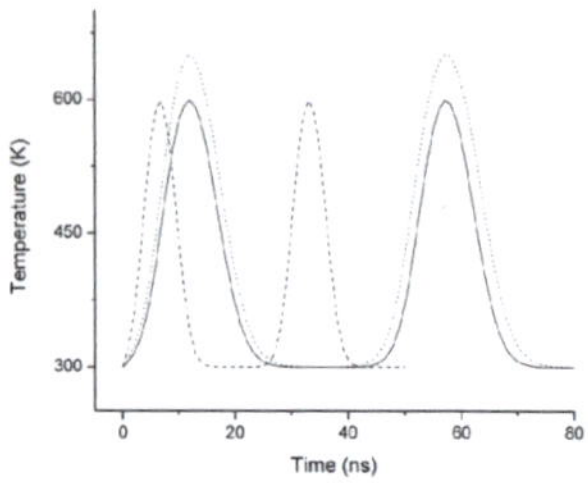

Figure. Multipulse time-temperature profiles of a gold nanoparticle for 10 MHz (solid), 37 MHz (dot) and 62.5 MHz (dash) pulse generation frequencies.

Slide 20.6 Multipulse time–temperature profiles of gold nanoparticles

to the large time gap (relative to the pulse width) between each incident pulse when using the 10 Hz pulse generation frequency. These profiles demonstrate visually that even at relatively high generation frequencies, multiple pulses do not produce any accumulative heating effect in metal nanoparticles over time (Slide 20.6). The particle reaches the peak temperature and falls to equilibrium (300 K) in roughly three times the pulse duration, but the time delay between pulses in the experiments exceeds the pulse duration by five to ten million times.

Several of the experimental papers show effects from multipulse laser heating that are not present in the single-pulse laser mode, even though the pulse generation frequency for the multipulse mode is well below the MHz generation frequency [1]. In experiments [4] it is found that bacterial killing at 3–5 J/cm^2 is observed for only 1–3 pulses, while at 0.5–1 J/cm^2, more than 100 pulses are required for bacterial damage. We feel the reason for these effects are that at 3–5 J/cm^2 the particles undergo phenomena beyond simple heating and ablation of the surrounding cell due to the extremely high temperature of the nanoparticle, and thus have a large area of effect. In the case of 0.5–1 J/cm^2 pulses, the nanoparticles can still surpass the melting threshold, but without extreme particle phenomena (such as explosion) being observed. In either case, each pulse is incident on ambient temperature particles, raising their temperature to the critical temperature for cell death. Additionally, the energy densities involved are well above the medical standard of 100 mJ/cm^2, so that such results are helpful to show the high-energy phenomena and potential treatment options. However, the methods of implementing treatment require refinement before direct medical applications can be developed. Our current investigation to determine the potential of accumulative heating effects in metal

Multipulse Time -Temperature Profiles of Gold Nanoparticles for High Pulse-Generation Frequency

- The *figure on the right* three different multipulse calculations for gold nanorods in water irradiated by *high pulse-generation frequencies* of 50 MHz, 66.6 MHz and 80 MHz.
- *Even at the high pulse-generation frequencies*, the 10 ns pulses *do not create an accumulative heating effect in the metal nanoparticles*.
- Furthermore, *changing the pulse duration* concurrently with high-frequency pulse-generation simulations only changes the magnitude of the *maximum particle temperature*.

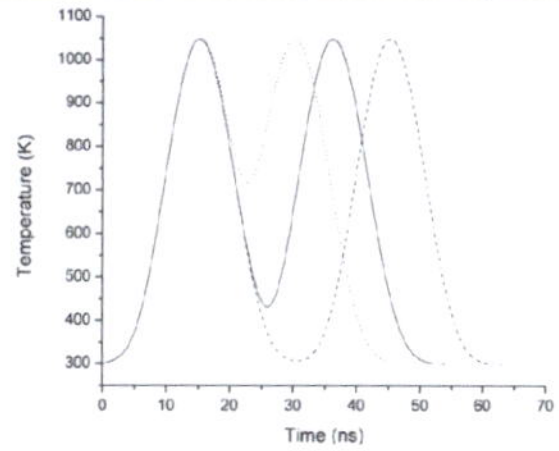

Figure. Time-temperature profile for multipulse laser heating of gold nanorods (r_{eff} = 11.8 nm, AR = 3.2, K_{abs} = 11.5 at 725 nm) irradiated by 0.01 J/cm^2/10 ns pulses at repetition rates of 50 MHz (dash), 66.6 MHz (solid) and 80 MHz (dot).

Slide 20.7 Multipulse time–temperature profiles of gold nanoparticles for high pulse-generation frequency

nanoparticles is an attempt at localizing damage to cancer cells by using low-energy density lasers while still killing the cancer cells [1–3].

After determining that most other experimental characteristics (pulse duration, particle size, etc.) have no ability to create an accumulative heating phenomenon of interest in metal nanoparticles, the pulse generation frequency used in multipulse scenarios was tested experimentally with a high pulse-generation frequency [9] (Slide 20.7). This experiment used two pulses of 10 ns duration that are separated by 20 and 150 ns time intervals –50 MHz and 11.8 MHz, respectively. Figure 7.2 shows three different multipulse calculations based on the experimental setup used in [9] for gold nanorods ($r_{\mathrm{eff}} = 11.8$ nm, AR = 3.2, $K_{\mathrm{abs}} = 11.5$ at 725 nm) in water irradiated by 0.01 J/cm^2/10 ns pulses with pulse-generation frequencies of 50, 66.6, and 80 MHz.

We exclude the case with pulse separation of 150 ns due to the rapid cooling of metal nanoparticles (32.8 ns for a 10 ns pulse in [1–3]) previously shown in Chap. 6, as well as the experiment [9], that there are no additional effects from two pulses separated by 150 ns over a single pulse. Even at the high-pulse generation frequencies shown, the 10 ns pulses do not create an accumulative heating effect in the metal nanoparticles (Slide 20.7). Furthermore, changing the pulse duration concurrently with high-frequency pulse generation simulations only changes the magnitude of the maximum particle temperature and does not enable the accumulative heating effect in the metal nanoparticles of interest to our investigation.

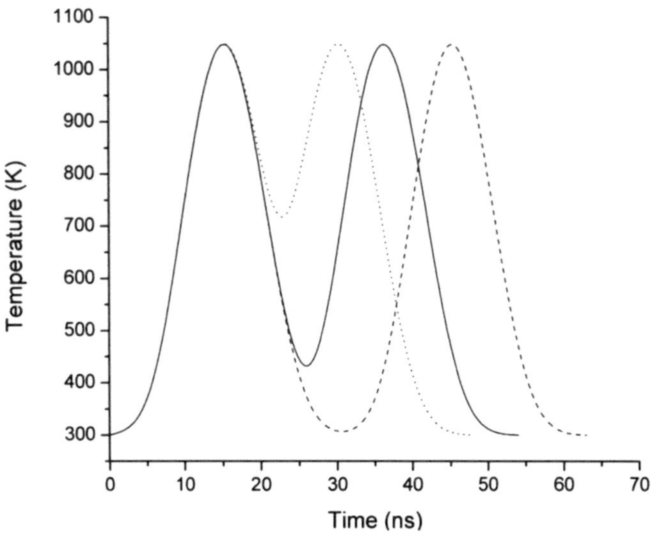

Fig. 7.2 Time–temperature profile for multipulse laser heating of gold nanorods ($r_{eff} = 11.8$ nm, AR = 3.2, $K_{abs} = 11.5$ at 725 nm) irradiated by 0.01 J/cm^2/10 ns pulses at repetition rates of 50 MHz (*dash*), 66.6 MHz (*solid*), and 80 MHz (*dot*). Reprinted from [2]. © With permission from Elsevier

7.1.4 Multipulse Heating of Healthy and Cancerous Cell Organelles

Using the multipulse form described before, we have also tested the heating and cooling kinetics of biological cell organelles to determine the ability to selectively destroy cancer cells [1] (Slide 20.8). Acquiring data for organelles (nucleus, mitochondrion, and ribosome) is difficult due to the wide variety of situations under which they must act; the added complexity of the large number of cancers makes finding very specific data difficult. Using a variety of cited dimensions and organelle properties (Tables 6.5, 7.2, and 7.3), we have calculated the "average" organelle size to determine the theoretical effects of the multipulse mode of laser irradiation on the organelle temperature with the same procedure as detailed earlier (Slides 20.8 and 20.9).

The cell organelles demonstrate the heat accumulation effect due to the mutipulse mode of heating as shown [2] in Figs. 7.3 and 7.4 (Slides 20.10 and 20.11). The ribosomal heating in Fig. 7.3 and Slide 20.10 contrasts with heating for the nuclei and mitochondria in Fig. 7.4 (Slide 20.11); the ribosomes do not experience accumulative heating to the extent that nuclei or mitochondria do, based primarily on the much smaller size. Even with the small size, within the first few pulses the accumulative heating effect is apparent, which contrasts with any nanoparticle simulations previously discussed (Slide 20.10). The much higher

Multipulse Heating of Healthy and Cancerous Cell Organelles

Using the multipulse form described above, we also can test the ***heating and cooling kinetics of biological cell organelles*** to determine the ability to selectively destroy cancer cells.

Table. Experimental data for healthy and tumor cells.

Source	Cancer	Organelle	Healthy diameter (μm)	Well differentiated diameter (μm)	Moderately differentiated diameter (μm)	Poorly differentiated diameter (μm)
[11]	Hepatocellular carcinoma	Nucleus	7.5	8.5	8.8	10.3
[11]	Hepatocellular carcinoma	Mitochondria	1.1	0.9	0.9	0.66
[12]	Colorectal adenocarcinoma	Nucleus	5.2	7.2	-	-
Source	Cancer	Organelle	Benign diameter (μm)	*In situ* diameter (μm)	Invasive diameter (μm)	
[13]	Ductal breast carcinoma	Nucleus	7	11.6	8.33	
[14]	Intraductal breast cancer	Nucleus	8.58	-	8.97	
[15]	Breast cancer	Nucleus	5.6	-	5.8	

Slide 20.8 Multipulse heating of healthy and cancerous cell organelles

values of specific heat in the organelles and the thermal conductivity of the medium is what causes the organelles to undergo accumulative heating, whereas metal nanoparticles do not experience accumulative heating, even at high-frequency pulse generation rates. This gives an advantage for organelles in using the mutipulse mode vs. single pulse, since the laser energy density per pulse can be reduced dramatically below the safety standard for medical usage of lasers. In this case, any desired final temperature of the organelle can be reached easily by adjusting the number of pulses at low energy density per pulse (Slide 20.11).

While the organelles demonstrate a similarity in the heat accumulation effect, both the mitochondria and nuclei respond differently to the pulsed mode of heating [1]. As shown in Fig. 7.4, healthy and cancerous mitochondria do not respond differently in a significant way to laser heating, while the healthy nucleus is heated 40 K higher than the cancerous nucleus.

The difference in heating characteristics is due to several behavioral properties [1]: the size of the nucleus is much larger than that of the mitochondrion; mitochondrion has been shown to shrink when in cancerous cells; and the nucleus of cancerous cells is known to grow relative to their healthy cell size (Slide 20.11). The dynamics of size on temperature results in the mitochondria being heated to a higher temperature, but with very little specificity. Alternatively, nuclei are heated to a lower temperature at a higher specificity level, though the healthy nuclei are the first to get damaged.

Table 7.2 Experimental data for healthy and tumor cells. Reprinted from [2]. © With permission from Elsevier

Source	Cancer	Organelle	Healthy diameter (μm)	Well differentiated diameter (μm)	Moderately differentiated diameter (μm)	Poorly differentiated diameter (μm)
[11]	Hepatocellular carcinoma	Nucleus	7.5	8.5	8.8	10.3
[11]	Hepatocellular carcinoma	Mitochondria	1.1	0.9	0.9	0.66
[12]	Colorectal adenocarcinoma	Nucleus	5.2	7.2	–	–
Source	Cancer	Organelle	Benign diameter (μm)	In situ diameter (μm)	Invasive diameter (μm)	
[13]	Ductal breast carcinoma	Nucleus	7	11.6	8.33	
[14]	Intraductal breast cancer	Nucleus	8.58	–	8.97	
[15]	Breast cancer	Nucleus	5.6	–	5.8	

Table 7.3 Computational input data for OTM calculations. Reprinted from [2]. © With permission from Elsevier

Source	Cancer	Organelle	Organelle state	Diameter (μm)	Absorption efficiency (at $\lambda = 700$ nm)[a]
[11]	Hepatocellular carcinoma	Nucleus	Healthy	7.5	0.944183362
			Well differentiated	8.5	0.964865712
			Moderately differentiated	8.8	0.969903343
			Poorly differentiated	10.3	0.989209877
[11]	Hepatocellular carcinoma	Mitochondria	Healthy	1.1	0.134234082
			Well differentiated	0.9	0.110949860
			Moderately differentiated	0.9	0.110949860
			Poorly differentiated	0.66	0.082130836
[13]	Ductal breast carcinoma	Nucleus	Benign	7.0	0.930908259
			In situ	11.6	1.000194442
			Invasive	8.33	0.961677234
[14]	Intraductal breast cancer	Nucleus	Benign	8.58	0.966255375
			Invasive	8.97	0.972557278
[15]	Breast cancer	Nucleus	Benign	5.6	0.880181004
			Invasive	5.8	0.889019082

[a]700 nm was chosen because it is the wavelength with the greatest absorption efficiency difference between cancerous and healthy organelles for wavelengths in the range 400–1000 nm

7.1.5 *Conclusions and Summary*

In this section we have performed time-dependent simulations and detailed analyses of different nonstationary laser–nanoparticle interaction modes to determine the accumulative heating potential of metal nanoparticles and biological particles irradiated by multipulse lasers (Slide 20.12). Our simulations have shown that at the instant when the laser pulse stops, thc cooling rate reaches almost the same values as the heating rate in metal nanoparticles due to the large surface area of the nanoparticle and high thermal conductivity of the surrounding aqueous medium. This becomes crucial for the mutipulse mode of heating the metal nanoparticles. From our analysis of multipulse metal nanoparticle heating, we have determined that the multipulse mode of heating does not create accumulative heating effects in metal nanospheres or nanorods (diameters of 5–40 nm) due to the rapid cooling kinetics of metal nanoparticles in an aqueous environment [2] (Slide 20.12).

However, changing the target material to biological particles, like cell organelles with much higher values of specific heat, causes the biological particles to undergo accumulative heating, whereas metal nanoparticles do not experience accumulative

Multipulse Heating of Healthy and Cancerous Organelles (continued)

Source	Cancer	Organelle	Organelle State	Diameter (μm)	Absorption Efficiency (at λ = 700 nm)
[11]	Hepatocellular carcinoma	Nucleus	Healthy	7.5	0.944183362
			Well differentiated	8.5	0.964865712
			Moderately differentiated	8.8	0.969903343
			Poorly differentiated	10.3	0.989209877
[11]	Hepatocellular carcinoma	Mitochondria	Healthy	1.1	0.134234082
			Well differentiated	0.9	0.110949860
			Moderately differentiated	0.9	0.110949860
			Poorly differentiated	0.66	0.082130836
[13]	Ductal breast carcinoma	Nucleus	Benign	7.0	0.930908259
			In situ	11.6	1.000194442
			Invasive	8.33	0.961677234
[14]	Intraductal breast cancer	Nucleus	Benign	8.58	0.966255375
			Invasive	8.97	0.972557278
[15]	Breast cancer	Nucleus	Benign	5.6	0.880181004
			Invasive	5.8	0.889019082

Slide 20.9 Multipulse heating of healthy and cancerous cell organelles (continued)

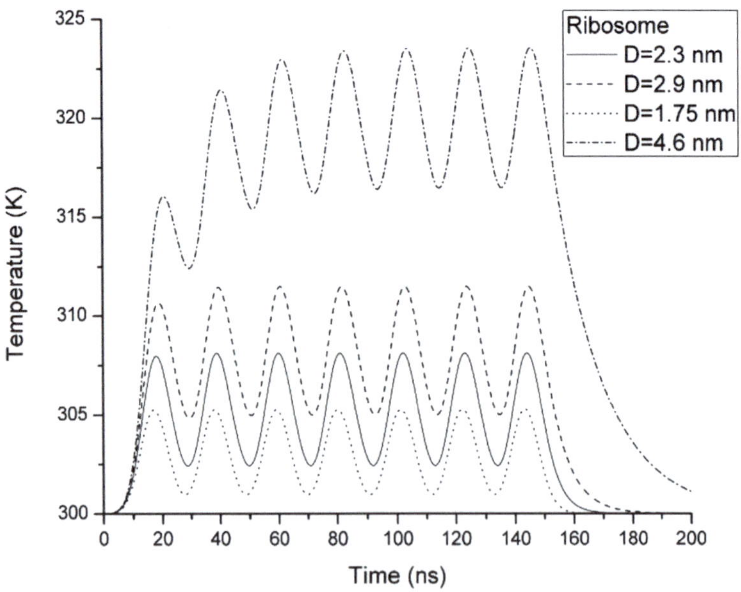

Fig. 7.3 Ribosomal time–temperature profiles at various diameter values. Multipulse laser parameters: 12 ns duration, 0.1 J/cm^2 at $\lambda = 700$ nm; $b_1 = 2.5$, $b_2 = 6$, $b_3 = 9.5$, $b_4 = 13$, $b_5 = 16.5$, $b_6 = 20$, $b_7 = 23.5$. Reprinted from [2]. © With permission from Elsevier

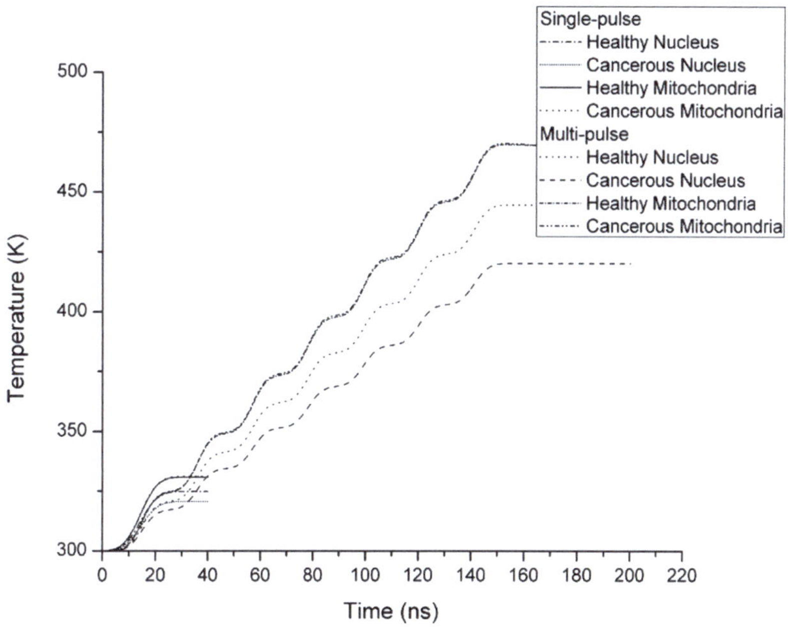

Fig. 7.4 Both single- and multipulse laser heating for "average" organelles. Average values: healthy nucleus—7.17 μm; cancerous nucleus—8.97 μm; healthy mitochondria—1.1 μm); cancerous mitochondria—0.82 μm. Single-pulse laser parameters include a 0.1 J/cm^2 pulse at 700-nm wavelength. Reprinted from [2]. © With permission from Elsevier

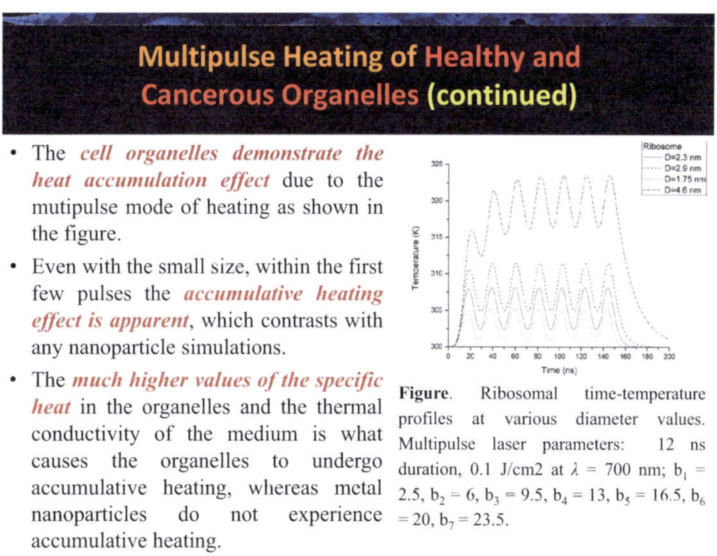

Slide 20.10 Multipulse heating of healthy and cancerous cell organelles (continued)

Multipulse Heating of Healthy and Cancerous Organelles (continued)

- ***This gives an advantage*** for organelles in using the mutipulse mode *vs.* single pulse, since the laser energy density per pulse can be reduced dramatically below the safety standard for medical usage of lasers.
- In this case, any ***desired final temperature*** of the organelle can be reached easily ***by adjusting the number of pulses at low energy density per pulse***.
- While the organelles demonstrate a similarity in the heat accumulation effect, both the ***mitochondria and nuclei respond differently*** to the pulsed mode of heating.
- The difference in heating characteristics is ***due to the size*** of the nucleus being much larger than mitochondrion.

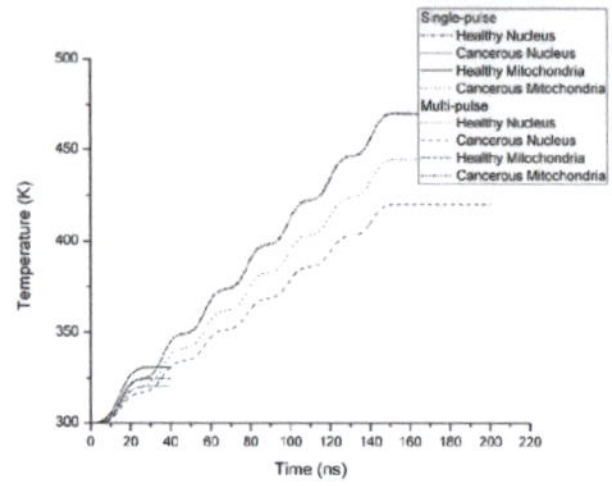

Figure. Multipulse laser heating for 'average' organelles. Average values: healthy nucleus – 7.17 μm;, cancerous nucleus – 8.97 μm; healthy mitochondria – 1.1 μm; cancerous mitochondria – 0.82 μm.

Slide 20.11 Multipulse heating of healthy and cancerous cell organelles (continued)

Conclusions and Summary

- ***In this lecture*** we have demonstrated time-dependent simulations and conducted detailed analyses of different non-stationary laser-nanoparticle interaction modes to determine the ***accumulative heating potential of metal nanoparticles and biological particles*** irradiated by multipulse lasers.
- Our simulations have shown that at the instant when the laser pulse stops, the ***cooling rate reaches almost the same values as the heating rate in metal nanoparticles*** due to the large surface area of the nanoparticle and high thermal conductivity of the surrounding aqueous medium.
- This becomes crucial for the mutipulse mode of heating the metal nanoparticles. From our analysis of multipulse metal nanoparticle heating, we have determined that the ***multipulse mode of heating does not create accumulative heating effects in metal nanospheres or nanorods*** (diameters of 5-40 nm) due to the rapid cooling kinetics of metal nanoparticles in an aqueous environment.

Slide 20.12 Conclusions and summary

heating, even at high frequency generation rates (Slide 20.13). The accumulative heating effect provides an advantage for organelles and biological tissue in using the mutipulse mode vs. single pulse, since the laser energy density per pulse can be reduced dramatically below the safety standard for medical usage of lasers. In this case, any desired final temperature of the biotissue can be reached easily by adjusting the number of pulses at low energy density per pulse [2] (Slide 20.13).

Conclusions and Summary (continued)

- However, ***changing the target material to biological particles***, like cell organelles with much higher values of specific heat, causes the biological particles to ***undergo accumulative heating***, whereas metal nanoparticles do not experience accumulative heating, even at high frequency generation rates.
- The ***accumulative heating effect provides an advantage*** for organelles and biological tissue in using the mutipulse mode *vs.* single pulse, since the laser energy density per pulse can be reduced dramatically below the safety standard for medical usage of lasers.
- In this case, any ***desired final temperature*** of the biotissue can be reached easily ***by adjusting the number of pulses at low energy density per pulse.***

Slide 20.13 Conclusions and summary (continued)

7.2 Computer Practicum 4 (Lecture 21): Multipulse Mode of Heating Healthy and Cancerous Cell Organelles (Slide 21.1)

- Introduction
- Time Profile of the Laser Pulse
- Practice Problem 6: Thermal Dynamics of Healthy and Cancerous Lysosomes in Cytoplasm Media Using Multipulsed Radiation
  - *Abstract*
  - *Model and Input Data*
  - *Discussion and Results*
  - *Conclusion*

7.2.1 Introduction

In Sect. 7.1 we showed that heat accumulation is a phenomenon that happens in a particle heated in a multipulse mode when energy does not have enough time to completely diffuse outside of the material between the pulses, so that some of it ends up being stored inside the particle (Slide 21.2). Therefore, if a pulse of radiation is applied again, heat will begin to build up inside the material. As shown in Sect. 7.1, this phenomenon is observed for biological particles, like cell

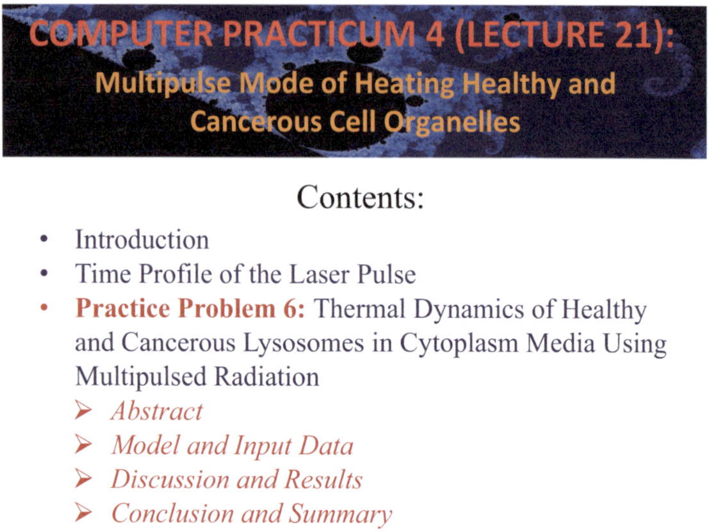

Slide 21.1 Contents

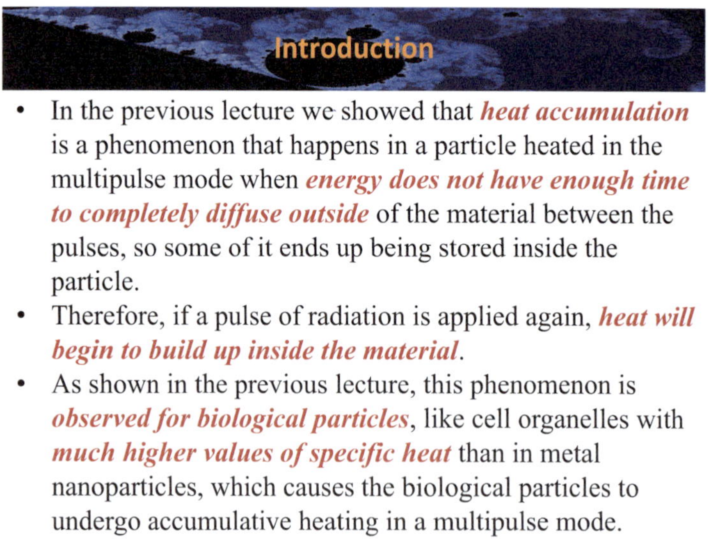

Slide 21.2 Introduction

organelles with much higher values of specific heat than in metal nanoparticles, which causes the biological particles to undergo accumulative heating in a multipulse mode.

The main purpose of this *Computer Practicum* is to determine the feasibility of destroying cancerous cell organelles using pulsed radiation (Slide 21.3). Once the

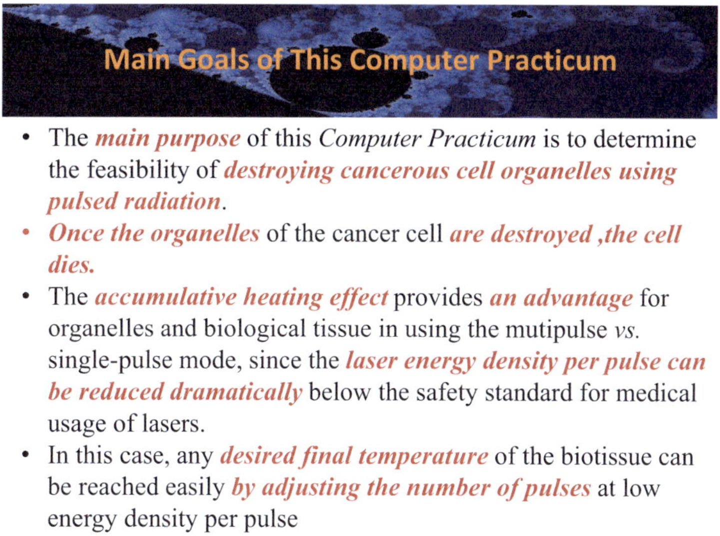

Slide 21.3 Main goals of this computer practicum

organelles of the cancer cell are destroyed, the cell dies. The accumulative heating effect provides an advantage for organelles and biological tissue in using a multipulse vs. single-pulse mode, since the laser energy density per pulse can be reduced dramatically below the safety standard for medical usage of lasers. In this case, any desired final temperature of the biotissue can be reached easily by adjusting the number of pulses at low energy density per pulse [1] (Slide 21.3).

7.2.2 *Time Profile of Multipulsed Radiation*

We use here same one-temperature model (OTM) introduced in Chap. 6 and Maple code given in Appendix B, but both modified for a multipulse mode (Slide 21.4). A main modification to the Maple code is made to compute multiple pulses as a periodic Gaussian function given by the following equation:

$$f(t) = \mathrm{e}^{-\sin(a*t-b)^2 + c}. \tag{7.2}$$

Here, a, b, and c are constants which are chosen depending on what the pulse width should be for a particular simulation and how many pulses are desired. The periodic Gaussian function $f(t)$ and *plot* of the multipulse shape in a defined time interval are shown in Fig. 7.5 and Slide 21.5. A modified Maple code for a multipulse mode of heating healthy and cancerous cell organelles is provided in Appendix D.

- We use here same ***one-temperature model*** (OTM) introduced in Chapter 6 and ***Maple code given in Appendix B***, but both ***modified to multipulsed radiation***.
- A main modification to the Maple code is made ***to compute multiple pulses as a periodic Gaussian function*** given by the following equation:

$$f(t) = e^{-\sin(a*t-b)^2 + c}.$$

- The constants a, b and c are chosen depending on what the pulse width should be for a particular simulation and how many pulses are desired.
- A ***modified Maple code for a multipulse mode*** of heating healthy and cancerous cell organelles is provided in ***Appendix D***.

Slide 21.4 Time profile of the multipulsed radiation

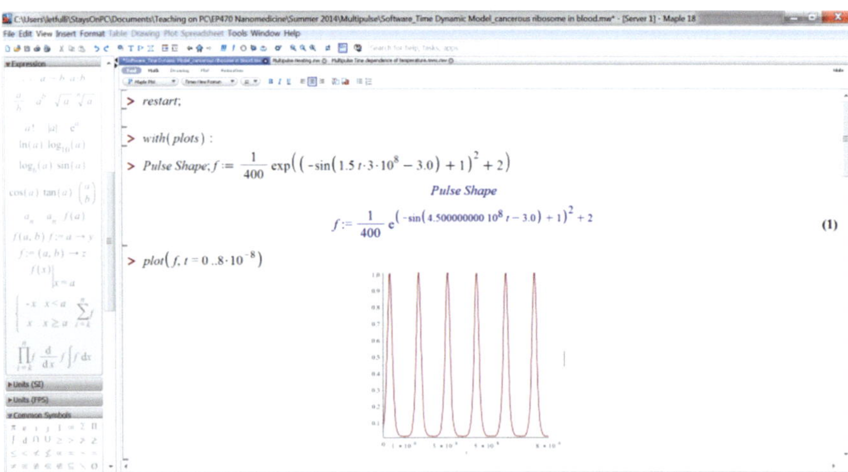

Fig. 7.5 Modeling a multipulse profile

The results of the simulations as shown in Fig. 7.6 and Slide 21.6 provide the time dynamics of heating and cooling of healthy and cancerous cell organelles in cytoplasm in a multipulse mode of heating, allowing us to draw conclusion whether it is possible to use multipulsed radiation as a means of cancerous cell destruction while leaving the healthy cells unharmed. The following *Practice Examples* present the output results of these simulations in detail for specific normal and abnormal cell organelles.

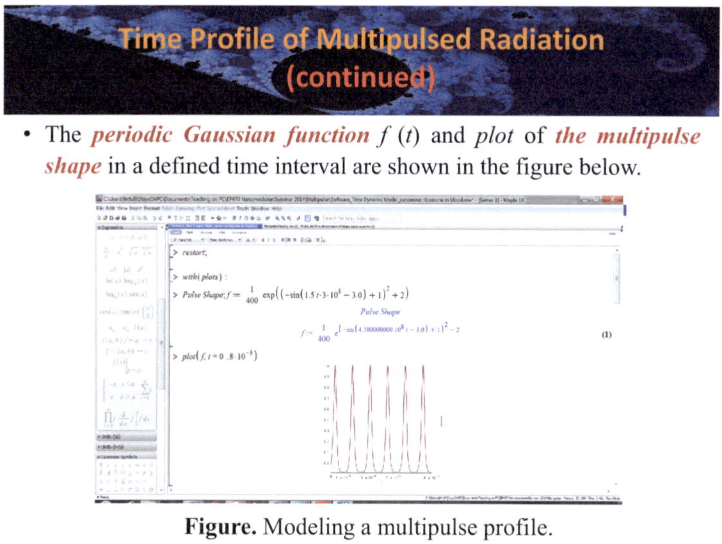

Slide 21.5 Time profile of the multipulsed radiation (continued)

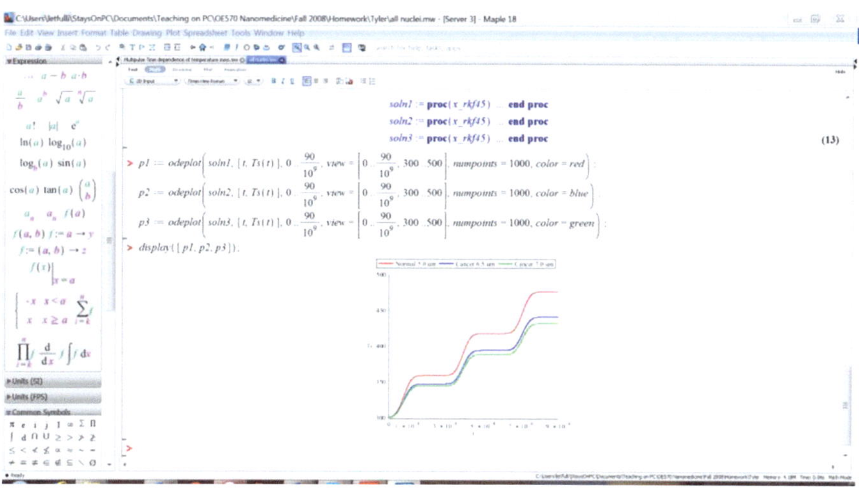

Fig. 7.6 Plots of the temperature profiles of healthy and cancerous cell nuclei in cytoplasm heated by multipulsed radiation

The results of the simulations

- The ***results*** of the simulations as shown in the figure below provide the time dynamics of ***heating and cooling of healthy and cancerous cell organelles*** in cytoplasm ***in a multipulse mode of heating***, allowing us to draw conclusiona as to whether it is possible to use ***multipulsed radiation as a means of cancerous cell destruction*** while leaving the healthy cells unharmed.

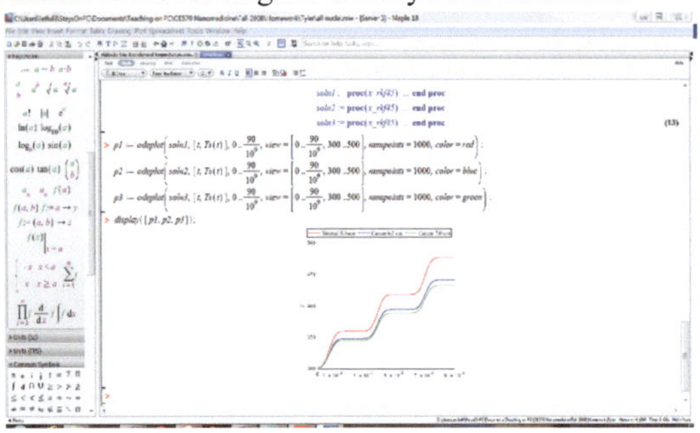

Slide 21.6 The results of the simulations

7.2.3 *Practice Problem 6: Thermal Dynamics of Healthy and Cancerous Lysosomes in Cytoplasm Media Using Multipulsed Radiation*

7.2.3.1 **Abstract** (Slide 21.7)

The purpose of this practice is to determine the feasibility of destroying cancerous cell lysosome using a pulsed laser. Using more than one pulse will help to heat the lysosome without damaging the surrounding tissue, since the laser energy density per pulse can be reduced dramatically. The lysosome itself may actually act as a therapeutic particle, which would mean cancer cells could be destroyed without inserting any nanoparticles into the body. This reduces the risk to the patient and creates an easier treatment process.

7.2.3.2 **Model and Input Data**

The one-temperature model is used, which is described in detail in Chap. 6 and given by the equation

$$\frac{dT_{\mathrm{s}}}{dt} = \frac{3Q_{\mathrm{abs}}I_0 f(t)}{4r_0 C_{\mathrm{s}}(T_{\mathrm{s}})\rho_{\mathrm{s}}} - \frac{3\mu_{\infty}T_{\mathrm{s}}}{(\alpha+1)r_0^2 C_{\mathrm{s}}(T_{\mathrm{s}})\rho_{\mathrm{s}}}\left[\left(\frac{T_{\mathrm{s}}}{T_{\infty}}\right)^{\alpha+1} - 1\right], \tag{7.3}$$

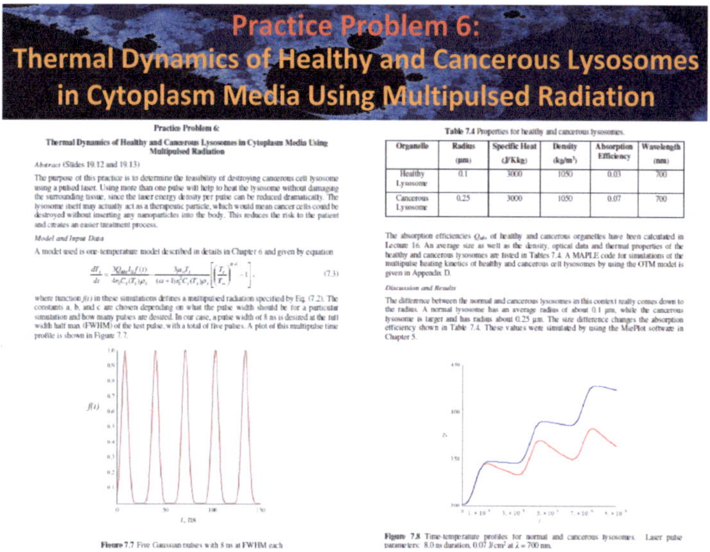

Slide 21.7 Practice Problem 6: thermal dynamics of healthy and cancerous lysosomes in cytoplasm media using multipulsed radiation

Where the function $f(t)$ in these simulations defines multipulsed radiation specified by Eq. (7.2). The constants a, b, and c are chosen depending on what the pulse width should be for a particular simulation and how many pulses are desired. In our case, a pulse width of 8 ns is desired at the full width at half maximum (FWHM) of the test pulse, with a total of five pulses. A plot of this multipulse time profile is shown in Fig. 7.7.

The absorption efficiencies Q_{abs} of healthy and cancerous organelles have been calculated in Lecture 16. An average size as well as the density, optical data, and thermal properties of the healthy and cancerous lysosomes are listed in Tables 7.4. A Maple code for simulations of the multipulse heating kinetics of healthy and cancerous cell lysosomes by using OTM is given in Appendix D.

7.2.3.3 Discussion and Results

The difference between normal and cancerous lysosomes in this context really comes down to the radius. A normal lysosome has an average radius of about 0.1 μm, while the cancerous lysosome is larger with a radius of about 0.25 μm. The size difference changes the absorption efficiency shown in Table 7.4. These values are simulated by using the MiePlot software in Chap. 5.

Greater values of Q_{abs} affect the temperature of the cancerous lysosome when struck by the multiple ultrashort laser pulses. Figure 7.8 shows the time–temperature profiles for heating healthy and cancerous lysosomes in the cell cytoplasm by multipulsed radiation. The laser is set to an energy density of 0.07 J/cm^2 with a

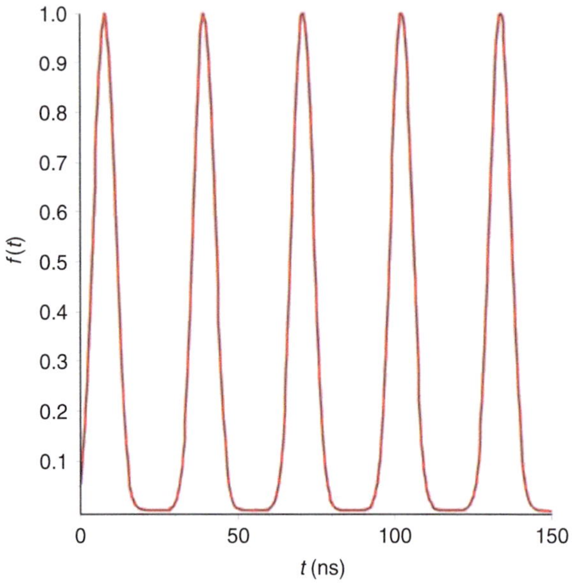

Fig. 7.7 Five Gaussian pulses with 8 ns at FWHM for each

Table 7.4 Properties for healthy and cancerous lysosomes

Organelle	Radius (μm)	Specific heat (J/K kg)	Density (kg/m^3)	Absorption efficiency	Wavelength (nm)
Healthy lysosome	0.1	3000	1050	0.03	700
Cancerous lysosome	0.25	3000	1050	0.07	700

pulse duration of 8.0 ns. It is evident from this figure that the cancer lysosomes heat up much faster than the normal lysosomes, reaching the critical temperature for ablation of 425 K, while a healthy lysosome temperature is well below this critical value. With the dramatic increase in temperature of the cancerous lysosome, the cancer cells undergo thermal ablation after the third pulse of radiation. Thus, multipulsed radiation with a low energy density of 0.07 J/cm^2 is capable of producing a significant temperature change and thermal damage in a cancerous lysosome, while a healthy organelle would be kept safe.

In conclusion (Slide 21.8), we have evaluated the heating dynamics of healthy and cancerous lysosomes in a multipulse mode of heating. It is found that the lysosomes demonstrate the heat accumulation effect due to the mutipulse mode of heating. A cancerous lysosome reaches a higher temperature and has a longer cooling time than a healthy lysosome. Such heating dynamics means that incident multipulsed radiation with low energy density could cause damage to the unhealthy lysosomes after a few pulses, while keeping the healthy lysosomes unharmed.

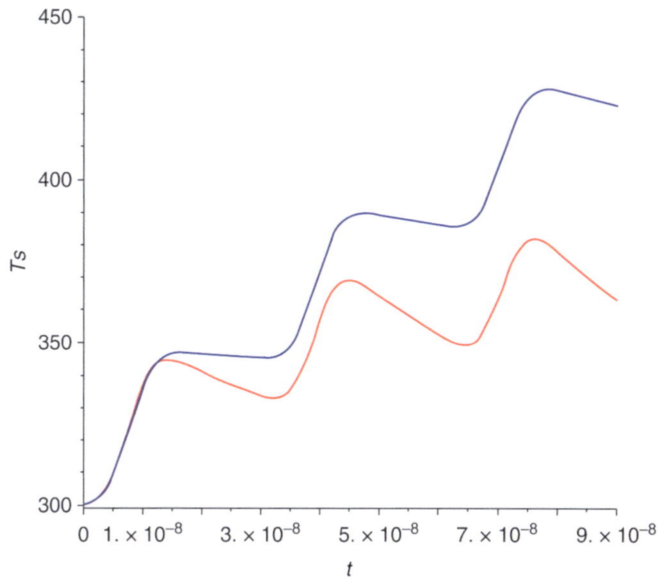

Fig. 7.8 Time–temperature profiles for normal and cancerous lysosomes. Laser pulse parameters: 8.0 ns duration, 0.07 J/cm^2 at $\lambda = 700$ nm

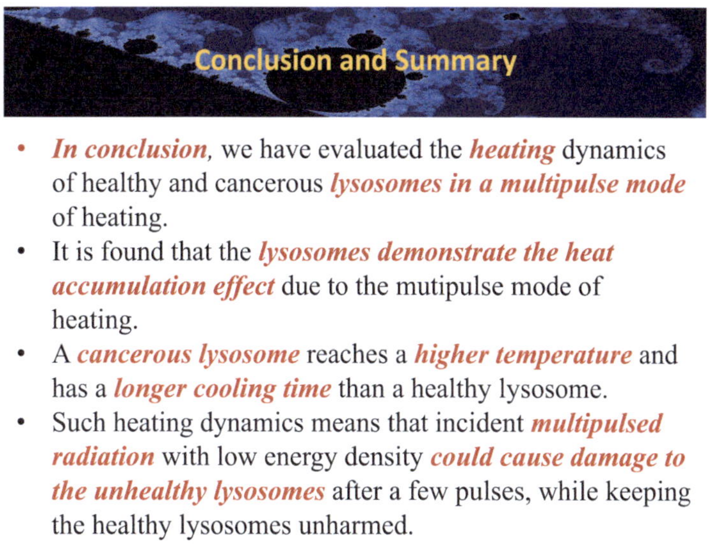

Slide 21.8 Conclusions and summary

7.3 Lecture 22: X-Ray and RF Heating of Nanoparticles (Slide 22.1)

- Introduction
- Nanoparticle-Enhanced X-Ray Therapy of Cancer
  - *Therapeutic Effects of X-Ray-Activated Nanoparticles*
  - *Thermal Modeling and Spectra Modeling*
  - *Discussion and Results*
  - *Conclusions*
- RF Activation of Nanoparticles in Single-Pulse and Multipulse Modes of Heating
  - *Single-Pulse RF Heating Mode*
  - *Multipulse RF Heating Mode*
  - *Conclusions*

7.3.1 Introduction

In Chap. 3 we discussed nanodrug design and methods of activation (Slide 22.2). We showed that nanodrug design depends on the *methods of nanodrug activation*, i.e., using different types of radiation for a total body scan. Under activation of nanoparticles, we mean heating them to initiate cell ablation as well as charged particle generation from the surface of nanoparticles caused by the applied radiation. The ideal radiation for nanophotodynamic therapy involves deep penetration of the human body as well as enough power to activate the nanoparticles without harming

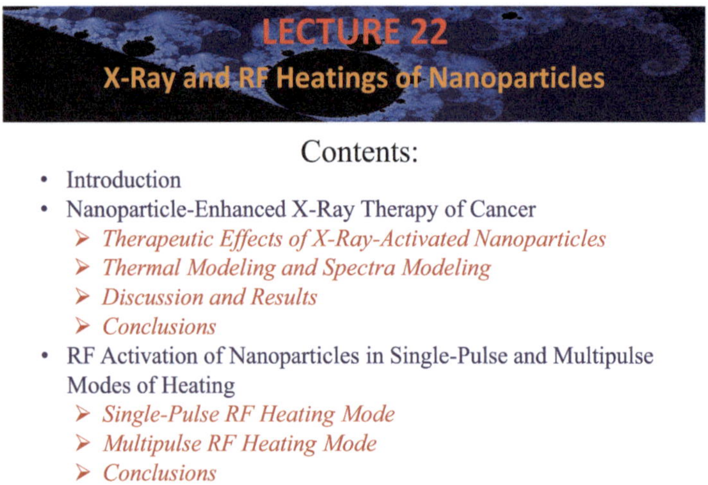

Slide 22.1 Contents

Introduction

- In Chapter 3 we discussed ***nanodrug design and methods of activation***.
- We showed that nanodrug design depends on the ***methods of nanodrug activation***, i.e., using different types of radiation for a total body scan.
- ***Under activation of nanoparticles***, we mean heating them to initiate cell ablation as well as charged particle generation from the surface of nanoparticles caused by the applied radiation.
- The ***ideal radiation*** for nanophotodynamic therapy involves deep penetration of the human body as well as enough power to activate the nanoparticles without harming the patient.
- In Chapter 3 we investigated a ***wide range of the electromagnetic spectrum***, from the very short wavelengths of gamma and x-rays radiation to the very long wavelengths of RF radiation.

Slide 22.2 Introduction

Introduction (continued)

- The search for ***a universal tool for nanoparticle activation*** shows that the very ends of the electromagnetic spectrum – ***x-rays and RF waves*** – are the best candidates for a total body scan to activate the nanoparticles inside the body.
- ***Both radiation types***, short x-rays and long RF waves, have excellent ability to penetrate into the human body, giving a great opportunity to activate/heat nanoparticles delivered inside the body as a contrast agent for diagnosis and treatment purposes.
- This lecture discusses ***heating nanoparticles*** by electromagnetic radiation, specifically ***in the x-ray and RF wave portions of the spectrum***.
- ***Time-dependent simulations*** are carried out for different nonstationary x-ray and RF- nanoparticle interactions for single-pulse and multipulse modes of heating.

Slide 22.3 Introduction (continued)

the patient. In Sect. 3.2 (Lecture 10) of Chap. 3, we investigated a wide range of the electromagnetic spectrum, from the very short wavelengths of gamma and X-ray radiation to the very long wavelengths of RF radiation. This section contains material adopted from our publications [16, 17].

The search for a universal tool for nanoparticle activation showed that the very ends of the electromagnetic spectrum—X-rays and RF waves—are the best candidates for a total body scan to activate the nanoparticles inside the body (Slide 22.3). Both radiation types, short X-rays and long RF waves, have excellent ability to penetrate into the human body, giving great opportunity to activate/heat nanoparticles delivered inside the body as a contrast agent for diagnosis and treatment purposes [16, 17]. This lecture discusses heating nanoparticles by electromagnetic radiation, specifically in the X-ray and RF wave portions of the spectrum. Time-dependent simulations are carried out for different nonstationary X-ray and RF–nanoparticle interactions for single-pulse and multipulse modes of heating. We also discuss the specific features of the heating kinetics of nanoparticles, such as (1) a short time scale for heating and cooling of nanoparticles in a liquid biological environment with high thermal conductivity properties; and (2) the effect of the radiation field structure (pulse shape, pulse duration, repetition rate, time interval between pulses) on the kinetics of heating and cooling by single-pulse and multipulse modes of heating.

7.3.2 Nanoparticle-Enhanced X-Ray Therapy of Cancer

Photothermal therapies of nanophotohyperthermia and nanophotothermolysis utilize the light absorptive properties of nanoparticles to create heat and free radicals in a small localized region [16] (Slide 22.4). Conjugating nanoparticles with various biomolecules allow for targeted delivery to specific tissues or even specific cells, cancerous cells being of particular interest. Previous studies have investigated nanoparticles at visible and infrared wavelengths where a surface plasmon resonance leads to unique absorption characteristics. However, issues such as poor

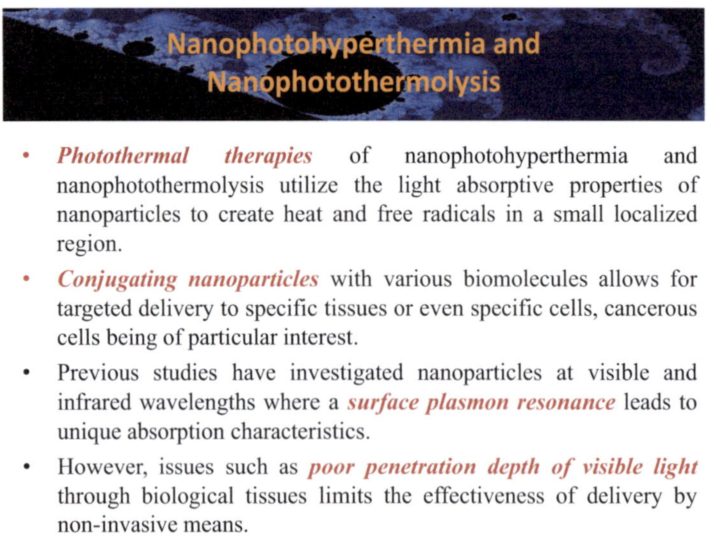

Slide 22.4 Nanophotohyperthermia and nanophotothermolysis

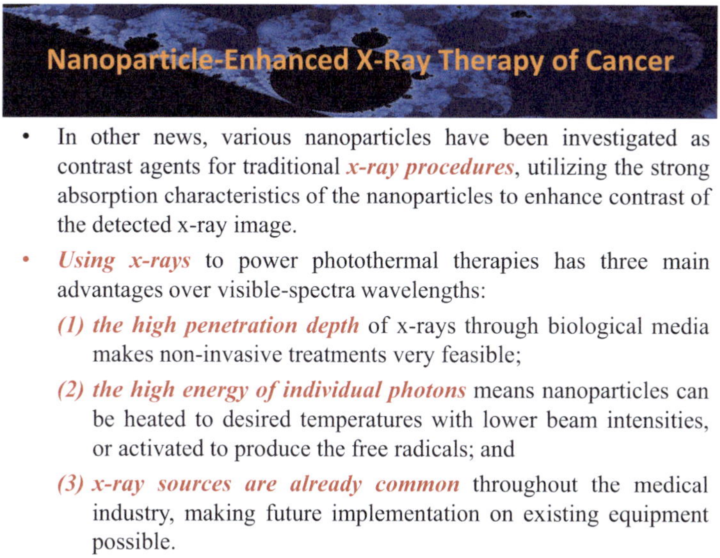

Slide 22.5 Nanoparticle enhanced X-ray therapy of cancer

penetration depth of visible light through biological tissues limit the effectiveness of delivery by noninvasive means.

In other news, various nanoparticles have been investigated as contrast agents for traditional X-ray procedures, utilizing the strong absorption characteristics of the nanoparticles to enhance contrast of the detected X-ray image (Slide 22.5). Using X-rays to power photothermal therapies has three main advantages over visible spectra wavelengths [16]: (1) the high penetration depth of X-rays through biological media makes noninvasive treatments very feasible; (2) the high energy of individual photons means nanoparticles can be heated to desired temperatures with lower beam intensities, or activated to produce the free radicals; and (3) X-ray sources are already common throughout the medical industry, making future implementation on existing equipment possible.

Therapeutic effects of localized activation of nanoparticles by X-ray radiation are achieved by the emission of charged particles from the surface of the nanoparticles in photoelectric events, Compton scattering events, and thermionic emission, or heating the nanoparticles to temperatures high enough for cell ablation (Slide 22.6). Applying X-rays to heat nanoparticles in order to induce heat transfer into adjacent unhealthy cells may seem convoluted (compared to radiating the cells directly with X-rays), but this extra step allows for two important improvements: (1) selective targeting of the cancer cells and (2) significantly reducing the therapeutic doses and exposure of X-ray radiation to the diagnostic level. Nanoparticles can be combined with cancer-targeting molecules, allowing the nanoparticles to be selectively accumulated on the unhealthy cells (Slide 22.7). When the X-ray is turned on, these nanoparticles, which are thousands of times stronger absorbers than

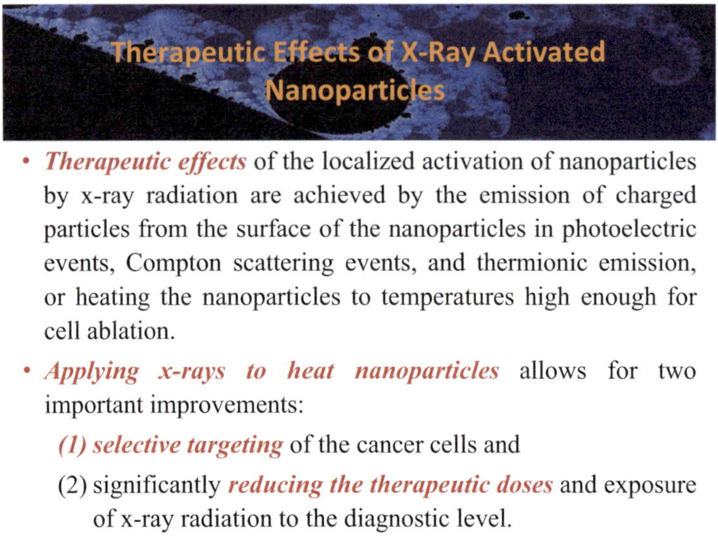

Slide 22.6 Therapeutic effects of X-ray-activated nanoparticles

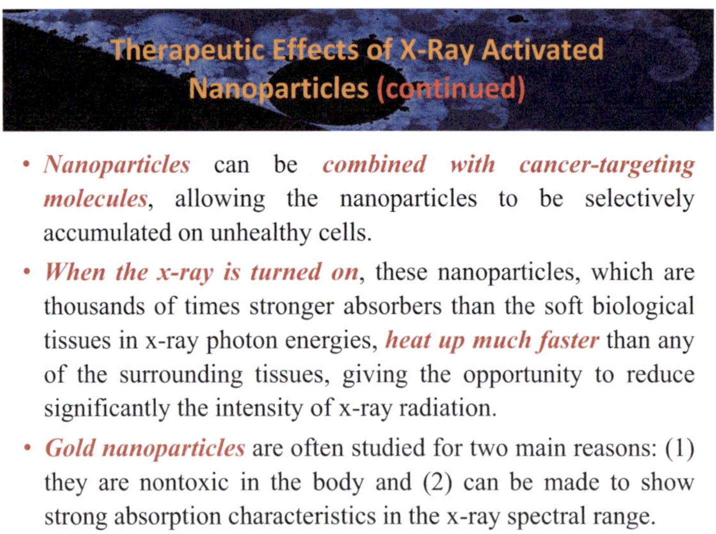

Slide 22.7 Therapeutic effects of X-ray-activated nanoparticles (continued)

the soft biological tissues in X-ray photon energies, heat up much faster than any of the surrounding tissues, giving the opportunity to reduce significantly the intensity of X-ray radiation. Gold nanoparticles are often studied for two main reasons: (1) they are nontoxic in the body and (2) can be made to show strong absorption characteristics in the X-ray spectral range [16].

In this section we use the results of Lorenz–Mie simulations of the absorption properties of various size gold nanoparticles over photon energies in the 1–100 keV range performed in Sect. 4.2 (Lecture 13) of Chap. 4. These absorption values are then plugged into a thermal model to determine the temperatures reached by the nanoparticles for X-ray exposures of differing time and intensity. The results of these simulations are discussed in relation to the effective implementation of nanophotohyperthermia and nanophotothermolysis treatments.

7.3.2.1 Thermal Modeling and Spectra Modeling

In Chap. 6 we introduced the one-temperature model (OTM) for time-dependent thermal modeling of nanoparticle heating kinetics shown by the following equation (Slide 22.8):

$$\frac{dT_{\mathrm{s}}}{dt} = \frac{3Q_{\mathrm{abs}}I_0 f(t)}{4r_0 C_{\mathrm{s}}(T_{\mathrm{s}})\rho_{\mathrm{s}}} - \frac{3\mu_{\infty}T_{\mathrm{s}}}{(\alpha+1)r_0^2 C_{\mathrm{s}}(T_{\mathrm{s}})\rho_{\mathrm{s}}}\left[\left(\frac{T_{\mathrm{s}}}{T_{\infty}}\right)^{\alpha+1} - 1\right]. \quad (7.4)$$

In the previous equation, the first term on the right-hand side deals with the heat generation by light absorption, and the second term describes heat lost to the surrounding medium by conduction. Absorption efficiency values for gold nanoparticles in the X-ray portion of the radiation spectrum have been calculated by using Lorenz–Mie theory in Sect. 4.2 and shown in Fig. 13.4 (Chap. 4).

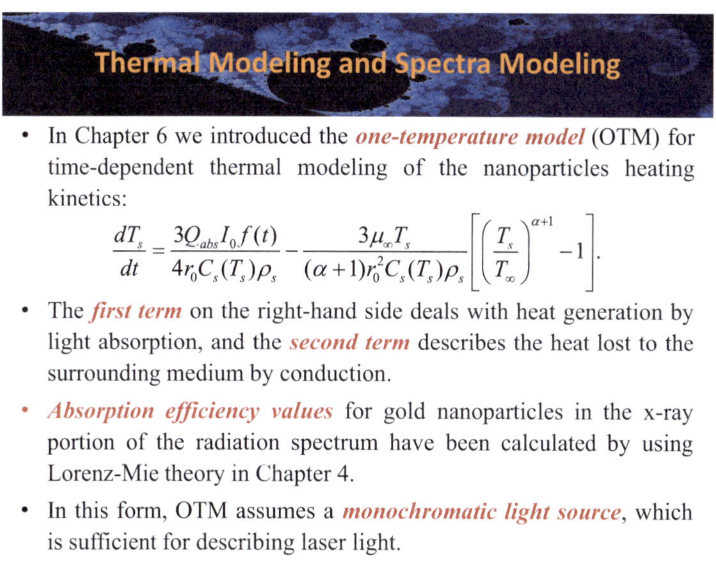

Slide 22.8 Thermal modeling and spectra modeling

Thermal Modeling and Spectra Modeling (continued)

- However, the ***output from most medical diagnostic x-ray devices*** follows a wide spectrum of photon energies as shown in the figure below from Chapter 4.

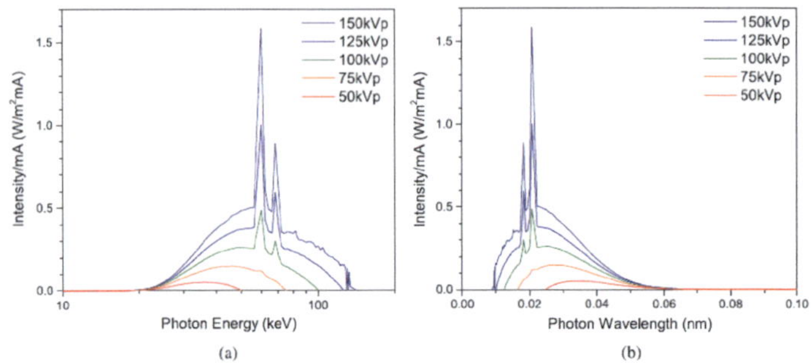

Slide 22.9 Thermal modeling and spectra modeling (continued)

Thermal Modeling and Spectra Modeling (continued)

- Because of this, it is better to ***integrate over all spectral energies*** when trying to determine the amount of heat generated:

$$\frac{dT_s}{dt}=\frac{3f(t)}{4r_0C_s(T_s)\rho_s}\left[\int_0^{\infty}Q_{abs}(\varepsilon)I_0(\varepsilon)d\varepsilon\right]-\frac{3\mu_\infty T_s}{(\alpha+1)r_0^2C_s(T_s)\rho_s}\left[\left(\frac{T_s}{T_\infty}\right)^{\alpha+1}-1\right].$$

- Rather than involving the energy integration during every numerical computation for $T_s(t)$, we introduced in Chapter 4 an ***"effective" value*** for the combined terms $Q_{abs}I_0$:

$$\frac{dT_s}{dt}=\frac{3Q_{eff}I_{eff}f(t)}{4r_0C_s(T_s)\rho_s}-\frac{3\mu_\infty T_s}{(\alpha+1)r_0^2C_s(T_s)\rho_s}\left[\left(\frac{T_s}{T_\infty}\right)^{\alpha+1}-1\right],$$

where $Q_{eff}I_{eff}=\int_0^{\infty}Q_{abs}(\varepsilon)I_0(\varepsilon)d\varepsilon.$

Slide 22.10 Thermal modeling and spectra modeling (continued)

In this form, OTM assumes a monochromatic light source, which is sufficient for describing laser light. However, the output from most medical diagnostic X-ray devices follows a wide spectrum of photon energies as shown in Fig. 13.5 in Sect. 4.2 (Lecture 13) of Chap. 4 (Slide 22.9). Because of this, it is better to integrate over all spectral energies when trying to determine the amount of heat generated, as shown as follows [16] (Slide 22.10):

$$\frac{dT_s}{dt} = \frac{3f(t)}{4r_0 C_s(T_s)\rho_s}\left[\int_0^\infty Q_{abs}(\varepsilon) I_0(\varepsilon) d\varepsilon\right] - \frac{3\mu_\infty T_s}{(\alpha+1) r_0^2 C_s(T_s)\rho_s}\left[\left(\frac{T_s}{T_\infty}\right)^{\alpha+1} - 1\right]. \quad (7.5)$$

Rather than involving the energy integration during every numerical computation for $T_s(t)$, we introduced in Sect. 4.2 an "effective" value for the combined terms $Q_{abs} I_0$. In this way, the effects of refinements to the spectral model can be seen separately as a change in the effective absorption efficiency before it is plugged into the OTM. Also, it reduces the equation to the equivalent of the original OTM (Slide 22.10),

$$\frac{dT_s}{dt} = \frac{3Q_{eff} I_{eff} f(t)}{4r_0 C_s(T_s)\rho_s} - \frac{3\mu_\infty T_s}{(\alpha+1) r_0^2 C_s(T_s)\rho_s}\left[\left(\frac{T_s}{T_\infty}\right)^{\alpha+1} - 1\right], \quad (7.6)$$

where $Q_{eff} I_{eff} = \int_0^\infty Q_{abs}(\varepsilon) I_0(\varepsilon) d\varepsilon$

In Sect. 4.2 (Lecture 13) of Chap. 4, we computed the effective intensity-absorption efficiency for gold nanoparticles by using the output spectrum of a typical X-ray source (Slide 22.11). The true spectrum will vary from machine to machine and depends greatly on the materials used in the X-ray tube construction

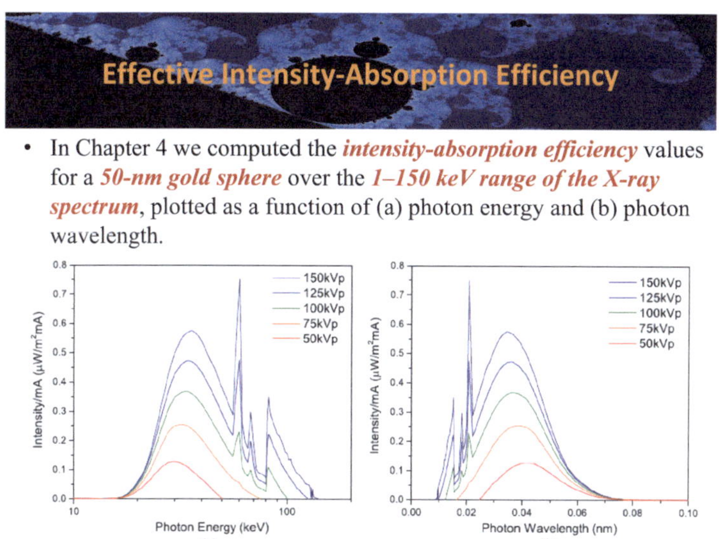

Slide 22.11 Effective intensity-absorption efficiency

Effective Intensity-Absorption Efficiency Values

- The table below from Chapter 4 lists the ***effective intensity-absorption efficiency values*** calculated ***for nanoparticles*** of radii varying ***from 5 to 50 nm*** absorbing radiation generated by ***tungsten anode x-ray tubes*** with kilovolt potentials ranging ***from 50 to 150 kVp***.

kVp	$Q_{eff} \cdot I_{eff}$ (10^{-6} W/m²)[a]										Exposure
	5 nm	10nm	15nm	20nm	25nm	30nm	35nm	40nm	45nm	50nm	mR[a]
50	0.229	0.459	0.688	0.917	1.146	1.375	1.604	1.833	2.062	2.290	1.3314
75	0.666	1.331	1.996	2.661	3.326	3.991	4.656	5.320	5.984	6.648	3.9734
100	1.257	2.514	3.771	5.027	6.283	7.539	8.795	10.050	11.306	12.561	7.5995
125	2.014	4.027	6.040	8.053	10.065	12.077	14.089	16.100	18.111	20.122	12.0014
150	2.794	5.588	8.381	11.174	13.966	16.758	19.550	22.341	25.132	27.922	16.8100

- The 50 nm Au particles have $Q_{eff} I_{eff}$ values ranging from ***2.290 to 27.922 ×10⁻⁶ W/m²***, with corresponding ***exposures*** ranging from ***1.33 to 16.81 mR/mA***.
- Typically, ***a single chest x ray*** will give an ***exposure of 6-25 mrem***, and standard ***x-ray computed tomography scans*** can range from ***300 to 2700 mrem***.

Slide 22.12 Effective intensity-absorption efficiency values

and filters used to harden the beam. X-rays produced by X-ray tubes are generated by bremsstrahlung and X-ray fluorescence. Most commonly, X-ray tubes use tungsten anodes, giving characteristic peaks at 59.3 and 67.2 keV above the general bremsstrahlung output [16]. In Sect. 4.2 we used the program called Spektr to produce realistic X-ray spectra shown in Fig. 13.5. Table 4.1 of Chap. 4 lists the effective intensity-absorption efficiency values calculated for nanoparticles of radii varying from 5 to 50 nm, absorbing radiation generated by tungsten anode X-ray tubes with kilovolt potentials ranging from 50 to 150 kVp (Slide 22.12).

7.3.2.2 Discussion and Results

To reach the desired final temperature, the incoming radiation intensity must be sufficiently large to induce heating [16]. To find a minimum value for this intensity, the model has been solved for the steady-state condition to achieve the maximum temperature (Slide 22.13):

$$0 = \frac{\partial T_{\mathrm{s,max}}}{\partial t} = \frac{3Q_{\mathrm{eff}} I_{\mathrm{eff}} f(t)}{4 r_0 C_{\mathrm{s}}(T_{\mathrm{s}}) \rho_{\mathrm{s}}} - \frac{3\mu_{\infty} T_{\mathrm{s}}}{(\alpha+1) r_0^2 C_{\mathrm{s}}(T_{\mathrm{s}}) \rho_{\mathrm{s}}} \left[\left(\frac{T_{\mathrm{s,max}}}{T_{\infty}} \right)^{\alpha+1} - 1 \right]. \tag{7.7}$$

We solve the previous equation for the effective intensity absorption efficiency:

$$[Q_{\mathrm{abs}} I_0 f(t)]_{\mathrm{eff}} = \frac{4\mu_{\infty} T_{\mathrm{s}} C_{\mathrm{s}} \rho_{\mathrm{s}}}{3(\alpha+1) r_0^2 C_{\infty} \rho_{\infty}} \left[\left(\frac{T_{\mathrm{s,max}}}{T_{\infty}} \right)^{\alpha+1} - 1 \right]. \tag{7.8}$$

Simulation Results

- *To reach the desired final temperature*, the incoming radiation intensity must be sufficiently large to induce heating.
- *To find a minimum value for this intensity*, the model was solved for the steady-state condition to achieve the maximum temperature:

$$0 = \frac{\partial T_{s,\max}}{\partial t} = \frac{3Q_{eff} I_{eff} f(t)}{4 r_0 C_s(T_s)\rho_s} - \frac{3\mu_\infty T_s}{(\alpha+1) r_0^2 C_s(T_s)\rho_s}\left[\left(\frac{T_{s,\max}}{T_\infty}\right)^{\alpha+1} - 1\right].$$

- We solve the above equation *for the effective intensity absorption efficiency*:

$$\left[Q_{abs} I_0 f(t)\right]_{eff} = \frac{4\mu_\infty T_s C_s \rho_s}{3(\alpha+1) r_0^2 C_\infty \rho_\infty}\left[\left(\frac{T_{s,\max}}{T_\infty}\right)^{\alpha+1} - 1\right].$$

- *Using the material properties* values listed in Tables 7.5 and 7.6, the value of $[Q_{abs} I_0 f(t)]_{\text{eff}}$ required to heat nanoparticles from body temperature, $T_0 = 310$ K, to hyperthermia, $T_{max} = 319$ K (46°C), is approximately 2.1×10^4 W/cm^2.

Slide 22.13 Simulation results

Table 7.5 Material properties of bone (Slide 22.14)

Symbol	Property	Values
μ	Thermal conductivity	5.60×10^{-1} W/m K
ρ	Density	1.85×10^3 kg/m^3
C	Specific heat	1.33×10^3 J/kg K
χ	Thermal diffusivity	2.276×10^{-7} m^2/s

Table 7.6 Material properties of gold (Slide 22.14)

Symbol	Property	Values
μ	Thermal conductivity	3.18×10^2 W/m K
ρ	Density	1.93×10^4 kg/m^3
C	Specific heat	1.29×10^2 J/kg K
χ	Thermal diffusivity	1.27×10^{-4} m^2/s

The right-hand side of the previous equation is solved using the material properties values listed in Tables 7.5 and 7.6 (for $r_0 = 30$ nm) to find $[Q_{\text{abs}} I_0 f(t)]_{\text{eff}}$ that would be required to heat nanoparticles from body temperature, $T_0 = 310$ K, to hyperthermia, $T_{\text{max}} = 319$ K (46 °C). Using the reference data from Tables 7.5 and 7.6, the required value of $[Q_{\text{abs}} I_0 f(t)]_{\text{eff}}$ is seen to be approximately 2.1×10^4 W/cm^2 (Slide 22.13).

The OTM is then solved using this input energy, and the time evolution of temperature for a 50-nm diameter gold nanoparticle is shown in Fig. 7.9 (Slide 22.15). The calculated effective intensity absorption efficiency values for a 50-nm particle are on the order of 10^{-6} W/m^2 per amp [16], as shown in the two rightmost columns of Table 4.1 in Sect. 4.2 of Chap. 4. The 50-nm Au particles

Table 7.5. Material properties of bone.

Symbol	Property	Values
μ	Thermal conductivity	$5.60\cdot10^{1}$ W/m·K
ρ	Density	$1.85\cdot10^{+3}$ kg/m^3
C	Specific heat	$1.33\cdot10^{+3}$ J/kg·K
χ	Thermal diffusivity	$2.276\cdot10^{-7}$ m^2/s

Table 7.6. Material properties of gold.

Symbol	Property	Values
μ	Thermal conductivity	$3.18\cdot10^{+2}$ W/m·K
ρ	Density	$1.93\cdot10^{+4}$ kg/m^3
C	Specific heat	$1.29\cdot10^{+2}$ J/kg·K
χ	Thermal diffusivity	$1.27\cdot10^{-4}$ m^2/s

Slide 22.14 Material properties used in simulations

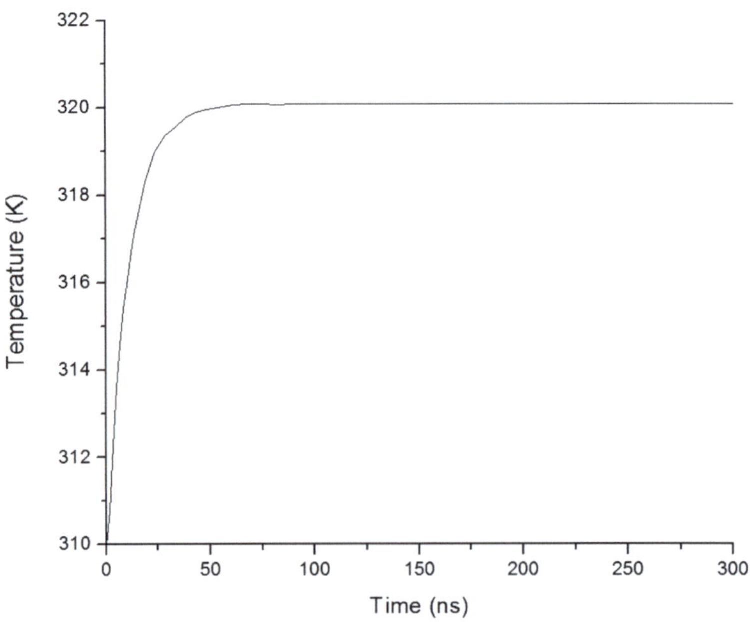

Fig. 7.9 Time evolution of temperature for a 50-nm diameter Au NP heated to 46 °C, the lower limit for thermal damage to bone tissues

Time Evolution of Nanoparticle Temperature Exposed to X-Rays

- The OTM is then solved using this input data, and the *time evolution of temperature* for a 50-nm diameter gold nanoparticle is shown in the figure.
- *The 50-nm Au particles* have $[Q_{abs}I_0f(t)]_{\text{eff}}$ ranging from $(2.290 \text{ to } 27.922)\times10^{-6}$ W/m^2, with the corresponding *exposures ranging from 1.33 to 16.81 mR/mA*.
- Typically, a single *chest x-ray* will give an *exposure of 6 to 25 mrem*, and standard *CT scans* can range from *300 to 2,700 mrem*.
- *Radiation therapy* treatments use much stronger radiation, as much as 5 orders of magnitude greater than standard chest x-rays, or roughly *10^6 mrem.*

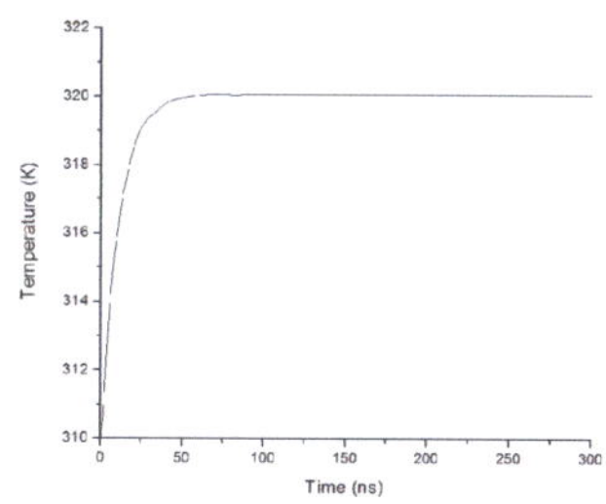

Figure. Time evolution of temperature for a 50-nm diameter Au NP heated to 46°C, the lower limit for thermal damage to bone tissues.

Slide 22.15 Time evolution of nanoparticle temperature exposed to X-rays

have $[Q_{\text{abs}}I_0f(t)]_{\text{eff}}$ in the range $2.290–27.922 \times 10^{-6}$ W/m^2, with the corresponding exposures ranging from 1.33 to 16.81 mR/mA. Typically, a single chest X-ray will give an exposure of 6–25 mrem, and standard CT scans can range from 300 to 2700 mrem. Radiation therapy treatments use much stronger radiation, as much as five orders of magnitude greater than standard chest X-rays, or roughly 10^6 mrem (Slide 22.15).

The output from Spektr assumes a standardized exposure time of 1 s at 100 cm from the focal spot. By doubling the amperage while cutting the exposure time in half, the overall exposure should be unchanged, while the light intensity has doubled. Looking closely at Fig. 7.9 it can be seen that the maximum heat is attained in less than 10^{-7} s due to the very small size of the nanoparticle. Decreasing the exposure time down to this extreme, the effective intensity can be increased to the extent that nanophotothermolysis of cancer cells could be done at diagnostic radiation equivalent exposures. Recent research on laser-generated X-rays hints at a promising future for X-ray light sources [16]. Output from these new X-ray emitters already produces pulses on the nanosecond timescale, and additionally, the output is much more directional and focused, limiting unnecessary exposure to healthy tissues.

In conclusion we have discussed the use of X-ray radiation as a photon source for nanophotothermolysis and nanophotohyperthermia treatments [16] (Slide 22.16). X-rays are much more capable of penetrating deep into biological tissues than visible light. Incorporation of nanoparticles into cancer cells leads to synergistic effects such as localized heating, photoelectron emission, thermionic emission of charged particles, or producing free radicals when used with standard

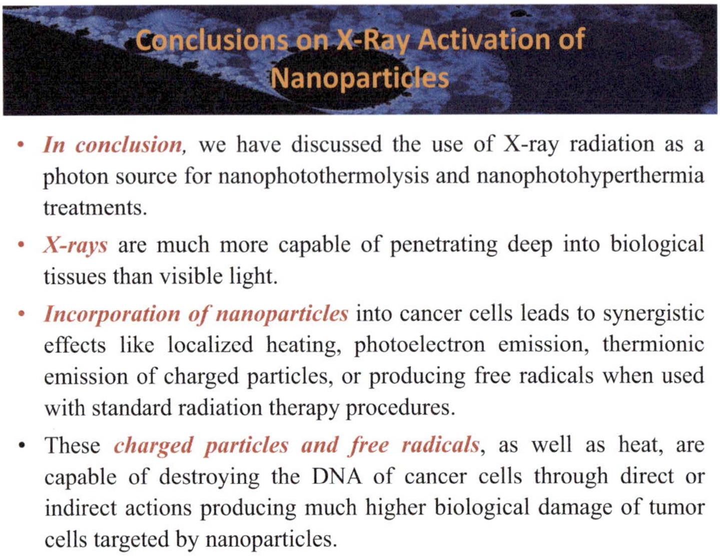

Slide 22.16 Conclusions on X-ray activation of nanoparticles

radiation therapy procedures. These charged particles and free radicals, as well as heat, are capable of destroying the DNA of cancer cells through direct or indirect actions producing much higher biological damage of tumor cells targeted by nanoparticles.

7.3.3 RF Activation of Nanoparticles in Single-Pulse and Multipulse Modes of Heating

The idea of using radiofrequency (RF) waves to heat nanoparticles (NPs: nanospheres, nanoshells, nanorods, nanoclusters) has emerged from nanomedicine applications (Slide 22.17). The human body, consisting of 60 % water, has negligible absorption of RF waves with penetration depth of 20–200 cm at 13.56 MHz, as shown in the Sect. 4.3 (Lecture 14) of Chap. 4. This gives a great opportunity to activate/heat nanoparticles injected in the human body with RF waves for diagnosis and treatment of diseases at any place inside the body harmlessly. This section contains material adopted from our publication [17].

The major concept of the heating kinetics of nanoparticles is often misunderstood (Slide 22.18). The key idea is that due to their small nanosize, particles gain and lose heat almost instantaneously, taking only nanoseconds to heat up and cool down, which is why throughout this section a nanosecond time scale is considered. We use here again the well-proven OTM thermal model for heating the nanoparticles over the time given by Eq. (7.4). The absorption efficiencies for

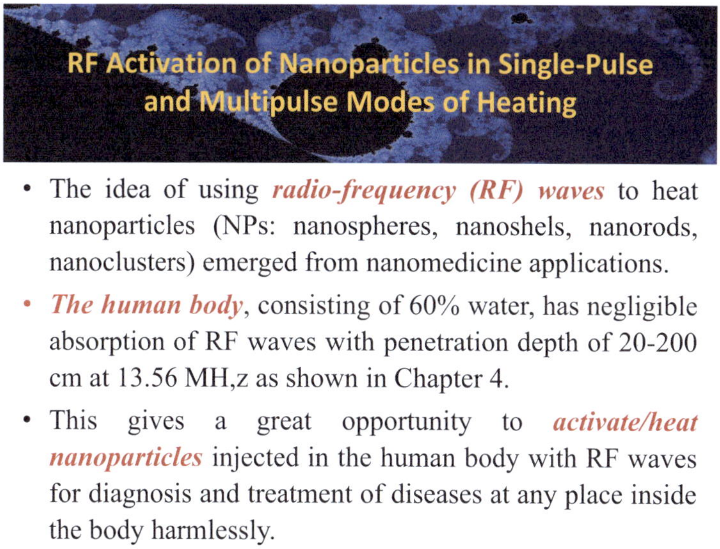

Slide 22.17 RF activation of nanoparticles in single-pulse and multipulse modes of heating

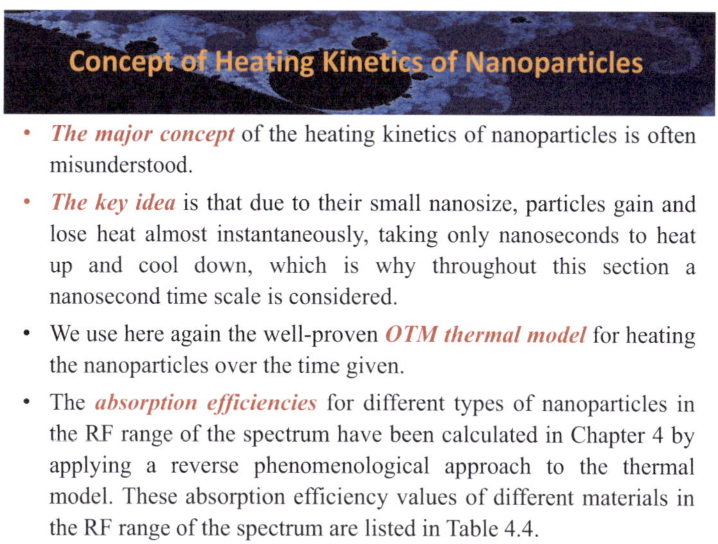

Slide 22.18 Concept of heating kinetics of nanoparticles

different types of nanoparticles in the RF range of the spectrum have been calculated in Sect. 4.3 by applying a reverse phenomenological approach to the thermal model. These absorption efficiency values of different materials in the RF range of the spectrum are listed in Table 4.4 (Slide 22.19).

RF Absorption Efficiency of Nanoparticles

- Results for the ***absorption efficiency in the RF spectral range*** along with the highest temperature achieved ***for metal and dielectric nanoparticles***.

Material	Absorption efficiency	Maximum temperature (K)
Metal (aluminum oxide, gold, magnesium oxide, nickel oxide, silver) sphere, 2.5-nm radius	0.8×10^{-9}	366.5
Metal sphere, 5-nm radius	1.15×10^{-9}	349.1
Metal sphere, 20-nm radius	6.10×10^{-9}	360.1
Metal sphere, 50-nm radius	2.73×10^{-8}	360.2
Metal sphere, 100-nm radius	1.47×10^{-7}	360.5
Dielectric (carbon, fullerene, glass, polystyrene) sphere, 5-nm radius	1.21×10^{-11}	300.6
Dielectric sphere, 50-nm radius	1.01×10^{-8}	326.0
Dielectric sphere, 100-nm radius	7.62×10^{-8}	339.0
Dielectric sphere, 150-nm radius	26×10^{-8}	344.6
Dielectric sphere, 250-nm radius	112×10^{-8}	344.8
Dielectric sphere, 500-nm radius	870×10^{-8}	345.2

Slide 22.19 RF absorption efficiency of nanoparticles

As we saw earlier, the pulse shape, pulse duration, and time interval between pulses are described by the function $f(t)$, which determines the time structure of the radiation delivery, using either single-pulse or multipulse modes. The single-pulse shape is described by a Gaussian curve,

$$f(t) = \mathrm{e}^{-(0.2*t-2)^2}, \tag{7.9}$$

whereas the multipulse mode is described by a periodic Gaussian function,

$$f(t) = \mathrm{e}^{-2*\sin(0.167*t-2)^2}. \tag{7.10}$$

Both of these pulse shapes generated by these equations can be viewed in Fig. 7.10 (Slide 22.20). Depending on the particular nanoparticle material, a multipulse mode can lead to the heat accumulation effect as we observed in the previous Sects. 7.1 and 7.2. Heat accumulation in dielectric nanoparticles happens because the particle doesn't have enough time to cool off from the first pulse before the second pulse heats it up again. The heat accumulation effect which occurs in a multipulse mode of heating is explained in detail in Sects. 7.1 and 7.2.

The OTM simulates the temperature change of the nanoparticle over time at a given RF pulse function and energy density for both heating modes—single-pulse and multipulse modes. The single-pulse and multipulse modes of heating are both studied in this section. The input data used in the OTM simulations are listed in Table 7.7 (Slide 22.20).

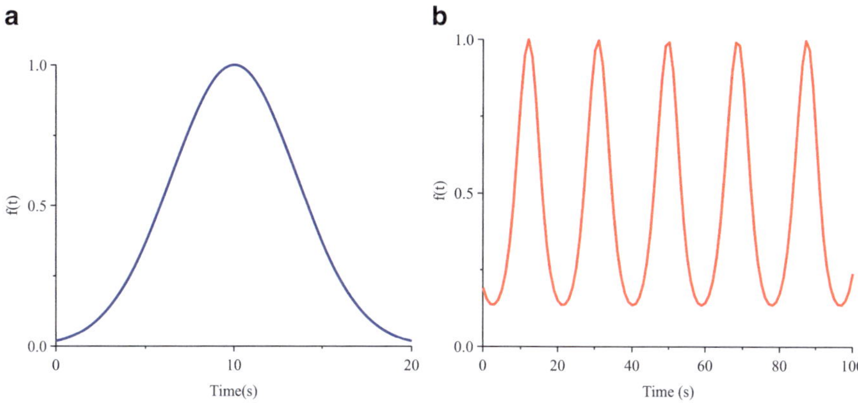

Fig. 7.10 (**a**) Single-pulse shape generated by the Gaussian curve function Eq. (7.9). (**b**) Multipulse function generated by Eq. (7.10)

Input Data for RF Heating Simulations

- The *OTM simulates* the temperature change of the nanoparticle over time at a given RF pulse function and energy density for both heating modes – *single-pulse and multipulse modes*. The input data used in the OTM simulations are listed in the table.

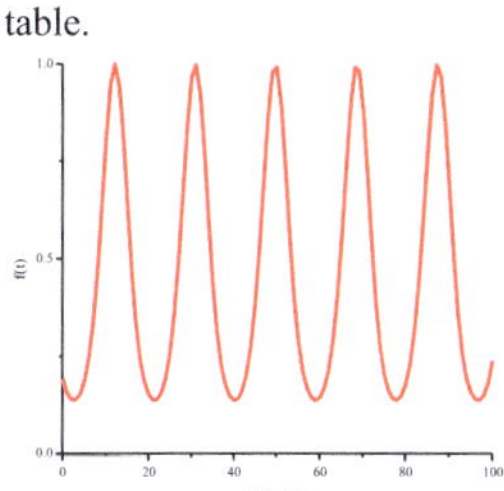

Single-pulse function	$e^{-(0.2\times t-2)^2}$
Multipulse function	$e^{-2\times\sin(0.167\times t-2)^2}$
Radius of NP	5 nm
Initial temperature	300 K
Duration of pulse	10 ns
Energy density	0.005 J/cm^2
Frequency of radiation	13.56 MHz
Surrounding medium	Deionized water
Power	600 W

Slide 22.20 Input data for RF heating simulations

7.3.3.1 Single-Pulse RF Heating Mode

The curves achieved through the OTM simulations in the single-pulse mode of heating by using a Maple code provided in Appendix B can be viewed in Fig. 14.1 in Chap. 4 for metal and dielectric nanoparticles (Slide 22.21). The metals examined were silver, magnesium oxide, aluminum oxide, and NiO, and their absorption efficiency is estimated to be that of gold. The dielectrics examined were

Table 7.7 Input data used in the OTM simulations. Reprinted from [17]. © With permission from Elsevier

Single-pulse function	$e^{-(0.2\times t-2)^2}$
Multipulse function	$e^{-2\times \sin(0.167\times t-2)^2}$
Radius of NP	5 nm
Initial temperature	300 K
Duration of pulse	10 ns
Energy density	0.005 J/cm^2
Frequency of radiation	13.56 MHz
Surrounding medium	Deionized water
Power	600 W

Single-Pulse RF Heating Mode

- The curves achieved through the ***OTM simulations*** can be viewed in the figure below.

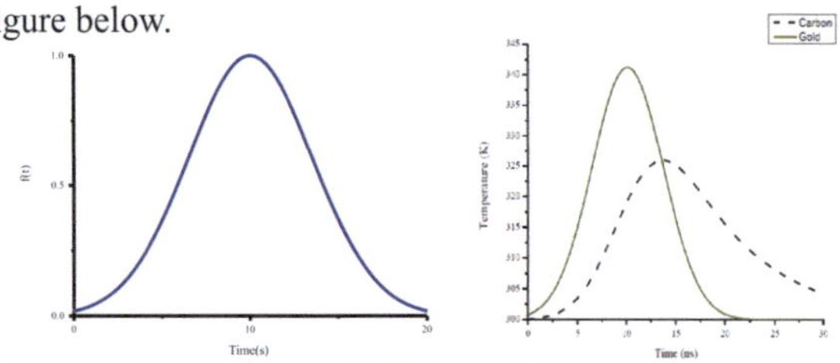

- The simulations show that ***all of the metals*** have the ***same heating kinetics*** and almost identical time-temperature profiles ***for the same conditions of RF heating***.
- The heating of different ***dielectric nanoparticles***, like in the metal case, has ***similar curves*** as well for the same conditions of RF heating.

Slide 22.21 Single-pulse RF heating mode

polystyrene, fullerene, and glass, and their absorption efficiency, as well as their experimental behavior, was modeled after carbon, while metal nanoparticles were modeled after gold. In order to compare these materials, the initial conditions were kept the same for every type of nanoparticle. These conditions are similar to those used in experiments. The results for the absorption efficiency in the RF spectral range along with the highest temperature achieved for metal and dielectric nanoparticles studied are summarized in Table 4.4. Our simulations have shown that all of the metals have the same heating kinetics and almost identical time–temperature profiles for the same conditions of RF heating.

Since the simulation curves for the dielectric materials are very similar to each other, they can be represented by only one dielectric material, such as carbon. The single-pulse temperature graphs for carbon nanoparticles of various sizes are shown in Fig. 7.11 (Slide 22.22). It follows from the figure that during the pulse lifetime,

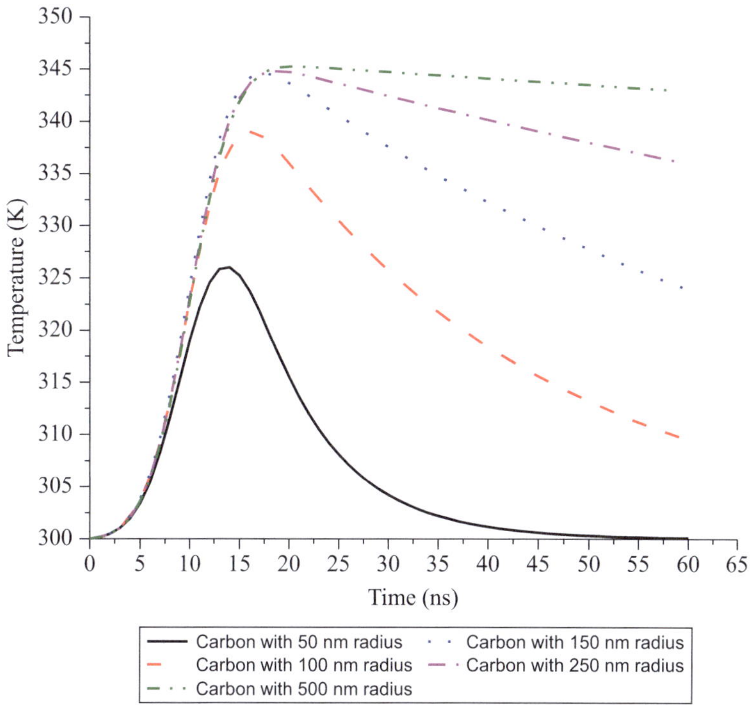

Fig. 7.11 Single-pulse heating of carbon nanoparticles of various sizes. Reprinted from [17]. © With permission from Elsevier

Single-Pulse RF Heating Mode (continued)

- ***Single-pulse temperature graphs*** for carbon nanoparticles of various sizes are shown in the figure.
- During the pulse lifetime, the ***transfer of heat*** from the nanoparticle into the surrounding medium is slight, and the particles rapidly ***reach the maximum temperature.***
- ***After the RF pulse degrades***, the transfer of heat from the particle to the surrounding medium becomes increasingly important, since the energy source is no longer present in the system.

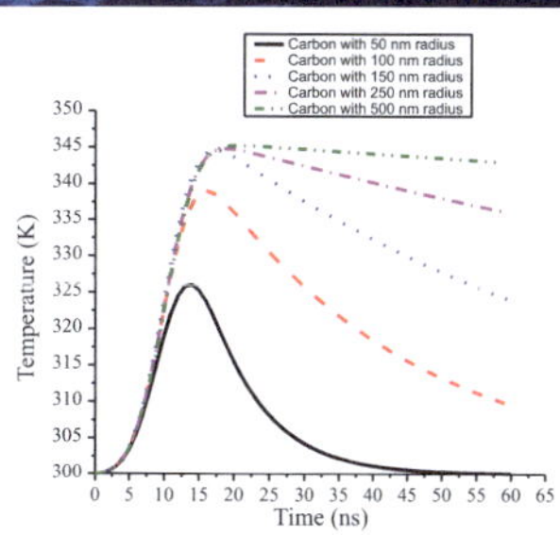

Figure. Single-pulse heating of carbon nanoparticles of various sizes.

- ***The cooling rate*** greatly depends on the size of the carbon nanoparticle and keeping a high temperature of the nanoparticle and surrounding medium (≥ 335 K) for a relatively long time, reaching a microsecond time scale for particles larger than 500 nm in radius.

Slide 22.22 Single-pulse RF heating mode (continued)

the transfer of heat from the nanoparticle into the surrounding medium is slight, and the particles rapidly reach the maximum temperature. The highest heating rate, about 2.1×10^{10} K/s, is observed for the carbon nanoparticle of 50-nm radius. The maximum temperature for the given energy density of 0.005 J/cm^2 is reached in the first 15–20 ns, depending on the nanoparticle size. The heating time for a particle is longer than the pulse duration, 8–10 ns, because the heating kinetics involves first the transfer of energy from the irradiative source to the electron subsystem, and then further to the phonon subsystem. This creates the time delay for the particle to reach the maximum temperature.

After the RF pulse degrades, the transfer of heat from the particle to the surrounding medium becomes increasingly important, since the energy source is no longer present in the system (Slide 22.22). As evident from Fig. 7.11, the cooling rate greatly depends on the size of the carbon nanoparticle and keeping a high temperature of the nanoparticle and surrounding medium ($\geq$335 K) for a relatively long time, reaching a microsecond time scale for particles larger than 500 nm in radius. The total time for one heating cycle (heating from the initial temperature 300 K to maximum temperature, followed by cooling back to the initial temperature) strongly depends on the nanoparticle size as well.

7.3.3.2 Multipulse RF Heating Mode

It is interesting to investigate the effect of the multipulse mode of heating metal and dielectric nanoparticles to determine what would enable accumulative heating of the nanoparticles in the RF range of the spectrum (Slide 22.23). Here we

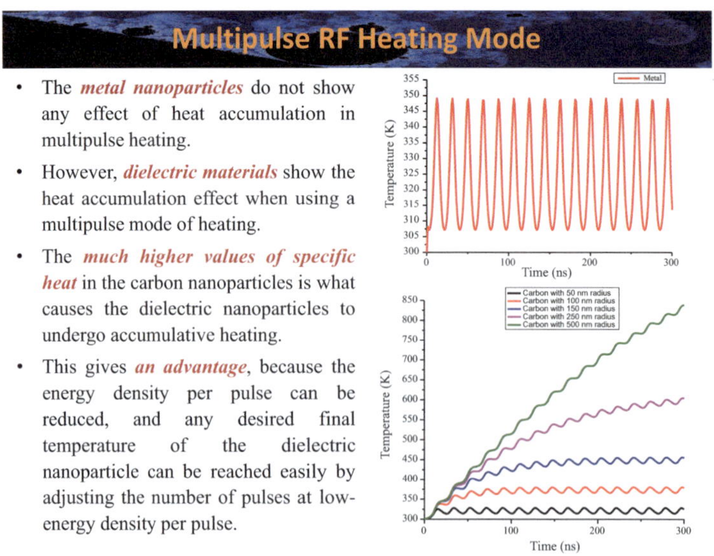

Slide 22.23 Multipulse RF heating mode

reproduce our results published in [17]. The multiple pulses are modeled by Gaussian curves shown in Fig. 7.10b with changeable repetition rate, which depends on several values listed in Table 7.7. The results of these simulations of temperature profiles by using a Maple code provided in Appendix D for both metal and dielectric nanoparticles exposed to the multipulse mode of heating are shown in Fig. 7.12. As we showed in Sect. 7.1 and can be seen from Fig. 7.12a, even for high frequency of pulse repetition, the metal nanoparticles do not show any effect of heat accumulation in multipulse heating. This is due to the extremely quick cooling kinetics of metal nanoparticles in the surrounding aqueous medium. The particle reaches the peak temperature and rapidly falls to equilibrium (300 K) in a time less than the time delay between the pulses, even for high-frequency pulse generation.

However, dielectric materials generally show the heat accumulation effect and an increase of the highest temperature when using a multipulse mode of heating [17], as demonstrated in Fig. 7.12b for the example of heating carbon nanoparticles (Slide 22.23). Even with the small size, within the first few pulses, the accumulative heating effect is noticeable, in contrast with any metal nanoparticle heating. The much higher values of specific heat in the carbon nanoparticles is what causes the dielectric nanoparticles to undergo accumulative heating, whereas metal nanoparticles do not experience accumulative heating, even at high frequency of pulse repetition. This gives an advantage for dielectric nanoparticles in using the multipulse mode vs. single pulse, because the energy density per pulse can be reduced, and any desired final temperature of the dielectric nanoparticle can be reached easily by adjusting the number of pulses at low energy density per pulse (Slide 22.23).

In conclusion we have performed time-dependent simulations of different nonstationary RF–nanoparticle interaction modes (Slides 22.24 and 22.25). The study of heating kinetics in the RF range of spectrum shows the following:

- The maximum temperature of metal nanoparticles is reached in the first 10–15 ns of heating with a heating rate of 3.4×10^{10} K/s.
- The maximum temperature of dielectric nanoparticles is reached in the first 15–20 ns of heating with a heating rate of 2.1×10^{10} K/s.
- The cooling rate of dielectric nanoparticles greatly depends on the particle size and keeping a high temperature of the nanoparticle and surrounding medium ($\geq$335 K) for a relatively long time, reaching a microsecond time scale for particles larger than 500 nm in radius.
- The multipulse mode of heating does not create accumulative heating effects in metal nanoparticles because of their rapid cooling kinetics in an aqueous environment.
- Changing the target material of nanoparticles to carbon with a much higher value of specific heat causes the dielectric NPs to undergo accumulative heating while in a multipulse mode.
- The highest temperature of 834.5 K is reached by a carbon nanosphere with radius 500 nm using a multipulse heating mode.

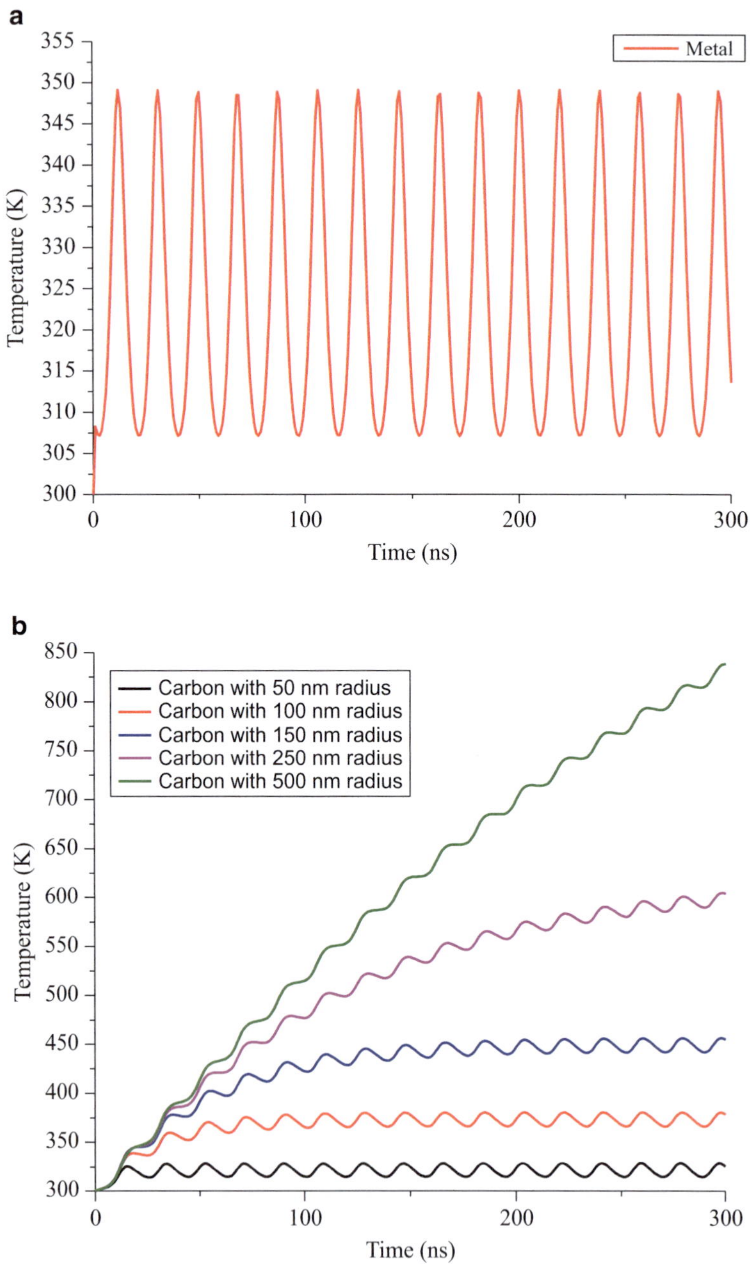

Fig. 7.12 (**a**) Temperature curve of metal nanoparticles using multipulse heating. (**b**) Multipulse heating of carbon nanoparticles of different sizes. Reprinted from [17]. © With permission from Elsevier

Conclusions on RF Wave Activation of Nanoparticles

In conclusion, we have performed time-dependent simulations of different nonstationary RF-nanoparticle interaction modes. The study of heating kinetics in the RF range of the spectrum has shown the following:

- The ***maximum temperature of metal nanoparticles*** is reached in the first 10-15 ns of heating with a heating rate of 3.4×10^{10} K/s.
- The ***maximum temperature of dielectric nanoparticles*** is reached in the first 15-20 ns of heating with a heating rate of 2.1×10^{10} K/s.
- The ***cooling rate*** of dielectric nanoparticles greatly depends on the particle size and keeping a high temperature of the nanoparticle and surrounding medium ($\geq$ 335 K) for a relatively long time, reaching a microsecond time scale for large particles greater than 500 nm in radius.

Slide 22.24 Conclusions on RF waves activation of nanoparticles

Conclusions on RF Wave Activation of Nanoparticles (continued)

- The ***multipulse mode of heating*** does not create accumulative heating effects in metal nanoparticles because of their rapid cooling kinetics in an aqueous environment.
- ***Changing the target material*** of nanoparticles to carbon with a much higher value of specific heat causes the dielectric NPs to undergo accumulative heating while in a multipulse mode.
- The ***highest temperature*** of 834.5 K is reached by a carbon nanosphere with radius 500 nm using a multipulse heating mode.

Slide 22.25 Conclusions on RF waves activation of nanoparticles (continued)

Appendix D: Maple Code for Bioparticle Heating in a Multipulse Mode

This appendix contains a Maple code designed for using OTM to find the heating kinetics of healthy and cancerous cell organelles heated in cytoplasm by multipulsed radiation (Slide 21.9). The comments within the Maple code should explain the variables, procedures, and functions that are used.

Note: This code comes directly from the Maple file and is therefore formatted to word-wrap according to Maple standard character-per-line limits.

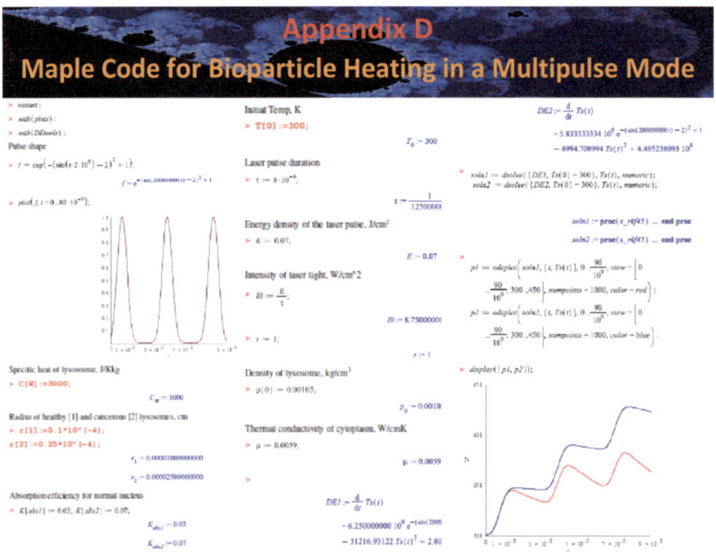

Slide 21.9 Appendix D: maple code for bioparticle heating in a multipulse mode

```
> restart :
> with(plots) :
> with(DEtools) :
```

Pulse shape

> $f := \exp\left(-\left(\sin\left(t\cdot 2\cdot 10^{8}\right) - 2\right)^{2} + 1\right);$

$$f := e^{-(\sin(200000000\, t) - 2)^2 + 1}$$

> $plot\left(f, t = 0\,..80\cdot 10^{-9}\right);$

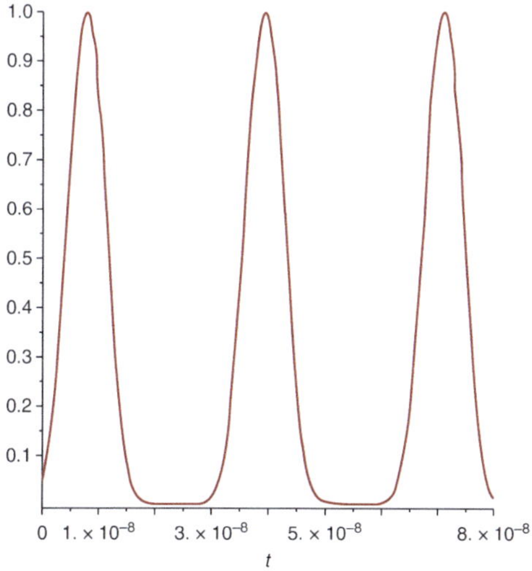

Specific heat of lysososme, J/Kkg

```
> C[W]:=3000;
```

$$C_W := 3000$$

Radius of healthy [1] and cancerous [2] lysosomes, cm

```
> r[1]:=0.1*10^(-4);
r[2]:=0.25*10^(-4);
```

$$r_1 := 0.00001000000000$$

$$r_2 := 0.00002500000000$$

Absorption efficiency for normal nucleus

> $K[abs1] := 0.03;\ K[abs2] := 0.07;$

$$K_{abs1} := 0.03$$

$$K_{abs2} := 0.07$$

Initial Temp, K

```
> T[0]:=300;
```

$$T_0 := 300$$

Laser pulse duration

> $\tau := 8 \cdot 10^{-9};$

$$\tau := \frac{1}{125000000}$$

Energy density of the laser pulse, J/cm^2

> $E := 0.07;$

$$E := 0.07$$

Intensity of laser light, W/cm^2

> $I0 := \frac{E}{\tau};$

$$I0 := 8.75000000\ 10^6$$

> $s := 1;$

$$s := 1$$

Density of lysosome, kg/cm^3

> $\rho[0] := 0.00105;$

$$\rho_0 := 0.00105$$

Thermal conductivity of cytoplasm, W/cmK

> $\mu := 0.0059;$

$$\mu := 0.0059$$

>

$$DE1 := diff(Ts(t), t) = \frac{3 \cdot K[abs1] \cdot I0 \cdot f}{4 \cdot r[1] \cdot C[W] \cdot \rho[0]} - \frac{\mu \cdot T[0]}{(s+1) \cdot r[1]^2 \cdot C[W] \cdot \rho[0]} \cdot \left(\left(\frac{Ts(t)}{T[0]} \right)^{s+1} - 1 \right);$$

$$DE2 := diff(Ts(t), t) = \frac{3 \cdot K[abs2] \cdot I0 \cdot f}{4 \cdot r[2] \cdot C[W] \cdot \rho[0]} - \frac{\mu \cdot T[0]}{(s+1) \cdot r[2]^2 \cdot C[W] \cdot \rho[0]} \cdot \left(\left(\frac{Ts(t)}{T[0]} \right)^{s+1} - 1 \right);$$

$$DE1 := diff(Ts(t), t) = \frac{3 \cdot K[abs1] \cdot I0 \cdot f}{4 \cdot r[1] \cdot C[W] \cdot \rho[0]} - \frac{\mu \cdot T[0]}{(s+1) \cdot r[1]^2 \cdot C[W] \cdot \rho[0]} \cdot \left(\left(\frac{Ts(t)}{T[0]} \right)^{s+1} - 1 \right);$$

$$DE2 := diff(Ts(t), t) = \frac{3 \cdot K[abs2] \cdot I0 \cdot f}{4 \cdot r[2] \cdot C[W] \cdot \rho[0]} - \frac{\mu \cdot T[0]}{(s+1) \cdot r[2]^2 \cdot C[W] \cdot \rho[0]} \cdot \left(\left(\frac{Ts(t)}{T[0]} \right)^{s+1} - 1 \right);$$

$$DE1 := \frac{d}{dt} Ts(t) = 6.250000000\ 10^9\ e^{-(\sin(200000000\,t) - 2)^2 + 1} - 31216.93122\ Ts(t)^2 + 2.809523810\ 10^9$$

$$DE2 := \frac{d}{dt} Ts(t) = 5.833333334\ 10^9\ e^{-(\sin(200000000\,t) - 2)^2 + 1} - 4994.708994\ Ts(t)^2 + 4.495238095\ 10^8$$

\> $soln1 := dsolve(\{DE1, Ts(0) = 300\}, Ts(t), numeric);$
$soln2 := dsolve(\{DE2, Ts(0) = 300\}, Ts(t), numeric);$

$soln1 :=$ **proc**(x_rkf45) ... **end proc**

$soln2 :=$ **proc**(x_rkf45) ... **end proc**

\> $$p1 := odeplot\left(soln1, [t, Ts(t)], 0 .. \frac{90}{10^9}, view = \left[0 .. \frac{90}{10^9}, 300 .. 450 \right], numpoints = 1000, color = red \right):$$

$$p2 := odeplot\left(soln2, [t, Ts(t)], 0 .. \frac{90}{10^9}, view = \left[0 .. \frac{90}{10^9}, 300 .. 450 \right], numpoints = 1000, color = blue \right):$$

> *display*([*p1*, *p2*]);

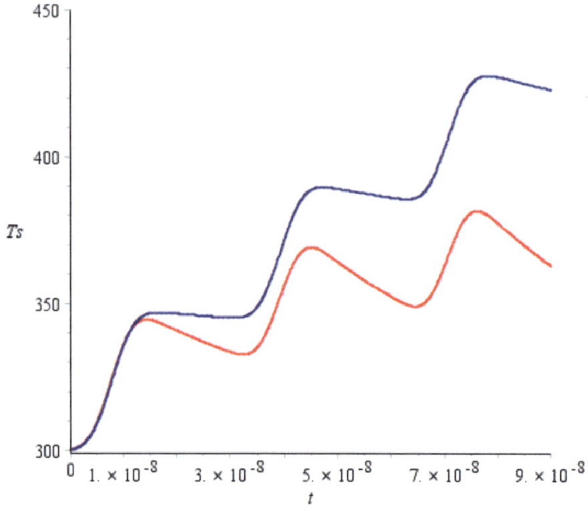

Homework Exercises

Section 7.1: Modeling Photothermal Heating for Multipulsed Radiation

7.1. List the potential advantages of photothermal sensitizers delivered to the tumor.
7.2. What are three main categories of factors defining the heating of the nanoparticle?
7.3. True or false? Particle heating depends on the properties of the surrounding medium.
7.4. What factors stay relatively constant across many nanoparticle heating experiments?
7.5. What pulse durations are often used in nanoparticle heating experiments?
7.6. Why are gold nanoparticles often used in nanoparticle heating experiments?
7.7. Why is water employed as a surrounding medium?
7.8. How many radiation pulses are used in nanoparticle heating experiments?
7.9. What size range of gold nanoparticles is often used in nanoparticle heating experiments?
7.10. What frequency range of pulse generation is often used in nanoparticle heating experiments?
7.11. What is a key goal of the multipulse mode of heating nanoparticles?
7.12. What function is used to simulate multiple pulses?
7.13. What is a heat accumulation effect?

7.14. True or false? Multiple pulses produce an accumulative heating effect in metal nanoparticles.
7.15. Describe multipulse heating of metal nanoparticles.
7.16. True or false? High pulse-generation frequencies do not create an accumulative heating effect in metal nanoparticles.
7.17. Why do cell organelles undergo a heat accumulation effect?
7.18. True or false? The cell organelles have much lower values of the specific heat than the metal ones.
7.19. What are the advantages of multipulse heating of cell organelles?
7.20. True or false? A multipulse mode of heating allows us to reduce the energy density of the radiation per pulse but still able to cause biological damage.
7.21. How is a desired final temperature of the organelle reached in a multipulse mode of heating?
7.22. Is there any difference in the heating kinetics of the mitochondria and cell nuclei? Explain.
7.23. Compare multipulse heating kinetics of metal and biological particles.
7.24. True or false? The cooling rate has the same values as the heating rate in metal nanoparticles.

Section 7.2: Computer Practicum: Multipulse Mode of Heating Healthy and Cancerous Cell Organelles

7.25. How is a thermal model modified to compute multiple pulses?
7.26. What conclusions can be made from simulation results of multipulse heating of healthy and cancerous organelles?
7.27. True or false? The healthy lysosome reaches a higher temperature and has a longer cooling time than cancerous lysosome.
7.28. What are the energy density of radiation and minimum number of pulses that could cause damage to unhealthy lysosomes while keeping healthy lysosomes unharmed?
7.29. Perform multipulse heating simulations for a silver nanoparticle of optimal radius in cytoplasm, fat, and tumor surrounding media by using thermal data from Tables 6.1 to 6.4 and the Maple code in Appendix D. Find the optimal radius of the nanoparticle, optimal wavelength of the radiation, absorption efficiency, and energy density of the radiation required to achieve a maximum temperature of the nanoparticle in each surrounding medium. Write a report on your simulation results. Make conclusions about the heating and cooling kinetics of the nanoparticle in a multipulse heating mode.
7.30. Perform multipulse heating simulations for a fullerene nanoparticle of optimal radius in cytoplasm, fat, and tumor surrounding media by using thermal data from Tables 6.1 to 6.4 and the Maple code in Appendix D. Find the optimal radius of the nanoparticle, optimal wavelength of the radiation, absorption efficiency, and energy density of the radiation required to achieve

a maximum temperature of the nanoparticle in each surrounding medium. Write a report on your simulation results. Make conclusions about the heating and cooling kinetics of the nanoparticle in a multipulse heating mode.

7.31. Perform multipulse heating simulations for a polystyrene nanoparticle of optimal radius in cytoplasm, fat, and tumor surrounding media by using thermal data from Tables 6.1 to 6.4 and the Maple code in Appendix D. Find the optimal radius of the nanoparticle, optimal wavelength of the radiation, absorption efficiency, and energy density of the radiation required to achieve a maximum temperature of the nanoparticle in each surrounding medium. Write a report on your simulation results. Make conclusions about the heating and cooling kinetics of the nanoparticle in a multipulse heating mode.

7.32. Perform multipulse heating simulations for a glass nanoparticle of optimal radius in cytoplasm, fat, and tumor surrounding media by using thermal data from Tables 6.1 to 6.4 and the Maple code in Appendix D. Find the optimal radius of the nanoparticle, optimal wavelength of the radiation, absorption efficiency, and energy density of the radiation required to achieve a maximum temperature of the nanoparticle in each surrounding medium. Write a report on your simulation results. Make conclusions about the heating and cooling kinetics of the nanoparticle in a multipulse heating mode.

7.33. Perform multipulse heating simulations for a carbon nanoparticle of optimal radius in cytoplasm, fat, and tumor surrounding media by using thermal data from Tables 6.1 to 6.4 and the Maple code in Appendix D. Find the optimal radius of the nanoparticle, optimal wavelength of the radiation, absorption efficiency, and energy density of the radiation required to achieve a maximum temperature of the nanoparticle in each surrounding medium. Write a report on your simulation results. Make conclusions about the heating and cooling kinetics of the nanoparticle in a multipulse heating mode.

7.34. Perform multipulse heating simulations for an aluminum oxide nanoparticle of optimal radius in cytoplasm, fat, and tumor surrounding media by using thermal data from Tables 6.1 to 6.4 and the Maple code in Appendix D. Find the optimal radius of the nanoparticle, optimal wavelength of the radiation, absorption efficiency, and energy density of the radiation required to achieve a maximum temperature of the nanoparticle in each surrounding medium. Write a report on your simulation results. Make conclusions about the heating and cooling kinetics of the nanoparticle in a multipulse heating mode.

7.35. Perform multipulse heating simulations for a magnesium oxide nanoparticle of optimal radius in cytoplasm, fat and tumor surrounding media by using thermal data from Tables 6.1 to 6.4 and the Maple code in Appendix D. Find the optimal radius of the nanoparticle, optimal wavelength of the radiation, absorption efficiency, and energy density of the radiation required to achieve a maximum temperature of the nanoparticle in each surrounding medium. Write a report on your simulation results. Make conclusions about the heating and cooling kinetics of the nanoparticle in a multipulse heating mode.

7.36. Perform multipulse heating simulations for a nickel oxidefullerene nanoparticle of optimal radius in cytoplasm, fat, and tumor surrounding media by using thermal data from Tables 6.1 to 6.4 and the Maple code in Appendix D. Find the optimal radius of the nanoparticle, optimal wavelength of the radiation, absorption efficiency, and energy density of the radiation required to achieve a maximum temperature of the nanoparticle in each surrounding medium. Write a report on your simulation results. Make conclusions about the heating and cooling kinetics of the nanoparticle in a multipulse heating mode.

7.37. Perform multipulse heating simulations for normal and cancerous cell nuclei in cytoplasm by using thermal data from Tables 6.1 to 6.5 and the Maple code in Appendix D. Write a report on your simulation results according to the Practice Example 6. Make conclusions about how multipulse laser heating would affect cancerous organelles in comparison to healthy organelles.

7.38. Perform multipulse heating simulations for normal and cancerous cell mitochondria in cytoplasm by using thermal data from Tables 6.1 to 6.5 and the Maple code in Appendix D. Write a report on your simulation results according to the Practice Example 6. Make conclusions about how multipulse laser heating would affect cancerous organelles in comparison to healthy organelles.

7.39. Perform multipulse heating simulations for normal and cancerous cell microtubules in cytoplasm by using thermal data from Tables 6.1 to 6.5 and the Maple code in Appendix D. Write a report on your simulation results according to the Practice Example 6. Make conclusions about how multipulse laser heating would affect cancerous organelles in comparison to healthy organelles.

7.40. Perform multipulse heating simulations for normal and cancerous cell ribosomes in cytoplasm by using thermal data from Tables 6.1 to 6.5 and the Maple code in Appendix D. Write a report on your simulation results according to the Practice Example 6. Make conclusions about how multipulse laser heating would affect cancerous organelles in comparison to healthy organelles.

7.41. Perform multipulse heating simulations for normal and cancerous cell cytoskeleton in cytoplasm by using thermal data from Tables 6.1 to 6.5 and the Maple code in Appendix D. Write a report on your simulation results according to the Practice Example 6. Make conclusions about how multipulse laser heating would affect cancerous organelles in comparison to healthy organelles.

Section 7.3: X-Ray and RF Heating of Nanoparticles

7.42. What are the properties of ideal radiation for nanophotodynamic therapy?

7.43. What parts of the electromagnetic spectrum satisfy the requirements of ideal radiation for nanophotodynamic therapy?

7.44. True or false? Short X-rays and long RF waves are universal tools for nanoparticle activation.
7.45. True or false? A surface plasmon resonance is observed in the visible and infrared range of spectrum.
7.46. True or false? Visible light has deep penetration into soft biological tissue.
7.47. List three main advantages of using X-rays to power photothermal therapies.
7.48. True or false? X-rays have low energy of individual photons.
7.49. What are therapeutic effects of X-ray-activated nanoparticles?
7.50. Select one or more. Therapeutic effects of the localized activation of nanoparticles by X-ray radiation can be achieved by:

(a) Surface plasmon resonance
(b) Heat generation
(c) Photoelectric effect
(d) Compton effect
(e) Thermionic emission
(f) Shock-wave generation
(g) Sound production
(h) All of the above
(i) None of the above

7.51 What improvements can nanoparticles bring to traditional X-ray radiation therapy?
7.52 True or false? Gold nanoparticles can dramatically reduce the therapeutic doses and exposure of X-ray radiation in cancer treatments.
7.53 True or false? The one-temperature model is designed for a monochromatic light source.
7.54 Select one or more. The output from most medical diagnostic X-ray devices has a:

(a) Narrow spectrum of photon energies
(b) Monochromatic radiation
(c) Wide spectrum of photon energies
(d) Polyenergetic beam
(e) Monoenergetic beam
(f) All of the above
(g) None of the above

7.55 What is an effective intensity-absorption efficiency?
7.56 What is the X-ray exposure range for heating a 50-nm gold nanoparticle?
7.57 What is the X-ray exposure range for a typical chest X-ray?
7.58 Select one or more. The exposure range for standard X-ray computed tomography scans is:

(a) 1.33–16. 81 mR/mA
(b) 6–25 mrem
(c) 300–2700 mrem

(d) All of the above
(e) None of the above

7.59 Select one or more. Incorporation of nanoparticles into cancer cells leads to synergistic effects like:

(a) Localized heating
(b) Photoelectron emission
(c) Thermionic emission
(d) Producing free radicals
(e) All of the above
(f) None of the above

7.60 True or false? Charged particles as well as free radicals are capable of repairing the DNA of cancer cells through direct or indirect actions.
7.61 What is the frequency range of RF waves?
7.62 True or false? The human body has a negligible absorption of RF waves.
7.63 Describe a major concept of the heating kinetics of nanoparticles.
7.64 True or false? All of the metals have the same heating kinetics and almost identical time–temperature profiles for the same conditions of RF heating.
7.65 True or false? A heating of different dielectric nanoparticles has similar time–temperature curves for the same conditions of RF heating.
7.66 Describe the single-pulse temperature graphs for carbon nanoparticles of various sizes.
7.67 True or false? The cooling rate of carbon nanoparticles doesn't depend on the nanoparticle size.
7.68 Explain why dielectric materials show the heat accumulation effect when using a multipulse mode of heating.
7.69 True or false? Dielectric nanoparticles have much higher values of specific heat than metal ones.
7.70 True or false? The maximum temperature of the nanoparticle is reached after degradation of the heating radiation pulse.
7.71 True or false? During the pulse lifetime, the transfer of heat from the nanoparticles into the surrounding medium is slight, and the particles rapidly reach the maximum temperature.

References

1. R.R. Letfullin, T.F. George, Plasmonic nanomaterials in nanomedicine, in *Springer Handbook of Nanomaterials*, ed. by R. Vajtai (Springer, Berlin, 2013), pp. 1063–1097
2. R.R. Letfullin, C.B. Iversen, T.F. George, Modeling nanophotothermal therapy: kinetics of thermal ablation of healthy and cancerous cell organelles and gold nanoparticles. Nanomed.: Nanotechnol. Biol. Med. **7**, 137–145 (2011)
3. R.R. Letfullin, T.F. George, New dynamic modes for selective laser cancer nanotherapy, in *Computational Studies of New Materials II: From Ultrafast Processes and Nanostructures to*

Optoelectronics, Energy Storage and Nanomedicine, ed. by T.F. George, D. Jelski, R.R. Letfullin, G.P. Zhang (World Scientific, Singapore, 2011), pp. 131–172
4. V.P. Zharov, K.E. Mercer, E.N. Galitovskaya, M.S. Smeltzer, Photothermal nanotherapeutics and nanodiagnostics for selective killing of bacteria targeted with gold nanoparticles. Biophys. J. **90**, 619–627 (2006)
5. C.M. Pitsillides, E.K. Joe, X. Wei, R.R. Anderson, C.P. Lin, Selective cell targeting with light-absorbing microparticles and nanoparticles. Biophys. J. **84**, 4023–4032 (2003)
6. V.P. Zharov, R.R. Letfullin, E.N. Galitovskaya, Microbubbles-overlapping mode for laser killing of cancer cells with absorbing nanoparticle clusters. J. Phys. D. Appl. Phys. **38**, 2571–2581 (2005)
7. Z. Peng, T. Walther, K. Kleinermanns, Influence of intense pulsed laser irradiation on optical and morphological properties of gold nanoparticle aggregates produced by surface acid-base reactions. Langmuir **21**, 4249–4253 (2005)
8. V.S. Kalambur, E.K. Longmire, J.C. Bischof, Cellular level loading and heating of superparamagnetic iron oxide nanoparticles. Langmuir **23**, 12329–12336 (2007)
9. E.Y. Hleb, D.O. Lapotko, Photothermal properties of gold nanoparticles under exposure to high optical energies. Nanotechnology **19**, 1–10 (2008)
10. H. Takahashi, T. Niidome, A. Nariai, Y. Niidome, S. Yamada, Gold nanorod-sensitized cell death: microscopic observation of single living cells irradiated by pulsed near-infrared laser light in the presence of gold nanorods. Chem. Lett. **35**, 500–501 (2006)
11. T. Torimura, T. Ueno, S. Inuzuka, Y. Kimura, P. Ko, M. Kin, T. Minetoma, T. Majima, M. Sata, H. Abe, K. Tanikawa, Ultrastructural observation on hepatocellular carcinoma: correlation of tumor grade and degree of atypia of cell organelles by morphometry. Med. Electron Microsc. **26**, 19–28 (1993)
12. G.T. Deans, P.W. Hamilton, P.C.H. Watt, M. Heatly, K. Williamson, C.C. Patterson, B.J. Rowlands, G. Parks, R. Spence, Morphometric analysis of colorectal cancer. Dis. Colon Rectum **36**, 450–456 (1993)
13. M.M. Radwan, K.A. Amer, N.M. Mokhtar, M.A. Kandil, A.M. El-Barbary, H.A. Aiad, Nuclear morphometry in ductal breast carcinoma with correlation to cell proliferative activity and prognosis. J. Egypt. Natl Cancer Inst. **15**, 169–182 (2003)
14. D. Ozaki, Y. Kondo, Comparative morphometric studies of benign and malignant intraductal proliferative lesions of the breast by computerized image analysis. Hum. Pathol. **26**, 1109–1113 (1995)
15. Y. Cui, E.A. Koop, P.J. van Diest, R.A. Kandel, T.E. Rohan, Nuclear morphometric features in benign breast tissue and risk of subsequent breast cancer. Breast Cancer Res. Treat. **104**, 103–107 (2007)
16. R.R. Letfullin, C.E.W. Rice, T.F. George, X-ray optics of gold nanoparticles. Appl. Opt. **53**, 7208–7214 (2014)
17. R.R. Letfullin, A.R. Letfullin, T.F. George, Absorption efficiency and heating kinetics of nanoparticles in the RF range for selective nanotherapy of cancer. Nanomed.: Nanotechnol. Biol. Med. **11**, 413–420 (2015)
18. R.R. Letfullin, T.F. George, Nanomaterials in nanomedicine, in *Computational Studies of New Materials II: From Ultrafast Processes and Nanostructures to Optoelectronics, Energy Storage and Nanomedicine*, ed. by T.F. George, D. Jelski, R.R. Letfullin, G.P. Zhang (World Scientific, Singapore, 2011), pp. 103–129

Chapter 8
Spatial Behavior of Temperature Fields at Nanoscale

Contents

Abstract This chapter involves the heating of a single nanoparticle that absorbs the radiation to generate heat, which is then transferred via conduction to the surrounding medium. This entails solving the heat flow equations for both the particle and surrounding fluid, while satisfying the boundary condition at the edge of the nanoparticle. In this chapter, we model 3D thermal fields within and around a plasmonic nanoparticle and cell organelles surrounded by biological medium heated by radiation. This chapter contains material adapted from our publications [2, 3]. A detailed list of references and reviews on a given topic of this chapter can be found in those original papers.

Electronic supplementary material: The online version of this chapter (doi:10.1007/978-3-319-43577-0_8) contains supplementary material, which is available to authorized users.

R.R. Letfullin, T.F. George, *Computational Nanomedicine and Nanotechnology*,
DOI 10.1007/978-3-319-43577-0_8

8.1 Lecture 23: Modeling Heat Diffusion (Slide 23.1)

- Heat Transfer Model
- Temperature Fields Around the Nanoparticle
- Material Properties, Optical Data, and Input Parameters
- Conclusions

8.1.1 *Heat Transfer Model*

For simulations of a space distribution of the temperature in and outside the nanoparticle, we solve the heat-mass transfer equation introduced in Sect. 6.1 (Lecture 17) of Chap. 6 in a quasistationary heating approximation, namely, $dT_s(r,t)/dt = 0$ (Slide 23.2). In this approximation, we neglect the time required to reach a maximum temperature (extremum in the $T_s(t)$ curve) of the nanoparticle and solve the heat diffusion equation for the space distribution for that particular temperature. The calculations are simplified by assuming the particle is spherical and rigid, and that no mass transfer takes place. This is assumed valid so long as the maximum temperature of the nanoparticle is kept below a certain threshold, which we will take as the melting temperature of the metal composing the nanoparticle, $T_{s,max} < T_{s,melt}$. Here we reproduce our results published in Refs. [2, 3].

Consider a homogeneous sphere of radius r_0 surrounded by an infinite homogeneous medium, with heat produced in the sphere for time $t > 0$ at the constant rate A per unit time per unit volume. Since the surrounding medium is transparent for the chosen wavelength, we can neglect thermally induced changes in the cellular refractive index during the action of a radiation pulse. Both the sphere and medium are initially at zero temperature. Let the suffix "s" refer to the sphere and "m" to the medium, and T, μ, and χ denote the temperature, thermal conductivity, and

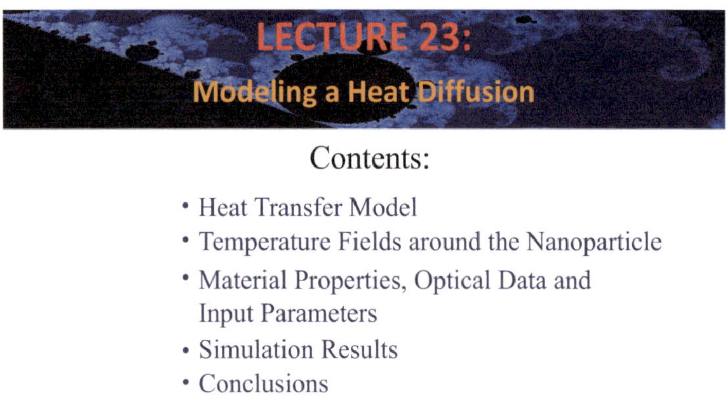

Slide 23.1 Contents

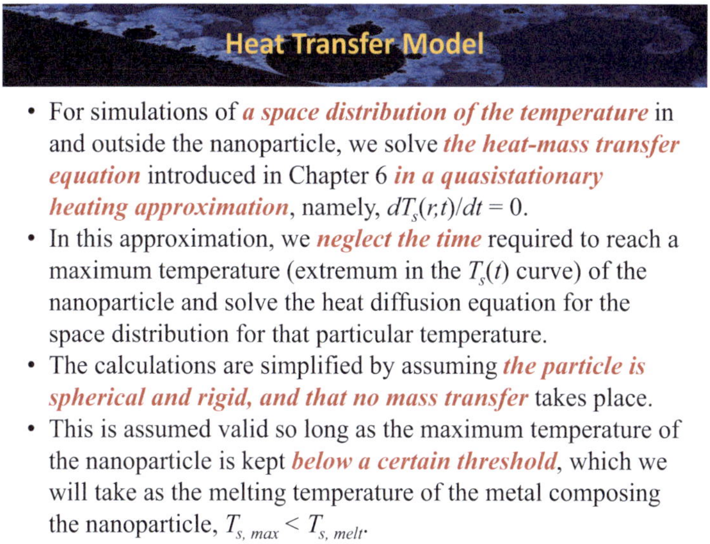

Slide 23.2 Modeling heat transfer

diffusivity, respectively. To describe the temperature changes created in three-dimensional space, we return to Eq. (6.1). This time the nanoparticle boundary is used to couple a pair of heat equations. Inside the boundary, the coupled equations are identical to Eq. (6.3), while outside the particle the heat generation term q_{gen} is removed. The two heat equations are coupled by two boundary conditions, requiring the temperature, when moving across the nanoparticle boundary, to be continuous and smooth [3] (Slide 23.3):

$$\left.\begin{aligned} \frac{1}{\chi_{\text{s}}}\frac{\partial T_{\text{s}}}{\partial t} &= \frac{1}{r^2}\frac{\partial}{\partial r}\left(r^2\frac{\partial T_{\text{s}}}{\partial r}\right) + \frac{A}{\mu_{\text{s}}}, \quad 0 \le r < r_0, \\ \frac{1}{\chi_{\text{m}}}\frac{\partial T_{\text{m}}}{\partial t} &= \frac{1}{r^2}\frac{\partial}{\partial r}\left(r^2\frac{\partial T_{\text{m}}}{\partial r}\right), \quad r > r_0, \end{aligned}\right\} t > 0, \tag{8.1}$$

$$\left.\begin{aligned} T_{\text{s}} &= T_{\text{m}} \\ \mu_{\text{s}}\frac{\partial T_{\text{s}}}{\partial r} &= \mu_{\text{m}}\frac{\partial T_{\text{m}}}{\partial r} \end{aligned}\right\} \quad \text{where } r = r_0, \tag{8.2}$$

$$T_{\text{s}} = T_{\text{m}} = 0 \quad \text{when } t = 0, \tag{8.3}$$

$$T_{\text{s}} \text{ finite as } r \to 0 \quad \text{and} \quad T_{\text{m}} \text{ finite as } r \to \infty. \tag{8.4}$$

Then, as done by Goldenberg and Tranter [1], the Laplace transform of T, defined by

$$\overline{T} = \int_0^\infty \mathrm{e}^{-pt} T(t)dt = L[T(t)], \tag{8.5}$$

Heat Transfer Model (continued)

- The ***two heat equations*** are coupled by ***two boundary conditions***, requiring that the temperature, when moving across the nanoparticle boundary, be continuous and smooth:

$$\left.\begin{aligned}\frac{1}{\chi_s}\frac{\partial T_s}{\partial t} &= \frac{1}{r^2}\frac{\partial}{\partial r}\left(r^2\frac{\partial T_s}{\partial r}\right)+\frac{A}{\mu_s}, \quad 0 \le r < r_0,\\ \frac{1}{\chi_m}\frac{\partial T_m}{\partial t} &= \frac{1}{r^2}\frac{\partial}{\partial r}\left(r^2\frac{\partial T_m}{\partial r}\right), \qquad r > r_0,\end{aligned}\right\} t>0$$

$$\left.\begin{aligned}T_s &= T_m\\ \mu_s\frac{\partial T_s}{\partial r} &= \mu_m\frac{\partial T_m}{\partial r}\end{aligned}\right\} \textit{where} \quad r = r_0$$

$$T_s = T_m = 0 \text{ when } t = 0$$

T_s finite as $r \to 0$ and T_m finite as $r \to \infty$.

Slide 23.3 Modeling heat transfer (continued)

is used to reduce Eqs. (8.1)–(8.4) to the following (Slide 23.4):

$$\left.\begin{aligned}\frac{\partial^2}{\partial r^2}\left(r\overline{T_s}\right) - rq_s^2\overline{T_s} &= -\frac{rA}{\mu_s p}, \quad \text{if } 0 \le r < r_0\\ \frac{\partial^2}{\partial r^2}\left(r\overline{T_m}\right) - rq_m^2\overline{T_m} &= 0, \quad \text{if } r > r_0\end{aligned}\right\}, \tag{8.6}$$

$$\left.\begin{aligned}\overline{T_s} &= \overline{T_m}\\ \mu_s\frac{\partial\overline{T_s}}{\partial r} &= \mu_m\frac{\partial\overline{T_m}}{\partial r}\end{aligned}\right\} \quad \text{when } r = r_0, \tag{8.7}$$

$$\overline{T_s} \text{ finite as } r \to 0 \quad \text{and} \quad \overline{T_m} \text{ finite as } r \to \infty \tag{8.8}$$

with ${q_s}^2 = p/\chi_s$ and ${q_m}^2 = p/\chi_m$.

The solutions $\overline{T_s}$ and $\overline{T_m}$ to the system of Eqs. (8.6)–(8.8) are then inverse transformed via the procedure described in [1–3], to find the following (Slide 23.5):

$$\begin{aligned}\overline{T_s} = \frac{r_0^2 A}{\mu_s}\Bigg\{&\frac{1}{3}\frac{\mu_s}{\mu_m}+\frac{1}{6}-\frac{2b}{\pi}\int_0^\infty \frac{\exp(-y^2 t/\gamma)}{y^2}\\ &\times\frac{(\sin\ y - y\ \cos\ y)\sin(ry/r_0)}{\left[(c\ \sin\ y - y\ \cos\ y)^2 + b^2y^2\sin^2 y\right]}dy\Bigg\}, r = 0,\end{aligned} \tag{8.9}$$

Heat Transfer Model (continued)

- The ***Laplace transform*** of T, defined by

$$\overline{T} = \int_0^\infty e^{-pt} T(t)dt = L[T(t)]$$

is used to ***reduce the previous equations*** to the following:

$$\left.\begin{aligned} &\frac{\partial^2}{\partial r^2}\left(r\overline{T_s}\right) - rq_s^2\overline{T_s} = -\frac{rA}{\mu_s p}, \quad if \quad 0 \le r < r_0, \\ &\frac{\partial^2}{\partial r^2}\left(r\overline{T_m}\right) - rq_m^2\overline{T_m} = 0, \qquad if \quad r > r_0, \end{aligned}\right\}$$

$$\left.\begin{aligned} &\overline{T_s} = \overline{T_m} \\ &\mu_s \frac{\partial \overline{T_s}}{\partial r} = \mu_m \frac{\partial \overline{T_m}}{\partial r} \end{aligned}\right\} \ when\ r = r_0,$$

with $q_s{}^2 = p/\chi_s$ and $q_m{}^2 = p/\chi_m$.

Slide 23.4 Modeling heat transfer (continued)

Heat Transfer Model (continued)

- The ***solutions*** for space distribution of the temperature of nanoparticle and medium are:

$$\overline{T_s} = \frac{r_0^2 A}{\mu_s}\left\{\frac{1}{3}\frac{\mu_s}{\mu_m} + \frac{1}{6} - \frac{2b}{\pi}\int_0^\infty \frac{\exp(-y^2t/\gamma)}{y^2}\frac{(\sin y - y\cos y)\sin(ry/r_0)}{[(c\sin y - y\cos y)^2 + b^2y^2\sin^2 y]}dy\right\}, \quad r = 0,$$

$$\overline{T_s} = \frac{r_0^2 A}{\mu_s}\left\{\frac{1}{3}\frac{\mu_s}{\mu_m} + \frac{1}{6}\left(1 - \frac{r^2}{r_0^2}\right) - \frac{2r_0 b}{\pi r}\int_0^\infty \frac{\exp(-y^2t/\gamma)}{y^2}\frac{(\sin y - y\cos y)\sin(ry/r_0)}{[(c\sin y - y\cos y)^2 + b^2y^2\sin^2 y]}dy\right\}, \quad 0 < r \le r_0,$$

$$\overline{T_m} = \frac{r_0^3 A}{r\mu_s}\left\{\frac{1}{3}\frac{\mu_s}{\mu_m} - \frac{2}{\pi}\int_0^\infty \frac{\exp(-y^2t/\gamma)}{y^3}\frac{(\sin y - y\cos y)[by\sin y\cos\sigma y - (c\sin y - y\cos y)\sin\sigma y]}{[(c\sin y - y\cos y)^2 + b^2y^2\sin^2 y]}dy\right\}, r > r_0,$$

where $b = \frac{\mu_m}{\mu_s}\sqrt{\frac{\chi_m}{\chi_s}}$; $c = 1 - \frac{\mu_m}{\mu_s}$; $\gamma = \frac{r^2}{\chi_s}$ and $\sigma = \left(\frac{r}{r_0} - 1\right)\sqrt{\frac{\chi_s}{\chi_m}}$.

Slide 23.5 Modeling heat transfer (continued)

$$\overline{T_s} = \frac{r_0^2 A}{\mu_s}\left\{\frac{1\mu_s}{3\mu_m} + \frac{1}{6}\left(1 - \frac{r^2}{r_0^2}\right) - \frac{2r_0 b}{\pi r}\int_0^\infty \frac{\exp(-y^2t/\gamma)}{y^2} \right.$$
$$\left. \times \frac{(\sin y - y\cos y)\sin(ry/r_0)}{[(c\sin y - y\cos y)^2 + b^2y^2\sin^2 y]}dy\right\}, \quad 0 < r \le r_0, \tag{8.10}$$

$$\overline{T_{\mathrm{m}}} = \frac{r_0^3 A}{r\mu_{\mathrm{s}}}\left\{\frac{1}{3}\frac{\mu_{\mathrm{s}}}{\mu_{\mathrm{m}}} - \frac{2}{\pi}\int_0^{\infty}\frac{\exp(-y^2 t/\gamma)}{y^3}\right.$$
$$\left.\times\frac{(\sin y - y\cos y)[by\sin y\cos\sigma y - (c\sin y - y\cos y)\sin\sigma y]}{[(c\sin y - y\cos y)^2 + b^2y^2\sin^2 y]}dy\right\},\quad r > r_0, \tag{8.11}$$

where

$$b = \frac{\mu_{\mathrm{m}}}{\mu_{\mathrm{s}}}\sqrt{\frac{\chi_{\mathrm{m}}}{\chi_{\mathrm{s}}}};\quad c = 1 - \frac{\mu_{\mathrm{m}}}{\mu_{\mathrm{s}}};\quad \gamma = \frac{r^2}{\chi_{\mathrm{s}}}\quad \text{and}\quad \sigma = \left(\frac{r}{r_0} - 1\right)\sqrt{\frac{\chi_{\mathrm{s}}}{\chi_{\mathrm{m}}}}.$$

The heat source in the earlier equations is characterized here by a constant energy rate A per unit time per unit volume with units in W/m^3 as

$$\int_0^{r_0} A(t)\cdot 4\pi r^2 dr = A\cdot V_0 = Q_{\mathrm{abs}}I_0S_0 \rightarrow A = \frac{3Q_{\mathrm{abs}}I_0}{r_0}. \tag{8.12}$$

If the earlier expression for the heat generation rate A is applied with the substitution $\chi = \mu/\rho C$, we can rewrite Eqs. (8.9)–(8.11) for T_{s} and T_{m} in terms similar to the one-temperature model terms given by Eq. (6.8). With these solutions, we can solve for the temperature as a function of the radial distance $T(r)$, both inside and outside the nanoparticle boundary, r_0 (Slide 23.6):

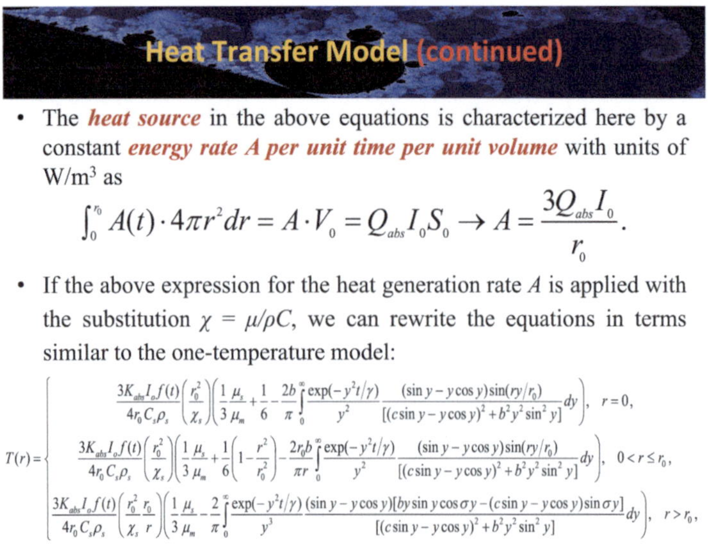

Slide 23.6 Modeling heat transfer (continued)

$$T(r)=\begin{cases}\dfrac{3K_{\mathrm{abs}}I_{\mathrm{o}}f(t)}{4r_0C_{\mathrm{s}}\rho_{\mathrm{s}}}\left(\dfrac{r_0^2}{\chi_{\mathrm{s}}}\right)\left(\dfrac{1}{3}\dfrac{\mu_{\mathrm{s}}}{\mu_{\mathrm{m}}}+\dfrac{1}{6}-\dfrac{2b}{\pi}\displaystyle\int_0^{\infty}\dfrac{\exp(-y^2t/\gamma)}{y^2}\right.\\ \qquad\left.\times\dfrac{(\sin y-y\cos y)\sin(ry/r_0)}{\left[(c\sin y-y\cos y)^2+b^2y^2\sin^2y\right]}dy\right), & r=0\\ \dfrac{3K_{\mathrm{abs}}I_{\mathrm{o}}f(t)}{4r_0C_{\mathrm{s}}\rho_{\mathrm{s}}}\left(\dfrac{r_0^2}{\chi_{\mathrm{s}}}\right)\left(\dfrac{1}{3}\dfrac{\mu_{\mathrm{s}}}{\mu_{\mathrm{m}}}+\dfrac{1}{6}\left(1-\dfrac{r^2}{r_0^2}\right)-\dfrac{2r_0b}{\pi r}\displaystyle\int_0^{\infty}\dfrac{\exp(-y^2t/\gamma)}{y^2}\right.\\ \qquad\left.\times\dfrac{(\sin y-y\cos y)\sin(ry/r_0)}{\left[(c\sin y-y\cos y)^2+b^2y^2\sin^2y\right]}dy\right), & 0<r\le r_0\\ \dfrac{3K_{\mathrm{abs}}I_{\mathrm{o}}f(t)}{4r_0C_{\mathrm{s}}\rho_{\mathrm{s}}}\left(\dfrac{r_0^2r_0}{\chi_{\mathrm{s}}r}\right)\left(\dfrac{1}{3}\dfrac{\mu_{\mathrm{s}}}{\mu_{\mathrm{m}}}-\dfrac{2}{\pi}\displaystyle\int_0^{\infty}\dfrac{\exp(-y^2t/\gamma)}{y^3}\right.\\ \qquad\left.\times\dfrac{(\sin y-y\cos y)\left[by\sin y\cos\sigma y-(c\sin y-y\cos y)\sin\sigma y\right]}{\left[(c\sin y-y\cos y)^2+b^2y^2\sin^2y\right]}dy\right) & r>r_0.\end{cases}\tag{8.13}$$

As can be seen from the earlier solutions, both the thermal and optical properties affect the behavior of the thermal field. The thermal properties of the material govern how easily a material can dissipate heat to the surrounding medium, while the optical properties govern how much heat is generated when exposed to an energy source [2].

- In these simulations we wish to examine the photothermal effect of a strongly-absorbing ***nanoparticle in hard tissue*** for applications to medical procedures. The bone tissues make laser light delivery more difficult by being trapped deep inside the body. Different approaches have been made, where one uses a hollow metal needle through which the optical fiber is placed, and another shapes and sharpens the glass of the silica fiber itself to create a needle.

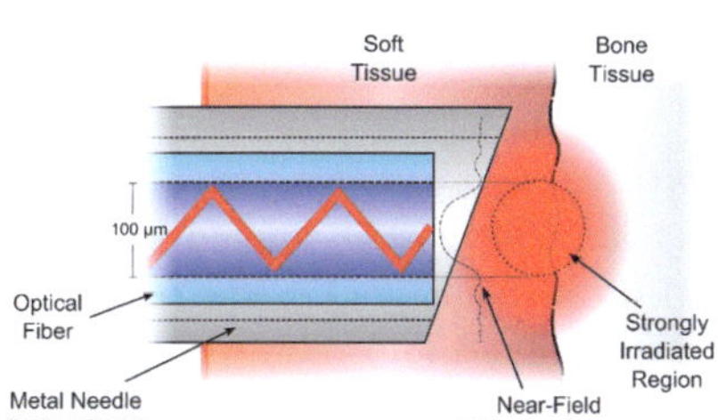

Figure. Depiction of the minimally-invasive fiber-optic needle system used to deliver laser light more directly to hard tissues by injecting a needle past most of the softer tissues.

Slide 23.7 Space distribution of temperature fields around the nanoparticle

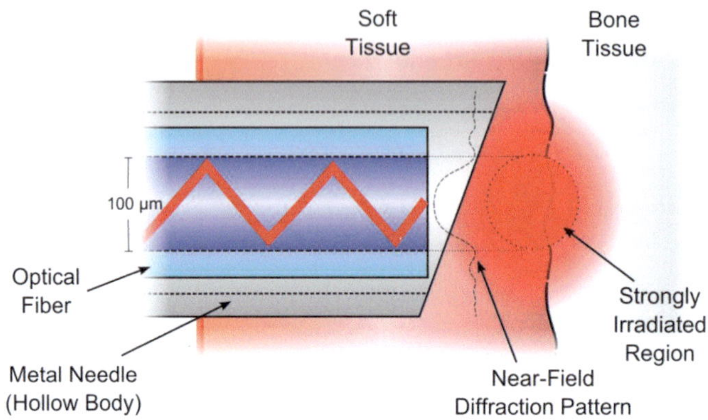

Fig. 8.1 Depiction of the minimally invasive fiber-optic needle system used to deliver laser light more directly to hard tissues by injecting a needle past most of the softer tissues. Reprinted with permission from [2]

8.1.2 Temperature Fields Around the Nanoparticle

In these simulations we wish to examine the photothermal effect of a strongly absorbing nanoparticle in hard tissue for applications to medical procedures (Slide 23.7). The bone tissues make laser light delivery more difficult by being trapped deep inside the body. In the visible spectra, the strong scattering properties of the biological tissues mean that only a small portion of the incident radiation will reach the desired particles. To remedy this, engineering attempts have been made to create fiber-optic needles capable of delivering light much closer to the targeted region [2]. Different approaches have been made, where one uses a hollow metal needle through which the optical fiber is placed, and another shapes and sharpens the glass of the silica fiber itself to create a needle. Figure 8.1 is a depiction of a fiber-needle system delivering light using 100 μm diameter optical fiber. Light waves undergo diffraction spreading on the optical fiber interface and deliver electromagnetic energy to the biological tissue containing nanoparticles (Slide 23.7).

8.1.3 Material Properties and Input Parameters

As discussed earlier [2, 3], gold nanoparticles are promising for near-future medical uses because their biological effects have been well studied (and shown to be noncytotoxic and almost completely inert in the body). For this reason, the material properties of bone and gold are listed in Tables 8.1 and 8.2. Also, to our advantage, plasmonic effects in gold nanoparticles have been widely investigated, and the absorption efficiencies of gold nanoparticles can be readily modified by changes to their size, shape, or structure. The silver nanoparticles are another example of

Table 8.1 Material properties of bone. Reprinted with permission from [2]

Property	Values
Thermal conductivity	5.60×10^{-1} W/m K
Density	1.85×10^{3} kg/m^3
Specific heat	1.33×10^{3} J/kg K
Thermal diffusivity	2.276×10^{-7} m^2/s

Table 8.2 Material properties of gold. Reprinted with permission from [2]

Property	Values
Thermal conductivity	3.18×10^{2} W/m K
Density	1.93×10^{4} kg/m^3
Specific heat	1.29×10^{2} J/kg K
Thermal diffusivity	1.27×10^{-4} m^2/s

Table 8.3 Material properties of silver. Reprinted with permission from [3]

Property	Values
Thermal conductivity	4.29×10^{2} W/m K
Density	1.05×10^{4} kg/m^3
Specific heat	2.35×10^{2} J/kg K
Thermal diffusivity	1.74×10^{-4} m^2/s

Table 8.4 Sample absorption efficiency values for different nanoparticle shapes and effective radii at various wavelengths. Reprinted with permission from [2]

Nanoparticle type	Dimension (nm)	Wavelength	Absorption efficiency
Gold nanospheres	35	532	4.02
	40	550	3.50
Silica-gold nanoshells	120	1100	3.50
	120	750	3.75
Gold nanorods	11.4	800	14.0
	17.9	825	14.5

plasmonic nanoparticles used in nanomedicine. The material properties of silver are listed in Table 8.3. A few examples of the optical properties of gold nanoparticles of different shapes are shown in Table 8.4, which includes nanosphere and nanorod shapes of multiple radii as well as silica–gold nanoshells (Slide 23.8).

The optical properties of nanoparticles can be much more varied, however, as the absorption efficiency of the particles is affected by the material, size, and geometry of the nanoparticle [2, 3]. To achieve sufficient heating with pulses of lower energy densities, it is important to find absorption efficiency peaks. These peaks occur due to the plasmon resonance of the nanoparticle at that particular wavelength. To determine the absorption characteristics of spherical particles in various biomedia, generalized Lorenz–Mie diffraction theory is used (discussed in Chaps. 4 and 5), which takes into account the plasmon-resonance absorption effect.

Material Properties

- As discussed earlier, gold nanoparticles are promising for near-future medical uses because their biological effects have been well studied. For this reason the material properties of ***bone and gold*** are listed below.

Property	Values
Thermal conductivity	$5.60 \cdot 10^{-1}$ W/m·K
Density	$1.85 \cdot 10^{3}$ kg/m^3
Specific heat	$1.33 \cdot 10^{3}$ J/kg·K
Thermal diffusivity	$2.276 \cdot 10^{-7}$ m^2/s

Property	Values
Thermal conductivity	$3.18 \cdot 10^{2}$ W/m·K
Density	$1.93 \cdot 10^{4}$ kg/m^3
Specific heat	$1.29 \cdot 10^{2}$ J/kg·K
Thermal diffusivity	$1.27 \cdot 10^{-4}$ m^2/s

Nanoparticle type	Dimension	Wavelength	Absorption efficiency
Gold nanospheres	35 nm	532 nm	4.02
	40 nm	550 nm	3.50
Silica-gold nanoshells	120 nm	1100 nm	3.50
	120 nm	750 nm	3.75
Gold nanorods	11.4 nm	800 nm	14.0
	17.9 nm	825 nm	14.5

Slide 23.8 Material properties

Using the complex indices of refraction for the medium and particle (which depend on the wavelength, as well as particle size), the theory allows us to determine the absorption efficiency of spherical particles at any wavelength within the bounds of the refractive index values provided. The results of these calculations for the absorption efficiency of gold, silver, and copper nanoparticles are shown in Figs. 8.2 and 8.3 (Slide 23.9).

Reduction in the size of the spherical nanoparticles lowers the absorption efficiency, as seen when comparing a 30-nm gold nanoparticle to a 5-nm Au nanoparticle, where Q_{abs} loses an entire order of magnitude. Moving into the IR and longer wavelengths, the absorption efficiency decreases continuously and without any noticeable peaks upon which to capitalize. This means larger energy densities would be required to heat the nanospheres to the critical temperature. Thus, utilization of silica–gold nanoshells and/or gold nanorods, which absorb IR wavelengths much better than solid nanospheres, is recommended for selective nanothermolysis of bone tissue when working in the longer wavelength range of the spectrum [2, 3].

The choices of nanoparticle material, size, and shape used in the model are listed in Tables 8.5 and 8.6 along with the choice of wavelength and corresponding absorption efficiency for that wavelength (Slide 23.10). The energy densities are chosen to bring the nanoparticle to the relatively high temperature of ~600 K in order to maintain the bone ablation temperature at a feasible distance around the nanoparticle.

The simulations are conducted through Eqs. (8.13) to see the general behavior of the spatial temperature distribution [2] (Slide 23.11). In the case of gold

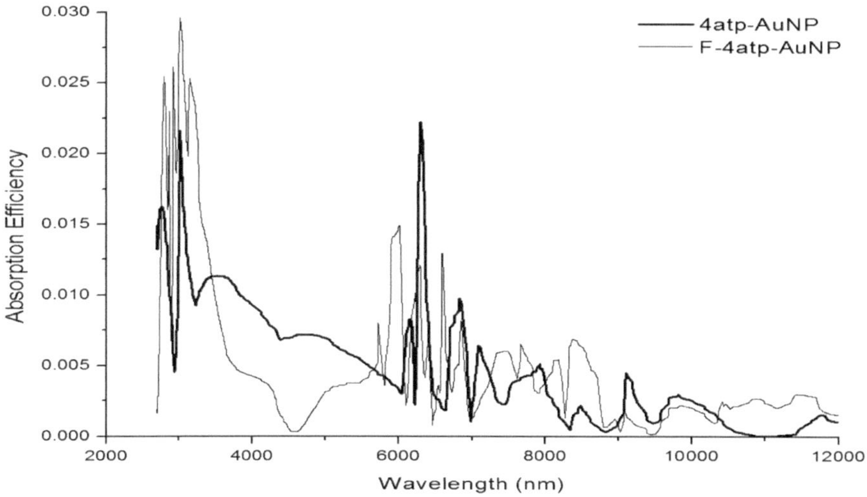

Fig. 8.2 Absorption efficiency for 5-nm aminothiophenol conjugated gold nanoparticles (4atp-AuNP) and folate-aminothiophenol conjugated gold nanoparticles (F-4atp-AuNP). The values are calculated with Mie theory using the complex index of refraction data for gold. Reprinted with permission from [2]

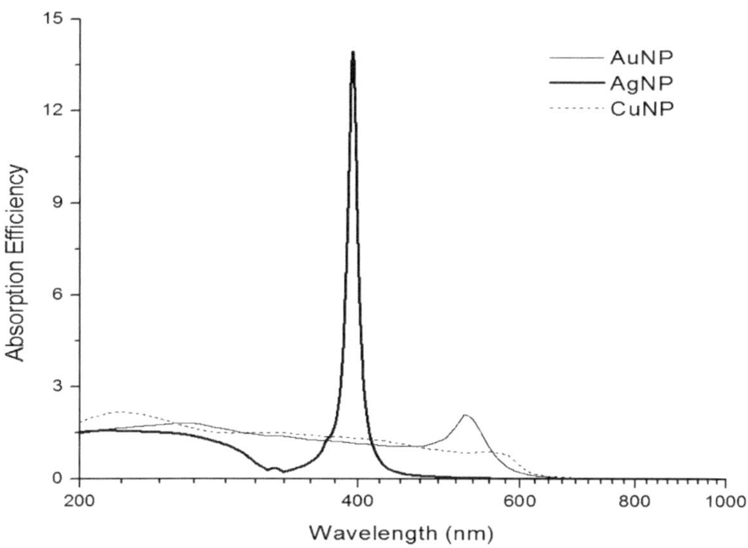

Fig. 8.3 Absorption efficiency for 30-nm diameter gold, silver, and copper nanoparticles (AuNP, AgNP, CuNP) surrounded by a water medium. These values are calculated with Mie theory using the complex index of refraction data for the listed materials. Reprinted with permission from [3]

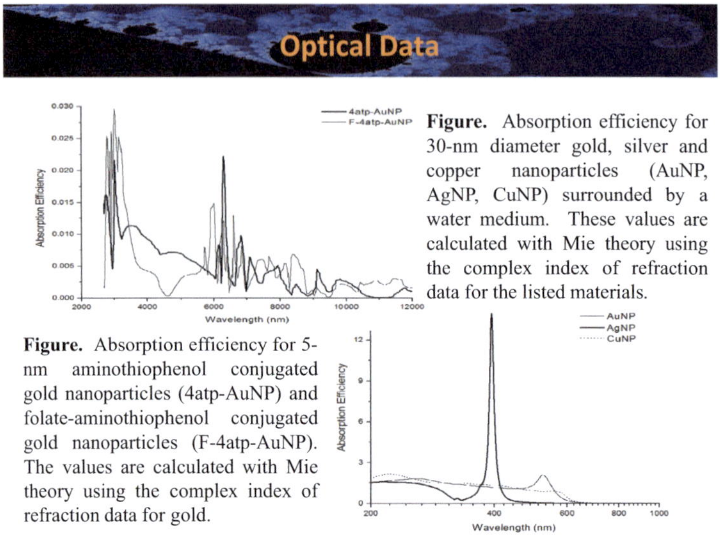

Slide 23.9 Optical data

Table 8.5 Parameters for gold nanoparticle heating simulation achieving the critical heat of vaporization for bone at the nanoparticle edge. Reprinted with permission from [3]

Symbol	Property	Values
D	Nanoparticle diameter	30 nm
E	Pulse energy density	0.788 J/cm^2
λ	Wavelength	520 nm
Q_{abs}	Absorption efficiency	2.1

Table 8.6 Parameters for silver nanoparticle heating simulation achieving the critical heat of vaporization for bone at the nanoparticle edge. Reprinted with permission from [3]

Symbol	Property	Values
D	Nanoparticle diameter	30 nm
E	Pulse energy density	0.119 J/cm^2
Λ	Wavelength	394 nm
Q_{abs}	Absorption efficiency	13.9

nanoparticles in bone tissue, the high thermal conductivity of gold relative to bone (about three orders of magnitude) means that temperature variations within the particle are much smaller than the overall temperature difference from the initial medium rest temperature (see Fig. 8.5). Temperatures inside the nanoparticle can therefore be considered essentially homogeneous, and the temperature solutions (Eqs. 8.13) for all $0 < r < r_0$ are replaced by the solution at $r = r_0$, resulting in the following solution [2]:

Input Parameters

- The choices of nanoparticle ***material, size and shape*** used in the model are listed in the tables below along with the choice of ***wavelength and corresponding absorption efficiency*** for that wavelength.
- The ***energy densities*** are chosen to bring the nanoparticle to the relatively high temperature of ~ 600 K in order to maintain the bone ablation temperature at a feasible distance around the nanoparticle.

Symbol	Property	Values
d	Nanoparticle diameter	30 nm
E	Pulse energy density	0.788 J/cm^2
λ	Wavelength	520 nm
Q_{abs}	Absorption efficiency	2.1

Symbol	Property	Values
d	Nanoparticle diameter	30 nm
E	Pulse energy density	0.119 J/cm^2
λ	Wavelength	394 nm
Q_{abs}	Absorption efficiency	13.9

Slide 23.10 Input parameters

Simulation Results

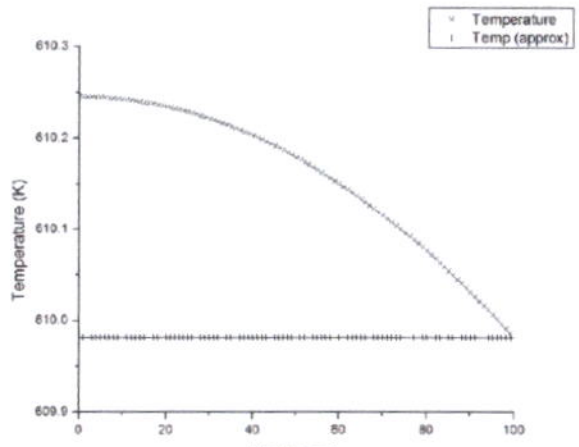

- The simulations are conducted for a ***gold nanoparticle in bone tissue***.
- The ***high thermal conductivity of gold*** relative to bone (about 3 orders of magnitude), means that temperature variations within the particle are much smaller than the overall temperature difference from the initial medium rest temperature. ***Temperatures inside the nanoparticle can therefore be considered essentially homogeneous***.

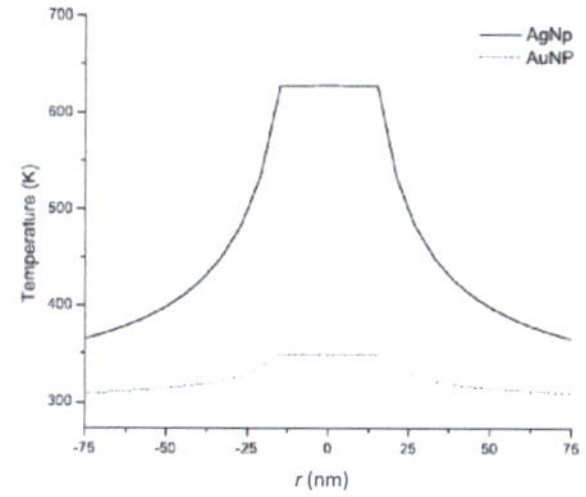

- The ***temperature distribution*** is mainly dependent upon ***two factors: the thermal conductivities***, which affect the slopes of the curves, and the ***absorption efficiency***, which effects the maximum temperature difference reached.

Slide 23.11 Simulation results

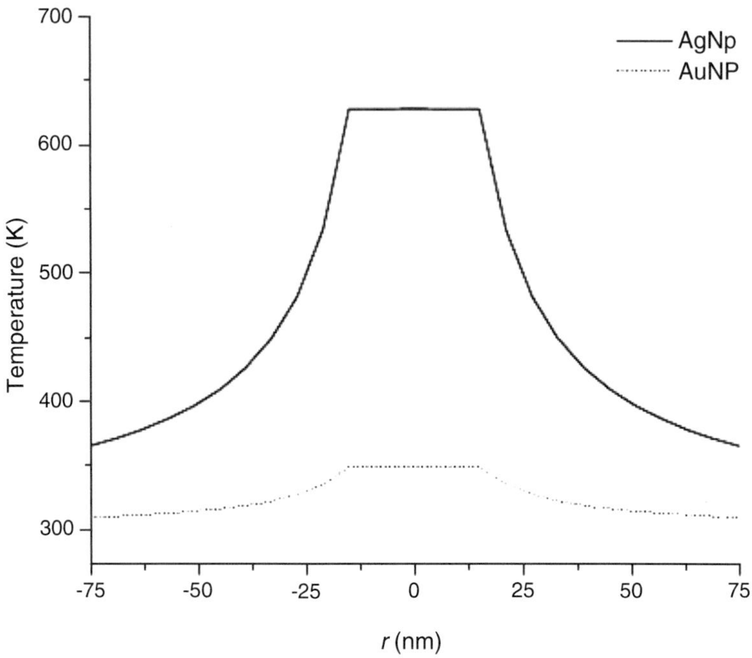

Fig. 8.4 Heat profiles for a 30-nm diameter silver nanoparticle heated by a pulse of $\lambda = 394$ nm and by a pulse of $\lambda = 520$ nm. Both pulses have equivalent energy densities, and both spheres are within an infinite medium of the steady-state temperature 300 K. Reprinted with permission from [3]

$$T(r) = \begin{cases} \dfrac{3K_{\text{abs}}I_0 f(t)}{4r_0 C_s \rho_s}\left(\dfrac{r_0^2}{\chi_s}\right)\left(\dfrac{1}{3}\dfrac{\mu_s}{\mu_m} - \dfrac{2b}{\pi}\displaystyle\int_0^\infty \dfrac{\exp(-y^2 t/\gamma)}{y^2} \right. \\ \left. \quad \times \dfrac{(\sin y - y\cos y)\sin(ry/r_0)}{[(c\sin y - y\cos y)^2 + b^2 y^2 \sin^2 y]} dy \right), \quad 0 < r \le r_0 \\ \dfrac{3K_{\text{abs}}I_0 f(t)}{4r_0 C_s \rho_s}\left(\dfrac{r_0^2 r_0}{\chi_s\, r}\right)\left(\dfrac{1}{3}\dfrac{\mu_s}{\mu_m} - \dfrac{2}{\pi}\displaystyle\int_0^\infty \dfrac{\exp(-y^2 t/\gamma)}{y^3} \right. \\ \left. \quad \times \dfrac{(\sin y - y\cos y)[by\sin y \ \ \cos\ \ \sigma y - (c\sin y - y\cos y)\sin\sigma y]}{[(c\sin y - y\cos y)^2 + b^2 y^2 \sin^2 y]} dy \right), \quad r > r_0. \end{cases} \tag{8.14}$$

Using the earlier simplification, plots have been generated to show the spatial temperature profile using 60-ns pulses at different laser intensities [2]. Figures 8.4, 8.5, and 8.6 show the spatial distribution of the temperature for a 1 mW/cm^2 intensity pulse (Slide 23.12). Next, simulations have been conducted with lower intensities to keep nanoparticle temperatures in the nanophotohyperthermia range of 40–44 °C (313.15–317.15 K). By selecting a pulse intensity of 24 μW/cm^2, the

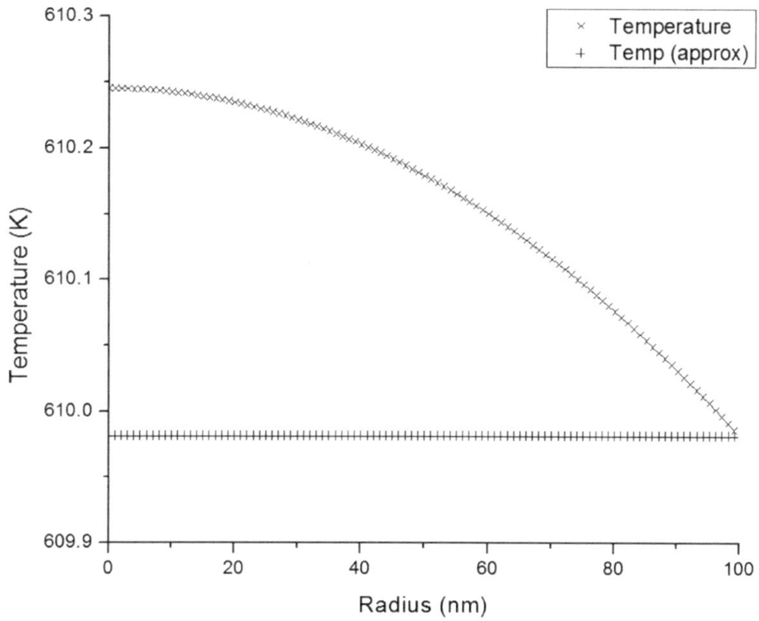

Fig. 8.5 Temperature distribution within a gold nanoparticle as a function of radial distance from the center of a nanoparticle with 100 nm radius. Reprinted with permission from [2]

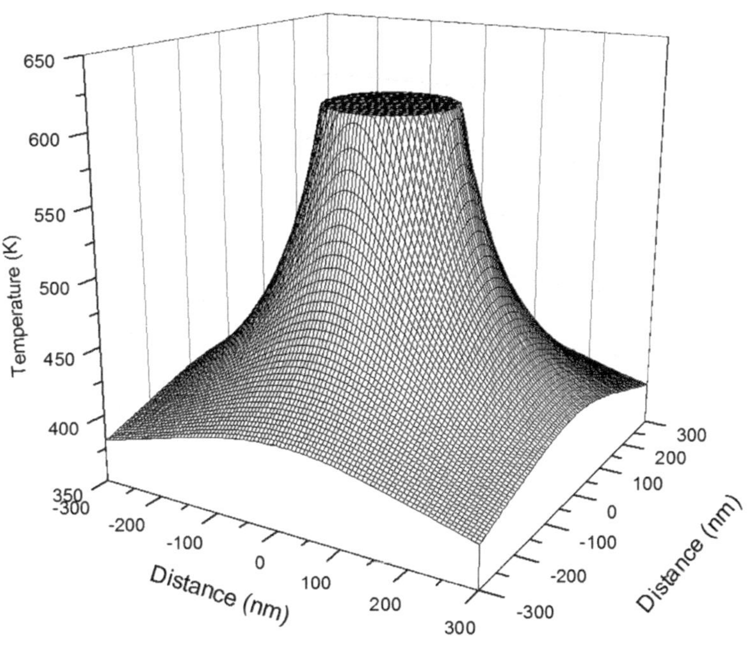

Fig. 8.6 Spatial temperature distribution for a gold nanoparticle with 100-nm radius heated by a 60-ns pulse of 1 mW/cm^2 intensity. Reprinted with permission from [2]

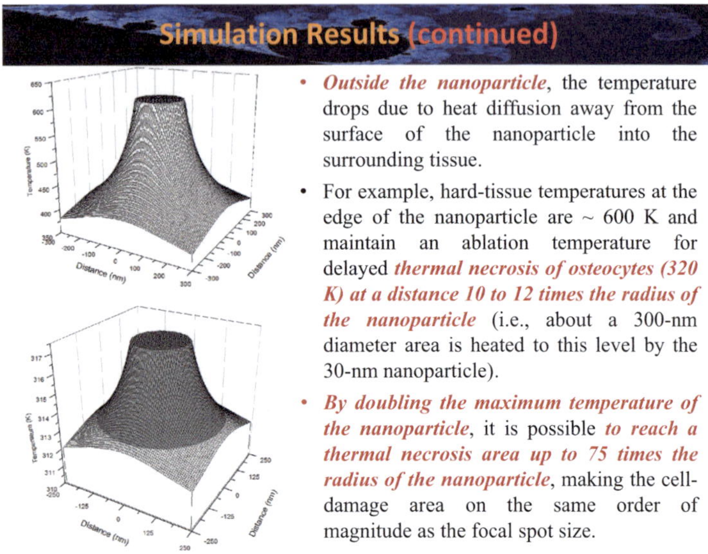

Slide 23.12 Simulation results (continued)

temperature within the nanoparticle reaches 317.15 K. The spatial temperature profile created by this lower intensity pulse is shown in Fig. 8.7 (Slide 23.12).

The temperature distribution is mainly dependent upon two factors [2]: the thermal conductivities, which affect the slopes of the curves, and the absorption efficiency, which affects the maximum temperature difference reached. In the case of noble metal nanoparticles, the thermal conductivities of the metals are much greater than that of bone. This leads to an essentially homogeneous temperature distribution within the gold nanoparticle, with an exponential cooling outside the particle as shown in Figs. 8.4, 8.6, and 8.7.

In Fig. 8.4, it is clear that temperature is around the same within the nanoparticle radius with significant cooling outside the radius. Closer inspection shows that the temperature distribution on the plateau varies slightly from the hottest point at the center, to the coolest temperature at the nanoparticle edge, as can be seen in Fig. 8.5. For silver nanoparticles, this is a variance from 628.1 to 627.8 K, and for nanoparticles it varies from 349.6 to 349.5 K. Both show less than 0.1 % variance from the maximum temperature, and thus the plateau regions can be considered homogeneously heated [2, 3].

Outside the nanoparticle, the temperature drops due to heat diffusion away from the surface of the nanoparticle into the surrounding tissue (Slide 23.12). In Figs. 8.4 and 8.6, for example, the hard-tissue temperatures at the edge of the nanoparticle are ~600 K and maintain an ablation temperature for delayed thermal necrosis of osteocytes (320 K) at a distance 10–12 times the radius of the nanoparticle (i.e., about a 300-nm diameter area is heated to this level by the 30-nm nanoparticle). By doubling the maximum temperature of nanoparticle, it is possible to reach

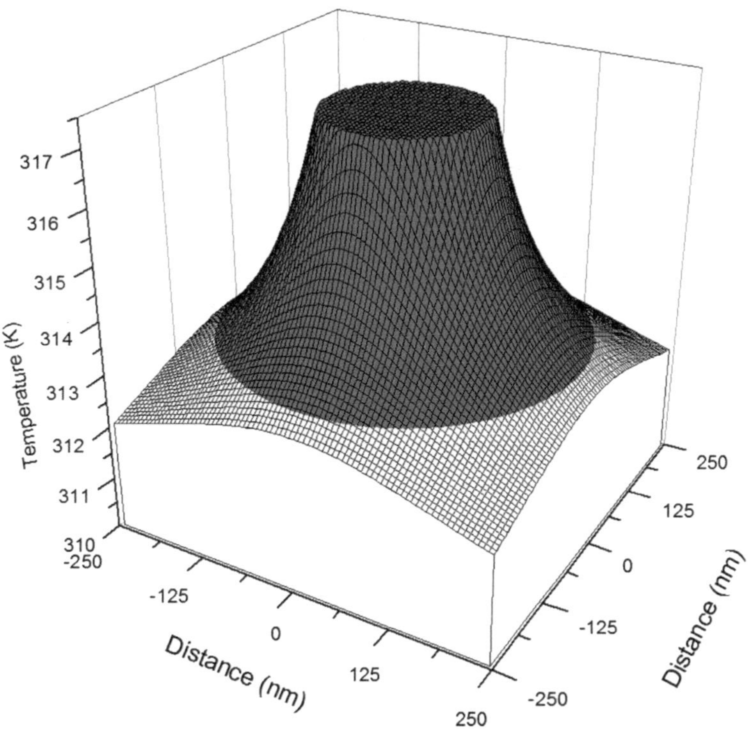

Fig. 8.7 Spatial temperature distribution for a gold nanoparticle with 100-nm radius heated by 60-ns pulse of 24 μW/cm^2 intensity. Reprinted with permission from [2]

a thermal necrosis area up to 75 times the radius of the nanoparticle, making the cell-damage area on the same order of magnitude as the focal spot size.

In conclusion, we have modeled 3D thermal fields of plasmonic nanoparticles surrounded by hard tissue heated by laser light (Slide 23.13). Our simulations show [2, 3] that these thermal fields can readily be controlled by careful consideration of nanoparticle morphological properties (size, shape, structure) as well as laser wavelength and laser intensity. For example, in the UV and visible ranges of the spectrum, silver and gold solid spherical nanoparticles exhibit a strong plasmonic effect in laser heating of nanoparticles. For longer wavelengths in the transparency window of most biotissues, we recommend utilization of silica–gold nanoshells and/or gold nanorods for selective nanothermolysis of bone tissue. Additionally, heat loss from the surface of a nanoparticle to the surrounding medium is observed, creating a large thermal damage area in the bone tissue up to 75 times the radius of the nanoparticle.

Our simulations show [2, 3] that thermal necrosis of osteocytes can be reached in most of these cases using laser pulses with intensities of less than 1 mW/cm^2 (Slide 23.14). This is a relatively low intensity as far as laser safety is concerned, and many commercially available Class 3b pulsed lasers could satisfy these

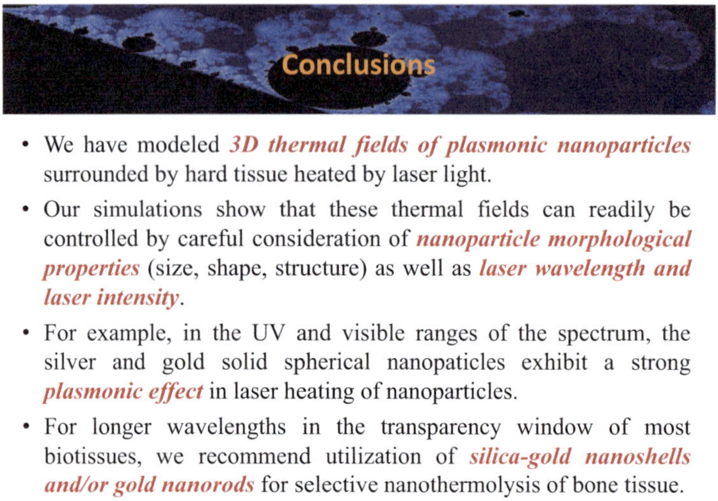

Slide 23.13 Conclusions

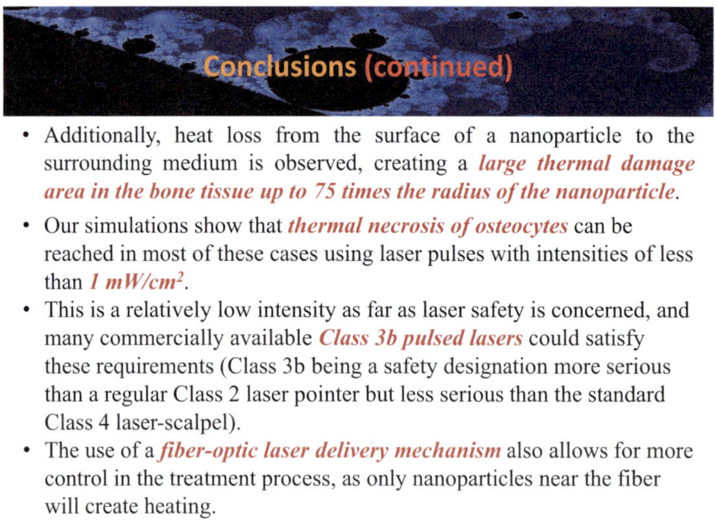

Slide 23.14 Conclusions (continued)

requirements (Class 3b being a safety designation more serious than a regular Class 2 laser pointer, but less serious than the standard Class 4 laser-scalpel). Because the pulse energies required for nanophotohyperthermia are much lower than those needed for nanophotohyperthermia, it should be even easier to prove them safe for these medical applications. The use of a fiber-optic laser delivery mechanism also allows for more control in the treatment process, as only nanoparticles near the fiber will create heating.

8.2 Computer Practicum 5 (Lecture 24): Simulating the Temperature Fields Within and Outside a Nanoparticle (Slide 24.1)

- Computer Code
- Practice Example 7: Thermal Fields Around a Gold Nanoparticle in Aqueous Solution
  - Abstract
  - Method, Model, and Input Data
  - Discussion and Results
  - Internal Temperature Distribution in a Single Gold Nanoparticle
  - Outside Temperature Distribution Around a Single Gold Nanoparticle
  - Conclusions
- Practice Example 8: Thermal Ablation of Cancerous Cell Nuclei in Cytoplasm
  - Abstract
  - Model and Input Data
  - Discussion and Results
  - Conclusions

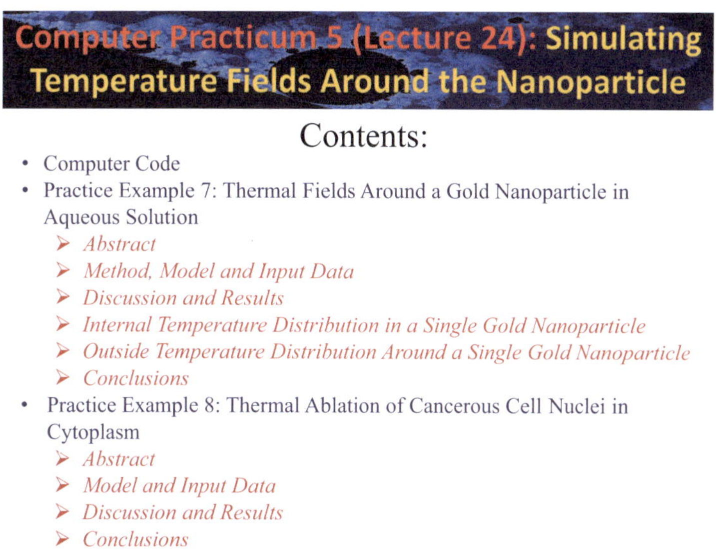

Slide 24.1 Contents

8.2.1 Computer Code

The model described in Sect. 8.1 of this chapter can be used for simulating the space distribution of the temperature fields within and outside a nanoparticle in a biological surrounding medium. *The research task* for this computer practicum (schematically shown in Slide 24.2) is to numerically solve the heat transfer equations (8.9)–(8.11) to determine the special behavior of temperature fields around a single nanoparticle. To perform the heat diffusion simulations at the nanoscale level, we have developed an effective Maple code, shown on the slides and Appendixes E and F. Appendix E contains a Maple code for 2D simulations of thermal fields (second "dimension" is a temperature), and Appendix F provides a code for 3D simulations of the temperature distributions around the nanoparticle (third "dimension" is a temperature).

A Maple code begins with introducing the energy density of the radiation, radius of the gold nanoparticle, and absorption efficiency of the gold nanoparticle at a given size, and using 530 nm wavelength, thermal conductivities and diffusivities of the particle's material and surrounding media (water and blood in the provided example), as shown in Fig. 8.8 and Slide 24.3. Then we define all constants and the rate of heat generation used in the thermal model (8.9)–(8.11) shown in Fig. 8.9 and Slide 24.4, and we introduce the integral solutions for the temperature distributions within and outside the nanoparticle as shown in Fig. 8.10 and Slide 24.5.

Next we perform integrations and plot the results of simulations of the temperature distribution around the nanoparticle in water and blood surrounding

Statement of The Research Task

- The ***research task*** for this computer practicum is to numerically solve the heat transfer equations below to determine the special behavior of temperature fields around a single nanoparticle.
- To perform the heat diffusion simulations at the nanoscale, we have developed an effective ***Maple code***, shown in Appendices E and F.
- ***Appendix E*** contains a Maple code for 2D simulations of thermal fields, and ***Appendix F*** provides a code for 3D simulations of the temperature distributions around the nanoparticle (third "dimension" is a temperature).

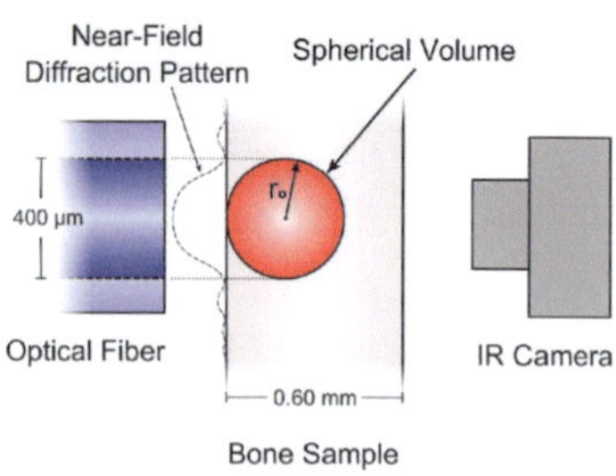

$$T(r)=\begin{cases}\dfrac{3K_{abs}I_o f(t)}{4r_0 C_s\rho_s}\left(\dfrac{r_0^2}{\chi_s}\right)\left(\dfrac{1}{3}\dfrac{\mu_s}{\mu_m}+\dfrac{1}{6}-\dfrac{2b}{\pi}\displaystyle\int_0^\infty\dfrac{\exp(-y^2t/\gamma)}{y^2}\dfrac{(\sin y-y\cos y)\sin(ry/r_0)}{[(c\sin y-y\cos y)^2+b^2y^2\sin^2 y]}dy\right), & r=0,\\ \dfrac{3K_{abs}I_o f(t)}{4r_0 C_s\rho_s}\left(\dfrac{r_0^2}{\chi_s}\right)\left(\dfrac{1}{3}\dfrac{\mu_s}{\mu_m}+\dfrac{1}{6}\left(1-\dfrac{r^2}{r_0^2}\right)-\dfrac{2r_0 b}{\pi r}\displaystyle\int_0^\infty\dfrac{\exp(-y^2t/\gamma)}{y^2}\dfrac{(\sin y-y\cos y)\sin(ry/r_0)}{[(c\sin y-y\cos y)^2+b^2y^2\sin^2 y]}dy\right), & 0<r\le r_0,\\ \dfrac{3K_{abs}I_o f(t)}{4r_0 C_s\rho_s}\left(\dfrac{r_0^2}{\chi_s}\dfrac{r_0}{r}\right)\left(\dfrac{1}{3}\dfrac{\mu_s}{\mu_m}-\dfrac{2}{\pi}\displaystyle\int_0^\infty\dfrac{\exp(-y^2t/\gamma)}{y^3}\dfrac{(\sin y-y\cos y)[by\sin y\cos\sigma y-(c\sin y-y\cos y)\sin\sigma y]}{[(c\sin y-y\cos y)^2+b^2y^2\sin^2 y]}dy\right), & r>r_0,\end{cases}$$

Slide 24.2 Statement of the research task

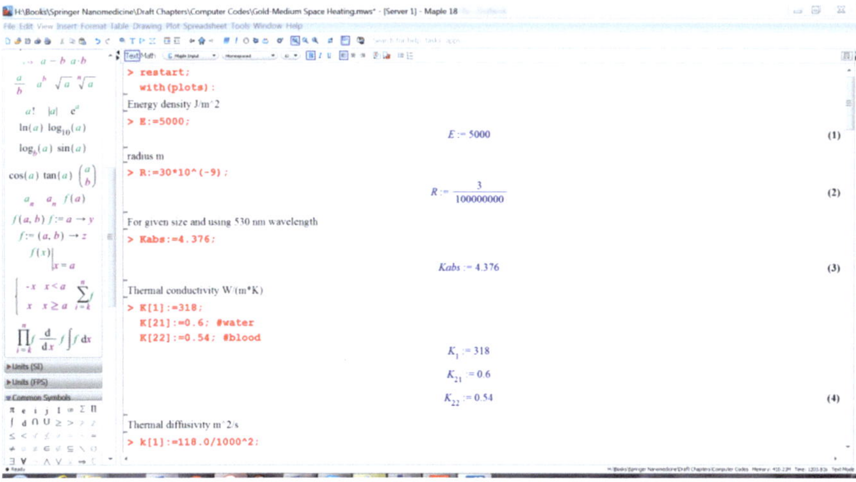

Fig. 8.8 Input data for a heat diffusion model

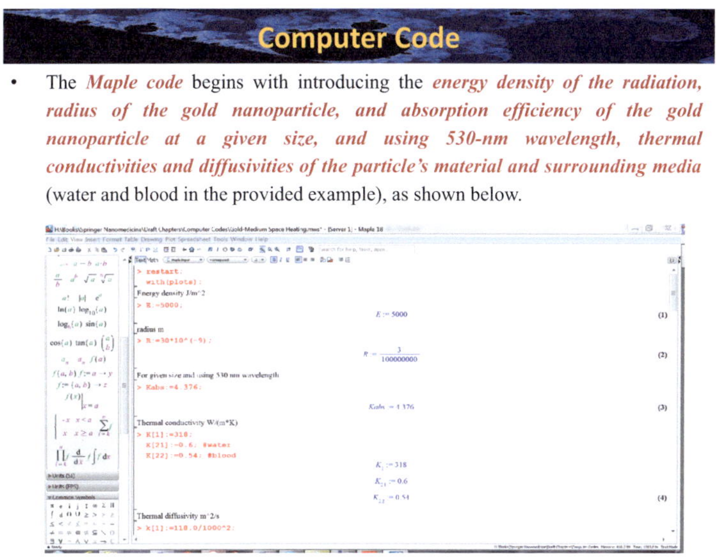

Slide 24.3 Computer code

media shown in Fig. 8.11 and Slide 24.6. Figure 8.12 and Slide 24.7 show the space temperature distributions inside the nanoparticle.

The following Practice Example presents the output results of these simulations in detail for gold nanoparticles in water and blood surrounding media.

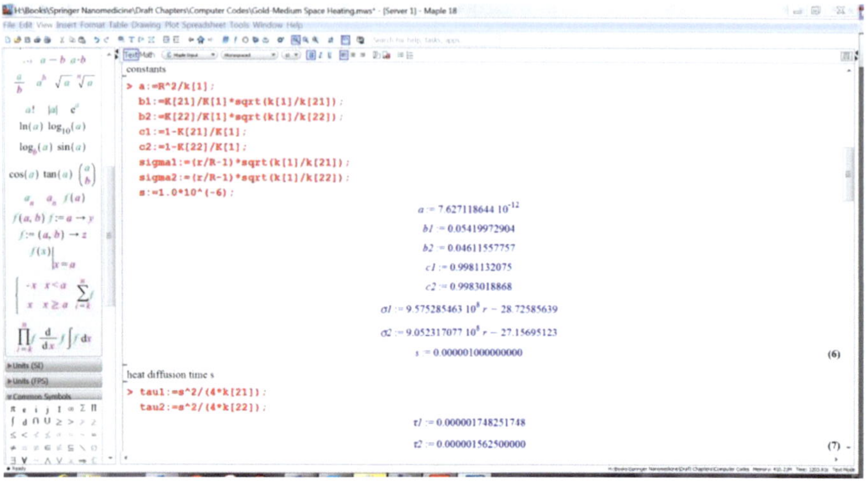

Fig. 8.9 Introducing the simulation constants for the heat diffusion model

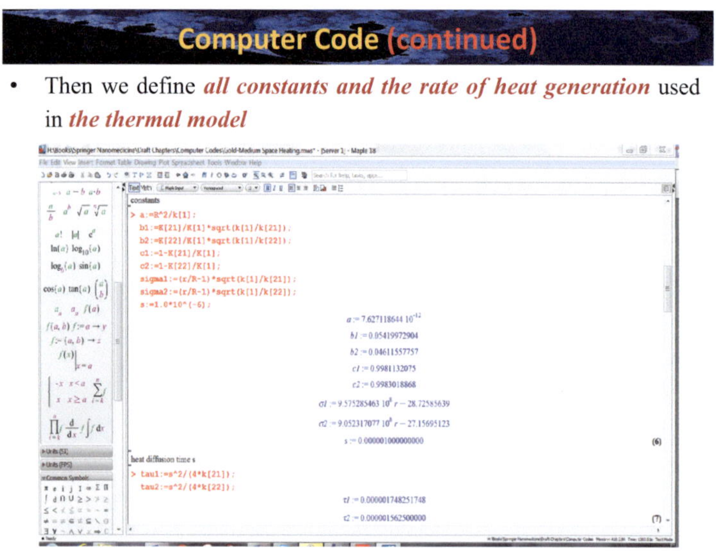

Slide 24.4 Computer code (continued)

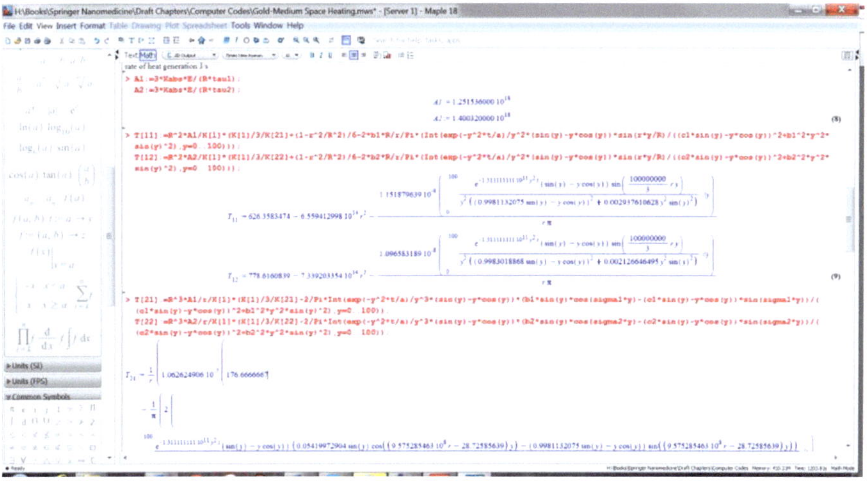

Fig. 8.10 Defining the heat generation rate and introducing the integral solutions for the temperature distributions around the nanoparticle

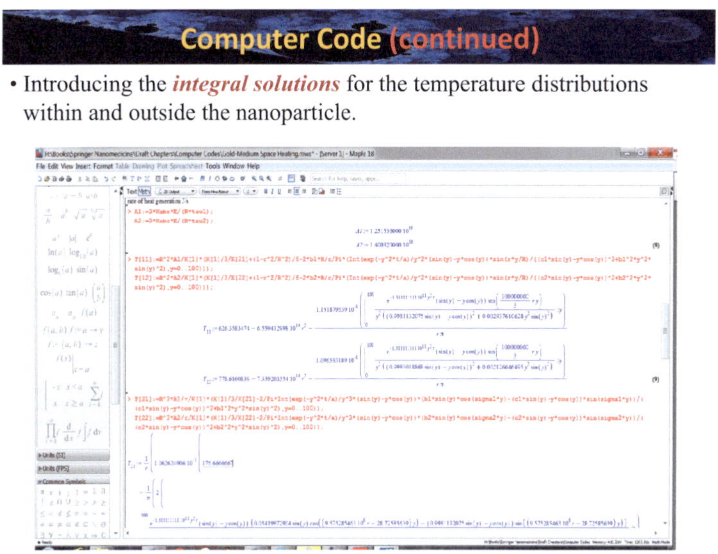

Slide 24.5 Computer code (continued)

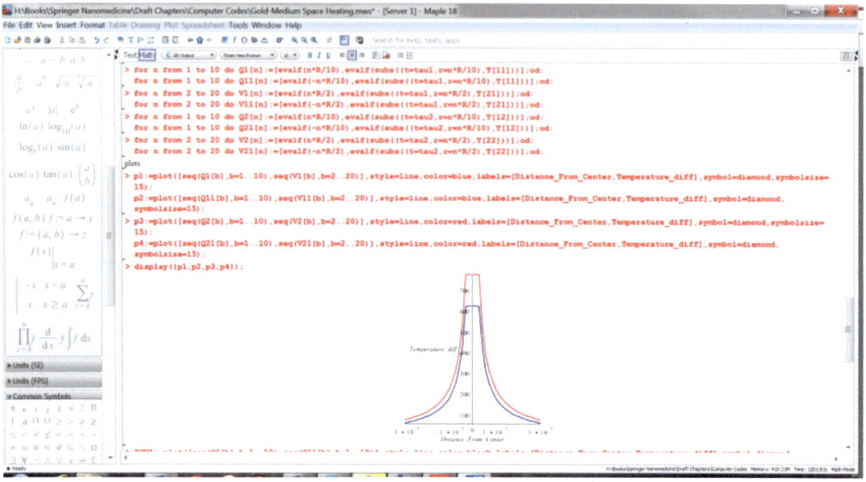

Fig. 8.11 Plots of the space distribution of temperature around the nanoparticle

Computer Code (continued)

- Defining the heat generation rate and introducing the integral solutions for the ***temperature distributions around the nanoparticle.***

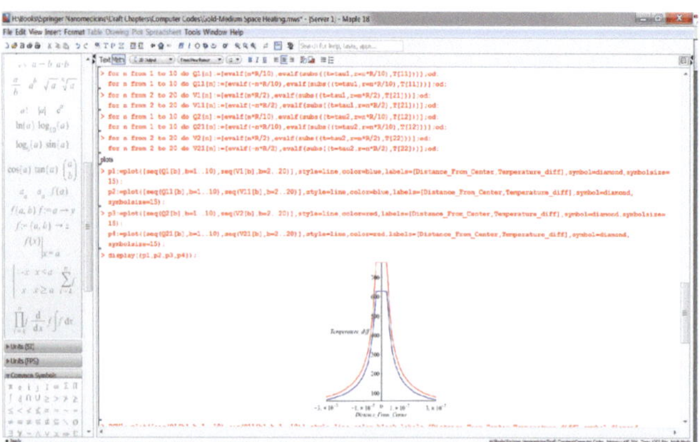

Slide 24.6 Computer code (continued)

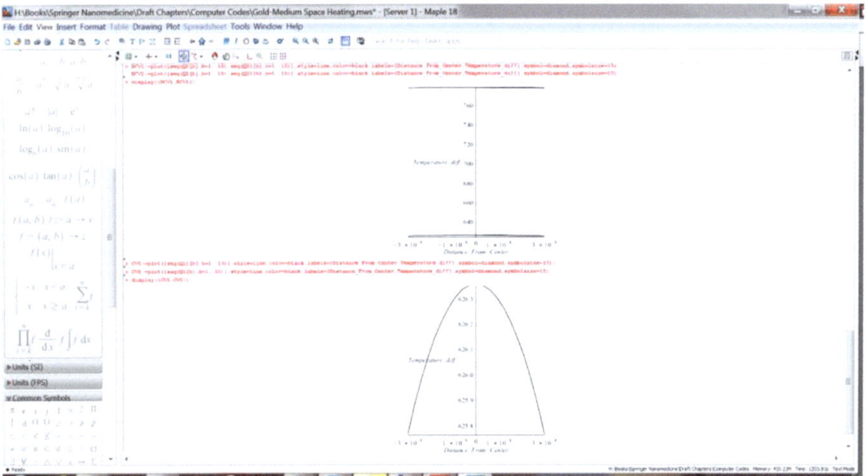

Fig. 8.12 The temperature distributions inside the nanoparticle

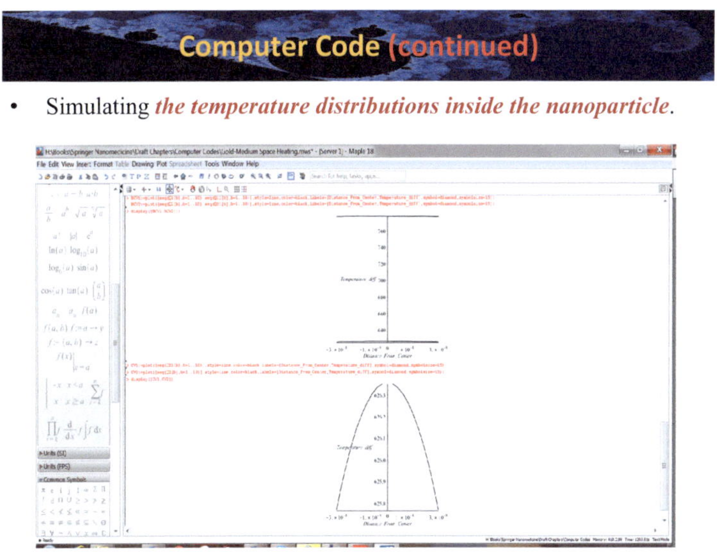

Slide 24.7 Computer code (continued)

8.2.2 *Practice Example 7: Thermal Fields Around a Gold Nanoparticle in Aqueous Solution*

8.2.2.1 Abstract

The objective of this practice example is to calculate the temperature distribution in a gold nanoparticle and the surrounding medium of water and blood as a function of position when heated by the radiation. These simulations will allow us to estimate the collateral thermal damage area produced by single nanoparticle in a biological medium.

8.2.2.2 Method, Model, and Input Data

Thermal ablation of cancerous cells is one of the primary therapeutic advancements that are made possible with the use of nanoparticles. After surrounding cancerous cells/growths with nanoparticles, they can be heated via radiation to kill the cells by concentrating high levels of heat, causing cell destruction by burning, evaporating, or bubbling. For therapeutic purposes, the temperature around the nanoparticle in the surrounding cellular medium being treated must reach a critical value of 425 K in order to initiate the cancer cell killing phenomena associated with thermal ablation, nanobubble formation, high-pressure waves, and sound generation. This practice example investigates the potential of using gold nanoparticles in therapeutic applications by calculating the temperature as a function of distance away from the particle.

The method used in this practice example consists of two steps (Slide 24.8):

1. Computing the optimal radius and maximum absorption efficiency of gold nanoparticles in surrounding water and blood media using the Mie diffraction theory.
2. Plugging these absorption values into the heat diffusion model to simulate the spatial distribution of temperature fields around the nanoparticle.

The absorption efficiency for the optimal radius of the gold nanoparticle is computed by using the MiePlot software demonstrated in the *Computer Practicum 1* of Chap. 5. The absorption values found from these simulations are summarized in Table 8.4 for different gold nanoparticle shapes and effective radii at various optimal wavelengths of radiation.

The next step is to numerically solve the heat transfer equations (8.9)–(8.11) by using a Maple code provided in Appendix E for 2D simulations of thermal fields (second "dimension" is a temperature), or Appendix F for 3D simulations of the temperature distributions around the nanoparticle (third "dimension" is a temperature). The material properties of the gold nanoparticle needed for simulations are listed in Table 8.2. The thermal properties of some selected surrounding biological media are given in Table 8.7 (Slide 24.9).

Practice Example 7:
Thermal Fields Around a Gold Nanoparticle in Aqueous Solution

- The *method* used in this practice example consists of *two steps*:
- (1) ***computing the optimal radius and maximum absorption efficiency*** of gold nanoparticles in surrounding water and blood media using the Mie diffraction theory, and
- (2) plugging these absorption values into the heat diffusion model to ***simulate the spatial distribution of temperature fields*** around the nanoparticle.
- The ***absorption efficiency*** for the optimal radius of the gold nanoparticle is computed by using the ***MiePlot software*** demonstrated in the *Computer Practicum 1*.

Slide 24.8 Practice example 7: thermal fields around a gold nanoparticle in aqueous solution

Table 8.7 Thermal properties of selected biological media

Medium	Density ρ (kg/m^3)	Specific heat C (J/K kg)	Interval of T (K)	Phase transition point T_{ph} (K)	Thermal conductivity μ (W/m K)	Thermal diffusivity χ (m^2/s)
Blood	1060	3645–3897	273–373 (323)	373	0.45–0.6 (0.525)	1.6×10^{-7}
Fat	900	2975	273–373 (323)		0.185–0.233 (0.209)	7.81×10^{-8}
Cytoplasm/ water	1000	4181.6–4215.6	273–373 (323)		0.597–0.682 (0.640)	1.52×10^{-7}
Cell membrane	900	3000	273–373 (323)		0.3	1.11×10^{-7}
Collagen	1900	440	273–373 (323)		0.16–0.34 (0.25)	2.99×10^{-7}

8.2.2.3 Discussion and Results

Let we calculate the laser heating of single gold nanoparticles with radii $r_0 = 30$–35 nm in surrounding water and blood biomedia when the incident laser pulse has an energy density of $E = 4.0$ J/cm^2 and pulse duration of $\tau_L = 8$ ns. Here we reproduce our results published in Refs. [2, 3]. The laser flux is chosen at the level of 4.0 J/cm^2 to provide cell lethality during a single laser pulse shot, and it is comparable to the laser fluence currently used in clinical treatments of pigmented skin lesions. The results of these simulations are presented in Figs. 8.13 and 8.14.

Practice Example 7: Input Data

Medium	Density ρ (kg/m³)	Specific heat C (J/K kg)	Interval of T (K)	Phase transition point T_{ph} (K)	Thermal Conductivity μ (W/mK)	Thermal Diffusivity χ (m²/s)
Blood	1060	3645 - 3897	273-373 (323)		0.45-0.6 (0.525)	1.6E-07
Fat	900	2975	273-373 (323)		0.185-0.233 (0.209)	7.81E-08
Cytoplasm/ water	1000	4181.6-4215.6	273-373 (323)	373	0.597-0.682 (0.640)	1.52E-07
Cell Membrane	900	3000	273-373			
Collagen	1900	440				

Property of Au	Values
Thermal conductivity	3.18×10^2 W/m·K
Density	1.93×10^4 kg/m³
Specific heat	1.29×10^2 J/kg·K
Thermal diffusivity	1.27×10^{-4} m²/s

Nanoparticle type	Dimension	Wavelength	Absorption efficiency
Gold nanospheres	35 nm	532 nm	4.02
	40 nm	550 nm	3.50
Silica-gold nanoshells	120 nm	1100 nm	3.50
	120 nm	750 nm	3.75
Gold nanorods	11.4 nm	800 nm	14.0
	17.9 nm	825 nm	14.5

Slide 24.9 Practice example 7: input data

Figure 8.13 illustrates the space behavior of the temperature inside the 30–35-nm gold particles heated and cooled in the surrounding water and blood media (Slide 24.10). Figure 8.14 shows the space distribution of the temperature outside these nanoparticles (Slide 24.11).

The events that take place after absorption of the laser pulse energy by small particles depend on the size of the locally heated region and duration of laser exposure [3]. Long pulses that exceed the thermal relaxation time in biological tissue, $\tau_r = l^2/4\chi_t$ (where χ_t is the thermal diffusivity of the biological tissue, and l is the minimum size of the locally heated region), cause heating of both the particle and the surrounding media. If the size of the locally heated region l is chosen to be equal to the size of the laser focal spot, which is about 1 μm, the thermal relaxation time for water is 1.75 μs and for blood is 1.56 μs, which is much longer than the laser pulse duration (~10 ns) used in our simulations. For ultrashort laser pulses when $\tau_L \ll \tau_r$, the absorbed energy can be thermally confined within the target, causing rapid heating of the absorber itself.

8.2.2.4 Internal Temperature Distribution in a Single Gold Nanoparticle

The integration of Eqs. (8.9) and (8.10) gives the temperature distribution inside the heated nanoparticle surrounded by the biomedia shown in Fig. 8.13 and Slide 24.10. The simulations are performed for the 30 nm gold particle heated

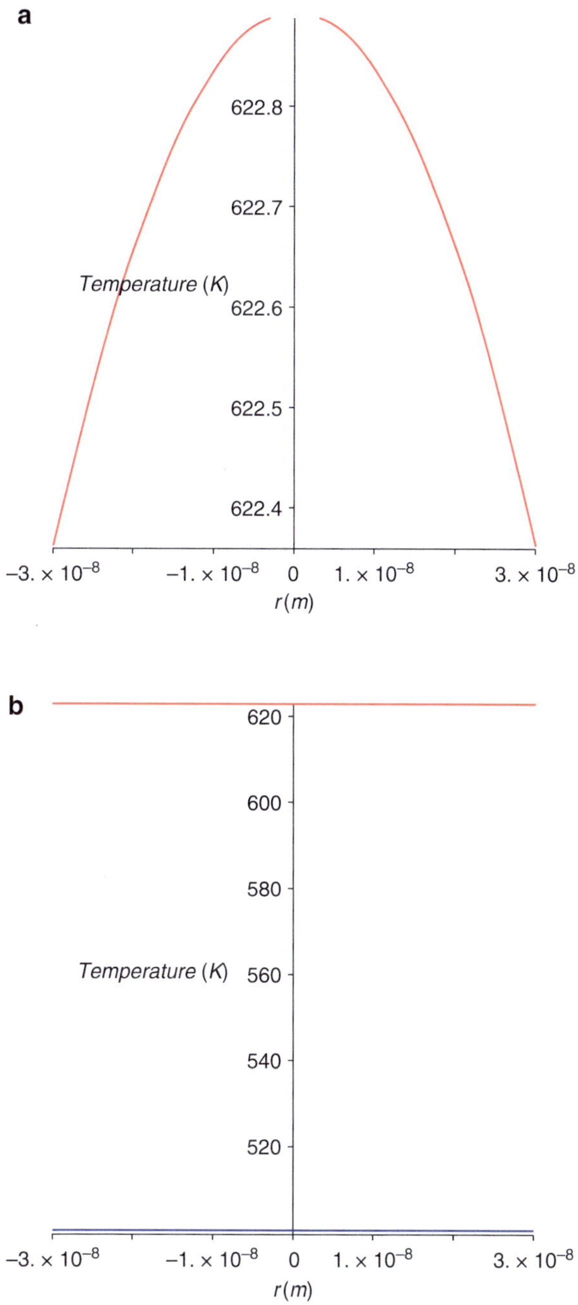

Fig. 8.13 Temperature distribution inside a 30-nm gold nanoparticle (**a**) in blood and (**b**) in blood (*red curve*) and water (*blue curve*), with the surrounding media irradiated by a single laser pulse of energy density 4 J/cm^2 and duration 8 ns in both cases

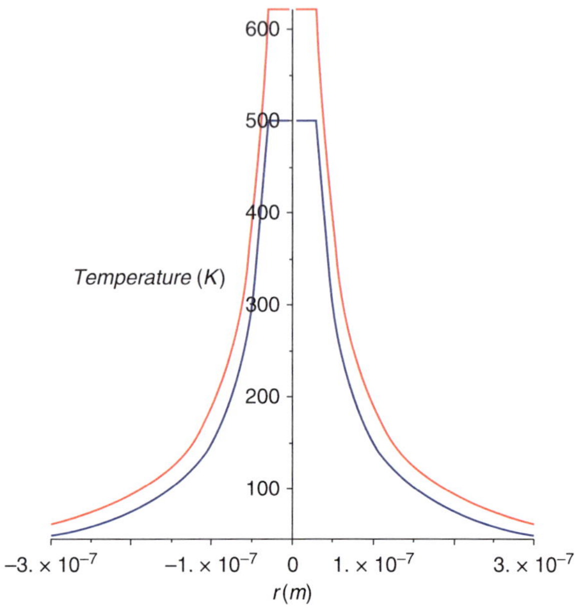

Fig. 8.14 Spatial distribution of the temperature outside the 30-nm gold particle in water (*blue curve*) and blood (*red curve*), with the surrounding media irradiated by a single laser pulse of energy density 4.0 J/cm^2 and duration 8 ns

by a 4.0 J/cm^2 laser pulse of 8 ns width in the water and blood surrounding media. As can be seen from Fig. 8.13, the gold nanoparticle reaches its maximum temperature $T_{\max} \sim 500$ K in water and $T_{\max} = 622.9$ K in blood at the center of the sphere. Then we observe a small decrease of the inside temperature by 0.6 K to the particle's boundaries, establishing the stationary surface temperature of 622.3 K in blood (Fig. 8.13a) and about 500 K in water. Since the temperature variation inside the nanoparticle occurs over a small range of 0.6 K, the internal particle's temperature can be assumed to be constant (Fig. 8.13b). Thus, the quasistationary and homogeneous approach can be applied to modeling the laser heating and evaporation of the nanoparticles in the biological media (Slide 24.10).

8.2.2.5 Outside Temperature Distribution Around a Single Gold Nanoparticle

As follows from Fig. 8.14 and Slide 24.11, there is a significant heat loss from the surface of a nanoparticle to the surrounding medium, suggesting that even lower energy pulses are enough to achieve true and large thermal damage of the biological surroundings [3]. If the temperature of the surrounding medium exceeds some thermal thresholds, secondary phenomena of cell killing effects may develop, such as photothermal ablation, sound generation, bubble formation, and so on. For most cancer cell killing effects, the threshold temperature is about

Internal Temperature Distribution in a Single Gold Nanoparticle

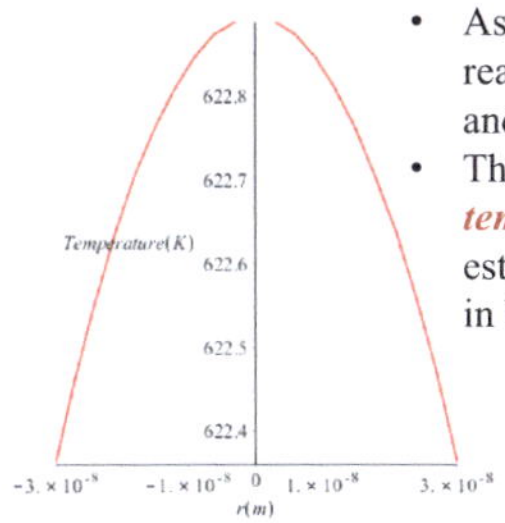

- As can be seen from the figure, the gold nanoparticle reaches its maximum temperature T_{max} ~ 500 K in water and T_{max} = 622.9 K in blood at the center of the sphere.
- Then we observe a ***small decrease of the inside temperature by 0.6 K*** to the particle's boundaries, establishing the stationary surface temperature of 622.3 K in blood and about 500 K in water.

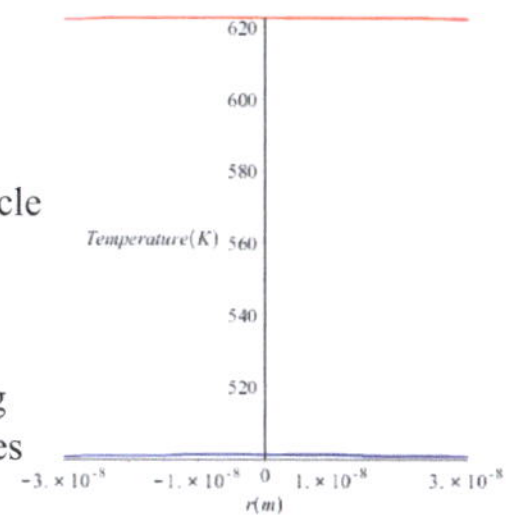

Since the temperature variation inside the nanoparticle occurs over a small range of 0.6 K, the ***internal particle's temperature can be assumed to be constant***. Thus, ***the quasistationary and homogeneous approach*** can be applied to modeling the laser heating and evaporation of the nanoparticles in the biological media.

Slide 24.10 Internal temperature distribution in a single gold nanoparticle

Outside Temperature Distribution Around a Single Gold Nanoparticle

- As follows from the figure, there is ***a significant heat loss from the surface of a nanoparticle to the surrounding medium***, suggesting that even lower-energy pulses are enough to achieve true and large thermal damage of the biological surroundings.
- The calculations show that at the level of the threshold temperature (425 K), the ***damage area produced by a heated 30-nm gold particle*** in blood irradiated by 5 J/cm^2 can reach a size of ~ ***0.5 μm in diameter***, which ***exceeds by 17 times the size of the nanoparticle.*** This size of a damage area requires the nanoparticle to be heated to over 1000 K.
- Thus, under the chosen conditions, ***the damage area*** produced by a 30-nm particle heated by a single laser pulse of 4-5 J/cm^2 ***is comparable to the size of the laser focal spot.***

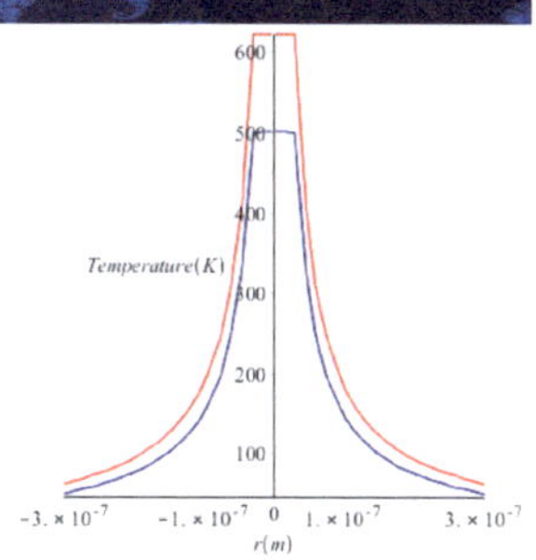

Figure. Spatial distribution of the temperature outside the 30-nm gold particle in the surrounding water (blue curve) and blood (red curve) media irradiated by a single laser pulse of energy density 4.0 J/cm^2 and duration 8 ns.

Slide 24.11 Outside temperature distribution around a single gold nanoparticle

150 °C (423 K). Our calculations show that at this level of the threshold temperature, the damage area produced by a heated 30-nm gold particle in blood irradiated by 5 J/cm^2 can reach a size of ~0.5 μm in diameter, which exceeds by 17 times the size of the nanoparticle [3]. This size of a damage area requires the nanoparticle to be heated to over 1000 K. Please note that Fig. 8.14 demonstrates heat diffusion at much lower temperature of the nanoparticle, and because of this it shows less damage area. Thus, under the chosen conditions, the damage area produced by a 30-nm particle heated by a single laser pulse of 4–5 J/cm^2 is comparable to the size of the laser focal spot (Slide 24.11). Decreasing the energy density of the incident single laser pulse leads to smaller damage volume around the nanoparticle.

8.2.2.6 **Conclusions** (Slide 24.12)

1. The temperature variation inside the nanoparticle during the laser heating by an 8-ns pulse occurs in a small temperature range and can be assumed to be constant. Thus, the quasistationary and homogeneous approach can be applied for modeling the laser heating and evaporation of the nanoparticles in the biological media [3].
2. The damage area produced by a 30-nm particle heated by a single laser pulse of 5 J/cm^2 is about 0.5 μm [3].

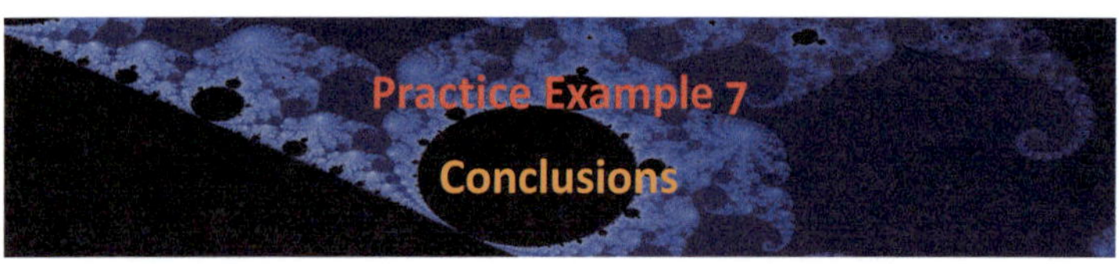

- The ***temperature variation inside the nanoparticle*** during the laser heating by a 8 ns pulse occurs in a small temperature range and ***can be assumed to be constant***.
- Thus, ***the quasistationary and homogeneous approach can be applied*** for modeling the laser heating and evaporation of the nanoparticles in the biological media.
- The ***damage area*** produced by a 30-nm particle heated by a single laser pulse of 5 J/cm^2 is about 0.5 μm.

Slide 24.12 Practice example 7: conclusions

8.2.3 Practice Example 8: Thermal Ablation of Cancerous Cell Nuclei in Cytoplasm

8.2.3.1 Abstract

In this computer practicum, we perform spatial simulations of the thermal fields around healthy and abnormal cell nuclei in cytoplasm. When a cell becomes cancerous, it swells to unregulated sizes. The morphological changes in the cancerous cells affect the heat distribution across an abnormal cell. Given the effects of cancer on cell organelles, here we investigate if it is possible to selectively heat a cancerous nucleus with laser radiation as a method of cancer treatment without using nanoparticles. By utilizing the physical changes experienced by nucleus organelles in abnormal cells, we can determine how laser irradiation would affect cancerous cells in comparison to healthy cells [3]. The critical temperature for organelle heating is about 425 K, as that results in protein denaturing and thermal ablation.

8.2.3.2 Model and Input Data

The main purpose of this practice example is to determine the feasibility of destroying cancerous cells using pulsed laser radiation (Slide 24.13). Specifically, we are exploring the possibility of heating the cancerous nucleus to the ablation temperatures, without damaging the surrounding tissue, by applying low-energy

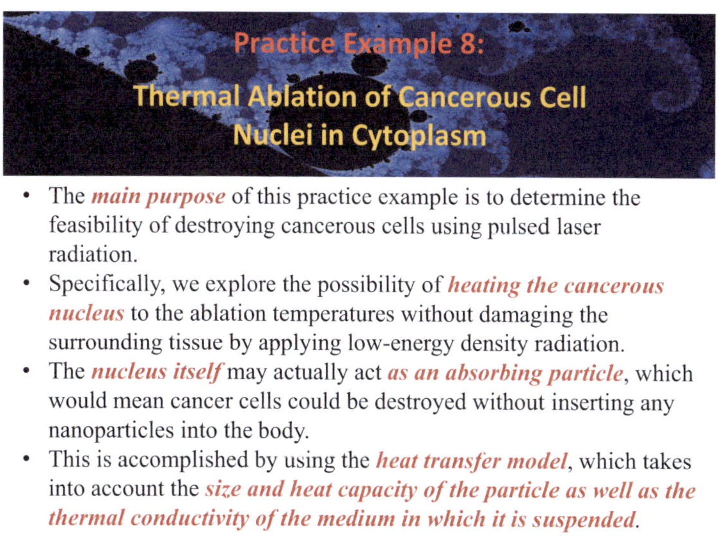

Slide 24.13 Practice example 8: thermal ablation of cancerous cell nuclei in cytoplasm

Practice Example 8: Input Data

- Working from the ***Maple worksheet*** presented in ***Appendix E or F***, the first step is plugging in the input data for healthy and cancerous nuclei and cytoplasm.
- The ***MiePlot software*** is also used to determine the absorption efficiencies of normal and cancerous nuclei along with the optimum wavelengths of radiation for treatment purposes.

Organelle and cytoplasm properties.

Organelle	Diameter, d (μm)	Specific heat, C (J/kg-K)	Thermal conductivity, μ (W/m-K)	Density, ρ (kg/m³)
Nucleus	5-10	3000	0.3	1050-1400
Mitochondria	0.85-3.57	3000	0.3	1050-1100
Ribosome	0.02-0.025	3000	0.3	1050-1600
Microtubules	0.02-2	3000	0.3	1050-1100
Lysosomes	0.1-0.25	3000	0.3	1050-1100
Cytoskeleton	0.004-0.012	3000	0.3	1050-1100
Cytoplasm	-	4180	0.59	

Nucleus	Q_{abs}	λ_{opt} (nm)
Normal (r_0 = 5.0 μm)	0.9859	942.18
Cancerous (r_0 = 6.5 μm)	1.008365	899.99
Cancerous (r_0 = 7.0 μm)	1.0125	899.99

Slide 24.14 Practice example 8: input data

density radiation. The nucleus itself may actually act as an absorbing particle, which would mean cancer cells could be destroyed without inserting any nanoparticles into the body, reducing risk to the patient and creating an easier treatment procedure. This is accomplished by using the heat transfer equations (8.9)–(8.11), which takes into account the size and heat capacity of the particle as well as the thermal conductivity of the medium in which it is suspended. Working from the Maple worksheet presented in Appendix E or F, the first step is plugging in the input data for healthy and cancerous nuclei and cytoplasm (Slide 24.14). The MiePlot software is also used to determine the absorption efficiencies of normal and cancerous nuclei along with the optimum wavelengths of radiation for treatment purposes.

The difference between the normal and cancerous nuclei in this context really comes down to the radius. A normal nucleus has a radius of about 5.0 μm, while the cancerous nucleus has a radius that can range from 6.5 to 10.0 μm, as shown in Tables 5.5 and 6.5. The size difference changes the absorption efficiency (Q_{abs}) values and the optimal wavelength for absorption presented in Table 8.8 and Slide 24.14. These values are simulated with the optical data in Table 5.6 using the MiePlot software. Thermal properties of cell organelles and cytoplasm are listed in Table 6.5. The thermal diffusivity χ of the cell organelles can be calculated from the specific heat C, thermal conductivity μ, and density ρ of each organelle given in Table 6.5 using the relation: $\chi = \mu/(C{\cdot}\rho)$.

Table 8.8 Absorption properties of normal and cancerous nuclei

Nucleus	Q_{abs}	λ_{opt} (nm)
Normal ($r_0 = 5.0$ μm)	0.9859	942.18
Cancerous ($r_0 = 6.5$ μm)	1.008365	899.99
Cancerous ($r_0 = 7.0$ μm)	1.0125	899.99

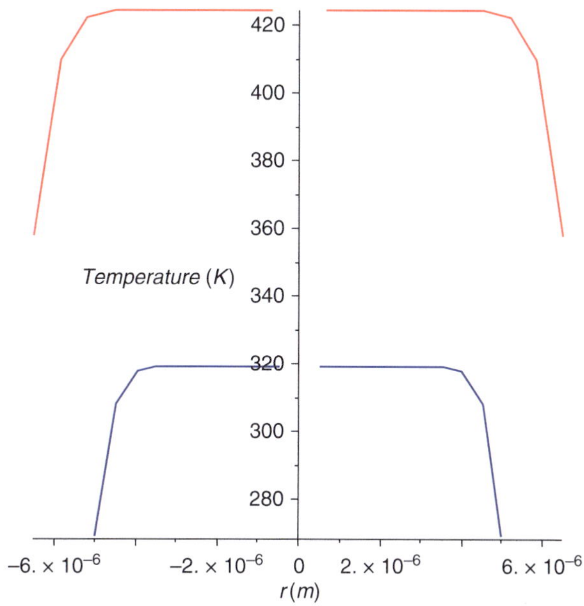

Fig. 8.15 Spatial distribution of the temperature around healthy (*blue curve*) and cancerous (*red curve*) cell nuclei

8.2.3.3 Discussion and Results

The simulations are performed for heating the healthy ($r_0 = 5.0$ μm) and cancerous ($r_0 = 7.0$ μm) nuclei in cytoplasm by using the thermal and optical properties from Tables 6.5 and 8.8. The results of these simulations are presented in Figs. 8.15 and Slide 24.15. The figure illustrates the space behavior of the temperature inside and outside the healthy and cancerous nuclei heated and cooled in the cytoplasm. It is evident from the figure that a cancerous nucleus heats up to much higher temperatures than a normal nucleus, reaching the critical temperature for ablation of 425 K. With the significant increase in temperature of the cancerous nucleus, the cancer cells undergo cell death via the thermal ablation of the nucleus. At the same time, the change in temperature for the normal nucleus in the cytoplasm is only 10 K above the normal body temperature of 310 K. Thus, a single laser pulse is capable of producing a significant temperature change and thermal damage in the cancerous nucleus, while a healthy organelle would be kept safe. The temperature distribution outside the nucleus drops off rapidly without causing any harmful effect on surrounding normal cells.

Practice Example 8
Discussion and Results

- Simulations have been performed for heating healthy ($r_0 = 5.0$ μm, ***blue curve***) and cancerous ($r_0 = 7.0$ μm, ***red curve***) nuclei in cytoplasm.
- It is evident from this figure that a ***cancerous nucleus*** heats up to much higher temperature than a normal nucleus, ***reaching the critical temperature for ablation of 425 K***.
- With the significant increase in temperature of the cancerous nucleus, the ***cancer cells undergo cell death via the thermal ablation of the nucleus***.
- At the same time, the change in temperature for the ***normal nucleus in the cytoplasm is only 10 K above the normal body temperature*** of 310 K.

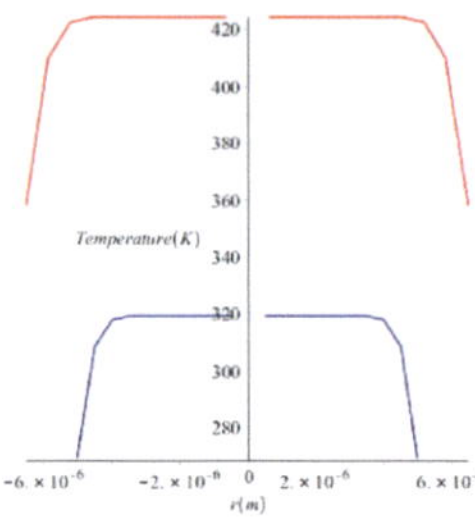

Figure. Spatial distribution of the temperature around healthy (***blue curve***) and cancerous (***red curve***) cell nuclei.

Slide 24.15 Practice example 8: discussion and results

Practice Example 8
Conclusions

- ***Heating the largest cell organelle***, i.e., the cancerous nucleus, can effectively ***kill a cancer cell***.
- The ***size difference*** between normal and abnormal nuclei ***provides a significant temperature difference*** with equal incident energy density which could be used to kill cancer cells while leaving healthy cells unharmed.
- In this simulation, the ***cancerous nucleus temperature*** reaches the ***damage threshold (425 K)***, while the ***healthy nucleus*** heats up ***to only 320 K***, which is a little above the normal body temperature.
- Thus, at high enough energy density, ***heating of the nuclei*** can be used ***to kill cancer cells*** which have enlarged nuclei while keeping the healthy cells undamaged.

Slide 24.16 Practice example 8: conclusions

8.2.3.4 Conclusions

Heating the largest cell organelle, i.e., the cancerous nucleus, can effectively kill a cancer cell (Slide 24.16). The size difference between normal and abnormal nuclei provides a significant temperature difference with equal incident energy density, which could be used to kill cancer cells while leaving healthy cells unharmed. In this simulation, the cancerous nucleus temperature reaches the damage threshold (425 K), while the healthy nucleus heats up to only 320 K, which is a little above normal body temperature. Thus, at high enough energy density, heating of the nuclei can be used to kill cancer cells which have enlarged nuclei, while keeping the healthy cells undamaged.

Appendix E: Maple Code for Simulating Heat Diffusion Around a Nanoparticle

This appendix contains a Maple code designed for 2D simulations of the temperature distributions (second "dimension" is a temperature) around the nanoparticle (Slide 24.17). The comments within the Maple code should explain the variables, procedures, and functions that are used.

Note: This code takes about 10–15 min to complete the integrations. The code comes directly from the Maple file and is therefore formatted to word-wrap according to Maple standard character-per-line limits.

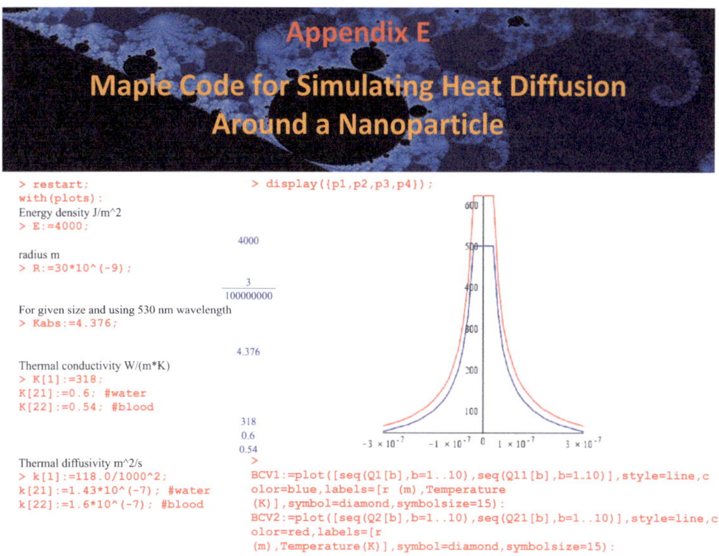

Slide 24.17 Appendix E: Maple Code for simulating heat diffusion around a nanoparticle

```
> restart;
with(plots):
```
Energy density (J/m^2)
```
> E:=4000;
```
$$4000$$

Radius (m)
```
> R:=30*10^(-9);
```
$$\frac{3}{100000000}$$

For given size and using 530-nm wavelength
```
> Kabs:=4.376;
```
$$4.376$$

Thermal conductivity W/(m K)
```
> K[1]:=318;
K[21]:=0.6; #water
K[22]:=0.54; #blood
```
$$318$$
$$0.6$$
$$0.54$$

Thermal diffusivity (m^2/s)
```
> k[1]:=118.0/1000^2;
k[21]:=1.43*10^(-7); #water
k[22]:=1.6*10^(-7); #blood
```
$$0.0001180000000$$
$$1.430000000\ 10^{-7}$$
$$1.600000000\ 10^{-7}$$

Constants
```
> a:=R^2/k[1];
b1:=K[21]/K[1]*sqrt(k[1]/k[21]);
b2:=K[22]/K[1]*sqrt(k[1]/k[22]);
c1:=1-K[21]/K[1];
c2:=1-K[22]/K[1];
sigma1:=(r/R-1)*sqrt(k[1]/k[21]);
sigma2:=(r/R-1)*sqrt(k[1]/k[22]);
s:=1.0*10^(-6);
```
$$7.627118644\ 10^{-12}$$
$$0.05419972904$$
$$0.04611557757$$
$$0.9981132075$$
$$0.9983018868$$
$$9.575285463\ 10^{8}\ r - 28.72585639$$
$$9.052317077\ 10^{8}\ r - 27.15695123$$
$$0.000001000000000$$

Heat diffusion time (s)

```
> tau1:=s^2/(4*k[21]);
tau2:=s^2/(4*k[22]);
```

$$0.000001748251748$$
$$0.000001562500000$$

Rate of heat generation (J/s)

```
> A1:=3*Kabs*E/(R*tau1);
A2:=3*Kabs*E/(R*tau2);
```

$$1.001228800\ 10^{18}$$
$$1.120256000\ 10^{18}$$

```
> T[11]:=R^2*A1/K[1]*(K[1]/3/K[21]+(1-r^2/R^2)/6-
2*b1*R/r/Pi*(Int(exp(-y^2*t/a)/y^2*(sin(y)-
y*cos(y))*sin(r*y/R)/((c1*sin(y)-
y*cos(y))^2+b1^2*y^2*sin(y)^2),y=0..100)));
T[12]:=R^2*A2/K[1]*(K[1]/3/K[22]+(1-r^2/R^2)/6-
2*b2*R/r/Pi*(Int(exp(-y^2*t/a)/y^2*(sin(y)-
y*cos(y))*sin(r*y/R)/((c2*sin(y)-
y*cos(y))^2+b2^2*y^2*sin(y)^2),y=0..100)));
```

$$501.0866779 - 5.247530398\ 10^{14}\ r^2 - \frac{1}{r\pi}\left(9.215037112\ 10^{-9}\left(\int_0^{100} \left(e^{-1.311111111\ 10^{11}\ y^2 t}\ (\sin(y) - y\cos(y))\ \sin\left(\frac{100000000}{3}\ r\ y\right)\right) \Big/ \left(y^2\ \left((0.9981132075\ \sin(y) - y\cos(y))^2 + 0.002937610628\ y^2\ \sin(y)^2\right)\right) \mathrm{d}y\right)\right)$$

$$622.8928672 - 5.871362683\ 10^{14}\ r^2 - \frac{1}{r\pi}\left(8.772665512\ 10^{-9}\left(\int_0^{100} \left(e^{-1.311111111\ 10^{11}\ y^2 t}\ (\sin(y) - y\cos(y))\ \sin\left(\frac{100000000}{3}\ r\ y\right)\right) \Big/ \left(y^2\ \left((0.9983018868\ \sin(y) - y\cos(y))^2 + 0.002126646495\ y^2\ \sin(y)^2\right)\right) \mathrm{d}y\right)\right)$$

```
> T[21]:=R^3*A1/r/K[1]*(K[1]/3/K[21]-2/Pi*Int(exp(-
y^2*t/a)/y^3*(sin(y)-y*cos(y))*(b1*sin(y)*cos(sigma1*y)-
(c1*sin(y)-y*cos(y))*sin(sigma1*y))/((c1*sin(y)-
y*cos(y))^2+b1^2*y^2*sin(y)^2),y=0..100));
T[22]:=R^3*A2/r/K[1]*(K[1]/3/K[22]-2/Pi*Int(exp(-
y^2*t/a)/y^3*(sin(y)-y*cos(y))*(b2*sin(y)*cos(sigma2*y)-
(c2*sin(y)-y*cos(y))*sin(sigma2*y))/((c2*sin(y)-
y*cos(y))^2+b2^2*y^2*sin(y)^2),y=0..100));
```

$$\frac{1}{r}\left(8.500999245\ 10^{-8}\left(176.6666667-\frac{1}{\pi}\left(2\left(\int_0^{100}\left(e^{-1.311111111\ 10^{11}\ y^2\ t}\left(\sin(y)-y\cos(y)\right)\left(0.05419972904\sin(y)\cos\left(\left(9.575285463\ 10^8\ r-28.72585639\right)y\right)-\left(0.9981132075\sin(y)-y\cos(y)\right)\sin\left(\left(9.575285463\ 10^8\ r-28.72585639\right)y\right)\right)\right)\Big/\left(y^3\left(\left(0.9981132075\sin(y)-y\cos(y)\right)^2+0.002937610628\,y^2\sin(y)^2\right)\right)dy\right)\right)\right)\right)$$

$$\frac{1}{r}\left(9.511607547\ 10^{-8}\left(196.2962963-\frac{1}{\pi}\left(2\left(\int_0^{100}\left(e^{-1.311111111\ 10^{11}\ y^2\ t}\left(\sin(y)-y\cos(y)\right)\left(0.04611557757\sin(y)\cos\left(\left(9.052317077\ 10^8\ r-27.15695123\right)y\right)-\left(0.9983018868\sin(y)-y\cos(y)\right)\sin\left(\left(9.052317077\ 10^8\ r-27.15695123\right)y\right)\right)\right)\Big/\left(y^3\left(\left(0.9983018868\sin(y)-y\cos(y)\right)^2+0.002126646495\,y^2\sin(y)^2\right)\right)dy\right)\right)\right)\right)$$

```
> for n from 1 to 10 do
Q1[n]:=[evalf(n*R/10),evalf(subs((t=tau1,r=n*R/10),T[11]))];od:
for n from 1 to 10 do Q11[n]:=[evalf(-
n*R/10),evalf(subs((t=tau1,r=n*R/10),T[11]))];od:
> for n from 2 to 20 do
V1[n]:=[evalf(n*R/2),evalf(subs((t=tau1,r=n*R/2),T[21]))];od:
for n from 2 to 20 do V11[n]:=[evalf(-
n*R/2),evalf(subs((t=tau1,r=n*R/2),T[21]))];od:
> for n from 1 to 10 do
Q2[n]:=[evalf(n*R/10),evalf(subs((t=tau2,r=n*R/10),T[12]))];od:
for n from 1 to 10 do Q21[n]:=[evalf(-
n*R/10),evalf(subs((t=tau2,r=n*R/10),T[12]))];od:
> for n from 2 to 20 do
V2[n]:=[evalf(n*R/2),evalf(subs((t=tau2,r=n*R/2),T[22]))];od:
for n from 2 to 20 do V21[n]:=[evalf(-
n*R/2),evalf(subs((t=tau2,r=n*R/2),T[22]))];od:
```

Plots

```
p1:=plot([seq(Q1[b],b=1..10),seq(V1[b],b=2..20)],style=line,colo
r=blue,labels=[r
(m),Temperature(K)],symbol=diamond,symbolsize=15):
p2:=plot([seq(Q11[b],b=1..10),seq(V11[b],b=2..20)],style=line,co
lor=blue,labels=[r
(m),Temperature(K)],symbol=diamond,symbolsize=15):
>
p3:=plot([seq(Q2[b],b=1..10),seq(V2[b],b=2..20)],style=line,colo
r=red,labels=[r
(m),Temperature(K)],symbol=diamond,symbolsize=15):
p4:=plot([seq(Q21[b],b=1..10),seq(V21[b],b=2..20)],style=line,co
lor=red,labels=[r
(m),Temperature(K)],symbol=diamond,symbolsize=15):
> display({p1,p2,p3,p4});
```

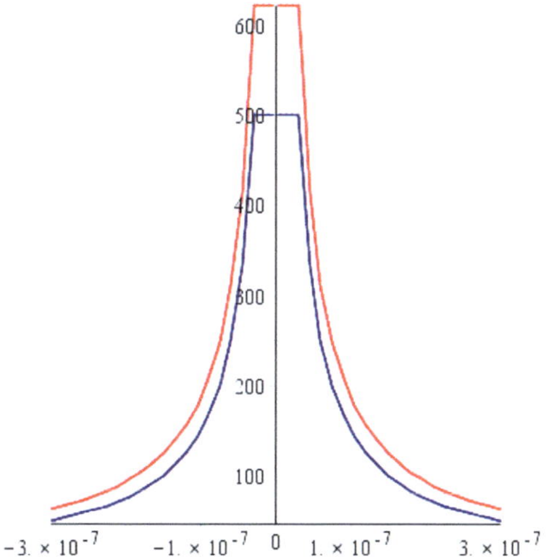

```
>
BCV1:=plot([seq(Q1[b],b=1..10),seq(Q11[b],b=1..10)],style=line,c
olor=blue,labels=[r (m),Temperature
(K)],symbol=diamond,symbolsize=15):
BCV2:=plot([seq(Q2[b],b=1..10),seq(Q21[b],b=1..10)],style=line,c
olor=red,labels=[r
(m),Temperature(K)],symbol=diamond,symbolsize=15):
> display({BCV1,BCV2});
```

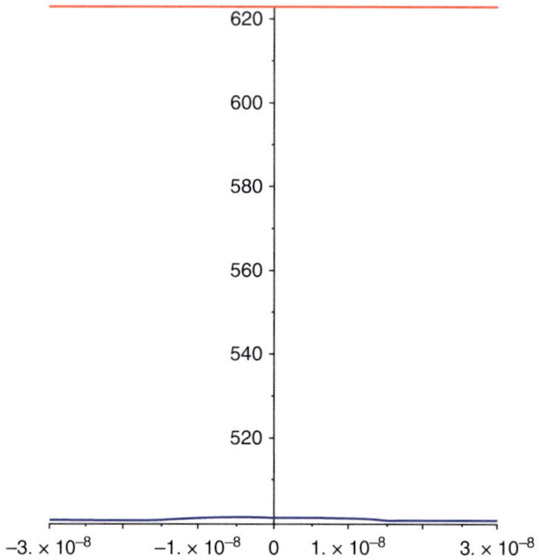


```
> CV1:=plot([seq(Q21[b],b=1..10)],style=line,color=red,labels=[r
(m),Temperature(K)],symbol=diamond,symbolsize=15):
> CV2:=plot([seq(Q2[b],b=1..10)],style=line,color=red,labels=[r
(m),Temperature(K)],symbol=diamond,symbolsize=15):
> display({CV1,CV2});
```

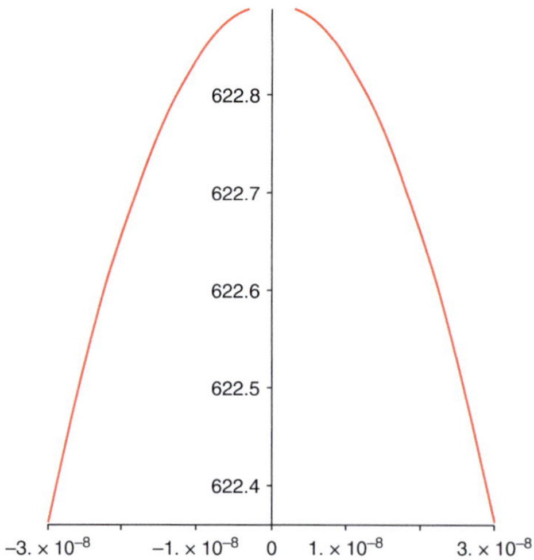

Appendix F: Maple Code for 3D Simulation of Heat Diffusion Around the Nanoparticle

This appendix contains a Maple code designed for 3D simulations of the temperature distributions (third "dimension" is a temperature) around the nanoparticle (Slide 24.18). The comments within the Maple code should explain the variables, procedures, and functions that are used.

Note: This code takes about 10–15 min to complete the integrations. The code comes directly from the Maple file and is therefore formatted to word-wrap according to Maple standard character-per-line limits.

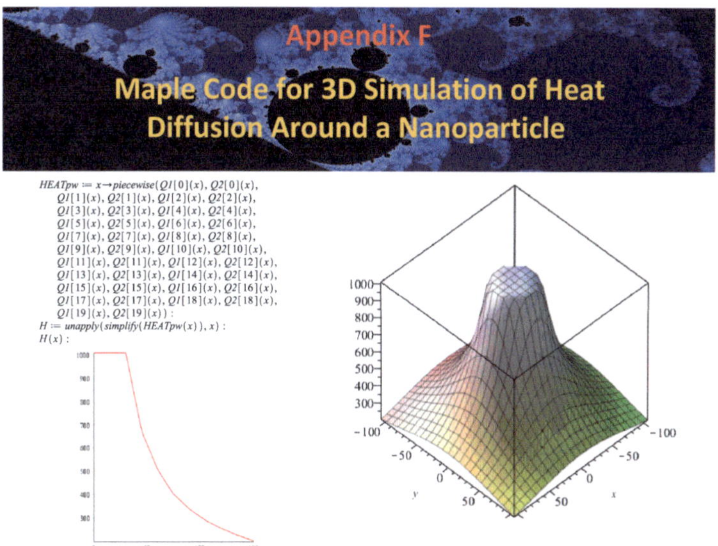

Slide 24.18 Appendix F: Maple Code for 3D simulation of heat diffusion around a nanoparticle

```
> restart;
> E:=5000;
```

$$E := 5000$$

Radius (m)

```
> R:=30*10^(-9);
```

$$R := \frac{3}{100000000}$$

```
> Kabs:=4.376;
```

$$Kabs := 4.376$$

Thermal conductivity (W/m K)

```
> K[1]:=318;
```

$$K_1 := 318$$

```
> K[2]:=0.58;
```

$$K_2 := 0.58$$

Thermal diffusivity (m^2/s)

```
> k[1]:=118.0/1000^2;
```

$$k_1 := 0.0001180000000$$

```
> k[2]:=2.276*(10^(-7));
```

$$k_2 := 2.276000000\ 10^{-7}$$

Constants

```
> a:=R^2/k[1];
```

$$a := 7.627118644\ 10^{-12}$$

```
> b:=K[2]/K[1]*sqrt(k[1]/k[2]);
```

$$b := 0.04152940878$$

```
> c:=1-K[2]/K[1];
```

$$c := 0.9981761006$$

```
> w:=(r/R-1)*sqrt(k[1]/k[2]);
```

$$w := 7.589857467\ 10^{8}\ r - 22.76957240$$

```
> s:=1.01*10^(-6);
```

$$s := 0.000001010000000$$

Heat diffusion time (s)

```
> tau:=s^2/(4*k[2]);
```

$$\tau := 0.000001120496485$$

```
>
```

Rate of heat generation (J/s)

```
> A:=3*Kabs*E/(R*tau);
```

$$A := 1.952705813\ 10^{18}$$

```
>
> T[1]:=R^2*A/K[1]*(K[1]/3/K[2]+(1-r^2/R^2)/6-
2*b*R/r/Pi*(Int(exp(-y^2*t/a)/y^2*(sin(y)-
y*cos(y))*sin(r*y/R)/((c*sin(y)-
y*cos(y))^2+b^2*y^2*sin(y)^2),y=0..100)));
```

$$T_1 := 1010.941336 - 1.023430720\ 10^{15}\ r^2 - \frac{1}{r\,\pi}\left(1.377080116\ 10^{-8}\left(\int_0^{100}\left(e^{-1.311111111\ 10^{11}\ y^2\ t}\ (\sin(y) - y\cos(y))\ \sin\left(\frac{100000000}{3}\ r\ y\right)\right)\Big/\left(y^2\left((0.9981761006\ \sin(y) - y\cos(y))^2 + 0.001724691794\ y^2\ \sin(y)^2\right)\right)\,dy\right)\right)$$

```
> T[2]:=R^3*A/r/K[1]*(K[1]/3/K[2]-2/Pi*Int(exp(-
y^2*t/a)/y^3*(sin(y)-y*cos(y)) *(b*sin(y)*cos(w*y) - (c*sin(y)-
y*cos(y))*sin(w*y))       /(    (c*sin(y)-
y*cos(y))^2+b^2*y^2*sin(y)^2    ) ,y=0..100));
```

$$T_2 := \frac{1}{r}\left(1.657957766\ 10^{-7}\left(182.7586207 - \frac{1}{\pi}\left(2\left(\int_0^{100}\left(e^{-1.311111111\ 10^{11}\ y^2\ t}\ (\sin(y) - y\cos(y))\ \left(0.04152940878\ \sin(y)\ \cos\left(\left(7.589857467\ 10^{8}\ r - 22.76957240\right) y\right) - (0.9981761006\ \sin(y) - y\cos(y))\ \sin\left(\left(7.589857467\ 10^{8}\ r - 22.76957240\right) y\right)\right)\right)\Big/\left(y^3\left((0.9981761006\ \sin(y) - y\cos(y))^2 + 0.001724691794\ y^2\ \sin(y)^2\right)\right)\,dy\right)\right)\right)\right)$$

```
> Nin := 10; Nout := 10;
```

$$Nin := 10$$

$$Nout := 10$$

```
> for n from 1 to Nin do
Q[n]:=[evalf(n*R/Nin),evalf(subs({t=tau,r=n*R/Nin},T[1]))];od:

> for n from 2 to Nout do
V[n]:=[evalf(n*R/2),evalf(subs({t=tau,r=n*R/2},T[2]))];od:
```

$>$
for n **from** 1 **to** 19 **do**
$HEAT[n] := \begin{cases} [Q[n][2], Q[n][1]\cdot 10^9] & n < 11 \\ [V[n-9][2], V[n-9][1]\cdot 10^9] & n \geq 11 \end{cases};$
od:
$HEAT[20] := [HEAT[19][1], \infty]:$

The above QV[n] returns data as [radius (nm), temperature difference (K)] – see below:

$>$ $HEAT[0] := [HEAT[1][1], 0]:$
for n **from** 1 **to** 19 **do** $HEAT[n]$;**od**:

$>$
for n **from** 1 **to** 19 **do**
$Q1[n] := unapply(x \geq HEAT[\text{n}][2]$ **and** $x < HEAT[n+1][2]):$
od:
$Q1[0] := unapply(x < HEAT[1][2]):$

$>$
for k **from** 1 **to** 19 **do**
$a[k-1] := \left(\frac{\text{HEAT[k][1]} - HEAT[\text{k}-1][1]}{\text{HEAT[k][2]} - HEAT[\text{k}-1][2]}\right):$
$b[k-1] := HEAT[\text{k}-1][2]:$
$c[k-1] := HEAT[\text{k}-1][1]:$

$Q2[k-1] := unapply(a[k-1]\cdot(x - b[k-1]) + c[k-1], x);$
od:
$Q2[0] := HEAT[0][1]:$
$Q2[19] := HEAT[19][1]:$

$>$
$HEATpw := x \rightarrow piecewise(Q1[0](x), Q2[0](x),$
$Q1[1](x), Q2[1](x), Q1[2](x), Q2[2](x),$
$Q1[3](x), Q2[3](x), Q1[4](x), Q2[4](x),$
$Q1[5](x), Q2[5](x), Q1[6](x), Q2[6](x),$
$Q1[7](x), Q2[7](x), Q1[8](x), Q2[8](x),$
$Q1[9](x), Q2[9](x), Q1[10](x), Q2[10](x),$
$Q1[11](x), Q2[11](x), Q1[12](x), Q2[12](x),$
$Q1[13](x), Q2[13](x), Q1[14](x), Q2[14](x),$
$Q1[15](x), Q2[15](x), Q1[16](x), Q2[16](x),$
$Q1[17](x), Q2[17](x), Q1[18](x), Q2[18](x),$
$Q1[19](x), Q2[19](x)):$
$H := unapply(simplify(HEATpw(x)), x):$
$H(x):$

> $plot(H(x), x = 0 ..\mathrm{HEAT}[19][2]);$

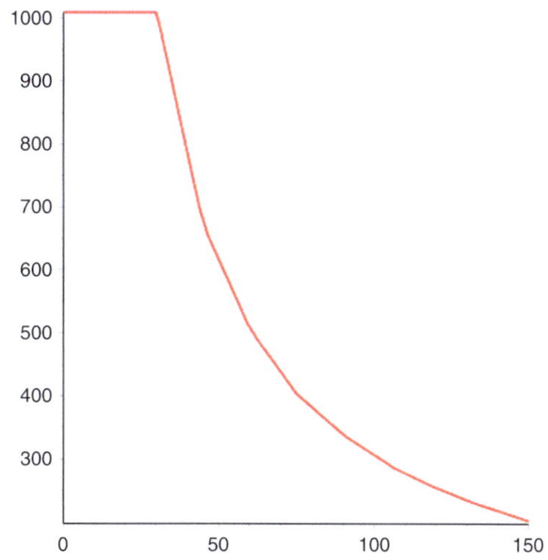

> $plot3d\left(\mathrm{H}\left(\left(x^2+y^2\right)^{\frac{1}{2}}\right), \mathrm{x}=-\left(\frac{\mathrm{sqrt}(2)}{2}\right) \cdot HEAT[19][2] .. \left(\frac{\mathrm{sqrt}(2)}{2}\right) \cdot \mathrm{HEAT}[19][2], y = -\left(\frac{\mathrm{sqrt}(2)}{2}\right) \cdot \mathrm{HEAT}[19][2] .. \left(\frac{\mathrm{sqrt}(2)}{2}\right) \cdot \mathrm{HEAT}[19][2], axes = boxed\right);$

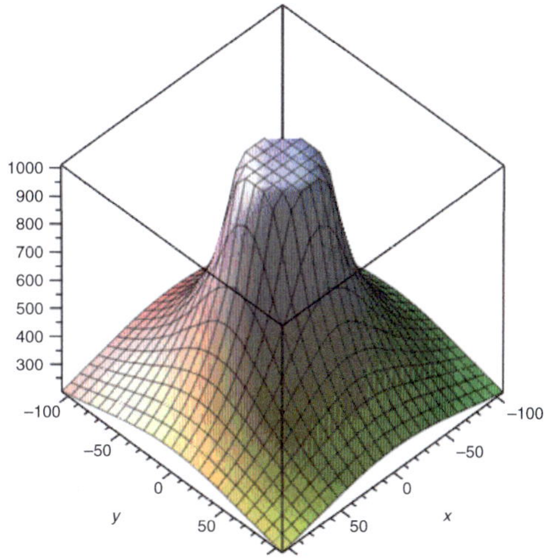

Homework Exercises

Section 8.1: Modeling Heat Diffusion

8.1. What is the quasistationary heating approximation?
8.2. True or false? In the quasistationary heating approximation, we take into account the time required to reach a maximum temperature by the nanoparticle.
8.3. True or false? In the quasistationary heating approximation, the heat diffusion equation gives the space distribution for the maximum temperature of the particle.
8.4. What solutions are found from a heat transfer model?
8.5. How is the heat source calculated in a heat transfer model?
8.6. Describe how radiation is delivered to bone tissue to activate/heat nanoparticles?
8.7. What input data are needed for solving a heat transfer model?
8.8. Why is the nanoparticle heated to the relatively high temperature of ~600 K?
8.9. True or false? The temperature variations within the heated gold nanoparticle are large.
8.10. True or false? The temperatures inside the nanoparticle can be considered homogeneous.
8.11. What factors define the temperature distribution around the nanoparticle?
8.12. How is the thermal damage area of the surrounding tissue around the nanoparticle determined?
8.13. By how many times can the thermal necrosis size in bone tissue exceed the diameter of a heated nanoparticle?
8.14. True or false? The morphological properties of a nanoparticle can affect the thermal field distribution around the particle.
8.15. What energy intensity of the laser pulse can cause thermal necrosis of osteocytes by heated nanoparticles in most cases?
8.16. True or false? The energy intensity of the laser pulse for thermal necrosis of osteocytes by heated nanoparticles is greater than the intensity for laser safety requirements.

Section 8.2: Computer Practicum 5: Simulating the Temperature Fields Within and Outside the Nanoparticle

8.17. What are the two steps used for simulations of thermal fields around the nanoparticle?
8.18. True or false? There is a significant heat loss from the surface of a heated nanoparticle to the surrounding medium.

8.19. Using the Maple code in Appendix E, perform spatial simulations of thermal fields distribution inside and around a silver nanoparticle of optimal radius in surrounding cytoplasm, fat, and cell membrane media by employing thermal data from Tables 6.3 and 8.7. Find the optimal radius of the nanoparticle, optimal wavelength of the radiation, absorption efficiency, and energy density of the radiation required to achieve the maximum temperature of the nanoparticle in each surrounding medium. Write a report on your simulation results similar to the practice example 7. Make conclusions about the spatial heating of the nanoparticle and thermal damage area produced by heated nanoparticle in a surrounding biomedia.

8.20. Using the Maple code in Appendix E, perform spatial simulations of thermal fields distribution inside and around a fullerene nanoparticle of optimal radius in surrounding cytoplasm, fat, and cell membrane media by employing thermal data from Tables 6.3 and 8.7. Find the optimal radius of the nanoparticle, optimal wavelength of the radiation, absorption efficiency, and energy density of the radiation required to achieve the maximum temperature of the nanoparticle in each surrounding medium. Write a report on your simulation results similar to the practice example 7. Make conclusions about the spatial heating of the nanoparticle and thermal damage area produced by heated nanoparticle in a surrounding biomedia.

8.21. Using the Maple code in Appendix E, perform spatial simulations of thermal fields distribution inside and around a polystyrene nanoparticle of optimal radius in surrounding cytoplasm, fat, and cell membrane media by employing thermal data from Tables 6.3 and 8.7. Find the optimal radius of the nanoparticle, optimal wavelength of the radiation, absorption efficiency, and energy density of the radiation required to achieve the maximum temperature of the nanoparticle in each surrounding medium. Write a report on your simulation results similar to the practice example 7. Make conclusions about the spatial heating of the nanoparticle and thermal damage area produced by heated nanoparticle in a surrounding biomedia.

8.22. Using the Maple code in Appendix E, perform spatial simulations of thermal fields distribution inside and around a glass nanoparticle of optimal radius in surrounding cytoplasm, fat, and cell membrane media by employing thermal data from Tables 6.3 and 8.7. Find the optimal radius of the nanoparticle, optimal wavelength of the radiation, absorption efficiency, and energy density of the radiation required to achieve the maximum temperature of the nanoparticle in each surrounding medium. Write a report on your simulation results similar to the practice example 7. Make conclusions about the spatial heating of the nanoparticle and thermal damage area produced by heated nanoparticle in a surrounding biomedia.

8.23. Using the Maple code in Appendix E, perform spatial simulations of thermal fields distribution inside and around a carbon nanoparticle of optimal radius in surrounding cytoplasm, fat, and cell membrane media by employing thermal data from Tables 6.3 and 8.7. Find the optimal radius of the nanoparticle, optimal wavelength of the radiation, absorption efficiency, and energy density of the radiation required to achieve the maximum temperature of the nanoparticle in each surrounding medium. Write a report on your simulation results similar to the practice example 7. Make conclusions about the spatial heating of the nanoparticle and thermal damage area produced by heated nanoparticle in a surrounding biomedia.

8.24. Using the Maple code in Appendix E, perform spatial simulations of thermal fields distribution inside and around an aluminum oxide silver nanoparticle of optimal radius in surrounding cytoplasm, fat, and cell membrane media by employing thermal data from Tables 6.3 and 8.7. Find the optimal radius of the nanoparticle, optimal wavelength of the radiation, absorption efficiency, and energy density of the radiation required to achieve the maximum temperature of the nanoparticle in each surrounding medium. Write a report on your simulation results similar to the practice example 7. Make conclusions about the spatial heating of the nanoparticle and thermal damage area produced by heated nanoparticle in a surrounding biomedia.

8.25. Using the Maple code in Appendix E, perform spatial simulations of thermal fields distribution inside and around a magnesium oxide nanoparticle of optimal radius in surrounding cytoplasm, fat, and cell membrane media by employing thermal data from Tables 6.3 and 8.7. Find the optimal radius of the nanoparticle, optimal wavelength of the radiation, absorption efficiency, and energy density of the radiation required to achieve the maximum temperature of the nanoparticle in each surrounding medium. Write a report on your simulation results similar to the practice example 7. Make conclusions about the spatial heating of the nanoparticle and thermal damage area produced by heated nanoparticle in a surrounding biomedia.

8.26. Using the Maple code in Appendix E, perform spatial simulations of thermal fields distribution inside and around a nickel oxide nanoparticle of optimal radius in surrounding cytoplasm, fat, and cell membrane media by employing thermal data from Tables 6.3 and 8.7. Find the optimal radius of the nanoparticle, optimal wavelength of the radiation, absorption efficiency, and energy density of the radiation required to achieve the maximum temperature of the nanoparticle in each surrounding medium. Write a report on your simulation results similar to the practice example 7. Make conclusions about the spatial heating of the nanoparticle and thermal damage area produced by heated nanoparticle in a surrounding biomedia.

8.27. What is a main purpose of the practice example 8?

8.28. True or false? The cell organelle itself may act as an absorbing particle which, on being heated, could destroy the cancer cells without inserting any nanoparticles into the body.

8.29. True or false? The cancerous nucleus heats up to a lower temperature than the normal nucleus.

8.30. True or false? With the significant increase in temperature of the cancerous nucleus, the cancer cells undergo cell death via thermal ablation of the nucleus.

8.31. What causes a significant temperature difference between heating the healthy and cancerous cell organelles?

8.32. Using the Maple code in Appendix E, perform spatial simulations of the thermal field distributions around normal and cancerous cell mitochondria in cytoplasm by employing thermal data from Tables 6.5 and 8.7. Find the optimal wavelength of the radiation, absorption efficiency of healthy and cancerous organelles, and energy density of the radiation required to achieve the critical temperature for cancerous cell organelle ablation. Write a report on your simulation results similar to the practice example 8. Make conclusions about the possibility of killing cancer cells by heating the cell organelle.

8.33. Using the Maple code in Appendix E, perform spatial simulations of the thermal field distributions around normal and cancerous cell microtubules in cytoplasm by employing thermal data from Tables 6.5 and 8.7. Find the optimal wavelength of the radiation, absorption efficiency of healthy and cancerous organelles, and energy density of the radiation required to achieve the critical temperature for cancerous cell organelle ablation. Write a report on your simulation results similar to the practice example 8. Make conclusions about the possibility of killing cancer cells by heating the cell organelle.

8.34. Using the Maple code in Appendix E, perform spatial simulations of the thermal field distributions around normal and cancerous cell ribosomes in cytoplasm by employing thermal data from Tables 6.5 and 8.7. Find the optimal wavelength of the radiation, absorption efficiency of healthy and cancerous organelles, and energy density of the radiation required to achieve the critical temperature for cancerous cell organelle ablation. Write a report on your simulation results similar to the practice example 8. Make conclusions about the possibility of killing cancer cells by heating the cell organelle.

8.35. Using the Maple code in Appendix E, perform spatial simulations of the thermal field distributions around normal and cancerous cell cytoskeletons in cytoplasm by employing thermal data from Tables 6.5 and 8.7. Find the optimal wavelength of the radiation, absorption efficiency of healthy and cancerous organelles, and energy density of the radiation required to achieve the critical temperature for cancerous cell organelle ablation. Write a report on your simulation results similar to the practice example 8. Make conclusions about the possibility of killing cancer cells by heating the cell organelle.

8.36. Using the Maple code in Appendix E, perform spatial simulations of the thermal field distributions around normal and cancerous cell lysosomes in cytoplasm by employing thermal data from Tables 6.5 and 8.7. Find the optimal wavelength of the radiation, absorption efficiency of healthy and cancerous organelles, and energy density of the radiation required to achieve the critical temperature for cancerous cell organelle ablation. Write a report on your simulation results similar to the practice example 8. Make conclusions about the possibility of killing cancer cells by heating the cell organelle.

References

1. H. Goldenberg, C.J. Tranter, Heat flow in an infinite medium heated by a sphere. Br J Appl Phys **3**, 296–301 (1952)
2. R.R. Letfullin, C.E.W. Rice, T.F. George, Bone cancer therapy by plasmonic nanoparticles. Ther. Deliv. **2**, 1259–1273 (2011)
3. R.R. Letfullin, T.F. George, Plasmonic nanomaterials in nanomedicine, in *Springer Handbook of Nanomaterials*, ed. by R. Vajtai (Springer, Berlin, 2013), pp. 1063–1097

Chapter 9
New Dynamic Modes in Selective Nanophotothermolysis

Contents

Electronic supplementary material: The online version of this chapter (doi:10.1007/978-3-319-43577-0_9) contains supplementary material, which is available to authorized users.

R.R. Letfullin, T.F. George, *Computational Nanomedicine and Nanotechnology*,
DOI 10.1007/978-3-319-43577-0_9

Abstract In this chapter, we perform optical, thermal, kinetic, and dynamic modeling to develop and test new ideas and new dynamic modes in selective nanophotothermolysis, involving nanocluster aggregation in living cells, nanobubble overlapping around heated intracellular nanoparticles/clusters, and the laser thermal explosion mode of single nanoparticles—*nanobombs*—delivered to the cells. This chapter contains material adapted from our publications listed at the end of the chapter. A detailed list of references and reviews on a given topic of this chapter can be found in those original papers.

9.1 Lecture 25: Nanocluster Aggregation Mode (Slide 25.1)

- Introduction
- Self-Assembling of Nanoclusters in Living Systems
- Criteria for Interparticle Distances and Nanocluster Concentration
- Thermal Field Simulations for Nanoclusters
  - *Temperature distribution inside the nanocluster*
  - *Temperature distribution outside the nanocluster*
- Conclusions

9.1.1 Introduction

Progress toward the development of selective nanophotothermolysis technology requires the investigation of new physical concepts and new approaches to the study of pulsed laser radiation interaction with biological systems containing nanostructures [1, 2] (Slide 25.2). The quick temperature rise in the nanoparticle/cell volume irradiated by short laser pulses is accompanied by nonlinear effects in the surrounding nanoparticle biological medium (e.g., protein denaturation, nanobubble formation, shock and sound wave generation, explosive evaporation,

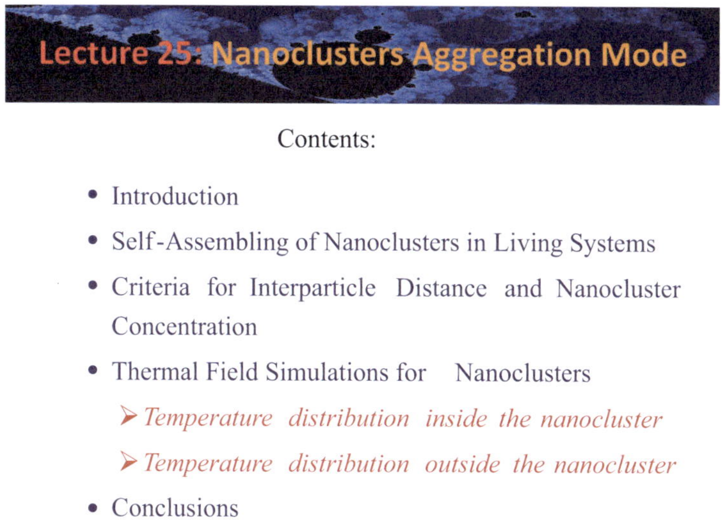

Slide 25.1 Contents

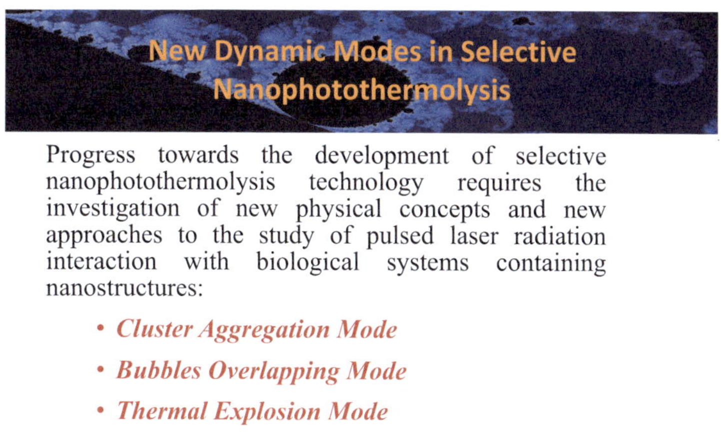

Slide 25.2 New dynamic modes in selective nanophotothermolysis

etc.) leading to irreparable target (e.g., abnormal cell) damage. By optimizing the laser pulse duration, particle size and shape, this technology can provide highly localized damage in a controlled manner, potentially from a few nanometers (e.g., for DNA nanosurgery) to tens of microns (the scale of single cancer cells) without harmful effects on the surrounding healthy tissue [2]. As discussed in this book, among different nanomaterials, gold nanoparticles in different modifications (spherical, rods, shells, etc.) are the most promising candidates as photothermal sensitizers (i.e., as effective converting laser energy into thermal energy), because they are strong absorbers; photostable; nontoxic; easily conjugated to antibodies,

Nanocluster Aggregation Mode

- A ***cluster*** is a self-assembled group of closely-located nanoparticles.
- The ***cluster aggregation mode*** of nanophotothermolysis can be realized when the ***incubation time*** for collecting nanoparticles at a tumor site is long enough.
- The ***incubation time*** depends on the nanoparticle sizes and coatings, which increase the effectiveness of penetration inside the cell. For example, the incubation time is 1-5 hours for 1-50 nm sized particles, and it is 6-12 hours for 50-100 nm sized particles.
- The ***effectiveness of cancer cell killing*** by activated nanoclusters is achieved due to the large damage area produced by a nanocluster in comparison to the single nanoparticle case at the relatively low energy density of the incident laser pulse, which is secure for healthy cells.

Slide 25.3 Nanocluster aggregation mode

proteins, or viruses; and have adjustable optical properties through changing their shape and composition [1].

It has been demonstrated [1] that an accumulation of gold nanoparticles on a cell membrane or inside the cell and, especially, formation of *gold nanoclusters*, lead to more severe cancer cell damage at relatively low radiation intensities. A *cluster* is a self-assembling group of closely located nanoparticles (Slide 25.3). The cluster aggregation mode of nanophotothermolysis can be realized when the incubation time for collecting the nanoparticles at a tumor site is long enough. The incubation time depends on the size of the nanoparticles and nanoparticle coatings, which increases the effectiveness of penetration inside the cell. For example, the incubation time is 1–5 h for 1–50 nm particles and 6–12 h for 50–100 nm particles.

The effectiveness of cancer cell killing by activated nanoclusters is achieved due to large damage area produced by a nanocluster in comparison to the single nanoparticle case at the relatively low energy density of the incident laser pulse, which is secure for the healthy cells (Slide 25.3). In addition, the therapeutic effect of gold nanoclusters is increased because of amplification of optical, thermal, bubble, acoustic, and shock wave generation phenomena around the nanocluster. In this section we study two basic mechanisms of cancer cell killing in clusters aggregation mode—thermal damage of abnormal cells produced by gold nanoclusters, and bubble generation in the cellular medium around the nanoclusters.

9.1.2 *Self-Assembling of Nanoclusters in Living Systems*

After delivery of an appropriate number of nanoparticles to the target cells, which depends on the incubation time, the nanoparticles are then self-assembled into larger nanoclusters (i.e., many adjacent nanoparticles) directly within live systems (e.g., cancer cells, bacteria, vascular legions, etc.). Self-assembled aggregation of gold nanoparticles into the nanoclusters has been observed experimentally [1] (Slide 25.4). For short incubation time less than 30 min, a relatively smooth distribution of single nanoparticles within cell volume is observed. In this short time interval, no nanoclusters are formed, and the average distance between single particles is about 0.5 μm. With increasing incubation time, the number of particles in a cell increases, and the distances between them decrease to 100–300 nm [2]. Further increase in the incubation time leads to aggregation of particles into clusters, as schematically shown in Fig. 9.1, consisting of closely located three to seven particles on the cell membrane or inside the cell or tumor volume (interparticle spacing becomes about 10–30 nm). A long incubation time leads to the creation of clusters of more significant size (up to 200–300 nm as experimentally observed) consisting of 10–20 particles each.

The presence of nanoclusters in cancer cells or tumors may provide further synergistic enhancement of selective nanophotothermolysis due to the following [2] (Slide 25.5): (1) increase of the nanocluster's average local absorption, (2) red shifting of absorption to the near-IR range (window of transparency in most biotissues), (3) achieving better heating efficiency, and (4) decrease in threshold energy for bubble formation around the nanocluster and cell damage.

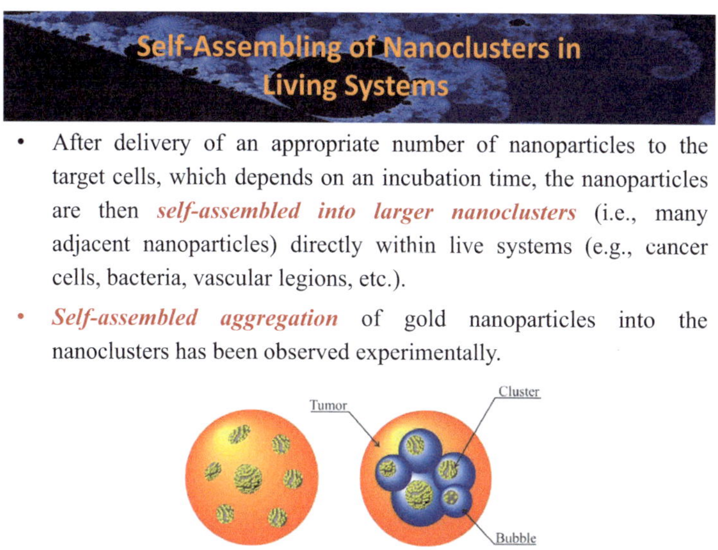

Slide 25.4 Self-assembling of nanoclusters in living systems

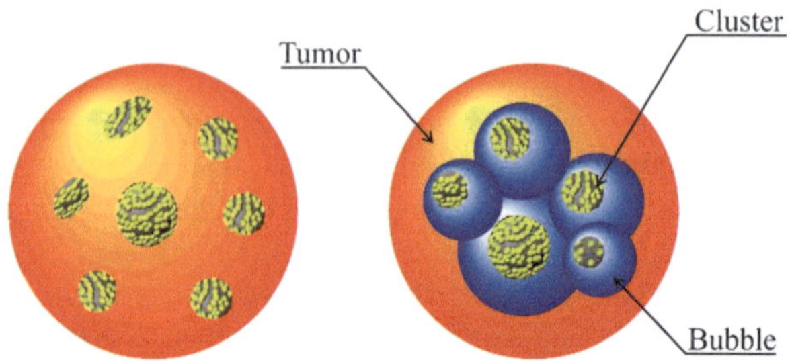

Fig. 9.1 Nanocluster aggregation mode in selective nanophotothermolysis of cancer. Reprinted from [2]. © With permission from World Scientific

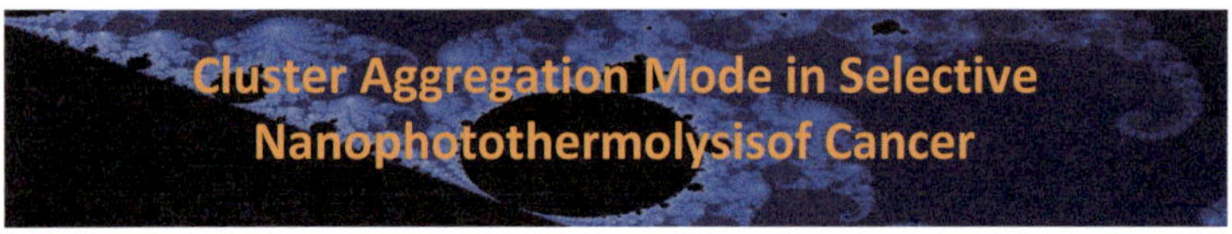

The presence of nanoclusters in cancer cells or tumors may provide further ***synergistic enhancement*** of selective nanophotothermolysis due to the following :

(1) increase of the nanocluster's average ***local absorption*** ;

(2) red shifting of absorption to the near-IR range (window of transparency in most biotissues);

(3) Achieving better heating efficiency ; and

(4) decrease in threshold energy for bubble formation around the nanocluster and cell damage .

Slide 25.5 Cluster aggregation mode in selective nanophotothermolysis of cancer

Laser-induced optical, acoustical, bubble formation, thermal ablation, and other phenomena in abnormal cells targeted with nanoclusters depend greatly on the space distribution of temperature fields inside and outside the clusters [2] (Slide 25.6). Due to close proximity of nanoparticles in a cluster, these fields around each nanoparticle may overlap to create linear as well as nonlinear—especially synergistic—effects. The linear effects are simply proportional to the number of particles in the cluster, whereas synergistic effects can be used as a significant amplification of the above-mentioned phenomena. To provide this synergistic enhancement, it is important to find the space distribution of temperature within a nanocluster and estimate the characteristic spatial dimensions of the accompanied effects compared to the distances between nanoparticles in the cluster.

In this section we develop a theoretical model for the laser space heating of nanoclusters, perform the computer simulations of the space distribution of thermal fields inside/outside the clusters, determine the thermal damage area of abnormal cells produced by gold nanocluster, and then compare it (damage) to the single

Synergetic Effects in the Cluster Aggregation Mode

- There is overlapping of ***laser-induced optical, acoustic, bubble formation, thermal ablation***, and other phenomena due to close proximity of nanoparticles in cluster, which creates linear as well as nonlinear – especially synergistic – effects.
- The ***linear effects*** are simply proportional to the number of particles in the cluster, whereas ***synergistic effects*** can be used as a significant amplification of the abovementioned phenomena.
- To provide this synergistic enhancement, it is important to find the ***space distribution of temperature*** within a nanocluster and estimate the characteristic spatial dimensions of the accompanied effects compared to the distances between nanoparticles in the cluster.

Slide 25.6 Synergetic effects in the cluster aggregation mode

nanoparticle case. The laser-induced overlapping thermal fields in nanoclusters depend on nanoparticle parameters (composition, size, shape), their number and location, laser performances (wavelength, fluence, pulse duration, number of pulses), and properties of the surrounding media [2]. Space overheating effects around nanoparticles may create many therapeutic actions through microbubble formation, schematically shown in Fig. 9.1, accompanied by mechanical stress, and laser-induced hydrodynamic pressures arising from a bubble's rapid expansion and collapse.

9.1.3 Criteria for Interparticle Distance and Nanocluster Concentration

In the cluster case, the optical, thermal, acoustical, and bubble-formation phenomena around each particle initiated by laser radiation can overlap well to create synergistic effects. The distance between nanoparticles in a cluster is crucial for the overlapping phenomena. We introduce the following criteria for interparticle spacing (Slide 25.7): *Effective overlap of thermal fields to provide considerable thermal damage to tumors or cells can be achieved if the temperature between the nanoparticles during heat transfer drops not less than the threshold temperature for most cancer cell killing effects.* For most biological tissues, including bone tissues, thermal ablation occurs at about 70 °C, and the threshold temperature for bubble formation is 150 °C (423 K). Thus, the surface temperature of the nanoparticles in a cluster and heat diffusion from the surface of nanoparticles

Criteria for Interparticle Distance and Nanocluster Concentration

- The ***distance between*** nanoparticles in the cluster is crucial for the overlapping phenomena.
- We introduce the following ***criteria*** for interparticle spacing: ***Effective overlapping of thermal fields to provide a considerable thermal damage to tumors or cells can be achieved if the temperature between the nanoparticles during heat transfer drops not less than the threshold temperature for most cancer cell killing effects.***
- For most biological tissues, including bone tissues, ***thermal ablation*** occurs at ***about 70°C***, and the ***threshold temperature for bubble formation is 150°C*** (423 K).

Slide 25.7 Criteria for interparticle distance and nanocluster concentration

Interparticle Distance and Nanocluster Concentration

- Thus, the ***surface temperature of the nanoparticles in a cluster*** and heat diffusion from the surface of nanoparticles into the cellular volume should provide a cell temperature rise ***in the range of 70-150°C*** to produce unreparable damage to the cancer cells.
- This intracellular temperature range can be reached by varying the energy density of the incident laser pulse for a given ***interparticle spacing***, which sets the lower limit for the ***nanoparticle concentration n in the cluster*** (i.e. the number of particles per unit volume) as

$$n = \left(\frac{1}{l^3}\right) \Rightarrow N = n \cdot volume$$

where N is an absolute number of nanoparticles in the cluster volume.

Slide 25.8 Interparticle distance and nanocluster concentration

into the cellular volume should provide a cell temperature rise in the range of 70–150 °C to produce unreparable damage to the cancer cells. This intracellular temperature range can be reached by varying the energy density of the incident laser pulse for a given interparticle spacing l, which sets the lower limit for the nanoparticle concentration n in the cluster (i.e., the number of particles per unit volume) as (Slide 25.8) [1, 2]

$$n = \left(\frac{1}{l^3}\right) \Rightarrow N = n \cdot \text{volume}, \quad (9.1)$$

where N is the absolute number of nanoparticles in the cluster volume.

9.1.4 Thermal Field Simulations for Nanoclusters

The heat transfer model (8.9)–(8.11) introduced in Sect. 8.1 of Chapter 8 can be used for simulations of the space distribution of the temperature produced by many heat sources in the case of a nanocluster (Slide 25.9). Figure 9.2 illustrates the temperature change (above the ambient) inside a gold nanocluster in the surrounding blood medium irradiated by a single laser pulse of various energy densities. As follows from these calculations, the temperature drop in the cellular medium between the nanoparticles inside the cluster strongly depends on the interparticle spacing. If the nanoparticles are spaced far from each other, then the temperature between particles drops below the critical value of 423 K for bubble formation and descends to a minimum temperature level for cancer cell killing as shown in Fig. 9.2a. Figure 9.2b illustrates the temperature change distribution between the nanoparticles when the particles are located close to each other. In this case, the upper temperature level for thermal ablation of the cancer cell is reached simply at the interparticle distance (i.e., between the centers of neighboring nanoparticles) $l \leq 2d_0$, where $d_0 = 2r_0$ is the diameter of the nanoparticle. As follows from these

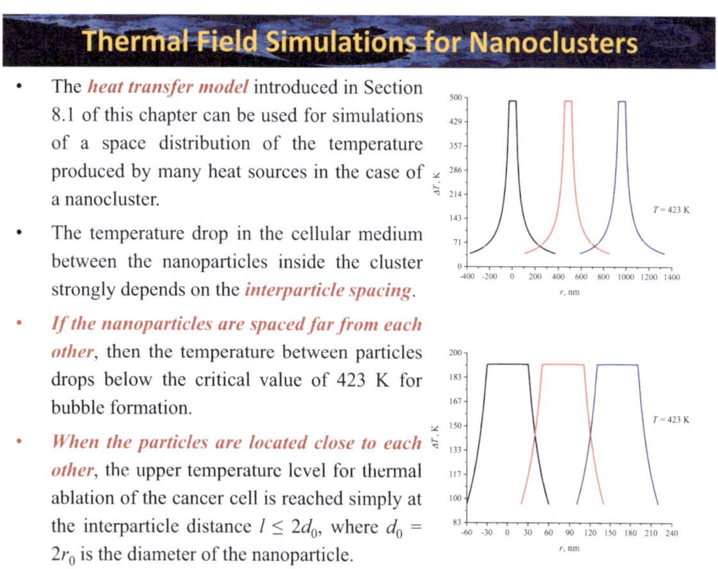

Slide 25.9 Thermal field simulations for nanoclusters

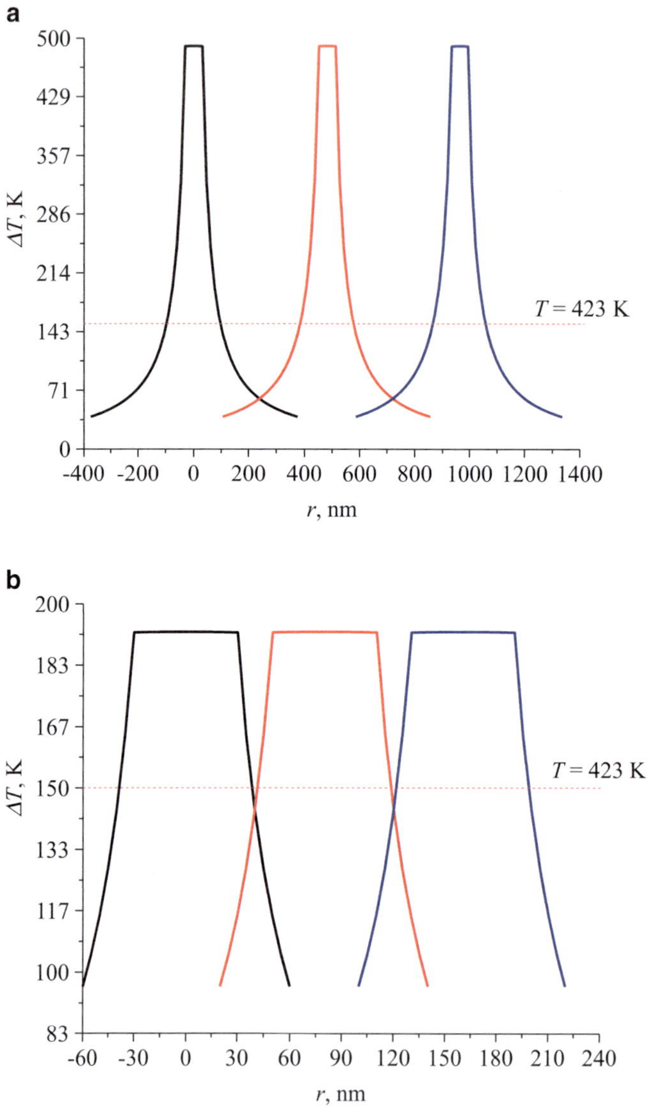

Fig. 9.2 Space distribution of the temperature change (above the ambient) produced inside a cluster consisting of 30-nm gold particles in the surrounding blood medium: (**a**) temperature distribution at far separation and (**b**) at close location of nanoparticles in the cluster. Reprinted from [2]. © With permission from World Scientific

calculations, to provide the threshold temperature range 70–150 °C between the particles everywhere in the cell volume (i.e., thermal damage in the cell), the interparticle distance l should not exceed $l \leq 480$ nm in a cluster made of 30–35 nm gold nanoparticles (Slide 25.10) [1, 2].

The previous chosen criteria can be reached by varying l and the energy density of the incident laser pulse as well. In order to find the optimal distance l where

Interparticle Spacing in a Cluster

- As follows from these calculations, to provide the threshold temperature range 70-150°C between the particles everywhere in the cell volume (i.e., thermal damage in the cell), the ***interparticle distance*** l should not exceed $l \leq 480$ nm in a cluster made of 30-35 nm gold nanoparticles.

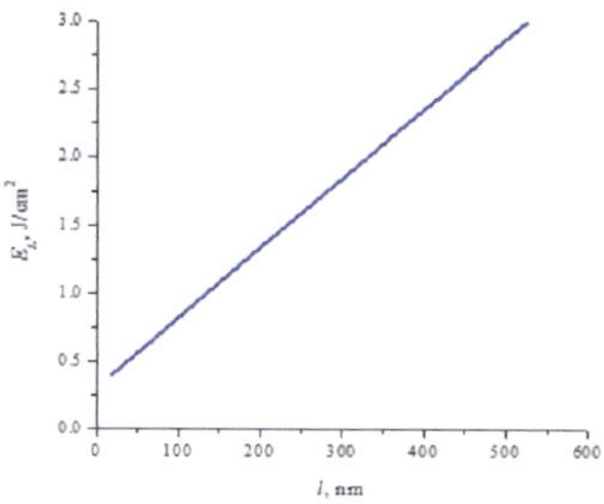

- The above chosen ***criteria*** can be reached by ***varying l and the energy density*** of the incident laser pulse as well.
- Results of these simulations presented in the figure show that an ***optimum l does not exist*** in the framework of the thermal calculations: l increases linearly increases with the energy density of the incident laser pulse.

Slide 25.10 Interparticle spacing in a cluster

the minimal temperature between the nanoparticles is equal to 423 K, we have conducted calculations of the interparticle temperature distribution for different values of the laser energy density. Results of these simulations presented in Fig. 9.3 show that an optimum l does not exist in the framework of the thermal calculations: l increases linearly with the energy density of the incident laser pulse (Slide 25.10) [2].

Let us now estimate the concentration n and absolute number N of nanoparticles which provide a high-temperature distribution inside the cancer cell volume (Slide 25.11). By using Eq. (9.1) we find that the concentration is $n = 9.0 \times 10^{12}$ cm^{-3} of 30-nm gold nanoparticles homogeneously distributed throughout the cell volume for interparticle spacing $l = 480$ nm. The absolute number of nanoparticles in cluster N then can be calculated as $N = nV_{\text{cell}}$, where V_{cell} is the cell volume. In the case of 10-μm diameter cells of the MDA-MB-231 breast cancer line, the number of 30-nm gold particles needed to provide the threshold temperature for cell death everywhere in the cell volume (thermal damage) is $N \approx 4710$ [2]. This is a relatively large number of particles per cell, requiring over 6 h of incubation time to collect them in the cell or cell membrane. We then calculate the space distribution of the temperature fields inside and outside the cluster formed by 30-nm particles conjugated with antibodies.

9.1.4.1 Temperature Distribution Inside the Nanocluster

In our experiments [1], the distance between the centers of neighboring particles in a cluster is about 80 nm, i.e., 20 nm between the surfaces of the 30-nm particles.

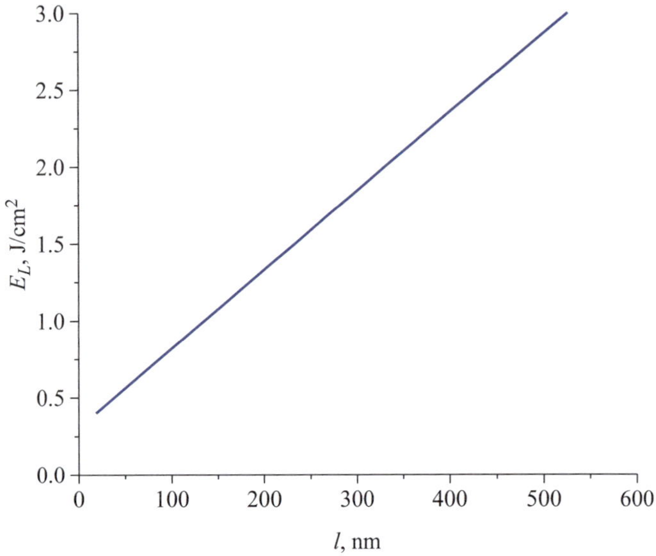

Fig. 9.3 Interparticle distance l as a function of the laser energy density. Reprinted from [2]. © With permission from World Scientific

- Let now estimate the ***concentration n*** and ***absolute number N of nanoparticles*** which provide a high-temperature distribution inside the cancer cell volume.
- By using the equation $n = \left(\frac{1}{l^3}\right) \Rightarrow N = n \cdot volume,$

  we can find that the concentration is ***$n = 9.0\times10^{12}$ cm^{-3} of 30-nm gold nanoparticles*** homogeneously distributed throughout the cell volume for ***interparticle spacing l = 480 nm***.
- The ***absolute number N of nanoparticles*** in the cluster then can be calculated as $N = nV_{cell}$, where V_{cell} is the cell volume.
- In the case of the ***10-µm diameter cells*** of the MDA-MB-231 breast cancer line, the number of 30-nm gold particles needed to provide the threshold temperature for cell death everywhere in the cell volume (thermal damage) is ***$N \approx 4{,}710$***.
- This is a relatively large number of particles per cell, requiring ***over 6 hours of incubation time*** to collect them in the cell or cell membrane.

Slide 25.11 Nanoparticle concentration in a cluster

The interparticle distance in the cluster cannot be less than 20 nm, since the particles are covered by antibodies with at least 10-nm thickness. To find the lowest energy level of an incident single laser pulse, we have conducted calculations of the space distribution of the temperature inside the cluster consisting of 30-nm gold

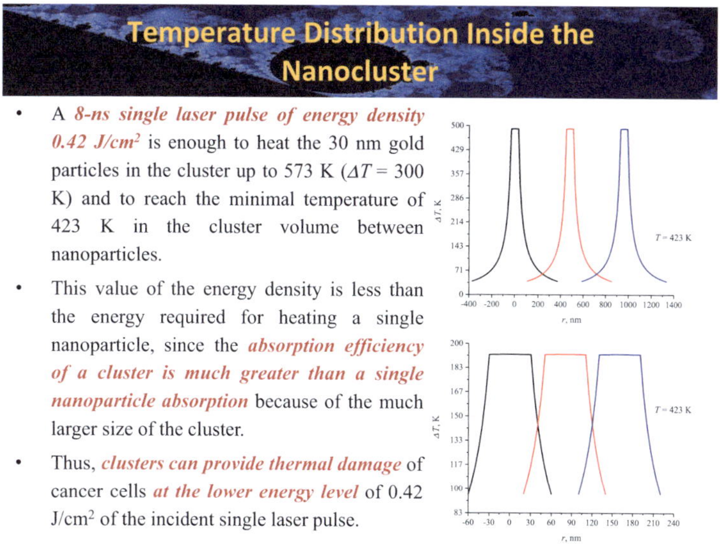

Slide 25.12 Temperature distribution inside the nanocluster

particles for different laser fluxes. The results of these simulations for the lowest energy, which provides a threshold temperature not less than 423 K inside the cluster (between nanoparticles), are illustrated in Fig. 9.2b. As follows from these calculations [2], a 8-ns single laser pulse of energy density 0.42 J/cm^2 is enough to heat the 30-nm gold particles in the cluster up to 573 K ($\Delta T = 300$ K) and to reach the minimal temperature of 423 K between nanoparticles in the cluster volume. This value of the energy density is less than the energy required for heating a single nanoparticle as calculated in Sect. 6.3, since the absorption efficiency of a cluster is much greater than a single nanoparticle absorption because of the much larger size of the cluster. Thus, clusters aggregated from small nanoparticles can provide thermal damage of cancer cells at the lower energy level of 0.42 J/cm^2 of the incident single laser pulse (Slide 25.12).

9.1.4.2 Temperature Distribution Outside the Nanocluster

We assume that the cluster is aggregated from 30-nm gold particles with 20-nm interparticle distances homogeneously compacted in the sphere. The cluster is located along the r-axis and surrounded by blood. It is irradiated by a single laser pulse of energy density 1 J/cm^2 and pulse duration of 8 ns at FWHM. Results of the simulations for the space distribution of the temperature field inside and outside the cluster in the surrounding blood medium are shown in Fig. 9.4a. As follows from these calculations [2], the temperature inside the cluster oscillates in the range of 500–560 K between the nanoparticles. So, the cluster heated by a single laser pulse of energy density 1 J/cm^2 has a high inside temperature which exceeds the threshold

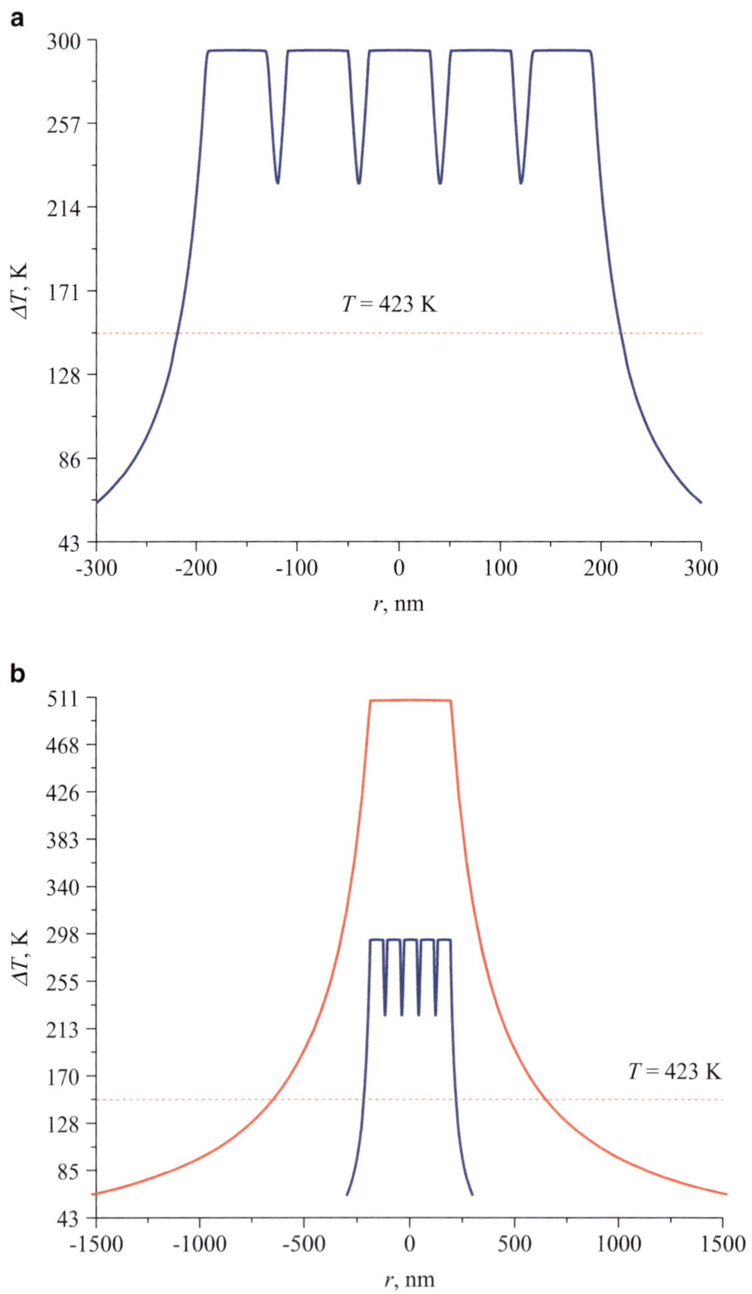

Fig. 9.4 Spatial distribution of the temperature change (above the ambient) (**a**) inside and outside a discrete cluster consisting of 30-nm gold particles and (**b**) outside the continuous "cluster" shown by the *red curve*, both surrounded by a blood medium and irradiated by a single laser pulse of energy density 1 J/cm^2

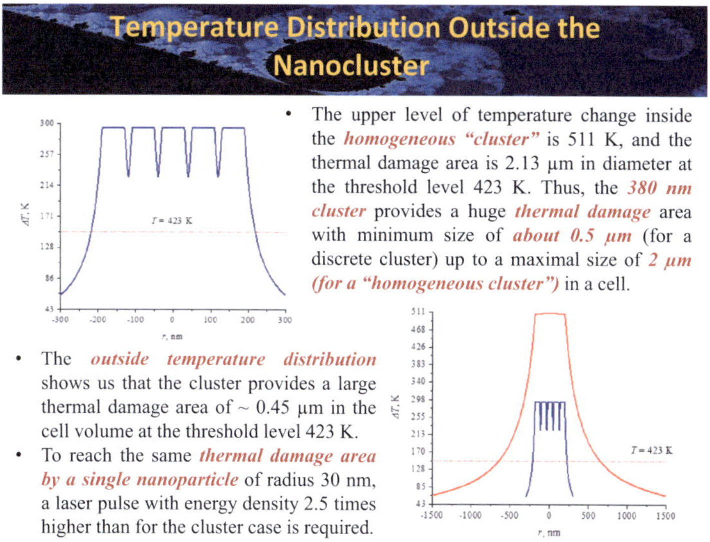

Slide 25.13 Temperature distribution outside the nanocluster

temperature for cell death. The outside temperature distribution shows us that the cluster provides a large thermal damage area of ~0.45 μm in the cell volume at the threshold level 423 K. To reach the same thermal damage area by a single nanoparticle of radius 30 nm, a laser pulse with energy density 2.5 times higher than for the cluster case is required (Slide 25.13).

We should note that since the cluster is treated here as a discrete medium (each particle considered separately), the earlier results obtained for the cluster temperature and thermal damage area have their lowest values [5]. The upper limit for these values can be found for a homogeneous (continuous) distribution of the medium in the cluster. To keep our notations here the same, we will label such a continuously distributed medium as a "homogeneous cluster." There are two reasons to support the homogeneous distribution in the cluster.

The first reason is the optical property of the cluster. The size of nanoparticles in a cluster is much smaller than the wavelength of laser radiation, $r_0 \ll \lambda$. As shown in Chap. 5 of this book, according to the Lorentz–Mie diffraction theory, the cross-section of absorption exceeds by four times the geometric cross-section of the gold nanoparticle, $\sigma_{abs} > \sigma_0$. Hence, the cross-sections of absorption of each nanoparticle are overlapping in the cluster. Because of this, the light does not distinguish the separate nanoparticles inside cluster, and the laser energy is distributed homogeneously among the particles [5].

The second reason is the heat transfer mechanism in the cluster. Since the temperature variation inside the nanocluster occurs over a relatively small temperature range in comparison to the huge temperature drop outside the cluster, we can assume that the internal cluster temperature is homogeneous and constant. This means that from a thermal calculation point of view, the cluster can be treated as a homogeneous (continuous) gold particle with the same size as a cluster [5].

Let us evaluate the damage area produced by continuous gold "clusters" in the blood medium (Slide 25.13). The temperature distribution for gold particles of radius 190 nm (cluster diameter 380 nm) in blood irradiated by a single laser pulse of the same energy density 1 J/cm^2 is illustrated in Fig. 9.4b (red curve). As follows from these simulations [2], the upper level of temperature change inside the homogeneous cluster is 511 K, and the thermal damage area is 2.13 μm in diameter at the threshold temperature level of 423 K. Thus, the 380-nm cluster provides a huge thermal damage area with minimum size of about 0.5 μm (for a discrete cluster) up to a maximal size of 2 μm (for a "homogeneous cluster") in a cell volume irradiated by a single laser pulse of energy density 1 J/cm^2.

9.1.4.3 Conclusions

The aggregation of nanoparticles into a cluster, i.e., cluster mode of nanophotothermolysis, raises the efficiency of cancer cell treatments due to a large damage area at the relatively low energy density of the incident single laser pulse (Slides 25.14 and 25.15). This conclusion is supported by the following results [5]:

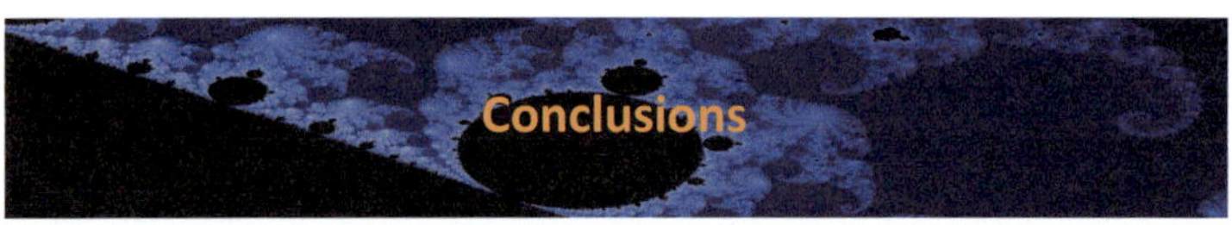

The aggregation of nanoparticles into a cluster, i.e., ***cluster mode of nanophotothermolysis***, raises the efficiency of cancer cell treatments due to a large damage area at the relatively low energy density of the incident single laser pulse:

- Clusters can produce ***thermal damage of cancer cells at a lower energy density*** level 0.42 J/cm^2 of the incident single laser pulse ***in comparison to the single nanoparticle case*** because of the higher absorption efficiency of a cluster.
- A clusters provides a larger ***thermal damage area*** with minimal size of ***0.5 μm (for the discrete cluster case) up to maximal size of 2 μm (for continuous "cluster") in a cell volume*** irradiated by a single laser pulse of energy density 1 J/cm^2 and duration 8 ns.

Slide 25.14 Conclusions

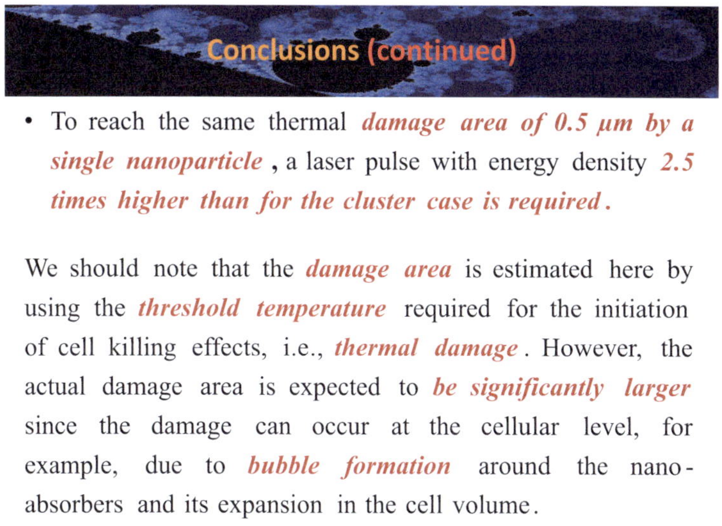

Slide 25.15 Conclusions (continued)

- Clusters can produce thermal damage to cancer cells at a lower energy density level 0.42 J/cm^2 of the incident single laser pulse in comparison to the single nanoparticle case because of the higher absorption efficiency of a cluster.
- A cluster provides a larger thermal damage area with minimal size of 0.5 μm (for the discrete cluster case) up to maximal size of 2 μm (for continuous "cluster") in a cell volume irradiated by a single laser pulse of energy density 1 J/cm^2 and duration 8 ns.
- To reach the same thermal damage area of 0.5 μm by a single nanoparticle, a laser pulse with energy density 2.5 times higher than for the cluster case is required.

We should note that the damage area is estimated here by using the threshold temperature required for the initiation of cell killing effects, i.e., thermal damage [5]. However, thc actual damage area is expected to be significantly larger since the damage can occur at the cellular level, for example, due to bubble formation around the nano-absorbers and its expansion in the cell volume. Section 9.3 is devoted to the investigation of the nanobubble dynamics around the nanoparticles/ nanoclusters in biological media.

9.2 Computer Practicum 6 (Lecture 26): Simulating Temperature Fields Around a Nanocluster (Slide 26.1)

- Computer Code
- Practice Problem 9: RF Heating of Nanoclusters for Cancer Therapy
  - *Abstract*
  - *Introduction*
  - *Theoretical model*
  - *RF heating of a single gold nanoparticle*
  - *RF heating of a single carbon nanoparticle*
  - *RF heating of a gold nanocluster*
  - *RF heating of a carbon nanocluster*
  - *Conclusions*
- Practice Example 10: Heating of Silver Nanoclusters by Laser Radiation
  - *Abstract*
  - *Method*
  - *Discussion and results*
  - *Conclusions*

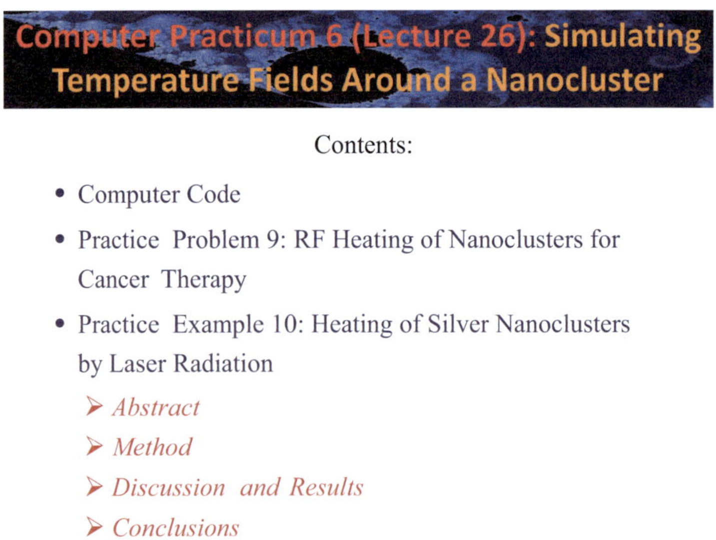

Slide 26.1 Contents

9.2.1 Computer Code

We have previously developed a model (see Eqs. (8.9)–(8.11)) that gives us the temperature distribution around a spherical particle in a medium (Sect. 8.1) (Slide 26.2). In Sect. 9.1 of this chapter, we showed how this model can be used for simulating the spatial distributions of temperature fields around a nanocluster in a surrounding biological medium. To determine the spatial behavior of the temperature fields produced by many heat sources, i.e., absorbing nanoparticles in the cluster volume, we use here the same Maple code which we designed for single nanoparticle heating listed in Appendix E, but modified to the nanocluster heating case. A main modification to the Maple code is to arrange the overlap of the temperature fields produced by each nanoparticle in the cluster, which is accomplished via the repetition statement (*for. . .from. . .to. . .do*) and separation distance as shown in Fig. 9.5 and Slide 26.3. The separation distance is the distance between the surfaces of neighboring particles in the cluster. Unlike the separation distance, the interparticle spacing l is the distance between the centers of neighboring particles in the cluster.

Next, we perform integrations and plot the results of simulations of the temperature distributions inside and outside the cluster consisting of five gold nanoparticles in the surrounding water media shown in Fig. 9.6 and Slide 26.4.

The modified Maple code adjusted for simulations of heating the nanoclusters is given in Appendix G. The Practice Examples 9 and 10 present the results of these simulations in detail for clusters made of gold, silver, and carbon nanoparticles heated by RF and optical radiation in various surrounding media.

Computer Code

- We have previously developed a ***model that gives us the temperature distribution around a spherical particle*** in a medium.
- In Section 9.1 of this chapter, we showed how ***this model*** can be used for simulating the spatial distributions of temperature fields ***around nanocluster*** in a biological surrounding medium.
- ***To determine the spatial behavior of temperature fields*** produced by many heat sources, i.e ., absorbing nanoparticles ***in the cluster volume*** , we use here the ***same Maple code*** which we designed for single nanoparticle heating listed ***in Appendix E*** , but modified to the nanocluster heating case.

Slide 26.2 Computer Code

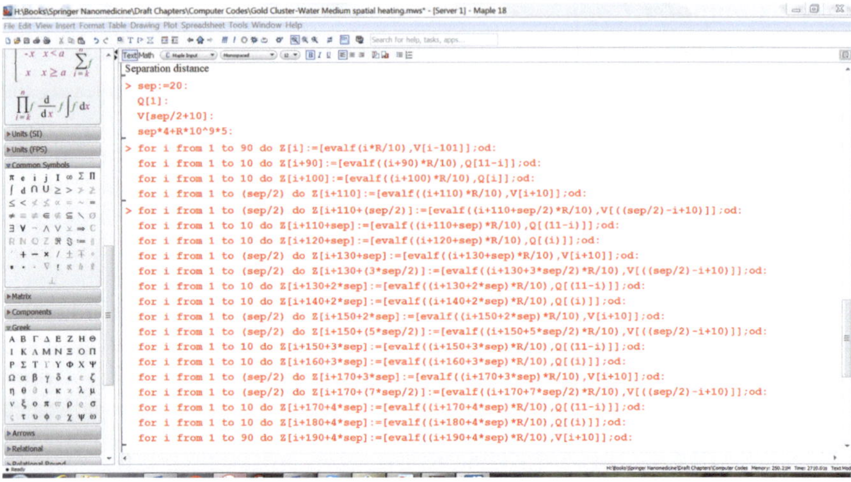

Fig. 9.5 Arranging the overlap of temperature fields produced by each nanoparticle in a cluster by using the repetition statement (*for...from...to...do*) and separation distance

Computer Code (continued)

- A ***main modification to the Maple code*** is to arrange the ***overlap of the temperature fields*** produced by each nanoparticle in cluster, which is accomplished ***via the repetition statement*** (***for..from..to..do***) and ***separation distance*** as shown in the figure below.
- ***The separation distance*** is the distance between the surfaces of neighboring particles in the cluster. Unlike the separation distance, the ***interparticle spacing l*** is the distance between the centers of neighboring particles in the cluster.

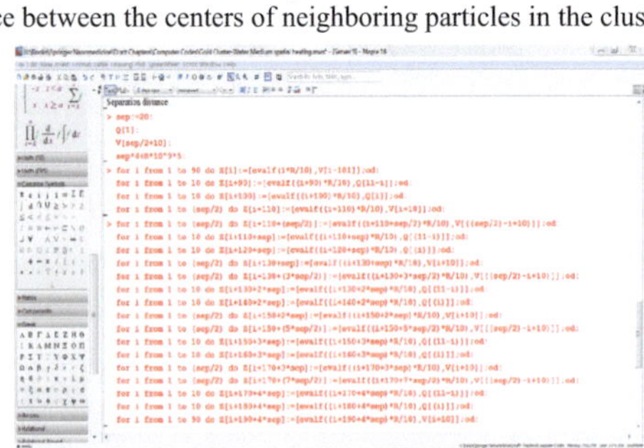

Slide 26.3 Computer code (continued)

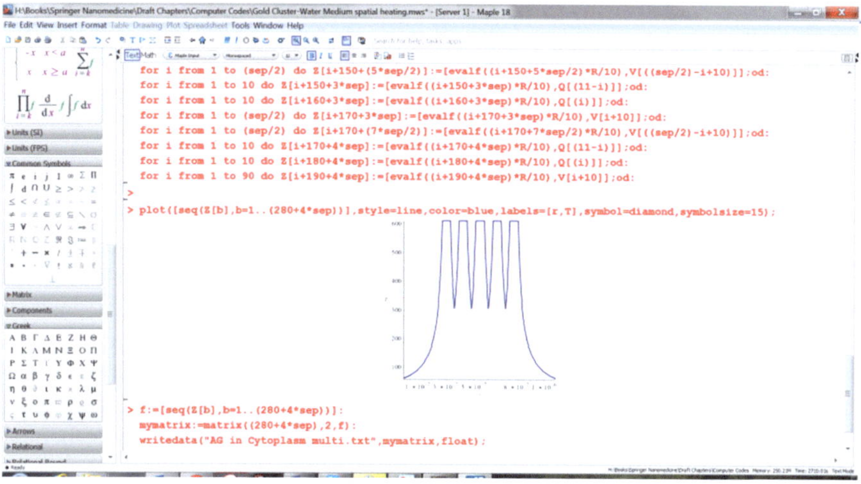

Fig. 9.6 Arranging the output results of simulations for the spatial distributions of the temperatures in and outside the nanocluster

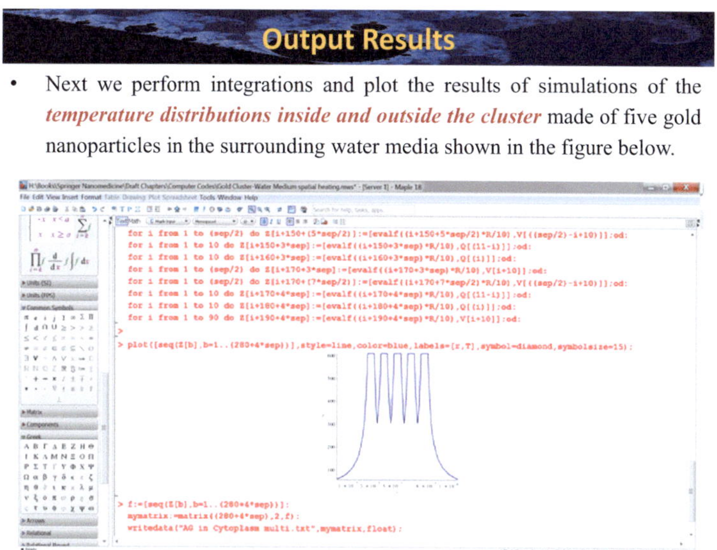

Slide 26.4 Output results

9.2.2 Practice Example 9: RF Heating of Nanoclusters for Cancer Therapy

9.2.2.1 Abstract

Radiofrequency (RF) waves have excellent ability to penetrate into the human body, providing a great opportunity to activate/heat nanoparticles inside the body for diagnosis and treatment purposes. However, the nanoparticles have low absorption efficiency in the RF range (Slide 26.5). To resolve this weakness in the RF activation of nanoparticles and dramatically increase the absorption of contrast agents in a tumor, we suggest aggregating the nanoclusters inside or on the surface of the cancer cells. We simulate the space distribution of temperature changes inside and outside metal and dielectric nanoparticles/nanoclusters, determine the number of nanoparticles needed to form a cluster, and estimate the thermal damage area produced in a surrounding medium by nanoparticles/nanoclusters heated in the RF field. This section contains material reproduced from our publication [7].

9.2.2.2 Introduction

As we stated earlier, the human body has negligible absorption of RF waves, which gives a great opportunity to heat nanoparticles injected in the body with RF waves

- ***Radio frequency*** (RF) waves have excellent ability to penetrate into the human body, providing a great opportunity to activate/heat nanoparticles inside the body for diagnosis and treatment purposes.
- However, the ***nanoparticles have low absorption efficiency in the RF range***.
- To dramatically increase the absorption of contrast agents in a tumor, ***we suggest aggregating the nanoclusters*** inside or on the surface of the cancer cells.
- We simulate the ***space distribution of temperature*** changes inside and outside metal and dielectric nanoparticles/nanoclusters, determine the ***number of nanoparticles*** needed to form a cluster, and estimate the ***thermal damage area*** produced in the surrounding medium by nanopraticles/nanoclusters heated in the RF field.

Slide 26.5 Practice Example 9: RF heating of nanoclusters for cancer therapy

Table 9.1 Input parameters used in the heat transfer model

Description	Values
Heat conductivity/thermal diffusivity of carbon	534 W/mK/0.000135 m^2/s
Heat conductivity/thermal diffusivity of gold	318 W/mK/0.000118 m^2/s
Heat conductivity/thermal diffusivity of deionized water (surrounding medium)	0.938 W/mK/1.43×10^{-7} m^2/s
Frequency of radiation	13.56 MHz
Q_{abs} of carbon for $r_0 = 20$–200 nm [3]	9.91×10^{-16}–1.09×10^{-13}
Q_{abs} of gold for $r_0 = 5$–100 nm [3]	1.15×10^{-9}–14.5×10^{-8}

at any place inside the body. However, the heating of nanoparticles in the RF spectral range is characterized by a low power load of RF waves and a low absorption efficiency of nanoparticles in the RF range. In Sect. 4.3 we have applied a reverse phenomenological approach to determine the absorption efficiency for different types of nanoparticles in the RF range of the spectrum presented in Table 9.1.

In order to improve the RF activation of nanoparticles and dramatically increase absorption of contrast agents in a tumor, we suggest aggregating the nanoclusters inside or on the surface of the cancer cells [7]. Nanoclusters could have much higher absorption efficiency than single nanoparticles, making the heating of nanoparticles by RF radiation feasible. We simulate the space distribution of the temperature around the single nanoparticles and nanoclusters, and estimate the thermal damage area produced in the surrounding medium by nanoparticles/clusters heated by RF radiation.

9.2.2.3 Theoretical Model

The temperatures reached in tissues heated by nanoparticles depend on a number of factors. The evolution of the thermal fields in and around the particle relies on an interplay of heat generation and thermal diffusion governed by material properties. As shown in Chap. 8, the changing temperature of a particle and its surroundings can be described by a set of continuity equations which track heat generation and heat diffusion across a boundary. The heat transfer model (Eqs. (8.9)–(8.11)) can be reduced to two equations (Sect. 8.1) (Slide 26.6): one representing the temperature change inside the nanoparticle, Eq. (9.2), and another representing temperature change outside the nanoparticle in surrounding medium, Eq. (9.3). This model uses several assumptions [7]:

- Both the surrounding medium and the nanoparticle are homogeneous.
- The surrounding medium is infinite.
- Heat in both the sphere and the surrounding medium is produced at a constant rate.
- Thermally induced changes in the refractive indices are neglected.

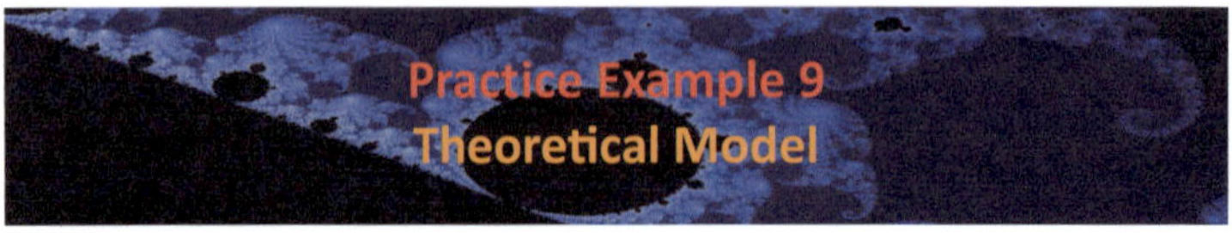

- As shown earlier, the ***heat transfer model*** can be reduced to ***two equations***: one representing the temperature change ***inside the nanoparticle***, and another representing temperature change ***outside the nanoparticle*** in surrounding medium.
- These equations are coupled by ***two boundary conditions*** requiring the temperature function to be continuous and smooth across the boundary.

$$T_1 = \frac{r_0^2 A}{\mu_1}\left\{\frac{1}{3}\frac{\mu_1}{\mu_2} + \frac{1}{6}\left(1 - \frac{r^2}{r_0^2}\right) - \frac{2r_0 b}{\pi r}\int_0^\infty \frac{\exp(-y^3 t/\gamma_1)}{y^2}\frac{(\sin y - y\cos y)\sin(ry/r_0)}{[(c\sin y - y\cos y)^2 + b^2y^2\sin^2 y]}dy\right\}$$

$$T_2 = \frac{r_0^3 A}{\mu_1}\left\{\frac{1}{3}\frac{\mu_1}{\mu_2} \pm \frac{2}{\pi}\int_0^\infty \frac{\exp(-y^3 t/\gamma_1)}{y^3}\frac{(\sin y - y\cos y)[by\sin y\cos\sigma y - \cos\sigma y - (c\sin y - y\cos y)\sin\sigma y]}{[(c\sin y - y\cos y)^2 + b^2y^2\sin^2 y]}dy\right\}$$

Slide 26.6 Practice Example 9: Theoretical model

$$T_1 = \frac{r_0^2 A}{\mu_1}\left\{\frac{1}{3}\frac{\mu_1}{\mu_2} + \frac{1}{6}\left(1 - \frac{r^2}{r_0^2}\right) - \frac{2r_0 b}{\pi r}\int_0^\infty \frac{\exp(-y^3 t/\gamma_1)}{y^2}\frac{(\sin y - y\cos y)\sin(ry/r_0)}{\left[(c\sin y - y\cos y)^2 + b^2y^2\sin^2 y\right]}dy\right\}, \quad (9.2)$$

$$T_2 = \frac{r_0^3 A}{\mu_1}\left\{\frac{1}{3}\frac{\mu_1}{\mu_2} \pm \frac{2}{\pi}\int_0^\infty \frac{\exp(-y^3 t/\gamma_1)}{y^3} \times \frac{(\sin y - y\cos y)[by\sin y\cos\sigma y - \cos\sigma y - (c\sin y - y\cos y)\sin\sigma y]}{\left[(c\sin y - y\cos y)^2 + b^2y^2\sin^2 y\right]}dy\right\}. \quad (9.3)$$

These equations are coupled by two boundary conditions requiring the temperature function to be continuous and smooth across the boundary. The input data used in the model is shown in Table 9.1 and Slide 26.7. The absorption efficiencies, Q_{abs}, for dielectric and metal nanoparticles in the RF range of spectrum are computed in Sect. 4.3 of Chap. 4.

9.2.2.4 RF Heating of a Single Gold Nanoparticle

We have performed simulations of the space distributions of temperature inside and outside gold nanoparticles heated by RF radiation by using a Maple code from Appendix E (Slide 26.8). Figure 9.7 contains the heat distribution curves, while

Practice Example 9
Input Data

- The ***input data*** used in the model is shown in the table below.
- The ***absorption efficiencies***, Q_{abs}, for dielectric and metal nanoparticles in the RF range of spectrum are computed in Section 4.3 of Chapter 4.

Description	Values
Heat conductivity / Thermal diffusivity of carbon	534 W/mK / 0.000135 m^2/s
Heat conductivity / Thermal diffusivity of gold	318 W/mK / 0.000118 m^2/s
Heat conductivity / Thermal diffusivity of deionized water (surrounding medium)	0.938 W/mK / 1.43 × 10^{-7} m^2/s
Frequency of radiation	13.56 MHz
Q_{abs} of carbon for r_0 = 20-200 nm [9]	9.91 × 10^{-16} - 1.09 × 10^{-13}
Q_{abs} of gold for r_0 = 5-100 nm [9]	1.15 × 10^{-9} - 14.5 × 10^{-8}

Slide 26.7 Practice Example 9: Input data

Practice Example 9
RF Heating of a Single Gold Nanoparticle

- As follow from the figure below, there is ***significant heat loss from the surface of a nanoparticle to the surrounding medium***, suggesting that even lower energy pulses are enough to achieve large thermal damage to the biological surroundings.

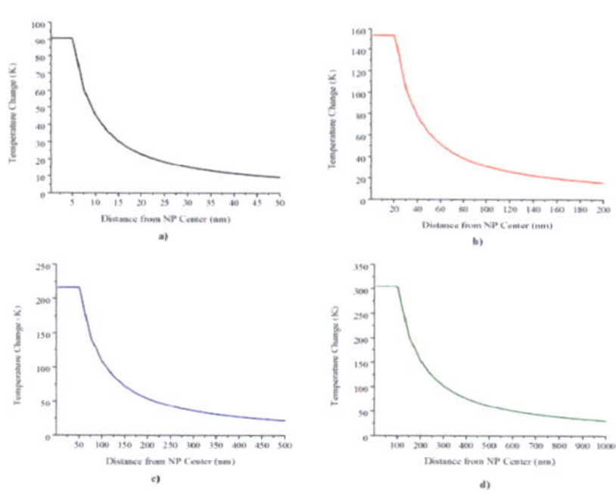

Slide 26.8 Practice Example 9: RF heating of a single gold nanoparticle

Table 9.2 provides the numerical results. As shown in Fig. 9.7, there is significant heat loss from the surface of a nanoparticle to the surrounding medium, suggesting that even lower energy pulses are enough to achieve large thermal damage to the biological surroundings.

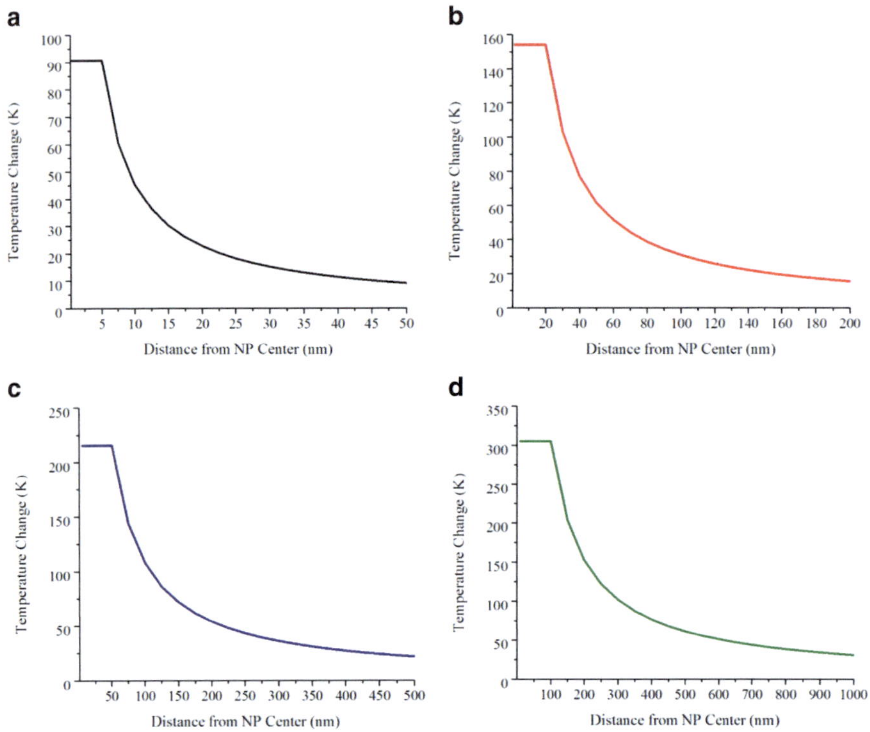

Fig. 9.7 Thermal fields inside and outside gold nanoparticle. The horizontal axis is the distance away from the center of the nanoparticle measured in nm, while the vertical axis represents the difference between the initial and final in K. The simulations are performed in a deionized surrounding water medium for different radii of gold nanoparticles: (**a**) 5 nm, (**b**) 20 nm, (**c**) 50 nm, and (**d**) 100 nm

Table 9.2 Results of simulations for heating gold nanoparticles

Radius (nm)	ΔT (K)	Q_{abs}	Power (W)	Heat duration (s)	Sample volume (mL)	Damage radius (nm)	Minimum concentration (NPs/cell)
5	90.84	1.15×10^{-9}	1000	1800	0.1	10	525
20	154.2	6.10×10^{-9}	600	1200	0.5	70	75
50	215.7	27.3×10^{-9}	600	150	0.5	225	23
100	305.4	14.5×10^{-8}	200	60	0.5	650	8

As shown in Eq. (9.2), the heat generated depends on the power of the RF field, the duration of the heating process, and the volume of the colloidal solution. In Table 9.2 we provide possible values for these parameters in order to achieve the ablation temperature of the surrounding medium around the nanoparticles [7]. As can be seen from the graphs and Table 9.2, size plays a crucial role in the heating process. If the temperature of the surrounding medium exceeds some thermal

Practice Example 9

RF Heating of a Single Gold Nanoparticle (continued)

- The ***damage radius*** around the nanoparticle is determined by computing the distance where curve's temperature change reaches a ***45 K difference above ambient***, since raising the cell's temperature by 45 K will cause ablation.
- ***The most efficient sizes*** of gold nanoparticles would be ***20 and 50 nm***.
- They can produce ***large damage areas of about 140 nm and 450 nm***, respectively, with a reasonable energy density of RF radiation.

Radius (nm)	ΔT (K)	Q_{abs}	Power (W)	Heat duration (s)	Sample volume (mL)	Damage radius (nm)	Minimum concentration (NPs/cell)
5	90.84	1.15×10^{-9}	1000	1800	0.1	10	525
20	154.2	6.10×10^{-9}	600	1200	0.5	70	75
50	215.7	27.3×10^{-9}	600	150	0.5	225	23
100	305.4	14.5×10^{-8}	200	60	0.5	650	8

Slide 26.9 Practice Example 9: RF heating of a single gold nanoparticle (continued)

thresholds, secondary phenomena of cell killing effect may develop, such as photothermal ablation, sound generation, bubble formation, and so on. The damage radius around the nanoparticle is determined by computing the distance where the curve's temperature reaches a 45 K difference above ambient, since raising the cell's temperature by 45 K will cause ablation. The damage area is also calculated, assuming that the area has a circular shape (Slide 26.9).

Next, we compute the minimum concentration of nanoparticles needed to ablate the cell (last column in Table 9.2). We estimate the cancer cell radius to be around 15 μm and calculate the minimum number of particles per cell, assuming that the cell dies once 70 % of its area has been damaged. All of the nanoparticle sizes used have achieved sufficient temperatures to damage the cell (second column in Table 9.2). However, the most efficient sizes would be 20 and 50 nm because they are small enough to move freely through the membrane of the cell, while they still can produce large damage areas of about 140 and 450 nm, respectively, with a reasonable energy density of RF radiation [7] (Slide 26.9). The number of 50-nm gold nanoparticles for effective killing of cancer cells is 23 per cell.

9.2.2.5 RF Heating of a Single Carbon Nanoparticle

We also model the temperature space curves for carbon nanoparticles of different sizes by using a Maple code from Appendix E. Results of these simulations are shown in Fig. 9.8 and summarized in Table 9.3 and Slide 26.10. Similar to the gold nanoparticles, the maximum temperature reached by carbon nanoparticles also

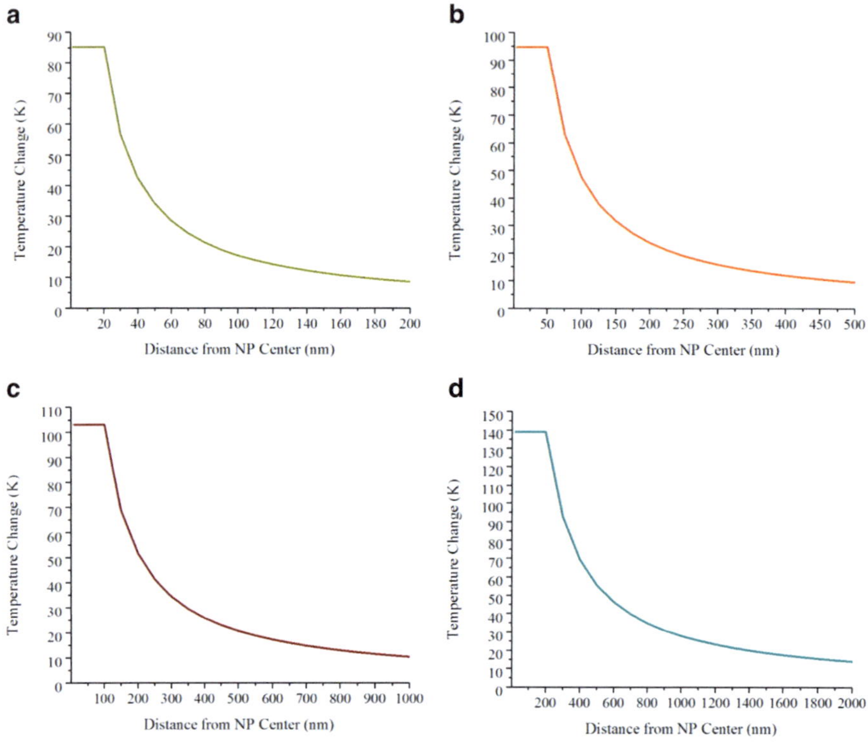

Fig. 9.8 Space distribution of thermal fields inside and outside carbon nanoparticles. The simulations are performed in a deionized surrounding water medium for different radii of carbon nanoparticles: (**a**) 20 nm, (**b**) 50 nm, (**c**) 100 nm, and (**d**) 200 nm

Table 9.3 Results of simulations for heating carbon nanoparticles

Radius (nm)	ΔT (K)	Q_{abs}	Sample volume (mL)	Damage radius (nm)	Minimum concentration (NPs/cell)
20	85.18	9.91×10^{-16}	0.001	30	175
50	94.66	1.44×10^{-14}	0.001	100	53
100	103.16	1.09×10^{-13}	0.001	200	26
200	138.81	8.24×10^{13}	0.001	600	9

increases with size. However, unlike gold nanoparticles much more energy is needed to reach ablation temperatures by carbon nanoparticles. This is due to the fact that carbon's absorption efficiency (Q_{abs}) is lower than gold's absorption efficiency, so that carbon nanoparticles need to have a greater size or be heated by a greater energy density of RF radiation in order to achieve the same ablation results as gold nanoparticles [7]. We use the same methods as described in the

Practice Example 9

RF Heating of Single Carbon Nanoparticle

- ***Carbon's absorption efficiency (Q_{abs}) is lower*** than gold's absorption efficiency, so that carbon nanoparticles need to have a greater size or be heated by a greater energy density of RF radiation in order to achieve same ablation results as gold nanoparticles.
- Overall, these results show that ***carbon nanoparticles are less efficient than gold nanoparticles***, producing half the damage area in the cell. For example, for a 50-nm particle, it takes over a twice more carbon nanoparticles and higher energy density of the radiation to kill a cell than golden nanoparticles.

Radius (nm)	ΔT (K)	Q_{abs}	Sample volume (mL)	Damage radius (nm)	Minimum concentration (np/cell)
20	85.18	9.91×10^{-16}	0.001	30	175
50	94.66	1.44×10^{-14}	0.001	100	53
100	103.16	1.09×10^{-13}	0.001	200	26
200	138.81	8.24×10^{13}	0.001	600	9

Slide 26.10 Practice Example 9: RF heating of a single carbon nanoparticle

previous section for gold nanoparticles to compute the damage radius produced in the surrounding medium and the minimum concentration of nanoparticles required for the ablation of cancer cells. Overall, these results show that carbon nanoparticles are less efficient than gold nanoparticles, producing half the damage area in the cell. For example, for a 50-nm particle, it takes over a twice more carbon nanoparticles and higher energy density of radiation to kill a cell than gold nanoparticles (Slide 26.10).

9.2.2.6 RF Heating of a Gold Nanocluster

The model (9.2)–(9.3) can be used for simulations of the space distribution of the temperature produced by many heat sources, i.e., nanocluster. For cluster simulations, we use a modified Maple code provided in Appendix G. Figure 9.9 and Slide 26.11 illustrate the temperature fields inside and outside gold nanoclusters of different sizes in a deionized surrounding water medium irradiated by 13.56 MHz RF radiation of various energy densities. The results of these simulations are summarized in Table 9.4. The following computations carried out using a nanocluster consisting of nanoparticles with radii of 5 and 20 nm. As follows from these computations [7], to provide the threshold temperature change of 45 K in the medium between the particles everywhere in the cell volume (i.e., thermal damage in the cell), the interparticle spacing l in the cluster (i.e., between the centers of neighboring nanoparticles) should not exceed $l \leq 20$ nm in case of a cluster made of 5-nm gold particles. In order to keep the minimum temperature

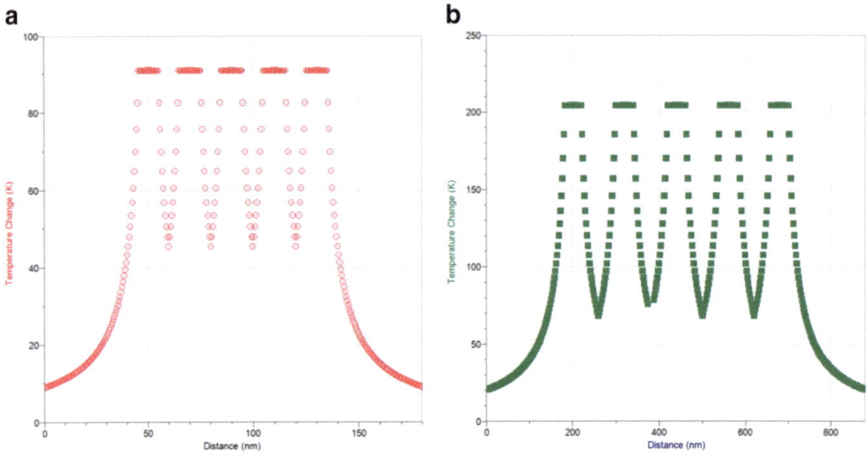

Fig. 9.9 Space distribution of thermal fields inside and outside gold nanoclusters. The simulations are performed in a deionized surrounding water medium for nanoclusters consisting of gold nanoparticles of different radii: (**a**) 5 nm and (**b**) 20 nm

change of 45 K in the cell medium between the nanoparticles, each particle is heated to the maximum temperature change of 90.87 K above the ambient in the cluster [7].

The interparticle spacing l sets the minimum number of particles per unit volume, n, needed to form a nanocluster (i.e., the number of particles per unit volume) as $n = \frac{1}{l^3}$. The minimum concentration of nanoparticles in a cluster is found to be 12.5×10^{16} particles/cm^3, which yields 48 nanoparticles (of 5 nm radius) per cluster [7].

The damage volume produced by a nanocluster can be found using the radius, which corresponds to the area where the ablation temperature is reached in the surrounding medium, as $V = \frac{4}{3}\pi r^3$. Based on Fig. 9.9, the damage radius produced by the nanocluster made of 5-nm gold particles is found to be 60 nm, which yields a damage volume of 9.04×10^{-16} cm^3. Now we can calculate how many particles are needed in order to kill one cancer cell. Let's assume that a cancer cell has a radius of 15 μm and that the cell dies after 70 % of its volume has been ablated. Then we find the number of nanoclusters needed to ablate the cell by dividing the 70 % of the cancer cell size by the damage area size. This means that each cancer cell needs to accumulate about 84 clusters made of 48 gold particles (5-nm) in order for ablation to occur [7].

Nanocluster simulation with-20 nm nanoparticles also have been performed. As expected, the nanocluster reaches a higher temperature change of 204 K, with a bigger particle separation of 120 nm, which leads to a concentration of 5.79×10^{14} particles/cm^3, or 43 particles per cluster. This cluster produces a damage radius of 329 nm and damage volume of 1.49×10^{-13} cm^3, which is about 160 times larger than the damage volume produced by a nanocluster consisting of 5-nm nanoparticles. If we assume a cancer cell of 15 μm and that an ablation of 70 %

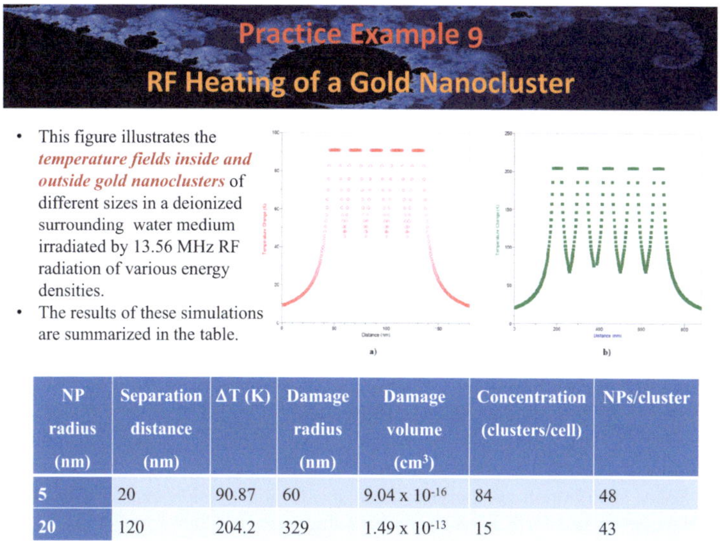

Slide 26.11 Practice Example 9: RF heating of a gold nanocluster

Table 9.4 Results of simulations for heating gold nanoclusters

NP radius (nm)	Separation distance (nm)	ΔT (K)	Damage radius (nm)	Damage volume (cm^3)	Concentration (clusters/cell)	NPs/ cluster
5	20	90.87	60	9.04×10^{-16}	84	48
20	120	204.2	329	1.49×10^{-13}	15	43

of the cell will result in its death, then each cell needs to accumulate 15 clusters per cell made of 20-nm gold particles, which is 5.6 times less than the number for the smaller cluster.

9.2.2.7 RF Heating of a Carbon Nanocluster

In this section we perform simulations for a carbon nanocluster using the same model and a Maple code from Appendix G. Figure 9.10 and Slide 26.12 illustrate the temperature fields inside and outside carbon nanoclusters of different sizes in a deionized surrounding water medium irradiated by 13.56 MHz RF radiation of various energy densities. The results of these simulations are summarized in Table 9.5. The following computations are obtained using a nanocluster consisting of carbon nanoparticles with radii of 20 and 50 nm. The minimum temperature change between the particles is 49 K above ambient, which is enough to cause ablation, while the maximum temperature change reaches 128.89 K [7].

To provide the minimum temperature change of 49 K between the particles, the separation distance between particles is 120 nm for the cluster consisting 20-nm

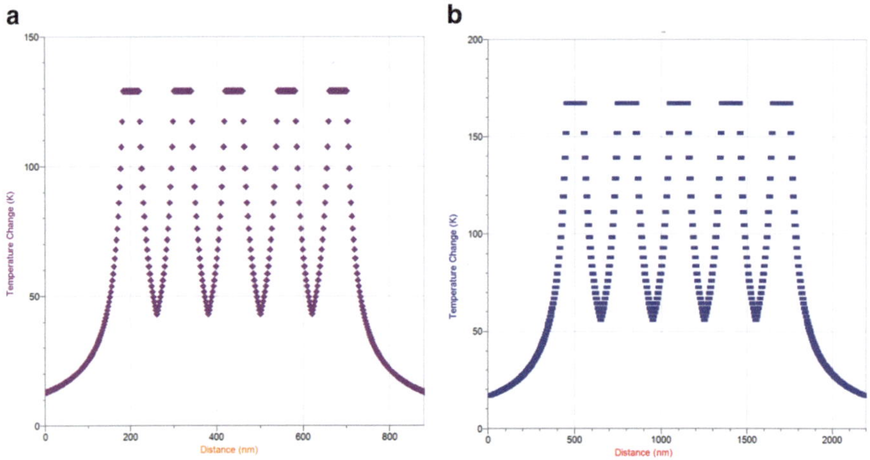

Fig. 9.10 Space distribution of thermal fields inside and outside carbon nanoclusters. The simulations are performed for a deionized surrounding water medium for nanoclusters consisting of carbon nanoparticles of different radii: (**a**) 20 nm and (**b**) 50 nm

carbon particles, which yields a concentration of 5.79×10^{14} particles/cm^3 and 43 particles per cluster. This concentration is identical to the gold nanocluster made of 20-nm particles. This is explained by the fact that the nanoparticle concentration in the nanocluster only depends on the radius of the nanoparticles used and the separation distance, which are identical between these two clusters. However, the carbon nanoparticle produces a smaller damage radius of 295 nm, which results in a 1.07×10^{-13} cm^3 damage volume. This means that a 15-μm cancer cell needs to accumulate 18 nanoclusters per cell in order to ablate 70 % of the cell [7].

We also have performed simulations for a nanocluster consisting of 50-nm carbon particles. As can be seen from Table 9.5, this cluster produces the largest damage area with a radius of about 1.6 μm. The reason it achieves such a large damage area is because of the larger size of the 50-nm particle and the larger separation of 300 nm from center to center. This separation yields a concentration of 3.73×10^{13} particles/cm^3 or 43 nanoparticles per cluster. For a 15-μm cancer cell, we compute that it takes only three nanoclusters per cell in order to ablate 70 % of its volume [7].

9.2.2.8 Conclusions

We have performed simulations of space distributions of the thermal fields around single and clustered nanoparticles heated by 13.56 MHz RF radiation (Slide 26.13). We find that the nanoclusters are more efficient than single particles. A single 5-nm gold nanoparticle produces only a 10-nm damage radius, while a cluster consisting

Practice Example 9
RF Heating of a Carbon Nanocluster

- Figure illustrates the *temperature fields inside and outside carbon nanoclusters* of different sizes in deionized water surrounding medium irradiated by 13.56 MHz RF radiation of various energy densities.
- The results of these simulations are summarized in table.

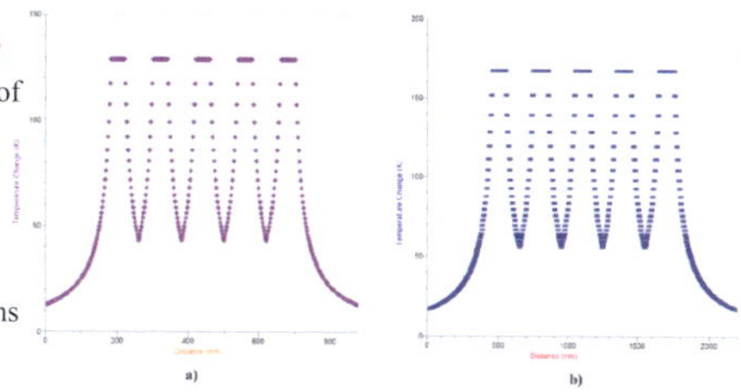

NP radius (nm)	Separation (nm)	ΔT (K)	Damage Radius (nm)	Damage Volume (cm^3)	Concentration (clusters/cell)	NPs/ cluster
20	120	128.9	295	1.07×10^{-13}	18	43
50	300	167.11	1565	1.61×10^{-11}	3	43

Slide 26.12 Practice Example 9: RF heating of a carbon nanocluster

Table 9.5 Results of simulations for heating carbon nanoclusters

NP radius (nm)	Separation (nm)	ΔT (K)	Damage radius (nm)	Damage volume (cm^3)	Concentration (clusters/cell)	NPs/ cluster
20	120	128.9	295	1.07×10^{-13}	18	43
50	300	167.11	1565	1.61×10^{-11}	3	43

Practice Example 9
Conclusions

- We have performed simulations of *space distributions of the thermal fields* around *single and clustered nanoparticles* heated by 13.56 MHz RF radiation.
- We find that the *nanoclusters are more efficient than single particles*.
- In comparison, a single 5-nm gold nanoparticle produces only a 10-nm *damage radius*, while a cluster consisting of 5-nm particles produces a 60-nm damage radius.
- In addition, the nanocluster requires *lower energy density* for its activation.
- The *highest temperature change* of 305 K is reached by a 100-nm gold nanoparticle, while the largest damage radius is reached by a carbon nanocluster consisting of 50-nm particles.
- Despite these results, we stand by our conclusion that *gold nanoclusters* are the most efficient candidates for nanodrugs because they require the least amount of radiation and the least amount of material.

Slide 26.13 Practice Example 9: Conclusions

of 5-nm particles produces a 60-nm damage radius. In addition, the nanocluster requires lower energy density for its activation [7].

The highest temperature change of 305 K is reached by a 100-nm gold nanoparticle, while the largest damage radius is reached by a carbon nanocluster consisting of 50-nm particles (Slide 26.13). Despite these results, we stand by our conclusion that gold nanoclusters are the most efficient candidates for nanodrugs because they require the least amount of radiation and the least amount of material [7].

9.2.3 Practice Example 10: Heating of Silver Nanoclusters by Laser Radiation

9.2.3.1 Abstract

The objective of this practice example is to calculate the temperature distribution in and outside a cluster made of silver nanoparticles of various sizes in different biological media (Slide 26.14). Based on these simulations, we determine the number of nanoparticles needed to form a cluster and estimate the thermal damage area produced in the surrounding biological medium by nanoclusters heated in the optical range of the spectrum [7].

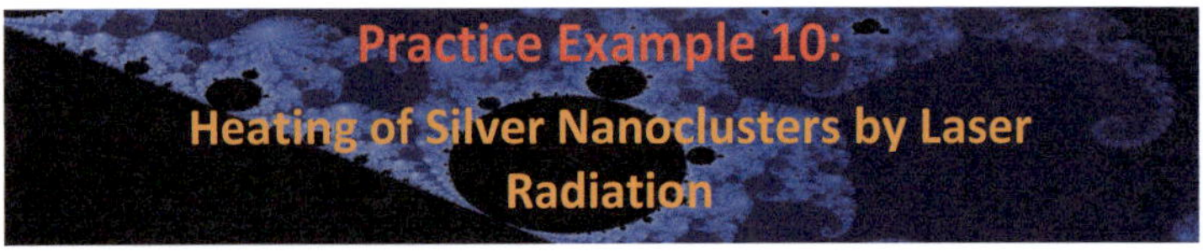

- ***The objective*** of this practice example is to calculate the temperature distribution in and outside a cluster made of silver nanoparticles of various sizes in different biological media.
- Based on these simulations, we determine the ***number of nanoparticles*** needed to form a cluster and estimate the ***thermal damage area*** produced in the surrounding biological medium by nanoclusters heated in the optical range of the spectrum.
- We use here the ***same heat transfer model and same methods*** as described in the previous example for RF heating of nanoclusters to compute the ***damage area*** produced by silver nanoclusters in the surrounding biological medium, and the minimum concentration of silver nanoparticles required for the ablation of cancer cells.

Slide 26.14 Practice Example 10: Heating of silver nanoclusters by laser radiation

9.2.3.2 Method

We use here the same model (9.2)–(9.3) and same methods as described in the previous section for RF heating of nanoclusters to compute the damage area produced by silver nanoclusters in the surrounding biological medium, and the minimum concentration of silver nanoparticles required for the ablation of cancer cells (Slide 26.14). According to the procedure described earlier, the interparticle spacing l sets the minimum number of particles per unit volume, n, needed to form a nanocluster (i.e., the number of particles per unit volume) as $n = \frac{1}{l^3}$. The damage volume produced by a nanocluster in the cancer cell can be found using the radius, which corresponds to the area where the ablation temperature is reached in the surrounding medium, as $V = \frac{4}{3}\pi r^3$. We assume that a cancer cell has a radius of 15 μm and that the cell dies after 70 % of its volume has been ablated. Then we can find the number of nanoclusters needed to ablate the cell by dividing the 70 % of the cancer cell size by the damage area size [7].

The spatial simulations of the temperature distributions inside and outside the silver nanocluster are performed by using a Maple code from Appendix G with input data listed in Table 9.6 and Slide 26.15. Please note that the optical values provided in Table 9.6 for silver nanoparticles are not optimized to reach the maximum absorption efficiency in terms of the optimal wavelength of radiation.

9.2.3.3 Discussion and Results

Simulations have been performed for a cluster made of silver nanoparticles heated in the optical range of the spectrum in various biological media: blood, cytoplasm, fat, intracellular components, and membrane. First we model the heating of the cluster at a fixed separation distance of 20 nm between the surfaces of neighboring silver nanoparticles and at a fixed damage radius of 1114 nm produced by the nanocluster, in order to investigate how each medium affects the amount of energy

Table 9.6 Input data used in the simulation

Surrounding biological medium				Optical properties of silver nanoparticle*		
Medium	Density (kg/m^3)	Heat conductivity W/(cm K)	Specific heat J/(kg K)	Radius (nm)	Wavelength (nm)	Q_{abs}
Blood	1060	0.6	3775	56	398	1.55
Cytoplasm	1010	0.59	4180	51	391	2.21
Fat	918	0.2	2975	45	418	2.12
Cell membrane	1050	0.3	3000	33	399	3.33
Intracellular medium	1100	0.3	2100	37	290	4.9

**Note*: The optical values for silver nanoparticles listed here are not optimized to reach the maximum absorption efficiency in terms of the optimal wavelength of radiation

- The ***spatial simulations of the temperature distributions inside and outside the silver nanocluster*** are performed by using a Maple code from Appendix G with input data listed in the table below.
- Please note that the optical values provided in the table below for silver nanoparticle are not optimized to reach the maximum absorption efficiency in terms of the optimal wavelength of radiation.

Surrounding Biological Medium				Optical Properties of Silver Nanoparticle		
Medium	Density kg/m^3	Heat conductivity W/(cm K)	Specific heat J/(kg K)	Radius (nm)	Wavelength (nm)	Q_{abs}
Blood	1060	0.6	3775	56	398	1.55
Cytoplasm	1010	0.59	4180	51	391	2.21
Fat	918	0.2	2975	45	418	2.12
Cell membrane	1050	0.3	3000	33	399	3.33
Intracellular medium	1100	0.3	2100	37	290	4.9

Slide 26.15 Practice Example 10: Input data

- The table shows the ***amount of energy*** required to achieve a 1114-nm damage radius by a silver cluster with a fixed separation distance of 20 nm between the surfaces of neighboring nanoparticles in various media.

Surrounding Biological Media	Energy density (J/cm^2)
Blood	0.9950
Cytoplasm	0.7360
Fat	0.4965
Intracellular	0.2675
Membrane	0.3430

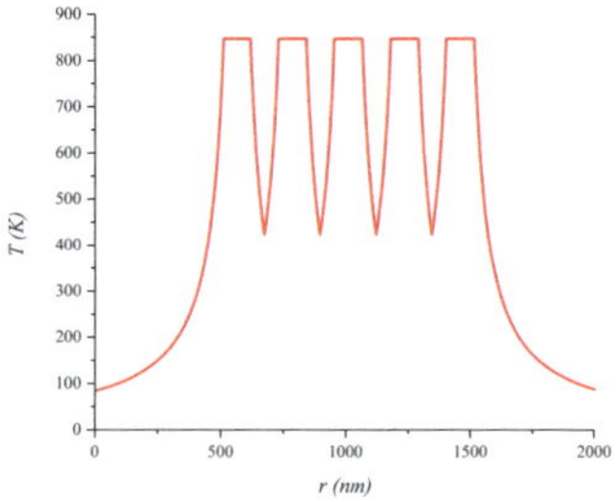

This figure demonstrates the temperature distributions produced by a cluster made of 56-nm silver particles in the blood surrounding medium heated by radiation of 398-nm wavelength and 0.9950-J/cm^2 energy density.

Slide 26.16 Practice Example 10: Discussion and results

needed to produce a specific damage area (Slide 26.16). The radius of the nanoparticles in cluster is chosen to provide the maximum absorption in the selected medium at the given wavelength of radiation, and it is varied according to Table 9.6. The results of these simulations are presented in Table 9.7

Table 9.7 Amount of energy required to achieve a 1114-nm damage radius by a silver cluster with a fixed separation distance of 20 nm between the surfaces of neighboring nanoparticles in various media

Surrounding biological media	Energy density (J/cm^2)
Blood	0.9950
Cytoplasm	0.7360
Fat	0.4965
Intracellular	0.2675
Membrane	0.3430

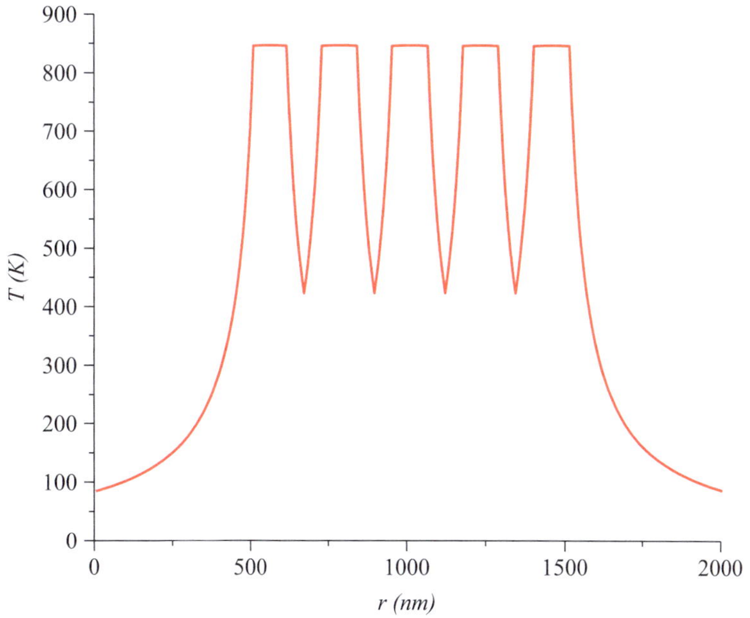

Fig. 9.11 Temperature distributions produced by a cluster made of 56-nm silver particles in the surrounding blood medium heated by radiation of 398-nm wavelength and 0.9950-J/cm^2 energy density

and Fig. 9.11. This figure demonstrates the temperature distributions produced by a cluster made of 56-nm silver particles in the surrounding blood medium heated by radiation of 398-nm wavelength and 0.9950-J/cm^2 energy density. The chosen separation distance of 20 nm sets the interparicle spacing l in cluster, i.e., the distance between the centers of neighboring particles, to be 136 nm in Fig. 9.11. Table 9.7 shows that for the fixed damage radius and fixed particle separation, blood is the least efficient medium, and the intracellular components are the most efficient media for heating the largest area with the minimum amount of energy.

Practice Example 10
Discussion and Results (continued)

- This table shows the results of simulations for the ***separation distance between the surfaces of nanoparticles*** in a cluster and thermal damage radius produced by a silver nanocluster heated by radiation of 0.21 J/cm^2 energy density in various biological media.

Surrounding biological media	Separation distance (nm)	Damage radius (nm)
Blood	20	1114
Cytoplasm	28	1219
Fat	44	1435
Intracellular	86	1957
Membrane	66	1419

Slide 26.17 Practice Example 10: Discussion and results (continued)

Table 9.8 Results of simulations for the separation distance between the surfaces of nanoparticles in a cluster and thermal damage radius produced by a silver nanocluster heated by radiation of 0.21 J/cm^2 energy density in various biological media

Surrounding biological media	Separation distance (nm)	Damage radius (nm)
Blood	20	1114
Cytoplasm	28	1219
Fat	44	1435
Intracellular	86	1957
Membrane	66	1419

Next, we investigate what is the maximum damage radius produced by nanoclusters for each medium heated at a fixed amount of energy density of 0.21 J/cm^2 (Slide 26.17). For each surrounding medium simulation, we adjust the nanoparticle separation distance until we find the spacing between particles that provides a minimum interparticle temperature change of 45 K above the ambient, which is our damage threshold. Table 9.8 shows the results of this simulation confirming again that the intracellular components are the most efficient as they have the largest damage radius at a given energy density.

Finally, we simulate a silver nanocluster in each medium at a fixed 20-nm particle separation distance and with a fixed energy density of 0.21 J/cm^2. Figure 9.12 and Slide 26.18 demonstrate how different media give different damage radii, even when each cluster has a constant nanoparticle spacing. In this case, blood

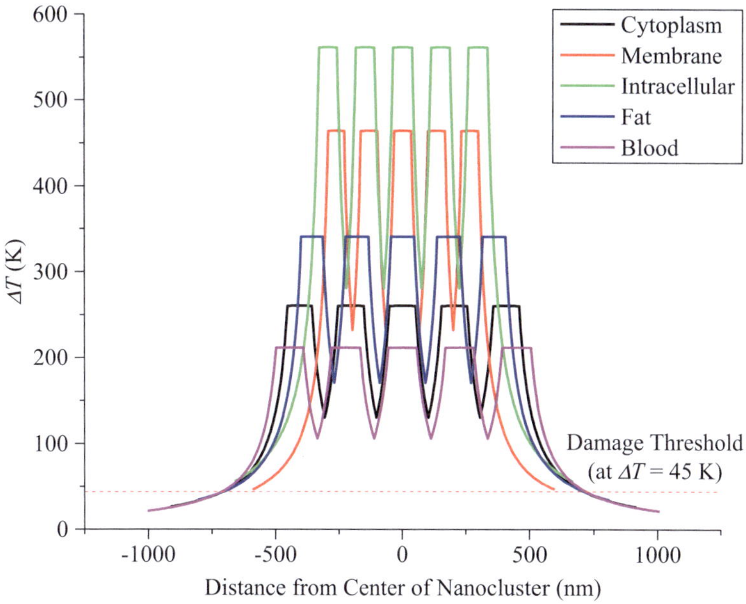

Fig. 9.12 Spatial temperature distributions produced by silver nanoclusters with fixed separation distance between the surfaces of the nanoparticles in various biological media heated by radiation of 0.21 J/cm^2 energy density

has the highest damage radius while the membrane clearly has the lowest. The silver nanoclusters optimized for use in the membrane and intracellular components perform rather poorly with separation distance of 20 nm, but when the spacing is increased to 66 and 86 nm, the damage radius becomes much higher as demonstrated in Table 9.8. This suggests that by tuning the nanoparticle size and spacing, and then actively targeting a particular medium in cancer cells, damage to healthy cells can be minimized.

9.2.3.4 **Conclusions** (Slide 26.19)

- Nanoparticle clusters produce significantly higher damage to cancer cells than single nanoparticles due to the increase in the total thermal damage area within the cell volume rather than heating many smaller areas.
- The damage radii of different nanoclusters in their respective media at a fixed separation distance are pretty similar. However, when the particles and separation distance between them are optimized for a particular medium, they can perform much better.

Practice Example 10
Discussion and Results

- Finally, we simulate a silver nanocluster ***in each medium*** at a fixed 20-nm particle separation distance and with a fixed energy density of 0.21 J/cm^2.
- In this case, ***blood*** has the highest damage radius while the membrane clearly has the lowest.
- By tuning the nanoparticle size and spacing, and then actively targeting a particular medium in cancer cells, damage to healthy cells can be minimized.

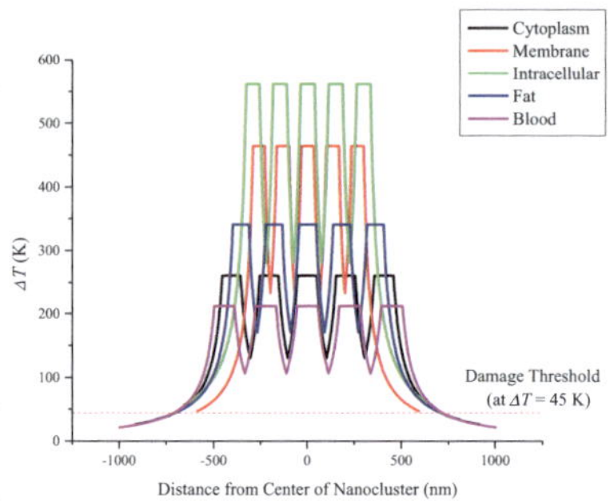

Slide 26.18 Practice Example 10: Discussion and results

Practice Example 10
Conclusions

- Nanoparticle ***clusters*** produce significantly ***higher damage*** to cancer cells than single nanoparticles due to increase in the total thermal damage area within the cell volume rather than heating many smaller areas.
- The ***damage radii of different nanoclusters*** in their respective media at a fixed separation distance are pretty similar.
- However, when the particles and separation distance between them are optimized for a particular medium, they can perform much better.

Slide 26.19 Practice Example 10: Conclusions

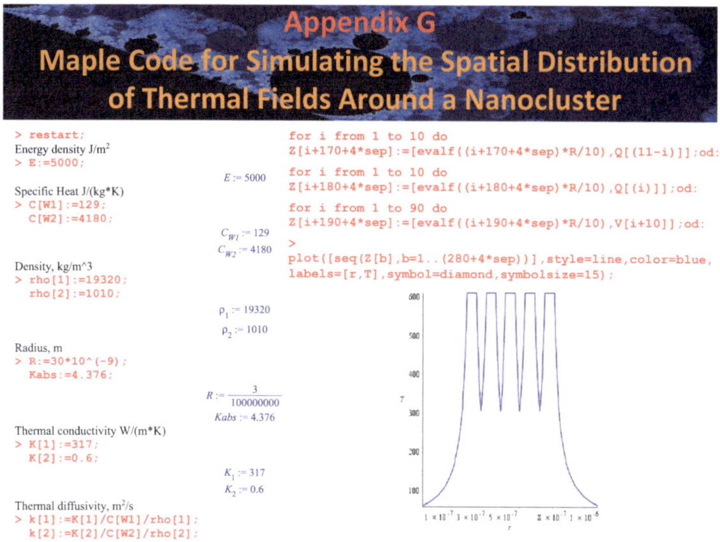

Slide 26.20 Appendix G: Maple code for simulating the spatial distribution of thermal fields around a nanocluster

9.3 Lecture 27: Nanobubble Overlapping Mode (Slide 27.1)

- Introduction
- Spherical Bubble Dynamics
- Nanobubble Overlapping
- Conditions for Nanoparticle Concentration
- Conditions for Nanoparticle Size
- Microbubble Generation Around Nanoparticles and Nanoclusters
- Conclusion

9.3.1 Introduction

As shown in the previous section, most of the radiation pulse energy remains localized at the absorbing nanoparticles/cluster whose temperatures rise high enough to cause vaporization of the immediate surrounding medium. This creates a bubble that then extends outward from the nanoparticle as shown in Slide 27.2. The heated nanoparticle/cluster with radius r_0 may be considered as a nucleation center. We now investigate bubble dynamics resulting from laser pulses with duration of 8 ns at FWHM. This section contains material adopted from our publications [1, 2, 5].

Lecture 27:

Nanobubbles Overlaping Mode

Contents:

Slide 27.1 Contents

Introduction

- As shown in the previous lectures, most of the radiation pulse energy remains localized at the absorbing ***nanoparticles/cluster whose temperature rises*** high enough to cause ***vaporization of the immediate surrounding medium***.
- This creates a ***bubble*** that then ***extends outward*** from the nanoparticle as shown in the picture on the right.
- The heated ***nanoparticle/cluster*** with radius r_0 may be considered as a ***nucleation center***.
- The experiments show that it is enough to ***heat*** the strongly absorbing nanoparticles above the ***boiling temperature of water*** only or above the critical temperature of the phase transition in water ***in order to generate the bubbles*** around the particles.

Slide 27.2 Introduction

The experiments [1] show that it is enough to heat the strongly absorbing nanoparticles above the boiling temperature of water only or above the critical temperature of the phase transition in water in order to generate the bubbles around the particles (Slide 27.2). So, for a relatively low laser flux, the nanoparticle will be surrounded by a rapidly expanding layer of vapor. Let us calculate the threshold

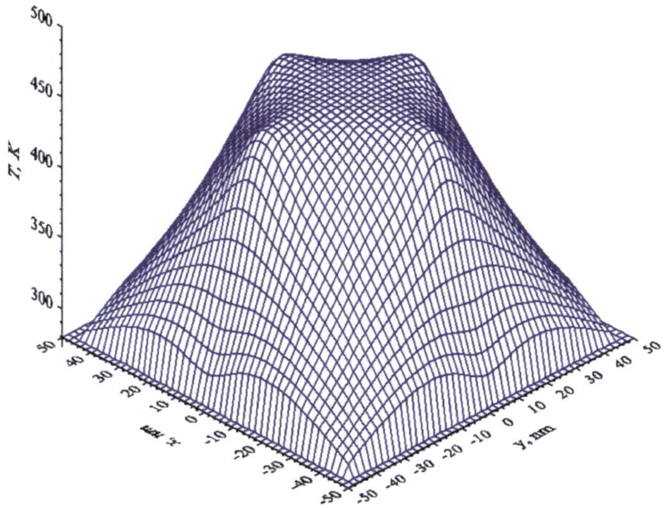

Fig. 9.13 Two-dimensional distribution of the temperature inside and outside a 30-nm gold nanoparticle in the surrounding blood medium irradiated by a single laser pulse of energy density 0.33 J/cm^2 and duration 8 ns. Reprinted from [2]. © With permission from World Scientific

energy density of the laser pulse deposited into the nanoparticle, which is required to raise its surface temperature to the level of the critical threshold temperature for the bubble generation.

An experimental threshold nucleation temperature of $T_{\text{thr}} = 150$ °C (423 K) for heterogeneous bubble generation on the melanosome's surface was determined in Ref. [4]. Our calculations demonstrate (see Fig. 9.13 and Slide 27.3) that a threshold energy density of 0.33 J/cm^2 for a single 8-ns laser pulse is required to reach a surface temperature of 450 K for 30-nm gold nanoparticles in a blood medium. Bubble nucleation around the nanoparticles and their rapid expansion in the cellular volume with high internal vapor pressure can destroy the cancer cells and tumors through mechanical damage or sound generation during sudden bubble collapsing. Thus, the bubble mode of photothermolysis can provide large damage area at the relatively low optical flux of 0.33 J/cm^2 ($\tau_{\text{L}} = 8$ ns) via rapid expanding nanobubbles around the nanoparticles.

9.3.2 *Spherical Bubble Dynamics*

Consider a spherical bubble of radius $R(t)$ (where t is time) in an infinite domain of liquid whose temperature and pressure far from the bubble are T_∞ and $p_\infty(t)$, respectively (Slide 27.4). We assume that the liquid density ρ_{L} and dynamic viscosity η_{L} are constant and uniform [2]. It will also be assumed that the contents of the bubble are homogeneous, and that its temperature $T(t)$ and pressure $p(t)$

Bubble Nucleation

- An experimental ***nucleation temperature*** for heterogeneous bubble generation on the melanosome's surface is T_{thr} ***= 150°C (423 K)***.
- Calculations demonstrated by the figure below show that a ***threshold energy density of 0.33 J/cm²*** for a single 8-ns laser pulse is required to ***reach a surface temperature of 450 K for 30-nm gold nanoparticles*** in a blood medium.
- ***Bubble nucleation*** around the nanoparticles and ***their rapid expansion*** in a cellular volume with high internal vapor pressure can ***destroy the cancer cells*** and tumors through ***mechanical damage or sound generation during sudden bubble collapsing***.
- Thus, the ***bubble mode*** of photothermolysis can ***provide large damage area at the relatively low optical flux*** of 0.33 J/cm² via rapid expanding nanobubbles.

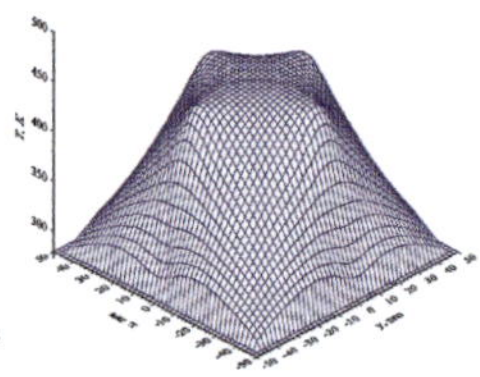

Slide 27.3 Bubble nucleation

Spherical Bubble Dynamics

- Consider a ***spherical bubble*** of radius $R(t)$ (where t is time) in an infinite domain of liquid whose temperature and pressure far from the bubble are T_∞ and $p_\infty(t)$, respectively .
- We assume that the ***liquid density*** ρ_L ***and dynamic viscosity*** η_L ***are constant and uniform .***
- It will also be assumed that the ***contents of the bubble are homogeneous, and that its temperature T(t) and pressure p(t) within the bubble are always uniform .***
- The ***radius of the bubble*** will be one of the primary results of the analysis .

Slide 27.4 Spherical bubble dynamics

within the bubble are always uniform. The radius of the bubble will be one of the primary results of the analysis. In the absence of mass transport across the boundary (evaporation or condensation) and with the earlier assumptions, the bubble dynamics can be described by the generalized Rayleigh–Plesset equation [4] (Slide 27.5):

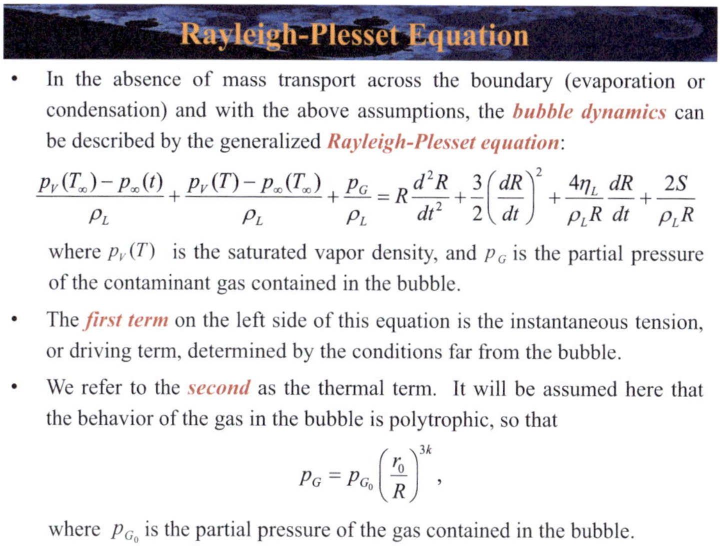

Slide 27.5 Rayleigh–Plesset equation

$$
\begin{aligned}
&\frac{p_{\mathrm{V}}(T_\infty)-p_\infty(t)}{\rho_{\mathrm{L}}}+\frac{p_{\mathrm{V}}(T)-p_\infty(T_\infty)}{\rho_{\mathrm{L}}}\\
&\quad+\frac{p_{\mathrm{G}}}{\rho_{\mathrm{L}}}=R\frac{d^2R}{dt^2}+\frac{3}{2}\left(\frac{dR}{dt}\right)^2+\frac{4\eta_{\mathrm{L}}}{\rho_{\mathrm{L}}R}\frac{dR}{dt}+\frac{2S}{\rho_{\mathrm{L}}R},
\end{aligned}
\tag{9.4}
$$

where $p_{\mathrm{V}}(T)$ is the saturated vapor density, and p_{G} is the partial pressure of the contaminant gas contained in the bubble. The first term on the left side of this equation is the instantaneous tension, or driving term, determined by the conditions far from the bubble. We refer to the second as the thermal term. It will be assumed here that the behavior of the gas in the bubble is polytrophic, so that [5] (Slide 27.5)

$$
p_{\mathrm{G}} = p_{\mathrm{G}_0}\left(\frac{r_0}{R}\right)^{3k},
\tag{9.5}
$$

where p_{G_0} is the partial pressure of the contaminant gas contained in the bubble at some reference size $R = r_0$, with k as approximately constant. Here, $k = 1$ implies a constant bubble temperature, and $k = \gamma$ would model adiabatic behavior, where γ is the ratio of the specific heat of the vapor at constant pressure to the specific heat at constant volumc.

Equation (9.4) can be readily integrated numerically to find $R(t)$ and the rate of bubble growth $v = \frac{dR}{dt}$, given the input $p_\infty(t)$, the temperature T_∞, and the other constants. Initial conditions are also required and, in the context of cavitating flows,

Rayleigh-Plesset Equation (continued)

- The generalized ***Rayleigh -Plesset equation*** can be readily integrated numerically to find $R(t)$ and the rate of bubble growth $v = \frac{dR}{dt}$ given the input $p_\infty(t)$ and the temperature T_∞.
- Initial conditions are also required and, in the context of cavitating flows, it is appropriate to assume that a nanobubble of radius $R = r_0$ is in equilibrium at $t = 0$ in the fluid at a pressure $p_\infty(0)$, so that

$$p_{G_0} = p_\infty(0) - p_V(T_\infty) + \frac{2S}{r_0}$$

and $\left(dR/dt\right)_{t=0} = 0$.

Slide 27.6 Rayleigh–Plesset equation (continued)

it is appropriate to assume that a nanobubble of radius $R = r_0$ is in equilibrium at $t = 0$ in the fluid at a pressure $p_\infty(0)$, so that [5] (Slide 27.6)

$$p_{G_0} = p_\infty(0) - p_V(T_\infty) + \frac{2S}{r_0} \tag{9.6}$$

and $(dR/dt)_{t=0} = 0$.

9.3.3 Nanobubble Overlapping

As discussed in Sect. 9.1, under certain delivery and particle accumulation conditions, the particles can locate on the membrane surface very close to each other and form clusters [2, 5], as shown in Fig. 9.1 and Slide 27.7. In this mode, the optical, thermal, acoustic, and bubble formation phenomena initiated by radiation around each particle can overlap well to create synergistic effects [1]. For example, the characteristic radii for these overlapping effects can be defined by an optical radius $R_{opt} = (\sigma_{abs}/\pi)^{1/2}$, thermal diffusion length $R_T = (kt)^{1/2}$, sound transfer distance $R_{ac} = c_s t$, and microbubble radius $R_{bubble} = v_b t$, where σ_{abs} is the cross-section of absorption, k is the heat diffusion coefficient, c_s is the speed of sound, and v_b is the bubble growth velocity (Slide 27.8). Since for a certain range of gold nanoparticle sizes the optical radius can exceed the geometrical particle radius r_0 at the given wavelength of laser radiation, a plasmon resonance at $R_{opt} > l$ (l is the interparticle spacing) can be achieved. This resonance may lead to "red-shifting" of the

Nanobubble Overlapping

- As discussed in the previous lecture, under certain delivery and ***particle accumulation*** conditions, the particles can locate on the membrane surface very close to each other and ***form clusters***, as shown in the figure below.
- In this mode, the ***optical, thermal, acoustic and bubble formation phenomena*** initiated by radiation around each particle can ***overlap well*** to create synergistic effects.

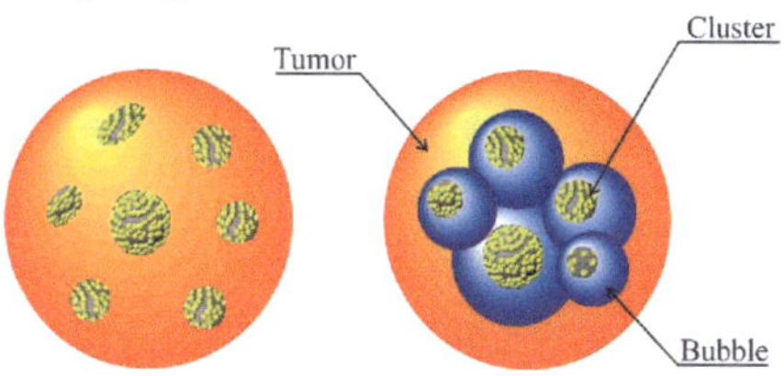

Slide 27.7 Nanobubble overlapping

Nanobubble Overlapping (continued)

- For example, the characteristic radii for these overlapping effects can be defined by an ***optical radius*** $R_{opt} = (\sigma_{abs}/\pi)^{1/2}$, ***thermal diffusion length*** $R_t = (kt)^{1/2}$, ***sound transfer distance*** $R_{ac} = c_s t$, and ***nanobubble radius*** $R_{bubble} = v_b t$, where σ_{abs} is the cross-section of absorption, k is the heat diffusion coefficient, c_s is the speed of sound, and v_b is the bubble growth velocity.
- Since for a certain range of gold nanoparticle sizes the ***optical radius*** can exceed the geometrical particle radius r_0 at the given wavelength of laser radiation, a ***plasmon resonance*** at $R_{opt} > l$ (l is the interparticle spacing) can be achieved. This resonance may lead to "red-shifting" of the maximum absorption and will potentially increase the integrated absorption efficiency of the particle's cluster.

Slide 27.8 Nanobubble overlapping (continued)

maximum absorption and will potentially increase the integrated absorption efficiency of the particle's cluster. The overlapping thermal fields at $R_T > l$ (during or after the laser pulse t_p) lead to a dramatic increase in thermal and accompanying effects (e.g., sudden appearance of different nonlinear effects such as bubble formation). The interaction of acoustic waves at $R_{ac} > l/2$ may significantly change

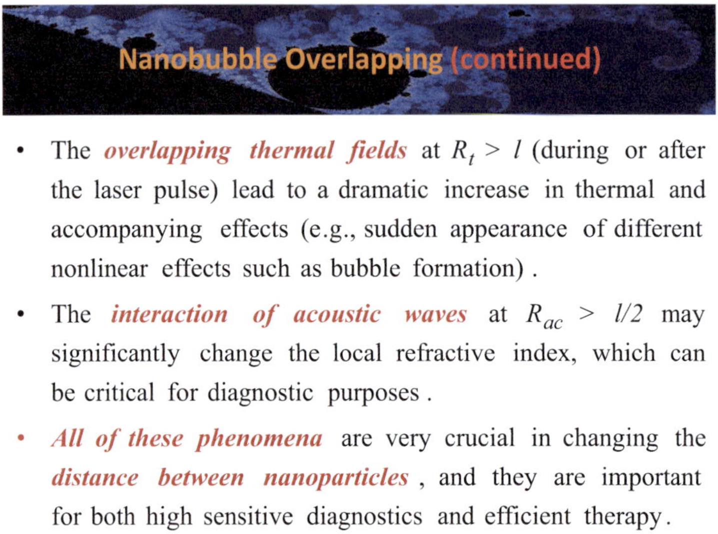

Slide 27.9 Nanobubble overlapping (continued)

the local refractive index, which can be critical for diagnostic purposes. All of these phenomena are very crucial in changing the distance between nanoparticles, and they are important for both high sensitive diagnostics and efficient therapy (Slide 27.9).

We focus here on therapeutic applications of this technology [5] (Slide 27.10). Indeed, laser-induced overheating effects around nanoparticles may create many therapeutic actions through the microbubble formation phenomena, accompanied by acoustic and cavitation effects, mechanical stress, and laser-induced hydrodynamic pressure arising from rapid bubble expansion and collapse. In this section, we discuss a dynamic mode for selective cancer treatments by laser-activated nanoheaters, involving the situation where the bubbles are overlapping inside the cell volume. The bubble overlapping mode (BOM) may dramatically increase the efficiency of a cancer treatment by laser-heated nanoparticles as the result of a large damage range.

Since bubbles do not all appear simultaneously, there are few possibilities for bubble overlapping [5]. The bubbles can overlap at different nucleation times and spreading velocities as shown in Fig. 9.14a, or simultaneously with the same spreading velocity when they reach each other midway between the neighboring gold particles (Slide 27.11). In our calculations, we choose stronger criteria than overlapping at midpoints (Slide 27.12). Effective overlapping with considerable damage of tumor cells can be achieved if laser-induced bubbles around each nanoparticle during expansion reach the other particles located in the neighborhood, as shown in Fig. 9.14b, not just the midpoints between nanoparticles. Also, bubbles can be created around the nanoclusters, as shown in Fig. 9.1.

- Laser-induced overheating effects around nanoparticles may create many therapeutic actions through the ***microbubble formation phenomena***, accompanied by ***acoustic and cavitation effects, mechanical stress, and laser-induced hydrodynamic pressure arising from bubble rapid expansion and collapse***.
- The ***bubble overlapping mode*** (BOM) may dramatically increase the efficiency of a ***cancer treatment*** by laser-heated nanoparticles as the result of a large damage range.

Slide 27.10 Nanobubble overlapping (continued)

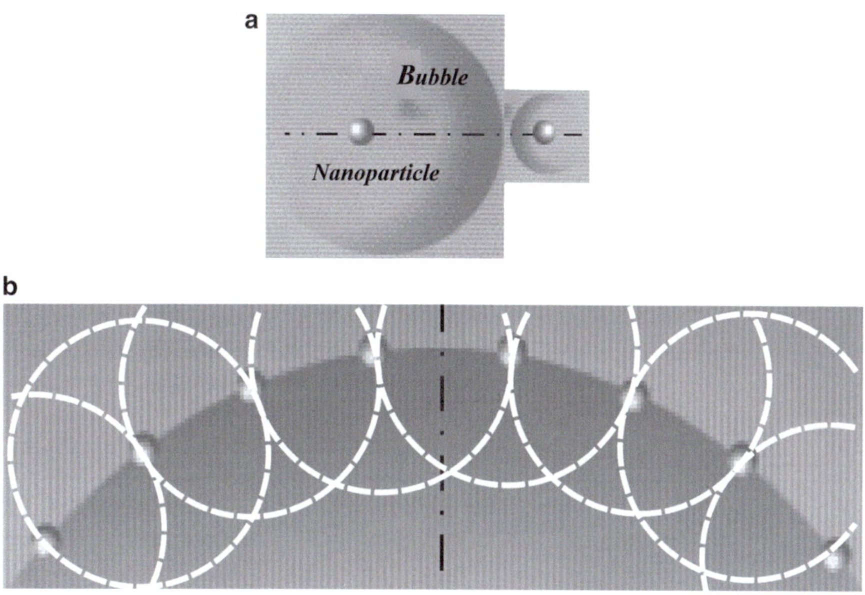

Fig. 9.14 Principle of the bubble overlapping mode: (**a**) appearance of bubbles at different nucleation time and (**b**) overlapping of bubbles appearing simultaneously with the same spreading velocity

Nanobubble Overlapping (continued)

- Since bubbles do not all appear simultaneously, there are few possibilities for ***bubble overlapping***.
- The bubbles can ***overlap at different nucleation times and spreading velocities***, as shown in figure, ***or simultaneously*** with the same spreading velocity when they reach each other midway between neighboring gold particles.

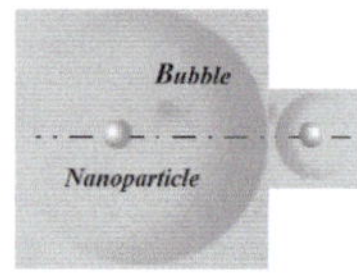

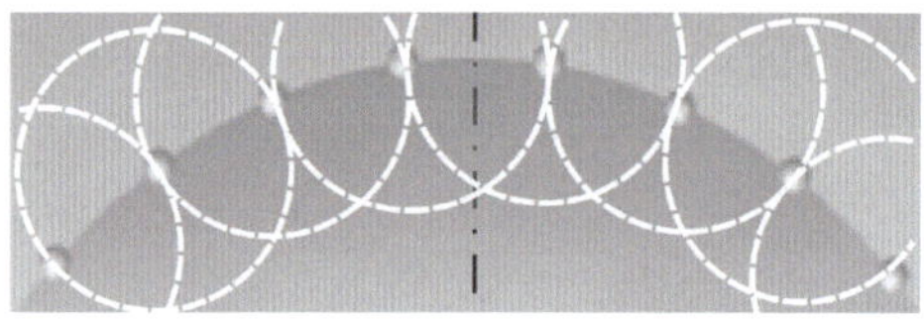

Slide 27.11 Nanobubble overlapping (continued)

Nanobubble Overlapping (continued)

- In our calculations, we choose stronger criteria than overlapping at midpoints. ***Effective overlapping*** with considerable damage of tumor cells can be achieved if laser-induced ***bubbles*** around each nanoparticle during expansion ***reach the other particles*** located in the neighborhood as shown in the left figure, not just the midpoints between nanoparticles. Also, the bubbles can be created around the nanoclusters as shown in right figure.

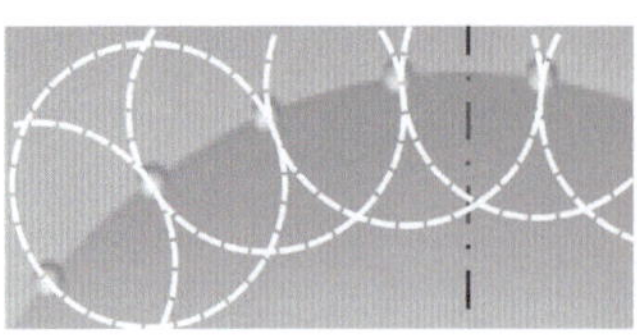

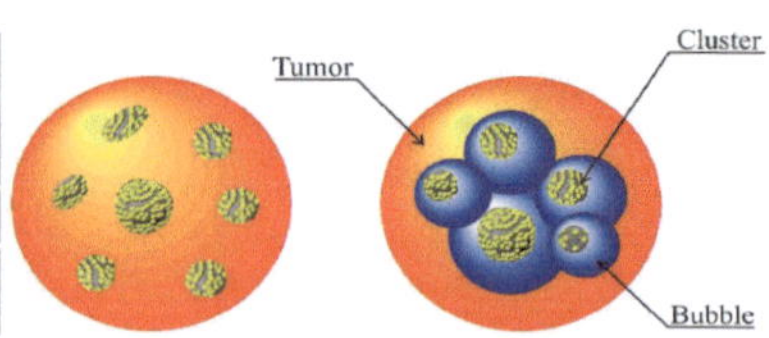

Slide 27.12 Nanobubble overlapping (continued)

9.3.4 Conditions for Nanoparticle Concentration

This section contains material reproduced from our publication [1]. The criteria for BOM introduced earlier assume a smooth distribution of nanoparticles on the cellular membrane (or in the cell volume) with a distance l between centers of two neighboring nanoparticles (Fig. 9.14b). Because a bubble must reach its neighboring particles during the growth time, τ_{growth}, the integration of the Rayleigh–Plesset equation (9.4) for bubble dynamics at the nucleation temperature T_{nuc} gives the requirement for interparticle spacing l as (Slide 27.13)

$$l \leq \sqrt{\frac{2}{3} \frac{P_{\mathrm{sat}}(T_{\mathrm{nuc}}) - P_{\infty}}{\rho(T_{\infty})}} \tau_{\mathrm{growth}}, \tag{9.7}$$

where $P_{\mathrm{sat}}(T_{\mathrm{nuc}})$ is the saturated vapor pressure at the nucleation temperature, P_{∞} is the ambient pressure, and $\rho(T_{\infty})$ is the water density. This sets the lower limit for concentration of nanoparticles required for the BOM, i.e., the number of particles per unit volume, as (Slide 27.13)

$$n = \left(1/l^3\right). \tag{9.8}$$

Let us now estimate the interparticle spacing, the nanoparticle concentration, and their absolute number collected into the tumor cell to provide BOM by laser-activated nanoheaters (Slide 27.14). The interparticle spacing l calculated by Eq. (9.7) for the experimentally measured nucleation temperature of $T_{\mathrm{nuc}} = 150$ °C is

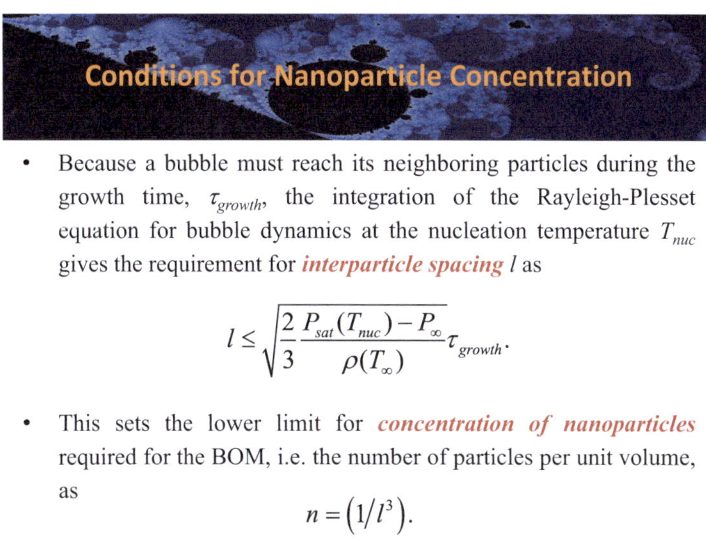

Slide 27.13 Conditions for nanoparticle concentration

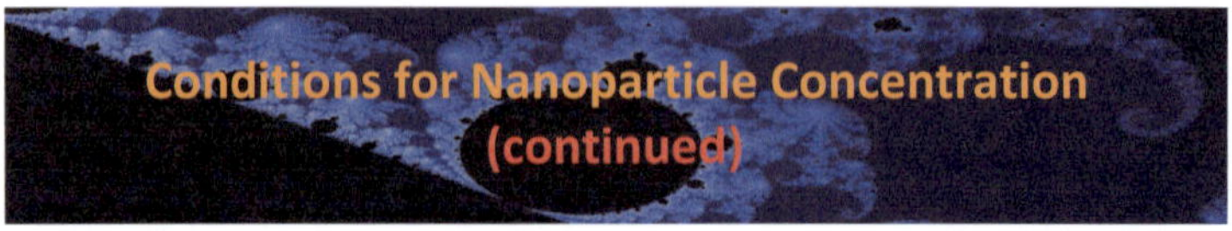

- Let us now estimate the ***interparticle spacing, the nanoparticle concentration and their absolute number*** collected into the tumor cell to provide BOM by laser-activated nanoheaters.
- The ***interparticle spacing*** l calculated for the experimentally measured nucleation temperature of T_{nuc} = 150°C is

$$l \approx 15.8\,\tau_{growth}.$$

- The ***bubble growth time*** τ_{growth} is limited by the bubble nucleation time t_{nuc} and bubble lifetime τ_{bubble}, where $t_{nuc} \leq \tau\ \tau_{growth} < \tau_{bubble}$. In our experiments [1], the nucleation time of the bubbles around isolated gold nanoparticles in aqueous suspension and irradiated by nanosecond laser pulses is $t_{nuc} \sim 100$ ns.

Slide 27.14 Conditions for nanoparticles concentration (continued)

$$l \approx 15.8\,\tau_{\text{growth}}. \tag{9.9}$$

The bubble growth time τ_{growth} is limited by the bubble nucleation time t_{nuc} and bubble lifetime τ_{buble}, where $t_{\text{nuc}} \leq \tau_{\text{growth}} < \tau_{\text{buble}}$ (Slide 27.14). In our experiments [1], the nucleation time of the bubbles around isolated gold nanoparticles in aqueous suspension and irradiated by nanosecond laser pulses is $t_{\text{nuc}} \sim 100$ ns. The bubble lifetime depends on the laser energy density and ranges from 100 ns to 2 μs. For the overwhelming majority of specific experimental cases, including ours, the bubble lifetime is in the range $\tau_{\text{bubble}} = 200$–$400$ ns (Slide 27.15). We can therefore conclude that the bubble growing time is close to the nucleation time, i.e., $\tau_{\text{growth}} \sim t_{\text{nuc}}$. Then, from the rate equation (9.9), we can estimate the interparticle spacing (microbubble radius) required for the BOM as $l = R_{\text{bubble}} \approx 1.6$ μm, which corresponds to the results of our experiments. Thus, Eq. (9.8) gives the lower limit of the concentration of nanoparticles to provide the BOM as $n \approx 2.44 \times 10^{11}$ cm^{-3}. The absolute number of particles homogeneously distributed inside a tumor cell for this mode can be found as $N = nV_{\text{cell}}$, where V_{cell} is the cell volume. In the case of 15-μm size cells of the MDA-MB-231 breast cancer line, the number of gold nanoparticles needed to produce BOM is $N \approx 430$, while in the case of a smooth distribution of nanoparticles on the cell membrane $N \approx 88$ [1] (Slide 27.15).

9.3.5 Conditions for Nanoparticle Size

The optimal range of nanoparticle sizes for effective laser initiation of BOM in tumor cells is governed by the nanoparticle optics described in Chaps. 4 and 5 of this book (Slide 27.16). The optical characteristics of spherical nanoparticles

Nanoparticle Concentration

- For the overwhelming majority of specific experimental cases, the ***bubble lifetime*** is in the range τ_{bubble} = 200-400 ns.
- We can therefore conclude that the ***bubble growing time*** is close to the ***nucleation time***, i.e., $\tau_{growth} \sim t_{nuc}$. Then, from the rate equation, we can estimate the ***interparticle spacing*** (***microbubble radius***) required for the BOM as $l = R_{bubble} \approx 1.6$ μm, which corresponds to the results of experiments.
- Then the ***absolute number of particles*** homogeneously distributed inside a tumor cell for this mode can be found as

$$N = nV_{cell},$$

where V_{cell} is the cell volume. ***In the case of 15-μm size cells*** of the MDA-MB-231 breast cancer line, *the* ***number of gold nanoparticles*** needed to produce BOM is N ≈ 430, while in the case of a smooth distribution of nanoparticles on the cell membrane N ≈ 88.

Slide 27.15 Nanoparticle concentration

Conditions for Nanoparticle Size

- ***The optimal range*** of nanoparticle sizes for effective laser initiation of BOM in tumor cells is governed by the ***nanoparticle optics***.
- The Lorentz-Mie diffraction theory along with bubble dynamics allows us to find the following conditions for BOM realization:
  - ➢ The ***maximal effect*** of laser heating of gold nanoparticles over the wavelength range λ = 400-580 nm can be achieved for particle ***radii of 10-45 nm***,
  - ➢ The ***optimal nanoparticle concentration*** is n ≈ 2.44×10^{11} cm^{-3}, which corresponds to an absolute number of nanoparticles in a breast cancer cell volume of 430 (88 on the cell surface).
  - ➢ The ***bubble radius*** providing BOM is $R_{bubble} \approx 1.6$ μm.

Slide 27.16 Conditions for nanoparticle size

dispersed in a biological medium at a given radiation wavelength λ can be calculated on the basis of Lorentz–Mie diffraction theory at the single-scattering approximation as shown in Sect. 4.1. In Chap. 5, we introduced the computer software and performed Mie simulations of absorption, scattering, and extinction efficiencies of different types of nanoparticles as a function of particle size and wavelength of radiation in different surrounding biological media. For the example of gold nanoparticle simulations we showed that gold particles are effective absorbers of radiation ($Q_{abs} > 1$, Q_{sca}) over a wide range of the optical spectrum, 400–580 nm, with a maximum efficiency $Q_{abs} = 4.02$ for 30-nm gold particles at the wavelength $\lambda = 538.3$ nm in the surrounding water medium. The optimal nanoparticle size range for effective radiation heating of nanoparticles is 10–45 nm, where $Q_{abs} \geq 1$ and $Q_{abs} > Q_{sca}$ in aqueous suspension (Figs. 5.9 and 5.11). The optical properties of the nanoparticles depend on the shape and internal structure of the particle as well. The highest absorption efficiency is observed for gold nanorods reaching values of $Q_{abs} = 14.5$. Silica–gold nanoshells have an advantage in being tunable over the longer wavelengths in the transparency window of most biotissues.

Thus, the Lorentz–Mie diffraction theory along with bubble dynamics, Eqs. (9.4)–(9.9), allows us to find the following conditions for BOM realization [2] (Slide 27.16):

- The maximal effect of laser heating of gold nanoparticles over the wavelength range $\lambda = 400$–580 nm can be achieved for particle radii of 10–45 nm.
- The optimal nanoparticle concentration is $n \approx 2.44 \times 10^{11}$ cm^{-3}, which corresponds to an absolute number of nanoparticles in a breast cancer cell volume of 430 (88 on the cell surface).
- The bubble radius providing BOM is $R_{bubble} \approx 1.6$ μm.

These theoretical predictions of the BOM conditions have been confirmed by our experimental results [1], which include the study of nanocluster-related phenomena as discussed in Sect. 9.1.

9.3.6 Microbubble Generation Around Nanoparticles and Nanoclusters

This section reproduces the results of simulations published in our papers [1, 2]. The model described earlier has been used to calculate the bubble growth in time around the 30-nm gold particles and homogeneous cluster with radius 190 nm surrounded by a blood biomedium when the incident laser pulse has an energy density of $E = 0.33$ J/cm^2 and pulse duration of $\tau_L = 8$ ns (Slide 27.17). Section 9.4 with Appendix H provides a Maple code used for bubble dynamic simulations, as well as discussing a specific example for bubble nucleation around the nanoparticles in different surrounding biological media. The physical properties of blood used in the calculations are listed in Table 9.9. The results of the simulations are presented in Figs. 9.15 and 9.16 (Slide 27.18). Figure 9.15 illustrates the time behavior of the radius and its rate for the bubble generated around the 30-nm particles heated and cooled in the blood. Figure 9.16 demonstrates a bubble expansion rate (growth velocity) caused by the same heated particle in the blood.

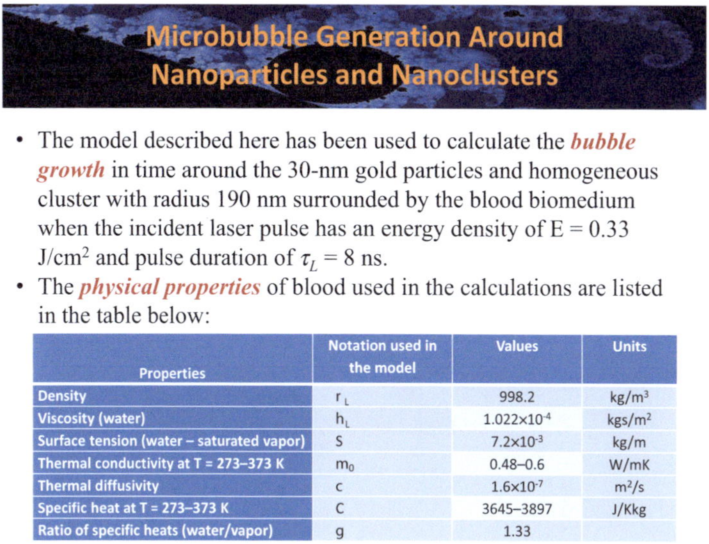

Slide 27.17 Microbubble generation around nanoparticles and nanoclusters

Table 9.9 Physical properties of blood at 1 atm pressure used in the simulations. Reprinted from [2]. © With permission from World Scientific

Properties	Notation used in the model	Values	Units
Density	ρ_L	998.2	kg/m^3
Viscosity (water)	η_L	1.022×10^{-4}	kg s/m^2
Surface tension (water–saturated vapor)	S	7.2×10^{-3}	kg/m
Thermal conductivity at $T = 273$–373 K	μ_0	0.48–0.6	W/mK
Thermal diffusivity	χ	1.6×10^{-7}	m^2/s
Specific heat at $T = 273$–373 K	C	3645–3897	J/K kg
Ratio of specific heats (water/vapor)	γ	1.33	

Calculations are performed for both cases at constant bubble temperature ($k = 1$) and for adiabatic conditions ($k = \gamma = 4/3$). Under such conditions, the time dependence of the bubble radius $R(t)$ is a better fit to our experimental results [1]. As follows from our calculations, the adiabatic expansion of the bubble provides a final radius of 3.5 μm for the bubble lifetime 200 ns, which agrees well with our experimental observations in [1].

The general feature of the simulation results is linear and smooth growth of the bubble radius with time when the bubble lifetime exceeds 30 ns [2]. The time dynamics of heating 30-nm nanoparticles calculated in Chap. 6 of this book shows that the total time for one cycle (heating the nanoparticles from an initial temperature 300 K to a maximum temperature, and then cooling back to the initial temperature) is about 30 ns. Thus, linear growth of the bubble radius with time after 30 ns can be explained by the absence of an active heat source in the medium (Slide 27.19). This result is confirmed also by the rate dynamics $v(t) = dR/dt$ for the bubble radius shown

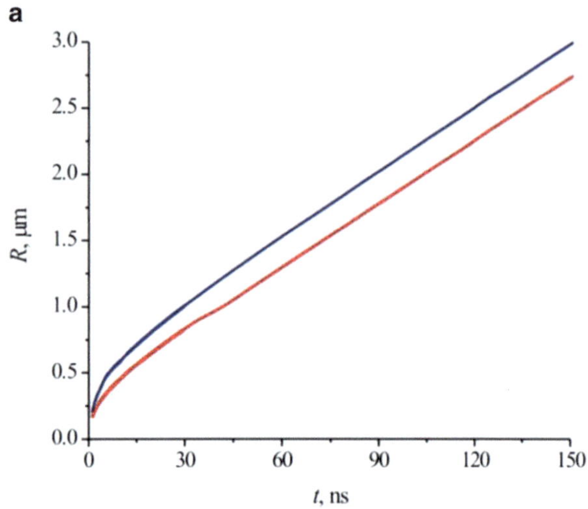

Fig. 9.15 (**a**) Bubble expansion around the 30-nm gold particle in blood irradiated by a laser pulse with energy density $E = 0.33$ J/cm^2 and pulse duration $\tau_L = 8$ ns at constant bubble temperature (*upper curve*) and under adiabatic conditions (*lower curve*). (**b**) Bubble expansion rate, $v(t) = dR/dt$. Reprinted from [2]. © With permission from World Scientific

in Fig. 9.15b. In the time range 0–30 ns, both curves $R(t)$ and $v(t)$ have strong nonlinearity due to active heat transfer from the nanoparticle to the surrounding medium. During the first several nanoseconds, the bubble rapidly expands with a high velocity of 40 m/s. Then, the expansion rate drops exponentially and reaches a constant saturation value of 16 m/s for $t > 30$ ns. Since there is no heating source in the system after 30 ns, the microbubble expands uniformly and linearly at a constant rate (Slide 27.19).

It is important to compare the bubble expansion rate to the thermal conduction rate in the surrounding medium (Slide 27.20). Using the thermal properties of blood given in Table 9.9 and the size of the locally heated region ($l \sim 1$ μm) as a focal spot of the laser radiation, we find that the approximate speed for heat conduction is [2]

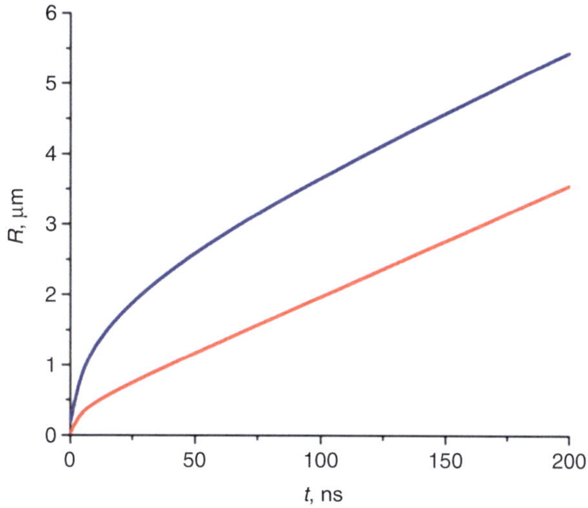

Fig. 9.16 Adiabatic expansion of the bubble produced by a single 30-nm particle (*lower curve*) and 190-nm continuous "cluster" (*upper curve*) irradiated by a laser pulse with energy density $E = 0.33$ J/cm^2 and pulse duration $\tau_L = 8$ ns. Reprinted from [2]. © With permission from World Scientific

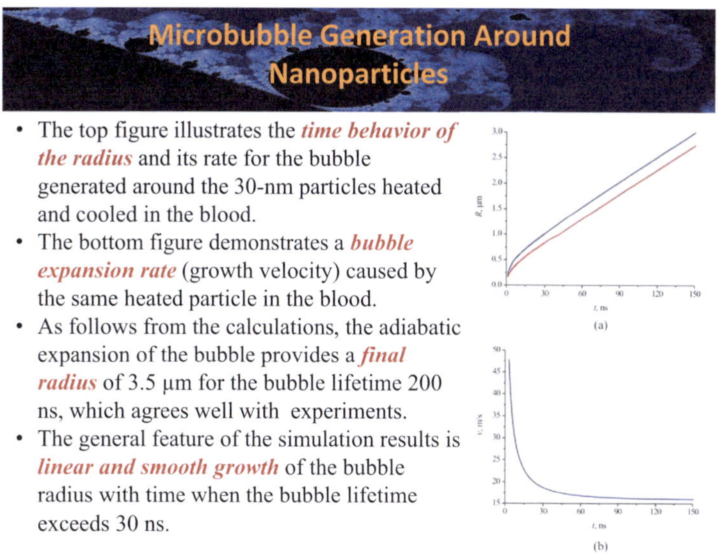

Slide 27.18 Microbubble generation around nanoparticles

$$v_{\text{thermal}} \approx \frac{\mu_0}{\rho C l} \approx 0.15\,\text{m/s}, \tag{9.10}$$

which is more than two orders of magnitude less than the bubble expansion rate. Here, ρ, C, and μ_0 are, respectively, the density, specific heat, and thermal conductivity of the surrounding medium. So, during a laser pulse of $\tau_L = 8$ ns, thermal damage occurs over a negligible distance from the nanoparticle in comparison to the bubble damage. Thus, bubble expansion occurs on a time scale much shorter than heat loss, which justifies the use of an adiabatic treatment during the expansion [2].

Microbubble Generation Around Nanoparticles (continued)

- The ***time dynamics of heating 30-nm nanoparticles*** calculated earlier shows that the total time for one cycle (heating of the nanoparticles from an initial temperature to maximum temperature, and then cooling back to the initial temperature) is about 30 ns.
- Thus, ***linear growth of the bubble radius*** with time after 30 ns can be explained by the absence of an active heat source in the medium.
- This result is confirmed also by the ***rate dynamics*** $v(t) = dR/dt$ for the bubble radius shown on previous slide.
- ***In the time range 0-30 ns***, both curves $R(t)$ and $v(t)$ have ***strong nonlinearity*** due to active heat transfer from the nanoparticle to the surrounding medium.
- ***During the first several nanoseconds***, the bubble rapidly expands with a high velocity of 40 m/s. Then, the ***expansion rate drops exponentially*** and reaches a constant saturation value of 16 m/s for t > 30 ns.

Slide 27.19 Microbubble generation around nanoparticles (continued)

Microbubble Generation Around Nanoparticles (continued)

- It is important to compare the ***bubble expansion rate*** to the ***thermal conduction rate*** in the surrounding medium.
- Using the thermal properties of blood and the size of the locally heated region ($l \sim 1$ μm) as a focal spot of the laser radiation, we find that the approximate ***speed for heat conduction*** is

$$v_{thermal} \approx \frac{\mu_0}{\rho C l} \approx 0.15 \text{ m/s},$$

which is more than ***two orders of magnitude less than the bubble expansion rate.*** So, during a laser pulse of $\tau_L = 8$ ns, thermal damage occurs over a negligible distance from the nanoparticle in comparison to the bubble damage.

- Thus, ***bubble expansion*** occurs on a time scale much shorter than heat loss, which justifies the use of an adiabatic treatment during the expansion.

Slide 27.20 Microbubble generation around nanoparticles (continued)

The thermal relaxation time is 1.75 μs for water and 1.56 μs for blood, which is much longer than the bubble lifetime of about 200 ns [5]. This means that for cell damage by bubble expansion, it is sufficient to heat only the nanoparticles up to the critical threshold temperature for the bubble generation (423 K), without having to

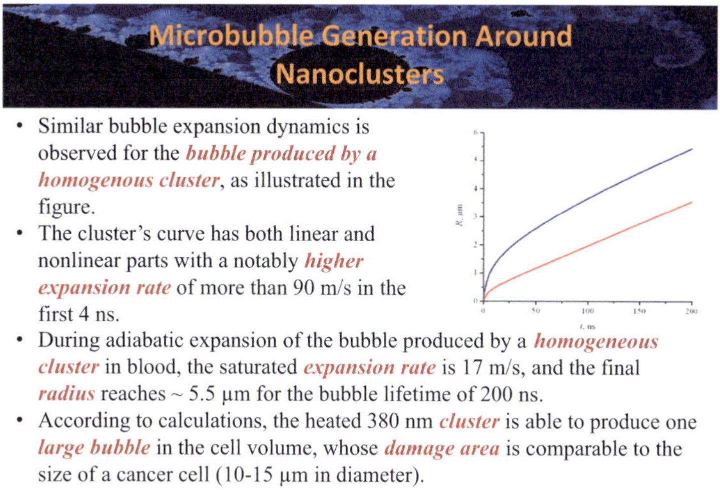

Slide 27.21 Microbubble generation around nanoclusters

heat the surrounding cellular medium. Therefore, the relatively low optical flux of 0.33 J/cm^2 ($\tau_L = 8$ ns) is enough for effective cell killing.

Similar bubble expansion dynamics is observed for the bubble produced by a homogenous cluster, as illustrated in Fig. 9.16 and Slide 27.21. The cluster's curve has both linear and nonlinear parts, with a notably higher expansion rate of more than 90 m/s in the first 4 ns. During adiabatic expansion of the bubble produced by a homogeneous cluster in blood, the saturated expansion rate is 17 m/s, and the final radius reaches ~5.5 μm for the bubble lifetime of 200 ns. According to our calculations, the heated 380 nm cluster is able to produce one large bubble in the cell volume, whose damage area is comparable to the size of a cancer cell (10–15 μm in diameter). Thus, the formation of just one large bubble around one large nanocluster aggregated in the cell volume is good enough for cancer cell killing [5].

In conclusion, the bubble mode of photothermolysis decreases the threshold optical flux up to 0.33 J/cm^2 ($\tau_L = 8$ ns) for cancer cell killing and dramatically increases the damage area due to rapid expanding nanobubbles around the nanoparticles and cluster [5] (Slides 27.22 and 27.23):

- The expansion of the bubble has strong nonlinearity in the time range 0–30 ns due to active heat transfer from nanoparticles to the surrounding medium. During the first several nanoseconds, the bubble produced around a 30-nm particle rapidly expands with a high velocity of 40 m/s. Then, the expansion rate drops exponentially and reaches a constant saturation value of 16 m/s for $t > 30$ ns. Since there is no heating source in the system after 30 ns, the microbubble expands uniformly and linearly at a constant rate [5].
- The adiabatic expansion of the bubble produced around a single 30-nm particle reaches a final radius of 3.5 μm for the bubble lifetime of 200 ns [5].
- The speed for heat conduction in the cellular medium ($v_{thermal} = 0.15$ m/s) is more than two orders of magnitude less than the bubble expansion rate

($v_{\text{saturated}} = 16$ m/s), so that during a laser pulse of $\tau_L = 8$ ns, thermal damage occurs over a negligible distance from the nanoparticle in comparison to the bubble damage [5].

- The heated 380 nm cluster is able to produce a single large bubble within the cell volume, whose damage area is comparable to the size of a cancer cell (10–15 μm in diameter) [5].

The ***bubble mode*** of photothermolysis decreases the threshold optical flux up to 0.33 J/cm^2 (τ_L = 8 ns) for cancer cell killing and ***dramatically increases the damage area*** due to rapid expanding nanobubbles around the nanoparticles and cluster:

- The ***expansion of the bubble*** has strong nonlinearity in the time range 0-30 ns due to active heat transfer from nanoparticles to the surrounding medium.
- ***During the first several nanoseconds***, the bubble produced around a 30-nm particle rapidly expands with a high velocity of 40 m/s.
- Then, the ***expansion rate*** drops exponentially and reaches a constant saturation value of 16 m/s for t > 30 ns. Since there is no heating source in the system after 30 ns, the microbubble expands uniformly and linearly at a constant rate.

Slide 27.22 Conclusions

- The adiabatic expansion of the bubble produced around a single 30-nm particle reaches a ***final radius*** of 3.5 μm for the bubble lifetime of 200 ns.
- The ***speed for heat conduction*** in the cellular medium ($v_{thermal}$ = 0.15 m/s) is more than ***two orders of magnitude less*** than the ***bubble expansion rate*** ($v_{saturated}$ = 16 m/s), so that during a laser pulse of τ_L = 8 ns, thermal damage occurs over a negligible distance from the nanoparticle in comparison to the bubble damage.
- The heated 380-nm ***cluster*** is able to produce a single ***large bubble*** within the cell volume, whose ***damage area*** is comparable to the size of a cancer cell (10-15 μm in diameter).

Slide 27.23 Conclusions (continued)

9.4 Computer Practicum 7 (Lecture 28): Simulating Nanobubble Dynamics (Slide 28.1)

- Computer Code
- Practice Problem 11: Bubble Formation and Growth
  - *Abstract*
  - *Model*
  - *Discussion and Results*
  - *Conclusions*

9.4.1 Computer Code

In Sect. 9.3 we introduced a spherical bubble dynamics model, Eqs. (9.4)–(9.6), based on the generalized Rayleigh–Plesset equation, allowing to simulate bubble nucleation around the nanoparticles and their rapid expansion within the cellular volume. In this section we develop a Maple code to numerically solve the Rayleigh–Plesset equation, and we give a practice example of nanobubble growth in time in different surrounding biological media (Slide 28.2).

Appendix H contains a Maple code for simulations of the nanobubble dynamics. The code begins with introducing the partial pressure of the contaminant gas contained in the bubble given by Eq. (9.5) and shown in Fig. 9.17 and Slide 28.2. Then we define all constants used in the bubble dynamic model, Eqs. (9.4)–(9.6), as shown in Fig. 9.17 and Slide 28.2.

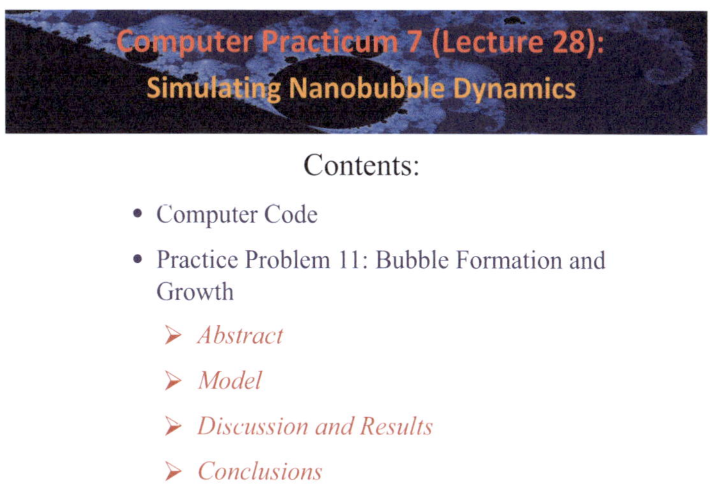

Slide 28.1 Contents

Computer Code

- *In this section* we develop a *Maple code* to numerically *solve the Rayleigh-Plesset equation*, and we give a *practice example* of nanobubble growth in time in different surrounding biological media.
- *Appendix H* contains a *Maple code* for simulations of the nanobubble dynamics.
- The *code* begins with introducing the partial pressure of the contaminant gas contained in the bubble given by Eq. (9.5) and shown below.
- Then we *define all constants* used in the *bubble dynamic model*, Eqs. (9.4)-(9.6).

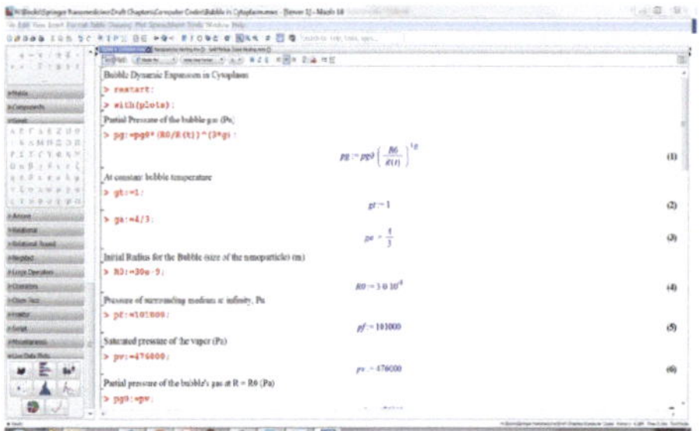

Slide 28.2 Computer code

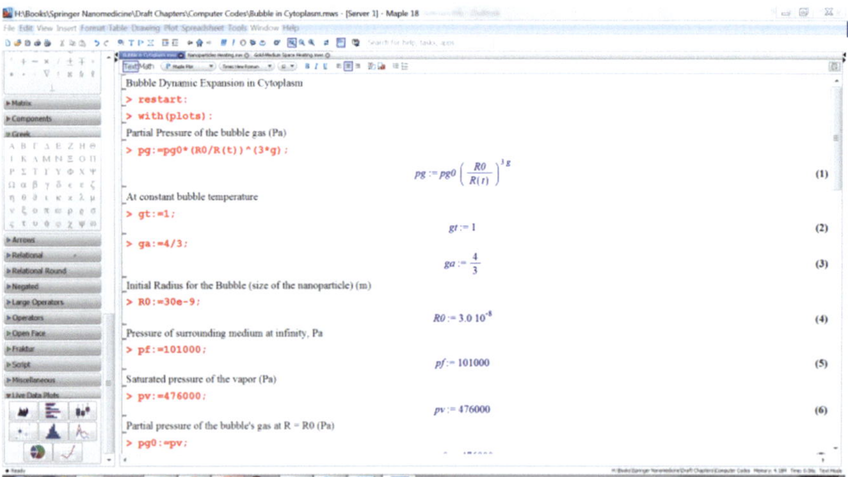

Fig. 9.17 Input data for the bubble dynamics model

Next, we solve the Rayleigh–Plesset equation for two different bubble growing modes as shown in Fig. 9.18 and Slide 28.3: (1) nonadiabatic expansion at constant bubble temperature when $k = 1$ and (2) adiabatic expansion when $k = \gamma$, where γ is the ratio of the specific heat of the vapor at constant pressure to the specific heat at constant volume. The solutions of second-order differential equations are derived using the *dsolve* function in Maple which relies on the numeric *Runge–Kutta* method.

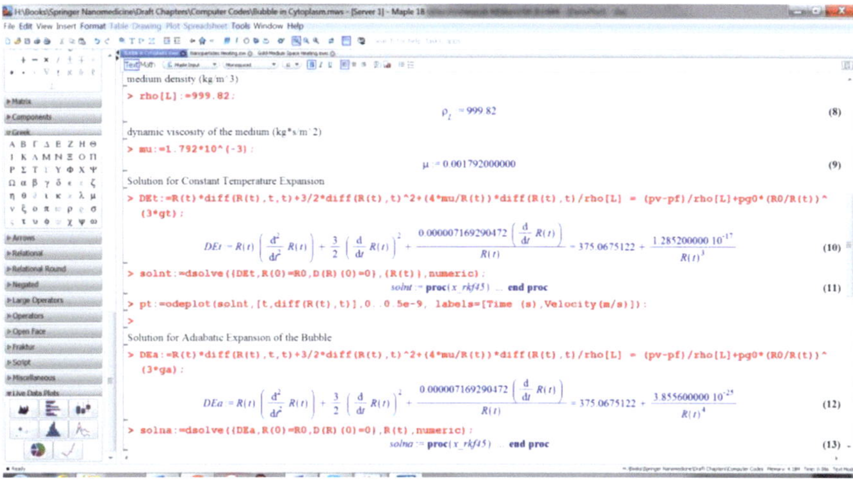

Fig. 9.18 Solving the Rayleigh–Plesset equation for two different bubble growing modes

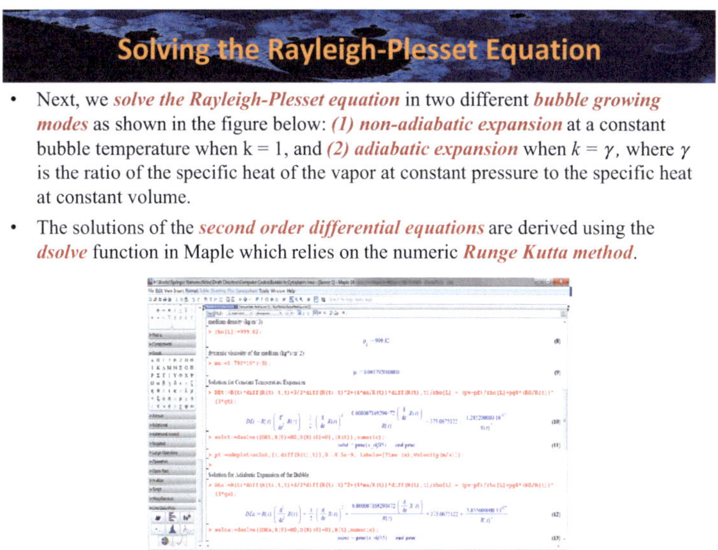

Slide 28.3 Solving the Rayleigh–Plesset equation

We arrange the results of the simulations in the form of plots for the bubble expansion rate (i.e., bubble growth velocity) and the radius of the bubble as a function of time shown in Fig. 9.19 and Slide 28.4. The Practice Example 11 presents the results of these simulations in detail.

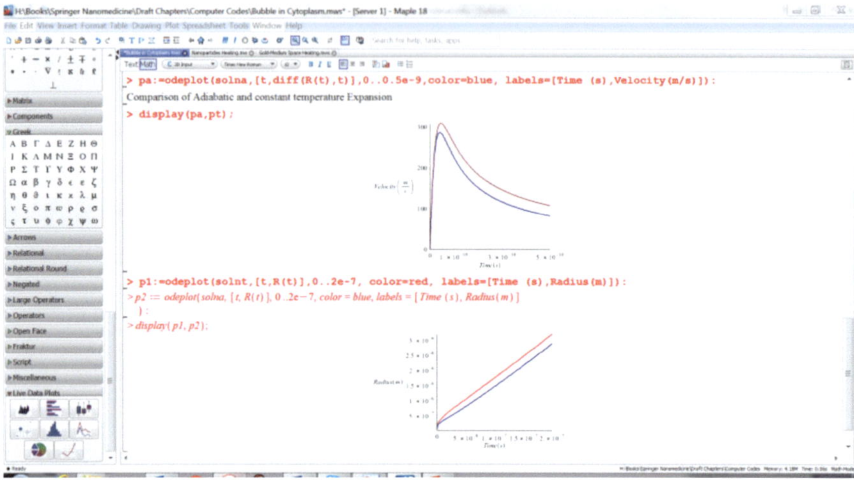

Fig. 9.19 Arranging the output results in the form of plots for the bubble expansion rate (i.e., bubble growth velocity) and the radius of the bubble as a function of time

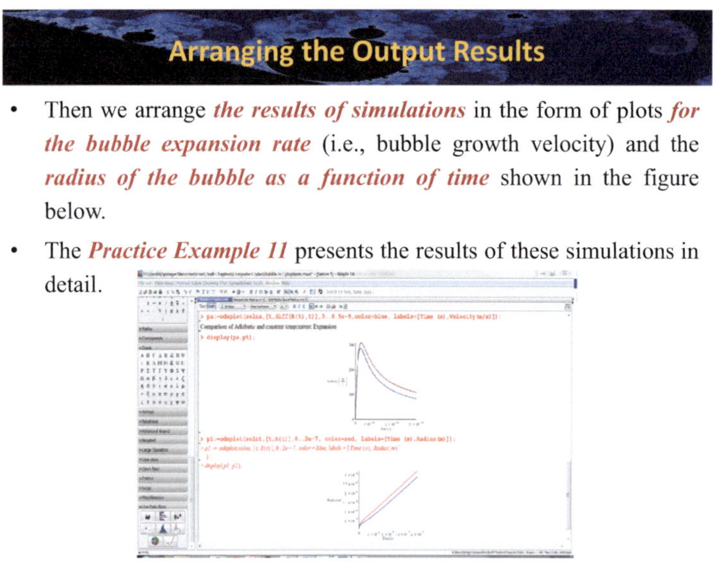

Slide 28.4 Arranging the output results

9.4.2 Practice Example 11: Bubble Formation and Growth

9.4.2.1 Abstract

In this simulation, we describe the behavior of a bubble formed by the vaporization of material surrounding a nanoparticle when the particle is heated to above 423 K (Slide 28.5). We use the generalized Rayleigh–Plesset equation to describe the bubble's expansion in time. In brief, the model takes into account forces due to outward pressure, viscosity of the surrounding medium, and ambient pressure of the surrounding medium. We model the bubble dynamics for a nonadiabatic expansion case with constant temperature and then take into account adiabatic expansion.

9.4.2.2 Model

One possible mode for causing damage in cancer cells is to use the nanoparticle heating to generate a nanobubble that rapidly expands to damage the cancer cell via mechanical force. As shown in Sect. 9.3, the expansion of the bubble causes the area of damage to be much larger than the area directly damaged by thermal ablation. The goal of this practice example is to determine the behavior of nanobubble formation around a nanoparticle. The behavior characteristics that will be determined are the time dynamics of the bubble expansion, specifically the bubble size and maximum expansion velocity. The media used for the calculations with density and dynamic viscosity properties are listed in Table 9.10

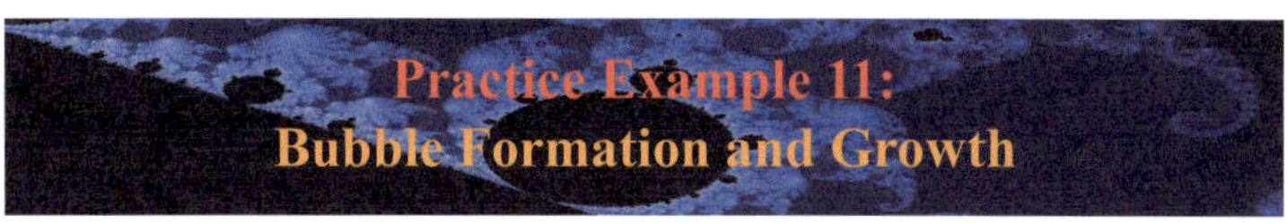

- In this simulation, we describe the ***behavior of a bubble*** formed by the ***vaporization of material surrounding a nanoparticle*** when the particle is heated to above 423 K.
- We use the ***generalized Rayleigh-Plesset equation*** to describe the bubble's expansion in time.
- In brief, the model takes into account ***forces due to outward pressure, viscosity of the surrounding medium, and ambient pressure of the surrounding medium***.
- We model the bubble dynamics for a ***non-adiabatic expansion*** case with constant temperature and then take into account ***adiabatic expansion***.
- ***The behavior characteristics*** that will be determined are the time dynamics of the bubble expansion, specifically the ***bubble size and maximum expansion velocity.***

Slide 28.5 Practice Example 11: Bubble formation and growth

Table 9.10 Density and dynamic viscosity of the surrounding media

Medium	Density (kg/m^3)	Dynamic viscosity (kg/s m)
Blood	998.2	1.022×10^{-4}
Cytoplasm	999.82	1.792×10^{-3}
Cell membrane	1100	8.90×10^{-4}
Fat	900	3.0×10^{-3}
Water	1000	1.022×10^{-4}

- The ***media*** used for the calculations ***with density and dynamic viscosity properties*** are listed in the table below.
- These media are the *most* common locations for nanoparticle therapy targets: ***cytoplasm, blood, cell membrane, fat and water***.

Medium	Density (kg/m^3)	Dynamic Viscosity (kg/s m)
Blood	998.2	1.022×10^{-4}
Cytoplasm	999.82	1.792×10^{-3}
Cell Membrane	1100	8.90×10^{-4}
Fat	900	3.0×10^{-3}
Water	1000	1.022×10^{-4}

Slide 28.6 Practice Example 11: Input data

and Slide 28.6. These media are the most common locations for nanoparticle therapy targets: cytoplasm, blood, cell membrane, fat, and water. We model the bubble formation around the nanoparticles and their expansion in those biological media using the Rayleigh–Plesset equation and Maple code given in Appendix H. The simulations are performed for two expansion modes: adiabatic bubble expansion and constant temperature expansion.

9.4.3 *Discussion and Results*

We are assuming that the energy from the radiation is sufficient to raise the temperature of the nanoparticle sufficiently to vaporize the surrounding medium and thus form the bubble (Slide 28.7). The Maple code written in Appendix H assumes that this radiation energy is present and does not take into account the

- We are assuming that the ***energy from the radiation*** is sufficient to raise the temperature of the nanoparticle sufficiently ***to vaporize the surrounding medium*** and thus form the bubble.
- The ***Maple code*** written in Appendix H assumes that this radiation energy is present and does not take into account the optical or thermal properties of the nanoparticles.
- The only ***contribution that the nanoparticle makes*** in the model is ***its size***, which is taken as the ***initial size of the nanobubble***.
- ***The nanoparticles*** serve the role of the ***bubble nucleation centers.***

Slide 28.7 Practice Example 11: Discussion and results

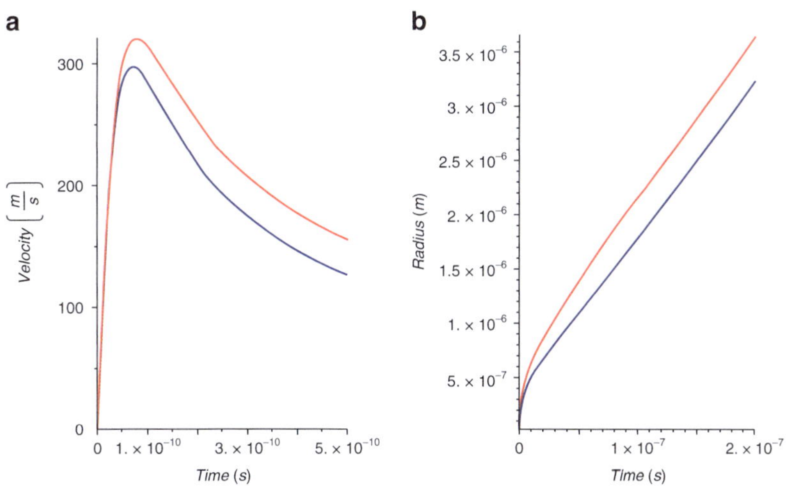

Fig. 9.20 Bubble expansion rate (**a**) and radius (**b**) as a function of time for adiabatic (*blue curve*) and nonadiabatic (*red curve*) expansion modes in cytoplasm around a heated 50-nm particle

optical or thermal properties of the nanoparticles. The only contribution that the nanoparticle makes in the model is its size, which is taken as the initial size of the nanobubble. The nanoparticles serve the role of the bubble nucleation centers.

First we study the effect of two different expansion modes on bubble formation and growth around a heated 50-nm particle in cytoplasm, where Fig. 9.20 and Slide 28.8 show the difference between the adiabatic and nonadiabatic modes at constant temperature. Clearly, the bubble expands more slowly for the

Practice Example 11:
Adiabatic and Non-Adiabatic Expansion

- First we study ***the effect of two different expansion modes*** on bubble formation and growth around a heated 50-nm particle.
- The ***figure shows the difference*** between the adiabatic mode and non-adiabatic mode (at constant temperature).
- Clearly the ***bubble expands more slowly for the adiabatic mode***.

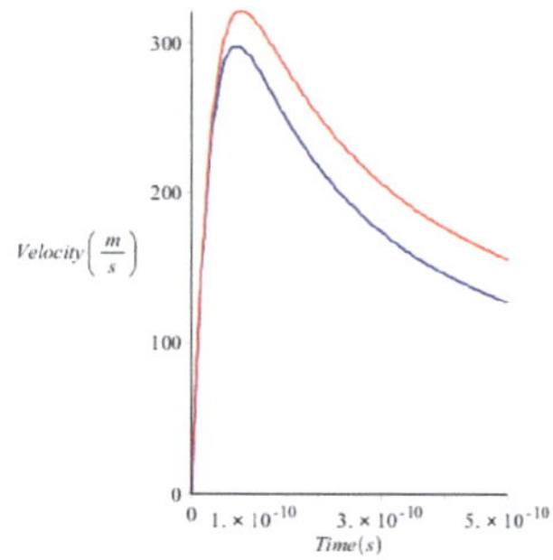

Slide 28.8 Practice Example 11: Adiabatic and nonadiabatic expansion

adiabatic mode. The nonadiabatic bubble growth at constant temperature has a slightly higher initial growth, resulting in a 0.5 μm bigger bubble, but both reach the same linear growth. Thus, the nonadiabatic expansion of the bubble at constant temperature provides an upper level for the expansion rate (velocity) and radius of the bubble at a given instant in the time, and the adiabatic expansion mode gives the lower values for the bubble growth rate (Slide 28.9).

These simulations also show that the initial nucleation of the bubble around an overheated nanoparticle happens in 0.1 ns with an explosive expansion rate of 300 m/s (Fig. 9.20a and Slide 28.10). After the bubble is generated, the expansion rate drops exponentially and reaches a constant saturation value of 16 m/s for time $t > 30$ ns. As might be expected, when formed, the vapor bubble quickly expands, and then as pressure decreases with increased volume, it slows dramatically as the internal pressure nears the external pressure. Overall, the linear growth spans a few microns with a steady rate of two orders of magnitude higher than that for the thermal effects discussed in the previous section (Slide 28.10).

After 200 ns (bubble lifetime), the adiabatic expansion bubble reaches the size of 3.0 μm (Fig. 9.20b and Slide 28.11), while the nonadiabatic constant temperature expansion bubble is 3.5 μm in radius (Fig. 9.20b and Slide 28.11). This is generated from an initial size of 50 nm for the nucleation center (nanoparticle), meaning the radius affected by the bubble is around 70 times larger than the radius of the nanoparticle that generated the bubble. Cell death from thermal ablation is only effective out to about 5 times the radius of the initial single particle. In addition, the laser pulse energy density required to heat a single nanoparticle for producing thermal damage over a significant area is much higher than used here to generate the bubble (Slide 28.11).

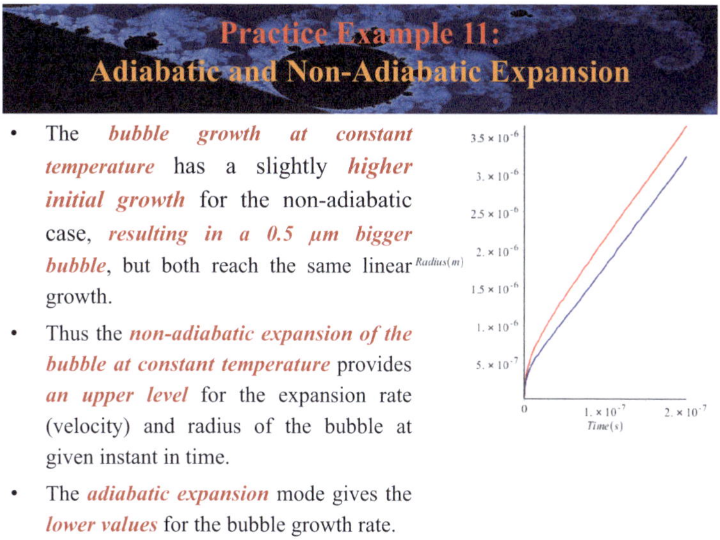

Slide 28.9 Practice Example 11: Adiabatic and nonadiabatic expansion (continued)

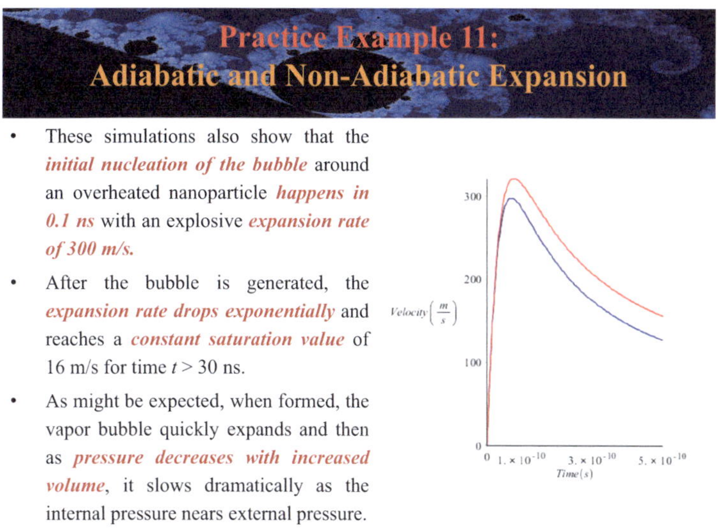

Slide 28.10 Practice Example 11: Adiabatic and nonadiabatic expansion (continued)

Next, we investigate the bubble formation and growth for varying nanoparticle sizes, demonstrated in Fig. 9.21 and Slide 28.12 for the nonadiabatic expansion case. Simulations are performed for 20-nm, 50-nm, and 100-nm particles in cytoplasm. Unexpectedly from the plots of varying nanoparticle sizes, the larger particle results in a much slower bubble generation and expansion rate (Fig. 9.21a), forming an overall slightly smaller bubble size for the same time interval of bubble growth (Fig. 9.21b).

Practice Example 11:
Adiabatic and Non-Adiabatic Expansion

- ***After 200 ns (bubble lifetime),*** the ***adiabatic expansion bubble reaches the size of 3.0 µm*** shown in figure, while the non-adiabatic ***constant temperature expansion bubble is 3.5 µm in radius.***
- This is generated from initial size of 50 nm for the nucleation center (nanoparticle), meaning the ***radius affected by the bubble is around 70 times larger*** than the ***radius of the nanoparticle*** that generated the bubble.

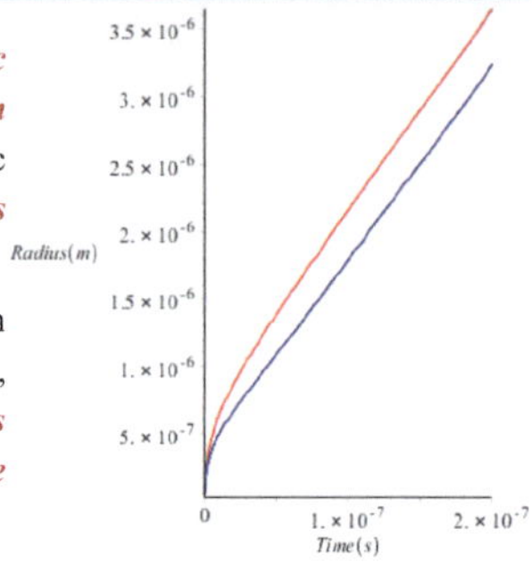

- Cell death from ***thermal ablation*** is only effective out to ***about 5 times the radius of the initial single particle***. In addition, the ***laser pulse energy density*** required ***to heat a single nanoparticle for producing thermal damage*** over a significant area ***is much higher*** than used here ***to generate the bubble***.

Slide 28.11 Practice Example 11: Adiabatic and nonadiabatic expansion (continued)

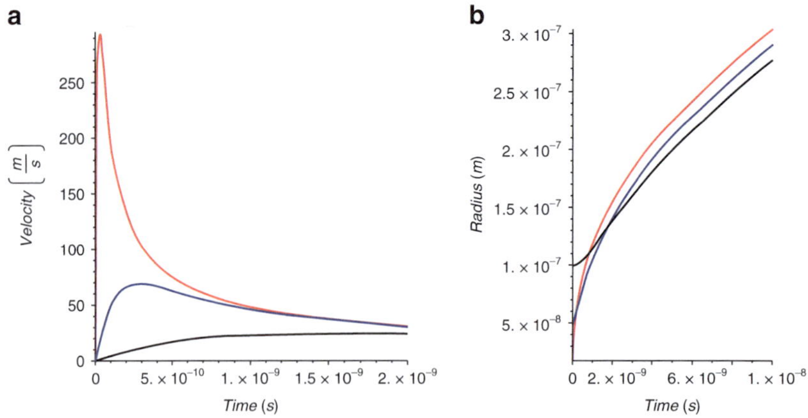

Fig. 9.21 Bubble expansion rate (**a**) and radius (**b**) as a function of time for the nonadiabatic expansion mode in cytoplasm. The curves in both graphs from top to bottom are for 20 nm (*red*), 50 nm (*blue*) and 100 nm (*black*) particles, providing the initial sizes for the nanobubbles

We also explore the effect of different biological media on nanobubble dynamics (Slide 28.13). The media used in the simulations are blood, cytoplasm, cell membrane, and fat. The results of calculations for bubble generation around a 20-nm particle and its expansion in those media at constant temperature are presented in Fig. 9.22 and Slide 28.13. Figure 9.22a shows the bubble expansion rate in four biological media, and Fig. 9.22b shows the bubble radius as a function of time. As follows from the plots, the bubble growth rate and expansion is distinctly different in the different media. For both the bubble radius and velocity, fat is not as

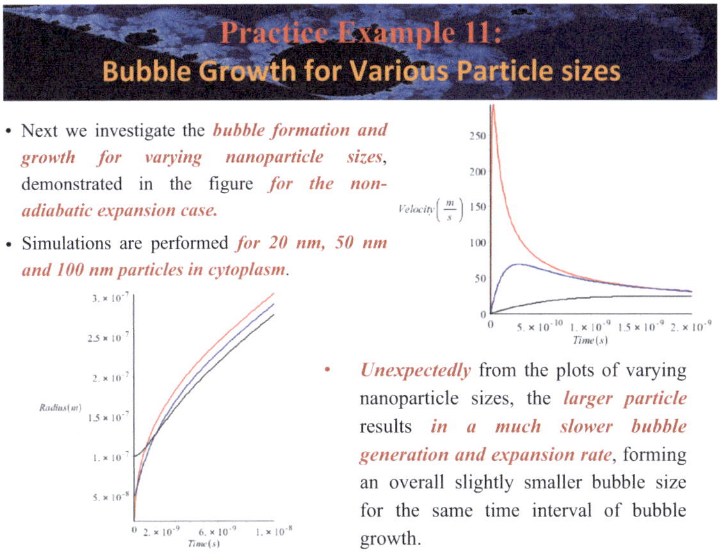

Slide 28.12 Practice Example 11: Bubble growth for various particle sizes

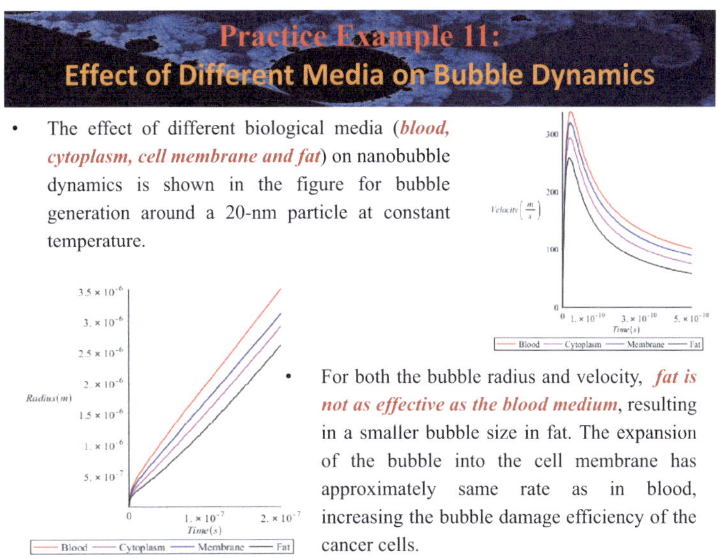

Slide 28.13 Practice Example 11: Effect of different media on bubble dynamics

effective as the blood medium, resulting in a smaller bubble size in fat. The expansion of the bubble into the cell membrane has approximately the same rate as in blood, increasing the bubble damage efficiency of the cancer cells.

The general feature of the simulation results is linear growth of the bubbles after the source of the radiation heating nanoparticle is turned off. For all biological

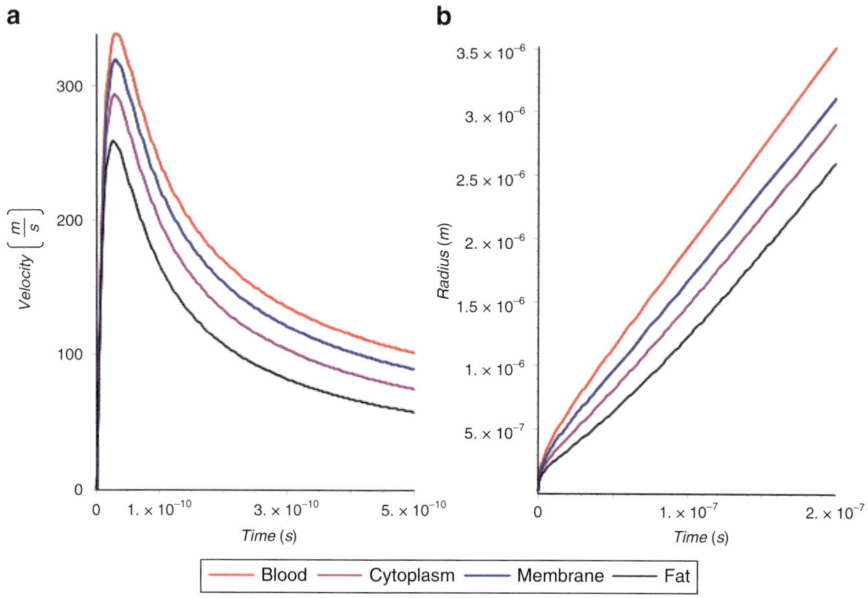

Fig. 9.22 (**a**) Nonadiabatic expansion rate and (**b**) radius of the bubble generated around a 20-nm particle in various biological media

media, the bubble radius greatly increases over time. From a 20-nm particle nucleation center, the bubbles can reach up to 3.5 μm (in blood) during a bubble lifetime of 200 ns (Fig. 9.22b). Even in fat, the final radius of the bubble is 2.5 μm, which is 125 times larger than the nanoparticle radius.

9.4.4 Conclusions (Slides 28.14 and 28.15)

- The bubble mode of photothermolysis decreases the threshold energy density of the radiation for cancer cell killing and significantly increases the damage area due to rapid expanding bubbles around the nanoparticles and cluster.
- The bubble generation mode for cancer cell killing can cause damage at distances around 70–120 times the radius of the initial size of the bubble at the instant of nucleation, i.e., the radius of the nanoparticle.
- In addition, only a few nanoparticles would be required to cause damage in all parts of a cancer cell, as the damage region from bubbles is on the order of microns. For thermal ablation by single nanoparticles, the damage area is hundreds of nanometers, meaning that many more particles would be required to cause death in all parts of the cell.
- Bubbles expand at constant temperature more rapidly than they expand adiabatically while still under the heat source, but both expand at the same rate after the source is removed. As follows from our calculations, the expansion

of a bubble at constant temperature provides the final radius of 3.5 μm for the bubble lifetime of 200 ns.

- Distinctly different behaviors are observed for bubbles in different biological media. The fat surrounding medium is less effective than the blood or cell membrane medium, resulting in a smaller bubble size in fat.

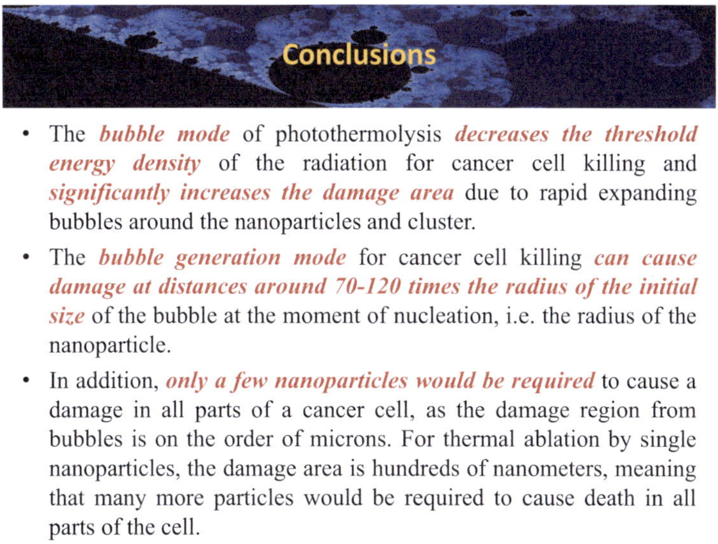

Slide 28.14 Conclusions

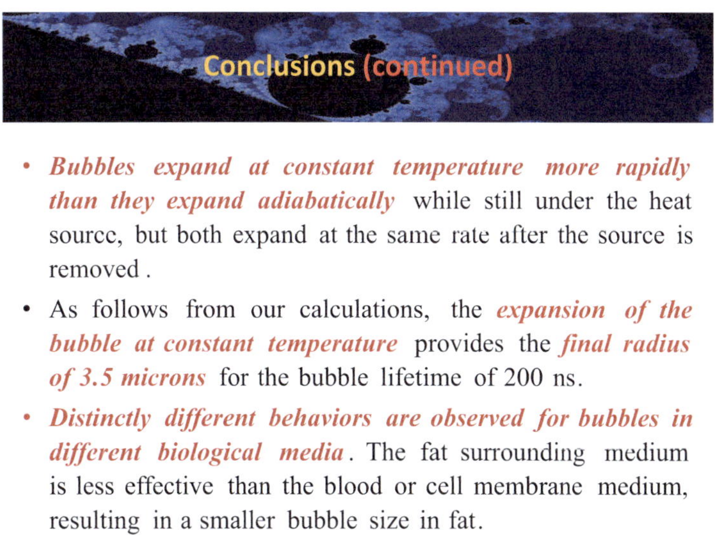

Slide 28.15 Conclusions (continued)

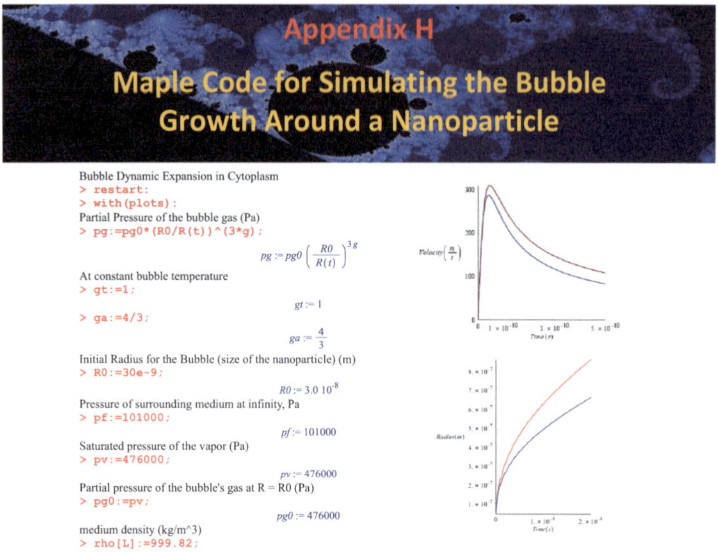

Slide 28.16 Appendix H: Maple code for simulating the bubble growth around a nanoparticle

9.5 Lecture 29: Laser-Induced Thermal Explosion Mode—"Nanobombs" (Slide 29.1)

- Introduction
- Thermal Explosion
- Coulomb Explosion
- Time Scale Approximations
- Threshold Intensity for Laser-Induced Thermal Explosion
- Cell Damage Effects
- Conclusions
- References

9.5.1 *Introduction*

In this section, we discuss a new mechanism for selective laser killing of abnormal cells by laser thermal explosion of single nanoparticles—"nanobombs"—delivered to the cells [5, 6] (Slide 29.2). Thermal explosion of nanoparticles is realized for ultrashort laser pulses when heat is generated within the strongly absorbing target more rapidly than the heat can diffuse away. Laser-induced rapid explosive evaporation of gold nanoparticles enables the generation of stress transients, shock waves, and high local pressure. A schematic picture of these complex

Lecture 29: Laser-Induced Thermal Explosion Mode – 'Nanobombs'

Contents:

- Introduction
- Thermal Explosion
- Coulomb Explosion
- Time Scale Approximations
- Threshold Intensity for Laser-Induced Thermal Explosion
- Cell Damage Effects
- Conclusions

Slide 29.1 Contents

Introduction

- In this lecture, we discuss a new mechanism for selective laser killing of abnormal cells by laser ***thermal explosion of single nanoparticles – 'nanobombs'*** – delivered to the cells.
- ***Thermal explosion of nanoparticles*** is realized ***for ultrashort laser pulses*** when heat is generated within the strongly-absorbing target more rapidly than the heat can diffuse away.

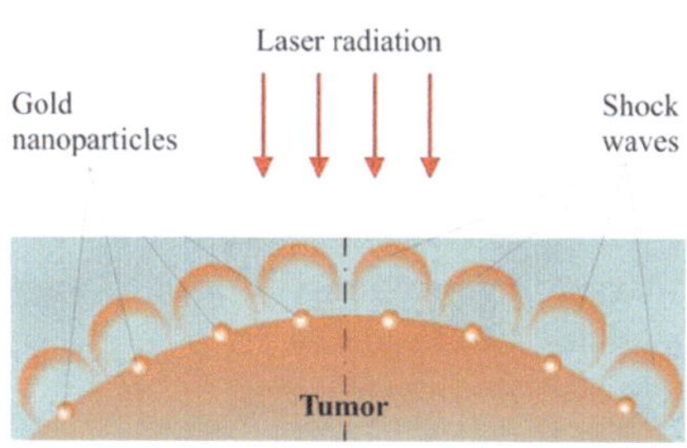

Slide 29.2 Introduction

physical effects is provided in Figs. 9.23 and 9.24, and Slide 29.3. There are two main physical mechanisms that can lead to the laser-induced explosion of nanoparticles: ***thermal explosion mode*** through electron–phonon excitation–relaxation and ***Coulomb explosion mode*** through multiphoton ionization. This section contains material adapted from our publications [2, 5, 6].

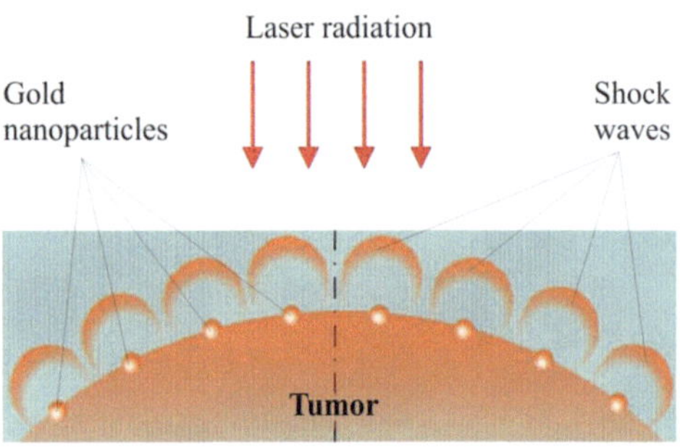

Fig. 9.23 Principle of the thermal explosion mode of nanoparticles—“nanobombs”—in selective nanophotothermolysis of cancer. Reprinted from [2]. © With permission from World Scientific

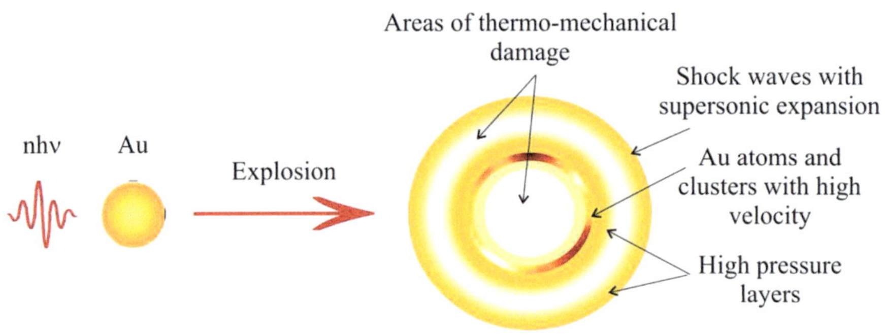

Fig. 9.24 Laser-induced thermal explosion of a gold nanoparticle. Reprinted from [2]. © With permission from World Scientific

9.5.2 *Thermal Explosion*

Under the action of ultrashort laser pulses in the spectral range of surface plasmon resonances (for a 30-nm solid spherical gold nanoparticle, the wavelength of radiation corresponding to the maximum absorption is ~530 nm), the atoms within a nanoparticle are excited to upper electronic states owing to the absorption of many photons (Slide 29.4). Through rapid relaxation (picosecond time scale), these atoms decay to their ground state with effective electron–phonon conversion of the absorbed photon energy into thermal energy, as shown in Chap. 6 of this book. Depending on the nanoparticle temperature, T, the following scenarios, individually or in combination, can occur [2, 6] (Slide 29.5):

1. For $T < T_{LV}$, where T_{LV} is the liquid vaporization temperature (water in many cases), ~150–350 °C, thermal expansion of a single gold nanoparticle and surrounding thin liquid layer is observed, which is accompanied by the generation of linear acoustic waves, known as the ‘classic’ photoacoustic effect.
2. If $T_{LV} \leq T < T_{GNM}$, where T_{GNM} is the gold nanoparticle melting point, ~1063 °C, a bubble is formed as discussed in Sects. 9.3 and 9.4. Bubble expansion and collapse is accompanied by the production of acoustic and shock waves as shown in Figs. 9.23 and 9.24.
3. For the temperature range $T_{GNM} \leq T < T_{GNB}$, where T_{GNB} is the gold nanoparticle boiling point of ~2710 °C, the nanoparticles are melting.

4. When the temperature exceeds a phase transition temperature, $T \geq T_{GNB}$, the particles are boiling with the formation of gold vapor around liquid gold drops.

The photothermal process of scenario 3, and especially scenario 4, can lead to gold nanoparticle fragmentation into smaller parts—"nanobullets"—and to thermal explosion of a gold nanoparticle into single atoms—"nanobombs" (Slide 29.5).

Introduction (continued)

- Laser-induced rapid ***explosive evaporation of gold nanoparticles*** enables the ***generation of stress transients, shock waves and high local pressure***, as shown chematically in the figure below.
- There are ***two main physical mechanisms*** that could lead to the laser-induced explosion of nanoparticles: ***thermal explosion mode*** through electron-phonon excitation-relaxation and ***Coulomb explosion mode*** through multiphoton ionization.

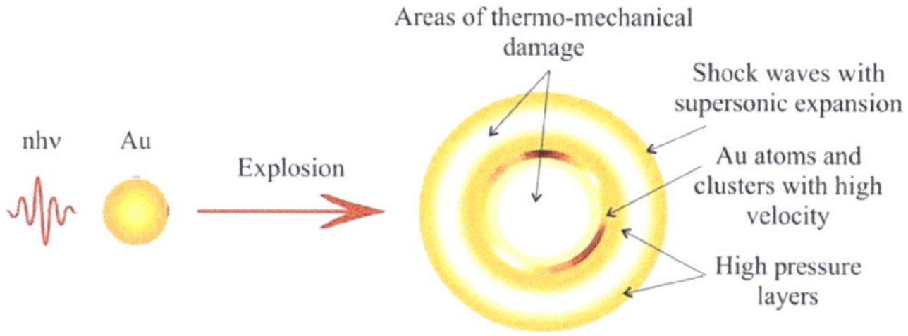

Slide 29.3 Introduction (continued)

Thermal Explosion

- Under the action of ultrashort laser pulses in the spectral range of the ***surface plasmon resonances*** (for a 30-nm solid spherical gold nanoparticle, the wavelength of radiation corresponding to the maximum absorption is ~ 530 nm), ***the atoms within a nanoparticle are excited to upper electronic states*** owing to the absorption of many photons.
- ***Through rapid relaxation*** (picosecond time scale), these atoms decay to their ground state with ***effective electron-phonon conversion*** of the absorbed photon energy ***into thermal energy***, as discussed earlier.
- ***Depending on the nanoparticle temperature***, *T*, the following scenarios, individually or in combination, can occur.

Slide 29.4 Thermal explosion

Thermal Explosion (continued)

1. ***For*** $T < T_{LV}$, where T_{LV} is the liquid vaporization temperature, ~ 150-350ºC, thermal expansion of a single gold nanoparticle and surrounding thin liquid layer is observed, which is accompanied by the generation of linear acoustic waves.
2. ***If*** $T_{LV} \leq T < T_{GNM}$, where T_{GNM} is the gold nanoparticle melting point, ~ 1,063°C, a bubble is formed as discussed earlier. Bubble expansion and collapse is accompanied by the production of acoustic and shock waves.
3. ***For the temperature range*** $T_{GNM} \leq T < T_{GNB}$, where T_{GNB} is the gold nanoparticle boiling point of ~ 2,710°C, the nanoparticles are melting.
4. ***When the temperatures exceeds a phase transition temperature*** $T \geq T_{GNB}$, the particles are boiling with the formation of gold vapor around liquid gold drops.

Scenario 3, and especially scenario 4, can lead to gold nanoparticle ***fragmentation*** into smaller parts – "***nanobullets***" – and to ***thermal explosion*** of a gold nanoparticles into single atoms – "***nanobombs***".

Slide 29.5 Thermal explosion (continued)

9.5.3 Coulomb Explosion

For ultrashort laser pulses (e.g., femtosecond time scale) comparable to the time scale of electron–electron interactions and shorter than the electron–phonon interaction time, a nonphotothermal mechanism of nanoparticle explosion can occur through multiphoton ionization when the absorbed photon energy is transferred directly to the electrons, leading to their ejection due to the Coulomb explosion mechanism [5] (Slide 29.6). High plasmon-resonance absorption of gold nanoparticles, discussed in Chap. 5 of this book, may facilitate this effect due to thermionic electron emission.

9.5.4 Time Scale Approximations

The goal of the theoretical modeling is to estimate the threshold laser energy density, E_{expl}, required for realization of the thermal explosion mode of gold nanoparticles, and to compare the calculated data with available experimental results [2] (Slide 29.7). This mode is realized through the rapid overheating of a strongly absorbing target during a short laser pulse when the influence of heat diffusion is minimal. Let us first estimate the time scale for thermal relaxation due to heat diffusion from the surface of a nanoparticle.

In the vicinity of the nanoparticle $\leq 5r_0$, where r_0 is the nanoparticle's radius, the thermal relaxation time for a spherical gold nanoparticle can be estimated as

Coulomb Explosion

- ***For ultrashort laser pulses*** (e.g., femtosecond time scale) comparable to the time scale of electron-electron interactions and shorter than the electron-phonon interaction time, a non-photothermal mechanism of nanoparticle explosion can occur through ***multiphoton ionization*** when the absorbed photon energy is transferred directly to the electrons, leading to their ejection due to the ***Coulomb explosion mechanism***.
- ***High plasmon-resonance absorption*** of gold nanoparticle, discussed in Chapter 5 of this book, may facilitate this effect due to thermionic electron emission.

Slide 29.6 Coulomb explosion

Time Scale Approximations

- ***The goal of the theoretical modeling*** is to estimate the ***threshold laser energy density***, E_{expl}, required ***for*** realization of the ***thermal explosion mode*** of gold nanoparticles, and to compare the calculated data with available experimental results.
- This mode is realized through the ***rapid overheating of a strongly-absorbing target*** during a short laser pulse when the influence of heat diffusion is minimal. Let us first estimate the ***time scale*** for thermal relaxation due to heat diffusion from the surface of the nanoparticle.
- In the vicinity of the nanoparticle $\leq 5r_0$, where r_0 is the nanoparticle's radius, the ***thermal relaxation time*** for a spherical gold nanoparticle can be estimated as $\tau_T = r_0^2/6.75k$, where k is the thermal diffusivity. For radii of $r_0 = 50$, 100 and 200 nm, the thermal relaxation times τ_T (for water, $k = 1.44 \times 10^{-3}$ cm^2/s) are approximately ***2.6, 10 and 41 ns***, respectively. For a laser pulse duration $\tau_L \leq \tau_T$, ***heat is generated within the gold nanoparticle more rapidly than can be diffused away***, which leads to ***explosion of the nanoparticle***. This condition for smaller gold nanoparticles is valid with picosecond and femtosecond laser pulses.

Slide 29.7 Time scale approximations

$\tau_T = r_0^2/6.75k$, where k is the thermal diffusivity. For radii of $r_0 = 50$, 100, and 200 nm, the thermal relaxation times τ_T (for water ($k = 1.44 \times 10^{-3}$ cm^2/s) are approximately 2.6, 10, and 41 ns, respectively. For a laser pulse duration $t_L \leq \tau_T$, heat is generated within the gold nanoparticle more rapidly than can be diffused away, which leads to explosion of the nanoparticle [2] (Slide 29.7). This condition for smaller gold nanoparticles is valid with picosecond and femtosecond laser pulses.

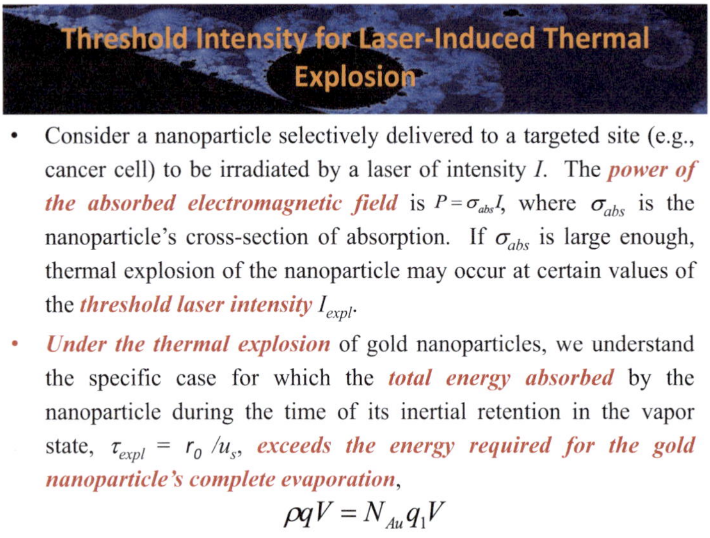

Slide 29.8 Threshold intensity for laser-induced thermal explosion

9.5.5 Threshold Intensity for Laser-Induced Thermal Explosion

Consider a nanoparticle selectively delivered to a targeted site (e.g., cancer cell) to be irradiated by a laser of intensity I (Slide 29.8). The power of the absorbed electromagnetic field is $P = \sigma_{\text{abs}} I$, where σ_{abs} is the nanoparticle's cross-section of absorption. If σ_{abs} is large enough, thermal explosion of the nanoparticle may occur at certain values of the threshold laser intensity I_{expl}, which is less than the threshold intensity for optical plasma formation in the surrounding medium [5, 6]. Under the thermal explosion of gold nanoparticles, we understand the specific case for which the total energy absorbed by the nanoparticle during the time of its inertial retention in the vapor state, $\tau_{\text{expl}} = r_0/u_{\text{s}}$, exceeds the energy required for the gold nanoparticle's complete evaporation [2] (Slide 29.8):

$$\rho q V = N_{\text{Au}} q_1 V. \tag{9.11}$$

Here, u_{s} is the sound velocity in Au vapor at the critical temperature $T_{\text{cr}} \sim T_{\text{GNB}}$; ρ is the volume density of nanoparticle; N_{Au} is the number of Au atoms per unit volume; q and q_1 are the particle's specific heat of evaporation per unit mass and per particle, respectively; and V is the nanoparticle's volume. We have a laser pulse duration of $\tau_{\text{L}} > \tau_{\text{expl}} = r_0/u_{\text{s}}$ (~1 ps for r_0 ~ 10 nm), where τ_{expl} is an explosive evaporation time.

Threshold Intensity for Laser-Induced Thermal Explosion (continued)

- To compute I_{expl}, we can write the following ***energy balance equation***:

$$\sigma_{abs} I_{\mathrm{expl}} \frac{r_0}{u_s} \approx \rho q V$$

- In terms of the absorption efficiency Q_{abs}, the ***threshold intensity*** of the laser radiation ***for thermal explosion*** of the spherical gold nanoparticle can be expressed as

$$I_{\mathrm{expl}}(r_0) \approx \frac{4\rho q u_s}{3Q_{abs}(r_0, \lambda)}.$$

- Thus, the solid nanoparticle ***in a relatively strong laser field*** with intensity $I_0 \geq I_{\mathrm{expl}}$ ***during the short time*** $\tau_{expl} = r_0/u_s$ ***transforms into a gas*** (vapor) sphere of radius ~ r_0, which has high temperature and high pressure.

Slide 29.9 Threshold intensity for laser-induced thermal explosion (continued)

To compute I_{expl}, we can write the following energy balance equation (Slide 29.9):

$$\sigma_{\mathrm{abs}} I_{\mathrm{expl}} \frac{r_0}{u_{\mathrm{s}}} \approx \rho q V. \tag{9.12}$$

In terms of the absorption efficiency, $Q_{\mathrm{abs}} = \sigma_{\mathrm{abs}}/\pi r_0^2$, the threshold intensity of the laser radiation for thermal explosion of the spherical gold nanoparticle can be expressed as [6] (Slide 29.9)

$$I_{\mathrm{expl}}(r_0) \approx \frac{4\rho q u_{\mathrm{s}}}{3Q_{\mathrm{abs}}(r_0, \lambda)}. \tag{9.13}$$

Thus, the solid nanoparticle in a relatively strong laser field with intensity of $I_0 \geq I_{\mathrm{expl}}$ during the short time $\tau_{\mathrm{expl}} \sim r_0/u_{\mathrm{s}}$ transforms into a gas (vapor) sphere of radius ~r_0, which has high temperature $T \sim T_{\mathrm{GNB}}$ and high pressure $P_{\mathrm{vap}} \gg P_{\infty}$, where P_{∞} is the ambient pressure (Slide 29.9).

We assume here that thermal explosion of a nanoparticle is accompanied by generation of shock waves that expand with a supersonic velocity [6] (Slide 29.10):

$$u_{\mathrm{s}} = \frac{r_0}{\tau_{\mathrm{expl}}} = \frac{I_{\mathrm{expl}}}{4\Delta H} \approx 10^5\,\mathrm{cm/s}. \tag{9.14}$$

Here, $\Delta H = \Delta H_{300-\mathrm{vap}} + \Delta H_{\mathrm{vap}}$, where $\Delta H_{300-\mathrm{vap}}$ is the enthalpy change per unit volume for heating the gold nanoparticle from the ambient temperature to the

Shock Waves Induced by Thermal Explosion

- We assume here that ***thermal explosion of a nanoparticle*** is accompanied by ***generation of shock waves*** that expand with a supersonic velocity

$$u_s = \frac{r_0}{\tau_{expl}} = \frac{I_{expl}}{4\Delta H} \approx 10^5 \text{ cm/s} .$$

- The ***shock waves*** could be waves of ***high acoustic or/and water vapor pressure***, which spread out over long distances around an epicenter of explosion and ***produce irreparable mechanical cell damage***.
- It should be noted that the ***pressure produced by*** vaporization of a ***gold nanoparticle*** itself (e.g., ~10^{-2} atm) ***is less than the water vapor pressure*** around the hot nanoparticles.

Slide 29.10 Shock waves induced by thermal explosion

vaporization temperature, and ΔH_{vap} is the vaporization enthalpy per unit volume. The shock waves could be waves of high acoustic or/and water vapor pressure, which spread out over long distances around an epicenter of explosion and produce irreparable mechanical cell damage. It should be noted that the pressure produced by vaporization of a gold nanoparticle itself (e.g., ~10^{-2} atm) is less than the water vapor pressure around the hot nanoparticles (Slide 29.10). Indeed, the number of atoms vaporized per unit of time and per unit of particle surface is given by [2]

$$\frac{N_{\text{vap}}}{dt} = u_s N_{\text{Au}} = \frac{I_{\text{expl}}}{4\Delta H} N_{\text{Au}}. \tag{9.15}$$

For example, the atomic density of bulk gold is 5.9×10^{22} cm^{-3}, while each gold nanoparticle contains approximately 10^9 atoms, leading to an average atom density of 10^{17} cm^{-3}. Higher gold vapor pressure can only be reached on the front end of the explosion, where explosion products are localized in a thin layer.

9.5.6 Cell Damage Effects

The therapeutic effect of laser-induced explosion of gold nanoparticles for cancer treatments can be realized through one or several phenomena, such as [2] (Slide 29.11):

Cell Damage Effects

The ***therapeutic effect*** of laser-induced explosion of gold nanoparticles for cancer treatments can be realized due to one or several phenomena, such as:

- ***bubble formation*** discussed earlier,
- ***protein inactivation*** (e.g., through its denaturation or coagulation) around hot gold nanoparticles, which will be discussed later,
- ***generation of acoustic and shock waves***, and
- interaction with nanoparticle fragments and ***products of explosion***, which fly with high kinetic energy and produce a mechanical damage to the cancer cell.

An important damage-related factor is the ***vapor pressure*** produced by both water and gold vapors, accompanied by cavitations and shock waves, with local pressure up to ***gigapascals***.

Slide 29.11 Cell damage effects

- Bubble formation discussed in Sects. 9.3 and 9.4 of this book,
- Protein inactivation (e.g., through its denaturation or coagulation) around hot gold nanoparticles, which is discussed in Chap. 10 of this book,
- Generation of acoustic and shock waves, and
- Interaction with nanoparticle fragments and products of explosion, which fly with high kinetic energy and produce mechanical damage to cancer cells.

An important damage-related factor is not only the temperature but the vapor pressure produced by both water and gold vapors, accompanied by cavitations and shock waves, with local pressure up to gigapascals [5, 6]. The damage of specific cellular structures (e.g., plasma membranes, nuclei, cytoskeletons, and other organelles) in the laser explosion mode depends on nanoparticle parameters (composition, size, and shape), their number and location, laser characteristics (wavelength, fluence, pulse duration, and number of pulses), and properties of the surrounding media.

The time required to destroy an abnormal cell of size d by supersonic expansion of shock waves can be roughly estimated as $\tau_{sw} \approx d/u_s$. In the case of 15-μm diameter cancer cells, this is $\tau_{sw} \approx 15$ ns, which is comparable to the lifetime of sound/pressure shock waves in an aqueous medium [2].

We can estimate the lower level of the threshold energy density of a laser pulse required for thermal explosion of the gold nanoparticle using $E_{expl}(R) \approx I_{expl}(R) \times \tau_{expl}$, where the threshold intensity, I_{expl}, depends on the absorption efficiency of the nanoparticle. Table 9.11 and Slide 29.12 contain the theoretical and experimental values of the threshold energy density for the laser-induced thermal explosion/fragmentation of the gold nanoparticles and cell damage by

Table 9.11 Theoretical and experimental values of the threshold energy density for laser-induced thermal explosion/fragmentation of gold nanoparticles and cell damage by ultrasound shock waves

Theoretical thresholds E_{expl} (mJ/cm^2)	Experimental thresholds E_{fragm} (mJ/cm^2)	Cell organelles damage thresholds by ultrasound shock waves
Solid gold sphere of $r_0 = 35$–40 nm: $E_{\text{expl}} = 38.5$ mJ/cm^2	Solid gold sphere of $r_0 = 45$ nm: $E_{\text{fragm}} = 30$ mJ/cm^2	Cell membrane: $E_{\text{damage}} = 12$ mJ/cm^2
		Vimentin: $E_{\text{damage}} = 21$ mJ/cm^2
Gold nanorod of $r_0 \approx 11$ nm: $E_{\text{expl}} = 11.2$ mJ/cm^2	Solid gold sphere of $r_0 = 25$ nm: $E_{\text{fragm}} = 23$ mJ/cm^2	Mitochondria: $E_{\text{damage}} = 33$ mJ/cm^2
		Nuclear membrane: $E_{\text{damage}} = 50$ mJ/cm^2

Cell Damage Effects (continued)

The ***damage*** of specific ***cellular structures*** (e.g,. plasma membranes, nuclei, cytoskeletons and other organelles) in the laser ***explosion mode*** depends on ***nanoparticle parameters*** (composition, size and shape), ***their number and location, laser characteristics*** (wavelength, fluence, pulse duration and number of pulses), and ***properties of the surrounding media***.

Theoretical Thresholds E_{expl} (mJ/cm^2)	Experimental Thresholds E_{fragm} (mJ/cm^2)	Cell Organelles Damage Thresholds by Ultrasound Shock Wave
Solid gold sphere nano-, shell of r_0 = 35- 40 nm E_{expl} = 38.5 mJ/cm^2	Solid gold sphere of r_0 = 45 nm E_{fragm} = 30 mJ/cm^2	Cell membrane: E_{damage} = 12 mJ/cm^2 Vimentin: E_{damage} = 21 mJ/cm^2
Gold nanorod of $r_0 \approx 11$ nm E_{expl} = 11.2 mJ/cm^2	Solid gold sphere of r_0 = 25 nm E_{fragm} = 23 mJ/cm^2	Mitochondria: E_{damage} = 33 mJ/cm^2 Nuclear membrane: E_{damage} = 50 mJ/cm^2

Slide 29.12 Cell damage effects (continued)

ultrasound shock waves. As can be seen from this table, for the laser pulse of $\tau_L = 8$–10 ns duration and wavelength $\lambda = 532$ nm often used in the experiments, the threshold energy density for thermal explosion of a solid gold nanosphere of size $r_0 = 35$ nm (absorption efficiency of $Q_{\text{abs}} = 4.02$) is $E_{\text{expl}} = 38.5$ mJ/cm^2 (Slide 29.13). This energy density strongly depends on the types of nanoparticles (solid gold nanospheres, nanoshells, and nanorods). For example, gold nanorods have a near-IR resonance absorption efficiency of approximately 14 for an effective radius of 11.43 nm. Due to the higher plasmon-resonance absorption efficiency of nanorods, the threshold energy density can be reduced by using gold nanorods up to $E_{\text{expl}} = 11.2$ mJ/cm^2. It is important to note that the estimated threshold energy densities for laser-induced thermal explosion of gold nanoparticles are 2.5–9 times less than the laser energy density of 100 mJ/cm^2 established as the safety standard for medical lasers [2] (Slide 29.13).

The calculated values of E_{expl} are in good agreement with some available experimental results presented in Table 9.11 and Slide 29.14. Indeed, for spherical

Cell Damage Effects (continued)

- As can be see from the table, for a laser pulse of τ_L = 8-10 ns duration and wavelength λ = 532 nm often used in the experiments, the ***threshold energy density*** for thermal explosion of a solid gold nanosphere of size r_0 = 35 nm (absorption efficiency of Q_{abs} = 4.02) is ***E_{expl} = 38.5 mJ/cm²***.
- This energy density strongly ***depends on the types of nanoparticles*** (solid gold nanospheres, nanoshells and nanorods). For example, ***gold nanorods have*** a near-IR resonance absorption efficiency of approximately 14 for an effective radius of 11.43 nm. Due to the ***higher plasmon-resonance absorption efficiency*** of nanorods, the ***threshold energy density can be reduced*** by using gold nanorods ***up to E_{expl} = 11.2 mJ/cm²***.
- It is important to note that the estimated ***threshold energy densities*** for laser-induced thermal explosion of gold nanoparticles are ***2.5 to 9 times less*** than the laser energy density of ***100 mJ/cm² established as the safety standard for medical lasers.***

Slide 29.13 Cell damage effects (continued)

Cell Damage Effects (continued)

- The ***calculated values for E_{expl}*** are ***in good agreement*** with some available ***experimental results*** presented in the table.
- Indeed, for spherical gold nanoparticles with sizes 45 nm and 25 nm irradiated with second-harmonic Nd:YAG laser pulses (λ = 532 nm for τ_L = 7 ns), the ***laser fluence thresholds for the nanoparticle fragmentation*** associated with particle melting and boiling phenomena ***are 30 and 23 mJ/cm²***, respectively.

Theoretical Thresholds E_{expl} (mJ/cm²)	Experimental Thresholds E_{fragm} (mJ/cm²)	Cell Organelles Damage Thresholds by Ultrasound Shock Wave
Gold solid sphere, nanoshell of r_0 = 35-40 nm E_{expl} = 38.5 mJ/cm² Gold nanorod of $r_0 \approx$ 11 nm E_{expl} = 11.2 mJ/cm²	Gold solid sphere of r_0 = 45 nm E_{fragm} = 30 mJ/cm² Gold solid sphere of r_0 = 25 nm E_{fragm} = 23 mJ/cm²	Cell membrane: E_{damage} = 12 mJ/cm² Vimentin: E_{damage} = 21 mJ/cm² Mitochondria: E_{damage} = 33 mJ/cm² Nuclear membrane: E_{damage} = 50 mJ/cm²

Slide 29.14 Cell damage effects (continued)

gold nanoparticles with sizes 45 nm and 25 nm irradiated by second-harmonic Nd: YAG laser pulses ($\lambda = 532$ nm for $\tau_L = 7$ ns), the laser fluence thresholds for the nanoparticle fragmentation associated with particle melting and boiling phenomena are 30 and 23 mJ/cm^2, respectively. The more intense gold nanoparticle fragmentation to small fragments of mainly 5–10 nm in size has been observed for laser

Cell Damage Effects (continued)

- The last column in the table below shows the ***threshold energy densities for cell organelle damage*** produced ***by shock waves***.
- The ***threshold energy densities for thermal explosion of nanoparticles*** are comparable to the ***energy fluences*** required ***to destroy the cancer cell organelles*** via, for example, shock waves generation. Thus, ***thermal explosion*** around nanoparticles may give ***significant contribution to cell damage*** alone or together with conventional water vapor bubble formation in nanophotothermolysis.

Theoretical Thresholds E_{expl} (mJ/cm^2)	Experimental Thresholds E_{fragm} (mJ/cm^2)	Cell Organelles Damage Thresholds by Ultrasound Shock Wave
Gold solid sphere, nanoshell of r_0 = 35- 40 nm E_{expl} = 38.5 mJ/cm^2 Gold nanorod of $r_0 \approx 11$ nm E_{expl} = 11.2 mJ/cm^2	Gold solid sphere of r_0 = 45 nm E_{fragm} = 30 mJ/cm^2 Gold solid sphere of r_0 = 25 nm E_{fragm} = 23 mJ/cm^2	Cell membrane: E_{damage} = 12 mJ/cm^2 Vimentin: E_{damage} = 21 mJ/cm^2 Mitochondria: E_{damage} = 33 mJ/cm^2 Nuclear membrane: E_{damage} = 50 mJ/cm^2

Slide 29.15 Cell damage effects (continued)

fluences in the range 30–140 mJ/cm^2 and higher. For a 30-ps pulse, 25-nm gold nanoparticle fragmentation has been observed at 16 mJ/cm^2, with a slight effect on changing the particle shape, even at 2–5 mJ/cm^2 [2].

The last column in Table 9.11 shows the threshold energy densities for cell organelle damage produced by the shock waves (Slide 29.15). As evident from these data, the threshold energy densities for laser-induced thermal explosion of nanoparticles are comparable to the energy fluences required to destroy the cancer cell organelles via, for example, shock waves generation. Thus, thermal explosion around nanoparticles may give significant contribution to cell damage alone or together with conventional water vapor bubble formation in nanophotothermolysis.

The laser-induced explosion effect can explain the experimental results discussed in [5, 6] with notable cancer damage with 1.4-nm gold nanoparticles inside viruses on a cell membrane with picosecond laser pulses (30 ps, 50–100 mJ/cm^2) when the probability of classic water bubble formation is very low. The explosion mode could be essential in selective nanophotothermolysis of DNA, and this mode is definitely becoming dominant in the absence of a sufficient amount of water around gold nanoparticles (Slide 29.16). This mode can be combined with a bubble-overlapping mode and nanocluster aggregation mode discussed earlier in this Chap. 9, where the explosion of few closely located gold nanoparticles in a cluster can produce one large bubble with enhanced killing efficiency. The explosion of nanoparticles on smaller particles or single atoms during a laser pulse may provide the condition for the interaction of the pulse with its atoms, leading to their ionization and plasma formation. However, these effects should appear at relatively high laser fluences that are not safe for normal cells.

Cell Damage Effects (continued)

- The ***thermal explosion mode*** could be essential in selective ***nanophotothermolysis of DNA***, and this mode is definitely becoming dominant ***in the absence of a sufficient amount of water*** around gold nanoparticles.
- ***This mode*** can be combined ***with a bubble-overlapping mode and nanocluster aggregation mode*** discussed earlier, where the explosion of few closely-located gold nanoparticles in a cluster can produce one large bubble with enhanced killing efficiency.
- The ***explosion of nanoparticles*** on smaller particles or single atoms during a laser pulse may provide the condition for the interaction of the pulse with its atoms, leading to their ionization and plasma formation. However, these effects should appear at relatively high laser fluences that are not safe for normal cells.

Slide 29.16 Cell damage effects (continued)

9.5.7 Conclusions (Slides 29.17 and 29.18)

We have considered a new mechanism for selective laser killing of abnormal cells by laser thermal explosion of single nanoparticles (nanobombs) delivered to the cells [2, 5, 6]. Thermal explosion is realized when heat is generated within a strongly absorbing target more rapidly than the heat can diffuse away. On the basis of simple energy balance, we have shown that the threshold energy density of a single laser pulse required for the thermal explosion of a solid gold nanosphere is 38.5 mJ/cm^2. The threshold can be reduced further (to 11.2 mJ/cm^2) by using gold nanorods (and probably nanoshells) whose optical plasmon resonance lies in the near-IR region, where the biological tissue transmissivity is the highest. Since the therapeutic effect of the "nanobomb" mode is based on thermal explosion of the nanoparticle itself, it can be applied to cancer treatments of dense solid tumors, bone tissues, atherosclerotic plaques, and other targets with lack of a sufficient amount of liquid for efficient bubble generation. Additionally, the effective therapeutic effect for cancer cell killing is achieved due to nonlinear phenomena that accompany the thermal explosion of the nanoparticles, such as the generation of nanoparticle explosion products with high kinetic energy (nanobullets), as well as strong shock waves with supersonic expansion of dense vapor within the cell volume, producing sound and high pressure waves. It is important that most of these phenomena can explain some published experimental results whose interpretation was performed without taking into account this effect. Thus, the thermal explosion mode can be recommended for selective laser killing of cancer cells.

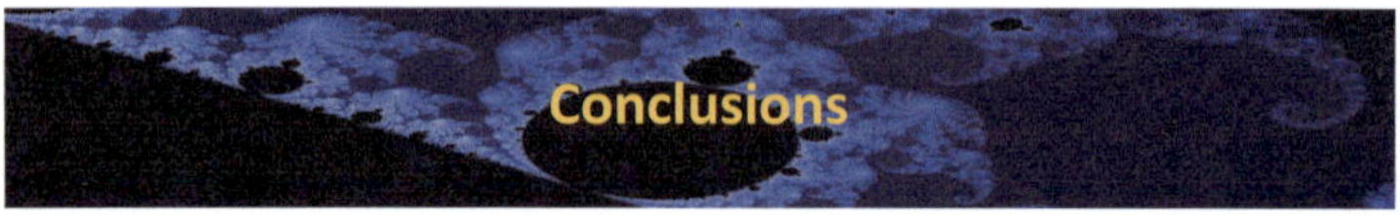

- We have considered a ***new mechanism*** for selective laser killing of abnormal cells by laser ***thermal explosion*** of single nanoparticles (nanobombs) delivered to the cells.
- ***Thermal explosion*** is realized when ***heat is generated*** within a strongly-absorbing target ***more rapidly than the heat can diffuse away***.
- On the basis of simple energy balance, we have shown that the ***threshold energy density*** of a single laser pulse required for the thermal explosion of a solid gold nanosphere ***is 38.5*** mJ/cm^2.
- The ***threshold*** can be reduced further (to ***11.2*** mJ/cm^2) by using ***gold nanorods*** (and probably nanoshells) whose optical plasmon resonance lies in the near-IR region, where the biological tissue transmissivity is the highest.

Slide 29.17 Conclusions

- Since the ***therapeutic effect of the 'nanobomb' mode*** is based on thermal ***explosion of the nanopatrticle itself***, it can be applied to cancer treatments of ***dense solid tumors, bone tissues, atherosclerotic plaques and other targets with lack of a sufficient amount of liquid*** for efficient bubbles generation.
- Additionally, the effective ***therapeutic effect*** for cancer cell killing is achieved due to ***nonlinear phenomena*** that accompany the thermal explosion of the nanoparticles, such as the generation of nanoparticle ***explosion products*** with high kinetic energy ***(nanobullets)***, as well as ***strong shock waves*** with supersonic expansion of dense vapor within the cell volume, producing ***sound and high pressure waves***.
- Thus, the ***thermal explosion mode*** can be recommended ***for selective laser killing of cancer cells.***

Slide 29.18 Conclusions (continued)

Appendix G: Maple Code for Simulating the Spatial Distribution of Thermal Fields Around a Nanocluster

This appendix contains a Maple code designed for spatial simulations of the temperature fields inside and outside a cluster made of gold nanoparticle in a surrounding water medium (Slide 26.20). The comments within the Maple code should explain the variables, procedures, and functions that are used.

Note: This code takes about 10–15 min to complete the integrations. The code comes directly from the Maple file and is therefore formatted to word-wrap according to Maple standard character-per-line limits.

```
> restart;
```

Energy density, J/m^2

```
> E:=5000;
```

$$E := 5000$$

Specific heat, J/(kg*K)

```
> C[W1]:=129;
  C[W2]:=4180;
```

$$C_{W1} := 129$$
$$C_{W2} := 4180$$

Density, kg/m^3

```
> rho[1]:=19320;
  rho[2]:=1010;
```

$$\rho_1 := 19320$$
$$\rho_2 := 1010$$

Radius, m

```
> R:=30*10^(-9);
  Kabs:=4.376;
```

$$R := \frac{3}{100000000}$$
$$Kabs := 4.376$$

Thermal conductivity, W/(m*K)

```
> K[1]:=317;
  K[2]:=0.6;
```

$$K_1 := 317$$
$$K_2 := 0.6$$

Thermal diffusivity, m^2/s

```
> k[1]:=K[1]/C[W1]/rho[1];
  k[2]:=K[2]/C[W2]/rho[2];
```

$$k_1 := \frac{317}{2492280}$$
$$k_2 := 1.421194751\ 10^{-7}$$

Constants

```
> a:=R^2/k[1];
  b:=K[2]/K[1]*sqrt(k[1]/k[2]);
  c:=1-K[2]/K[1];
  sigma:=(r/R-1)*sqrt(k[1]/k[2]);
  s:=1.01*10^(-6);
```

$$a := \frac{560763}{79250000000000000}$$

$$b := 0.05662346026$$

$$c := 0.9981072555$$

$$\sigma := 9.972020503\ 10^{8}\ r - 29.91606151$$

$$s := 0.000001010000000$$

Heat diffusion time, s

```
> tau:=s^2/(4*k[2]);
```

$$\tau := 0.000001794440908$$

Rate of heat generation, J/s

```
> A:=3*Kabs*E/(R*tau);
```

$$A := 1.219321288\ 10^{18}$$

```
> T[1]:=R^2*A/K[1]*(K[1]/3/K[2]+(1-r^2/R^2)/6-
2*b*R/r/Pi*(Int(exp(-y^2*t/a)/y^2*(sin(y)-
y*cos(y))*sin(r*y/R)/((c*sin(y)-
y*cos(y))^2+b^2*y^2*sin(y)^2),y=0..100)));
```

$$T_1 := 610.2376098 - 6.410732322\ 10^{14}\ r^2 - \frac{1}{r\,\pi}\left(1.176113024\ 10^{-8}\left(\int_0^{100}\left(\mathrm{e}^{-\frac{79250000000000000}{560763}y^2 t}\,(\sin(y) - y\cos(y))\sin\left(\frac{100000000}{3}\,r\,y\right)\right)\Big/\left(y^2\left((0.9981072555\sin(y) - y\cos(y))^2 + 0.003206216252\,y^2\sin(y)^2\right)\right)\mathrm{d}y\right)\right)$$

```
> T[2]:=R^3*A/r/K[1]*(K[1]/3/K[2]-2/Pi*Int(exp(-
y^2*t/a)/y^3*(sin(y)-y*cos(y))*(b*sin(y)*cos(sigma*y)-(c*sin(y)-
y*cos(y))*sin(sigma*y))/((c*sin(y)-
y*cos(y))^2+b^2*y^2*sin(y)^2),y=0..100));
```

$$T_2 := \frac{1}{r}\left(1.038538637\ 10^{-7}\left(176.1111111 - \frac{1}{\pi}\left(2\left(\int_0^{100}\left(e^{-\frac{79250000000000000}{560763} y^2 t}(\sin(y) - y\cos(y))(0.05662346026\sin(y)\cos((9.972020503\ 10^8 r - 29.91606151)y) - (0.9981072555\sin(y) - y\cos(y))\sin((9.972020503\ 10^8 r - 29.91606151)y))\right)\Big/\left(y^3\left((0.9981072555\sin(y) - y\cos(y))^2 + 0.003206216252\, y^2\sin(y)^2\right)\right) dy\right)\right)\right)\right)$$

```
> for n from 1 to 10 do
Q[n]:=evalf(subs((t=tau,r=n*R/10),T[1]));od:
for n from -10 to -1 do
Q[n]:=evalf(subs((t=tau,r=n*R/10),T[1]));od:
> for n from 10 to 100 do
V[n]:=evalf(subs((t=tau,r=n*R/10),T[2]));od:
for n from -100 to -10 do V[n]:=evalf(subs((t=tau,r=n*R/10),-
T[2]));od:
```

Separation distance

```
> sep:=20:
  Q[1]:
  V[sep/2+10]:
  sep*4+R*10^9*5:
> for i from 1 to 90 do Z[i]:=[evalf(i*R/10),V[i-101]];od:
for i from 1 to 10 do Z[i+90]:=[evalf((i+90)*R/10),Q[11-i]];od:
for i from 1 to 10 do Z[i+100]:=[evalf((i+100)*R/10),Q[i]];od:
for i from 1 to (sep/2) do
Z[i+110]:=[evalf((i+110)*R/10),V[i+10]];od:
> for i from 1 to (sep/2) do
Z[i+110+(sep/2)]:=[evalf((i+110+sep/2)*R/10),V[((sep/2)-
i+10)]];od:
for i from 1 to 10 do
Z[i+110+sep]:=[evalf((i+110+sep)*R/10),Q[(11-i)]];od:
for i from 1 to 10 do
Z[i+120+sep]:=[evalf((i+120+sep)*R/10),Q[(i)]];od:
for i from 1 to (sep/2) do
Z[i+130+sep]:=[evalf((i+130+sep)*R/10),V[i+10]];od:
for i from 1 to (sep/2) do
Z[i+130+(3*sep/2)]:=[evalf((i+130+3*sep/2)*R/10),V[((sep/2)-
i+10)]];od:
```

```
for i from 1 to 10 do
Z[i+130+2*sep]:=[evalf((i+130+2*sep)*R/10),Q[(11-i)]];od:
for i from 1 to 10 do
Z[i+140+2*sep]:=[evalf((i+140+2*sep)*R/10),Q[(i)]];od:
for i from 1 to (sep/2) do
Z[i+150+2*sep]:=[evalf((i+150+2*sep)*R/10),V[i+10]];od:
for i from 1 to (sep/2) do
Z[i+150+(5*sep/2)]:=[evalf((i+150+5*sep/2)*R/10),V[((sep/2)-
i+10)]];od:
for i from 1 to 10 do
Z[i+150+3*sep]:=[evalf((i+150+3*sep)*R/10),Q[(11-i)]];od:
for i from 1 to 10 do
Z[i+160+3*sep]:=[evalf((i+160+3*sep)*R/10),Q[(i)]];od:
for i from 1 to (sep/2) do
Z[i+170+3*sep]:=[evalf((i+170+3*sep)*R/10),V[i+10]];od:
for i from 1 to (sep/2) do
Z[i+170+(7*sep/2)]:=[evalf((i+170+7*sep/2)*R/10),V[((sep/2)-
i+10)]];od:
for i from 1 to 10 do
Z[i+170+4*sep]:=[evalf((i+170+4*sep)*R/10),Q[(11-i)]];od:
for i from 1 to 10 do
Z[i+180+4*sep]:=[evalf((i+180+4*sep)*R/10),Q[(i)]];od:
for i from 1 to 90 do
Z[i+190+4*sep]:=[evalf((i+190+4*sep)*R/10),V[i+10]];od:
>
plot([seq(Z[b],b=1..(280+4*sep))],style=line,color=blue,labels=[
r,T],symbol=diamond,symbolsize=15);
```

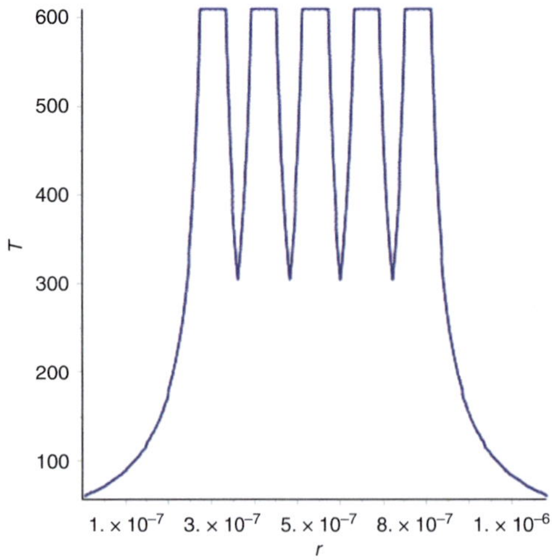


```
> f:=[seq(Z[b],b=1..(280+4*sep))]:
mymatrix:=matrix((280+4*sep),2,f):
writedata("AG in Cytoplasm multi.txt",mymatrix,float);
```

Appendix H: Maple Code for Simulating Bubble Growth Around a Nanoparticle

This appendix contains a Maple code designed for simulating the bubble dynamics in a surrounding cytoplasm medium (Slide 28.16). The comments within the Maple code should explain the variables, procedures, and functions that are used.

Note: The code comes directly from the Maple file and is therefore formatted to word-wrap according to Maple standard character-per-line limits.

```
> restart:
> with(plots):
```
Partial pressure of the bubble gas (Pa)
```
> pg:=pg0*(R0/R(t))^(3*g);
```
$$pg := pg0\left(\frac{R0}{R(t)}\right)^{3g}$$

At constant bubble temperature
```
> gt:=1;
```
$$gt := 1$$
```
> ga:=4/3;
```
$$ga := \frac{4}{3}$$

Initial radius for the bubble (size of the nanoparticle) (m)
```
> R0:=30e-9;
```
$$R0 := 3.0\ 10^{-8}$$

Pressure of surrounding medium at infinity, Pa
```
> pf:=101000;
```
$$pf := 101000$$

Saturated pressure of the vapor (Pa)
```
> pv:=476000;
```
$$pv := 476000$$

Partial pressure of the bubble's gas at R = R0 (Pa)
```
> pg0:=pv;
```
$$pg0 := 476000$$

Medium density (kg/m^3)
```
> rho[L]:=999.82;
```
$$\rho_L := 999.82$$

Dynamic viscosity of the medium (kg*s/m^2)
```
> mu:=1.792*10^(-3);
```
$$\mu := 0.001792000000$$

Solution for constant temperature expansion
```
>
DEt:=R(t)*diff(R(t),t,t)+3/2*diff(R(t),t)^2+(4*mu/R(t))*diff(R(t
),t)/rho[L] = (pv-pf)/rho[L]+pg0*(R0/R(t))^(3*gt);
```
$$DEt := R(t)\left(\frac{d^2}{dt^2}R(t)\right) + \frac{3}{2}\left(\frac{d}{dt}R(t)\right)^2 + \frac{0.000007169290472\left(\frac{d}{dt}R(t)\right)}{R(t)} = 375.0675122 + \frac{1.285200000\ 10^{-17}}{R(t)^3}$$

```
> solnt:=dsolve({DEt,R(0)=R0,D(R)(0)=0},{R(t)},numeric);
```

$$solnt := \mathbf{proc}(x\_rkf45) \ \ldots \ \mathbf{end\ proc}$$

```
> pt:=odeplot(solnt,[t,diff(R(t),t)],0..0.5e-9, labels=[Time
(s),Velocity(m/s)]):
>
```

Solution for adiabatic expansion of the bubble

```
>
DEa:=R(t)*diff(R(t),t,t)+3/2*diff(R(t),t)^2+(4*mu/R(t))*diff(R(t
),t)/rho[L] = (pv-pf)/rho[L]+pg0*(R0/R(t))^(3*ga);
```

$$DEa := R(t)\left(\frac{d^2}{dt^2}R(t)\right) + \frac{3}{2}\left(\frac{d}{dt}R(t)\right)^2 + \frac{0.000007169290472\left(\frac{d}{dt}R(t)\right)}{R(t)} = 375.0675122 + \frac{3.855600000\ 10^{-25}}{R(t)^4}$$

```
> solna:=dsolve({DEa,R(0)=R0,D(R)(0)=0},R(t),numeric);
```

$$solna := \mathbf{proc}(x\_rkf45) \ \ldots \ \mathbf{end\ proc}$$

```
> pa:=odeplot(solna,[t,diff(R(t),t)],0..0.5e-9,color=blue,
labels=[Time (s),Velocity(m/s)]):
```

Comparison of Adiabatic and constant temperature Expansion

```
> display(pa,pt);
```

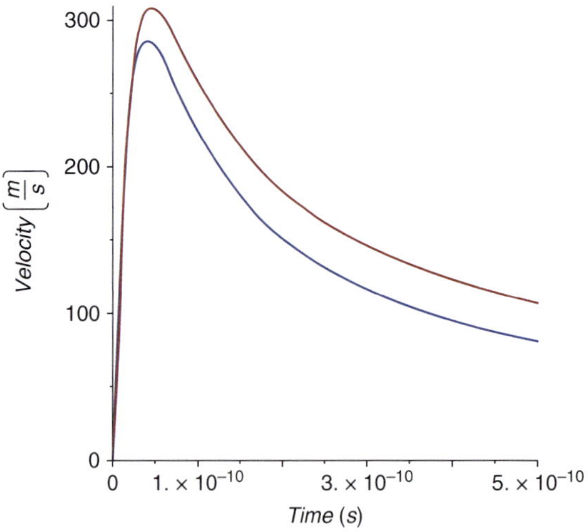

```
> p1:=odeplot(solnt,[t,R(t)],0..2e-7, color=red, labels=[Time
(s),Radius(m)]):
> p2:=odeplot(solna,[t,R(t)],0..2e-7, color=blue, labels=[Time
(s),Radius(m)]):
> display(p1,p2);
```

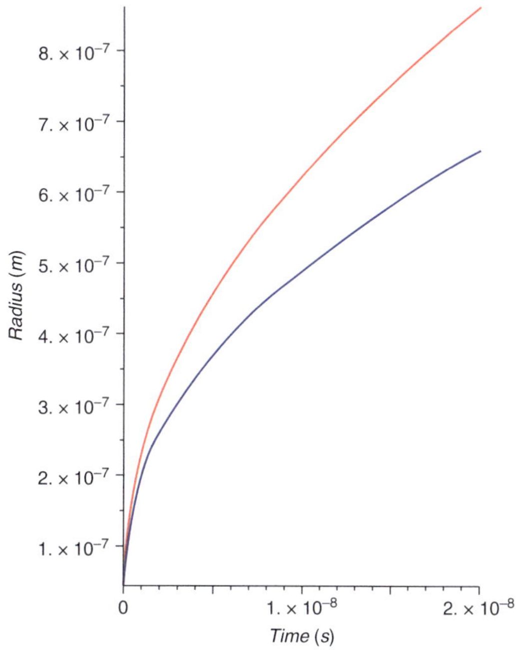

Homework Exercises

Section 9.1: Nanocluster Aggregation Mode

9.1. List three main new dynamic modes in selective nanophotothermolysis of cancer.
9.2. What is a nanocluster?
9.3. What is an incubation time?
9.4. True or false? The incubation time doesn't depend on the nanoparticle sizes and surface coatings.
9.5. What is an incubation time for 1–50 nm sized particles?
9.6. What is an incubation time for 50–100 nm sized particles?
9.7. Describe the nanocluster aggregation mode in selective nanophotothermolysis of cancer.
9.8. How may the nanocluster aggregation mode increase the effectiveness of cancer cell killing?

9.9. True or false? A large damage area can be produced by a nanocluster at higher energy density of the incident laser pulse as compared to the single nanoparticle case.
9.10. True or false? Nanoparticles are able to self-assemble into larger nanoclusters directly within live systems (e.g., cancer cells, bacteria, vascular legions, etc.).
9.11. What are synergetic enhancements of cancer nanophotothermolysis associated with nanoclusters?
9.12. What is the main difference between linear and synergetic effects related to nanoclusters in a tumor?
9.13. Define the interparticle spacing criteria used in the nanocluster aggregation mode.
9.14. What is the numerical value of the thermal ablation temperature for most biological tissues?
9.15. What is the numerical value of the threshold temperature for bubble formation in biological media?
9.16. Calculate the nanoparticle concentration in a cluster made of 20-nm radius gold particles with interparticle spacing $l = 2d_0$, where d_0 is the diameter of the nanoparticle.
9.17. Find the absolute number of nanoparticles in a 400-nm diameter cluster made of 30-nm radius silver particles with interparticle spacing $l = 2d_0$, where d_0 is the diameter of the nanoparticle.
9.18. If the concentration of 30-nm gold nanoshells in a cluster is $n = 4.5 \times 10^{12}\ \mathrm{cm}^{-3}$, what is the interparticle spacing?
9.19. The scanning electron microscope image shows 50 silica–gold nanoparticles in a 350-nm diameter cluster. (a) What is the interparticle spacing? (b) Estimate the radius of the nanoparticle in that cluster.
9.20. Describe qualitatively the temperature distribution in a cluster in two cases: (a) when the nanoparticles are spaced far from each other, and (b) when the particles are located close to each other.
9.21. What is an ideal arrangement for interparticle spacing in a cluster for effective cancer cell killing?
9.22. How do the interparticle spacing criteria in a nanocluster can be reached?
9.23. How many 30-nm radius gold particles are needed to destroy a 10-μm diameter cancer cell?
9.24. Why are individual nanoparticles in a cluster heated to temperatures higher than the critical temperature for cell death?
9.25. What is the minimal size of the damage area produced by a nanocluster in a cancer cell?
9.26. What is a typical energy density of radiation needed to heat a nanocluster to produce thermal damage of cancer cells?
9.27. True or false? To reach the same thermal damage area by a single nanoparticle, a laser pulse with energy density 2.5 times higher than for the cluster case is required.

Section 9.2: Computer Practicum 6: Simulating Temperature Fields Around a Nanocluster

9.28. What is the difference between the separation distance used in the Maple code and the interparticle spacing in a cluster?
9.29. What is a main modification made to the Maple code to simulate the spatial temperature distribution in a cluster?
9.30. Describe how the absorption efficiency of contrast agents can increase in the RF range of the spectrum.
9.31. True or false? The aggregation of nanoparticles in a cluster can dramatically increase the absorption efficiency of contrast agents in the RF range of the spectrum.
9.32. True or false? There is negligible heat loss from the surface of a heated nanoparticle to the surrounding medium.
9.33. What are the most efficient sizes of gold nanoparticles used for RF ablation of cancer cells?
9.34. What is the area size of thermal damage produced in a cancer cell by 20-nm gold nanoparticles?
9.35. What is the area size of thermal damage produced in a cancer cell by 50-nm gold nanoparticles?
9.36. What is the minimum concentration of 50-nm gold particles per cell needed to destroy cancer cells by thermal ablation?
9.37. True or false? Carbon's absorption efficiency is lower than that of gold in the RF range of the spectrum.
9.38. True or false? Carbon nanoparticles are more efficient than gold nanoparticles as contrast agents for RF activation.
9.39. What is the minimum concentration of 50-nm carbon particles per cell needed to destroy cancer cells by thermal ablation?
9.40. True or false? It takes over twice more carbon nanoparticles and higher energy density of radiation to kill a cell than for gold nanoparticles.
9.41. What is the area size (radius) of thermal damage produced in a cancer cell by 20-nm carbon nanoparticles?
9.42. What is the area size (radius) of thermal damage produced in a cancer cell by 50-nm carbon nanoparticles?
9.43. What is the minimum concentration of 20-nm carbon particles per cell needed to destroy cancer cells by thermal ablation?
9.44. What is the area size (radius) of thermal damage produced in a cancer cell by a cluster made of 20-nm gold nanoparticles?
9.45. How many 20-nm gold particles in a cluster are used to destroy cancer cells?
9.46. How many clusters per cell made of 20-nm gold nanoparticles are needed to destroy cancer cells by thermal ablation?
9.47. What is the separation distance between nanoparticles in a cluster made of 20-nm gold particles used to destroy cancer cells by thermal ablation?

9.48. Compare the cancer cell damage area produced by a single 5-nm gold particle to that by a cluster consisting of the same 5-nm gold particles.
9.49. What is the thermal damage area (radius) produced in a cancer cell by a cluster made of 50-nm carbon nanoparticles?
9.50. How many 50-nm carbon particles are in a cluster used to destroy cancer cells?
9.51. How many clusters per cell made of 50-nm carbon nanoparticles are needed to destroy cancer cells by thermal ablation?
9.52. What is the separation distance between nanoparticles in a cluster made of 50-nm carbon particles used to destroy cancer cells by thermal ablation?
9.53. Which clusters—made of gold or of carbon nanoparticles—are most efficient for cancer treatments?
9.54. What energy density of optical radiation is required to produce about a 1-μm damage area by a cluster made of 56-nm silver particles in a surrounding blood medium?
9.55. True or false? The damage area, produced by the same silver clusters and same amount of radiation energy density are different in different biological media.
9.56. In what media is the damage area produced by a silver nanocluster the highest?
9.57. In what media is the damage area produced by a silver nanocluster the lowest?
9.58. True or false? By tuning the nanoparticle size and spacing, and then actively targeting a particular medium in cancer cells, damage to healthy cells can be minimized.
9.59. Using the Maple code in Appendix G, perform spatial simulations of thermal field distributions inside and around a cluster made of gold nanoparticles of optimal radius in surrounding blood, cytoplasm, fat, and cell membrane media by employing thermal data from Tables 6.3, 8.7, and 9.6.

    (a) Use Mie simulations to find the optimal radius of a nanoparticle and optimal wavelength of radiation to achieve the maximum absorption efficiency of the nanoparticle in each surrounding medium.
    (b) Determine the number of nanoparticles needed to form a cluster.
    (c) Estimate the thermal damage area produced in the surrounding biological media by nanoclusters heated in the optical range of the spectrum.
    (d) Find how many clusters per cell are needed to destroy cancer cells by thermal ablation.
    (e) Write a report on your simulation results similar to the practice examples 9 and 10 in Sect. 9.2 of this chapter.

9.60. Using the Maple code in Appendix G, perform spatial simulations of thermal field distributions inside and around a cluster made of nickel oxide nanoparticles of optimal radius in surrounding blood, cytoplasm, fat, and cell membrane media by employing thermal data from Tables 6.3, 8.7, and 9.6.

(a) Use Mie simulations to find the optimal radius of a nanoparticle and optimal wavelength of radiation to achieve the maximum absorption efficiency of the nanoparticle in each surrounding medium.
(b) Determine the number of nanoparticles needed to form a cluster.
(c) Estimate the thermal damage area produced in the surrounding biological media by nanoclusters heated in the optical range of the spectrum.
(d) Find how many clusters per cell are needed to destroy cancer cells by thermal ablation.
(e) Write a report on your simulation results similar to the practice examples 9 and 10 in Sect. 9.2 of this chapter.

9.61. Using the Maple code in Appendix G, perform spatial simulations of thermal field distributions inside and around a cluster made of magnesium oxide nanoparticles of optimal radius in surrounding blood, cytoplasm, fat, and cell membrane media by employing thermal data from 6.3, 8.7, and 9.6.

(a) Use Mie simulations to find the optimal radius of a nanoparticle and optimal wavelength of radiation to achieve the maximum absorption efficiency of the nanoparticle in each surrounding medium.
(b) Determine the number of nanoparticles needed to form a cluster.
(c) Estimate the thermal damage area produced in the surrounding biological media by nanoclusters heated in the optical range of the spectrum.
(d) Find how many clusters per cell are needed to destroy cancer cells by thermal ablation.
(e) Write a report on your simulation results similar to the practice examples 9 and 10 in Sect. 9.2 of this chapter.

9.62. Using the Maple code in Appendix G, perform spatial simulations of thermal field distributions inside and around a cluster made of aluminum oxide nanoparticles of optimal radius in surrounding blood, cytoplasm, fat, and cell membrane media by employing thermal data from Tables 6.3, 8.7, and 9.6.

(a) Use Mie simulations to find the optimal radius of a nanoparticle and optimal wavelength of radiation to achieve the maximum absorption efficiency of the nanoparticle in each surrounding medium.
(b) Determine the number of nanoparticles needed to form a cluster.
(c) Estimate the thermal damage area produced in the surrounding biological media by nanoclusters heated in the optical range of the spectrum.
(d) Find how many clusters per cell are needed to destroy cancer cells by thermal ablation.
(e) Write a report on your simulation results similar to the practice examples 9 and 10 in Sect. 9.2 of this chapter.

9.63. Using the Maple code in Appendix G, perform spatial simulations of thermal field distributions inside and around a cluster made of carbon nanoparticles of optimal radius in surrounding blood, cytoplasm, fat, and cell membrane media by employing thermal data from Tables 6.3, 8.7, and 9.6.

(a) Use Mie simulations to find the optimal radius of a nanoparticle and optimal wavelength of radiation to achieve the maximum absorption efficiency of the nanoparticle in each surrounding medium.
(b) Determine the number of nanoparticles needed to form a cluster.
(c) Estimate the thermal damage area produced in the surrounding biological media by nanoclusters heated in the optical range of the spectrum.
(d) Find how many clusters per cell are needed to destroy cancer cells by thermal ablation.
(e) Write a report on your simulation results similar to the practice examples 9 and 10 in Sect. 9.2 of this chapter.

9.64. Using the Maple code in Appendix G, perform spatial simulations of thermal field distributions inside and around a cluster made of glass nanoparticles of optimal radius in surrounding blood, cytoplasm, fat, and cell membrane media by employing thermal data from Tables 6.3, 8.7, and 9.6.

(a) Use Mie simulations to find the optimal radius of a nanoparticle and optimal wavelength of radiation to achieve the maximum absorption efficiency of the nanoparticle in each surrounding medium.
(b) Determine the number of nanoparticles needed to form a cluster.
(c) Estimate the thermal damage area produced in the surrounding biological media by nanoclusters heated in the optical range of the spectrum.
(d) Find how many clusters per cell are needed to destroy cancer cells by thermal ablation.
(e) Write a report on your simulation results similar to the practice examples 9 and 10 in Sect. 9.2 of this chapter.

9.65. Using the Maple code in Appendix G, perform spatial simulations of thermal field distributions inside and around a cluster made of polystyrene nanoparticles of optimal radius in surrounding blood, cytoplasm, fat, and cell membrane media by employing thermal data from Tables 6.3, 8.7, and 9.6.

(a) Use Mie simulations to find the optimal radius of a nanoparticle and optimal wavelength of radiation to achieve the maximum absorption efficiency of the nanoparticle in each surrounding medium.
(b) Determine the number of nanoparticles needed to form a cluster.
(c) Estimate the thermal damage area produced in the surrounding biological media by nanoclusters heated in the optical range of the spectrum.
(d) Find how many clusters per cell are needed to destroy cancer cells by thermal ablation.
(e) Write a report on your simulation results similar to the practice examples 9 and 10 in Sect. 9.2 of this chapter.

9.66. Using the Maple code in Appendix G, perform spatial simulations of thermal field distributions inside and around a cluster made of fullerene nanoparticles of optimal radius in surrounding blood, cytoplasm, fat, and cell membrane media by employing thermal data from Tables 6.3, 8.7, and 9.6.

(a) Use Mie simulations to find the optimal radius of a nanoparticle and optimal wavelength of radiation to achieve the maximum absorption efficiency of the nanoparticle in each surrounding medium.
(b) Determine the number of nanoparticles needed to form a cluster.
(c) Estimate the thermal damage area produced in the surrounding biological media by nanoclusters heated in the optical range of the spectrum.
(d) Find how many clusters per cell are needed to destroy cancer cells by thermal ablation.
(e) Write a report on your simulation results similar to the practice examples 9 and 10 in Sect. 9.2 of this chapter.

Section 9.3: Nanobubble Overlapping Mode

9.67. Describe how nanobubbles are generated around a heated nanoparticle/nanocluster.
9.68. What is a bubble nucleation center?
9.69. Give a value of the critical threshold temperature for heterogeneous bubble generation in biological media.
9.70. What are therapeutic effects of the bubble mode in nanophotothermolysis of cancer cells?
9.71. True or false? The bubble mode of nanophotothermolysis can provide a large damage area in cancer cells at the relatively low energy density of radiation required to generate those bubbles around nanoparticles.
9.72. List the assumptions used in the spherical bubble dynamics model.
9.73. Describe the terms in the Rayleigh–Plesset equation.
9.74. What are the output results of the bubble dynamics simulations?
9.75. True or false? The optical, thermal, acoustic and bubble formation phenomena initiated by radiation around each nanoparticle can overlap well to create synergistic effects within the cellular volume.
9.76. Define the optical radius of a nanoparticle.
9.77. Define the thermal diffusion length.
9.78. Define the sound transfer distance.
9.79. Describe the overlapping of optical fields around nanoparticles.
9.80. Describe the overlapping of thermal fields around nanoparticles.
9.81. Describe the overlapping of acoustic waves around nanoparticles.
9.82. List the therapeutic effects produced in cancer cells by lascr-induced overheating of nanoparticles in a cell environment.
9.83. How can the bubble overlapping mode dramatically increase the efficiency of cancer treatments?
9.84. Describe how bubbles can overlap.

9.85. What criteria are used for effective overlapping of bubbles around heated nanoparticles?
9.86. True or false? Bubbles cannot be created around heated nanoclusters.
9.87. Describe how to find the interparticle spacing, the nanoparticle concentration, and their absolute number within the cell volume required for the bubble overlapping mode.
9.88. What is an experimental value for a bubble nucleation time?
9.89. What is an experimental value for a bubble lifetime?
9.90. What is a bubble growth time?
9.91. What is a typical final bubble radius?
9.92. Calculate the absolute number of nanoparticles required for bubbles overlapping in a 15-μm cancer cell.
9.93. What is the optimal range of nanoparticle radii for effective laser initiation of the bubble overlapping mode?
9.94. What is a general feature of bubble radius growth?
9.95. Which rate—the bubble expansion rate or thermal conduction rate—is greater in the liquid environment?
9.96. What is a typical value of the thermal conduction rate in blood?
9.97. What is the saturated value of the bubble's expansion rate?
9.98. True or false? Bubble expansion occurs on a time scale much greater than heat loss.
9.99. True or false? A heated 380-nm cluster is able to produce one large bubble within the cell volume, whose damage area is comparable to the size of a cancer cell (10–15 μm in diameter).
9.100. List the advantages of the bubble overlapping mode for cancer nanophototherapy.
9.101. Explain why the expansion of the bubble has strong nonlinearity in the time range 0–30 ns.
9.102. Explain why, after 30 ns, a microbubble expands uniformly and linearly at a constant rate.
9.103. True or false? Thermal damage occurs over a negligible distance from the nanoparticle in comparison to bubble damage.

Section 9.4: Computer Practicum 7: Simulating Nanobubble Dynamics

9.104. In what two modes can bubbles grow?
9.105. What contribution does a nanoparticle make to the bubble dynamics model?
9.106. In which expansion mode is bubble expansion quicker?
9.107. In which expansion mode is bubble growth slower?
9.108. What is the difference between the adiabatic and nonadiabatic bubble expansion modes?

9.109. True or false? The initial nucleation of a bubble around an overheated nanoparticle happens at an explosive expansion rate of 300 m/s.
9.110. Explain why after a bubble is generated, the expansion rate drops exponentially.
9.111. By how many times is the radius affected by the bubble greater than the radius of the nanoparticle that generated the bubble?
9.112. True or false? The laser pulse energy density required to heat a single nanoparticle for producing thermal damage over a significant area is much higher than used to generate the bubble.
9.113. Explain how the nanoparticle size effects the bubble growth dynamic.
9.114. True or false? A larger particle results in a much faster bubble generation and expansion rate, forming an overall larger bubble size for the same time interval of bubble growth.
9.115. Explain how different biological media affect the nanobubble dynamics.
9.116. True or false? As a medium, fat is not as effective as blood, resulting in a smaller bubble size in fat.
9.117. True or false? The expansion of the bubble into the cell membrane has approximately the same rate as in blood, increasing the bubble damage efficiency of the cancer cells.
9.118. True or false? The bubble mode of nanophotothermolysis decreases the threshold energy density of the radiation for cancer cell killing.
9.119. True or false? Bubbles expand at constant temperature more rapidly than they expand adiabatically.
9.120. True or false? Distinctly different behaviors are observed for bubbles in different biological media.
9.121. Using the Maple code in Appendix H and data from Table 9.10, perform simulations for adiabatic and nonadiabatic bubble expansions around a heated 20-nm particle in blood.

(a) Determine the bubble saturated expansion rate in both expansion modes.
(b) Estimate the thermal damage area produced in the surrounding biological media by bubble expansion in both expansion modes.
(c) Determine by how many times the final radius of a bubble is greater than the radius of the nanoparticle that generated the bubble.
(d) Write a report on your simulation results similar to the practice example 11 in Sect. 9.4 of this chapter.

9.122. Using the Maple code in Appendix H and data from Table 9.10, perform simulations for adiabatic and nonadiabatic bubble expansions around a heated 20-nm particle in a cell membrane.

(a) Determine the bubble saturated expansion rate in both expansion modes.
(b) Estimate the thermal damage area produced in the surrounding biological media by bubble expansion in both expansion modes.
(c) Determine by how many times the final radius of a bubble is greater than the radius of the nanoparticle that generated the bubble.

(d) Write a report on your simulation results similar to the practice example 11 in Sect. 9.4 of this chapter.

9.123. Using the Maple code in Appendix H and data from Table 9.10, perform simulations for adiabatic and nonadiabatic bubble expansions around a heated 20-nm particle in fat.

(a) Determine the bubble saturated expansion rate in both expansion modes.
(b) Estimate the thermal damage area produced in the surrounding biological media by bubble expansion in both expansion modes.
(c) Determine by how many times the final radius of a bubble is greater than the radius of the nanoparticle that generated the bubble.
(d) Write a report on your simulation results similar to the practice example 11 in Sect. 9.4 of this chapter.

9.124. Using the Maple code in Appendix H and data from Table 9.10, perform simulations for adiabatic and nonadiabatic bubble expansions around a heated 20-nm particle in water.

(a) Determine the bubble saturated expansion rate in both expansion modes.
(b) Estimate the thermal damage area produced in the surrounding biological media by bubble expansion in both expansion modes.
(c) Determine by how many times the final radius of a bubble is greater than the radius of the nanoparticle that generated the bubble.
(d) Write a report on your simulation results similar to the practice example 11 in Sect. 9.4 of this chapter.

9.125. Using the Maple code in Appendix H and data from Table 9.10, perform comparative simulations for adiabatic bubble expansion in blood for varying nanoparticle sizes: 20 nm, 50 nm, and 100 nm.

(a) Determine the bubble saturated expansion rate for the varying nanoparticle sizes.
(b) Estimate the thermal damage area produced in the surrounding biological media by the bubble expansion for the varying nanoparticle sizes.
(c) Determine by how many times the final radius of a bubble is greater than the radius of the nanoparticle that generated the bubble.
(d) Write a report on your simulation results similar to the practice example 11 in Sect. 9.4 of this chapter.

9.126. Using the Maple code in Appendix H and data from Table 9.10, perform comparative simulations for adiabatic bubble expansion in cytoplasm for varying nanoparticle sizes: 20 nm, 50 nm, and 100 nm.

(a) Determine the bubble saturated expansion rate for the varying nanoparticle sizes.
(b) Estimate the thermal damage area produced in the surrounding biological media by the bubble expansion for the varying nanoparticle sizes.

(c) Determine by how many times the final radius of a bubble is greater than the radius of the nanoparticle that generated the bubble.
(d) Write a report on your simulation results similar to the practice example 11 in Sect. 9.4 of this chapter.

9.127. Using the Maple code in Appendix H and data from Table 9.10, perform comparative simulations for adiabatic bubble expansion in a cell membrane for varying nanoparticle sizes: 20 nm, 50 nm, and 100 nm.

(a) Determine the bubble saturated expansion rate for the varying nanoparticle sizes.
(b) Estimate the thermal damage area produced in the surrounding biological media by the bubble expansion for the varying nanoparticle sizes.
(c) Determine by how many times the final radius of a bubble is greater than the radius of the nanoparticle that generated the bubble.
(d) Write a report on your simulation results similar to the practice example 11 in Sect. 9.4 of this chapter.

9.128. Using the Maple code in Appendix H and data from Table 9.10, perform comparative simulations for adiabatic bubble expansion in fat for varying nanoparticle sizes: 20 nm, 50 nm, and 100 nm.

(a) Determine the bubble saturated expansion rate for the varying nanoparticle sizes.
(b) Estimate the thermal damage area produced in the surrounding biological media by the bubble expansion for the varying nanoparticle sizes.
(c) Determine by how many times the final radius of a bubble is greater than the radius of the nanoparticle that generated the bubble.
(d) Write a report on your simulation results similar to the practice example 11 in Sect. 9.4 of this chapter.

9.129. Using the Maple code in Appendix H and data from Table 9.10, perform comparative simulations for adiabatic bubble expansion in water for varying nanoparticle sizes: 20 nm, 50 nm, and 100 nm.

(a) Determine the bubble saturated expansion rate for the varying nanoparticle sizes.
(b) Estimate the thermal damage area produced in the surrounding biological media by the bubble expansion for the varying nanoparticle sizes.
(c) Determine by how many times the final radius of a bubble is greater than the radius of the nanoparticle that generated the bubble.
(d) Write a report on your simulation results similar to the practice example 11 in Sect. 9.4 of this chapter.

9.130. Using the Maple code in Appendix H and data from Table 9.10, perform comparative simulations for adiabatic bubble expansion around 10-nm particles in different biological media: blood, cytoplasm, cell membrane, and fat.

(a) Determine the bubble saturated expansion rate in different biological media.
(b) Estimate the thermal damage area produced in the different surrounding biological media by the bubble expansion.
(c) Write a report on your simulation results similar to the practice example 11 in Sect. 9.4 of this chapter.

9.131. Using the Maple code in Appendix H and data from Table 9.10, perform comparative simulations for adiabatic bubble expansion around 20-nm particles in different biological media: blood, cytoplasm, cell membrane, and fat.

(a) Determine the bubble saturated expansion rate in different biological media.
(b) Estimate the thermal damage area produced in the different surrounding biological media by the bubble expansion.
(c) Write a report on your simulation results similar to the practice example 11 in Sect. 9.4 of this chapter.

9.132. Using the Maple code in Appendix H and data from Table 9.10, perform comparative simulations for adiabatic bubble expansion around 30-nm particles in different biological media: blood, cytoplasm, cell membrane, and fat.

(a) Determine the bubble saturated expansion rate in different biological media.
(b) Estimate the thermal damage area produced in the different surrounding biological media by the bubble expansion.
(c) Write a report on your simulation results similar to the practice example 11 in Sect. 9.4 of this chapter.

9.133. Using the Maple code in Appendix H and data from Table 9.10, perform comparative simulations for adiabatic bubble expansion around 40-nm particles in different biological media: blood, cytoplasm, cell membrane, and fat.

(a) Determine the bubble saturated expansion rate in different biological media.
(b) Estimate the thermal damage area produced in the different surrounding biological media by the bubble expansion.
(c) Write a report on your simulation results similar to the practice example 11 in Sect. 9.4 of this chapter.

9.134. Using the Maple code in Appendix H and data from Table 9.10, perform comparative simulations for adiabatic bubble expansion around 50-nm particles in different biological media: blood, cytoplasm, cell membrane, and fat.

(a) Determine the bubble saturated expansion rate in different biological media.

(b) Estimate the thermal damage area produced in the different surrounding biological media by the bubble expansion.
(c) Write a report on your simulation results similar to the practice example 11 in Sect. 9.4 of this chapter.

Section 9.5: Laser-Induced Thermal Explosion Mode—"Nanobombs"

9.135. Describe how to produce a thermal explosion of nanoparticles.
9.136. True or false? Thermal explosion of nanoparticles is realized for the continuous-wave mode of laser heating the nanoparticles.
9.137. True or false? In the thermal explosion mode, heat diffuses away from the surface of the nanoparticle more rapidly than heat is generated within the nanoparticle.
9.138. What phenomena accompany the thermal explosion of nanoparticles?
9.139. What are the two main physical mechanisms that can lead to laser-induced explosion of nanoparticles?
9.140. What is the Coulomb explosion mode?
9.141. What physical phenomena can be generated around a nanoparticle in the temperature range $T < T_{LV}$, where T_{LV} is the liquid vaporization temperature?
9.142. What physical phenomena can be generated around a nanoparticle in the temperature range $T_{LV} \leq T < T_{GNM}$, where T_{GNM} is the gold nanoparticle melting point?
9.143. What physical phenomena can be generated around a nanoparticle in the temperature range $T_{GNM} \leq T < T_{GNB}$, where T_{GNB} is the gold nanoparticle boiling point?
9.144. What physical phenomena can be generated around a nanoparticle in the temperature range $T \geq T_{GNB}$, where T_{GNB} is the gold nanoparticle boiling point?
9.145. Calculate the thermal relaxation time for 50-, 100-, and 200-nm radii gold nanoparticles in blood.
9.146. Calculate the thermal relaxation time for 50-, 100-, and 200-nm radii gold nanoparticles in cytoplasm.
9.147. Calculate the thermal relaxation time for 50-, 100-, and 200-nm radii gold nanoparticles in a cell membrane.
9.148. Calculate the thermal relaxation time for 50-, 100-, and 200-nm radii gold nanoparticles in fat.
9.149. What limitation is applied for a laser pulse duration to generate the explosion of nanoparticles?
9.150. Give the definition for thermal explosion of nanoparticles in terms of time, energy absorbed, and energy required for the complete particle evaporation.
9.151. Define the shock waves generated by thermal explosion of nanoparticles.

9.152. True or false? The pressure produced by vaporization of a nanoparticle itself is less than the water vapor pressure around hot nanoparticles.

9.153. List and describe the therapeutic effects produced by laser-induced explosion of nanoparticles in cancer cells.

9.154. True or false? The vapor pressure produced by nanoparticle explosion in a liquid environment can reach up to gigapascals.

9.155. List and describe what affects the damage of cellular structure in the thermal explosion mode.

9.156. Calculate the threshold energy density of the laser radiation for thermal explosion of silver nanoparticles in blood, cytoplasm, cell membrane, and fat.

(a) Use Mie simulations to find the optimal radius of the nanoparticle and optimal wavelength of radiation to achieve maximum absorption efficiency of the nanoparticle in each surrounding medium.
(b) Determine the threshold intensity of the laser radiation for thermal explosion of nanoparticles in each surrounding medium.
(c) Find the threshold energy density of the laser radiation for thermal explosion of nanoparticles in each surrounding medium.
(d) Compare the calculated threshold energy densities to the cell organelles damage thresholds given in Table 9.11.

9.157. Calculate the threshold energy density of the laser radiation for thermal explosion of nickel oxide nanoparticles in blood, cytoplasm, cell membrane, and fat.

(a) Use Mie simulations to find the optimal radius of the nanoparticle and optimal wavelength of radiation to achieve maximum absorption efficiency of the nanoparticle in each surrounding medium.
(b) Determine the threshold intensity of the laser radiation for thermal explosion of nanoparticles in each surrounding medium.
(c) Find the threshold energy density of the laser radiation for thermal explosion of nanoparticles in each surrounding medium.
(d) Compare the calculated threshold energy densities to the cell organelles damage thresholds given in Table 9.11.

9.158. Calculate the threshold energy density of the laser radiation for thermal explosion of magnesium oxide nanoparticles in blood, cytoplasm, cell membrane, and fat.

(a) Use Mie simulations to find the optimal radius of the nanoparticle and optimal wavelength of radiation to achieve maximum absorption efficiency of the nanoparticle in each surrounding medium.
(b) Determine the threshold intensity of the laser radiation for thermal explosion of nanoparticles in each surrounding medium.
(c) Find the threshold energy density of the laser radiation for thermal explosion of nanoparticles in each surrounding medium.

(d) Compare the calculated threshold energy densities to the cell organelles damage thresholds given in Table 9.11.

9.159. Calculate the threshold energy density of the laser radiation for thermal explosion of carbon nanoparticles in blood, cytoplasm, cell membrane, and fat.

(a) Use Mie simulations to find the optimal radius of the nanoparticle and optimal wavelength of radiation to achieve maximum absorption efficiency of the nanoparticle in each surrounding medium.
(b) Determine the threshold intensity of the laser radiation for thermal explosion of nanoparticles in each surrounding medium.
(c) Find the threshold energy density of the laser radiation for thermal explosion of nanoparticles in each surrounding medium.
(d) Compare the calculated threshold energy densities to the cell organelles damage thresholds given in Table 9.11.

9.160. Calculate the threshold energy density of the laser radiation for thermal explosion of glass nanoparticles in blood, cytoplasm, cell membrane, and fat.

(a) Use Mie simulations to find the optimal radius of the nanoparticle and optimal wavelength of radiation to achieve maximum absorption efficiency of the nanoparticle in each surrounding medium.
(b) Determine the threshold intensity of the laser radiation for thermal explosion of nanoparticles in each surrounding medium.
(c) Find the threshold energy density of the laser radiation for thermal explosion of nanoparticles in each surrounding medium.
(d) Compare the calculated threshold energy densities to the cell organelles damage thresholds given in Table 9.11.

9.161. Calculate the threshold energy density of the laser radiation for thermal explosion of polystyrene nanoparticles in blood, cytoplasm, cell membrane, and fat.

(a) Use Mie simulations to find the optimal radius of the nanoparticle and optimal wavelength of radiation to achieve maximum absorption efficiency of the nanoparticle in each surrounding medium.
(b) Determine the threshold intensity of the laser radiation for thermal explosion of nanoparticles in each surrounding medium.
(c) Find the threshold energy density of the laser radiation for thermal explosion of nanoparticles in each surrounding medium.
(d) Compare the calculated threshold energy densities to the cell organelles damage thresholds given in Table 9.11.

9.162. Calculate the threshold energy density of the laser radiation for thermal explosion of fullerene nanoparticles in blood, cytoplasm, cell membrane, and fat.

(a) Use Mie simulations to find the optimal radius of the nanoparticle and optimal wavelength of radiation to achieve maximum absorption efficiency of the nanoparticle in each surrounding medium.
(b) Determine the threshold intensity of the laser radiation for thermal explosion of nanoparticles in each surrounding medium.
(c) Find the threshold energy density of the laser radiation for thermal explosion of nanoparticles in each surrounding medium.
(d) Compare the calculated threshold energy densities to the cell organelles damage thresholds given in Table 9.11.

9.163. What is a value of laser energy density established as the safety standard for medical lasers?

9.164. Describe how the nanoparticle's explosion threshold energy density can be reduced.

9.165. List and describe the advantages of using the thermal explosion mode in nanotherapy.

References

1. V.P. Zharov, R.R. Letfullin, E.N. Galitovskaya, Microbubble-overlapping mode for laser killing of cancer cells with absorbing nanoparticle clusters. J. Phys. D Appl. Phys. **38**, 2571–2581 (2005)
2. R.R. Letfullin, T.F. George, New dynamic modes for selective laser cancer nanotherapy, in *Computational Studies of New Materials II: From Ultrafast Processes and Nanostructures to Optoelectronics, Energy Storage and Nanomedicine*, ed. by T.F. George, D. Jelski, R.R. Letfullin, G.P. Zhang (World Scientific, Singapore, 2011), pp. 131–172
3. R.R. Letfullin, A.R. Letfullin, T.F. George, Absorption efficiency and heating kinetics of nanoparticles in the RF range for selective nanotherapy of cancer. Nanomedicine **11**, 413–420 (2015)
4. C.E. Brennen, *Cavitation and Bubble Dynamics* (Oxford University Press, New York, 1995)
5. R.R. Letfullin, T.F. George, Plasmonic nanomaterials in nanomedicine, in *Springer Handbook of Nanomaterials*, ed. by R. Vajtai (-Verlag, Berlin, 2013), pp. 1063–1097
6. R.R. Letfullin, T.F. George, Plasmonic nanomaterials in nanomedicine, in *Springer Handbook of Nanomaterials*, ed. by R. Vajtai (-Verlag, Berlin, 2013), pp. 1063–1097
7. R.R. Letfullin, A.R. Letfullin, T.F. George, RF heating of nanoclusters for cancer therapy, in *Colloidal Nanoparticles for Biomedical Applications X (Photonics West 2015: BiOS—NanoBiophotonics)*, ed. By W.J. Parak, M. Osinski, X.J. Liang, SPIE Proc. **9338**, 93381G/1-9 (2015)

Chapter 10
Ablation of Biological Tissues by Plasmonic Nanoparticles

Contents

Abstract In this chapter, we perform modeling of soft and hard biological tissue ablation by plasmonic nanoparticles heated by short and ultrashort laser pulses. A goal of plasmonic laser nanosurgery/nanoablation is to deliver extremely precise heat blasts in programmed patterns to cancer cells, abnormal cell organelles, or

---

Electronic supplementary material: The online version of this chapter (doi:10.1007/978-3-319-43577-0_10) contains supplementary material, which is available to authorized users.

R.R. Letfullin, T.F. George, *Computational Nanomedicine and Nanotechnology*,
DOI 10.1007/978-3-319-43577-0_10

mutated DNA molecules in order to destroy them. Short and ultrashort laser pulses hold key interest in precise nanoablation as they can allow for deposition of thermal energy to the target area by many orders of magnitude faster than the heat diffusion process that leaks heat to adjacent healthy cells. For these reasons, time-dependent thermal ablation simulations are performed in this chapter for short and ultrashort single pulse laser–nanoparticle–tissue interactions, as well as simulation for multipulse plasmonic nanoparticle ablation that utilizes the repetition of ultrashort pulses. This chapter contains material adapted from our publications [1, 2]. The detailed list of references and reviews on a given topic of this chapter can be found in those original papers.

10.1 Lecture 30: Bone Cancer Ablation by Short and Ultrashort Pulses (Slide 30.1)

- Introduction
- Bone Targeting
- Radiation Delivery
- Plasmonic Nanoparticles
- Modeling
  - *Ablation Model*
  - *Modeling Heat Transfer*
- Materials Properties and Input Parameters
- Results and Discussion
  - *Heating of Nanoparticles in Hard Tissue*
  - *Heating of Bone Particles*
- Ablation of Bone Tissue
- Future Perspectives
- Conclusions

10.1.1 Introduction

Despite the popular image portraying the skeleton as a simple static frame for the soft tissues of the body, the truth is that bone too is a living, growing, and constantly changing tissue (Slide 30.2). Bones are not permanent; rather, the mineral frame is in constant renewal. Biological signals maintain a balance of reabsorption and mineralization, continuously recycling old bone and replacing it with new. Because only a small percentage of the bone volume is ever being destroyed or rebuilt at a given time, and because the destruction and construction processes are properly

Lecture 30: Bone Cancer Ablation by Short and Ultrashort Pulses

Contents:

Slide 30.1 Contents

Introduction

- ***Bone*** is a living, growing and constantly changing tissue.
- ***Bones are not permanent***; rather, the mineral frame is ***in constant renewal***.
- Biological signals maintain ***a balance of reabsorption and mineralization***, continuously recycling old bone and replacing it with new.
- If hard-tissue cells become ***cancerous***and begin producing normal bone-regulatory molecules in excess, then the natural ***balance between reabsorption and mineralization could be upset*** , leading to bone disorders.
- As shown earlier,***nanoparticles*** are being investigated as a possible tool***for therapeutic treatments***.
- In two different techniques, ***nanophotohyperthermia and nanophotothermolysis***, light wave radiation is used to irradiate a nanoparticle and heat the immediate surrounding material.

Slide 30.2 Introduction

balanced, this complex system can maintain the stable equilibrium we often take for granted as static. Only when this balance is displaced for significant periods of time will changes be noticeable on the macroscopic scale. This section contains material reproduced from our publication [1].

Cancers are caused by mutated cells that begin to reproduce uncontrollably as discussed in Chap. 2 of this book. As these unwanted cells multiply, they produce hormones, proteins, and other biological signals the same way as normal cells do, but as the number of cancerous cells increases, so does the strength of those rogue cell signals. Eventually, there comes a point when healthy cell signals are overcome by those from cancer, and the body's natural systems can be confused. If hard-tissue cells become cancerous and begin producing normal bone-regulatory molecules in excess, then the natural balance between reabsorption and mineralization could be upset, leading to bone disorders (Slide 30.2) [1].

Traditional oncology relies on three main forms of treatment: surgery, chemotherapy, and radiation therapy, as we discussed in Sect. 3.1 of Chap. 3. Surgery is able to remove large tumors from the body in a single treatment. However, this also carries the risk of dislodging some surviving cancerous cells into the bloodstream. Each of these cells has the potential to proliferate and develop into a new tumor elsewhere in the body, called metastasis. Chemotherapy uses chemicals and drugs to limit the ability of cancer cells to multiply, invade, and metastasize. However, the drugs are not always cancer cell specific and may hinder the normal functions of healthy cells as well. Radiation therapy uses radiation to damage cells; generally, X-rays are used to ionize the targeted tissue to create free radicals that can damage DNA. Cancer cells have a diminished ability to repair these DNA defects, and so after multiplication, the cancer cells will have diminished viability. Again, this process of creating free radicals is nonspecific to cancer tissues, and so creating localized damage requires proper aiming of the radiation to create a maximum of ionization in the cancerous region and a minimum to the healthy surrounding tissues. All three forms of therapy have benefits and detriments that must be weighed to create an effective, and safe, cancer treatment [1].

As shown in Chap. 3 nanoparticles are being investigated as a possible tool for therapeutic treatments (Slide 30.2). In two different techniques, nanophotohyperthermia and nanophotothermolysis, light wave radiation is used to irradiate a nanoparticle and heat the immediate surrounding material. In nanophotohyperthermia, temperatures in the target region are increased slightly above normal for an extended period of time (Slide 30.3). Because temperature affects protein shape, it can affect how well a protein carries out its intended functions. By disrupting these functions, cancers that rely on such functions can have reduced growth rates or die off entirely. In certain cancers, prolonged exposure to temperatures of only 40–44 °C has been shown to trigger cell death in the cancerous cells with little to no effect on surrounding healthy tissues [1, 2]. In nanophotothermolysis, however, high-intensity pulses are used to heat the nanoparticles very quickly, reaching thresholds for various nonlinear synergistic effects in the targeted region, including thermal coagulation, vaporization, sound production, bubble formation, and shock wave generation, which have shown great potential for selective damage to cancer cells, bacteria, viruses, and DNA (Slide 30.3). More importantly, these therapies are poised to remove many of the detrimental effects seen in current oncological treatments. Because the nanoparticles are not cytotoxic in and of themselves, they can be delivered into the body without harming

Introduction (continue)

- ***In nanophotohyperthermia***, ***temperatures*** in the target region are increased ***slightly above normal*** for an extended period of time.
- Because ***temperature affects protein*** shape, it can affect how well a protein carries out its intended functions.
- ***By disrupting these functions***, cancers that rely on such functions can have reduced growth rates or die off entirely.
- In certain cancers, prolonged exposure to ***temperatures*** of only ***40-44°C*** has been shown to trigger ***cell death in the cancerous cells*** with little to ***no effect on surrounding healthy tissue***.
- ***In nanophotothermolysis***, however, high-intensity pulses are used to heat the nanoparticles very quickly, reaching thresholds for various nonlinear synergistic effects in the targeted region, including ***thermal coagulation, vaporization, sound production, bubble formation and shock wave generation***, which have shown great potential for selective damage to cancer cells, bacteria, viruses and DNA.

Slide 30.3 Introduction (continued)

Introduction (continue)

- However, the successful use of such therapies in cancer treatment relies ***on two main factors: targeted delivery of the nanoparticles*** to the cancer cells, and ***controlled heating of the nanoparticles*** to induce the desired thermal damage mechanisms.
- Thus, ***a balance*** must be found between the ***laser energies*** required ***for*** effective ***ablation*** and the energies that ***cause excessive thermal damage*** to surrounding tissue.
- The ***use of plasmonic nanoparticles*** heated by short and ultrashort laser pulses discussed here seeks to meet the challenge of achieving safe and precision nanosurgery of the targeted areas.
- ***Bone biology*** will be further discussed because it applies not only to possible bone-targeting pathways but also to thermal diffusion and ablation models.
- The ***plasmonic properties of nanoparticles*** will then be considered as they apply to heat generation for thermal ablation of the surrounding biological medium.

Slide 30.4 Introduction (continued)

or hindering healthy cell functions; this is in contrast to chemotherapy, which can have detrimental effects that are pervasive to many types of healthy tissues. The damage mechanism, heat, does not kill cells by creating DNA damage in the way that radiation therapy does, and therefore removes the risk that healthy cells get DNA defects, resulting in reduced viability or mutation. Like chemotherapy or radiation therapy, this approach can be used in cases where the pervasiveness of metastases or the delicate nature of local tissues would make traditional surgery infeasible [1].

However, the successful use of such therapies in cancer treatment relies on two main factors: targeted delivery of the nanoparticles to the cancer cells, and controlled heating of the nanoparticles to induce the desired thermal damage mechanisms (Slide 30.4). Thus, a balance must be found between the laser energies required for effective ablation and the energies that cause excessive thermal damage to surrounding tissue. The use of plasmonic nanoparticles heated by short and ultrashort laser pulses discussed in this chapter seeks to meet the challenge of achieving safe and precision nanosurgery of the targeted areas. Bone biology will be further discussed because it applies not only to possible bone-targeting pathways but also to thermal diffusion and ablation models. The plasmonic properties of nanoparticles will then be considered as they apply to heat generation for thermal ablation of the surrounding biological medium [1].

10.1.2 Bone Targeting

The average human body is composed of about 60 % water by weight; of this, approximately 70 % is found within the cells, the remainder being extracellular. Tissues are made up of multiple cells, which may or may not be made up of different cell types, that ultimately work together to perform a specific function. Because of this, tissues are often classified according to their functions in the body. More broadly, tissues can be differentiated according to their composition. This chapter differentiates tissues into one of two groups: hard tissues, where a mineral component is the largest extracellular constituent; and soft tissues, being composed mainly of water [1].

Soft tissues, like those found in the heart and lungs, are 80–90 % water. As such, these tissues can generally be modeled as water, but with a few corrections to factor in changes due to the lipids, proteins, and any other organic molecules making up the remaining 10–20 % of the tissue weight (Slide 30.5). However, in hard tissues, a mineral structure made up of nanoscale calcium phosphate crystals accounts for about 70 % of total bone material [1]. Other biological molecules, notably collagen and lipids, are present as well, bringing the total water weight of bone tissues down to about 22 %. Because of this, hard tissues cannot be modeled in the same way as soft tissues.

There are three main cells to consider when investigating bone tissue: osteoblasts, which are responsible for mineralization of bone; osteocytes, which maintain

Bone Structure

- ***Soft tissues***, like those found in the heart and lungs, are 80-90% water.
- As such, these tissues can generally be ***modeled as water***, but with a few corrections to factor in changes due to the ***lipids, proteins and any other organic molecules*** making up the remaining ***10-20% of the tissue weight***.
- However, in ***hard tissues***, a mineral structure made up of nanoscale ***calcium phosphate crystals*** accounts for about ***70% of total bone material***.
- Other biological molecules, notably ***collagen and lipids***, are present as well, bringing the total water weight of ***bone tissues*** down to ***about 22%.***
- Because of this, ***hard tissues*** cannot be modeled in the same way as soft tissues.

Slide 30.5 Bone Structure

Bone Biology

- ***There are three main cells*** to consider when investigating bone tissue: ***osteoblasts***, which are responsible for mineralization of bone; ***osteocytes***, which maintain bone; and ***osteoclasts***, which are responsible for resorption.
- These cells are ***highly specialized*** and generally do not reproduce by replication.
- Rather, a ***nearby*** population of immature (less differentiated) cells stands ready, and in response to biochemical signaling molecules, these ***cells can*** be made to ***proliferate and differentiate*** into the more specialized cell types.
- These ***signaling molecules*** are activated during the remodeling process, but will also be released in response to a fracture in order to promote healing.

Slide 30.6 Bone biology

bone; and osteoclasts, which are responsible for resorption (Slide 30.6). These cells are highly specialized and generally do not reproduce by replication. Rather, a nearby population of immature (less differentiated) cells stands ready, and in response to biochemical signaling molecules, these cells can be made to proliferate

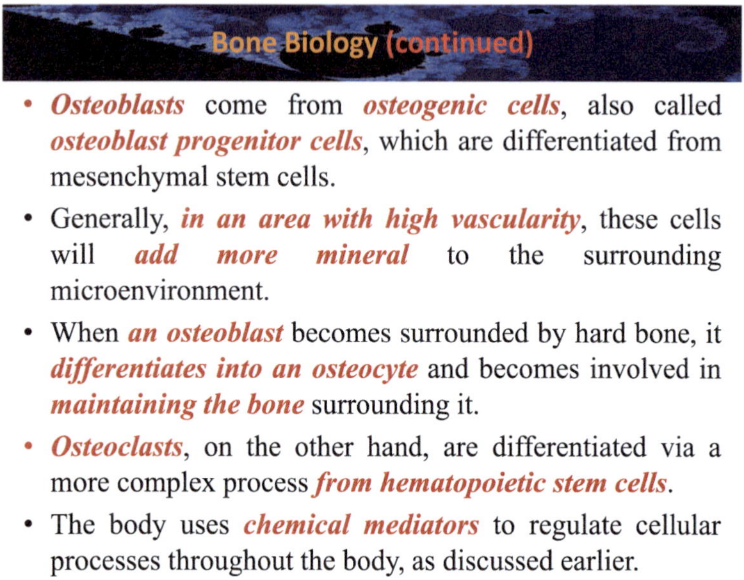

Slide 30.7 Bone biology (continued)

and differentiate into the more specialized cell types. These signaling molecules are activated during the remodeling process, but will also be released in response to a fracture in order to promote healing. Osteoblasts come from osteogenic cells, also called osteoblast progenitor cells, which are differentiated from mesenchymal stem cells (Slide 30.7). Generally, in an area with high vascularity, these cells will add more mineral to the surrounding microenvironment. When an osteoblast becomes surrounded by hard bone, it differentiates into an osteocyte and becomes involved in maintaining the bone surrounding it. Osteoclasts, on the other hand, are differentiated via a more complex process from hematopoietic stem cells [1].

The body uses chemical mediators to regulate cellular processes throughout the body, as discussed in Chap. 2. Growth factors are proteins or steroid hormones that regulate cell growth, function, or motility; typically they act as signaling molecules, binding to specialized receptors on target cells, which then signal the cell to increase or decrease a specific function (Slide 30.8). Bone growth factors can influence bone tissue in a number of ways. Growth factors sent to mesenchymal or hematopoietic stem cells can regulate the rate at which those stem cells proliferate, or can determine what types of specialized cells they differentiate into. Other growth factors can directly affect differentiated cells by regulating cell function, inducing motility, or by inducing differentiation of osteoblasts to osteocytes [1].

Many growth factors play roles in the mediation of bone formation and resorption processes. A large group of factors, known as the transforming growth factor beta (TGF-β) superfamily, regulate many different biological processes. More specifically related to bone are the bone morphogenic proteins (BMP 2-16) which belong to this superfamily (Slide 30.9). BMPs are the only known growth

Bone Targeting

- ***Growth factors are proteins or steroid hormones*** that ***regulate cell growth, function or motility***; typically they act as signaling molecules, binding to specialized receptors on target cells, which then signal the cell to increase or decrease a specific function.
- Growth factors sent to ***mesenchymal or hematopoietic stem cells*** can ***regulate the rate*** at which those ***stem cells proliferate***, or can determine what types of specialized cells they differentiate into.
- Other growth factors can directly affect differentiated cells by ***regulating cell function***, inducing ***motility***, or by inducing ***differentiation of osteoblasts to osteocytes***.
- Many ***growth factors*** play roles in the mediation of ***bone formation and resorption processes***. A large group of factors, known as the transforming growth factor beta (TGF-β) superfamily, regulate many different biological processes.

Slide 30.8 Bone targeting

Bone Targeting (continued)

- More specifically related to bone are the ***bone morphogenic proteins*** (BMP 2-16) which belong to this superfamily.
- BMPs are the only known growth factors to stimulate ***differentiation of mesenchymal stem cells into osteoblasts or chondroblasts***.
- A member of the ***tumor necrosis factor*** (TNF) superfamily, known as the receptor activator of NF-χβ ligand (RANKL), plays a major role in ***osteoclast formation***, and has important interplay with its receptor RANK and the decoy molecule ***osteoprotegerin*** (OPG).
- However, ***OPG*** also binds to the ***TNF-related apoptosis-inducing ligand (TRAIL), reducing*** TRAIL's ability to induce ***cell death***, which generally promotes cancer growth in several cancer types.

Slide 30.9 Bone targeting (continued)

factors to stimulate differentiation of mesenchymal stem cells into osteoblasts or chondroblasts. Different BMPs are more or less effective than others at producing bone formation, with propensities changed based on the presence of other mediator molecules, suggesting a rather complex synergy of multiple factors. A member of

Bone Targeting (continued)

- ***Silicon dioxide particles***, 10 nm and 50 nm in size, surface-doped with calcium (Ca-SiO2), have shown successful ***transfer*** of plasmid ***DNA to osteoblast cells*** and improved selectivity versus hepatocytes, giving yet another possibility for ***bone-targeting bio-conjugates***.
- A mixture of ***BMP-2 conjugated nanoparticles with mesenchemal stem cells*** (MSC) in fibrin hydrogel has shown ***increased bone regrowth rate*** versus controls, using the MCSs with nanoparticles alone or MSCs with BMP-2 alone.
- ***Biphosphonates*** (BPs) have high affinity for calcium ions and are therefore ***targeted to bone mineral***.
- After binding to the mineral bone, these BPs are ***internalized*** specifically ***by osteoclasts during resorption***, where they can either inhibit further resorption or trigger the death of the osteoclast (depending on the specific type of BP in use).

Slide 30.10 Bone targeting (continued)

the tumor necrosis factor (TNF) superfamily, known as the receptor activator of NF-χβ ligand (RANKL), plays a major role in osteoclast formation, and has important interplay with its receptor RANK and the decoy molecule osteoprotegerin (OPG). However, OPG also binds to the TNF-related apoptosis-inducing ligand (TRAIL), reducing TRAIL's ability to induce cell death, which generally promotes cancer growth in several cancer types. Silicon dioxide particles, 10 and 50 nm in size, surface-doped with calcium (Ca-SiO_2), have shown successful transfer of plasmid DNA to osteoblast cells and improved selectivity versus hepatocytes, giving yet another possibility for bone-targeting bioconjugates [1] (Slide 30.10). A mixture of BMP-2 conjugated nanoparticles with mesenchymal stem cells (MSC) in fibrin hydrogel has shown increased bone regrowth rate versus controls, using the MCSs with nanoparticles alone or MSCs with BMP-2 alone.

Bisphosphonates (BPs) have high affinity for calcium ions and are therefore targeted to bone mineral (Slide 30.10). After binding to the mineral bone, these BPs are internalized specifically by osteoclasts during resorption, where they can either inhibit further resorption or trigger the death of the osteoclast (depending on the specific type of BP in use). Following the success of BPs ability to deliver themselves to osteoclasts, many research groups have conjugated various payload molecules to the BPs and tested the delivery effectiveness (Slide 30.11). Radium-223 has been investigated in nuclear medicine as an α-emitting isotope, showing selective absorption to bone resorption regions when conjugated to BPs; it has also shown targeting potential when administered as radium-223 chloride ($^{223}RaCl_2$), or as a stand-alone Ra^{2+} ion (due to the natural bone-seeking properties of heavy

Bone Targeting (continued)

- Following the success of BPs ability to deliver themselves to osteoclasts, many research groups have ***conjugated*** various ***payload molecules to the BPs*** and tested the delivery effectiveness.
- ***Radium-223*** has been investigated in nuclear medicine as an α-emitting isotope, showing selective absorption to bone resorption regions when ***conjugated to BPs***; it has also shown targeting potential when administered as radium-223 chloride ($^{223}RaCl_2$), or as a stand-alone Ra^{2+} ion (due to the natural bone-seeking properties of heavy alkaline earth element cations).
- Currently, $^{223}RaCl_2$ is being investigated in Phase III ***clinical trials*** under the trade name Alpharadin in the ALSYMPCA study (ALpharadin in SYMptomatic Prostate CAncer patients).
- Phase II results show approximately a ***40% survival rate over placebos***, demonstrating the effectiveness of the bone-targeting molecule.

Slide 30.11 Bone targeting (continued)

alkaline earth element cations). Currently, $^{223}RaCl_2$ is being investigated in Phase III clinical trials under the trade name Alpharadin in the ALSYMPCA study (ALpharadin in SYMptomatic Prostate CAncer patients). The patients in the trial all present castration-resistant prostate cancer with bone metastases, and for this specific type of prostate cancer, bone metastases are known to be the main cause of disability and death. Phase II results show approximately a 40 % survival rate over placebos, demonstrating the effectiveness of the bone-targeting molecule (Slide 30.11) [1].

Additionally, a radiometal bifunctional chelating agent, known as 1,4,7,10-tetraazacyclododecane-1,4,7,10-tetraacetic acid (DOTA), has been conjugated to a bone-targeting biphosphate to form a new molecule, dubbed DO3A-BP, which is able to deliver a ^{111}In labeling agent to osteoclasts (Slide 30.12). Another interesting delivery mechanism utilizes BPs, but designs the bond connecting the attached molecules to hydrolyze within a given time period; in this way, the drugs can be targeted locally to bone tissues but do not require resorption by osteoclasts, making the drugs somewhat less osteoclast selective and allowing delivery to neighboring cells. Similarly, a tetracycline-derived compound, which (like BPs) has high affinity for the mineral hydroxyapatite (HAP), is conjugated with estradiol to produce a bone-targeting estrogen compound (BTE2-A1) for application to osteoporosis (Slide 30.12) [1].

Bone Targeting (continued)

- Additionally, a ***radiometal bifunctional chelating agent***, known as ***1,4,7,10-tetraazacyclododecane-1,4,7,10-tetraacetic acid*** (DOTA), has been conjugated to a bone-targeting biphosphate to form a new molecule, dubbed DO3A-BP, which is able to ***deliver a ^{111}In labeling agent to osteoclasts***.
- Another interesting delivery mechanism utilizes ***BPs***, but designs the bond connecting the attached molecules to hydrolyze within a given time period; in this way, the ***drugs can be targeted locally to bone tissues*** but do not require resorption by osteoclasts, making the drugs somewhat less osteoclast selective and allowing delivery to neighboring cells.
- Similarly, a ***tetracycline-derived compound***, which (like BPs) has high affinity for the mineral hydroxyapatite (HAP), is ***conjugated with estradiol*** to ***produce a bone-targeting estrogen compound*** (BTE2-A1) for application to osteoporosis.

Slide 30.12 Bone targeting (continued)

10.1.3 Radiation Delivery

In addition to the challenge of finding suitable conjugates for specific types of cancer, biological tissues are not ideal for transmitting light radiation (Slide 30.13). Such tissues have strong light scattering properties that diffuse the light and increase the time photons spend in the tissue, therefore increasing the probability that these photons are absorbed by tissue. This is not ideal, because it leads to heat generation throughout the tissue rather than locally at the nanoparticle. As discussed in Sect. 3.2 of Chap. 3, for biological tissue in general, there is a spectral region of reduced absorption, called an "optical window" or "window of transparency" for biological media. This region is bounded between the strong absorption effects of hemoglobin for wavelengths below 600 nm and the strong absorption effects of water at wavelengths above 1300 nm and is a region of increased penetration depth of light through the tissue (Slide 30.13) [1].

Current light-therapy methods look at light in or near this spectrum because of the relatively stronger absorption characteristics of nanoparticles in this region as well as the fact that these wavelengths can utilize laser light sources to allow greater control over the light intensity. We should note, as discussed in Sect. 3.2 of Chap. 3, the body's transparency in the X-ray and radio-wave spectrum would enable ideal biological transparency windows [1].

Although the visible light-scattering issue is a problem for both soft tissue and hard tissues, the problem is more pronounced for hard tissues because they are generally surrounded by a significant amount of light-scattering soft tissue (Slide 30.14).

Radiation Delivery

- In addition to the ***challenge*** of finding suitable conjugates for specific types of cancer, biological tissues are not ideal for ***transmitting light radiation***.
- Such ***tissues have strong light scattering properties*** that diffuse the light and increase the time photons spend in the tissue, therefore increasing the probability that these photons are absorbed by tissue.
- It leads to ***heat generation throughout the tissue*** rather than locally at the nanoparticle.
- As discussed earlier, for biological tissue in general, there is a ***spectral region of reduced absorption***, called an '***optical window***' or '***window of transparency***' for biological media.
- ***This region*** is bounded ***between*** the strong ***absorption*** effects ***of hemoglobin*** for wavelengths below 600 nm ***and*** the strong absorption effects of ***water*** at wavelengths above 1300 nm, and is a ***region of increased penetration depth*** of light through the tissue.

Slide 30.13 Radiation delivery

Radiation Delivery (continued)

- The problem is more pronounced for ***hard tissues*** because they are generally ***surrounded by a significant amount of light-scattering soft tissue***. To address this issue, we use the ***fiber-optic needle system***, which is capable of delivering light much closer to the targeted region as demonstrated in the figure below.
- ***Light*** waves undergo ***diffraction spreading*** on the optical fiber interface and ***deliver electromagnetic energy*** to the biological tissue containing nanoparticles.

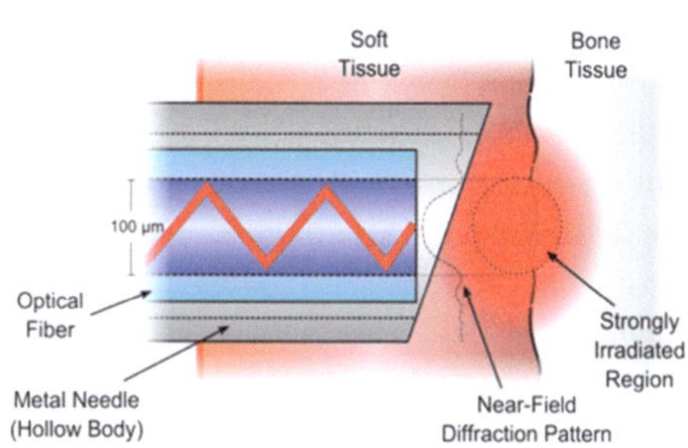

Slide 30.14 Radiation delivery (continued)

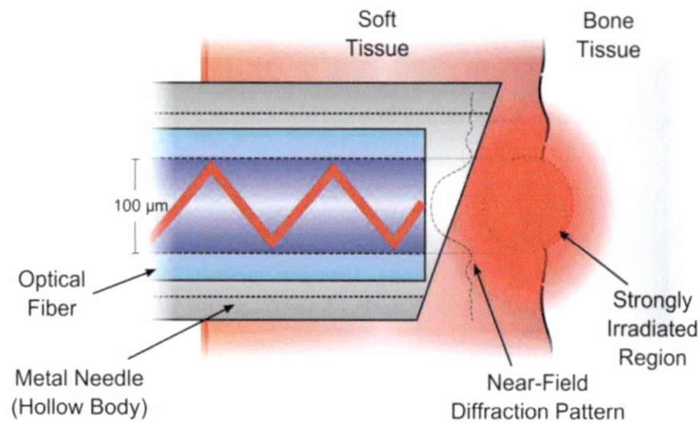

Fig. 10.1 Depiction of the minimally invasive fiber-optic needle system used to deliver laser light more directly to hard tissues by injecting a needle past most of the softer tissues. Reprinted with permission from [1]

To address this issue, we have considered the fiber-optic needle system in Sect. 8.1 of Chap. 8, which is capable of delivering light much closer to the targeted region as demonstrated in Fig. 10.1 and Slide 30.14. An experimental study on stimulation of bone formation in osteopenic mice shows that less than 15.9 % of light can penetrate to the femur without the needle device, but using the needle-delivery method, significant bone growth improvements are apparent with energy densities of only 3 J/cm^2, using 100-μm diameter fiber operating at 10-mW output power [1].

10.1.4 Plasmonic Nanoparticles

Light waves undergo diffraction spreading on the optical fiber interface and deliver electromagnetic energy to the biological tissue containing nanoparticles. As we saw in Chaps. 4 and 5, an electromagnetic wave drives electron charge oscillations on the particle surface, but the frequency and strength of these oscillations depends upon the differences in the complex refractive index between a particle and surrounding medium, in addition to geometric considerations (Slide 30.15). Because of this, the choice of material, as well as the size and shape of the nanoparticle, affects the resonance frequency for the electron oscillations. Additional factors may affect the plasmon resonance, such as coupling between electron oscillations by particles in close proximity, which has shown the ability to redshift the surface plasmon resonance frequency of gold nanoparticle clusters as discussed in Chap. 9 [1].

Our simulations performed in Chap. 5 show that gold nanoparticles provide an excellent opportunity for capitalizing on surface plasmon effects within the body (Slide 30.16). First, gold nanoparticles have been well studied in the context of their production, making it possible to produce nanoparticles of almost any size (from tiny 5-nm spheres to the micron scale and beyond) and shape (spheres, rods, and shells have all been produced). From a chemical standpoint, gold is also well understood in the context of its surface chemistry, making the addition of a variety of conjugates to the nanoparticle surface very straightforward. Additionally, gold is

Plasmonic Nanoparticles

- As we studied earlier, an ***electromagnetic wave drives electron charge oscillations*** on the particle surface, but the frequency and strength of these oscillations depends upon the differences in the ***complex refractive index*** between a particle and surrounding medium, in addition to geometric considerations.
- Because of this, the choice of ***material***, as well as the ***size and shape*** of the nanoparticle, ***affect the resonance frequency*** for the electron oscillations.
- Additional factors may affect the ***plasmon resonance***, such as coupling between electron oscillations by particles ***in close proximity***, which has shown the ability to redshift the surface plasmon resonance frequency ***of gold nanoparticle clusters***.

Slide 30.15 Plasmonic nanoparticles

Plasmonic Nanoparticles (continued)

- Our simulations show that ***gold nanoparticles*** provide an excellent opportunity for ***capitalizing on surface plasmon effects*** within the body.
- First, ***gold nanoparticles*** have been well studied in the context of their ***production***, making it possible to produce nanoparticles of most ***any size*** (from tiny 5-nm spheres to the micron scale and beyond) and ***shape*** (spheres, rods and shells have all been produced).
- From a chemical standpoint, ***gold*** is also well understood in the context of its surface chemistry, ***making*** the addition of a variety of ***conjugates*** to the nanoparticle surface very straightforward.
- Additionally, ***gold*** is relatively ***biocompatible*** because it is inert in the human body.

Slide 30.16 Plasmonic nanoparticles (continued)

relatively biocompatible because it is inert in the human body (Slide 30.16). Since biomedical applications of gold nanoparticles have been studied in great detail recently, and have been found safe, these particles could be more quickly developed into future medical trials. Other, less studied materials would require further cytotoxicity studies before being able to be used in any medical trials. All of these factors make gold a very promising material for near-future nanoparticle medical applications [1].

10.1.5 Modeling

10.1.5.1 Ablation Model

Thermal ablation leads to irreversible changes in the biological medium that cause cell death. Irreversible damage in biotissues can happen for a specific threshold temperature. For example, protein denaturation starts at temperatures as low as 45 °C (318 K) (Slide 30.17).

Let us consider a simple Arrhenius-type model to describe photothermal ablation of biological tissue [2]. Photothermal ablation refers to a process in which the absorption of photons causes a rise in temperature of the biological medium, typically resulting in a phase change of the material, i.e., vaporization in most cases. This empirical model describes the ablation kinetics according to the equation (Slide 30.17)

Ablation Model

- ***Thermal ablation*** leads to ***irreversible changes*** in the biological medium that cause cell death.
- ***Irreversible damage in biotissues*** can be happen for a specific ***threshold temperature***. For example, protein denaturation starts at temperatures as low as ***45 °C (318 K)***.
- Let us consider a simple ***Arrhenius-type model*** to describe photothermal ablation of biological tissue,

$$v = v_0 e^{-\frac{\rho\Omega}{CT}},$$

where ρ is the density of biological tissue, Ω is the vaporization heat, C is the specific heat of the target media, and v is the velocity of the evaporation front at a given temperature T, with the constant v_0 depending on the type of biological material.

Slide 30.17 Ablation model

$$v = v_0 \mathrm{e}^{-\frac{\rho \Omega}{CT}}, \tag{10.1}$$

where ρ is the density of biological tissue, Ω is the vaporization heat, C is the specific heat of the target media, and v is the velocity of the evaporation front at a given temperature T, with the constant v_0 depending on the type of biological material (e.g., types of soft or hard tissues).

When the laser pulse is very short, the heat generation due to the absorption of photons and its diffusion in the surrounding biological medium can be approximated by a one-dimensional temperature distribution in the z direction as (Slide 30.18)

$$T(z) \approx \frac{I_0 \alpha}{C} \mathrm{e}^{-\alpha z}, \tag{10.2}$$

where I_0 is the laser intensity, and α is the absorption coefficient of the surrounding medium. Substituting this into the earlier Arrhenius-type model and solving for the depth, z, gives the simple expression (Slide 30.18) [2]

$$L = \frac{1}{\alpha} \ln \frac{I_0}{I_{\mathrm{thr}}}, \tag{10.3}$$

where $I_{\mathrm{thr}} = \rho\Omega/\alpha$ is the threshold intensity for ablation, and L represents the final depth of the thermal ablation created by one full pulse.

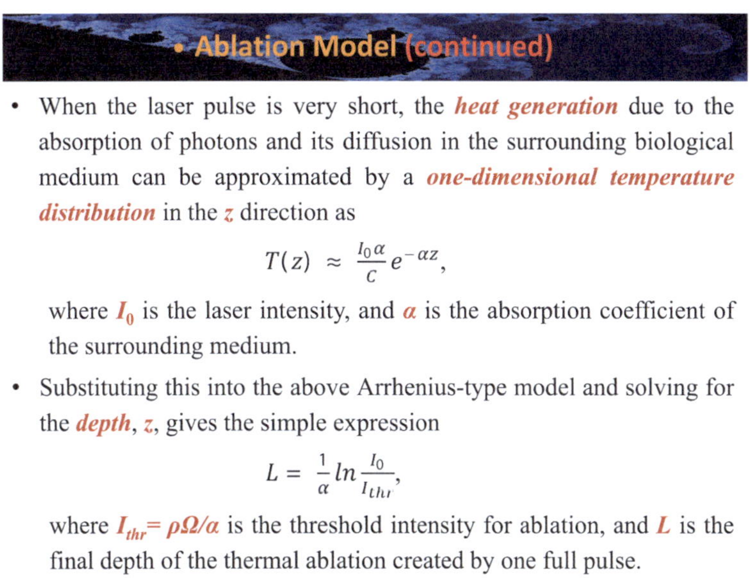

Slide 30.18 Ablation model (continued)

To model the time dependence of the ablation process, we assume that the number of molecules of the undamaged tissue is proportional to the original tissue number density, and that the rate of change is temperature dependent (Slide 30.19):

$$\frac{\partial N}{\partial t} = N_0 A \mathrm{e}^{-\frac{\Delta E}{RT}}. \tag{10.4}$$

Here, N_0 is the number of molecules in the surrounding medium, R is the ideal gas constant, ΔE is the change in energy required to transition from the nonablated to the ablated state, A is an experimentally determined constant, and T is the temperature of the medium obtained by solving Eq. (10.2). When the temperature of the medium rises to the level of $T(z) = \frac{\Delta E}{R}$, then as follows from Eq. (10.4) a major portion (63 %) of the tissue is ablated, which is considered as complete tissue damage (Slide 30.19).

Let us now estimate the threshold energy density required for thermal ablation of the biological tissue. There is a phenomenological relationship between the total number of ablated molecules in the medium and energy density of the laser pulse (Slide 30.20),

$$N_{\mathrm{total}} = N_0 L \ln\left[\frac{E_{\mathrm{L}}}{L(\Delta E - CT_0)}\right], \tag{10.5}$$

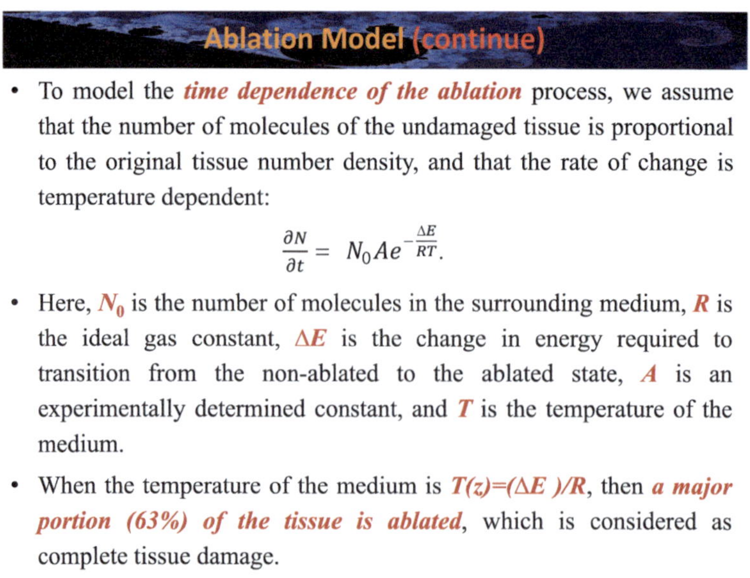

Slide 30.19 Ablation model (continued)

Ablation Model (continue)

- There is a phenomenological relationship between the ***total number of ablated molecules*** in the medium and ***energy density of the laser*** pulse,

$$N_{total} = N_0 L \, ln\left[\frac{E_L}{L(\Delta E - CT_0)}\right],$$

where E_L is the laser energy density, and T_0 is the initial temperature of the medium.

- We can conclude that the ***photothermal ablation threshold energy density*** is

$$E_{thr} = L(\Delta E - CT_0).$$

- For the case $E_L >> E_{thr}$, the ***total number of ablated molecules*** in the medium is

$$N_{total} = N_0 L \, ln\left[\frac{E_L}{E_{thr}}\right].$$

Slide 30.20 Ablation model (continued)

where E_{L} is the laser energy density, and T_0 is the initial temperature of the medium. We can conclude from Eq. (10.5) that the photothermal ablation threshold energy density is (Slide 30.20)

$$E_{\mathrm{thr}} = L(\Delta E - CT_0). \tag{10.6}$$

For the case $E_{\mathrm{L}} >> E_{\mathrm{thr}}$, the total number of ablated molecules in the medium is (Slide 30.20)

$$N_{\mathrm{total}} = N_0 L \, \ln\left[\frac{E_{\mathrm{L}}}{E_{\mathrm{thr}}}\right]. \tag{10.7}$$

For precise thermal ablation and nanosurgery of biological tissue/molecules, we use the heat generated in highly absorbing plasmonic nanoparticles due to the absorption of the radiation pulse by the nanoparticles (Slide 30.21). When the heat diffuses away from the surface of the nanoparticles, it causes a rise in temperature of the biological medium as discussed in detail in Chap. 8. Thus, the nanoparticles serve the role as a heat source for thermal ablation of the surrounding biological medium.

10.1.5.2 Modeling Heat Transfer

The biotissue ablation model discussed earlier involves the heating of a nanoparticle that absorbs laser radiation to generate heat, which is then transferred via conduction to the surrounding medium [2] (Slide 30.21). In Chap. 6 we introduced a

Modeling Heat Transfer

- For precise thermal ablation and nanosurgery of biological tissue/molecules, we use the ***heat generated in highly-absorbing plasmonic nanoparticles*** due to the absorption of the radiation pulse by the nanoparticles.
- When the ***heat diffuses away*** from the surface of the nanoparticles, it causes a ***rise in temperature of the biological medium***.
- Thus, the ***nanoparticles*** serve the role as a ***heat source for thermal ablation*** of the surrounding biological medium.
- Earlier we introduced a ***heat transfer equation*** describing heat flow in both the particle and surrounding fluid. This equation is then solved in a uniform heating approximation throughout the particle volume to investigate the ***time dependence of the temperature*** within the nanoparticle. Alternatively, to describe the ***spatial behavior of the thermal fields*** around the nanoparticle in the surrounding biological medium, we solved the heat transfer equation in a quasistationary heating approximation.

Slide 30.21 Modeling heat transfer

heat transfer equation (6.1) describing heat flow in both the particle and surrounding fluid. This equation is then solved in a uniform heating approximation throughout the particle volume to investigate the time dependence of the temperature within the nanoparticle (Sect. 6.1). Alternatively, to describe the spatial behavior of the thermal fields around the nanoparticle in the surrounding biological medium, in Chap. 8 we solved the heat transfer equation in a quasi-stationary heating approximation (Slide 30.21).

10.1.6 Materials Properties and Input Data

Both thermal and optical properties affect the behavior of the thermal field and ablation of the tissue. The thermal properties of a material govern how easily the material can dissipate heat to the surrounding medium, while the optical properties govern how much heat is generated when exposed to an energy source [1].

In these simulations, we wish to examine the photothermal effect of a strongly absorbing gold nanoparticle for hard-tissue ablation (Slide 30.22). The material properties of bone and gold are listed in Tables 8.1 and 8.2 of Chap. 8. Optical properties of the bone tissue are discussed in detail in Sect. 5.2 of Chap. 5. Examples of Mie calculations for the absorption efficiency of bone and water particles of radius 200 μm in water and surrounding bone media are shown in Fig. 10.2 and Slide 30.22. The radius of the bone and water particles is chosen to be 200 μm to fit the laser exposure area, which is determined by the diffraction

Material Properties and Input Data

- In these simulations, we wish to examine the ***photothermal effect of*** a strongly-absorbing gold nanoparticle for ***hard-tissue ablation***.
- Examples of ***Mie calculations for the absorption efficiency of bone and water*** particles of radius 200 μm in water and surrounding bone media are shown in the figure below.
- ***Bone tissues*** have a much ***weaker absorption than water***, but also follow the trend of ***increased absorption*** for wavelengths moving from the visible ***into the near-infrared***.

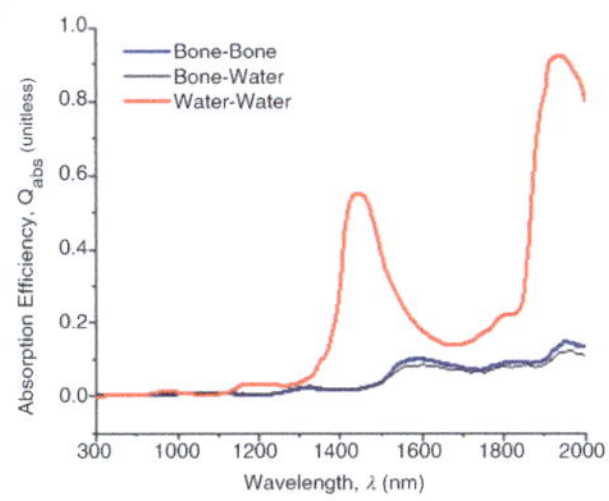

Slide 30.22 Materials properties and input data

spreading of the light on the optical fiber interface as shown in Fig. 10.1. These Mie calculations show that while water may be a strong absorber of light of wavelengths 1400 nm and up, it does not dominate the absorption behavior of the complex bone tissue. Instead, bone tissues have a much weaker absorption than water, but also follow the trend of increased absorption for wavelengths moving from the visible into the near-infrared (Slide 30.22).

Sample absorption efficiency values for different gold nanoparticle shapes and effective radii at various wavelengths can be found in Table 8.4 of Chap. 8. For illustrative purposes, the absorption efficiencies of gold nanoparticles calculated for nanoparticle radii in the range 5–100 nm over a wide range of wavelengths are shown in Fig. 10.3 and Slide 30.23.

10.1.7 Results and Discussion

10.1.7.1 Heating of Nanoparticles in Hard Tissue

This section reproduces the results of simulations published in our papers [1, 2]. The temperature kinetics seen in the one-temperature model (OTM) presented in Sect. 6.1 are both r_0 and Q_{abs} dependent. However, because the plasmon resonance effects which govern Q_{abs} are also size dependent, it can be difficult to discern the changes in heat dynamics due solely to particle size (Slide 30.24). For this reason, a

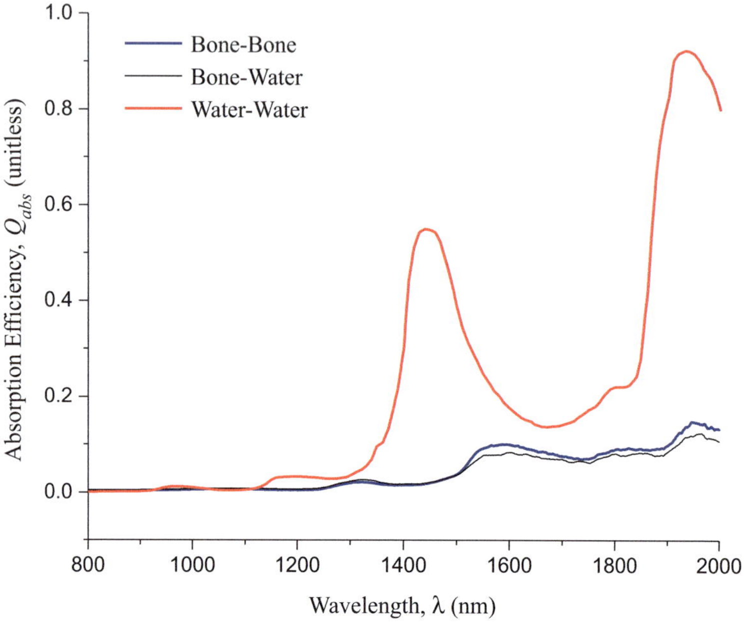

Fig. 10.2 Results of Mie calculations for the absorption efficiency of bone and water particles of radius 200 μm

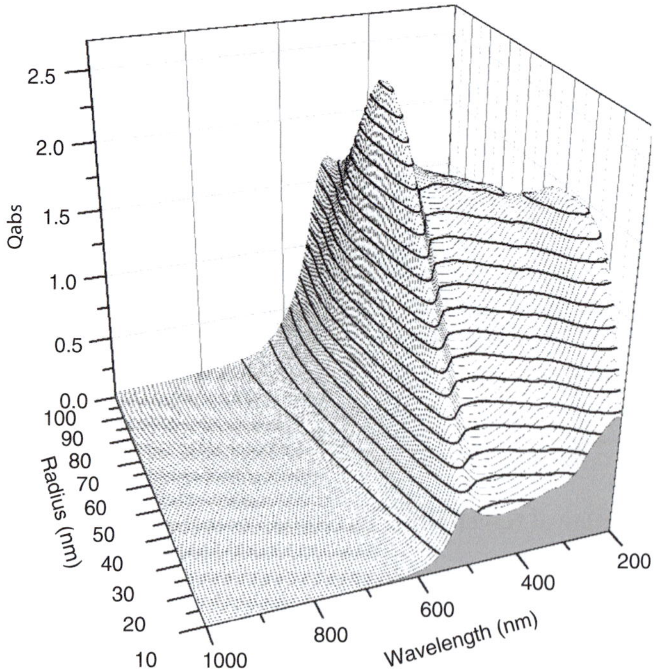

Fig. 10.3 Absorption efficiency of gold nanoparticles as a function of both nanoparticle radius and light wavelength. Reprinted with permission from [1]

Materials Properties and Input Data (continue)

- Sample ***absorption efficiency*** values ***for gold nanoparticles*** calculated for nanoparticle ***radii in the range 5-100 nm*** over ***a wide range of wavelengths*** are shown in the figure below.

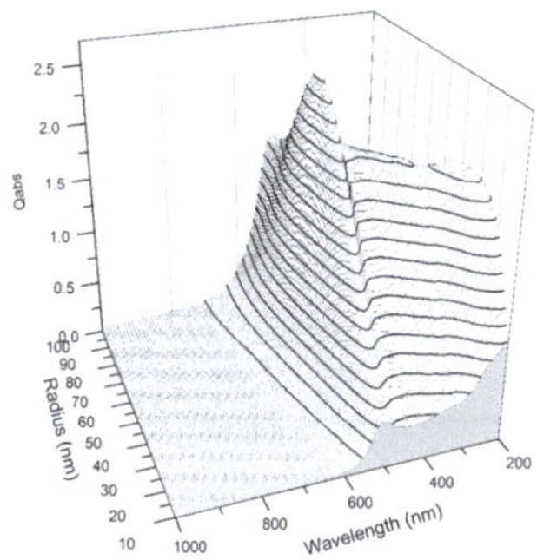

Slide 30.23 Materials properties and input data (continued)

Heating of Nanoparticles in Hard Tissue

- A first set of simulations was run for the simplified case of ***keeping Q_{abs} constant*** and varying only the radius of the nanoparticle.
- ***Two distinct size-dependent*** behaviors can be seen.
- ***Very small nanoparticles*** gain and lose heat very quickly; the maximum temperature is gained at essentially the same time as the maximum intensity of the incoming laser pulse, and most of the thermal energy is dissipated soon after the pulse is over.
- ***Larger nanoparticles***, which lose heat more slowly, take the entire lifetime of the pulse to reach their maximum temperature.

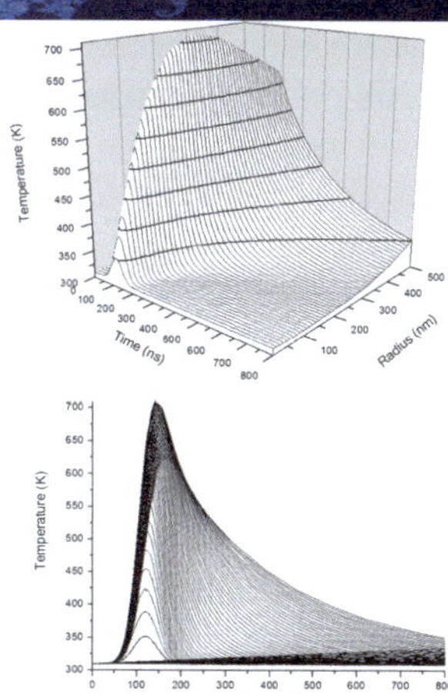

Slide 30.24 Heating of nanoparticles in hard tissue

first set of simulations was run for the simplified case of keeping Q_{abs} constant and varying only the radius of the nanoparticle. This shows how the particle radius affects the interplay between heat generation, proportional to the volume of the particle, and heat lost to the surrounding media, proportional to the surface area of the particle. The results of these simulations are shown in Fig. 10.4 and Slide 30.24, where two distinct size-dependent behaviors can be seen. Very small nanoparticles gain and lose heat very quickly; the maximum temperature is gained at essentially the same time as the maximum intensity of the incoming laser pulse, and most of the thermal energy is dissipated soon after the pulse is over. On the other hand, larger nanoparticles, which lose heat more slowly, take the entire lifetime of the pulse to reach their maximum temperature. Because the heat generated is dissipated much more slowly, the cooling dynamics take much longer than the lifetime of the pulse, allowing for pulse-train heating to reach higher maximum temperatures than single-pulse heating modalities. Particles of intermediate size fall somewhere in between, showing cooling tail broadening with increasing radius. Also of note is the fact that the maximum peak temperatures are realized for intermediate-sized particles, as peak temperatures decrease for particles beyond a critical size [1].

The OTM was then tested using actual, size-dependent, Q_{abs} values computed for spherical gold nanoparticles at a few visible spectral wavelengths. These plots again were carried out for a 60-ns pulse with an intensity of $I_0 = 1$ mW/cm^2 incident on the nanoparticle. The results are shown in Figs. 10.5, 10.6, and 10.7 and Slides 30.25, 30.26, and 30.27 [1].

Two main therapies being investigated, nanophotohyperthermia and nanophotothermolysis, occur at very different temperatures (Slide 30.28). The former simply elevates temperatures to a point where the cancers no longer thrive, previously stated to be 45 °C for protein denaturation. Nanophotothermolysis, on the other hand, requires temperatures high enough to cause local destructive effects. In general, other ablative procedures currently in clinical use operate at ablative temperatures in the range of 65–85 °C, where the damage mechanism in the heated region is coagulation necrosis. For nanophotothermolysis, the damage mechanism is different; rather than heating the whole region to some elevated temperature, the nanoparticles are heated to cause violent reaction at the cell membrane. This effectively breaks apart the cell membrane in the vicinity of the nanoparticle (Slide 30.28). Because nanoparticles are so much smaller than the cell (in the range 10–100 nm versus 8 μm for blood cells or 30 μm for of skin cells), the nanoparticle can give off this heat to the surrounding media without significantly increasing the overall temperature of that media. The actual temperature elevation in the media would have to be calculated by determining the amount of heat energy generated per nanoparticle, and then evaluating the heat gain in the tissue based on the thermal capacity of the tissue and the nanoparticle concentrations in the area, and should in fact fall below the ablative temperatures using the typical damage mechanism [1].

The bone tissues, if heated to 320 °C (in a sufficiently short time period), can boil the water in bones with enough pressure to overcome the tensile strength of the bone itself (Slide 30.29). For this reason, 320 °C will be used in this chapter as the

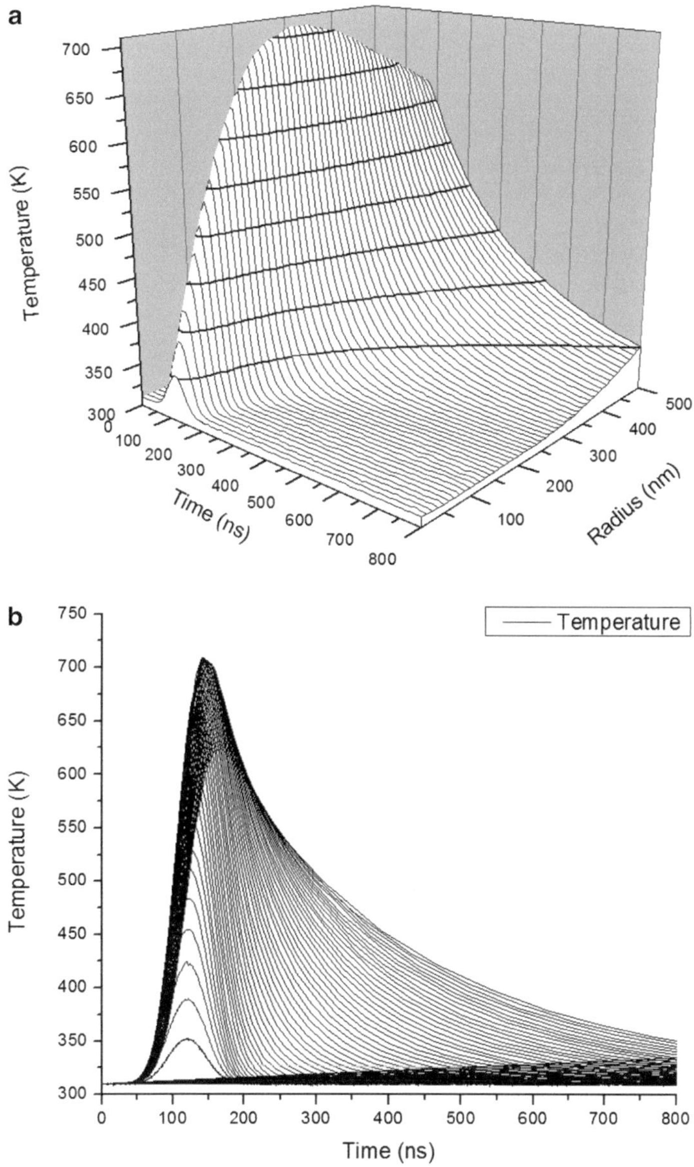

Fig. 10.4 Time dynamics of the heating and cooling process using the one-temperature model. A 60-ns pulse with intensity $I_0 = 1$ mW/cm^2 and constant absorption efficiency $Q_{abs} = 1$ is used in these simulations, while the nanoparticle radius is varied over the range 10–500 nm. (**a**) Temperature versus time as a function of the nanoparticle radius. (**b**) The same heat curves in 2D, projected onto the time–temperature axes. Reprinted with permission from [1]

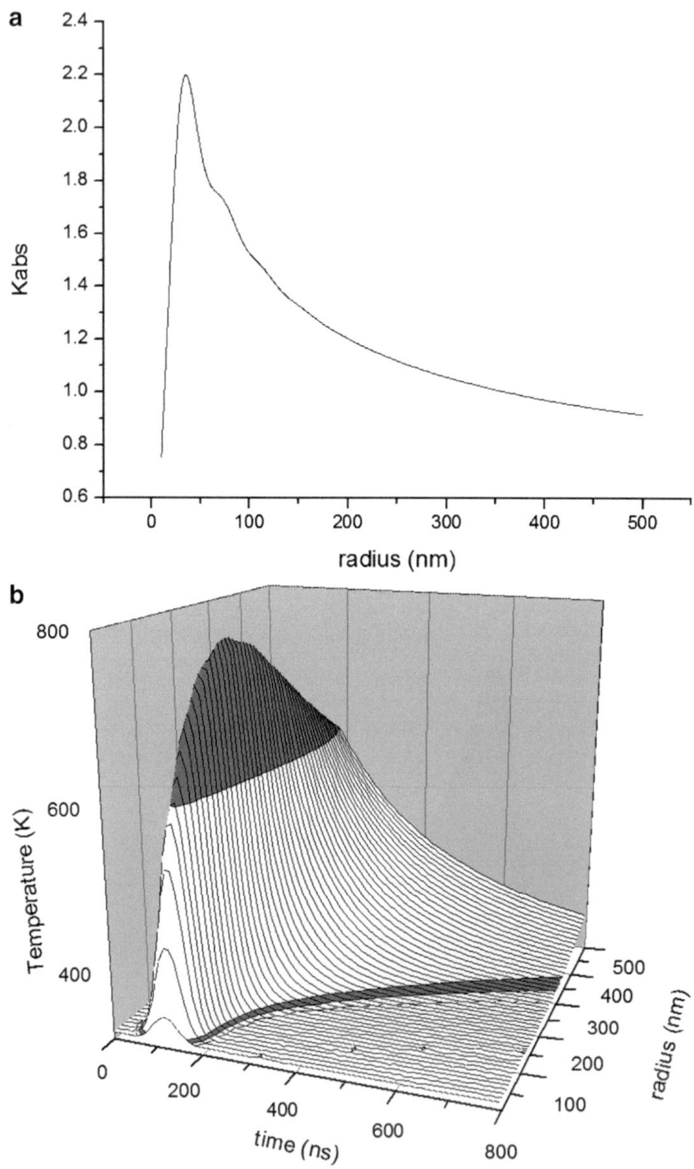

Fig. 10.5 Time dynamics of the heating and cooling process using the one-temperature model. A 60-ns pulse with intensity $I_0 = 1$ mW/cm^2 and absorption efficiency for gold nanoparticles at a laser wavelength of 400 nm are used. (**a**) Absorption efficiency $K_{abs}(\lambda, r)$ evaluated at the laser wavelength 400 nm. (**b**) One-temperature model results using these K_{abs} parameters: in dark gray are the temperature bands $T = 40$–44 °C ($T = 313$–317 K on the vertical axis) corresponding to nanophotohyperthermia therapies and $T > 320$ °C ($T > 593$ K) to nanophotothermolysis. Reprinted with permission from [1]

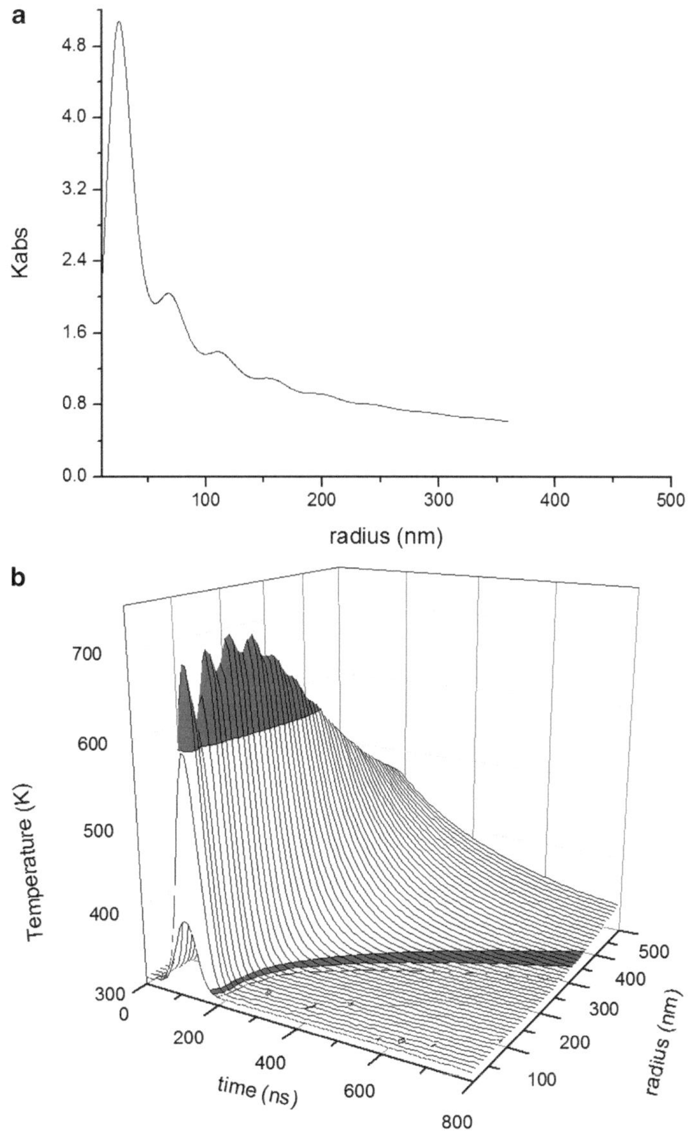

Fig. 10.6 Time dynamics of the heating and cooling process using the one-temperature model. A 60-ns pulse with intensity $I_0 = 1$ mW/cm^2 and absorption efficiency for gold nanoparticles at a laser wavelength of 550 nm are used. (**a**) Absorption efficiency $K_{\text{abs}}(\lambda, r)$ evaluated at the laser wavelength 550 nm. (**b**) One-temperature model results using these K_{abs} parameters: in dark gray are the temperature bands $T = 40$–44 °C corresponding to nanophotohyperthermia therapies and $T > 320$ °C to nanophotothermolysis. Reprinted with permission from [1]

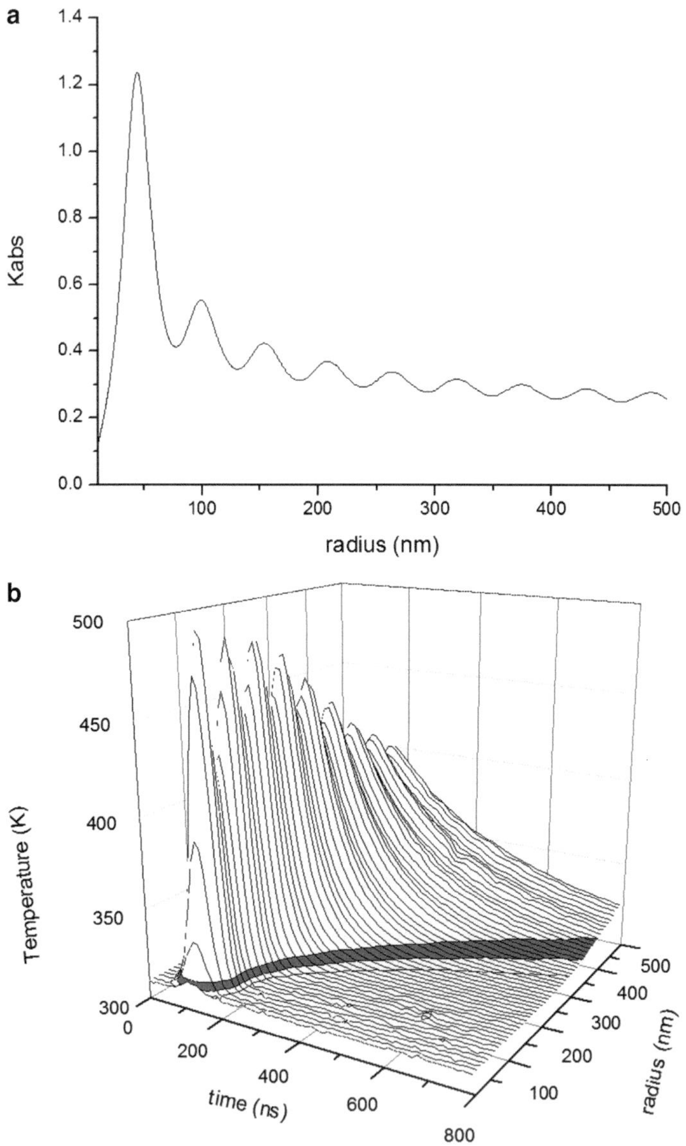

Fig. 10.7 Time dynamics of the heating and cooling process using the one-temperature model. A 60-ns pulse with intensity $I_0 = 1$ mW/cm^2 and absorption efficiency for gold nanoparticles at a laser wavelength of 633 nm are used. (**a**) Absorption efficiency $K_{abs}(\lambda, r)$ evaluated at the laser wavelength 633 nm. (**b**) One-temperature model results using these K_{abs} parameters: in dark gray are the temperature bands $T = 40$–44 °C corresponding to nanophotohyperthermia therapies and $T > 320$ °C to nanophotothermolysis. Reprinted with permission from [1]

Heating of Nanoparticles in Hard Tissue (continue)

The one-temperature model is employed to determine the time dynamics of the heating and cooling process. A ***60-ns pulse with intensity $I_0 = 1$ mW/cm²*** and ***absorption efficiency for gold nanoparticles*** at a laser ***wavelength of 400 nm*** are used. (Left figure) ***Absorption efficiency $K_{abs}(\lambda,r)$*** evaluated at the laser ***wavelength 400 nm***. (Right figure) ***One-temperature model results*** using these ***K_{abs}*** parameters: in dark gray are the temperature bands ***T = 40-44°C (T = 313-317 K*** on the vertical axis) corresponding to **nanophotohyperthermia** therapies and ***T > 320°C (T > 593 K)*** to ***nanophotothermolysis***.

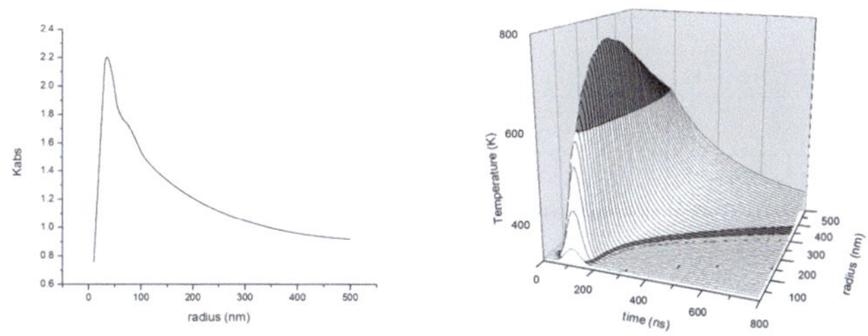

Slide 30.25 Heating of nanoparticles in hard tissue (continued)

Heating of Nanoparticles in Hard Tissue (continue)

The one-temperature model is employed to determine the time dynamics of the heating and cooling process. A ***60-ns pulse with intensity $I_0 = 1$ mW/cm²*** and ***absorption efficiency for gold nanoparticles*** at a laser ***wavelength of 550 nm*** are used. (Left figure) ***Absorption efficiency $K_{abs}(\lambda,r)$*** evaluated at the laser ***wavelength 550 nm***. (Right figure) ***One-temperature model results*** using these ***K_{abs}*** parameters: in dark gray are the temperature bands ***T = 40-44°C (T = 313-317 K*** on the vertical axis) corresponding to **nanophotohyperthermia** therapies and ***T > 320°C (T > 593 K)*** to ***nanophotothermolysis***.

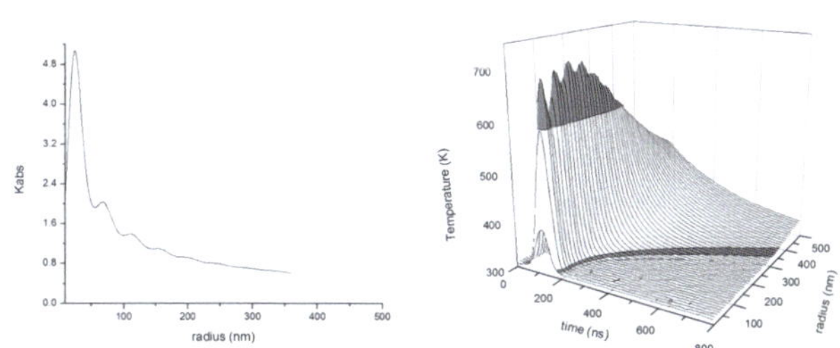

Slide 30.26 Heating of nanoparticles in hard tissue (continued)

The one-temperature model is employed to determine the time dynamics of the heating and cooling process. A ***60-ns pulse with intensity $I_0 = 1$ mW/cm^2*** and ***absorption efficiency for gold nanoparticles*** at a laser ***wavelength of 633 nm*** are used. (Left figure) ***Absorption efficiency $K_{abs}(\lambda,r)$*** evaluated at the laser ***wavelength 633 nm***. (Right figure) ***One-temperature model results*** using these ***K_{abs}*** parameters: in dark gray are the temperature bands ***T = 40-44°C (T = 313-317 K*** on the vertical axis) corresponding to **nanophotohyperthermia** therapies and ***T > 320°C (T > 593 K)*** to ***nanophotothermolysis***.

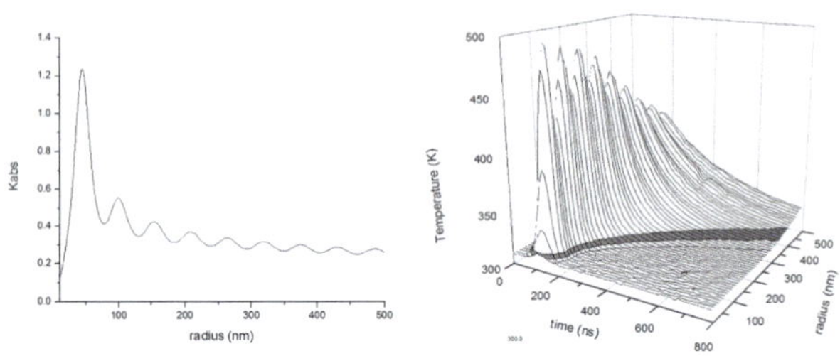

Slide 30.27 Heating of nanoparticles in hard tissue (continued)

- ***Two main therapies*** being investigated, ***nanophotohyperthermia*** and ***nanophotothermolysis***, occur at very different temperatures.
- ***Nanophotohyperthermia*** simply elevates temperatures to a point where the cancers no longer thrive, previously stated to be ***45°C for protein denaturation***.
- ***Nanophotothermolysis***, requires temperatures high enough to cause local destructive effects. In general, other ablative procedures currently in clinical use ***operate at ablative temperatures in the range of 65-85°C***, where the damage mechanism in the heated region is coagulation necrosis.
- For ***nanophotothermolysis***, the damage mechanism is different; rather than heating the whole region to some elevated temperature, the ***nanoparticles are heated to cause violent reaction at the cell membrane.***
- This effectively ***breaks apart the cell membrane*** in the vicinity of the nanoparticle.

Slide 30.28 Heating of nanoparticles in hard tissue (continued)

Heating of Nanoparticles in Hard Tissue (continue)

- The ***bone*** tissues, if ***heated to 320°C*** (in a sufficiently short time period), can ***boil the water in bones*** with enough pressure to overcome the tensile strength of the bone itself.
- For this reason, ***320°C*** will be used here as the ***nanophotothermolysis temperature*** cutoff ***for*** the violent local ***damage ablation mechanism***.
- These results show that, even for intensities of $I_0 = 1\ mW/cm^2$, ***nanoparticles*** have the potential to reach ***high temperatures***.
- In the cases ***of 400-nm and 550-nm wavelength*** incident light, the ***ablation temperatures*** required for nanophotothermolysis ***are*** readily ***achieved*** due to the high absorption efficiencies for smaller wavelengths.
- This is ***advantageous***, as it allows for a great range of ***thermal control***, even using low intensities.

Slide 30.29 Heating of nanoparticles in hard tissue (continued)

nanophotothermolysis temperature cutoff for the violent local damage ablation mechanism. These results show that, even for intensities of $I_0 = 1$ mW/cm^2, nanoparticles have the potential to reach high temperatures. In the cases of 400- and 550-nm wavelength incident light, the ablation temperatures required for nanophotothermolysis are readily achieved due to the high absorption efficiencies for smaller wavelengths. This is advantageous, as it allows for a great range of thermal control, even using low intensities (Slide 30.29) [1].

In Sect. 8.1 of Chap. 8, we have performed spatial simulations for heat transfer from the surface of nanoparticle to the surrounding bone tissue by solving the heat diffusion equation in a quasi-stationary heating approximation (Slide 30.30). Figures 8.13, 8.14, and 8.15 show the spatial distribution of the temperature for a 1 mW/cm^2 intensity pulse. The simulations show that heat loss from the surface of a nanoparticle to the hard tissue is observed, creating a large thermal damage area in the bone tissue up to 75 times the radius of the nanoparticle. Thus, the thermal necrosis of osteocytes can be reached in most of these cases using laser pulses with intensities of about or less than 1 mW/cm^2. This is a relatively low intensity as far as laser safety is concerned.

For nanophotohyperthermia we see extremely low pulse intensities compared to other cancer ablation techniques, but bear in mind that the destruction mechanism is not evaporation of the surrounding medium for hyperthermia. These small pulses cause the nanoparticles to reach the temperature target, but this heat is quickly lost to the surrounding medium, so that these pulses will have to be repeated again and again until sufficient thermal energy is deposited in the medium to bring the overall

Heating of Nanoparticles in Hard Tissue (continue)

- Earlier we have performed ***spatial simulations for heat transfer*** from the surface of nanoparticle to the surrounding bone tissue by solving the heat diffusion equation in a quasistationary heating approximation.
- The simulations show that ***heat loss from the surface of a nanoparticle to the hard tissue*** is observed, creating a large thermal damage area in the bone tissue ***up to 75 times the radius of the nanoparticle***.
- Thus, the ***thermal necrosis of osteocytes*** can be reached in most of these cases using laser pulses with intensities of about or less than $I_0 = 1\ mW/cm^2$. This is a ***relatively low intensity*** as far as ***laser safety*** is concerned.

Slide 30.30 Heating of nanoparticles in hard tissue (continued)

temperature of the tissue (not just the nanoparticle itself) to the 40–44 °C operating temperature. The repetition rate of the pulses is therefore also dependent on the thermal capacity of the specific tissue being warmed and the concentration of nanoparticles in the area [1].

10.1.7.2 Heating of Bone Particles

With the earlier absorption efficiency values for hard tissue, we have performed one-temperature model simulations to describe the temperature kinetics caused by short and ultrashort laser-pulse heating of bone particles on the nanosecond, picosecond, and femtosecond time scales (Slide 30.31). The kinetics due to a multipulse mode of heating, described in Chap. 7, is also modeled here for the bone particles. The thermal conductivity, specific heat, and density of bone tissue, as well as input data used in simulations based on the one-temperature model, are listed in Table 10.1. The laser pulse is modeled by a Gaussian curve.

The results of the bone particle heating by using a single-pulse mode are shown in Fig. 10.8 and Slide 30.32. As seen in Fig. 10.8a–c, the heating kinetics of the bone particle is very fast. The maximum temperature is achieved at the end of the pulse's lifetime. As shown in Fig. 10.8d, the cooling kinetics are relatively slow; temperatures fall to the half-maximum temperature approximately 0.03 s after the pulse. This cooling time (rate) is six orders of magnitude greater (less) than the heating time for even the nanosecond pulse. Substantial cooling is realized only approximately 0.1–0.2 s after the pulse. This slow rate of heat diffusion would

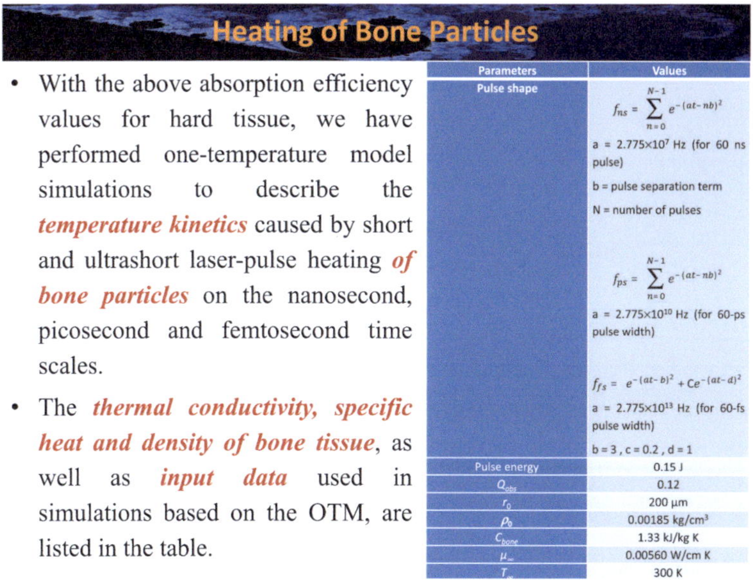

Slide 30.31 Heating of bone particles

Table 10.1 Input data used in the one-temperature model heating simulations. Reprinted from [2]. © With permission from World Scientific

Parameters	Values
Pulse shape	$f_{\mathrm{ns}} = \sum_{n=0}^{N-1} \mathrm{e}^{-(at-nb)^2}$
	$a = 2.775 \times 10^7$ Hz (for 60 ns pulse)
	$b =$ pulse separation term
	$N =$ number of pulses
	$f_{\mathrm{ps}} = \sum_{n=0}^{N-1} \mathrm{e}^{-(at-nb)^2}$
	$a = 2.775 \times 10^{10}$ Hz (for 60-ps pulse width)
	$f_{\mathrm{fs}} = \mathrm{e}^{-(at-b)^2} + C\mathrm{e}^{-(at-d)^2}$
	$a = 2.775 \times 10^{13}$ Hz (for 60-fs pulse width)
	$b = 3$, $c = 0.2$, $d = 1$
Pulse energy	0.15 J
Q_{abs}	0.12
r_0	200 μm
ρ_0	0.00185 kg/cm^3
C_{bone}	1.33 kJ/kg K
μ_{∞}	0.00560 W/cm K
T_{∞}	300 K

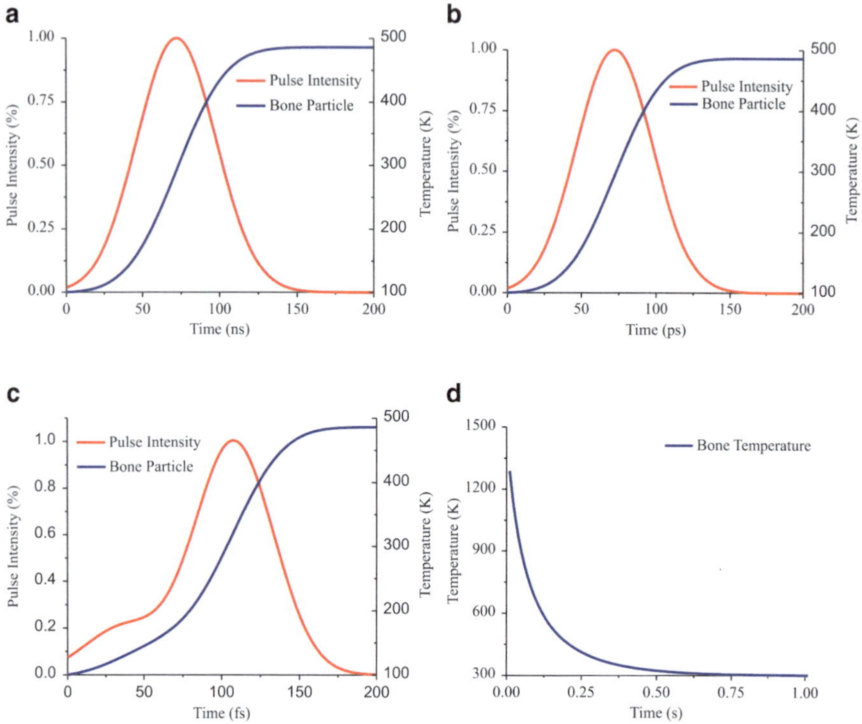

Fig. 10.8 Single-pulse heating of bone tissue. The pulse shape is shown by the *red line* and the heating kinetics by the blue line for bone tissue using a single 0.15-J pulse with pulse width (**a**) 60 ns, (**b**) 60 ps, and (**c**) 60 fs. (**d**) Shows the cooling kinetics of the same bone particle

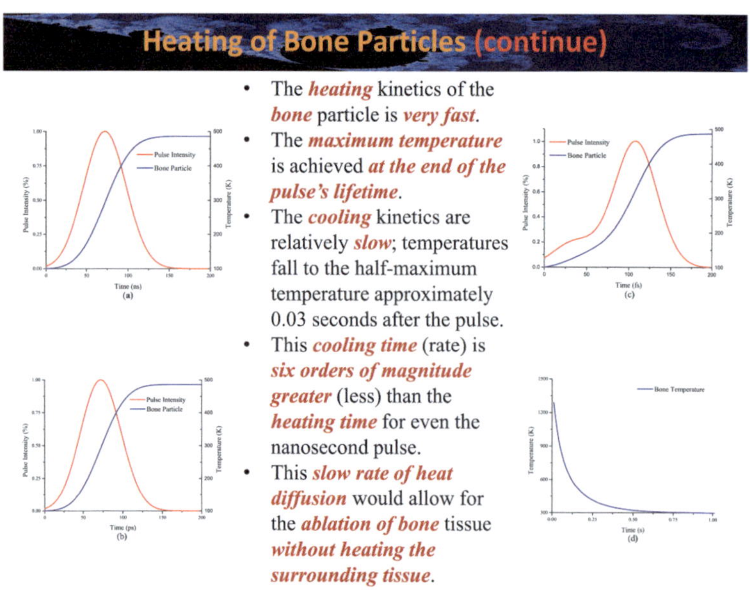

Slide 30.32 Heating of bone particles (continued)

allow for the ablation of bone tissue without heating the surrounding tissue (Slide 30.32). Given a pulse of sufficiently high energy density, the bone particle could be heated beyond its ablation temperature, and due to the slow rate of heat diffusion, very little heat could be accumulated by the surrounding medium before the ablated materials are expelled [2].

This result has been confirmed by experiments, demonstrating that the complex microstructure of dentine is exposed after ablation without applying any etchant, showing no signs of melting or resolidification of the surrounding tissues. Also it shows that protein denaturation does not occur in living cortical bone adjacent to the removed tissue for femtosecond pulses, which is quite impressive given the fact that the protein being monitored denatures at a temperature of only 45 °C [2].

This slow rate of cooling also allows for accumulation of heat by a multipulse heating mode as seen in Fig. 10.9 and Slide 30.33, where five 0.06-J pulses reach a maximum temperature about 750 K, approximately equal to that obtained by a single 0.15-J pulse. This method could be used to attain the higher temperatures required for tissue ablation using many pulses with low individual energies. Because the time scale for cooling is many orders of magnitude longer, it is possible to use energy pulses that are smaller by ten to a hundred times (or even less) to reproduce virtually any single-pulse energy. This gives a large amount of control over the individual pulse energies, which is especially important in medical and dental laser applications where regulations may limit the maximum pulse energies for different procedures (Slide 30.33) [2].

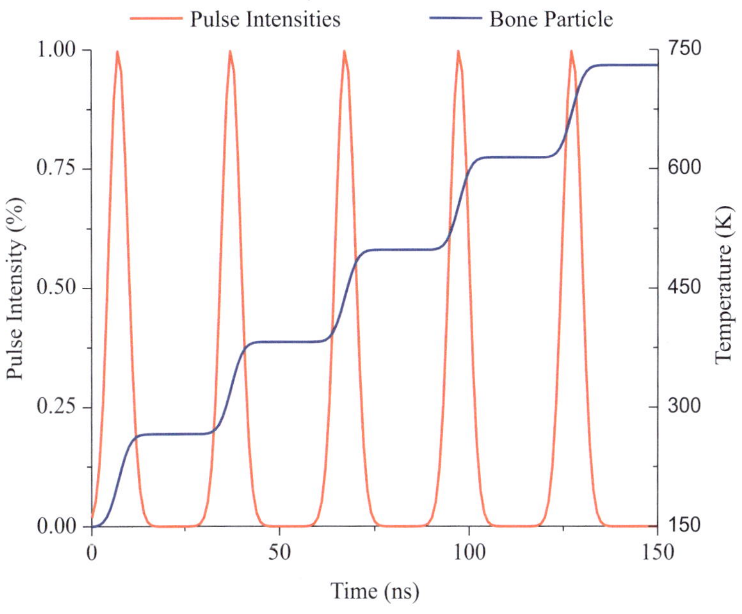

Fig. 10.9 Multipulse mode of heating of bone tissue. A representative multipulse function is drawn with a *red line* and the heating kinetics with a *blue line*, showing five 60-mJ pulses of 60-ns pulse width repeating every 300 ns

Heating of Bone Particles (continue)

- This slow rate of cooling also allows for accumulation of heat by a multipulse heating mode as seen in the figure, where five 0.06-J pulses reach a maximum temperature of about 750 K, approximately equal to that obtained by a single 0.15-J pulse.
- This method could be used to attain the higher temperatures required for tissue ablation using many pulses with low individual energies.
- Because the time scale for cooling is many orders of magnitude longer, it is possible to use energy pulses that are smaller by ten to a hundred times (or even less) to reproduce virtually any single-pulse energy.
- This gives a large amount of control over the individual pulse energies, which is especially important in medical and dental laser applications.

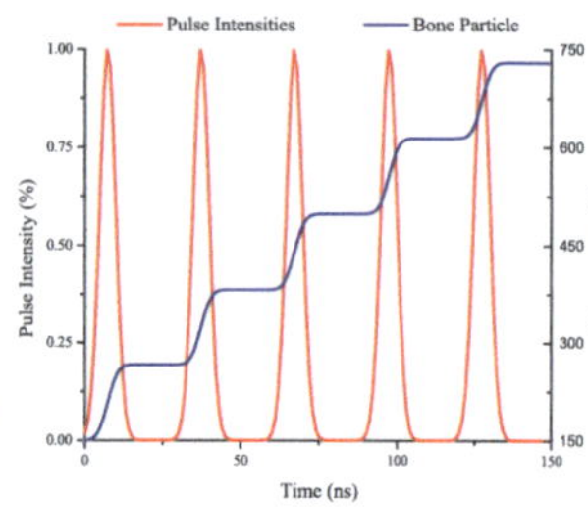

Slide 30.33 Heating of bone particles (continued)

10.1.8 Ablation of Bone Tissue

Here we reproduce the results of simulations published in our papers [2]. To simulate the ablation rate of bone tissue, the ablation model, Eqs. (10.1)–(10.7), with thermal and optical properties of hydroxyapatite (HAP) is applied (Slide 30.34). The vaporization heat for bone tissue Ω is calculated to be 456 J/g using the molar density and molar heat of vaporization from the literature. Bone tissues are composed of 70 % inorganic material, mostly HAP, and 30 % organic material. To account for the change in composition, we multiply I_0 by an accumulation factor k. The ablation rates and fluencies obtained from the experimental work [3] performed on human and bovine compacta using Yb:glass laser pulses, at a wavelength of $\lambda = 1040$ nm, and on bovine cortical bone [4] using Ho:YAG laser pulses, at a wavelength $\lambda = 2150$ nm, are compared to the rates for pure HAP. Fitting a curve to the experimental data yields the absorption coefficient α and the threshold fluence I_{thr}/k for bone tissue shown in Fig. 10.10a, b and Slide 30.35. Using the relationship $I_{thr} = \rho\Omega/\alpha$ and dividing by the density of bone, $\rho = 0.00185$ kg/cm^3, yields $\Omega/k = 3910$ J/g for the Ho:YAG laser and $\Omega/k = 3780$ J/g for the Yb:glass laser. Using the calculated value of $\Omega = 456$ J/g, for pure HAP, yields the accumulation factors of $k = 0.117$ (Ho:YAG) and $k = 0.121$ (Yb:glass) [2].

The differences in the fluencies and material absorption coefficients can be attributed the different laser wavelengths used and the differences in the experimental procedures (Slide 30.35). The optical properties could vary considerably from 1040 to 2150 nm in bone tissue. Unlike experiments in Ref. [4], the

Ablation of Bone Tissue

- To simulate the ablation rate of bone tissue, the ***ablation model with thermal and optical properties of hydroxyapatite*** (HAP) is applied.
- The ***vaporization heat for bone*** tissue ***Ω*** is calculated to be ***456 J/g*** using the molar density and molar heat of vaporization from the literature.
- ***Bone*** tissues are composed of ***70% inorganic material***, mostly HAP, and ***30% organic material***.
- To account for the ***change in composition***, we multiply I_0 by an ***accumulation factor k***.
- Using the relationship $I_{thr} = \rho\Omega/\alpha$ and dividing by the ***density of bone***, $\rho = 0.00185\ kg/cm^3$, yields $\Omega/k = 3910\ J/g$ for the Ho:YAG laser and $\Omega/k = 3780\ J/g$ for the Yb:glass laser.
- Using the calculated value of $\Omega = 456\ J/g$, for pure HAP, yields the accumulation factors of $k = 0.117$ (Ho:YAG) and $k = 0.121$ (Yb:glass).

Slide 30.34 Ablation of bone tissue

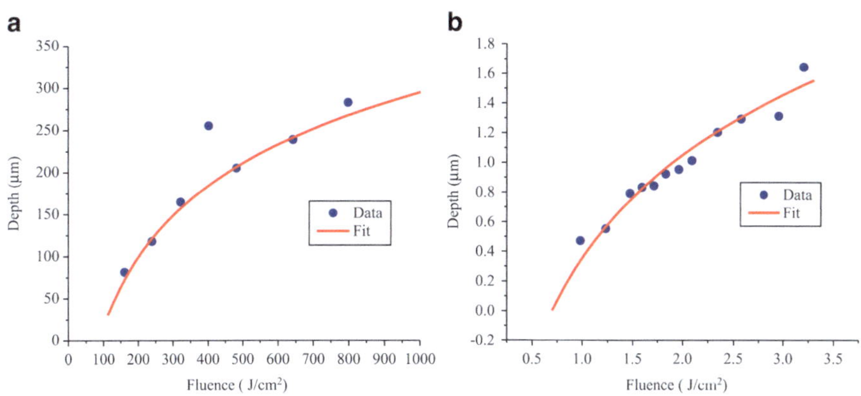

Fig. 10.10 Ablation data and fit curves for (**a**) Ho:YAG laser pulses (2150 nm) on bovine cortical bone and (**b**) Yb:glass laser pulses (1040 nm) on bovine compacta. Reprinted from [2]. © With permission from World Scientific

experiments in Ref. [3] did not use single pulses, but rather many small pulses at a repetition rate of 1 kHz, giving the average depth per pulse in the data. However, it has been shown that multipulse interactions can have a dramatic effect on lowering the threshold fluence for ablation (Slide 30.35) [2].

For the time progression of the ablation of particles due to a single pulse, we use the one-temperature model to simulate the temperature kinetics for a 0.15-J pulse with a pulse width of 60 ns (Slide 30.36). This temperature function $T(t)$ is then used

Ablation of Bone Tissue (continued)

- The ***differences in the fluencies and material absorption coefficients*** can be attributed the different laser wavelengths used and the differences in the experimental procedures.
- The ***optical properties*** could vary considerably from 1040 to 2150 nm in bone tissue.
- It has been shown that ***multipulse interactions*** can have a dramatic effect on ***lowering the threshold fluence for ablation***.

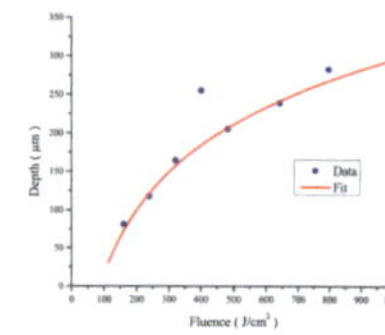

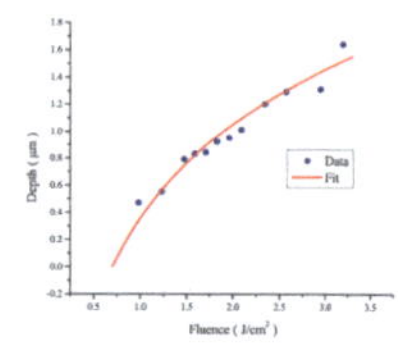

Parameters	Values
ΔE	458 kJ/mol
R	8.31 J/molK
N_0	1.9×10^{27} m^{-3}
A	10^{23}

Slide 30.35 Ablation of bone tissue (continued)

Ablation of Bone Tissue (continued)

- For the ***time progression of the ablation*** of particles due to a single pulse, we ***use the OTM*** to simulate the temperature kinetics for a 0.15-J pulse with a pulse width of 60 ns.
- This ***temperature function T(t)*** is then used ***in the ablation model***, along with the parameters found in the table, for simulation of the number density change for unablated particles plotted in the figure.
- The ***constant A*** is experimentally determined to be on the order of magnitude 10^{23}, as this results in the majority of particles being ablated on the 0-500 ns time scale as observed experimentally.

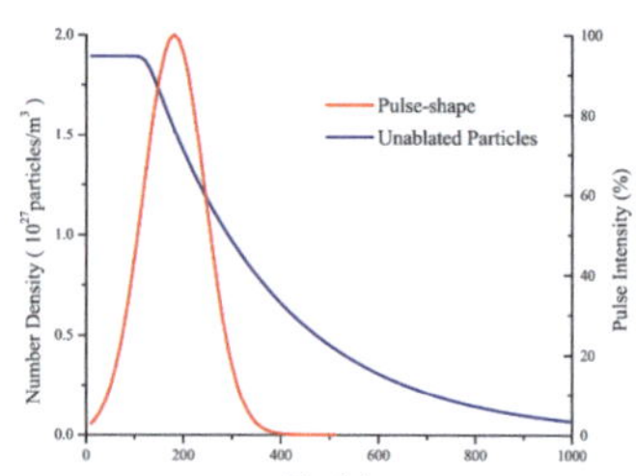

Parameters	Values
ΔE	458 kJ/mol
R	8.31 J/molK
N_0	1.9×10^{27} m^{-3}
A	10^{23}

Slide 30.36 Ablation of bone tissue (continued)

Table 10.2 Input data used in the ablation simulations

Parameters	Values
ΔE	458 kJ/mol
R	8.31 J/mol K
N_0	1.9×10^{27} m^{-3}
A	10^{23}

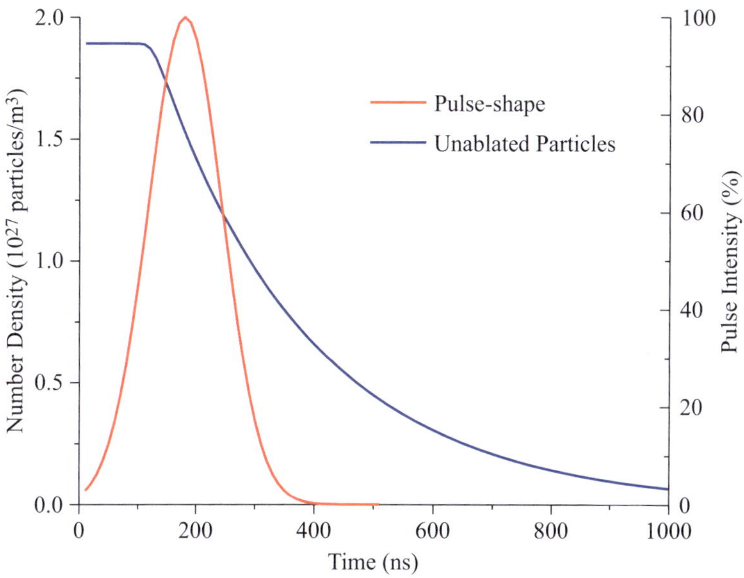

Fig. 10.11 Pulse shape and ablation dynamics of bone tissue. Reprinted from [2]. © With permission from World Scientific

in Eq. (10.4), along with the parameters found in Table 10.2, for a simulation of the number density change for unablated particles at the ablation site, as plotted in Fig. 10.11 and Slide 30.36. The constant A is experimentally determined to be on the order of magnitude 10^{23}, as this results in the majority of particles being ablated on the 0–500 ns time scale as observed experimentally [2].

10.1.9 Future Perspectives

To experimentally test the validity of these simulations it will be necessary to test temperature elevation in water and tissue phantoms (Slide 30.37). Although the instantaneous temperatures described by the equations would be all but impossible to determine, the total heat deposition caused by nanoparticle light absorption could be determined by measuring the bulk tissue temperature in the

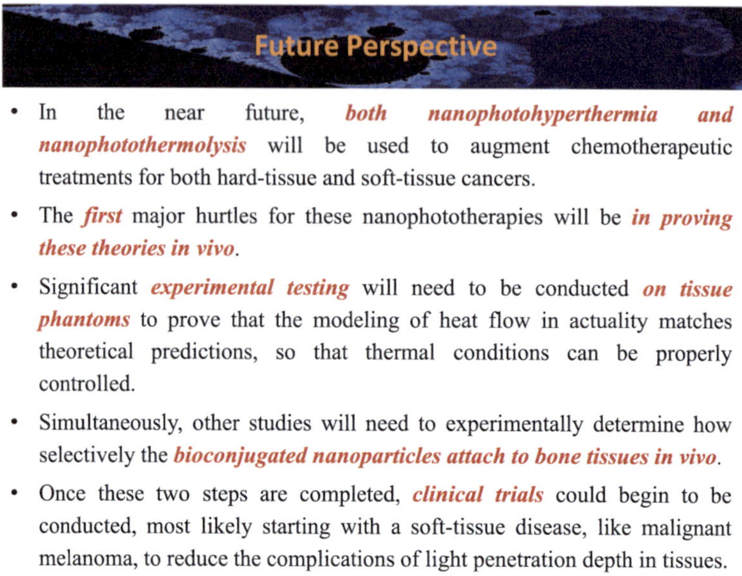

Slide 30.37 Future perspectives

nanoparticle-filled region after thermal equilibrium is reached (which from Figs. 10.4, 10.5, 10.6, and 10.7 appears to be in the μs timeframe for hard tissues). These experiments would have to be conducted with many different concentrations of nanoparticles in the tissue and may require higher pulse energies or the use of repeated pulses in order to deposit enough thermal energy to be measured in a practical way [1].

In the near future both nanophotohyperthermia and nanophotothermolysis will be used to augment chemotherapeutic treatments for both hard-tissue and soft-tissue cancers (Slide 30.37). The first major hurdles for these nanophototherapies will be in proving these theories in vivo. Significant experimental testing will need to be conducted on tissue phantoms to prove that the modeling of heat flow in actuality matches theoretical predictions, so that thermal conditions can be properly controlled. Simultaneously, other studies will need to experimentally determine how selectively the bioconjugated nanoparticles attach to bone tissues in vivo. Once these two steps are completed, clinical trials could begin to be conducted, most likely starting with a soft-tissue disease, like malignant melanoma, to reduce the complications of light penetration depth in tissues. The first bone-tissue nanophototherapies will likely use bisphosphonates as bone-targeting conjugates in cases that minimize penetration depth problems, e.g., metastasis on the bones of the hands, feet, or anywhere else that the depth of fat and muscle tissue between the skin and tumor is minimal. The efficacy of the results in these trials will then determine the rate at which other bioconjugates are tested for application of these treatments to target other specific forms of cancer [1].

10.1.10 Conclusions

We have modeled the thermal behavior of gold nanoparticles in biological hard tissue heated by laser light [1] (Slide 30.38). Our simulations show that these thermal fields can readily be controlled by careful consideration of nanoparticle morphological properties (size, shape, structure) as well as laser wavelength and intensity. Even the more substantial temperatures required for nanophotothermolysis can be reached in most of these cases using laser pulses with intensities of less than 1 mW/cm^2. This is a relatively low intensity as far as laser safety is concerned, and many commercially available Class 3b pulsed lasers could satisfy these requirements. Because the pulse energies required for nanophotohyperthermia are much lower than those needed for nanophotohyperthermia, it should be even easier to prove them safe for these medical applications.

We have also discussed the utility of bioconjugates as delivery vectors to cancerous tissues, specifically allowing for bone-tumor targeted delivery of nanoparticles [1]. The literature surveyed mostly uses these conjugates as mechanisms for drug delivery of medicines related to osteoporosis, and some in nuclear medicine to deliver α-emitting particles and label radioisotopes to the bone tissues. Changing the payload molecule from drugs to radioisotopes does not change the effectiveness of the conjugate targeting technique; therefore, replacing the payload with biologically inert nanoparticles should theoretically work as well. Of particular interest is the time-dependent hydrogen bond scheme can be utilized to bring drugs locally to the bone tissue and then release the drugs to diffuse in a very localized region.

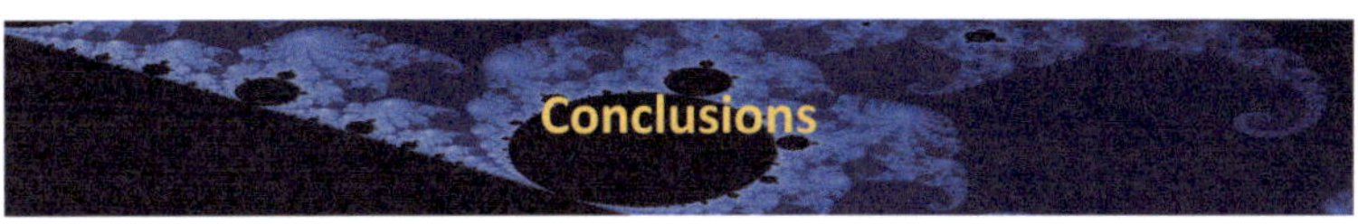

- We have modeled the ***thermal behavior of gold nanoparticles in*** biological ***hard tissue*** heated by laser light.
- Our simulations show that these ***thermal fields*** can readily be ***controlled by*** careful consideration of ***nanoparticle morphological properties*** (size, shape, structure) as well as ***laser wavelength and intensity***.
- Even the more ***substantial temperatures*** required for nanophotothermolysis ***can be reached*** in most of these cases using laser pulses ***with intensities*** of less than ***1 mW/cm^2***.
- This is a relatively ***low intensity*** as far as ***laser safety*** is concerned.
- We have also discussed the utility of ***bioconjugates*** as delivery vectors to cancerous tissues, specifically allowing for ***bone-tumor targeted delivery of nanoparticles***.

Slide 30.38 Conclusions

Conclusions (continued)

- The use of a ***fiber-optic laser delivery mechanism*** also allows for more control in the treatment process, as only nanoparticles near the fiber will create heating.
- Simulations using the one-temperature model to describe the heating of bone tissue with laser pulses in the nanosecond, picosecond and femtosecond regimes show that:
  - ➢ The ***mineral component of bone tissues reduces the absorption efficiency*** in comparison to that of water-based soft-tissues, especially when compared to the near-infrared absorption peaks of water.
  - ➢ The ***cooling kinetics is much slower*** for short and ultrashort pulses than the heating process, even for nanosecond pulses, by several orders of magnitude.
  - ➢ ***Slow cooling rates allow for the use of multipulse modes of heating*** without causing collateral damage to surrounding tissues.
  - ➢ Use of ***multipulse techniques*** is recommended, because this allows ***greater flexibility*** in the choice of ***individual pulse energies*** without sacrificing the desired maximum temperature.

Slide 30.39 Conclusions (continued)

The use of a fiber-optic laser delivery mechanism also allows for more control in the treatment process, as only nanoparticles near the fiber will create heating (Slide 30.39). This means that should a single conjugate have other targeted organs, e.g., both the prostate and bone, treatment can still be localized to the bone tissue by keeping the fiber optics in close proximity to only the bone tissues [1].

Simulations using the one-temperature model to describe the heating of bone tissue with laser pulses in the nanosecond, picosecond, and femtosecond regimes show that [2] (Slide 30.39):

- The mineral component of bone tissues reduces the absorption efficiency in comparison to that of water-based soft tissues, especially when compared to the near-infrared absorption peaks of water.
- The cooling kinetics is much slower for short and ultrashort pulses than the heating process, even for nanosecond pulses, by several orders of magnitude.
- Slow cooling rates allow for the use of multipulse modes of heating without causing collateral damage to surrounding tissues.
- Multipulse modes of heating can reach the same high temperatures as single-pulse heating at much lower individual pulse energies. Therefore, the use of multipulse techniques for both laser heating and laser ablation of bone tissue is recommended, because the multipulse mode of heating allows greater flexibility in the choice of individual pulse energies without sacrificing the desired maximum temperature, and the threshold fluence required for ablation is reduced in multipulse ablation modes, maximizing etch depths with minimal thermal accumulation.

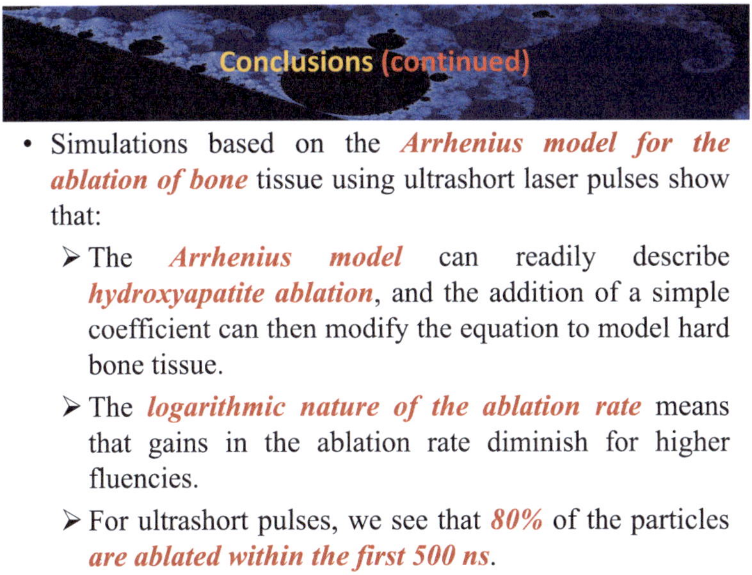

Slide 30.40 Conclusions (continued)

Simulations based on the Arrhenius model for the ablation of bone tissue using ultrashort laser pulses show that [2] (Slide 30.40):

- The Arrhenius model can readily describe hydroxyapatite ablation, and the addition of a simple coefficient can then modify the equation to model hard bone tissue.
- The logarithmic nature of the ablation rate means that gains in the ablation rate diminish for higher fluencies.
- For ultrashort pulses, we see that 80 % of the particles are ablated within the first 500 ns.

10.2 Lecture 31: Soft-Tissue Ablation by Short and Ultrashort Pulses (Slide 31.1)

- Introduction
- Hepatocellular Carcinoma Ablation
- Ductal Breast Carcinoma Ablation
- Ribosomal Ablation
- Conclusions

Lecture 31:

Soft-Tissue Ablation by Short and Ultrashort Pulses

Contents:

- Introduction
- Hepatocellular Carcinoma Ablation
- Ductal Breast Carcinoma Ablation
- Ribosomal Ablation
- Conclusions

Slide 31.1 Contents

Introduction

- Soft-tissue ***laser surgery*** is now common practice; many specific laser treatments and procedures have been developed.
- The use of ***laser scalpels*** in surgery has shown a reduced risk of infection as well as improved patient recovery time.
- This has been attributed to the laser's ability to ***seal small blood vessels and nerve endings*** at the cutting site, thus reducing pain and blood loss.
- ***LASIK surgery*** utilizes excimer lasers to vaporize small layers of corneal stroma and thus reshape the lens of the eye.
- Many applications have been found for laser resurfacing, from the treatment of ***sun damage and stretch marks*** to the removal of scars and tattoos.
- Lasers used for soft-tissue medical applications are chosen mainly for the ***strong absorption of their radiation by water***, the primary constituent of soft-tissues.

Slide 31.2 Introduction

10.2.1 Introduction

This section reproduces the material and results of simulations published in our paper [2]. Soft-tissue laser surgery is now common practice; many specific laser treatments and procedures have been developed (Slide 31.2). The use of laser scalpels in surgery has shown a reduced risk of infection as well as improved patient recovery time. This has been attributed to the laser's ability to seal small blood vessels and nerve endings at the cutting site, thus reducing pain and blood

Introduction (continued)

- In this section, we perform time-dependent simulations of ***soft-tissue laser ablation*** to determine the potential for cancer treatments via organelle specific ablation.
- Simulating both ***single- and multipulse modes of heating***, we calculate the time-temperature profiles for a variety of cancerous and healthy cell organelles.
- ***Each form of cancer*** is in its own way ***distinct***, so having a common treatment for all forms of cancer requires methods that also harm healthy cells, such as radiation or chemotherapy.
- Biologically speaking, ***cancer is*** a result of ***abnormal cells*** in the body that rapidly and ***uncontrollably replicate***.
- Normally ***cells undergo*** several ***checkpoints*** before division, as discussed earlier; these checkpoints are used to determine whether or not the cell should replicate its DNA and divide, and whether it should undergo apoptosis due to faulty DNA.

Slide 31.3 Introduction (continued)

loss. LASIK surgery (laser-assisted in situ keratomileusis) utilizes excimer lasers to vaporize small layers of corneal stroma, and thus reshape the lens of the eye. Many applications have been found for laser resurfacing, from the treatment of sun damage and stretch marks to the removal of scars and tattoos. Lasers used for soft-tissue medical applications are chosen mainly for the strong absorption of their radiation by water, the primary constituent of soft tissues. The success of soft-tissue laser surgery has promoted research for laser ablation of hard tissues discussed in Sect. 10.1 of this chapter [2].

In this section, we perform time-dependent simulations of soft-tissue laser ablation to determine the potential for cancer treatments via organelle-specific ablation. Simulating both single- and multipulse modes of heating, we calculate the time–temperature profiles for a variety of cancerous and healthy cell organelles (Slide 31.3) [2].

Technological improvements over the past several decades have led to rapid development of new equipment and techniques for all areas of medicine. Today's scientists have the opportunity to develop more effective and efficient techniques for cancer diagnosis and treatment with these new technologies and accompanying theories. This is still a difficult feat due to the complexity of the human body, which lends itself to a wide variety of "abnormalities" that can cause cancer. Each form of cancer is in its own way distinct, so having a common treatment for all forms of cancer requires methods that also harm healthy cells, such as radiation or chemotherapy [2].

Biologically speaking, cancer is a result of abnormal cells in the body that rapidly and uncontrollably replicate (Slide 31.3). Normally cells undergo several checkpoints before division, as discussed in Chap. 2; these checkpoints are used to

Introduction (continued)

- By utilizing the ***physical changes*** experienced by the ***organelles of abnormal cells,*** we can determine how laser irradiation would affect the cancerous cells in comparison to healthy cells.
- The ***critical temperature*** for organelle ablation is ***425 K***, as that results in ***protein denaturing*** and ***alters membrane permeability***.
- Thus, the ***thermal ablation*** of the soft tissues is defined here as ***irreversible changes*** in the biological medium ***due to the protein denaturation*** at the critical temperature, which causes cell death.
- Our procedure consists of two primary steps: ***(1) determining the absorption characteristics*** of the organelle and ***(2) numerically solving the heat transfer equation*** to determine the temperature of the organelle over time.
- For our analysis, we look at ***several specific cases*** from which we extrapolate the result to other forms of cancer.

Slide 31.4 Introduction (continued)

determine whether or not the cell should replicate its DNA and divide, whether it should undergo apoptosis due to faulty DNA, or if it should continue normal functioning. Inhibitory mechanisms will stop healthy cells from division when further growth would cause nutrient deprivation or stress the surrounding biological system in any other way. When the inhibitory mechanisms for cell replication malfunction, a cell can undergo rapid and uncontrolled replication regardless of the effects on the system. Such cell propagation can result in the starvation of the surrounding healthy cells as well as physical expansion into other biological systems, disturbing system function on multiple levels [2].

By utilizing the physical changes experienced by the organelles of abnormal cells, we can determine how laser irradiation would affect cancerous cells in comparison to healthy cells (Slide 31.4). The critical temperature for organelle ablation is 425 K, as that results in protein denaturing and alters membrane permeability [2]. Thus, the thermal ablation of soft tissues is defined here as irreversible changes in the biological medium due to the protein denaturation at the critical temperature that causes cell death [2].

Our procedure consists of two primary steps: (1) determining the absorption characteristics of the organelle and (2) numerically solving the heat transfer equation to determine the temperature of the organelles over time (Slide 31.4). To determine the absorption characteristics using the generalized Lorenz–Mie diffraction theory, we must assume the organelles are spherical and determine complex index of refraction data for the organelles and media. This data paired with the organelle size allows us to determine the absorption efficiency of the spherical organelle at a given wavelength [2].

Due to the complexity of the human biological system, it is not possible to make sweeping claims in terms of intercellular changes at the level of organelles. Furthermore, the morphometric analysis of organelles can produce a wide range of results for the same cancer depending on the developmental stage, which makes calculations tentative and case specific. For our analysis, we look at several specific cases from which we extrapolate the result to other forms of cancer (Slide 31.4) [2].

10.2.2 Hepatocellular Carcinoma Ablation

Hepatocellular carcinoma (HCC) is a cancer located in the liver that is normally a secondary effect of hepatitis or cirrhosis (Slide 31.5). It is historically difficult to treat due to the late recognition of tumors as well as difficulty removing the entire tumor upon discovery. HCC is not effectively treated by either chemotherapy or radiation therapy, and stands to be an area where a new cancer treatment would be invaluable [2].

Torimura et al. [5] have used electron microscopy to observe 31 cases of hepatocellular carcinoma in an attempt to determine the usefulness of organelle morphometry via electron microscopy in cancer diagnosis, concluding that as a stand-alone method, it is ineffective. They have determined the mitochondrial and nuclear area for 4 cases of well-differentiated HCC, 17 cases of moderately differentiated HCC, and 10 cases of poorly differentiated HCC. Using this area, we determine the effective radius of the mitochondria and nucleus, which is the

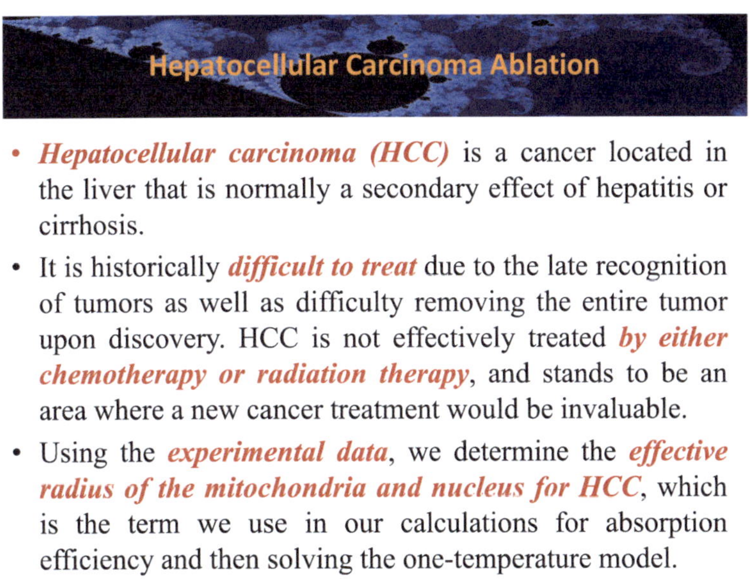

Slide 31.5 Hepatocellular carcinoma ablation

term used in our calculations for absorption efficiency performed in Sect. 5.2 of Chap. 5, and then solving the one-temperature model introduced in Chap. 6 of this book to describe the thermal ablation of the cancerous cell organelles. The results of the calculations for the absorption efficiency of normal and cancerous cell organelles are listed in Tables 5.7 and 5.8 and presented in Fig. 10.12 and Slide 31.6 for each of the three levels of differentiation for both the mitochondria and nuclei. Using these values and organelle-specific data given in Table 6.5, we then calculate time–temperature profiles for both normal and cancerous cell organelles at the different differentiation levels.

Looking at the single-pulse mode of heating (Fig. 10.13 and Slide 31.7), the healthy nuclei are heated to the highest temperature of all the organelles. As the cell passes from well- to poorly differentiated, the maximum nucleus temperature decreases. Alternatively, the healthy mitochondria reach a maximum temperature slightly below the well-differentiated and moderately differentiated cell mitochondria, while the poorly differentiated cell mitochondria reach the highest mitochondrial temperature [2].

This result indicates that trying to target hepatocellular carcinoma cells with single-pulse laser irradiation may not be an effective treatment due to the probable damage of healthy cell nuclei. Our multipulse calculations based on the model described in Chap. 7 produce a similar result (Fig. 10.14 and Slide 31.8), though in the multipulse mode of heating, the mitochondria reach a higher temperature than

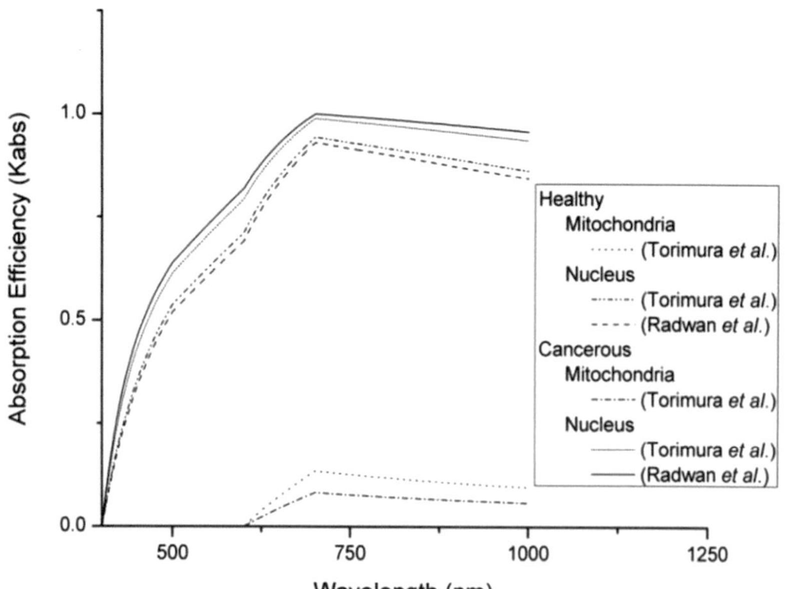

Fig. 10.12 Absorption efficiency (K_{abs}) profiles for mitochondria and nucleus. The organelle diameters are as follows: 1.1 nm (*dotted line*), 7 nm (*dashed line*), 7.5 nm (*dash-dot-dot*), 0.66 nm (*dash-dot*), 10.3 nm (*short-dot*), 11.6 nm (*solid*). Reprinted from [2]. © With permission from World Scientific

Hepatocellular Carcinoma Ablation (continued)

- The results of the calculations for the ***absorption efficiency of normal and cancerous cell organelles*** are presented in the figure for each of the three levels of differentiation for both the mitochondria and nuclei.
- Using these values and organelle specific data, we then ***calculate time-temperature profiles*** for both normal and cancerous cell organelles at the different differentiation levels.

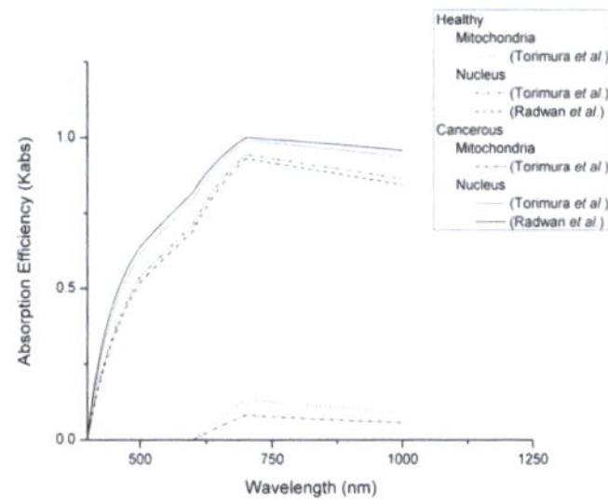

Slide 31.6 Hepatocellular carcinoma ablation (continued)

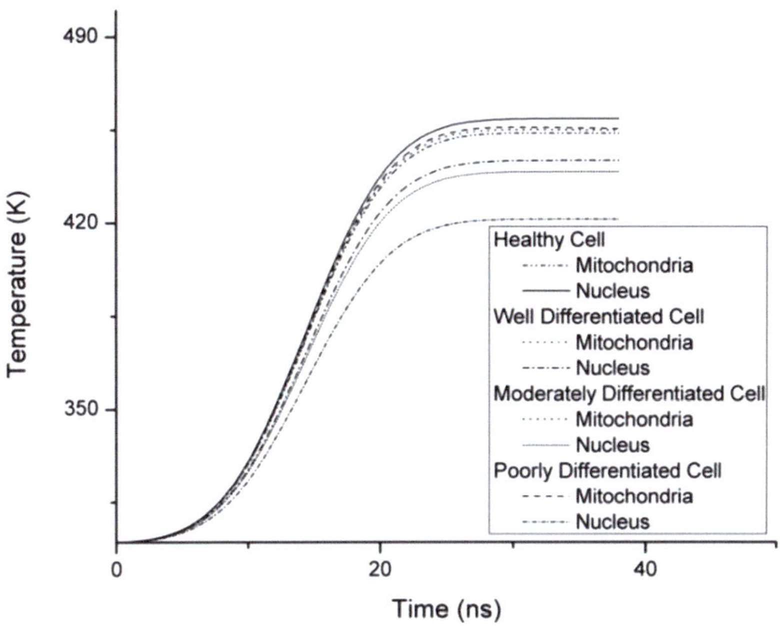

Fig. 10.13 Time–temperature profiles for single-pulse laser irradiation of the mitochondria and nuclei. Laser parameters: 12-ns duration, 0.5 J/cm^2 at $\lambda = 700$ nm. Reprinted from [2]. © With permission from World Scientific

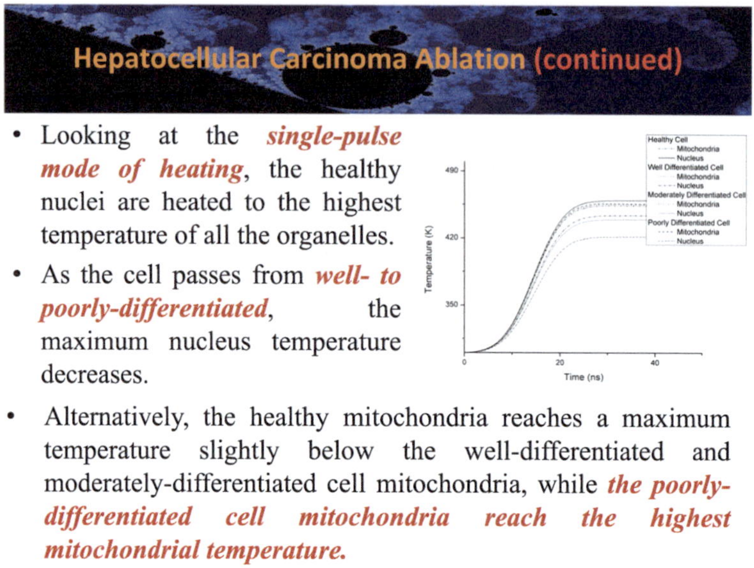

Slide 31.7 Hepatocellular carcinoma ablation (continued)

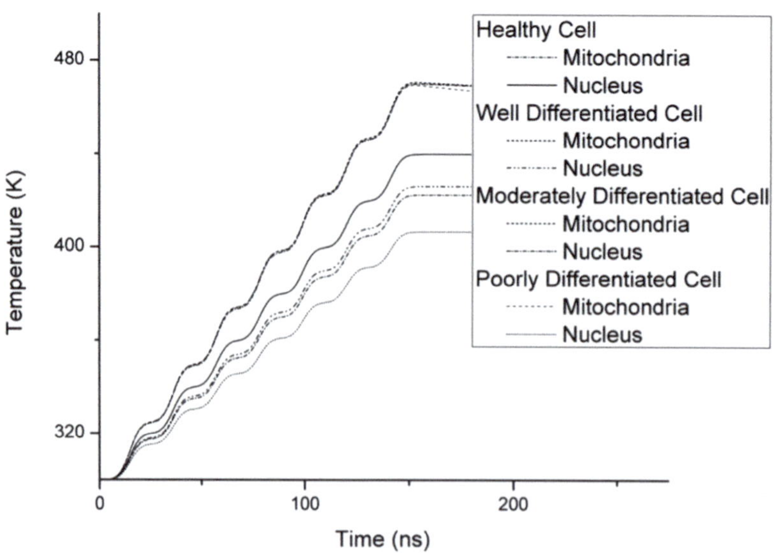

Fig. 10.14 Time–temperature profiles for multipulse laser irradiation of the mitochondria and nucleus based on data from Tables 10.1 and 10.2. Laser parameters: 12-ns duration, 0.1 J/cm^2 at $\lambda = 700$ nm; $b_1 = 2.5$, $b_2 = 6$, $b_3 = 9.5$, $b_4 = 13$, $b_5 = 16.5$, $b_6 = 20$, $b_7 = 23.5$. Reprinted from [2]. © With permission from World Scientific

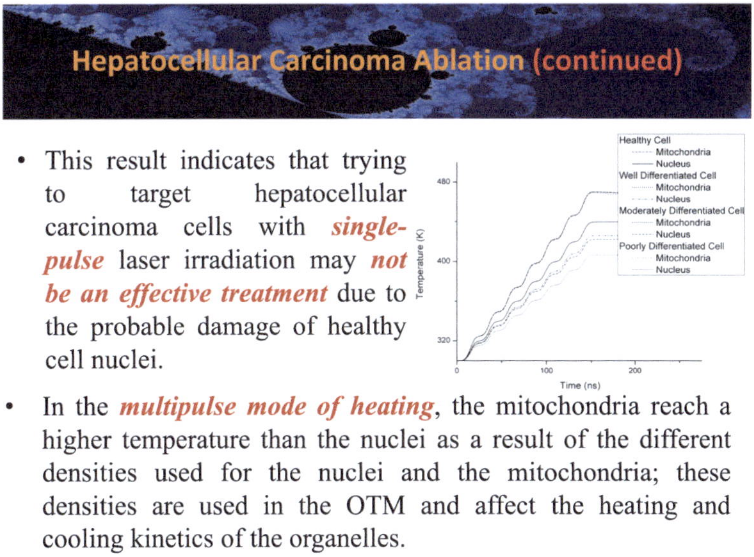

Slide 31.8 Hepatocellular carcinoma ablation (continued)

the nuclei, which results from the different densities used for the nuclei and the mitochondria; these densities are used in the one-temperature model and affect the heating and cooling kinetics of the organelle. Keeping that in mind, due to, which the semipermeable membranes for both organelles, determining a density value for the organelle is difficult; the density is a dynamic quantity [2].

Anson and Chivers [6] used 1050 kg/m^3 for the density of all organelles, which would bring the maximum and mitochondria and nuclei temperatures comparable when considering a multipulse heating mode. As well as the variable nature of organelle density due to standard system function, mitochondrial cristae have been shown to be affected in tumor cells, decreasing in number or almost disappearing completely. This alteration of internal structure of the mitochondria could contribute to permeability and density changes experienced by tumor mitochondria [2].

10.2.3 Ductal Breast Carcinoma Ablation

Ductal breast carcinoma is a prevalent form of breast cancer that has the therapeutic advantage of being near the surface, allowing it to be surgically removed or treated with radiation therapy (Slide 31.9). To test the results of nuclear heating in hepatocellular carcinoma against another form of cancer, we are using data acquired from the work of Radwan et al. [7] on the morphometric analysis of

Ductal Breast Carcinoma Ablation

- ***Ductal breast carcinoma*** is a prevalent form of breast cancer that has the therapeutic advantage of being near the surface, allowing it to be surgically removed or treated with radiation therapy.
- To test the results of nuclear heating in hepatocellular carcinoma against another form of cancer, we are using data acquired from the ***experimental work on the morphometric analysis of benign***, *in situ* and invasive ductal breast carcinoma.
- *In situ* ductal breast carcinoma has a high prevalence of becoming invasive, so ***early treatment via laser irradiation*** would allow ***safe prohibitive therapy*** that could decrease cancer incidence rates for those that are diagnosed early with *in situ* ductal carcinoma.

Slide 31.9 Ductal breast carcinoma ablation

benign, in situ, and invasive ductal breast carcinoma. In situ ductal breast carcinoma has a high prevalence of becoming invasive, so early treatment via laser irradiation would allow safe prohibitive therapy that could decrease cancer incidence rates for those that are diagnosed early with in situ ductal carcinoma. For their analysis, Radwan et al. measured the area and perimeter, along with other characteristics, of the varying cell nuclei. In our calculations, we determine an effective radius based on the area as well as the perimeter values given for each of the three divisions (benign, in situ, and invasive), and then average the two values. Additionally, we compare the heating for these cancerous nuclei to two breast cancer studies that attempt to determine breast cancer progression from benign to invasive by looking at nuclear characteristics. Data for our calculations can be found in Tables 5.7 and 6.5, while Fig. 10.15 and Slide 31.10 show time–temperature profiles for multipulse ablation of the varying cell nuclei. Once again, the calculations show progressively higher temperatures as the tumor becomes more benign, which is due to the increased size of the nucleus that follows from cancer development. As Table 5.7 shows, the larger the organelle, the larger the absorption efficiency; in the cases analyzed, the increase in absorption efficiency is not enough to offset the increase in size (resulting in decreased overall temperature of the organelle) [2].

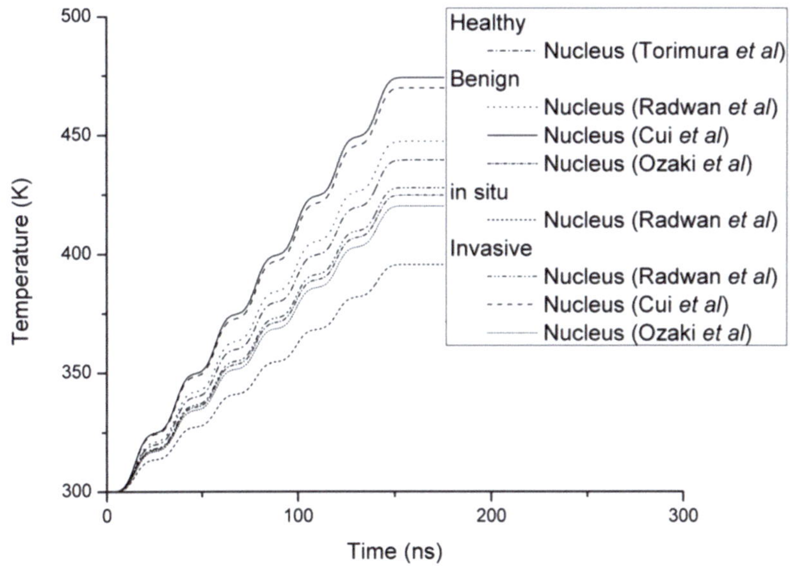

Fig. 10.15 Time–temperature profiles for multipulse laser irradiation of the nucleus based on data from Tables 10.1 and 10.2. Laser parameters: 12 ns duration, 0.1 J/cm^2 at $\lambda = 700$ nm; $b_1 = 2.5$, $b_2 = 6$, $b_3 = 9.5$, $b_4 = 13$, $b_5 = 16.5$, $b_6 = 20$, $b_7 = 23.5$. Reprinted from [2]. © With permission from World Scientific

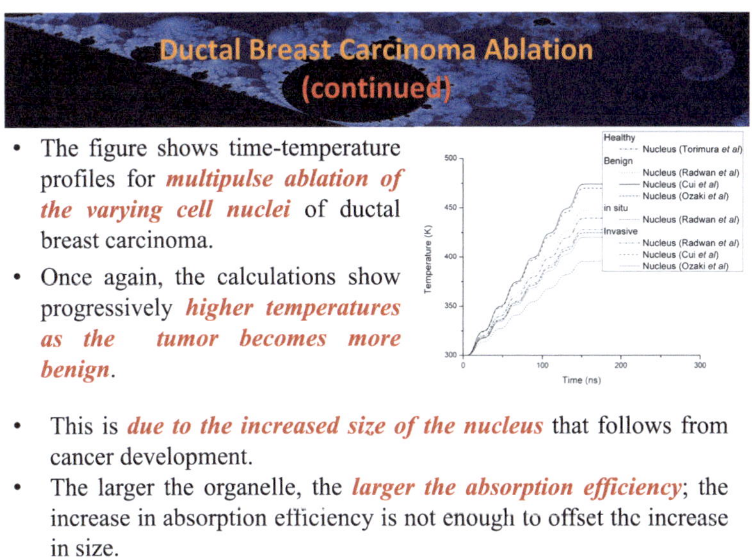

Slide 31.10 Ductal breast carcinoma ablation (continued)

10.2.4 Ribosomal Ablation

Ribosomes are key contributors to healthy cells, along with cells that have become cancerous (Slide 31.11). The optical properties of ribosomes are examined in Sect. 5.2 of Chap. 5 in order to find the differences in absorption and scattering properties of normal and cancerous ribosomes to design the ablation therapies. As shown in Sect. 5.2, the ribosomes in cancer cells have much greater scattering efficiency than normal cells in the spectral range of 200–500 nm, meaning that they might be useful organelles for cancer detection. Strong absorption of light for cancerous ribosomes is observed in the wavelength range 600–1000 nm, which will be optimal for ablation of cancer. Since the optical properties are considerably different between normal and cancer cells, the use of ribosomes holds much potential to facilitate diagnosis and therapy of cancer.

Heating of healthy and cancerous ribosomes in the single-pulse mode is discussed in Sect. 6.3 of Chap. 6. Because ribosomes have considerably less data published, we take some small diameters to demonstrate the effect of size on ribosome ablation. Figure 10.16 and Slide 31.12 demonstrate a time–temperature profiles for heating ribosomes of various small sizes. For the single-pulse mode of heating, the moderately large ribosome ($D = 2.9$ nm) is heated for most of all the organelles. Alternatively, in the multipulse mode of heating discussed in Sect. 7.1 of Chap. 7 and demonstrated in Fig. 7.3, the largest ribosome ($D = 4.6$ nm) exhibits the highest temperature. This change shows a similar dynamic between size (absorption efficiency) and maximum temperature (Slide 31.13). These results indicate that with a given increase in organelle size, the single- or multipulse

Ribosomal Ablation

- ***Ribosomes*** are key contributors to healthy cells, along with cells that have become cancerous.
- As shown earlier, the ***ribosomes in cancer cells*** have ***much greater scattering efficiency*** than normal cells ***in the spectral range 200-500 nm***, meaning that they might be useful organelles for cancer detection.
- ***Strong absorption of light for cancerous ribosomes*** is observed ***in the wavelength range 600-1000 nm***, which will be optimal for ablation of cancer.
- Since the ***optical properties are considerably different between normal and cancer cells***, the use of ribosomes holds much potential to facilitate diagnosis and therapy of cancer.

Slide 31.11 Ribosomal ablation

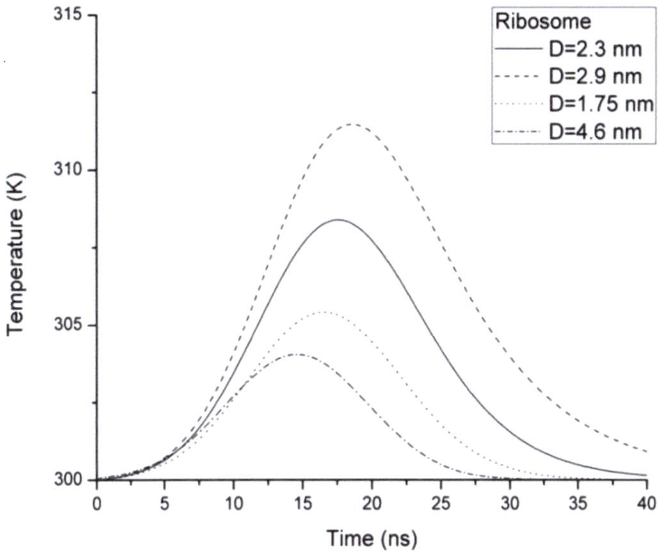

Fig. 10.16 Ribosomal time–temperature profiles at various diameter values. Laser parameters: 12 ns duration, 0.1 J/cm^2 at $\lambda = 700$ nm. Reprinted from [2]. © With permission from World Scientific

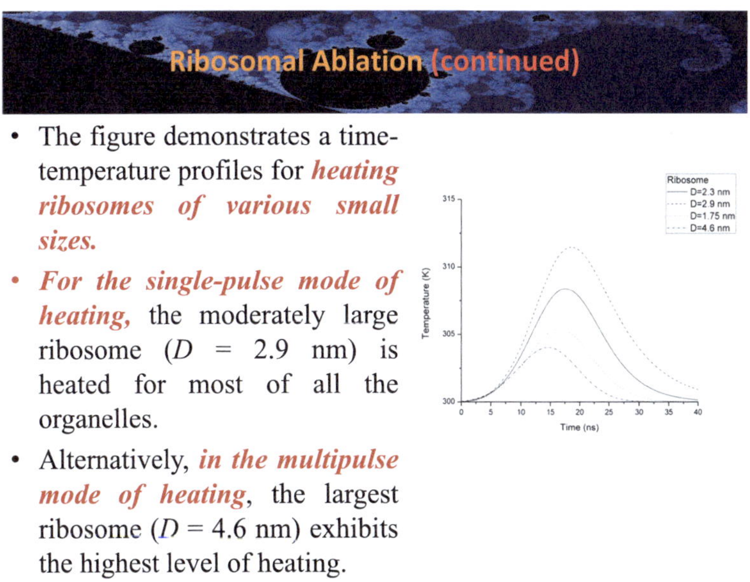

Slide 31.12 Ribosomal ablation (continued)

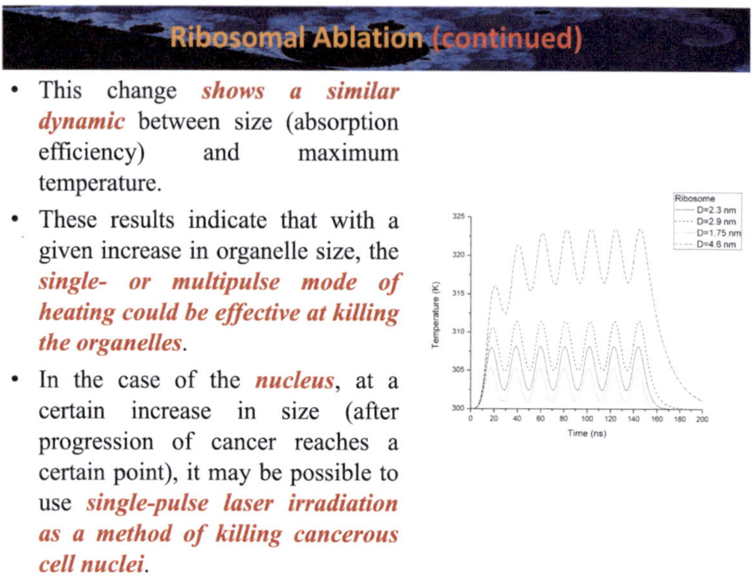

Slide 31.13 Ribosomal ablation (continued)

mode of heating could be effective for killing the organelles. In the case of the nucleus, at a certain increase in size (after progression of cancer reaches a certain point), it may be possible to use single-pulse laser irradiation as a method of killing cancerous cell nuclei [2].

10.2.5 Conclusions

Due to biological variability, the availability of organelle characteristics is minimal, even for specific cases. While it is hard to be fully concrete, several generalities can be made when considering tumor cells versus healthy cells [2] (Slide 31.14):

- A cancerous nucleus increases in size, and the cell has an increased nucleus-to-cytoplasm ratio.
- Mitochondrial activity fluctuates while mitochondrial cristae decrease, and the total number of mitochondria decreases.

Using an average value for organelles in healthy and cancerous forms, we can determine the relative single- and multipulse heating characteristics that produce thermal ablation in individual cases. For example, due to the heating and cooling kinetics of mitochondria, we find that for both single-pulse and multipulse modes of heating, tumor mitochondria will reach a higher maximum temperature than healthy mitochondria. A comparative analysis of mitochondrial and nuclear ablation shows that all types of mitochondria will heat to a temperature well above the nucleus (Slide 31.15) [2].

Conclusions

- Due to biological variability, the ***availability of organelle characteristics are minimal***, even for specific cases.
- While it is hard to be concrete, ***several generalities*** can be made when considering tumor cells versus healthy cells:
  - ➢ A ***cancerous nucleus increases in size***, and the cell has an increased nucleus-to-cytoplasm ratio.
  - ➢ ***Mitochondrial activity fluctuates*** while mitochondrial cristae decrease, and the total number of mitochondria decreases.
- ***Using an average value for organelles in healthy and cancerous forms***, we can determine the relative single- and multipulse heating characteristics that produce ***thermal ablation*** in individual cases.

Slide 31.14 Conclusions

Conclusions (continued)

- For example, due to the heating and cooling kinetics of mitochondria, we find that for both single-pulse and multipulse modes of heating, ***tumor mitochondria will reach a higher maximum temperature than healthy mitochondria.***
- A comparative analysis of mitochondrial and nuclear ablation shows that ***all types of mitochondria will heat to a temperature well above the nucleus***.
- The ***differences in cancer organelle morphology*** could all result from the numerous ***stages, grades and origins of cancer***.
- Investigations into cancer ***organelle morphology*** would need to be done under specific conditions, with the intent and purpose to provide more data for ***thermal ablation*** calculations.

Slide 31.15 Conclusions (continued)

The differences in cancer organelle morphology could all result from the numerous stages, grades, and origins of cancer. Before more concrete heating analysis can be made, investigations into cancer organelle morphology would need to be done under specific conditions, with the intent and purpose to provide more data for thermal ablation calculations (Slide 31.15) [2].

10.3 Computer Practicum 8 (Lecture 32): Simulating Biological Tissue Ablation (Slide 32.1)

- Computer Code
- Practice Example 12: Cell Ablation Through Nanoparticle Heating
  - *Abstract*
  - *Method*
  - *Results*
  - *Conclusions*

10.3.1 Computer Code

In Sect. 10.1 of this chapter, we introduced a simple Arrhenius-type model to describe photothermal ablation of biological tissue (see Eqs. (10.1)–(10.7)). Then we applied this model to study hard-tissue ablation through nanoparticle heating. In this section we develop a Maple code to simulate thermal ablation of biological tissues through plasmonic nanoparticles heated by short and ultrashort

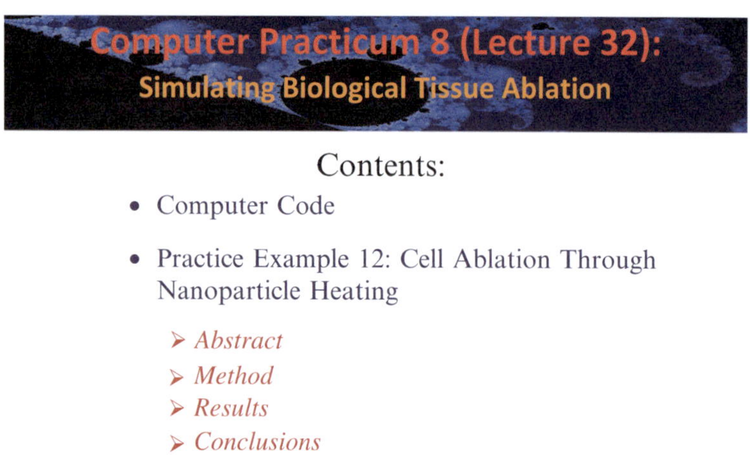

Slide 32.1 Contents

Computer Code

- We have introduced a simple ***Arrhenius-type model*** to describe ***photothermal ablation of biological tissue***.
- Here we develop a ***Maple code to simulate thermal ablation of biological tissues*** through plasmonic nanoparticles heated by short and ultrashort laser pulses.
- We also provide a ***practice example*** demonstrating simulation results ***for thermal ablation of soft biological cells***.
- ***A Maple code***, listed ***in Appendix I***, consists of two parts.
- ***The first*** solves a differential equation describing the time evolution of the number of unablated molecules of the biological tissue.
- ***The second part*** simulates the number of ablated molecules as a function of energy density of the incident laser pulse.

Slide 32.2 Computer code

laser pulses (Slide 32.2). We also provide a practice example demonstrating the simulation results for thermal ablation of soft biological cells.

A Maple code, listed in Appendix I, consists of two parts (Slide 32.2). The first solves a differential equation (10.4) describing the time evolution of a number of unablated molecules of the biological tissue. The second part simulates the number of ablated molecules as a function of energy density of the incident laser pulse by solving Eq. (10.7). The Maple code begins with introducing all the constants used in the ablation model simulations for the case of soft-tissue ablation as shown in Fig. 10.17 and Slide 32.3. The thermal ablation of soft tissues is defined here as irreversible changes in the biological medium due to the protein denaturation at the critical temperature that causes cell death.

Next, we solve a differential equation (10.4) by using the *dsolve* function in Maple which relies on the numeric *Runge–Kutta* method Slide 32.4. We arrange the results of the simulations in the form of a plot for the number of unablated molecules vs. time as shown in Fig. 10.18 and Slide 32.4. In the second part of the Maple code, we solve Eq. (10.7) and plot the number of ablated molecules vs. energy density of the laser pulse as shown in Fig. 10.19 and Slide 32.5.

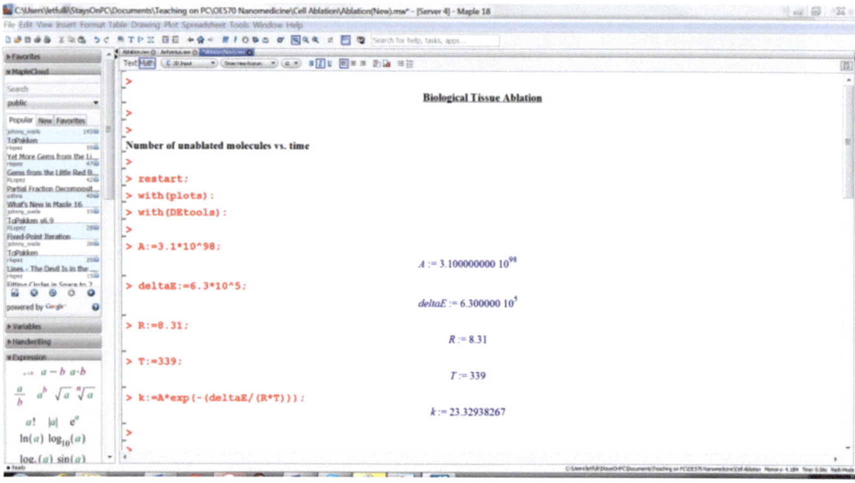

Fig. 10.17 Introducing the input data for the ablation model simulations

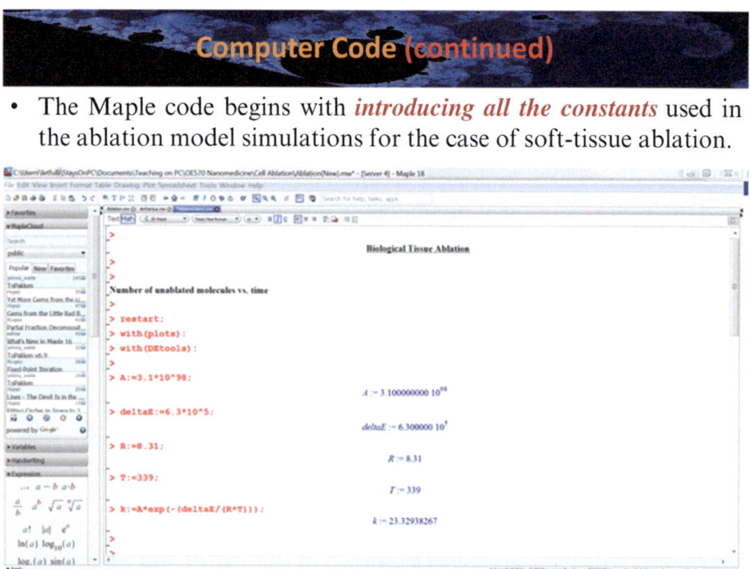

Slide 32.3 Computer code (continued)

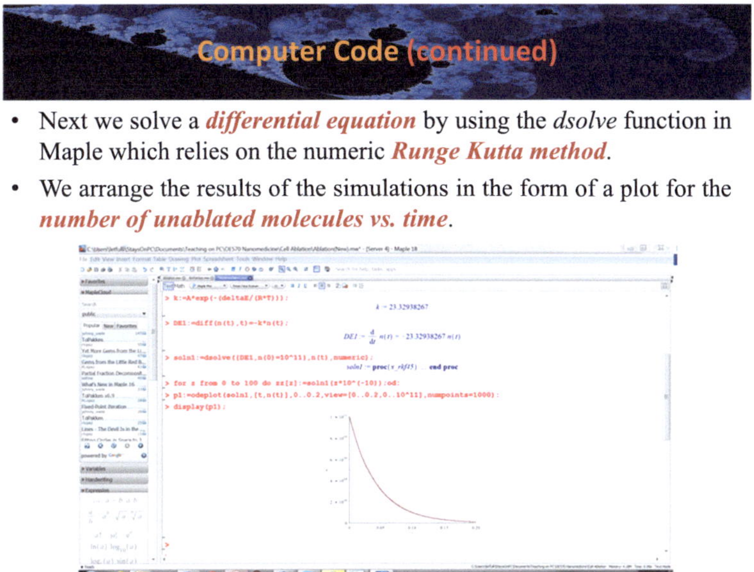

Slide 32.4 Computer code (continued)

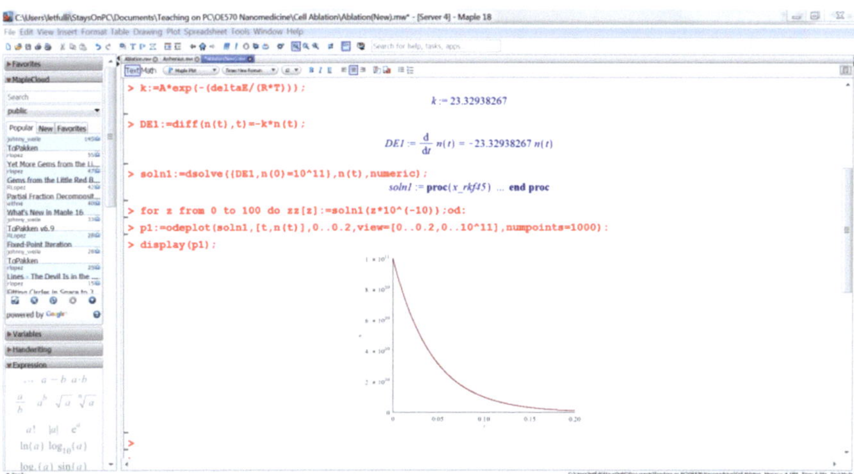

Fig. 10.18 Solving a differential equation and arranging the output results of simulations for the number of unablated molecules as a function of time

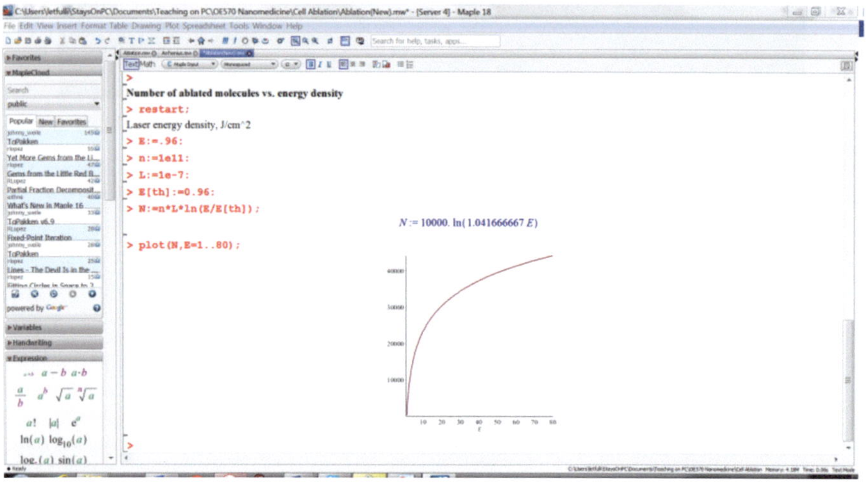

Fig. 10.19 Simulating the number of ablated molecules as a function of energy density of the laser pulse

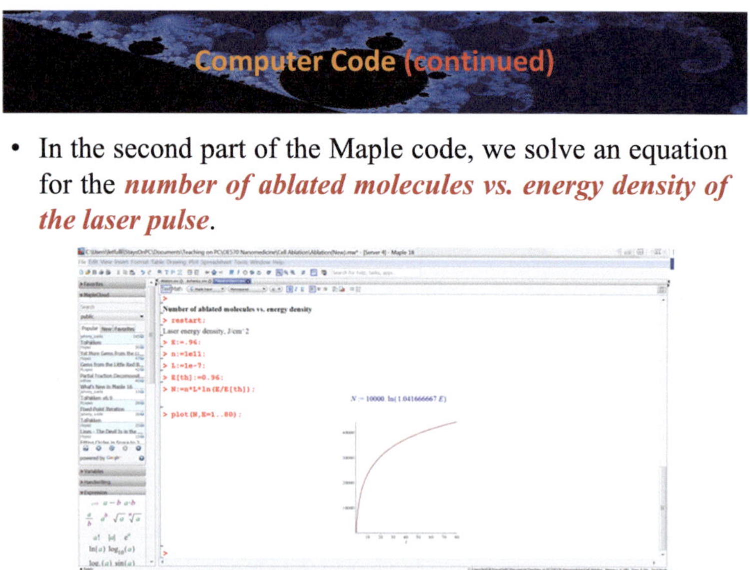

Slide 32.5 Computer code (continued)

10.3.2 Practice Example 12: Cell Ablation Through Nanoparticle Heating

10.3.2.1 Abstract

The interaction of laser radiation with targeted nanoparticles results in quasi-instantaneous conversion of laser pulse energy to thermal energy within the particles and further heat diffusion to the surrounding biological medium (Slide 32.6). The process of laser ablation of the soft surrounding tissue begins when the nanoparticle temperature achieves the threshold ablation temperature. The critical temperature for cancer cell ablation is in the range 319–425 K, as that results in protein denaturing and alters membrane permeability. Thus, the thermal ablation of soft tissues is defined here as irreversible changes in the biological medium due to the protein denaturation at the critical (threshold) temperatures that causes cell death. In this practice example, we model the ablation of molecules within cytoplasm in response to energy applied by heating nanoparticles.

10.3.2.2 Method

We use here a simple Arrhenius-type model (Eqs. (10.1)–(10.7)) and Maple code from Appendix I to simulate photothermal ablation of soft biological tissue (Slide 32.7). In the example of the simulations, we assume that the number of

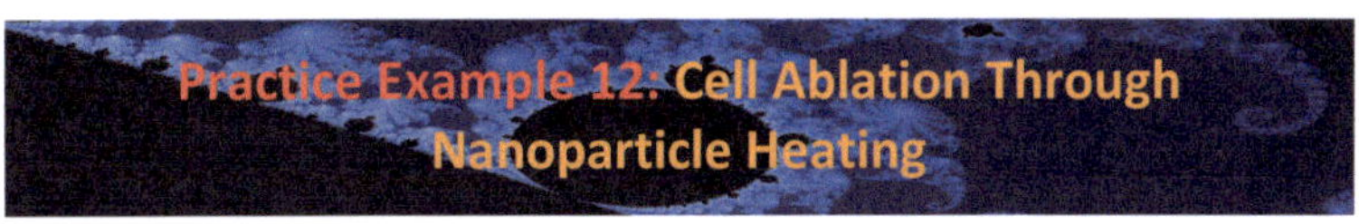

- The interaction of laser radiation with targeted nanoparticles results in quasi-instantaneous ***conversion of laser pulse energy to thermal energy within the particles*** and further heat diffusion to the surrounding biological medium.
- The process of ***laser ablation of the soft*** surrounding ***tissue*** begins when the nanoparticle temperature achieves the ***threshold ablation temperature***.
- The ***critical temperature*** for cancer cell ablation is in the range ***319-425 K***, as that results in ***protein denaturing*** and alters membrane permeability.
- Thus, ***thermal ablation*** of soft tissues is defined here as irreversible changes in the biological medium due to ***protein denaturation*** at the critical (threshold) temperatures that causes cell death.
- In this practice example, we model the ***ablation*** of molecules within ***cytoplasm*** in response to the energy applied by heating nanoparticles.

Slide 32.6 Practice Example 12: Cell ablation through nanoparticle heating

Practice Example 12: Method

- We use here a simple ***Arrhenius-type model and Maple code from Appendix I*** to simulate photothermal ablation of soft biological tissue.
- In the example of the simulations, we assume that the number of molecules in cytoplasm is $N_0 = 10^{11}\ cm^{-3}$; the threshold energy density for protein denaturation is $E_{threshold} = 0.96\ J/cm^2$.
- First, we model the ***number of molecules ablated*** in response to the particular ***energy density*** applied.
- Then we simulate ***the evolution of the number of unablated molecules in time*** by solving a differential equation.
- This model takes into account the ***heating caused by nanoparticles*** which then ablate surrounding molecules through heat diffusion.

Slide 32.7 Practice Example 12: Method

molecules in cytoplasm is $N_0 = 10^{11}$ cm^{-3}; the threshold energy density for protein denaturation is $E_{\text{thr}} = 0.96$ J/cm^2.

First, we use the simple relationship in Eq. (10.7) to model the number of molecules ablated in response to the particular energy density applied. Then we simulate the evolution of the number of unablated molecules in time by solving the differential equation (10.4). This model takes into account the heating caused by nanoparticles, which then ablate surrounding molecules through heat diffusion (Slide 32.7).

10.3.2.3 Results

The results of the simulations for ablated molecules as a function of laser energy density presented in Fig. 10.20 and Slide 32.8 show that the number of ablated molecules does not depend strongly on the energy applied to them. Even at high laser energies up to 80 J/cm^2, only about 45,000 out of 10^{11} molecules are denatured—an insignificant effect from which the cell could easily recover. These results demonstrate that direct ablation of the soft tissue by laser radiation is not efficient as ablation by heated nanoparticles, which absorb the incident radiation and transfer heat to the surrounding biological medium much more efficiently. It is, in fact, the heat generated by the nanoparticles' absorption of laser energy that causes the primary ablation within the cells (Slide 32.8).

Figure 10.21 and Slide 32.9 show the evolution of the number of ablated molecules in time for different temperatures simulated by using a differential

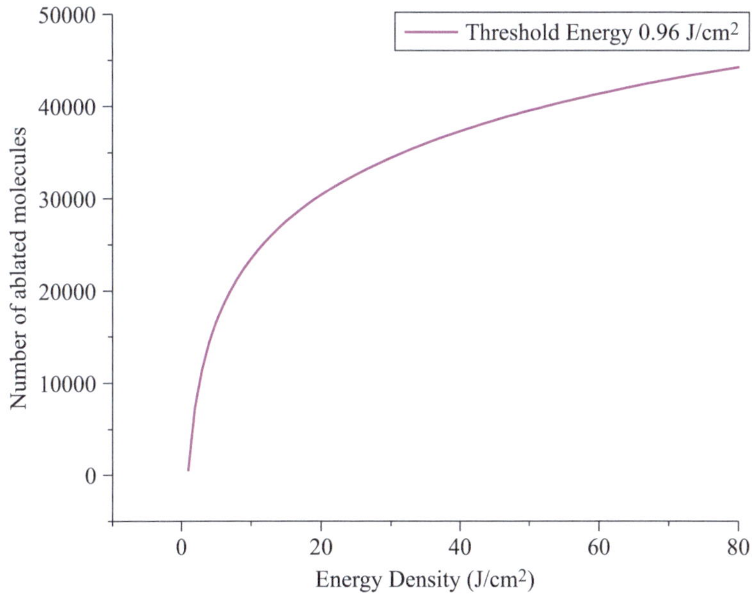

Fig. 10.20 Number of ablated molecules as a function of laser energy density. Note that the total number of molecules is 10^{11} cm^{-3}

Practice Example 12: Results

- The figure shows that the ***number of ablated molecules*** does not depend strongly on the amount of ***energy applied*** to them.
- Even at high laser energies up to 80 J/cm^2, only about ***45,000*** out of 10^{11} molecules ***are denatured*** – an insignificant effect from which the cell could easily recover.

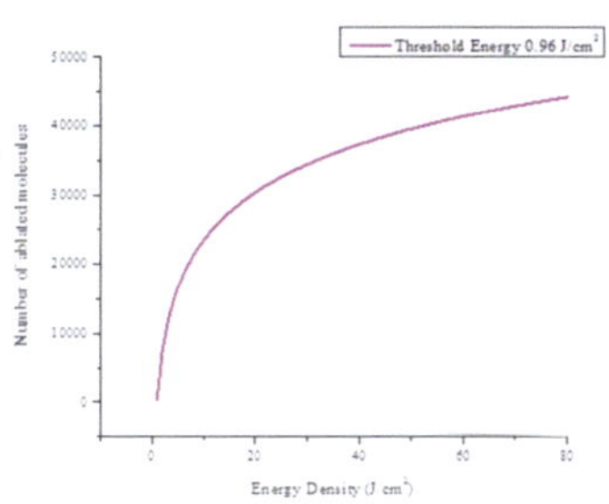

- These results demonstrate that the ***direct ablation*** of soft tissue by laser radiation ***is not as efficient*** as the ablation by heated nanoparticles, which absorb the incident radiation and transfer heat to the surrounding biological medium much more efficiently.
- It is, in fact, the ***heat generated by the nanoparticles***' absorption of laser energy that ***causes*** the primary ***ablation*** within the cells.

Slide 32.8 Practice Example 12: Results

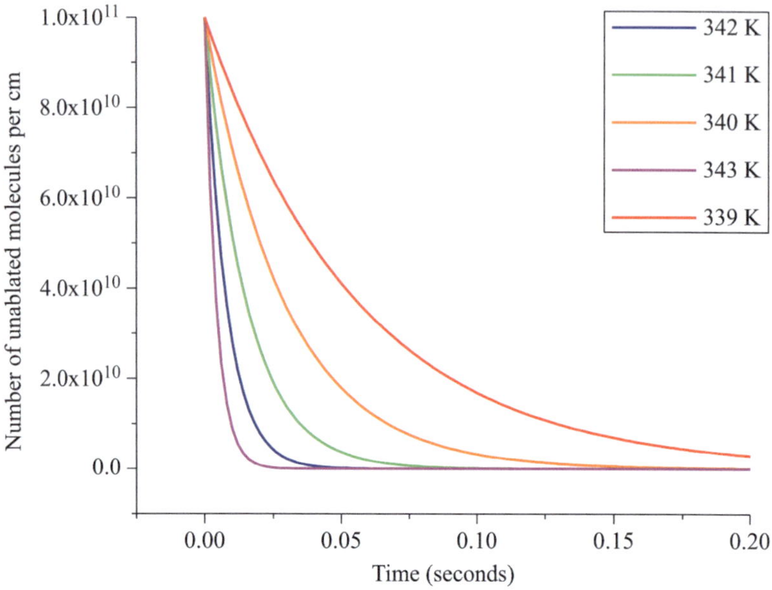

Fig. 10.21 Time evolution of the number of unablated molecules surrounding a heated nanoparticle

Practice Example 12: Results

- The figure shows the ***evolution of the number of ablated molecules in time*** for different temperatures.
- The ***lower end of the critical temperature*** range for ablation is only 319 K. However, it ***takes a long time (hours) to ablate 90% of the molecules*** at that temperature, so that higher temperatures are used in the simulations.

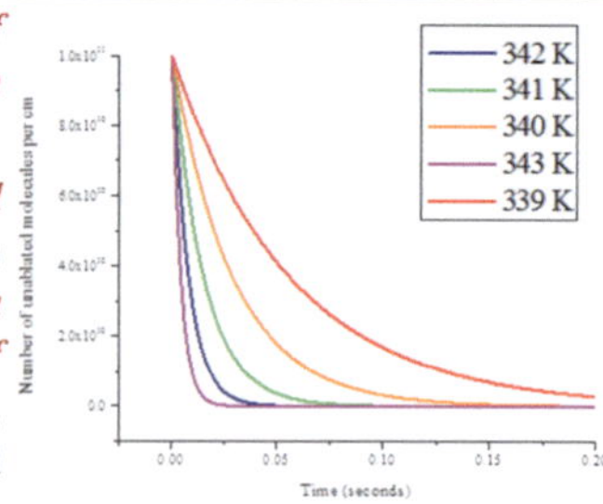

- This model definitively demonstrates that ***temperature has a huge effect on the ablation*** of molecules in vivo.
- The calculations show that an ***increase by only five degrees of the ablation temperature*** results in ***a tenfold decrease of the time*** necessary ***to ablate 95% of the molecules***.
- Thus, it is demonstrated that ***higher temperatures*** will produce ***much higher ablation rates*** with ***shorter ablation times***.

Slide 32.9 Practice Example 12: Results (continued)

equation (10.4). The lower end of the critical temperature range for ablation is only 319 K. However, it takes a long time (hours) to ablate 90 % of the molecules at that temperature, so that higher temperatures are used in simulations to achieve shorter ablation times on the millisecond to second time scale as shown in Fig. 10.21 and Slide 32.9. This model definitively demonstrates that temperature has a huge effect on the ablation of molecules in vivo. The calculations show that an increase by only 5° of the ablation temperature results in a tenfold decrease of the time necessary to ablate 95 % of the molecules. This relationship is consistent for other temperature ranges. Thus, it is demonstrated that higher temperatures will produce much higher ablation rates with shorter ablation times.

10.3.2.4 Conclusions

Our simulation showed that the laser energy absorbed directly by molecules does not have a significant effect on the ablation of those molecules (Slide 32.10). Heat diffusion from the surface of the nanoparticles into the surrounding biological tissue is the leading cause of thermal ablation in cells. An increase in ablation temperature will decrease the time necessary for full ablation of soft tissue by about 10 times for every 5°.

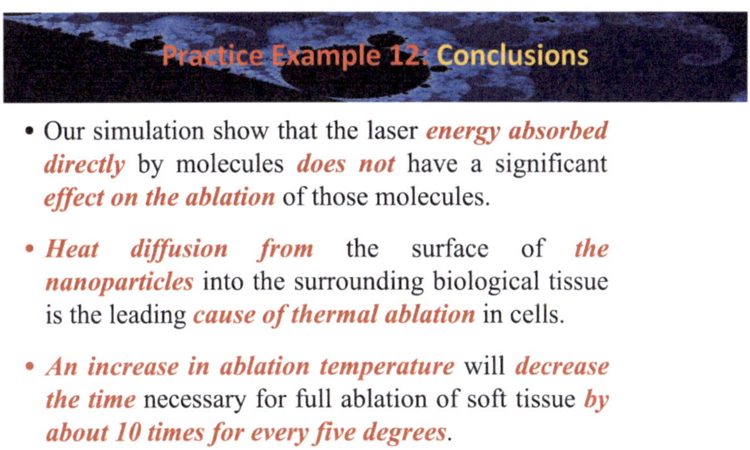

Slide 32.10 Practice Example 12: Conclusions

Appendix I: Maple Code for Simulating the Ablation of Biological Tissue

This appendix contains a Maple code designed to simulate photothermal ablation of soft biological tissue (Slide 32.11). The comments within the Maple code should explain the variables, procedures, and functions that are used.

Note: The code comes directly from the Maple file and is therefore formatted to word-wrap according to Maple standard character-per-line limits.

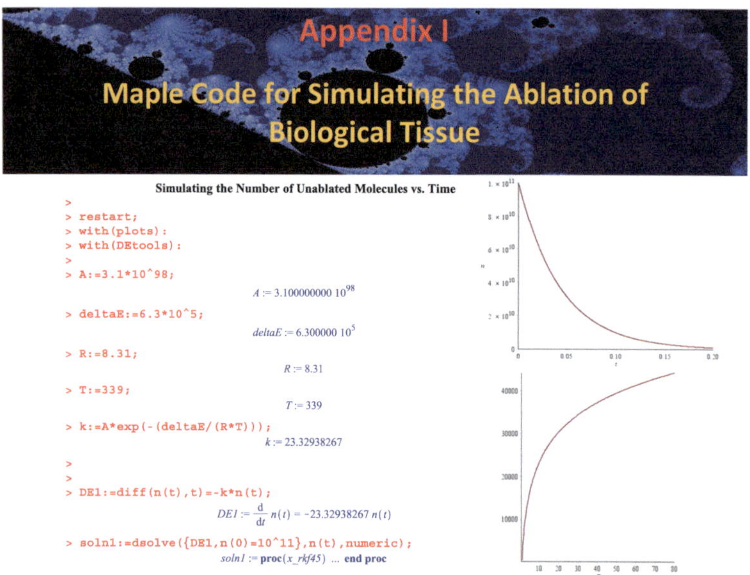

Slide 32.11 Appendix I: Maple Code for Simulating the Ablation of Biological Tissue

Biological Tissue Ablation

```
>
```

Simulating the Number of Unablated Molecules vs. Time

```
>
> restart;
> with(plots):
> with(DEtools):
>
> A:=3.1*10^98;
```

$$A := 3.100000000\ 10^{98}$$

```
> deltaE:=6.3*10^5;
```

$$deltaE := 6.300000\ 10^{5}$$

```
> R:=8.31;
```

$$R := 8.31$$

```
> T:=339;
```

$$T := 339$$

```
> k:=A*exp(-(deltaE/(R*T)));
```

$$k := 23.32938267$$

```
>
>
> DE1:=diff(n(t),t)=-k*n(t);
```

$$DE1 := \frac{d}{dt}\,n(t) = -23.32938267\,n(t)$$

```
> soln1:=dsolve({DE1,n(0)=10^11},n(t),numeric);
```

$$soln1 := \textbf{proc}(x\_rkf45)\ \ldots\ \textbf{end proc}$$

```
> for z from 0 to 100 do zz[z]:=soln1(z*10^(-10));od:
>
p1:=odeplot(soln1,[t,n(t)],0..0.2,view=[0..0.2,0..10^11],numpoin
ts=1000):
> display(p1);
```

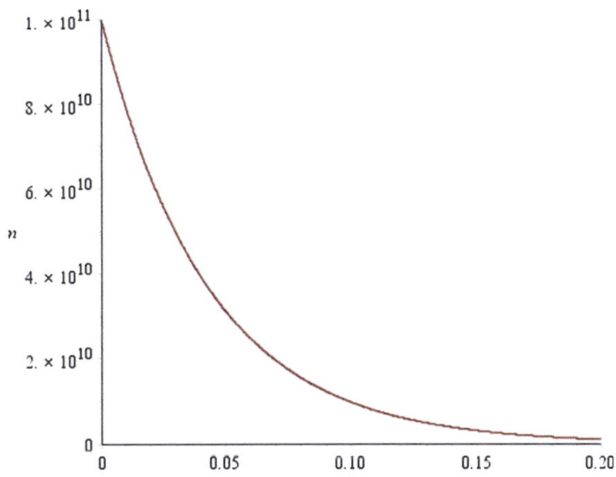


```
>
>
```

Simulating the Number of Ablated Molecules vs. Laser Energy Density

```
> restart;
```

Laser energy density, J/cm^2

```
> E:=.96;
```

$$E := 0.96$$

```
> n:=1e11;
```

$$n := 1.\ 10^{11}$$

```
> L:=1e-7;
```

$$L := 1.\ 10^{-7}$$

```
> E[th]:=0.96;
```

$$E_{th} := 0.96$$

```
> N:=n*L*ln(E/E[th]);
```

$$N := 10000.\ \ln(1.041666667\ E)$$

```
> plot(N,E=1..80);
```

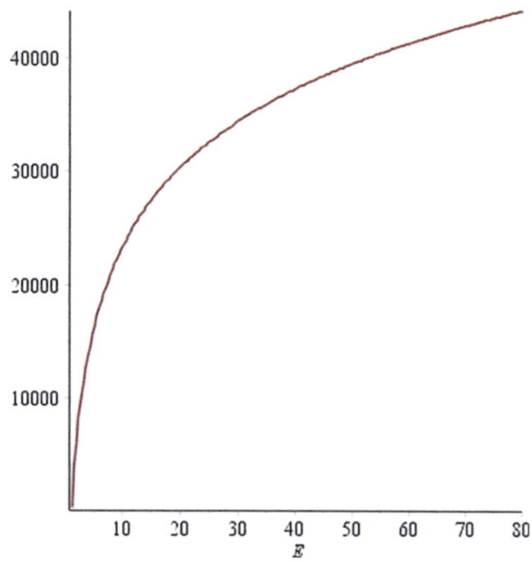


```
>
```

Homework Exercises

Section 10.1: Bone Cancer Ablation by Short and Ultrashort Pulses

10.1. Describe how bone tissue regenerates itself.
10.2. What is a nanophotohyperthermia?
10.3. How does high temperature effect biological tissue?
10.4. What temperatures cause cancerous cell death?
10.5. What is nanophotothermolysis?

10.6. List five nonlinear synergistic effects produced by nanophotothermolysis in tumors.
10.7. What are two main factors essential for nanophotothermal therapies?
10.8. True or false? Soft tissues, like those found in the heart and lungs, are 80–90 % water.
10.9. What is a hard tissue constituent?
10.10. True or false? In hard tissues, a mineral structure made up of nanoscale calcium phosphate crystals accounts for about 70 % of total bone material.
10.11. What biological molecules are present in bone tissue?
10.12. True or false? Hard tissues can be modeled in the same way as soft tissues.
10.13. List three main cells a bone tissue made of.
10.14. True or false? Osteoblasts, osteocytes, and osteoclasts are reproduced by replication.
10.15. Describe how osteoblast, osteocyte, and osteoclast cells are produced.
10.16. From what cells are osteoblasts differentiated from?
10.17. What is a main function of the osteoblasts?
10.18. True or false? When an osteoblast becomes surrounded by hard bone, it differentiates into an osteocyte and becomes involved in maintaining the bone surrounding it.
10.19. What are growth factors made of?
10.20. What are the main functions of growth factors?
10.21. True or false? Many growth factors play roles in the mediation of bone formation and resorption processes.
10.22. What are bone morphogenic proteins?
10.23. What is the tumor necrosis factor?
10.24. Give examples of bone-targeting bioconjugates.
10.25. What are bisphosphonates?
10.26. Describe a bone tissue targeting mechanism.
10.27. What is the spectral region for the window of transparency for biological media?
10.28. What are two main absorbers of light in soft tissue?
10.29. Describe the fiber-optic needle system for delivering light to bone tissues.
10.30. True or false? The choice of material, as well as the size and shape of a nanoparticle, affect the resonance frequency for the electron oscillations.
10.31. List several advantages of using gold nanoparticles for nanotherapy.
10.32. Give a definition of thermal ablation.
10.33. What is the protein denaturation temperature?
10.34. Describe an Arrhenius-type photothermal ablation model.
10.35. What role do nanoparticles play in thermal ablation of cancer cells?
10.36. Describe the mechanism of thermal ablation of biological tissues by plasmonic nanoparticles.
10.37. True or false? Bone tissue has a much weaker absorption than water in the optical range of the spectrum.
10.38. Describe the kinetics of heating nanoparticles by a laser pulse in hard tissue for small and larger nanoparticles.

10.39. What temperature range corresponds to nanophotohyperthermia therapies?
10.40. What temperature range corresponds to nanophotothermolysis?
10.41. What is the size of the thermal damage area produced in bone tissue by nanoparticles in comparison to the particle size?
10.42. What laser intensity is capable of producing thermal necrosis of osteocytes by absorbing nanoparticles?
10.43. Describe the heating and cooling kinetics of bone particles irradiated by short and ultrashort pulses.
10.44. True or false? The cooling time of bone particles is six orders of magnitude greater than the heating time.
10.45. Explain why a heat accumulation is observed in bone particles when heated by multipulse radiation.
10.46. Describe the advantages of using the multipulse mode for heating bone tissue particles.
10.47. What material is used to model the thermal and optical properties of bone tissue?
10.48. What is an accumulation factor in bone composition?

Section 10.2: Soft-Tissue Ablation by Short and Ultrashort Pulses

10.49. Give examples of laser usage in soft-tissue treatments.
10.50. What is the critical temperature for cell organelle ablation?
10.51. How is the thermal ablation of the soft tissue defined?
10.52. What is hepatocellular carcinoma?
10.53. Describe the absorption properties of cancerous nuclei and mitochondria.
10.54. Describe the kinetics of single-pulse heating of cancerous nuclei and mitochondria for well- and poorly differentiated cells.
10.55. Describe multipulse heating of cancerous nuclei and mitochondria.
10.56. What is ductal breast carcinoma?
10.57. True or false? The ribosomes in cancer cells have much greater scattering efficiency than normal cells in the spectral range 200–500 nm.
10.58. In what range of the spectrum do cancerous ribosomes effectively absorb radiation?
10.59. Describe the kinetics of heating normal and cancerous ribosome in the single- and multipulse modes.
10.60. True or false? Tumor mitochondria will reach a higher maximum temperature than healthy mitochondria when heated by laser radiation.
10.61. True or false? All types of mitochondria will heat to a temperature well above the nucleus when heated by laser radiation.
10.62. What are the origins of differences in cancer organelle morphology?

Section 10.3: Simulating Biological Tissue Ablation

10.63. Describe the two parts of the Maple code (Appendix I) designed to simulate thermal ablation of biological tissues.

10.64. How does an ablation model take into account the heating caused by nanoparticles?

10.65. True or false? The number of ablated molecules depends strongly on the amount of energy applied to them.

10.66. True or false? The direct ablation of soft tissue by laser radiation is not as efficient as the ablation by heated nanoparticles.

10.67. True or false? Temperature has a huge effect on the ablation of molecules in vivo.

10.68. How is the ablation time affected by the change in ablation temperature?

10.69. True or false? Higher temperatures will produce much higher ablation rates with shorter ablation times.

10.70. For the ablation temperature 335 K, find the time required to ablate 90 % of the molecules of soft biological tissue with initial number of molecules $N_0 = 10^{11}$ cm^{-3}.

10.71. How many molecules of soft biological tissue are ablated if the incident laser energy density is 50 J/cm^2, initial number of molecules is $N_0 = 10^{13}$ cm^{-3}, and the threshold energy density for protein denaturation is $E_{\mathrm{thr}} = 0.96$ J/cm^2?

10.72. For the ablation temperature 337 K, find the time required to ablate 90 % of the molecules of soft biological tissue with initial number of molecules $N_0 = 10^{13}$ cm^{-3}.

10.73. What is the laser energy density required to ablate 25 % of biological molecules with initial number of molecules $N_0 = 10^{11}$ cm^{-3} and threshold energy density for protein denaturation $E_{\mathrm{thr}} = 0.96$ J/cm^2?

10.74. What time is needed to ablate a 50 % of biological molecules with initial number of molecules $N_0 = 10^{11}$ cm^{-3} at the ablation temperature 338 K?

10.75. What is the laser energy density required to ablate 30 % of biological molecules with initial number of molecules $N_0 = 10^9$ cm^{-3} and threshold energy density for protein denaturation $E_{\mathrm{thr}} = 0.96$ J/cm^2?

References

1. R.R. Letfullin, C.E.W. Rice, T.F. George, Bone cancer therapy by plasmonic nanoparticles. Ther. Deliv. **2**, 1259–1273 (2011)
2. R.R. Letfullin, T.F. George, Laser ablation of biological tissue by short and ultrashort pulses, in *Computational Studies of New Materials II: From Ultrafast Processes and Nanostructures to Optoelectronics, Energy Storage and Nanomedicine*, ed. by T.F. George, D. Jelski, R.R. Letfullin, G.P. Zhang (World Scientific, Singapore, 2011), pp. 191–218
3. M. Strassl, V. Wieger, D. Brodoceanu, F. Beer, A. Moritz, E. Wintner, Ultrashort pulse laser ablation of biological hard tissue and biocompatibles. J. Laser Micro Nanoeng. **3**, 30–40 (2008)

4. T.G. Barton, H.J. Foth, M. Christ, K. Hormann, Interaction of holmium laser radiation and cortical bone: ablation and thermal damage in a turbid medium. Appl. Optics **36**, 32–43 (1997)
5. T. Torimura, T. Ueno, S. Inuzuka, Y. Kimura, P. Ko, M. Kin, T. Minetoma, T. Majima, M. Sata, H. Abe, K. Tanikawa, Ultrastructural observation on hepatocellular carcinoma: correlation of tumor grade and degree of atypia of cell organelles by morphometry. Med. Electron Microsc. **26**, 19–28 (1993)
6. L.W. Anson, R.C. Chivers, Ultrasonic propagation in mammalian cell suspensions based on a shell model. Phys. Med. Biol. **34**, 1153–1167 (1989)
7. M.M. Radwan, K.A. Amer, N.M. Mokhtar, M.A. Kandil, A.M. El-Barbary, H.A. Aiad, Nuclear morphometry in ductal breast carcinoma with correlation to cell proliferative activity and prognosis. J. Egypt. Natl. Canc. Inst. **15**, 169–182 (2003)

MIX
Papier aus verantwortungsvollen Quellen
Paper from responsible sources
FSC® C105338

If you have any concerns about our products,
you can contact us on
ProductSafety@springernature.com

In case Publisher is established outside the EU,
the EU authorized representative is:
Springer Nature Customer Service Center GmbH
Europaplatz 3, 69115 Heidelberg, Germany

Printed by Libri Plureos GmbH
in Hamburg, Germany